ADAPTIVE
WIRELESS
TRANSCEIVERS

*This book is dedicated to the numerous contributors of this field,
many of whom are listed in the Author Index.*

ADAPTIVE WIRELESS TRANSCEIVERS

Turbo-Coded, Turbo-Equalized and Space-Time Coded TDMA, CDMA and OFDM Systems

L. Hanzo
University of Southampton, UK

C. H. Wong
Multiple Access Communications Ltd, UK

M. S. Yee
University of Southampton, UK

IEEE PRESS

IEEE Communications Society, Sponsor

JOHN WILEY & SONS, LTD

Other Wiley Editorial Offices

John Wiley & Sons, Inc., 605 Third Avenue,
New York, NY 10158-0012, USA

WILEY-VCH Verlag GmbH
Pappelallee 3, D-69469 Weinheim, Germany

John Wiley & Sons Australia Ltd, 33 Park Road, Milton,
Queensland 4064, Australia

John Wiley & Sons (Canada) Ltd, 22 Worcester Road
Rexdale, Ontario, M9W 1L1, Canadá

John Wiley & Sons (Asia) Pte Ltd, 2 Clementi Loop #02-01,
Jin Xing Distripark, Singapore 129809

IEEE Communications Society, Sponsor
COMM-S Liaison to IEEE Press, Mostafa Hashem Sherif

British Library Cataloguing in Publication Data

A catalogue record for this book is available from the British Library

ISBN 0470 84689 5

Printed and bound by Antony Rowe Ltd, Eastbourne

Contents

Contributors of the book:
Chapter 5: C.H. Wong, S.X. Ng, L. Hanzo
Chapter 6: B.J. Choi, L. Hanzo
Chapter 12: E.L. Kuan, L. Hanzo
Chapter 13: T. Keller, L. Hanzo
Chapter 14: T.H. Liew, L. Hanzo

Chapter 1

Prologue

1.1 Motivation of the Book

In recent years the concept of intelligent multi-mode, multimedia transceivers (IMMT) has emerged in the context of wireless systems [1–6]. The range of various existing solutions that have found favour in already operational standard systems was summarised in the excellent overview by Nanda *et al.* [3]. *The aim of these adaptive transceivers is to provide mobile users with the best possible compromise amongst a number of contradicting design factors, such as the power consumption of the hand-held portable station (PS), robustness against transmission errors, spectral efficiency, teletraffic capacity, audio/video quality and so forth [2].*

The fundamental limitation of wireless systems is constituted by their time- and frequency-domain channel fading, as illustrated in Figure 14.39 in terms of the Signal-to-Noise Ratio (SNR) fluctuations experienced by a modem over a dispersive channel. The violent SNR fluctuations observed both versus time and versus frequency suggest that over these channels no fixed-mode transceiver can be expected to provide an attractive performance, complexity and delay trade-off. Motivated by the above mentioned performance limitations of fixed-mode transceivers, IMMTs have attracted considerable research interest in the past decade [1–6]. Some of these research results are collated in this monograph.

In Figure 1.1 we show the instantaneous channel SNR experienced by the 512-subcarrier OFDM symbols for a single-transmitter, single-receiver scheme and for the space-time block code G_2 [7] using one, two and six receivers over the shortened WATM channel. The average channel SNR is 10 dB. We can see in Figure 1.1 that the variation of the instantaneous channel SNR for a single transmitter and single receiver is severe. The instantaneous channel SNR may become as low as 4 dB due to deep fades of the channel. On the other hand, we can see that for the space-time block code G_2 using one receiver the variation in the instantaneous channel SNR is slower and less severe. Explicitly, by employing multiple transmit antennas as shown in Figure 1.1, we have reduced the effect of the channels' deep fades significantly. This is advantageous in the context of adaptive modulation schemes, since higher-order modulation modes can be employed, in order to increase the throughput of the system. However,

1

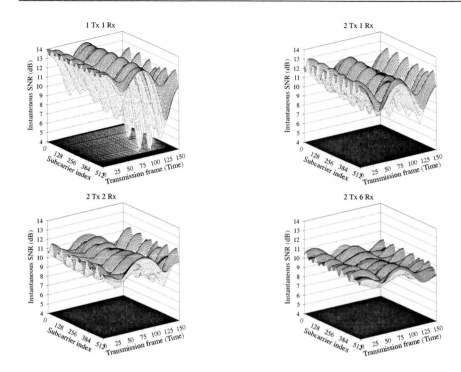

Figure 1.1: Instantaneous channel SNR versus time and frequency for a 512-subcarrier OFDM modem in the context of a single-transmitter single-receiver as well as for the space-time block code $\mathbf{G}_2$ [7] using one, two and six receivers when communicating over an indoor wireless channel. The average channel SNR is 10 dB. ©IEEE, Liew and Hanzo [8], 2001

as we increase the number of receivers, i.e. the diversity order, we observe that the variation of the channel becomes slower. Effectively, by employing higher-order diversity, the fading channels have been converted to AWGN-like channels, as evidenced by the scenario employing the space-time block code $\mathbf{G}_2$ using six receivers. Since adaptive modulation only offers advantages over fading channels, we argue that using adaptive modulation might become unnecessary, as the diversity order is increased. Hence, adaptive modulation can be viewed as a lower-complexity alternative to space-time coding, since only a single transmitter and receiver is required.

Our intention with the book is multifold:

1. Firstly, to pay tribute to all researchers, colleagues and valued friends, who contributed to the field. Hence this book is dedicated to them, since without their quest for better transmission solutions for wireless communications this monograph could not have been conceived. They are too numerous to name here, hence they appear in the author index of the book.

2. Although the potential of adaptive modulation and transmission was recognised some 30 years ago by Cavers [9] and during the nineties the associated research efforts in-

tensified, to date there is no monograph on the topic. Hence it is our hope that the conception of this monograph on the topic will provide an adequate portrayal of the last decade of research and fuel this innovation process.

3. As argued above, adaptive modulation only offers advantages when communicating over fading wireless channels. However, since the space-time coding assisted employment of transmit and receive diversity mitigates the effects of fading, we would like to portray adaptive modulation as a lower-complexity alternative to space-time coding, since only a single transmitter and receiver is required.

4. We expect to stimulate further research by exposing not only the information theoretical limitations of such IMMTs, but also by collating a range of practical problems and design issues for the practitioners. The coherent further efforts of the wireless research community is expected to lead to the solution of the vast range of outstanding problems, ultimately providing us with flexible wireless transceivers exhibiting a performance close to information theoretical limits.

The above mentioned calamities inflicted by the wireless channel can be mitigated by contriving a suite of near-instantaneously adaptive or Burst-by-Burst Adaptive (BbBA) wideband single-carrier [4], multi-carrier or Orthogonal Frequency Division Multiplex [4] (OFDM) as well as Code Division Multiple Access (CDMA) transceivers. The aim of these IMMTs is to communicate over hostile mobile channels at a higher integrity or higher throughput, than conventional fixed-mode transceivers. A number of existing wireless systems already support some grade of adaptivity and future research is likely to promote these principles further by embedding them into the already existing standards. For example, due to their high control channel rate and with the advent of the well-known Orthogonal Variable Spreading Factor (OVSF) codes the third-generation UTRA/IMT2000 systems are amenable to not only long-term spreading factor reconfiguration, but also to near-instantaneous reconfiguration on a 10ms transmission burst-duration basis.

With the advent of BbBA QAM, OFDM or CDMA transmissions it becomes possible for mobile stations (MS) to invoke for example in indoor scenarios or in the central propagation cell region - where typically benign channel conditions prevail - a high-throughput modulation mode, such as 4 bit/symbol Quadrature Amplitude Modulation (16QAM). By contrast, a robust, but low-throughput modulation mode, such as 1 bit/symbol Binary Phase Shift Keying (BPSK) can be employed near the edge of the propagation cell, where hostile propagation conditions prevail. The BbBA QAM, OFDM or CDMA mode switching regime is also capable of reconfiguring the transceiver at the rate of the channel's slow- or even fast-fading. This may prevent premature hand-overs and - more importantly - unnecessary powering up, which would inflict an increased interference upon co-channel users, resulting in further potential power increments. This detrimental process could result in all mobiles operating at unnecessarily high power levels.

A specific property of these transceivers is that their bit rate fluctuates, as a function of time. This is not an impediment in the context of data transmission. However, in interactive speech [5] or video [6] communications appropriate source codecs have to be designed, which are capable of promptly reconfiguring themselves according to the near-instantaneous bitrate budget provided by the transceiver.

The expected performance of our BbBA transceivers can be characterized with the aid of a whole plethora of performance indicators. In simple terms, adaptive modems outperform their individual fixed-mode counterparts, since given an average number of transmitted bits per symbol (BPS), their average BER will be lower than that of the fixed-mode modems. From a different perspective, at a given BER their BPS throughput will be always higher. In general, the higher the tolerable BER, the closer the performance to that of the Gaussian channel capacity. Again, this fact underlines the importance of designing programmable-rate, error-resilient source codecs - such as the Advanced Multi-Rate (AMR) speech codec to be employed in UMTS - which do not expect a low BER.

Similarly, when employing the above BbBA or AQAM principles in the frequency domain in the context of OFDM [4] or in conjunction with OVSF spreading codes in CDMA systems, attractive system design trade-offs and a high over-all performance can be attained [6]. However, despite the extensive research in the field by the international community, there is a whole host of problems that remain to be solved and this monograph intends to contribute towards these efforts. The signal processing techniques used are ambitious, but a range of emerging enabling technologies based on the design philosophy of Software-Defined Radio [10] (SDR) architectures have already been documented in the literature. The capabilities of the SDR technolgy are expected to evolve further over the forthcoming years.

1.2 Adaptation Principles

AQAM is suitable for duplex communication between the MS and BS, since the AQAM modes have to be adapted and signalled between them, in order to allow channel quality [1] estimates and signalling to take place. The AQAM mode adaptation is the action of the transmitter in response to time–varying channel conditions. In order to efficiently react to the changes in channel quality, the following steps have to be taken:

- *Channel quality estimation:* In order to appropriately select the transmission parameters to be employed for the next transmission, a reliable estimation of the channel transfer function during the next active transmit timeslot is necessary.

- *Choice of the appropriate parameters for the next transmission:* Based on the prediction of the channel conditions for the next timeslot, the transmitter has to select the appropriate modulation and channel coding modes for the subcarriers.

---

[1]Throughout the book we will be studying the effects of the multipath-induced channel quality fluctuations. However, the associated principles are equally applicable in the context of mitigating the co-channel interference effects imposed by the time-variant fluctuations of the number of users supported. Naturally, these co-channel interference effects can be mitigated with the aid of interference cancellation techniques, but in case of employing low-complexity single-user detection an adaptive scheme may simply activate a more robust transmission mode. The associated network-layer benefits of using adaptive transmission schemes have been quantified in [11].

We note furthermore that the multipath-induced channel quality fluctuations may be mitigated also with the aid of multiple transmitter and multiple receiver assisted space-time coding arrangements [12], if the associated higher complexity is affordable. In a multiple transmitter and multiple receiver assisted space-time coded scenario the performance benefits of AQAM erode, since the the multipath-induced channel quality fluctuations are mitigated by the space-time coding schemes [12] used. We note, however that fixed-mode modulation based space-time codecs are expected to be less efficient in terms of mitigating the effects of the time-variant co-channel interference fluctuations, than their adaptive counterparts, especially, if no interference cancellation is employed.

- *Signalling or blind detection of the employed parameters:* The receiver has to be informed, as to which demodulator parameters to employ for the received packet. This information can either be conveyed within the OFDM symbol itself, at the cost of loss of effective data throughput, or the receiver can attempt to estimate the parameters employed by the remote transmitter by means of blind detection mechanisms [4].

1.3 Channel Quality Metrics

The most reliable channel quality estimate is the bit error rate (BER), since it reflects the channel quality, irrespective of the source or the nature of the quality degradation. The BER can be estimated invoking a number of approaches.

Firstly, the BER can be estimated with a certain granularity or accuracy, provided that the system entails a channel decoder or - synonymously - Forward Error Correction (FEC) decoder employing algebraic decoding [13].

Secondly, if the system contains a soft-in-soft-out (SISO) channel decoder, the BER can be estimated with the aid of the Logarithmic Likelihood Ratio (LLR), evaluated either at the input or the output of the channel decoder. A particularly attractive way of invoking LLRs is employing powerful turbo codecs, which provide a reliable indication of the confidence associated with a particular bit decision in the context of LLRs.

Thirdly, in the event that no channel encoder / decoder (codec) is used in the system, the channel quality expressed in terms of the BER can be estimated with the aid of the mean-squared error (MSE) at the output of the channel equalizer or the closely related metric of Pseudo-Signal-to-Noise-Ratio (Pseudo-SNR) [6]. The MSE or pseudo-SNR at the output of the channel equalizer have the important advantage that they are capable of quantifying the severity of the inter-symbol-interference (ISI) and/or Co-channel Interference (CCI) experienced, in other words quantifying the Signal to Interference plus Noise Ratio (SINR).

As an example, let us consider OFDM. In OFDM modems [4] the bit error probability in each subcarrier can be determined by the fluctuations of the channel's instantaneous frequency domain channel transfer function H_n, if no co-channel interference is present. The estimate $\hat{H}_n$ of the channel transfer function can be acquired by means of pilot–tone based channel estimation [4]. For CDMA transceivers similar techniques are applicable, which constitute the topic of this monograph.

The delay between the channel quality estimation and the actual transmission of a burst in relation to the maximal Doppler frequency of the channel is crucial as regards to the adaptive system's performance. If the channel estimate is obsolete at the time of transmission, then poor system performance will result [6].

1.4 Transceiver Parameter Adaptation

Different transmission parameters - such as the modulation and coding modes - of the AQAM single- and multi-carrier as well as CDMA transceivers can be adapted to the anticipated channel conditions. For example, adapting the number of modulation levels in response to the anticipated SNR encountered in each OFDM subcarrier can be employed, in order to achieve a wide range of different trade–offs between the received data integrity and throughput. Corrupted subcarriers can be excluded from data transmission and left blank or used for example for Crest–factor reduction. A range of different algorithms for selecting the appropriate modulation modes have to be investigated by future research. **The adaptive channel coding parameters entail code rate, adaptive interleaving and puncturing for convolutional and turbo codes, or varying block lengths for block codes [4].**

Based on the estimated frequency–domain channel transfer function, **spectral pre–distortion at the transmitter of one or both communicating stations can be invoked, in order to partially of fully counteract the frequency–selective fading of the time–dispersive channel**. Unlike frequency–domain equalization at the receiver — which corrects for the amplitude– and phase–errors inflicted upon the subcarriers by the channel, but which cannot improve the SNR in poor quality OFDM subchannels — spectral pre–distortion at the OFDM transmitter can deliver near–constant signal–to–noise levels for all subcarriers and can be viewed as power control on a subcarrier–by–subcarrier basis.

In addition to improving the system's BER performance in time–dispersive channels, spectral pre–distortion can be employed in order to perform all channel estimation and equalization functions at only one of the two communicating duplex stations. Low–cost, low power consumption mobile stations can communicate with a base station that performs the channel estimation and frequency–domain equalization of the uplink, and uses the estimated channel transfer function for pre–distorting the down–link OFDM symbol. This setup would lead to different overall channel quality on the up– and downlink, and the superior pre-equalised downlink channel quality could be exploited by using a computationally less complex channel decoder, having weaker error correction capabilities in the mobile station than in the base station.

If the channel's frequency–domain transfer function is to be fully counteracted by the spectral pre-distortion upon adapting the subcarrier power to the inverse of the channel transfer function, then the output power of the transmitter can become excessive, if heavily faded subcarriers are present in the system's frequency range. In order to limit the transmitter's maximal output power, hybrid channel pre–distortion and adaptive modulation schemes can be devised, which would de–activate transmission in deeply faded subchannels, while retaining the benefits of pre–distortion in the remaining subcarriers.

BbBA mode signalling plays an important role in adaptive systems and the range of signalling options is summarised in Figure 1.2 for **closed–loop signalling**. If the channel quality estimation and parameter adaptation have been performed at the transmitter of a particular link, based on open–loop adaptation, then the resulting set of parameters has to be communicated to the receiver in order to successfully demodulate and decode the OFDM symbol. Once the receiver determined the requested parameter set to be used by the remote transmitter, then this information has to be signalled to the remote transmitter in the reverse link. If this signalling information is corrupted, then the receiver is generally unable to correctly decode the OFDM symbol corresponding to the incorrect signalling information, yielding an OFDM symbol error.

Unlike adaptive serial systems, which employ the same set of parameters for all data symbols in a transmission packet [4], adaptive OFDM systems [4] have to react to the frequency selective nature of the channel, by adapting the modem parameters across the subcarriers. The resulting signalling overhead may become significantly higher than that for serial modems, and can be prohibitive for example for subcarrier–by–subcarrier based modulation mode adaptation. In order to overcome these limitations, efficient and reliable signalling techniques have to be employed for practical implementation of adaptive OFDM modems.

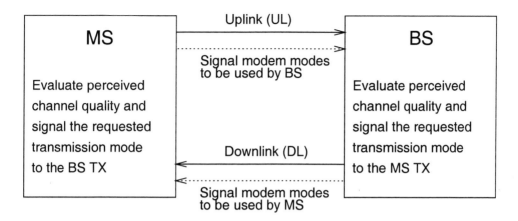

Figure 1.2: Parameter signalling in BbBA OFDM, CDMA and AQAM modems, IEEE Press-John Wiley, 2000, Hanzo, Webb, Keller [4].

If some flexibility in choosing the transmission parameters is sacrificed in an adaptation scheme, like in sub–band adaptive OFDM schemes [4], then the amount of signalling can be reduced. Alternatively, blind parameter detection schemes can be devised, which require little or no OFDM mode signalling information, respectively [4].

In conclusion, fixed mode transceivers are incapable of achieving a good trade-off in terms of performance and complexity. The proposed BbB adaptive system design paradigm is more promising in this respect. A range of problems and solutions were highlighted in conceptual terms with reference to an OFDM-based example, indicating the areas, where substantial future research is required. A specific research topic, which raised substantial research interest recently is invoking efficient channel quality prediction techniques [14]. Before we commence our indepth discourse in the forthcoming chapters, in the next section we provide a brief historical perspective on adaptive modulation.

1.5 Milestones in Adaptive Modulation History

1.5.1 Adaptive Single- and Multi-carrier Modulation

As we noted in the previous sections, mobile communications channels typically exhibit a near-instantaneously fluctuating time-variant channel quality [13] and hence conventional fixed-mode modems suffer from bursts of transmission errors, even if the system was designed for providing a high link margin. *An efficient approach to mitigating these detrimental effects is to adaptively adjust the modulation and/or the channel coding format as well as a range of other system parameters based on the near-instantaneous channel quality information perceived by the receiver, which is fed back to the transmitter with the aid of a feedback channel [15].* This plausible principle was recognised by Hayes [15] as early as 1968.

It was also shown in the previous sections that these near-instantaneously adaptive schemes require a reliable feedback link from the receiver to the transmitter. However, the channel quality variations have to be sufficiently slow for the transmitter to be able to adapt its modulation and/or channel coding format appropriately. The performance of these schemes can potentially be enhanced with the aid of *channel quality prediction techniques [14]*. As an efficient fading counter-measure, Hayes [15] proposed the employment of transmission power adaptation, while *Cavers [9] suggested invoking a variable symbol duration scheme* in response to the perceived channel quality at the expense of a variable bandwidth requirement. A disadvantage of the variable-power scheme is that it increases both the average transmitted power requirements and the level of co-channel interference imposed on other users, while requiring a high-linearity class-A or AB power amplifier, which exhibit a low power-efficiency. As a more attractive alternative, *the employment of AQAM was proposed by Steele and Webb*, which circumvented some of the above-mentioned disadvantages by employing various star-QAM constellations [16, 17].

With the advent of *Pilot Symbol Assisted Modulation (PSAM)* [18–20], Otsuki *et al.* [21] employed square-shaped AQAM constellations instead of star constellations [4], as a practical fading counter measure. With the aid of analysing the channel capacity of Rayleigh fading channels [22], *Goldsmith* et al. *[23] and Alouini* et al. *[24] showed that combined variable-power, variable-rate adaptive schemes are attractive* in terms of approaching the capacity of the channel and characterized the achievable throughput performance of variable-power AQAM [23]. However, they also found that *the extra throughput achieved by the additional variable-power assisted adaptation over the constant-power, variable-rate scheme is marginal* for most types of fading channels [23, 25].

In 1996 Torrance and Hanzo [26] proposed a set of *mode switching levels* s designed for achieving a high average BPS throughput, while maintaining the target average BER. Their method was based on defining a specific combined BPS/BER cost-function for transmission over narrowband Rayleigh channels, which incorporated both the BPS throughput as well as the target average BER of the system. *Powell's optimization was invoked for finding a set of mode switching thresholds, which were constant*, regardless of the actual channel Signal to Noise Ratio (SNR) encountered, i.e. irrespective of the prevalent instantaneous channel conditions. However, in 2001 Choi and Hanzo [27] noted that a *higher BPS throughput can be achieved, if under high channel SNR conditions the activation of high-throughput AQAM modes is further encouraged by lowering the AQAM mode switching thresholds. More explicitly, a set of SNR-dependent AQAM mode switching levels was proposed [27]*, which keeps the average BER constant, while maximising the achievable throughput. We note furthermore that the set of switching levels derived in [26, 28] is based on Powell's multidimensional optimization technique [29] and hence the optimization process may become trapped in a local minimum. This problem was overcome by Choi and Hanzo upon deriving an *optimum set of switching levels [27], when employing the Lagrangian multiplier technique*. It was shown that this set of switching levels results in the global optimum in a sense that the corresponding AQAM scheme obtains the maximum possible average BPS throughput, while maintaining the target average BER. An important further development was Tang's contribution [30] in the area of contriving an *intelligent learning scheme for the appropriate adjustment of the AQAM switching thresholds*.

These contributions demonstrated that *AQAM exhibited promising advantages*, when compared to fixed modulation schemes in terms of spectral efficiency, BER performance and robustness against channel delay spread, etc. Various systems employing AQAM were also characterized in [4]. The *numerical upper bound performance of narrow-band BbB-AQAM* over slow Rayleigh flat-fading channels was evaluated by Torrance and Hanzo [31], while over *wide-band channels* by Wong and Hanzo [32,33]. Following these developments, adaptive modulation was also studied *in conjunction with channel coding and power control techniques by Matsuoka et al. [34] as well as Goldsmith and Chua [35, 36].*

In the early phase of research more emphasis was dedicated to the system aspects of adaptive modulation in a narrow-band environment. A reliable method of transmitting the modulation control parameters was proposed by Otsuki *et al.* [21], where the parameters were embedded in the transmission frame's mid-amble using Walsh codes. Subsequently, at the receiver the Walsh sequences were decoded using maximum likelihood detection. Another technique of *signalling the required modulation mode used was proposed by Torrance and Hanzo [37], where the modulation control symbols were represented by unequal error protection 5-PSK symbols.* Symbol-by-Symbol (SbS) adaptive, *rather than BbB-adaptive systems were proposed by Lau and Maric in [38], where the transmitter is capable of transmitting each symbol in a different modem mode, depending on the channel conditions. Naturally, the receiver has to synchronise with the transmitter in terms of the SbS-adapted mode sequence, in order to correctly demodulate the received symbols and hence the employment of BbB-adaptivity is less challenging, while attaining a similar performance to that of BbB-adaptive arrangements under typical channel conditions.*

The adaptive modulation philosophy was then extended to wideband multi-path environments *amongst others for example by Kamio et al. [39] by utilizing a bi-directional Decision Feedback Equalizer (DFE)* in a micro- and macro-cellular environment. This equalization technique employed both forward and backward oriented channel estimation based on the pre-amble and post-amble symbols in the transmitted frame. Equalizer tap gain interpolation across the transmitted frame was also utilized for reducing the complexity in conjunction with space diversity [39]. The authors concluded that the cell radius could be enlarged in a macro-cellular system and a higher area-spectral efficiency could be attained for micro-cellular environments by utilizing adaptive modulation. The data transmission latency effect, which occurred when the input data rate was higher than the instantaneous transmission throughput was studied and solutions were formulated using *frequency hopping [40] and statistical multiplexing, where the number of Time Division Multiple Access (TDMA) timeslots allocated to a user was adaptively controlled [41].*

In reference [42] *symbol rate adaptive modulation* was applied, where the symbol rate or the number of modulation levels was adapted by using $\frac{1}{8}$-rate 16QAM, $\frac{1}{4}$-rate 16QAM, $\frac{1}{2}$-rate 16QAM as well as full-rate 16QAM and the criterion used for adapting the modem modes was based on the instantaneous received signal to noise ratio and channel delay spread. The slowly varying channel quality of the uplink (UL) and downlink (DL) was rendered similar by utilizing short frame duration Time Division Duplex (TDD) and the maximum normalized delay spread simulated was 0.1. A *variable channel coding rate* was then introduced by Matsuoka *et al.* in conjunction with adaptive modulation in reference [34], where the transmitted burst incorporated an outer Reed Solomon code and an inner convolutional code in order to achieve high-quality data transmission. The coding rate was varied according to the prevalent channel quality using the same method, as in adaptive modulation in order to achieve a certain target BER performance. A so-called *channel margin* was introduced in this contribution, which effectively increased the switching thresholds for the sake of preempting *the effects of channel quality estimation errors*, although this inevitably reduced the achievable BPS throughput.

In an effort to improve the achievable performance versus complexity trade-off in the context of AQAM, Yee and Hanzo [43] studied the design of various *Radial Basis Function (RBF) assisted neural network based schemes, while communicating over dispersive channels.* The advantage of these RBF-aided DFEs is that they are capable of delivering error-free decisions even in scenarios, when the received phasors cannot be error-freely detected by the conventional DFE, since they cannot be separated into decision classes with the aid of a linear decision boundary. In these so-called linearly non-separable decision scenarios the RBF-assisted DFE still may remain capable of classifying the received phasors into decision classes without decision errors. A further improved turbo BCH-coded version of this RBF-aided system was characterized by Yee *et al.* in [44], while a turbo-equalised RBF arrangement was the subject of the investigation conducted by Yee, Liew and Hanzo in [45, 46]. The RBF-aided AQAM research has also been extended to the turbo equalization of a convolutional as well as space-time trellis coded arrangement proposed by Yee, Yeap and Hanzo [47, 48]. The same authors then endeavoured to reduce the associated implementation complexity of an RBF-aided QAM modem with the advent of employing a separate in-phase / quadrature-phase turbo equalization scheme in the quadrature arms of the modem.

As already mentioned above, the performance of *channel coding in conjunction with adaptive modulation in a narrow-band environment* was also characterized by Chua and Goldsmith [35]. In their contribution trellis and lattice codes were used without channel interleaving, invoking a feedback path between the transmitter and receiver for modem mode control purposes. Specifically, the simulation and theoretical results by Goldsmith and Chua showed that a 3dB coding gain was achievable at a BER of 10^{-6} for a 4-sate trellis code and 4dB by an 8-state trellis code in the context of the adaptive scheme over Rayleigh-fading channels, while a 128-state code performed within 5dB of the Shannonian capacity limit. The *effects of the delay in the AQAM mode signalling feedback path* on the adaptive modem's performance were studied and this scheme exhibited a higher spectral efficiency, when compared to the non-adaptive trellis coded performance. Goeckel [49] also contributed in the area of adaptive coding and employed realistic outdated, rather than perfect fading estimates. Further research on adaptive multidimensional coded modulation was also conducted by Hole *et al.* [50] for transmissions over flat fading channels. *Pearce, Burr and Tozer [51] as well as Lau and Mcleod [52] have also analysed the performance trade-offs associated with employing channel coding and adaptive modulation or adaptive trellis coding,* respectively, as efficient fading counter measures. In an effort to provide a fair comparison of the various coded modulation schemes known at the time of writing, Ng, Wong and Hanzo have also studied Trellis Coded Modulation (TCM), Turbo TCM (TTCM), Bit-Interleaved Coded Modulation (BICM) and Iterative-Decoding assisted BICM (BICM-ID), where TTCM was found to be the best scheme at a given decoding complexity [53].

Subsequent contributions by Suzuki *et al.* [54] incorporated *space-diversity and power-adaptation* in conjunction with adaptive modulation, for example in order to combat the effects of the multi-path channel environment at a 10Mbits/s transmission rate. *The maximum tolerable delay-spread was deemed to be one symbol duration for a target mean BER performance of* 0.1%. This was achieved in a TDMA scenario, where the channel estimates were predicted based on the extrapolation of previous channel quality estimates. As mentioned above, variable transmitted power was applied in combination with adaptive modulation in reference [36], where the transmission rate and power adaptation was optimized for the sake of achieving an increased spectral efficiency. In their treatise a slowly varying channel was assumed and the instantaneous received power required for achieving a certain upper bound performance was assumed to be known prior to transmission. *Power control in conjunction with a pre-distortion type non-linear power amplifier compensator* was studied in the context of adaptive modulation in reference [55]. This method was used to mitigate the non-linearity effects associated with the power amplifier, when QAM modulators were used.

Results were also recorded concerning the performance of adaptive modulation in conjunction with *different multiple access schemes in a narrow-band channel environment.* In a TDMA system, *dynamic channel assignment* was employed by Ikeda *et al.*, where in addition to assigning a different modulation mode to a different channel quality, priority was always given to those users in their request for reserving time-slots, which benefitted from the best channel quality [56]. The performance was compared to fixed channel assignment systems, where substantial gains were achieved in terms of system capacity. Furthermore, a *lower call termination probability was recorded.* However, the probability of intra-cell hand-off increased as a result of the associated dynamic channel assignment (DCA) scheme, which constantly searched for a high-quality, high-throughput time-slot for supporting the actively communicating users. The application of adaptive modulation in packet transmission was introduced by Ue, Sampei and Morinaga [57], where the results showed an improved BPS throughput. The performance of adaptive modulation was also characterized in conjunction with an *automatic repeat request* (ARQ) system in reference [58], where the transmitted bits were encoded using a cyclic redundant code (CRC) and a convolutional punctured code in order to increase the data throughput.

A further treatise was published by Sampei, Morinaga and Hamaguchi [59] on *laboratory test results* concerning the utilization of adaptive modulation in a TDD scenario, where the modem mode switching criterion was based on the signal to noise ratio and on the normalized delay-spread. *In these experimental results, the channel quality estimation errors degraded the performance* and consequently - as laready alluded to earlier - a channel estimation error margin was introduced for mitigating this degradation. Explicitly, the channel estimation error margin was defined as the measure of how much extra protection margin must be added to the switching threshold levels for the sake of minimising the effects of the channel estimation errors. The delay-spread also degraded the achievable performance due to the associated irreducible BER, which was not compensated by the receiver. However, the performance of the adaptive scheme in a delay-spread impaired channel environment was better, than that of a fixed modulation scheme. *These experiments also concluded that the AQAM scheme can be operated for a Doppler frequency of $f_d = 10Hz$ at a normalized delay spread of 0.1 or for $f_d = 14Hz$ at a normalized delay spread of 0.02, which produced a mean BER of 0.1% at a transmission rate of 1 Mbits/s.*

Lastly, the *data buffering-induced latency and co-channel interference aspects* of AQAM modems were investigated in [60, 61]. Specifically, the latency associated with storing the information to be transmitted during severely degraded channel conditions was mitigated by frequency hopping or statistical multiplexing. As expected, the latency is increased, when either the mobile speed or the channel SNR are reduced, since both of these result in prolonged low instantaneous SNR intervals. It was demonstrated that as a result of the proposed measures, typically more than 4dB SNR reduction was achieved by the proposed adaptive modems in comparison to the conventional fixed-mode benchmark modems employed. However, the achievable gains depend strongly on the prevalant co-channel interference levels and hence interference cancellation was invoked in [61] on the basis of adjusting the demodulation decision boundaries after estimating the interfering channel's magnitude and phase.

The associated principles can also be invoked in the context of *multicarrier Orthogonal Frequency Division Multiplex (OFDM) modems [4]*. This principle was first proposed by Kalet [62] and was then further developed for example by Czylwik *et al.* [63] as well as by Chow, Cioffi and Bingham [64]. The associated concepts were detailed for example in [4] and will be also augmented in this monograph. Let us now briefly review the recent history of the BbB adaptive concept in the context of CDMA in the next section.

1.5.2   Adaptive Code Division Multiple Access

The techniques described in the context of single- and multi-carrier modulation are conceptually similar to multi-rate transmission [65] in CDMA systems. However, in BbB adaptive CDMA the transmission rate is modified according to the near-instantaneous channel quality, instead of the service required by the mobile user. BbB-adaptive CDMA systems are also useful for employment in arbitrary propagation environments or in hand-over scenarios, such as those encountered, when a mobile user moves from an indoor to an outdoor environment or in a so-called 'birth-death' scenario, where the number of transmitting CDMA users changes frequently [66], thereby changing the interference dramatically. Various methods of multi-rate transmission have been proposed in the research literature. Below we will briefly discuss some of the recent research issues in multi-rate and adaptive CDMA schemes.

Ottosson and Svensson compared various multi-rate systems [65], including multiple spreading factor (SF) based, multi-code and multi-level modulation schemes. According to the multi-code philosophy, the SF is kept constant for all users, but multiple spreading codes transmitted simultaneously are assigned to users requiring higher bit rates. In this case - unless the spreading codes's perfect orthogonality is retained after transmission over the channel - the multiple codes of a particular user interfere with each other. This inevitably reduces the system's performance.

Multiple data rates can also be supported by a variable SF scheme, where the chip rate is kept constant, but the data rates are varied, thereby effectively changing the SF of the spreading codes assigned to the users; at a fixed chip rate the lower the SF, the higher the supported data rate. Performance comparisons for both of these schemes have been carried out by Ottosson and Svensson [65], as well as by Ramakrishna and Holtzman [67], demonstrating that both schemes achieved a similar performance. Adachi, Ohno, Higashi, Dohi and Okumura proposed the employment of multi-code CDMA in conjunction with pilot symbol-assisted channel estimation, RAKE reception and antenna diversity for providing multi-rate capabilities [68, 69]. The employment of multi-level modulation schemes was also investigated by Ottosson and Svensson [65], where higher-rate users were assigned higher-order modulation modes, transmitting several bits per symbol. However, it was concluded that the performance experienced by users requiring higher rates was significantly worse, than that experienced by the lower-rate users. The use of M-ary orthogonal modulation in providing variable rate transmission was investigated by Schotten, Elders-Boll and Busboom [70]. According to this method, each user was assigned an orthogonal sequence set, where the number of sequences, M, in the set was dependent on the data rate required – the higher the rate required, the larger the sequence set. Each sequence in the set was mapped to a particular combination of $b = (\log_2 M)$ bits to be transmitted. The M-ary sequence was then spread with the aid of a spreading code of a constant SF before transmission. It was found [70] that the performance of the system depended not only on the MAI, but also on the Hamming distance between the sequences in the M-ary sequence set.

Saquib and Yates [71] investigated the employment of the decorrelating detector in conjunction with the multiple-SF scheme and proposed a modified decorrelating detector, which utilized soft decisions and maximal ratio combining, in order to detect the bits of the different-rate users. Multi-rate transmission schemes involving interference cancellation receivers have previously been investigated amongst others by Johansson and Svensson [72, 73], as well as by Juntti [74]. Typically, multiple users transmitting at different bit rates are supported in the same CDMA system invoking multiple codes or different spreading factors. SIC schemes and multi-stage cancellation schemes were used at the receiver for mitigating the MAI [72–74], where the bit rate of the users was dictated by the user requirements. The performance comparison of various multiuser detectors in the context of a multiple-SF transmission scheme was presented for example by Juntti [74], where the detectors compared were the decorrelator, the PIC receiver and the so-called group serial interference cancellation (GSIC) receiver. It was concluded that the GSIC and the decorrelator performed better than the PIC receiver, but all the interference cancellation schemes including the GSIC, exhibited an error floor at high SNRs due to error propagation.

The bit rate of each user can also be adapted according to the near-instantaneous channel quality, in order to mitigate the effects of channel quality fluctuations. Kim [75] analysed the performance of two different methods of combating the near-instantaneous quality variations of the mobile channel. Specifically, Kim studied the adaptation of the transmitter power or the switching of the information rate, in order to suit the near-instantaneous channel conditions. Using a RAKE receiver [76], it was demonstrated that rate adaptation provided a higher average information rate, than power adaptation for a given average transmit power and a given BER [75]. Abeta, Sampei and Morinaga [77] conducted investigations into an adaptive packet transmission based CDMA scheme, where the transmission rate was modified by varying the channel code rate and the processing gain of the CDMA user, employing the carrier to interference plus noise ratio (CINR) as the switching metric. When the channel quality was favourable, the instantaneous bit rate was increased and conversely, the instantaneous bit rate was reduced when the channel quality dropped. In order to maintain a constant overall bit rate, when a high instantaneous bit rate was employed, the duration of the transmission burst was reduced. Conversely, when the instantaneous bit rate was low, the duration of the burst was extended. This resulted in a decrease in interference power, which translated to an increase in system capacity. Hashimoto, Sampei and Morinaga [78] extended this work also to demonstrate that the proposed system was capable of achieving a higher user capacity with a reduced hand-off margin and lower average transmitter power. In these schemes the conventional RAKE receiver [76] was used for the detection of the data symbols. A variable-rate CDMA scheme – where the transmission rate was modified by varying the channel code rate and, correspondingly, the M-ary modulation constellations – was investigated by Lau and Maric [38]. As the channel code rate was increased, the bit-rate was increased by increasing M correspondingly in the M-ary modulation scheme. Another adaptive system was proposed by Tateesh, Atungsiri and Kondoz [79], where the rates of the speech and channel codecs were varied adaptively [79]. In their adaptive system, the gross transmitted bit rate was kept constant, but the speech codec and channel codec rates were varied according to the channel quality. When the channel quality was low, a lower rate speech codec was used, resulting in increased redundancy and thus a more powerful channel code could be employed. This resulted in an overall coding gain, although the speech quality dropped

with decreasing speech rate. A variable rate data transmission scheme was proposed by Okumura and Adachi [80], where the fluctuating transmission rate was mapped to discontinuous
transmission, in order to reduce the interference inflicted upon the other users, when there
was no transmission. The transmission rate was detected blindly at the receiver with the
help of cyclic redundancy check decoding and RAKE receivers were employed for coherent
reception, where pilot-symbol-assisted channel estimation was performed.

The information rate can also be varied in accordance with the channel quality, as it will
be demonstrated shortly. However, in comparison to conventional power control techniques
- which again, may disadvantage other users in an effort to maintain the quality of the links
considered - the proposed technique does not disadvantage other users and increases the
network capacity [81]. The instantaneous channel quality can be estimated at the receiver
and the chosen information rate can then be communicated to the transmitter via explicit
signalling in a so-called closed-loop controlled scheme. Conversely, in an open-loop scheme
- provided that the downlink and uplink channels exhibit a similar quality - the information
rate for the downlink transmission can be chosen according to the channel quality estimate
related to the uplink and vice versa. The validity of the above channel reciprocity issues in
TDD-CDMA systems have been investigated by Miya *et al.* [82], Kato *et al.* [83] and Jeong
et al. [84].

1.6   Outline of the book

In order to mitigate the impact of dispersive multi-path fading channels, equalization techniques are introduced, which are subsequently incorporated in a wideband adaptive modulation scheme. The performance of various wideband adaptive transmission scheme was then
analysed in different environments, resulting in the following outline:

- **Chapter 2**: Square Quadrature Amplitude Modulation (QAM) schemes are introduced and their corresponding performance is analysed over Gaussian and narrow-
  band Rayleigh fading channels. This is followed by an introduction to equalization
  techniques with an emphasis on the Minimum Mean Square Error (MMSE) Decision
  Feedback Equalizer (DFE). The performance of the DFE is then characterized using
  BPSK, 4QAM, 16QAM and 64QAM modems.

- **Chapter 3**: The recursive Kalman algorithm is formulated and employed in an adaptive channel estimator and adaptive DFE in order to combat the time-variant dispersion
  of the mobile propagation channel. In this respect, the system parameters of the algorithm are optimized for each application by evaluating the convergence speed of the
  algorithm. Finally, two receiver structures utilizing the adaptive channel estimator and
  DFE are compared.

- **Chapter 4**: The concept of AQAM is introduced, where the modulation mode is
  adapted based on the prevalent channel conditions. Power control is then implemented
  and analysed in conjunction with AQAM in a narrow-band environment. Subsequently,
  a wideband AQAM scheme - which incorporates the DFE - is jointly constructed in
  order to mitigate the effects of the dispersive multi-path fading channel. A numerical
  upper bound performance is derived for this wideband AQAM scheme, which is subsequently optimized for a certain target BER and transmission throughput performance.

Lastly, a comparison is made between the constituent fixed or time-invariant modulation modes and the wideband AQAM scheme in terms of their transmission throughput performance.

- **Chapter 5**: The performance of the wideband channel coded AQAM scheme is presented and analysed. Explicitly, turbo coding techniques are invoked, where each modulation mode was associated with a certain code rate and turbo interleaver size. Consequently, an adaptive code rate scheme is incorporated into the wideband AQAM scheme. The performance of such a scheme is compared to the constituent fixed modulation modes as well as the uncoded AQAM scheme, which was presented in Chapter 3. Furthermore, the concept of turbo equalization is introduced and applied in a wideband AQAM scheme. The iterative nature of the turbo equalizer is also exploited in estimating the channel impulse response (CIR). The chapter is concluded with a comparative study of various joint coding and adaptive modulation schemes, including Trellis Coded Modulation (TCM), turbo TCM (TTCM), Bit Interleaved Coded Modulation (BICM) and its iteratively detected (ID) version, namely BICM-ID.

- **Chapter 6**: Closed form expressions were derived for the average BER, the average BPS throughput and the mode selection probability of various adaptive modulation schemes, which were shown to be dependent on the mode-switching levels as well as on the average SNR experienced. Furthermore, a range of techniques devised for determining the adaptive mode-switching levels are studied comparatively. The optimum switching levels achieving the highest possible BPS throughput while maintaining the average target BER were developed based on the Lagrangian optimization method. The chapter is concluded with a brief comparison of space-time coding and adaptive modulation in the context of OFDM and MC-CDMA.

- **Chapter 7**: This chapter presents the practical aspects of implementing wideband AQAM schemes, which includes the effects of error propagation inflicted by the DFE and the more detrimental channel quality estimation latency impact of the scheme. The impact of latency is studied under different system delay and normalized Doppler frequencies. The impact of Co-Channel Interference (CCI) on the wideband AQAM scheme is also analysed. In this aspect, joint detection techniques and a more sophisticated switching regime is utilized, in order to mitigate the impact of CCI.

- In **Chapter 8** we cast channel equalization as a classification problem. We briefly give an overview of neural network and present the design of some neural network based equalizers. In this chapter we opted for studying a neural network structure referred to as the Radial Basis Function (RBF) network in more detail for channel equalization, since it has an equivalent structure to the so-called optimal Bayesian equalization solution [85]. The structure and properties of the RBF network is described, followed by the implementation of a RBF network as an equalizer. We will discuss the computational complexity issues of the RBF equalizer with respect to that of conventional linear equalizers and provide some complexity reduction methods. Finally, performance comparisons between the RBF equalizer and the conventional equalizer are given over various channel scenarios.

- **Chapter 9** commences by summarising the concept of adaptive modulation that adapts the modem mode according to the channel quality in order to maintain a certain target bit error rate and an improved bits per symbol throughput performance. The RBF based equalizer is introduced in a wideband Adaptive Quadrature Amplitude Modulation (AQAM) scheme in order to mitigate the effects of the dispersive multipath fading channel. We introduce the short-term Bit Error Rate (BER) as the channel quality measure. Lastly, a comparative study is conducted between the constituent fixed mode, the conventional DFE based AQAM scheme and the RBF based AQAM scheme in terms of their BER and throughput performance.

- In **Chapter 10** we incorporate turbo channel coding in the proposed wideband AQAM scheme. A novel reduced-complexity RBF equalizer utilizing the so-called Jacobian logarithmic relationship [44] is proposed and the turbo-coded performance of the Jacobian RBF equalizer is presented for the various fixed QAM modes. Furthermore, we investigate using various channel quality measures – namely the short-term BER and the average Log-Likelihood Ratio (LLR) magnitude of the data burst generated either by the RBF equalizer or the turbo decoder – in order to control the modem mode-switching regime for our adaptive scheme.

- **Chapter 11** introduces the principles of iterative, joint equalization and decoding techniques known as turbo equalization. We present a novel turbo equalization scheme, which employs a RBF equalizer instead of the conventional trellis-based equalizer. The structure and computational complexity of both the RBF equalizer and trellis-based equalizer are compared and we characterize the performance of these RBF and trellis-based turbo-equalizers. We then propose a reduced-complexity RBF assisted turbo equalizer, which exploits the fact that the RBF equalizer computes its output on a symbol-by-symbol basis and the symbols of the decoded transmission burst, which are sufficiently reliable need not be equalised in the next turbo equalization iteration. This chapter is concluded with the portrayal and characterization of RBF-based turbo equalised space-time coded schemes.

- In **Chapter 12** the recent history of smart CDMA MUDs is reviewed and the most promising schemes have been comparatively studied, in order to assist in the design of third- and fourth-generation receivers. Future transceivers may become BbB-adaptive, in order to be able to accommodate the associated channel quality fluctuations without disadvantageously affecting the system's capacity. Hence the methods reviewed in this chapter are advantageous, since they often assist in avoiding powering up, which may inflict increased levels of co-channel interference and power consumption. Furthermore, the techniques characterized in the chapter support an increased throughput within a given bandwidth and will contribute towards reducing the constantly increasing demand for more bandwidth. Both successive interference cancellation (SIC) and Parallel Interference Cancellation (PIC) receivers are investigated in the context of AQAM/CDMA schemes, along with joint-detection assisted schemes.

- In **Chapter 13** we provide a brief historical perspective on Orthogonal Frequency Division Multiplex (OFDM) transmissions with reference to the literature of the past 30 years. The advantages and disadvantages of various OFDM techniques are considered briefly and the expected performance is characterized for the sake of illustration in the context of indoor wireless systems. Our discussions will deepen, as we approach the subject of adaptive subcarrier modem mode allocation and turbo channel coding. Our motivation is that of quantifying the performance benefits of employing adaptive channel coded OFDM modems.

- In **Chapter 14** we provide an introduction to the subject of space-time coding combined with adaptive modulation and various channel coding techniques. A performance study is conducted in the context of both fixed-mode and adaptive modulation schemes, when communicating over dispersive wideband channels. We will demonstrate that in conjunction with space-time coding the advantages of employing adaptive modulation erode, since the associated multiple transmitter, multiple receiver assisted diversity scheme efficiently mitigates the channel quality fluctuations of the wireless channel.

∗   ∗

Having reviewed the historical developments in the field of AQAM, in the rest of this monograph we will consider wideband AQAM assisted single- and multi-carrier, as well as CDMA transceivers, communicating over dispersive wideband channels. We will also demonstrate that the potential performance gains attained by AQAM erode, as the diversity order of the systems is increased, although this is achieved at the cost of an increased complexity. We will demonstrate that this is particularly true in conjunction with space-time coding assisted transmitter diversity, since Multiple-Input, Multiple-Output (MIMO) systems substantially mitigate the effects of channel quality fluctuations. Hence if the added complexity of MIMOs has to be avoided, BbB-adaptive transceivers constitute powerful wideband fading counter-measures. By contrast, there is no need for the employment of BbB-adaptive transceivers, if the higher complexity of MIMOs is affordable, since MIMOs substantially mitigate the effects of channel quality fluctuations, rendering further fading counter-measures superfluous.

L. Hanzo, C.H. Wong, M.S. Yee
Department of Electronics and Computer Science
University of Southampton

Part I

Near-instantaneously Adaptive Modulation and Filtering Based Equalisation

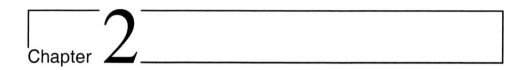

Chapter 2

Introduction To Equalizers

In most digital data transmission systems the dispersive linear channel encountered exhibits amplitude and phase distortion. As a result, the received signal is contaminated by Intersymbol Interference (ISI). In a system, which transmits a sequence of pulse-shaped information symbols, the time domain full response signalling pulses are smeared by the hostile dispersive channel, resulting in intersymbol interference. At the receiver, the linearly distorted signal has to be equalized in order to recover the information.

The equalizers that are utilized to compensate for the ISI can be classified according to their structure, the optimising criterion and the algorithms used to adapt the equalizer coefficients, which are summarized in Figure 2.1. On the basis of their structures, the equalizers can be classified as linear or decision feedback equalizers. Each of these structures will be discussed at a later stage with more emphasis on the Decision Feedback Equalizer (DFE).

Equalizers can also be distinguished on the basis of the criterion used to optimise their coefficients. The optimization is governed by the performance criteria used. For example, when applying the mean square error criterion (MSE), the equalizer is optimised such that the mean squared error between the distorted signal and the actual transmitted signal is minimized. Various optimization criteria will be discussed in the following sections, with more emphasis on the MSE criterion.

A range of adaptive algorithms can be invoked, in order to provide the equalizer the means of adapting its coefficients to the time-varying dispersive channels. We will not elaborate on the fine details of these algorithms until the next chapter, where attention will be given to the well known Least Squares (LS) algorithms, in particular to the Recursive Kalman Algorithm [86].

Before we proceed to highlight the different techniques of channel equalization, we will introduce the multilevel modulation schemes that are utilized throughout this treatise. This is necessary in order to study the performance of the equalizers in multilevel modulation environment, since the choice of the modulation mode will affect the performance of the equalizer.

Equalizer Structures

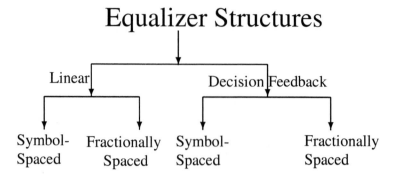

Optimizing Criteria

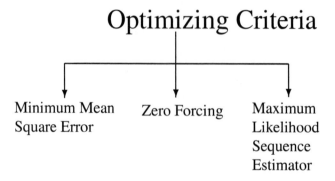

Adaptive Algorithms

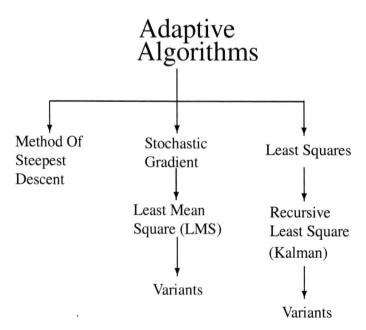

Figure 2.1: Classification of conventional equalizers on the basis of their structure, optimization criteria and coefficient adaptation algorithms.

2.1 Coherent Demodulation of Square-Constellation Quadrature Amplitude Modulation

In this section, various multilevel modulation schemes are presented and their performances are analysed in a Gaussian environment. The well known shaped Square Quadrature Amplitude Modulation (QAM) schemes are defined mainly by their phasor constellation arrangements. Throughout this treatise, the modulation schemes used are BPSK, 4QAM, 16QAM and 64QAM, representing the square-shaped constellations associated with one, two, four and six bits per symbol. The corresponding square-shaped phasor constellations are depicted in Figure 2.2 with their corresponding assigned bit sequences.

The phasor constellations shown in Figure 2.2 provide the location of each constellation point in terms of their in-phase (I) and quadrature-phase (Q) components, where each point is assigned a particular bit sequence. Gray-coding is applied in order to assign the bit sequence to their respective constellation points, ensuring that the nearest neighbour constellation points have a Hamming distance of one. Thereby the bit assignments are optimised in terms of minimising the Bit Error Rate (BER).

Gray-coding of the bit sequence gives rise to the formation of a number of different integrity subchannels [4]. These subchannels are formed in order to distinguish the different level of noise protection experienced by the bits in different subchannels. In the QAM constellations mentioned above, each subchannel consists of two bits. Thus for 16QAM, there are two different integrity subchannels, labelled as C1 and C2, where the C1 subchannel possesses a higher noise protection distance than the C2 subchannel. The same applies to the 64QAM mode having three subchannels. For a more in-depth understanding of QAM techniques, the reader is referred to Hanzo *et al.* [4].

2.1.1 Performance of Quadrature Amplitude Modulation in Gaussian Channels

The BER performance of QAM in a Gaussian environment is presented below, both theoretically and using simulations. The simulation results presented here assumed coherent detection and perfect timing synchronisation between the transmitter and receiver.

2.1.2 Bit Error Rate Performance in Gaussian Channels

The theoretical solutions for the BER performance of QAM in Gaussian channels have been quantified and published. According to Proakis [87], the closed form theoretical solution to the BPSK and 4QAM BER performance, P_{BPSK}^b and P_{4QAM}^b, respectively, are

$$P_{\text{BPSK}}^b = Q(\sqrt{2.\bar{\gamma}}), \tag{2.1}$$

$$P_{\text{4QAM}}^b = Q(\sqrt{\bar{\gamma}}), \tag{2.2}$$

where $\bar{\gamma}$ in the above equations represents the average channel signal to noise ratio (SNR) and the Q function is defined by Proakis as [87]

$$Q(x) = \frac{1}{\sqrt{2\pi}} \int_x^\infty e^{-y^2/2} \ dy. \tag{2.3}$$

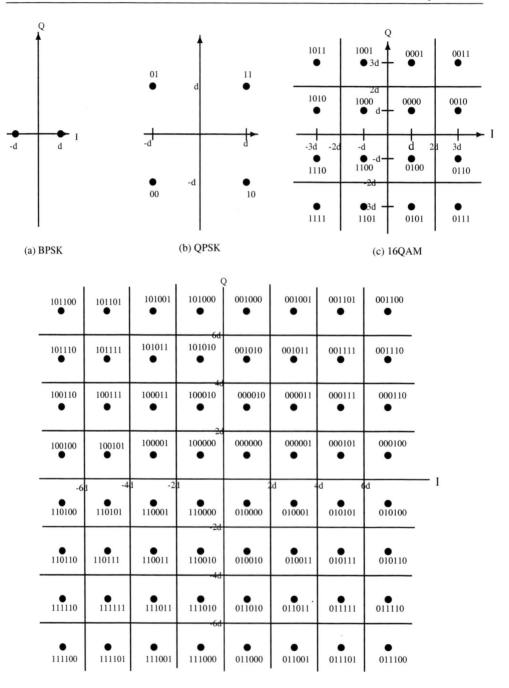

(a) BPSK   (b) QPSK   (c) 16QAM

(d) 64QAM

───── DECISION BOUNDARIES

Figure 2.2: QAM phasor constellations.

Following the approach outlined by Hanzo et al. [4], the closed form solution of the 16QAM and 64QAM modulation mode in terms of its individual subchannel BER performance can be written as :

$$P^b_{C1_{16QAM}} = Q\left(\sqrt{\frac{\bar{\gamma}}{5}}\right) \tag{2.4}$$

$$P^b_{C2_{16QAM}} = Q\left(3.\sqrt{\frac{\bar{\gamma}}{5}}\right) \tag{2.5}$$

$$P^b_{av_{16QAM}} = \frac{1}{2}(P^b_{C1_{16QAM}} + P^b_{C2_{16QAM}}) \tag{2.6}$$

$$P^b_{C1_{64QAM}} = \frac{1}{4}Q\left(\sqrt{\frac{\bar{\gamma}}{21}}\right) + \frac{1}{4}Q\left(3.\sqrt{\frac{\bar{\gamma}}{21}}\right) + \frac{1}{4}Q\left(5.\sqrt{\frac{\bar{\gamma}}{21}}\right) + \frac{1}{4}Q\left(7.\sqrt{\frac{\bar{\gamma}}{21}}\right) \tag{2.7}$$

$$P^b_{C2_{64QAM}} = \frac{1}{2}Q\left(\sqrt{\frac{\bar{\gamma}}{21}}\right) + \frac{1}{2}Q\left(3.\sqrt{\frac{\bar{\gamma}}{21}}\right) + \frac{1}{4}Q\left(5.\sqrt{\frac{\bar{\gamma}}{21}}\right) + \frac{1}{4}Q\left(7.\sqrt{\frac{\bar{\gamma}}{21}}\right) \tag{2.8}$$

$$P^b_{C3_{64QAM}} = Q\left(\sqrt{\frac{\bar{\gamma}}{21}}\right) + \frac{3}{4}Q\left(3.\sqrt{\frac{\bar{\gamma}}{21}}\right) - \frac{3}{4}Q\left(5.\sqrt{\frac{\bar{\gamma}}{21}}\right) - \frac{1}{2}Q\left(7.\sqrt{\frac{\bar{\gamma}}{21}}\right)$$
$$+ \frac{1}{2}Q\left(9.\sqrt{\frac{\bar{\gamma}}{21}}\right) + \frac{1}{4}Q\left(11.\sqrt{\frac{\bar{\gamma}}{21}}\right) - \frac{1}{4}Q\left(11.\sqrt{\frac{\bar{\gamma}}{21}}\right) \tag{2.9}$$

$$P^b_{av_{64QAM}} = \frac{1}{3}(P^b_{C1_{64QAM}} + P^b_{C2_{64QAM}} + P^b_{C3_{64QAM}}). \tag{2.10}$$

The notations $P^b_{Cx_{yQAM}}$ represent the BER of subchannel x of the y-point QAM, while $P^b_{av_{yQAM}}$ denotes the average BER of y-point QAM, assuming that the probability of transmission over any of the subchannels is equal. The theoretical BERs from the expressions above were compared to the experimental BERs and the results are presented graphically in Figure 2.3. The term - experimental - will be used throughout our discourse in the context of results generated by computer simulation.

In this section, the performance results of coherent QAM in a non-dispersive Gaussian channel were presented. In the next section, the performance of the QAM modems in a Rayleigh flat fading environment is evaluated by assuming perfect channel estimation and the employment of a matched filter at the receiver.

2.1.3   Bit Error Rate Performance in a Rayleigh Flat Fading Environment

In this section the performance of the multilevel modulation schemes outlined in Section 2.1 is presented and analysed in a Rayleigh flat fading environment [88]. Theoretical closed form solutions are presented and compared with the experimental results for each multilevel scheme utilized. The theoretical solution for BPSK followed the approach outlined by

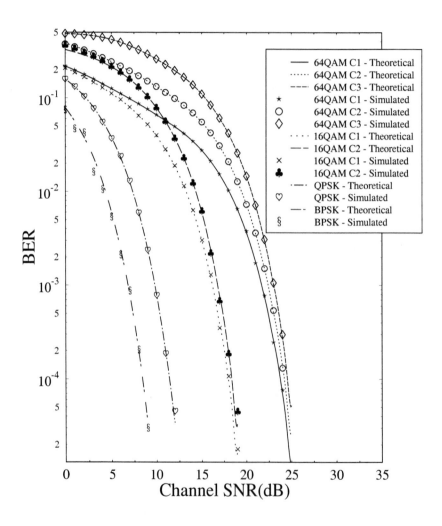

Figure 2.3: Theoretical and simulated BER performance of a BPSK, 4QAM, 16QAM and 64QAM over a Gaussian channel.

Proakis [87] , where perfect channel estimation and the employment of a matched filter [89] was assumed at the receiver. Since the matched filter was matched to the channel variations for all symbols transmitted, its transfer function $K^{MF}(z)$, can be written as:

$$K^{MF}(z) = \sum_{i=0}^{i=L_c-1} h_i^* z^{-i}, \tag{2.11}$$

where $*$ denotes complex conjugation and for a single-path flat Rayleigh fading channel, we have:

$$K^{MF}(z) = h_0^*. \tag{2.12}$$

Here h_0 is the single-path Channel Impulse Response (CIR) in a Rayleigh fading environment, which can be represented by Proakis [87] :

$$h_0 = \varepsilon e^{-j\theta}, \tag{2.13}$$

where ε represents the Rayleigh-distributed attenuation factor for the received signal and θ denotes the random phase variation, which is distributed uniformly between π and $-\pi$.

For a fixed attenuation ε, we can express the instantaneous SNR γ, at the receiver as $\gamma = \varepsilon^2 \gamma_G$, where γ_G is the channel SNR in a Gaussian environment. Upon utilizing Equation 2.1, we can characterize the BER performance of BPSK, assuming a fixed attenuation ε as [87]:

$$P_{\text{BPSK}}^b(\bar{\gamma}) = Q(\sqrt{2\gamma}). \tag{2.14}$$

For a random attenuation ε, we can obtain the theoretical BER performance by integrating $P_{BPSK}^b(\gamma)$ over the probability density function (PDF) of γ, $p(\gamma)$ for all possible values of γ, yielding :

$$P_{\text{BPSK}_{\text{FR}}}^b(\bar{\gamma}) = \int_0^\infty P_{\text{BPSK}}^b(\gamma) p(\gamma) d\gamma, \tag{2.15}$$

where $p(\gamma)$ can be expressed for Rayleigh fading as [87]:

$$p(\gamma) = \frac{1}{\bar{\gamma}} e^{-\frac{\gamma}{\bar{\gamma}}}, \tag{2.16}$$

where $\bar{\gamma}$, the average received SNR can be written as :

$$\bar{\gamma} = \gamma E(\varepsilon^2). \tag{2.17}$$

Now, by exploiting that $Q(x) = 0.5(1 - \text{erf}(\frac{x}{\sqrt{2}}))$ and employing Equation 2.16 and 2.14, we rewrite Equation 2.15 as :

$$P_{\text{BPSK}_{\text{FR}}}^b(\bar{\gamma}) = \frac{1}{2\bar{\gamma}} \int_0^\infty e^{-\frac{\gamma}{\bar{\gamma}}} (1 - \text{erf}(\sqrt{\gamma}) d\gamma. \tag{2.18}$$

We can exploit the fact stated by Gradshteyn *et al.* [90]:

$$\int_0^\infty e^{bx}(1 - \text{erf}(\sqrt{ax}))dx = \frac{1}{b}\left[\sqrt{\frac{a}{a-b}} - 1\right].$$ (2.19)

Upon comparing Equation 2.19 and 2.18 we can assign values for a and b in Equation 2.19 as :

$$a = 1, \quad b = -\frac{1}{\bar{\gamma}}.$$ (2.20)

Finally, by substituting Equation 2.20 into Equation 2.19, we arrive at the analytical bit error probability of BPSK in a Rayleigh flat fading environment :

$$P^b_{\text{BPSK}_{\text{FR}}}(\bar{\gamma}) = 0.5\left(1 - \sqrt{\frac{\bar{\gamma}}{1+\bar{\gamma}}}\right).$$ (2.21)

The same derivation can be extended to the higher-order modulation schemes discussed in Section 2.1. Therefore the theoretical BER performance of 4QAM, 16QAM and 64QAM over flat Rayleigh-fading channels can be written as :

$$P^b_{\text{4QAM}_{\text{FR}}}(\bar{\gamma}) = 0.5(1 - \sqrt{\frac{\bar{\gamma}}{2+\bar{\gamma}}})$$ (2.22)

$$P^b_{\text{16QAM}_{\text{FR}}}(\bar{\gamma}) = 0.5\left[0.5(1 - \sqrt{\frac{\bar{\gamma}}{10+\bar{\gamma}}}) + 0.5(1 - \sqrt{\frac{9\bar{\gamma}}{10+9\bar{\gamma}}})\right]$$ (2.23)

$$P^b_{\text{64QAM}_{\text{FR}}}(\bar{\gamma}) = \frac{1}{3}(P^b_{\text{64QAM}_1} + P^b_{\text{64QAM}_2} + P^b_{\text{64QAM}_3}),$$ (2.24)

where $P^b_{\text{64QAM}_1}$, $P^b_{\text{64QAM}_2}$ and $P^b_{\text{64QAM}_3}$ are defined as :

$$P^b_{\text{64QAM}_1} = \frac{1}{4}\left[0.5(1 - \sqrt{\frac{\bar{\gamma}}{42+\bar{\gamma}}}) + 0.5(1 - \sqrt{\frac{9\bar{\gamma}}{42+9\bar{\gamma}}})\right.$$
$$\left. +0.5(1 - \sqrt{\frac{25\bar{\gamma}}{42+25\bar{\gamma}}}) + 0.5(1 - \sqrt{\frac{49\bar{\gamma}}{42+49\bar{\gamma}}})\right]$$ (2.25)

$$P^b_{\text{64QAM}_2} = \frac{1}{2}\left[0.5(1 - \sqrt{\frac{\bar{\gamma}}{42+\bar{\gamma}}}) + 0.5(1 - \sqrt{\frac{9\bar{\gamma}}{42+9\bar{\gamma}}})\right.$$
$$\left. +0.25(1 - \sqrt{\frac{25\bar{\gamma}}{42+25\bar{\gamma}}}) + 0.25(1 - \sqrt{\frac{49\bar{\gamma}}{42+49\bar{\gamma}}})\right]$$ (2.26)

$$P^b_{64QAM_3} = \frac{1}{2}\left[(1 - \sqrt{\frac{\bar{\gamma}}{42 + \bar{\gamma}}}) + 0.75(1 - \sqrt{\frac{9\bar{\gamma}}{42 + 9\bar{\gamma}}})\right.$$

$$-0.75(1 - \sqrt{\frac{25\bar{\gamma}}{42 + 25\bar{\gamma}}}) - 0.5(1 - \sqrt{\frac{49\bar{\gamma}}{42 + 49\bar{\gamma}}})$$

$$+0.5(1 - \sqrt{\frac{81\bar{\gamma}}{42 + 81\bar{\gamma}}}) + 0.25(1 - \sqrt{\frac{121\bar{\gamma}}{42 + 121\bar{\gamma}}}) -$$

$$\left. 0.25(1 - \sqrt{\frac{169\bar{\gamma}}{42 + 169\bar{\gamma}}})\right]. \tag{2.27}$$

The results of the theoretical Rayleigh flat fading solutions and the experimental simulations are shown in Figure 2.4, where a close correspondence was observed between the analytical and experimental solutions. In the next section, we will categorize the effects of the dispersive channel, termed as (ISI).

2.2   Intersymbol Interference

Due to the linear, dispersive channel, the received linearly distorted instantaneous signal can be visualised as the superposition of several information symbols in the past and in the future. The so-called post-cursor ISI and pre-cursor ISI can be explained by using a hypothetical channel impulse response, as shown in Figure 2.5.

The channel impulse response shown in Figure 2.5 can be viewed as constituted by three distinct parts. The tap which possesses the highest relative amplitude h_2, is termed the main tap. The taps that exist before the main tap h_0 and h_1, are called pre-cursors and the taps that follow the main tap, namely h_3 and h_4, are referred to as post-cursors. The energy of the wanted signal is conveyed mainly by the contribution of the main channel tap. In addition to that, the received signal will also contain energy contributed by the convolution of the pre-cursors with the future transmitted symbols (relative to the channel main tap) and the convolution of the post-cursors with the past transmitted symbols, which are termed pre-cursor ISI and post-cursor ISI, respectively. Thus the received signal is distorted due to the superposition of the wanted signal, pre-cursor ISI and post-cursor ISI. Having defined the effects of a dispersive channel in terms of the post- and pre-cursor ISI, a range of equalization techniques used to combat the effects of this dispersive channel is introduced in the next section.

2.3   Basic Equalizer Theory

This section provides a rudimentary introduction to channel equalizers. A few equalizer structures are presented along with the criteria used to optimise their coefficients. Subsequently, the theoretical impact of these structures and their optimising criteria are highlighted. Before we proceed to discuss the structure and operation of the equalizers, their evolution history is briefly presented here.

In the context of linear equalizers, the pioneering work was achieved mainly byTufts [91], where the design of the transmitter and receiver was jointly optimised. The optimization was

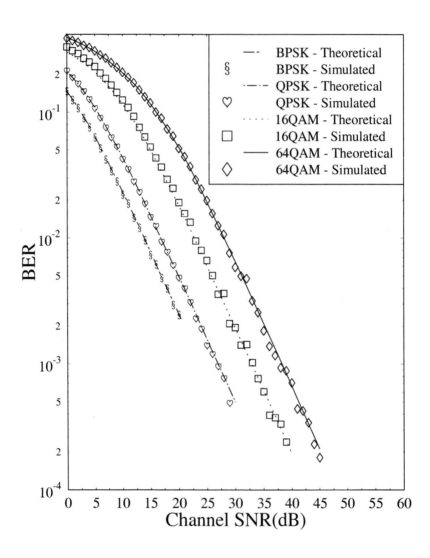

Figure 2.4: Theoretical and simulated BER performance of a single tap optimum DFE for BPSK, 4QAM, 16QAM and 64QAM over a flat fading Rayleigh channel using perfect channel estimation.

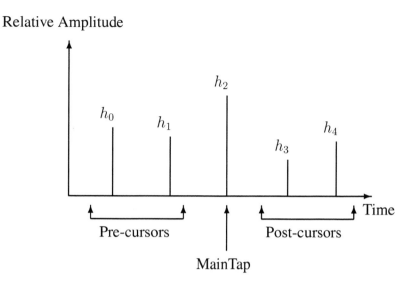

Figure 2.5: Channel Impulse Response (CIR) having pre-cursors, main tap and post-cursors.

based on the minimisation of the MSE between the transmitted signal and the equalized signal. This was achieved under the Zero Forcing (ZF) condition, where the ISI was completely mitigated at the sampling instances. Subsequently, Smith [92] introduced a similar optimization criterion with and without applying the ZF condition. Similar works as a result of these pioneering contributions were achieved by amongst others Hänsler [93], Ericson [94] and Forney [95].

The development of the DFE was initiated by the idea of using previous detected symbols to compensate for the ISI in a dispersive channel, which was first proposed by Austin [96]. This idea was adopted by Monsen [97], who managed to optimise the DFE based on minimizing the MSE between the equalized symbol and the transmitted symbol. The optimization of the DFE based on joint minimization of both the noise and ISI was undertaken by Salz [98], which was subsequently extended to QAM systems by Falconer and Foschini [99]. At about the same time, Price [100] optimised the DFE by utilizing the so-called ZF criterion, where all the ISI was compensated by the DFE. The pioneering work achieved so far assumed perfect decision feedback and that the number of taps of the DFE was infinite. A more comprehensive history of the linear equalizer and the DFE can be found in the classic papers by Lucky [101] or by Belfiore [102] and a more recent survey was produced by Qureshi [103].

In recent years, there has not been much development on the structure of the linear and decision feedback equalizers. However considerable effort has been given to the investigation of adaptive algorithms that are used to adapt the equalizers according to the prevalent CIR. These contributions will be elaborated in the next chapter. Nevertheless, some interesting work on merging the MLSE detectors with the DFE has been achieved by Cheung *et al.* [104, 105], Wu *et al.* [106, 107] and Gu *et al.* [108]. In these contributions, the structure of the MLSE and DFE was merged in order to yield an improved BER performance, when compared to the DFE, albeit at the cost of increased complexity. However, the complexity

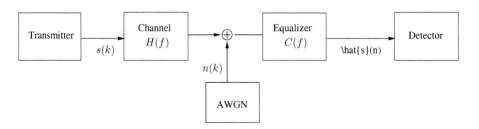

Figure 2.6: Schematic of the transmission system.

incurred was less, when compared to that of the MLSE.

In the context of error propagation in the DFE, which will be explained in Section 7.1, this phenomenon has been reported and researched in the past by Duttweiler *et al.* [109] and more recently by Smee *et al.* [110] and Altekar *et al.* [111]. In this respect some solutions have been proposed by amongst others, Tomlinson [112], Harashima [113], Russell *et al.* [114] and Chiani [115], in reducing the impact of error propagation.

In our subsequent discussion, the linear equalizer and the DFE are investigated using the Zero Forcing (ZF) and Minimum Mean Square Error (MMSE) criterion, with more emphasis on the DFE structure. In order to highlight the difference between the MMSE and ZF criteria, the linear equalizers based on these criteria are defined next.

2.3.1   Zero Forcing Equalizer

In Figure 2.6, the basic schematic of our system is shown. It simply consists of a pulse transmitter, the channel, an equalizer and a detector. The transmitted bits are labelled as $s(k)$, $n(k)$ represents the Additive White Gaussian Noise (AWGN) samples, while the output of the equalizer is denoted as $\hat{s}(k)$. The channel and equalizer transfer functions are denoted by $H(f)$ and $C(f)$, respectively.

The so-called zero forcing equalizer was devised and optimised by using the criterion. This essentially implied forcing all the impulse response contributions of the concatenated transmitter, channel and equalizer to zero at the signalling instants nT for $n \neq 0$, where T was the signalling interval duration. According to the zero ISI constraint, in the frequency domain the ZF criterion ensured the following relationship as stated by Lee *et al.* [89] when assuming zero delay:

$$H(f)C(f) = 1, \tag{2.28}$$

yielding the equalizer transfer function as the inverse of the channel transfer function :

$$C(f) = \frac{1}{H(f)}. \tag{2.29}$$

Consequently, the equalizer was reduced to a Finite Impulse Response (FIR) filter having an impulse response, which mimicked the inverse of the channel impulse response. A good insight into the behaviour of the equalizer can be obtained by studying the mean square error

(MSE) of the equalizer output [89]. The MSE is defined as :

$$E[|e(k)^2|] = E[|\hat{s}(k) - s(k)|^2], \tag{2.30}$$

$$e(k) = \hat{s}(k) - s(k), \tag{2.31}$$

where E[(.)] represents the expected value of (.) and e denotes the error signal at the equalizer output. Referring to Figure 2.6 we can write:

$$\hat{s}(k) = s(k) * h(k) * c(k) + n(k) * c(k), \tag{2.32}$$

where $*$ denotes the convolution process. As a result taking the Fourier Transform (FT) of $\hat{s}(k)$, we arrive at :

$$\hat{S}(f) = S(f)H(f)C(f) + N(f)C(f), \tag{2.33}$$

and upon applying Equation 2.28 the following can be obtained:

$$\hat{S}(f) - S(f) = N(f)C(f) = \frac{N(f)}{H(f)} = E(f), \tag{2.34}$$

where $E(f)$ is the FT of $e(n)$, leading to :

$$|E(f)|^2 = |N(f)|^2 |C(f)|^2. \tag{2.35}$$

The term $|E(f)|^2$ represents the power spectral density of the error signal and utilizing the Parseval Theorem [116], the expectation of the time domain mean square error is given by

$$E[|e^2|] = T \int_{-\frac{1}{2T}}^{\frac{1}{2T}} |E(f)|^2 df = T \int_{-\frac{1}{2T}}^{\frac{1}{2T}} |N(f)|^2 |C(f)|^2 df. \tag{2.36}$$

By noting that $|N(f)|^2$ is the single sided noise power spectral density of $\frac{N_o}{2}$, the mean square error can be written as :

$$E[|e^2|] = \frac{TN_o}{2} \int_{-\frac{1}{2T}}^{\frac{1}{2T}} \frac{1}{|H(f)|^2} df, \tag{2.37}$$

where f varies from $-\frac{1}{2T}$ to $\frac{1}{2T}$.

Upon exploiting Equation 2.37, the characteristics of the ZF equalizer can be analysed. The basic operation of the ZF equalizer is to provide gain at frequencies where the channel transfer function experiences attenuation and vice versa. However, in doing so, both the signal and noise are enhanced simultaneously. This results in noise enhancement, which degrades the performance of the equalizer. The MSE at the output of the ZF equalizer consists of only the effective noise component, since the ISI is completely removed. Thus, the MSE at the equalizer output can be used as a measure of the noise enhancement inflicted by the ZF equalizer. Referring again to Equation 2.37, at frequencies of the channel transfer function, where the gain is low, the MSE increases significantly, yielding high noise enhancement.

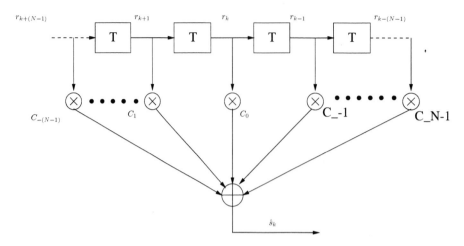

Figure 2.7: Linear mean square error equalizer schematic, where r_k and C_n denote the received signal and the equalizer coefficients, respectively.

Moreover, for a channel transfer function that exhibits a spectral null, the MSE is infinite. This implies an infinite gain, which is not realizable in an equalizer. As a result, no ZF equalizer exists for channels that exhibit a spectral null.

In mobile communications we often encounter spectral nulls and hence the ZF equalizer is rendered ineffective against this type of channel. Thus, the performance of the ZF equalizer is degraded by its noise enhancement and its inability to combat channels exhibiting spectral nulls. Consequently, in the next section, the linear minimum mean square equalizer is introduced, which can mitigate the disadvantages of the ZF equalizer to a certain extent.

2.3.2   Linear Mean Square Error Equalizer

The structure of the Linear Minimum Mean Square Error Equalizer (LE-MMSE) is shown in Figure 2.7, which was modelled as a FIR filter having tap delays equal to the symbol period T. The equalizer coefficients C_n, are determined by the specific optimization criterion utilized. Before considering the derivations and the criterion used to optimise this equalizer, the notations used in Figure 2.7 have to be introduced.

The equalizer consists of $2N + 1$ coefficient or taps. The taps are labelled as $C_{-(N-1)}$ $\rightarrow C_{N-1}$, where C_0 is the central tap of the equalizer. The equalizer input consists of past, present and future received samples,$(r_{k-1} \rightarrow r_{k-(N-1)})$, r_k and $(r_{k+1} \rightarrow r_{k+(N-1)})$, respectively. The employment of both past and future inputs underlines the fact that the equalizer is designed to combat both the so-called pre-cursor ISI and post-cursor ISI introduced by the dispersive channel. The presence of past inputs also implies the non-causality of the linear equalizer. Hence in order to facilitate causality in the equalizer, a delay is introduced.

The criterion used to optimise the equalizer coefficients was based on the error between the transmitted signal s_k and the estimate of the transmitted signal, i.e the equalizer output

$\hat{s}_k$, yielding :

$$e_k = s_k - \hat{s}_k, \tag{2.38}$$

where e_k was the error component at the equalizer output at time k. Here the MMSE criterion was invoked in order to minimise the mean square value of this error, termed as MSE, which was formulated as :

$$MSE = E[|s_k - \hat{s}_k|^2]. \tag{2.39}$$

In contrast to the ZF criterion, where the equalizer eliminated the ISI, the MMSE criterion allowed the joint minimization of both the noise and the ISI. Thus, when using the MMSE criterion, the total MSE was typically lower than that of the ZF criterion, which is made evident in the next section.

2.3.3 Derivation of the Linear Equalizer coefficients

The optimum coefficients are obtained by minimising the Mean Square Error (MSE) at the output of the equalizer. This can be accomplished in two different ways. The first technique obtains the first derivative of MSE with respect to the coefficients C_m, of the equalizer, equates this expression with zero, subsequently solving for C_m. Another simple technique is to invoke the so-called orthogonality principle [117], which states that the MSE is minimised, when the residual error of the equalizer is set orthogonal to the input signal of the equalizer. Thus, in order to arrive at the minimum MSE, we have :

$$E[e_k r^*_{k+l}] = 0, \tag{2.40}$$

where the superscript $*$ denotes complex conjugation. With the aid of Figure 2.7, the equalized signal is given by the following convolution :

$$\hat{s}_k = \sum_{m=-(N-1)}^{N-1} C_m r_{k+m} \tag{2.41}$$

and the received signal can be written as

$$r_k = \sum_{i=0}^{L_c-1} h_i s_{k-i} + n_k \tag{2.42}$$

where h_i is the ith tap of the CIR and L_c is the length of the CIR. Using Equations 2.38, 2.41 and 2.42, it can be shown that

$$
\begin{aligned}
E[e_k r^*_{k+l}] &= E\left[s_k \sum_{i=0}^{L_c-1} (h^*_i s^*_{k+l-i} + n^*_{k+l}) \right] - \\
&\quad E\left[\left(\sum_{m=-(N-1)}^{N-1} C_m \sum_{j=0}^{L_c-1} (h_j s_{k+m-j} + n_{k+m}) \right. \right. \\
&\quad \left. \left. \cdot \sum_{i=0}^{L_c-1} (h^*_i s^*_{k+l-i} + n^*_{k+l}) \right) \right].
\end{aligned} \tag{2.43}
$$

Assuming that the transmitted bits are wide sense stationary [118], we can write:

$$E[s_i s_j^*] = \begin{cases} \sigma_S^2 & for \ i = j \\ 0 & for \ i \neq j, \end{cases}$$

where σ_S^2 is the transmitted signal power and since the noise is uncorrelated, we arrive at :

$$E[n_i n_j^*] = \begin{cases} N_o & for \ i = j \\ 0 & for \ i \neq j. \end{cases}$$

Upon considering the first term at the right hand side of Equation 2.43 and using the above assumptions we can write:

$$E\left[s_k \sum_{i=0}^{L_c-1} h_i^* s_{k+l-i}^* + n_{k+l}^* \right] = \sigma_S^2 h_l^*. \tag{2.44}$$

We can also simplify the second term of the right hand side of Equation 2.43 to yield :

$$\left[\left(\sum_{m=-(N-1)}^{N-1} C_m \sum_{j=0}^{L_c-1} (h_j s_{k+m-j} + n_{k+m}) \cdot \sum_{i=0}^{L_c-1} h_i^* s_{k+l-i} + n_{k+l} \right) \right] = $$

$$\sum_{m=-(N-1)}^{N-1} C_m \sum_{i=0}^{L_c-1} \left[h_i^* h_{m+i-l} \sigma_S^2 + N_o \delta_{m-l} \right], \tag{2.45}$$

where δ is the Dirac delta function. When substituting Equation 2.44 and 2.45 into Equation 2.43, we arrive at

$$\sum_{m=-(N-1)}^{N-1} C_m \sum_{i=0}^{L_c-1} \left[h_i^* h_{m+i-l} \sigma_S^2 + N_o \delta_{m-l} \right] = \sigma_S^2 h_l^* \ for \ l = -(N-1)....N-1.$$

$$\tag{2.46}$$

Observe that Equation 2.46 constitutes a set of $2N + 1$ linear equations. Solving these equations simultaneously leads to the optimum equalizer coefficients corresponding to the MSE criterion.

The achievable minimum MSE (MMSE) can also be derived in terms of the equalizer coefficients and the CIR, which was given by Proakis [87] as:

$$MMSE = \sigma_S^2 \left(1 - \sum_{m=-(N-1)}^{N-1} C_m h_m \right). \tag{2.47}$$

In deriving this form of the MMSE, an important assumption was made in that the error signal between the transmitted signal and its estimate at the output of the equalizer had to be orthogonal to the equalizer's input signal. In other words, Equation 2.47 was only valid for the optimum coefficients corresponding to the MMSE criterion. This equation was useful in deriving numerical solutions, albeit it did not provide insight into the behaviour of the equalizer. A frequency domain representation of the MSE provided a deeper appreciation of its behaviour. Upon referring to Figure 2.6, we have:

$$E(f) = S(f) - \hat{S}(f) = S(f)[1 - H(f)C(f)] - N(f)C(f). \tag{2.48}$$

By noting that $|N(f)|^2 = \frac{N_o}{2}$, $|S(f)|^2 = \sigma_S^2$ and that the noise and the transmitted samples are uncorrelated, the power spectral density of the error signal between the transmitted signal and its estimate at the output of the equalizer is obtained as follows :

$$|E(f)|^2 = \sigma_S^2 |1 - H(f)C(f)|^2 + \frac{N_o}{2}|C(f)|^2. \tag{2.49}$$

Upon evaluating the square of the above equation, $|E(f)|^2$ can be written as :

$$|E(f)|^2 = \left(|H(f)|^2\sigma_S^2 + \frac{N_o}{2}\right)\left|C(f) - \frac{H^*(f)}{|H(f)|^2 + \frac{N_o}{2\sigma_S^2}}\right|^2 + \frac{N_o}{2}\left[\frac{1}{|H(f)|^2 + \frac{N_o}{2\sigma_S^2}}\right]. \tag{2.50}$$

Subsequently, by exploiting Parseval's Theorem [116], the MSE can be written as :

$$E[|e^2|] = T\int_{-\frac{1}{2T}}^{\frac{1}{2T}} |E(f)|^2 df. \tag{2.51}$$

The MSE, characterized by Equation 2.51 was minimised by ensuring that the squared term in Equation 2.50 was equal to zero. This yielded the optimum transfer function $C(f)$ of the equalizer, corresponding to the minimum MSE. Thus for the squared term of Equation 2.50, the optimum transfer function of the equalizer $C_{opt}(f)$, is given by :

$$C_{opt}(f) = \frac{H^*(f)}{|H(f)|^2 + \frac{N_o}{2\sigma_S^2}}. \tag{2.52}$$

Subsequently the MSE is obtained by using Equation 2.52, 2.51 and the non-zero term 2.50 yielding :

$$E[|e^2|] = \frac{TN_o}{2}\int_{-\frac{1}{2T}}^{\frac{1}{2T}} \frac{1}{|H(f)|^2 + \frac{N_o}{2\sigma_S^2}} df. \tag{2.53}$$

The frequency domain representation of the MSE is given by Equation 2.53, where several observations can be made. Firstly, the MMSE is less than that of the ZF criterion of Equation 2.37 due to the extra second term in the denominator of Equation 2.53. This follows from the initial statement made that the MSE criterion jointly minimises the noise and the ISI in order to yield a lower total MSE. This also implies that the noise enhancement in the linear MSE equalizer is lower than that of the ZF criterion. However, at high average channel SNRs, when the noise contribution is low, the MSE equation of the linear MSE equalizer approaches that of the ZF MMSE, causing both the linear MSE and ZF equalizer to exhibit similar characteristics. Consequently, the linear MSE equalizer experiences the same disadvantages as the ZF equalizer at high average channel SNRs. The other inference that

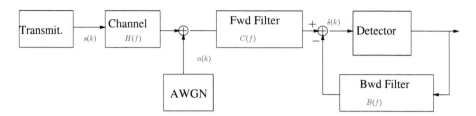

Figure 2.8: Schematic of the transmission system depicting the feedforward and feedback filter of the DFE, where $C(f)$ and $B(f)$ are their transfer functions, respectively.

Equation 2.53 provides is that, unlike Equation 2.37 of the ZF equalizer, Equation 2.53 is always integrable, even when the channel exhibits a spectral null. Consequently, the linear MSE equalizer exists for all types of channels.

In conclusion, the noise enhancement produced by the LE-MMSE was less than that produced by the ZF equalizer. Furthermore, the LE-MMSE existed, even when the channel exhibited a spectral null. In the next section, we will concentrate on the Decision Feedback Equalizer (DFE).

2.3.4   Decision Feedback Equalizer

As implied by the terminology, the decision feedback equalizer (DFE) employs a feedforward filter and feedback filter in order to combat the ISI inflicted by the dispersive channels. The non-linear function manifested by the decision device is introduced at the input of the feedback filter. The general block diagram of the DFE is shown in Figure 2.8. In general, as in the linear MSE equalizer, the forward filter partially eliminates the ISI introduced by the dispersive channel. The feedback filter, in the absence of decision errors, is fed with the error-free signal in order to further eliminate the ISI.

Explicitly, in Figure 2.8 the feedback filter - labelled as the Bwd Filter - receives the detected symbols. It then subtracts its output from the estimate made by the forward filter in order to yield the input signal to the detector. Again, since the feedback filter uses a "clean" signal as its input, the feedback loop mitigates the ISI without introducing noise into the system. However, the disadvantage is that when a wrong decision is fed back into the feedback loop, "error propagation" is inflicted and subsequently reduces the BER performance of the equalizer.

The detailed feedforward filter of Figure 2.9 is constituted by the coefficient taps labelled $C_0 \rightarrow C_{N_f-1}$, where N_f is the number of taps in the feed forward filter. The causal feedback filter consists of N_b feedback taps, denoted by $b_1 \rightarrow b_{N_b}$. The feedforward filter is fed only with present and future received signal samples, which implies that no latency is inflicted. As a result of this, the feedforward filter eliminates only the pre-cursor ISI, but not the post-cursor ISI, both of which were defined in Section 2.2. While the feedforward filter eliminates most of the pre-cursor ISI, the feedback filter mitigates the ISI caused by the past data symbols, i.e the post-cursor ISI. Since the feedforward filter only eliminates the pre-cursor ISI and not the post-cursor ISI, noise enhancement in the DFE is less significant, when compared to the linear MSE equalizer.

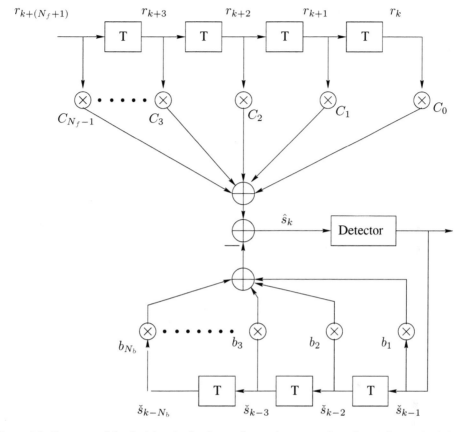

Figure 2.9: Structure of the decision feedback equalizer, where r_k and $\check{s}_k$ denote the received signal and detected symbol, respectively. The notations C_m and b_q, represent the coefficient taps of the feedforward and feedback filters, respectively.

The MSE criterion can be used to derive the optimum coefficients of the feedforward section of the DFE. The derivation of the feedforward filter is similar to that of the linear MSE equalizer given in Equation 2.46. Following the approach of Cheung [104], the optimum taps of the feedforward section of the DFE can be determined from the following set of N_f equations :

$$\sum_{m=0}^{N_f-1} C_m \left[\sum_{v=0}^{l} h_v^* h_{v+m-l} \sigma_S^2 + N_o \delta_{m-l} \right] = h_l^* \sigma_S^2 \quad l = 0.....N_f - 1. \tag{2.54}$$

This assumes that the CIR and the noise power was known. For the feedback filter, using Cheung's approach [104], we arrive at the following set of N_b equations :

$$b_q = \sum_{m=0}^{N_f-1} C_m h_{m+q} \qquad q = 1.....N_b. \tag{2.55}$$

Thus using Equation 2.54 and 2.55, the DFE coefficients can be determined. Having derived and studied the characteristics of the various equalizers, the signal to noise ratio loss incurred by the DFE will be determined in the next section with the aim of producing the analytical BER solution for the performance of the DFE.

2.4   Signal to Noise Ratio Loss of the DFE

The signal to noise ratio (SNR) loss of the equalizer was defined by Cheung [104] as:

$$SNR_{loss} = SNR_{input} - SNR_{output}, \tag{2.56}$$

where SNR_{input} is the SNR measured at the input of the equalizer, given by :

$$SNR_{input} = \frac{\sigma_s^2}{2\sigma_N^2}, \tag{2.57}$$

with σ_s^2 being the average transmitted power, assuming wide sense stationary conditions.

The calculation of SNR_{output} was slightly more complicated, since the equalizer output contained the wanted signal, the effective Gaussian noise, the residual ISI and the ISI caused by the past data symbols. In order to simplify our approach, correct bits were fed back into the DFE feedback filter, thus effectively eliminating all the post-cursor ISI from the equalizer output. Following Cheung's approach [104], we can write:

$$SNR_{output} = \frac{\text{Wanted Signal Power}}{\text{Residual ISI Power + Effective Noise Power}}. \tag{2.58}$$

In the above equation, the residual ISI was assumed to be extra noise possessing a Gaussian distribution, allowing us to evaluate the required terms as follows [104]:

$$\text{Wanted Signal Power} = E\left[\left|s_k \sum_{m=0}^{N_f-1} C_m h_m\right|^2\right], \tag{2.59}$$

$$\text{Effective Noise Power} = N_o \sum_{m=0}^{N_f-1} |C_m|^2, \tag{2.60}$$

$$\text{Residual ISI Power} = \sum_{q=-(N_f-1)}^{-1} E\left[|f_q s_{k-q}|^2\right], \tag{2.61}$$

where $f_q = \sum_{m=0}^{N_f-1} C_m h_{m+q}$ and h_i represents the ith path of the CIR. The remaining notations were given in Figure 2.9. Thus, using Equations 2.58, 2.59, 2.60 and 2.61, the final form of SNR_{output} was given by :

$$SNR_{output} = \frac{E\left[\left|s_k \sum_{m=0}^{N_f-1} C_m h_m\right|^2\right]}{\sum_{q=-(N_f-1)}^{-1} E\left[|f_q s_{k-q}|^2\right] + N_o \sum_{m=0}^{N_f-1} |C_m|^2}. \tag{2.62}$$

Having defined the signal to noise ratio output of the DFE, we can utilize it in order to create an analytical approach to quantifying the BER performance of the DFE.

2.4.1 Bit Error Rate Performance

An analytical approach to be outlined in Equation 2.63 was invoked, in order to quantify the theoretical BER performance of the DFE. The analytical approach simply used the calculated SNR_{output} characterized by Equation 2.62, where the theoretical BER for a BPSK modem in a multi-path channel environment was approximated as :

$$P^b_{BPSK_{mc}}(SNR_{input}) = P^b_{BPSK}(SNR_{output}), \qquad (2.63)$$

where P^b_{BPSK} and $P^b_{BPSK_{mc}}$ represent the BER probability in a Gaussian and in a multi-path channel environment, respectively. Again, when using this form of approximation, the residual ISI was considered to be Gaussian distributed and the closeness of the theoretical and experimental results to be highlighted in the context of Figure 2.10 confirmed this assumption to a certain degree. Let us now concentrate our attention on the performance of the DFE using multi-level modems.

2.5 Equalization in Multi-level Modems

In our forthcoming experiments the BER performance of the DFE was investigated using the channel models summarized in Table 2.1 in conjunction with BPSK, 4QAM, 16QAM and 64QAM. The corresponding results are shown in Figure 2.10, which portrayed both our experimental and theoretical results. There was a good correspondence between the simulated and theoretical results, as evidenced by Figure 2.10. The theoretical results were calculated using the approach outlined in Section 2.4.1 under the assumption that the residual ISI of Equation 2.61 was Gaussian distributed. Hence it was possible to account for it as additional noise, which degraded the output SNR of the DFE. Explicitly the theoretical solutions in a multi-path channel environment for each individual modulation mode can be formulated as :

$$P^b_{BPSK_{mc}}(SNR_{input})(SNR_{input}) = P^b_{BPSK}(SNR_{output}) \qquad (2.64)$$

$$P^b_{4QAM_{mc}}(SNR_{input})(SNR_{input}) = P^b_{av_{4QAM}}(SNR_{output}) \qquad (2.65)$$

$$P^b_{16QAM_{mc}}(SNR_{input}) = P^b_{av_{16QAM}}(SNR_{output}) \qquad (2.66)$$

$$P^b_{64QAM_{mc}}(SNR_{input}) = P^b_{av_{64QAM}}(SNR_{output}), \qquad (2.67)$$

where the notations SNR_{input} and SNR_{output} were defined in Equations 2.57 and 2.62, respectively. The average BER probabilities of the 4QAM, 16QAM and 64QAM in a Gaussian channel environment are represented by $P^b_{av_{4QAM}}$ $P^b_{av_{16QAM}}$ $P^b_{av_{64QAM}}$, respectively. Furthermore, perfect channel estimation and error-free decision feedback for the backward filter of the DFE was assumed at the receiver.

In the results shown in Figure 2.10, the performance degradation incurred over Channel 3 in comparison to the non-dispersive AWGN channel was the highest, followed by Channel 2 and finally Channel 1. This was consistent with our expectations, since Channel 3 possessed

Channel Label	Channel Impulse Response
Channel 1	$0.707 + 0.707z^{-1}$
Channel 2	$0.407 + 0.818z^{-1} + 0.407z^{-2}$
Channel 3	$0.227 + 0.460z^{-1} + 0.688z^{-2} + 0.460z^{-3} + 0.227z^{-4}$

Table 2.1: Static dispersive Gaussian CIRs used in our experiments.

the highest degree of linear distortion, as it was underlined by Proakis [87]. Hence, Channel 3 was expected to generate the most ISI, which resulted in significant degradation of the BER performance. In the context of the results shown in Figure 2.10 the channel SNR was related to $\frac{E_b}{N_o}$ by the expression

$$\frac{E_b}{N_o} = \frac{\text{Channel SNR}}{\text{number of bits per QAM symbol}} \qquad (2.68)$$

Having characterized the performance of the DFE in the context of multi-level QAM schemes, we will now review our results and formulate several conclusions.

2.6   Review and Discussion

The utilization of linear and decision feedback equalizers in conjunction with BPSK, 4QAM, 16QAM, and 64QAM was explored in terms of their behaviour and performance under a variety of conditions using the channel models of Table 2.1. Experimental results were generated based on the BER performance as shown in Figure 2.10, where the analytical approach exhibited close correspondence with the experimental results. The analytical approach was characterized by Equations 2.64 - 2.67, where the output SNR of the DFE governed the expected BERs.

The superiority of the MSE criterion over the ZF criterion, when applied to a linear equalizer was also highlighted analytically in Sections 2.3.1 and 2.3.2. However, it was also shown in Figure 2.10 that even in conjunction with the more powerful DFE, the error rate performance was degraded substantially, when a severe dispersive multi-path channel was used.

In general, the performance of the DFE was mainly degraded by two factors. Firstly, the feedforward filter produced noise enhancement at the output of the equalizer as a result of combating the pre-cursors. This led to the degradation of the SNR performance. The other potential factor - which was not discussed - in this chapter was due to incorrect decision feedback into the backward filter, which can lead to the error propagation phenomenon. This will be discussed in more detail in Section 7.1.

Another optimistic assumption in our simulations was that the (CIR) was estimated perfectly at the receiver. Moreover, the channels used were time invariant. In reality the CIR can only be imperfectly estimated and the quality of the estimation varies according to different estimation techniques. The CIR is in practice time-variant and experiences fading in a mobile environment. All these factors can contribute to further performance degradations. In the next chapter, the assumptions made regarding perfect knowledge of the channel and the usage of time invariant channels will be eliminated. This will lead to the introduction of time-variant channels and to the employment of methods in order to estimate these channels.

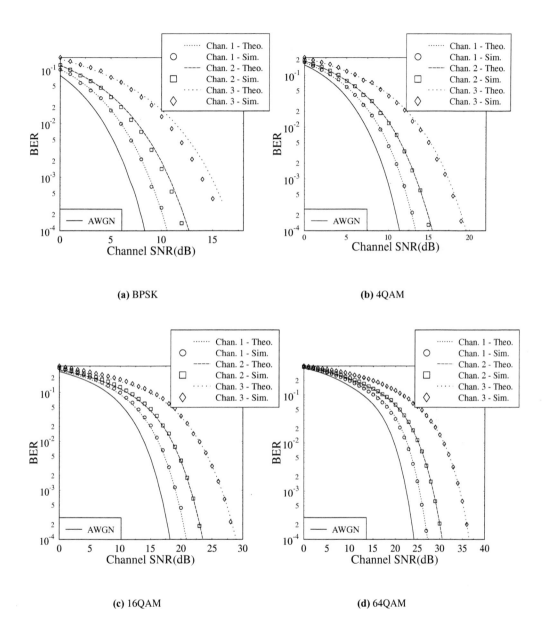

Figure 2.10: Theoretical and simulated BER performance over the channel models of Table 2.1 using **BPSK**, **4QAM**, **16QAM** and **64QAM** schemes. The DFE utilized seven forward taps and four feedback taps. Perfect channel estimation and error-free feedback were assumed at the receiver and the theoretical solutions were plotted using Equations 2.64, 2.65, 2.66 and 2.67, respectively.

The time-variant CIR estimation requires adaptive techniques in which the equalizer co-efficients are updated constantly in response to the varying CIR. There are various algorithms used to adaptively update the coefficients. Some of the notable ones are the steepest descent or the gradient algorithm, the least mean square (LMS) [119–121] and the Kalman algorithms [122, 123]. The adaptive scheme used in the equalizers or channel estimators have to converge rapidly and be able to track the channel variations adequately. As such, the algorithms listed above can be used to adapt the equalizer with the Kalman algorithm providing the best convergence speed. Hence, in the next chapter the recursive Kalman algorithm is analysed, in order to create an attractive adaptive equalizer and channel estimator.

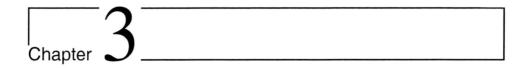

Chapter **3**

Adaptive Equalization

In Chapter 2 the equalizers operated under the assumption of perfect channel estimation, where the receiver always had perfect knowledge of the CIR. However, the CIR is typically time variant and consequently the receiver has to estimate the CIR or the coefficients of the equalizer, in order to compensate for the ISI induced by the channel.

Algorithms have been developed in order to automatically adapt the coefficients of the equalizer directly [118] or by utilizing the estimated CIR [124, 125]. These algorithms can be generally classified into three categories, which were depicted in Figure 2.1. The first category involves the steepest descent methods, where a considerable amount of the pioneering work was achieved by Lucky [126,127]. The second class of adaptive algorithms incorporates the stochastic gradient method, which is more commonly known as the Least Mean Square (LMS) algorithm that was widely documented by Widrow *et al.* [119–121]. The third and final category includes the Least Square (LS) algorithms. In this section, we shall concentrate on the LS algorithms, in particular on the well known Recursive LS (RLS) or Kalman algorithm [86].

The Kalman algorithm was first formulated by Kalman [86] in 1961. This was followed by the application of the algorithm for adaptive equalizers [104, 128–132]. Lee and Cunningham [133] extended the adaptive equalizer to QPSK modems, which invoked the complex version of the adaptive Kalman equalizer. Adaptive channel estimators utilizing the recursive Kalman algorithm were also researched by Cheung [104], Godard [123], Messe *et al.* [134], Harun *et al.* [124] and Shukla *et al.* [125]. In order to ensure stability and to reduce the complexity, variants of the Kalman algorithm have been developed, by amongst others, by Hsu [135], which was referred to as the Square Root Kalman algorithm and by Falconer *et al.* [122] termed as the Fast Kalman algorithm. A good comparison and survey of these variants of the recursive Kalman algorithm can be found in references by Haykin [118], Richards [136], Mueller [137] and Sayed *et al.* [138].

In this chapter, the Kalman algorithm is derived, applied and studied in the context of an adaptive channel estimator [124, 125] and a directly implemented adaptive DFE [118] in a multilevel modem scenario. Subsequently, the complexity incurred by this algorithm is discussed and compared in the above two scenarios. Let us now commence the derivation of the Kalman algorithm.

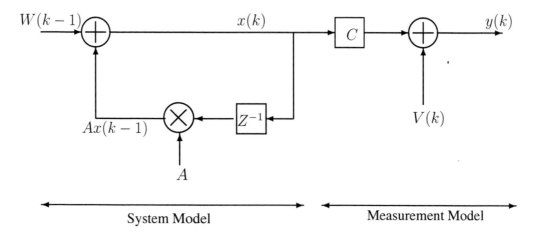

System Model   Measurement Model

Figure 3.1: Signal-flow representation of the Kalman system and measurement models, which are characterized by Equations 3.1 and 3.4.

3.1 Derivation of the Recursive Kalman Algorithm

In deriving the Kalman algorithm, we will utilize the approach outlined by Bozic [139], where the one-dimensional algorithm is derived and subsequently extended it to the multi-dimensional Kalman algorithm. Let us now proceed with the derivation of the one-dimensional recursive Kalman algorithm. This algorithm will then be invoked in the context of channel equalization and estimation in Section 3.2.

3.1.1 Derivation of the One-dimensional Kalman Algorithm

In estimating an unknown parameter $x(k)$, the Kalman algorithm is formulated based on two models, a system model and a measurement model, which are depicted in Figure 3.1. The purpose of the system model is to characterize the behaviour of an unknown parameter, which in generic terms, obeys a first order recursive system model, as stated below:

$$x(k) = Ax(k-1) + W(k-1). \tag{3.1}$$

In the above system model $x(k)$, is the unknown parameter, which is modelled as a first order recursive filter driven by a zero mean white noise process denoted by $W(k-1)$, as seen in Figure 3.1 and A relates the state of the system at times k and $k-1$ to each other. In the context of a channel estimator $x(k)$, represents the CIR, where its time varying characteristics are governed by the state transition contribution A, and a random perturbation based on a zero mean white noise process $W(k-1)$. The zero mean uncorrelated noise process $W(k-1)$, has the following characteristics :

$$E[W(k)] = 0, \tag{3.2}$$

$$E[W(i)W(j)^*] = \begin{cases} \sigma_w^2 & for \ i = j \\ 0 & for \ i \neq j. \end{cases} \tag{3.3}$$

The noisy measurement model, from which the unknown parameter $x(k)$ has to be extracted, is stated as :

$$y(k) = Cx(k) + V(k), \tag{3.4}$$

which is depicted in Figure 3.1, where $y(k)$ is the present observed signal C, represents the known measurement variable used to measure $x(k)$ and $V(k)$ is the zero mean white noise. This measurement model represents the observation of the time varying CIR by the received signal $y(k)$. This signal includes the contribution by the convolution of the CIR, $x(k)$, and a particular training signal C, as well as the random noise $V(k)$. The properties of $V(k)$ are the same as those of $W(k)$, although its variance is different:

$$E[V(k)] = 0, \tag{3.5}$$

$$E[V(i)V(j)^*] = \begin{cases} \sigma_v^2 & for \ i = j \\ 0 & for \ i \neq j. \end{cases} \tag{3.6}$$

In estimating $x(k)$, a recursive estimator utilizing $y(k)$ can be formulated as [139]:

$$\hat{x}(k) = \alpha\hat{x}(k-1) + \beta y(k). \tag{3.7}$$

The recursive estimator equation above stipulates that the estimate of $x(k)$, which is denoted by $\hat{x}(k)$, is dependent on two factors. The first term on the right hand side of Equation 3.7 represents the system model term based on the previous estimate $x(k-1)$, which is weighted by α. The second term in Equation 3.7 represents the measurement model based on the observed data signal $y(k)$, which is weighted by β. The factors α and β in Equation 3.7 can be determined by minimizing the mean square error (MSE) $P(k)$, between the estimated parameter $\hat{x}(k)$ and the actual parameter $x(k)$. The mean square error $P(k)$, and the error term $e(k)$ can be written as :

$$P(k) = E[e(k)e^*(k)], \tag{3.8}$$

$$e(k) = \hat{x}(k) - x(k). \tag{3.9}$$

The minimization of the MSE can be achieved by applying the previous orthogonality equations, which are formulated according to Bozic as [139]:

$$E[e(k)\hat{x}^*(k-1)] = 0, \tag{3.10}$$

$$E[e(k)y^*(k)] = 0. \tag{3.11}$$

We can substitute Equations 3.7 and 3.9 into Equation 3.10 yielding:

$$E\big[\alpha\hat{x}(k-1)\hat{x}^*(k-1)\big] = E\big[(x(k) - \beta y(k))\hat{x}^*(k-1)\big], \tag{3.12}$$

Adding and subtracting $\alpha x(k-1)\hat{x}^*(k-1)$ in the above equation will result in :

$$E\big[\alpha\hat{x}^*(k-1)[\hat{x}(k-1)+x(k-1)-x(k-1)]\big] = E\big[(x(k)-\beta y(k))\hat{x}^*(k-1)\big].$$
(3.13)

Equations 3.9 and 3.4 are then substituted in Equation 3.13 in order to yield :

$$\alpha E\big[e(k-1)\hat{x}^*(k-1)+x(k-1)\hat{x}^*(k-1)\big] = E\big[[(1-\beta C)x(k)-\beta V(k)]\hat{x}^*(k-1)\big].$$
(3.14)

The first term on the left-hand side $E\big[e(k-1)\hat{x}^*(k-1)\big] = 0$. This is true, since we can rewrite the term $\hat{x}^*(k-1) = \alpha\hat{x}^*(k-2)+\beta y^*(k-1)$ and use the orthogonality equations of 3.10 and 3.11 for the previous time instant of $k-1$. We also note that $V(k)$ and $\hat{x}^*(k-1)$ are independent of each other, resulting in the following:

$$\alpha E\big[x(k-1)\hat{x}^*(k-1)\big] = (1-\beta C)E\big[x(k)\hat{x}^*(k-1)\big].$$
(3.15)

We can then substitute Equation 3.1 into the above equation and noting that $W(k-1)$ is independent of $\hat{x}^*(k-1)$, we obtain the following expression for α:

$$\alpha E\big[x(k-1)\hat{x}^*(k-1)\big] = (1-\beta C)E\big[(Ax(k-1)+W(k-1))\hat{x}^*(k-1)\big],$$
(3.16)

leading to :

$$\alpha = A(1-C\beta).$$
(3.17)

Therefore by substituting Equation 3.17 into Equation 3.7, we can rewrite the recursive estimator relationship as :

$$\hat{x}(k) = A\hat{x}(k-1)+\beta[y(k)-\hat{y}(k)].$$
(3.18)

where $\hat{y}(k)$ is defined as :

$$\hat{y}(k) = CA\hat{x}(k-1).$$
(3.19)

The first term of Equation 3.18 denotes the prediction based on the past estimate $\hat{x}(k-1)$, while the second term represents a correction term, which consists of the difference between the actual observed sample $y(k)$ and the observation estimate $\hat{y}(k)$, which is formed by the past estimated parameter $\hat{x}(k-1)$ according to Equation 3.19.

In deriving an expression for β, we can use Equation 3.8 and the fact that $e^*(k) = \hat{x}^*(k) - x^*(k)$, in order to write the mean square error term in the following form :

$$P(k) = E[e(k)(\hat{x}^*(k)-x^*(k))].$$
(3.20)

We then proceed to substitute Equation 3.7 into the above equation, yielding :

$$P(k) = E[e(k)(\alpha\hat{x}^*(k-1)+\beta y^*(k)-x^*(k))].$$
(3.21)

This equation can be further simplified by applying the relationship stated in Equation 3.11 and 3.10, giving :

$$P(k) = -E[e(k)x^*(k)].$$
(3.22)

Proceeding further, we can substitute Equation 3.4 into Equation 3.11 to yield:

$$C^* E[e(k)x^*(k)] = -E[e(k)V^*(k)].$$ (3.23)

By substituting Equations 3.22, 3.9 and 3.7 into Equation 3.23 we can write the MSE as follows :

$$P(k) = \frac{1}{C^*} E[e(k)V^*(k)],$$ (3.24)

$$= \frac{1}{C^*} E[(\alpha \hat{x}(k-1) + \beta y(k) - x(k))V^*(k)].$$ (3.25)

Since $V^*(k)$ is independent of $\hat{x}(k-1)$ and $x(k)$, Equation 3.25 can be rewritten as :

$$P(k) = \frac{1}{C^*} E[\beta y(k)V^*(k)],$$ (3.26)

This can be further simplified by applying Equation 3.4, yielding the MSE as follows :

$$P(k) = \frac{1}{C^*} \beta E[(Cx(k) + V(k))V^*(k)],$$ (3.27)

$$P(k) = \frac{1}{C^*} \beta \sigma_v^2,$$ (3.28)

$$\beta = \frac{C^* P(k)}{\sigma_v^2}.$$ (3.29)

In determining β, we then substitute Equation 3.9 into Equation 3.8 in order to obtain the following expression for the MSE $P(k)$:

$$P(k) = E[(\hat{x}(k) - x(k))(\hat{x}(k) - x(k))^*].$$ (3.30)

We expand the above equation by substituting Equations 3.18, 3.4, 3.1 and 3.19 into it, yielding the MSE expression of :

$$P(k) = E\Big[\Big(A[1 - \beta C][e(k-1)] - [1 - \beta C]W(k-1) + \beta V(k)\Big) \times$$ (3.31)

$$\Big(A[1 - \beta C][e(k-1)] - [1 - \beta C]W(k-1) + \beta V(k)\Big)^*\Big].$$ (3.32)

Since the cross-products of the above equation averages to zero, because $e(k-1)$, $W(k-1)$ and $V(k)$ are independent of each other, the equation above can be simplified to yield :

$$P(k) = |A|^2 |[1 - C\beta]|^2 P(k-1) + |[1 - C\beta]|^2 \sigma_w^2 + |\beta|^2 \sigma_v^2.$$ (3.33)

Consequently, by substituting Equation 3.28 into Equation 3.33, we arrive at an expression for β :

$$\beta[\sigma_v^2 + |C|^2(|A|^2 P(k-1) + \sigma_w^2)] = \dot{C}^*(|A|^2 P(k-1) + \sigma_w^2),$$ (3.34)

Recursive estimator :	$\hat{x}(k) = A\hat{x}(k-1) + \beta[y(k) - CA\hat{x}(k-1)]$
Kalman gain :	$\beta = \dfrac{P(k,k-1)C^*}{\left[\sigma_v^2 + |C|^2 P(k,k-1)\right]}$
where	$P(k,k-1) = |A|^2 P(k-1) + \sigma_w^2$
Mean square error :	$P(k) = P(k,k-1) - C\beta P(k,k-1)$

Table 3.1: One-dimensional Kalman recursive equations.

$$\beta = \frac{C^* P(k,k-1)}{\left[\sigma_v^2 + |C|^2 P(k,k-1)\right]}, \tag{3.35}$$

where $P(k,k-1)$ is given by :

$$P(k,k-1) = |A|^2 P(k-1) + \sigma_w^2. \tag{3.36}$$

Finally, in order to obtain a recursive equation for $P(k)$, we can substitute Equation 3.35 into Equation 3.28 to give the MSE in the form of :

$$
\begin{aligned}
P(k) &= \frac{1}{C^*}\sigma_v^2 \frac{C^* P(k,k-1)}{\left[\sigma_v^2 + |C|^2 P(k,k-1)\right]} \tag{3.37}\\[2mm]
&= \frac{\sigma_v^2 P(k,k-1)}{\sigma_v^2 + |C|^2 P(k,k-1)} \tag{3.38}\\[2mm]
&= \frac{\sigma_v^2 P(k,k-1) - |C|^2 [P(k,k-1)]^2 + |C|^2 [P(k,k-1)]^2}{\sigma_v^2 + |C|^2 P(k,k-1)} \tag{3.39}\\[2mm]
&= \frac{P(k,k-1)(|C|^2 P(k,k-1) + \sigma_v^2) - |C|^2 P(k,k-1)P(k,k-1)}{\sigma_v^2 + |C|^2 P(k,k-1)}.
\end{aligned}
$$

$$\tag{3.40}$$

This can be further simplified to provide the MSE expression of :

$$P(k) = P(k,k-1) - C\beta P(k,k-1). \tag{3.41}$$

In conclusion, Equations 3.18, 3.35, 3.41 and 3.36 form the one-dimensional recursive Kalman equations, which are summarized in Table 3.1.

In this section we have derived the Kalman algorithm for the one-dimensional case. We are now equipped to derive the multi-dimensional or vector based Kalman algorithm in the next section, which will allow us to update the channel and equalizer coefficient vector.

3.1.2 Derivation of the Multi-dimensional Kalman Algorithm

In deriving the multi-dimensional or vector Kalman algorithm, we can apply the one-dimensional algorithm derived in Section 3.1.1. The one-dimensional algorithm can be extended to its vector form by replacing the one-dimensional Kalman variables and the corresponding operations with their matrix equivalents. This transformation of the associated operations from the one-dimensional to multi-dimensional space is summarized in Table 3.2, where a

Scalar	Matrix
$a + b$	$\mathbf{A} + \mathbf{B}$
ab	$\mathbf{AB}$
$a^2 b$	$\mathbf{ABA^{*T}}$
$\frac{1}{a+b}$	$[\mathbf{A} + \mathbf{B}]^{-1}$

Table 3.2: Transformation of scalar operations to matrix operations.

Recursive estimator :
$$\hat{\mathbf{x}}(k) = \mathbf{A}\hat{\mathbf{x}}(k-1) + \mathbf{K}(k)[\mathbf{y}(k) - \mathbf{CA}\hat{\mathbf{x}}(k-1)]$$

Estimator gain :
$$\mathbf{K}(k) = \mathbf{C}^*\mathbf{P}(k, k-1)[\mathbf{R}(k) + \mathbf{CP}(k, k-1)\mathbf{C}^{*T}]^{-1}$$

where $\mathbf{P}(k, k-1) = \mathbf{AP}(k-1)\mathbf{A}^{*T} + \mathbf{Q}(k-1)$

Error covariance matrix:
$$\mathbf{P}(k) = \mathbf{P}(k, k-1) - \mathbf{K}(k)\mathbf{CP}(k, k-1)$$

Table 3.3: Multi-dimensional Kalman recursive equations based on the scalar equations of Table 3.1.

bold face letter denotes a matrix. The superscripts $\mathbf{T}$ and -1 represent the transpose of a matrix and the inverse of a matrix, respectively.

We can now proceed to transform the one-dimensional Kalman equations to their vector equivalents in order to generate the multi-dimensional recursive Kalman equations listed in Table 3.3. The following system and measurement vector models - which are known collectively as the state space models - are used :

$$\mathbf{x}(k) = \mathbf{A}\mathbf{x}(k-1) + \mathbf{W}(k-1), \tag{3.42}$$

$$\mathbf{y}(k) = \mathbf{C}\mathbf{x}(k-1) + \mathbf{V}(k), \tag{3.43}$$

where $\mathbf{A}$ denotes the state transition matrix, which relates the various states of the system model at different times to each other, while $\mathbf{W}(k)$ and $\mathbf{V}(k)$ represent the system noise matrix and measurement noise matrix, respectively.

In the equations listed in Table 3.3 we have also introduced two new terms labelled as $\mathbf{R}(k)$ and $\mathbf{Q}(k)$, which represent the measurement noise covariance matrix and the system noise covariance matrix, respectively that are defined as :

$$\mathbf{R}(k) = E[\mathbf{V}(k)\mathbf{V}^{*T}(k)] \tag{3.44}$$

$$\mathbf{Q}(k) = E[\mathbf{W}(k)\mathbf{W}^{*T}(k)]. \tag{3.45}$$

We can now summarize the multi-dimensional Kalman recursive Equations in Table 3.4, where each variable is described and its matrix dimension is stated.

Description	Variable	Matrix Dimension (Row × Column)
Measurement Vector	$\mathbf{y}(k)$	$M \times 1$
Estimated Parameter Vector	$\hat{\mathbf{x}}(k)$	$N \times 1$
State Transition Matrix	$\mathbf{A}$	$N \times N$
Kalman Gain Matrix	$\mathbf{K}(k)$	$N \times M$
Measurement Matrix	$\mathbf{C}$	$M \times N$
Predicted State Error Covariance Matrix	$\mathbf{P}(k, k-1)$	$N \times N$
Measurement Noise Covariance Matrix	$\mathbf{R}(k)$	$M \times M$
System Noise Covariance Matrix	$\mathbf{Q}(k)$	$N \times N$
State Error Covariance Matrix	$\mathbf{P}(k)$	$N \times N$

Table 3.4: Summary of the Kalman vector and matrix variables, where M is the dimension of the measurement vector, while $\mathbf{y}(k)$ and N represents the dimension of the estimated parameter vector $\hat{\mathbf{x}}(k)$.

System parameters	$\mathbf{y}(k)$	$\mathbf{A}$	$\mathbf{C}$	$\mathbf{R}(k)$	$\mathbf{Q}(k)$

Table 3.5: System parameters in the Kalman algorithm.

3.1.3   Kalman Recursive Process

In this section, the recursive Kalman process is elaborated on and an overview of the process is provided, which is shown in Figure 3.2, depicting the recursive mechanism of the Kalman equations listed in Table 3.3. Explicitly, the recursive Kalman process can be separated into 5 stages, labelled as Stage $0-4$, which are portrayed in Figure 3.2.

In the Initialization Process, all the system parameters listed in Table 3.5 are defined. The unbiased initialization of the estimated parameter matrix $\hat{\mathbf{x}}(k)$, and that of the predicted state error covariance matrix $\mathbf{P}(k, k-1)$, ensue according to Haykin as follows [118]:

$$\hat{\mathbf{x}}(0) = \mathbf{0}. \tag{3.46}$$

$$\mathbf{P}(0) = \delta\mathbf{I}, \tag{3.47}$$

where $\mathbf{I}$ is the $N \times N$ Identity matrix and δ is a constant.

After the initialization stage, the Kalman gain matrix $\mathbf{K}(k)$, the estimated parameter matrix $\hat{\mathbf{x}}(k)$, the state error covariance matrix $\mathbf{P}(k)$ and the predicted state error covariance

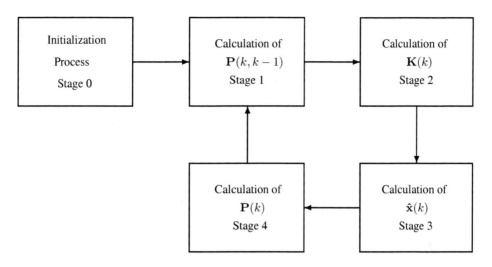

Figure 3.2: Recursive Kalman process.

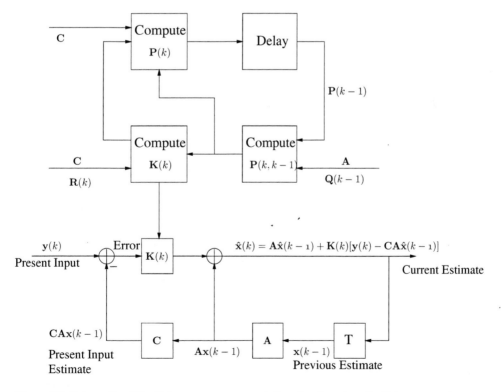

Figure 3.3: Schematic of the Kalman process computation steps based on the Kalman recursive equations of Table 3.3.

matrix $\mathbf{P}(k, k - 1)$ are computed according to the recursive Kalman equations listed in Table 3.3. The sequence of operations is portrayed in Figure 3.3, which displays the generation of the current estimated parameter matrix $\hat{\mathbf{x}}(k)$ based on the recursive Kalman equations of Table 3.3.

In summary, in this section we have derived the Kalman algorithm's recursive equations and described how the equations fit in a recursive process. The next section will introduce the application of the Kalman algorithm in the context of adaptive channel estimation and adaptive channel equalization.

3.2   Application of the Kalman Algorithm

The application of the Kalman algorithm to channel estimation and equalization is explored here. In this section we shall manipulate the recursive Kalman equations in order to form the channel estimator and equalizer. The convergence capability of the algorithm in both of these applications will also be investigated in this section. Finally, the performance of the adaptive DFE using the Kalman algorithm is quantified in the context of dispersive Rayleigh fading multi-path channels.

3.2.1   Recursive Kalman Channel Estimator

In a fading channel environment the receiver of a communications system has to optimally detect the corrupted symbols without the actual knowledge of the *exact* CIR encountered. The channel estimator is used to provide the receiver with an *estimate of the CIR* and thus to assist the equalizer in order to compensate for the dispersive multi-path fading environment. The channel estimator can estimate the channel with the aid of a training sequence in the transmitted burst. The training sequence is a string of symbols that is known both to the transmitter and receiver. Consequently, at the receiver, the channel estimator has access to the transmitted training sequence and to the corrupted received training sequence. By comparing these two sequences, the channel estimator can reconstruct the CIR that caused the corruption of the received training sequence. For more information on the training sequences and their properties, please refer to Milewski [140] and Steele [13].

In this estimation procedure, the Kalman algorithm can be employed in order to adaptively estimate the CIR based on the received and transmitted training sequence. The operation of the adaptive Recursive Kalman Channel Estimator (RKCE) is portrayed in Figure 3.4. In Figure 3.4, the adaptive **RKCE** consists of a channel estimator, which is a FIR filter and a Kalman process shown in Figure 3.2. The FIR filter stores the estimated CIR $\hat{\mathbf{h}}(k)$, and it generates the convolution of the transmitted training sequence $\mathbf{T_x}(k)$, and the estimated CIR. The output $\mathbf{T_x}(k) * \hat{\mathbf{h}}(k)$, is then subtracted from the received training sequence $\mathbf{R_x}(k)$, in order to form the error signal $\mathbf{e}(k)$. Subsequently the Kalman process applies the error signal $\mathbf{e}(k)$, and the transmitted training sequence $\mathbf{T_x}(k)$, in order to form the next best estimate of the CIR. This provides the channel estimator with a new CIR estimate in order to begin a new iteration. This process is then repeated, until convergence is achieved i.e until the MSE $(\mathbf{E}[|\mathbf{e}(k)|^2])$, is at its minimum value [119]. However, usually the number of iterations the algorithm can invoke is restricted by the length of the training sequence, since the algorithm can only generate a new error and a new CIR estimate with the aid of the training sequence

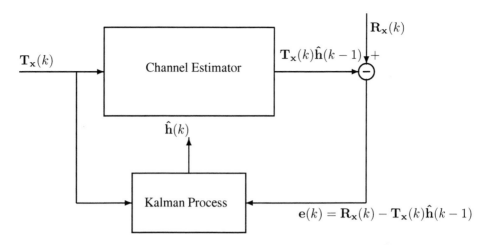

Figure 3.4: Schematic of the adaptive recursive Kalman channel estimator.

upon receiving a new symbol and hence the number of iterations is limited by the length of the training sequence. However, in the so-called decision-directed mode, the past symbol decisions can also be used to generate the error signal and hence to estimate the CIR. Consequently the entire transmission burst can be used to drive the algorithm, thus increasing the number of iterations.

The state space model, which was defined by Equations 3.42 and 3.43 can now be rewritten for the **RKCE** as :

$$\mathbf{h}(k) = \mathbf{h}(k-1) + \mathbf{W}(k-1), \tag{3.48}$$

$$\mathbf{R_x}(k) = \mathbf{T_x}(k)\mathbf{h}(k-1) + \mathbf{V}(k). \tag{3.49}$$

Physically Equation 3.48 implies that the previous and current CIR $\mathbf{h}(k-1)$ and $\mathbf{h}(k)$, differ from each other by a random vector $\mathbf{W}(k-1)$. Similarly, the received training sequence $\mathbf{R_x}(k)$, and the convolution of the transmitted one $\mathbf{T_x}(k)$ with the CIR $\mathbf{h}(k-1)$ differ also by a random vector $\mathbf{V}(k)$ in Equation 3.49. In forming these equations, the state transition matrix $\mathbf{A}$, of Equation 3.42 was assigned to be an identity matrix, since the receiver cannot anticipate the channel variation from state to state. We can now rewrite the Kalman recursive equations listed in Table 3.3 as :

$$\hat{\mathbf{h}}(k) = \hat{\mathbf{h}}(k-1) + \mathbf{K}(k)[\mathbf{R_x}(k) - \mathbf{T_x}(k)\hat{\mathbf{h}}(k-1)], \tag{3.50}$$

$$\mathbf{K}(k) = \mathbf{P}(k, k-1)\mathbf{T_x}(k)^*[\mathbf{T_x}(k)\mathbf{P}(k, k-1)\mathbf{T_x}(k)^{*\mathbf{T}} + \mathbf{R}(k)]^{-1}, \tag{3.51}$$

$$\mathbf{P}(k, k-1) = \mathbf{P}(k-1) + \mathbf{Q}(k-1), \tag{3.52}$$

$$\mathbf{P}(k) = \mathbf{P}(k, k-1) - \mathbf{K}(k)\mathbf{T_x}(k)\mathbf{P}(k, k-1). \tag{3.53}$$

	δ Variation	$\mathbf{R}(k)$ Variation	$\mathbf{Q}(k)$ Variation
Modulation	BPSK	BPSK	BPSK
No. of transm. bursts	100	100	100
Non-fading CIR	$0.707 + 0.707z^{-1}$	$0.707 + 0.707z^{-1}$	$0.707 + 0.707z^{-1}$
Channel SNR	30dB	30dB	30dB
No. of taps in **RKCE**	2	2	2
$\mathbf{R}(k)$	**I**	-	**I**
$\mathbf{Q}(k)$	0.1**I**	$0.1 \times \mathbf{I}$	–
δ (see Equation 3.47)	-	300	300
$\hat{\mathbf{h}}(\text{o})$	0	0	0

Table 3.6: Simulation parameters used to quantify the effects of δ, $\mathbf{R}(k)$ and $\mathbf{Q}(k)$ on the convergence of the **RKCE** of Figure 3.4.

Again, the recursive Kalman process cast in the context of the **RKCE** was depicted in Figure 3.2, where the estimated parameter is the CIR estimate $\hat{\mathbf{h}}(k)$, instead of $\hat{\mathbf{x}}(k)$. Now that we have formally presented the application of the Kalman algorithm in the context of recursive CIR estimation, we will proceed to investigate its convergence performance.

3.2.2 Convergence Analysis of the Recursive Kalman Channel Estimator

In this section, the convergence behaviour of the **RKCE** is investigated in relation to two different performance measures. The first measure is based on the so-called classic learning curve analysis, where the ensemble squared *a priori* estimation error $E[|\mathbf{e}(k)|^2]$, is observed with respect to the number of iterations. The second performance measure uses the CIR estimation error ϱ, which was defined by Cheung [104] as:

$$\varrho(dB) = 10log_{10}\left[\frac{\sum_{i=0}^{L_c-1} |\hat{h}_i - h_i|^2}{\sum_{i=0}^{L_c-1} |h_i|^2}\right], \tag{3.54}$$

where L_c represents the length of the CIR and $\hat{h}_i$ is the estimate of the actual CIR h_i at each iteration. These two performance measures are used to gauge the convergence performance of the **RKCE**, when the system parameters listed in Table 3.5 are varied. The parameters involved are the measurement noise covariance matrix $\mathbf{R}(k)$, the system noise covariance matrix $\mathbf{Q}(k)$ and δ, the constant initialization variable defined in Equation 3.47. We can now proceed to investigate the effects of varying δ on the convergence performance of the **RKCE**.

3.2.2.1 Effects of Varying δ in a Recursive Kalman Channel Estimator

In this section, simulations were performed in order to highlight the effects of the initialization variable δ defined in Equation 3.47. The matrix dimension of all the Kalman variables corresponds to $M = 1$ and $N = 2$, when applied to Table 3.4. The simulation parameters in this experiment are defined in Table 3.6.

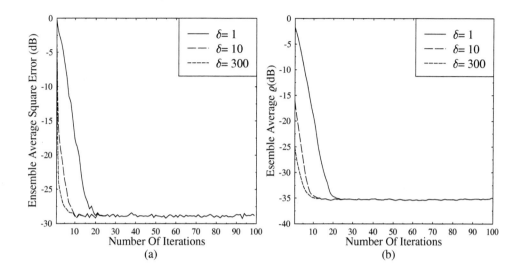

Figure 3.5: Convergence analysis of the **RKCE**, when δ of Equation 3.47 was varied using the simulation parameters listed in Table 3.6. (a) Ensemble average square error $E|e(k)|^2$ versus number of iterations. (b) Ensemble average CIR estimation error of Equation 3.54, ϱ versus number of iterations.

The results of our experiments are shown in Figure 3.5. The curves demonstrate that as δ was increased, the number of iterations needed for the algorithm to converge decreased and the final MSE was the same for all investigated values of δ simulated. This was shown in Figure 3.5(a), where at $\delta = 1$ the number of iterations needed to attain convergence was approximately 20, while at $\delta = 300$ the number of required iterations was approximately 10. The same trend was also observed in the CIR estimation error analysis of Figure 3.5(b). This correspondence was expected, since as the algorithm converged, the CIR estimate $\hat{h}_i$, approached that of the actual CIR h_i.

We can explain the trend upon varying δ by studying the Kalman recursive equations of Table 3.3 for the **RKCE** and by deriving a new mathematical formulation for the Kalman gain $\mathbf{K}(k)$. We commence the derivation by substituting Equation 3.51 into Equation 3.53, yielding:

$$\mathbf{P}(k) = \mathbf{P}(k, k-1) - \mathbf{P}(k, k-1)\mathbf{T_x}(k)^{*\mathbf{T}} \cdot$$
$$[\mathbf{R}(k) + \mathbf{T_x}(k)\mathbf{P}(k, k-1)\mathbf{T_x}(k)^{*\mathbf{T}}]^{-1}\mathbf{T_x}(k)\mathbf{P}(k, k-1). \quad (3.55)$$

We can apply the well-known matrix inversion lemma, which was defined by Scharf [141] as:

$$(\mathbf{A} + \mathbf{BCD})^{-1} = \mathbf{A}^{-1} - \mathbf{A}^{-1}\mathbf{B}(\mathbf{DA}^{-1}\mathbf{B} + \mathbf{C}^{-1})^{-1}\mathbf{DA}^{-1}, \quad (3.56)$$

where upon assigning $\mathbf{A}^{-1} = \mathbf{P}(k, k-1)$, $\mathbf{B} = \mathbf{T_x}(k)^*$, $\mathbf{C} = \mathbf{R}(k)$ and $\mathbf{D} = \mathbf{T_x}(k)$, we obtain :

$$\mathbf{P}(k) = [\mathbf{P}(k, k-1)^{-1} + \mathbf{T_x}(k)^{*\mathbf{T}}\mathbf{R}(k)^{-1}\mathbf{T_x}(k)]^{-1}. \qquad ' \qquad (3.57)$$

Referring again to Equation 3.51, we utilize $\mathbf{P}(k)\mathbf{P}(k)^{-1} = \mathbf{I}$ and $\mathbf{R}(k)^{-1}\mathbf{R}(k) = \mathbf{I}$ to give:

$$
\begin{aligned}
\mathbf{K}(k) &= \mathbf{P}(k)\mathbf{P}(k)^{-1}\mathbf{P}(k, k-1)\mathbf{T_x}(k)^*\mathbf{R}(k)^{-1}\mathbf{R}(k) \cdot \\
&\quad [\mathbf{T_x}(k)\mathbf{P}(k, k-1)\mathbf{T_x}(k)^{*\mathbf{T}}\mathbf{R}(k)^{-1}\mathbf{R}(k) + \mathbf{R}(k)]^{-1} \\
&= \mathbf{P}(k)\mathbf{P}(k)^{-1}\mathbf{P}(k, k-1)\mathbf{T_x}(k)^*\mathbf{R}(k)^{-1} \cdot \\
&\quad [\mathbf{T_x}(k)\mathbf{P}(k, k-1)\mathbf{T_x}(k)^{*\mathbf{T}}\mathbf{R}(k)^{-1} + \mathbf{I}]^{-1}. \qquad (3.58)
\end{aligned}
$$

We can now re-substitute Equation 3.57 into the above equation, in order to yield the final expression for the Kalman gain as :

$$
\begin{aligned}
\mathbf{K}(k) &= \mathbf{P}(k)[\mathbf{P}(k, k-1)^{-1} + \mathbf{T_x}(k)^{*\mathbf{T}}\mathbf{R}(k)^{-1}\mathbf{T_x}(k)]\mathbf{P}(k, k-1)\mathbf{T_x}(k)^*\mathbf{R}(k)^{-1} \cdot \\
&\quad [\mathbf{T_x}(k)\mathbf{P}(k, k-1)\mathbf{T_x}(k)^{*\mathbf{T}}\mathbf{R}(k)^{-1} + \mathbf{I}]^{-1} \\
&= \mathbf{P}(k)\mathbf{T'_x}(k)\mathbf{R}(k)^{-1}. \qquad (3.59)
\end{aligned}
$$

By observing Equation 3.59, we can deduce that the Kalman gain $\mathbf{K}(k)$, is proportional to the parameter error covariance matrix $\mathbf{P}(k)$, of Table 3.4 and inversely proportional to the measurement error covariance matrix $\mathbf{R}(k)$, of Table 3.4. This relationship is used to explain the trend shown in Figure 3.5. As δ increased, the initial value of $\mathbf{P}(k)$ also increased as a result of Equation 3.47. Therefore the initial Kalman gain was high according to Equation 3.59. Consequently, the second term of Equation 3.50, which was the measurement term, was dominant at the initial stage due to the high Kalman gain. This resulted in a faster convergence, since it reached the minimum MSE faster.

However at a later stage, the Kalman gain reduced, since $\mathbf{P}(k)$ decreased as dictated by Equation 3.53. Thus, at the this stage the Kalman gain was stabilized, irrespective of the initial value of δ, yielding the same steady-state MSE, as evidenced by Figure 3.5 for different values of δ. In this simple exercise, we observed that the convergence performance of the algorithm pivoted upon the Kalman gain, where the gain value was adapted automatically according to the current MSE with the aim of achieving the minimum MSE.

We have quantified and interpreted the convergence effects due to different initializing values of δ, hence we can now concentrate on the effects of the next Kalman variable, which is the measurement noise covariance matrix $\mathbf{R}(k)$.

3.2.2.2 Effects of Varying $\mathbf{R}(k)$ in a Recursive Kalman Channel Estimator

In this section, the effect of the measurement error covariance matrix $\mathbf{R}(k)$, on the convergence of the algorithm is investigated. The simulation parameters for this experiment were defined in Table 3.6 and the matrix $\mathbf{R}(k)$ was defined as :

$$\mathbf{R(k)} = g\mathbf{I}, \qquad (3.60)$$

where g is a positive real constant. The effects of varying $\mathbf{R}(k)$ are shown in Figure 3.6. As

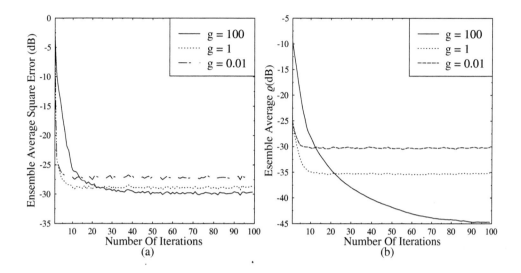

Figure 3.6: Convergence analysis of the **RKCE** parameterized with g of Equation 3.60, when $\mathbf{R}(k)$ of Equation 3.44 was varied using the simulation parameters listed in Table 3.6. (a) Ensemble average square error $E|e(k)|^2$ versus number of iterations. (b) Ensemble average CIR estimation error of Equation 3.54, ϱ versus number of iterations.

g was increased, the algorithm converged slower, where according to Figure 3.6, for $g = 100$ approximately 40 iterations were needed in order to converge, while for $g = 1$ approximately 10 iterations were required. The other significant effect was that the value of the final steady-state MSE decreased, when g was increased, as evidenced by Figure 3.6.

The different convergence rate for different values of $\mathbf{R}(k)$ was explained by invoking Equation 3.59, where the Kalman gain was inversely proportional to $\mathbf{R}(k)$. As the values in $\mathbf{R}(k)$ increased, the Kalman gain decreased, yielding a slower convergence rate according to the arguments of Section 3.2.2.1.

However, as the algorithm converged towards its optimum MSE value, higher values of $\mathbf{R}(k)$ resulted in lower Kalman gain. Since there was no controlling mechanism for reducing $\mathbf{R}(k)$, the algorithm having a lower Kalman gain possessed an increased resolution in the search for a lower MSE, compared to having a higher Kalman gain. Therefore the **RKCE** having the higher values of $\mathbf{R}(k)$ converged to a lower MSE, but at the expense of a lower convergence speed.

3.2.2.3   Effects of Varying $\mathbf{Q}(k)$ in a Recursive Kalman Channel Estimator

The system error covariance matrix $\mathbf{Q}(k)$, is a parameter that influences the adaptivity of the **RKCE**. Observing Equations 3.45 and 3.48, $\mathbf{Q}(k)$ predicts the time varying characteristics of the system, which in this case, is the transmission channel and subsequently adapts the

Kalman recursive process in order to track the CIR by applying Equation 3.52. The physical significance of Equation 3.52 is that the parameter $\mathbf{Q}(k)$ prompts the recursive Kalman process to always search for the optimum MSE. This is extremely important for improving the adaptivity of the **RKCE** in a time varying channel. Another way of explaining the significance of $\mathbf{Q}(k)$ is that this parameter matrix provides the ability for the **RKCE** to deduce the CIR by relying more on the current received signals than on past signals received. Therefore, for a fast adapting **RKCE**, the system's error covariance matrix $\mathbf{Q}(k)$, has to contain high values and vice-versa.

The other function of $\mathbf{Q}(k)$ is to provide the algorithm with a measure of stability. The main source of instability is when the parameter $\mathbf{P}(k-1)$ becomes singular and hence it does not have an inverse. Without the presence of $\mathbf{Q}(k)$ in Equation 3.52, the parameter $\mathbf{P}(k, k-1)$ would then also become singular. This is catastrophic, since Equation 3.51 cannot be evaluated, because it requires the inverse of $\mathbf{P}(k, k-1)$. This will lead to a phenomenon referred to as divergence [142]. However, by introducing the parameter matrix $\mathbf{Q}(k)$ in Equation 3.52, $\mathbf{P}(k, k-1)$ is prevented from becoming singular and this stabilises the algorithm as stated by Haykin [118] and Brown *et al.* [142]. This stability can be further improved by composing $\mathbf{Q}(k)$ as a diagonal matrix with positive real constants on its diagonal as it was demonstrated by Harun *et al.* [124].

For a stationary environment, where there is no difference between the current channel impulse response and the previous response, Equation 3.48 is reduced to :

$$\mathbf{h}(k) = \mathbf{h}(k-1), \tag{3.61}$$

where there is no need for the matrix $\mathbf{Q}(k)$. However, in order to ensure stability, $\mathbf{Q}(k)$ is needed and it is usually assigned as :

$$\mathbf{Q}(k-1) = q\mathbf{I}, \tag{3.62}$$

where q is a positive real constant.

Having highlighted the significance of $\mathbf{Q}(k)$, we can now proceed to investigate its effects on the convergence of the algorithm. The simulation parameters used in this investigation were those defined in Table 3.6.

Upon observing Figure 3.7 the number of iterations needed for the algorithm to converge was approximately 10 and it was about the same for the different values of q in Equation 3.62. However, the final MSE and CIR estimation error ϱ, were higher for larger values of q. At the initial stage, the rate of convergence was the same for different values of q, since the values in the matrix $\mathbf{P}(k-1)$ of Equation 3.47 were dominated by the initializing variable δ, which was set to 300, as seen in Table 3.6 and highlighted in Section 3.2.2.1. However, at a later stage, $\mathbf{Q}(k)$ of Equation 3.62 began to dominate, since the contribution of $\mathbf{P}(k-1)$ was reduced according to Equation 3.53. In contrast to $\mathbf{P}(k-1)$, $\mathbf{Q}(k)$ was constant and not controlled by any mechanism. As a result, the Kalman gain was dependent on the values in $\mathbf{Q}(k)$ where if q was low, the algorithm possessed an increased resolution in its search for a lower MSE, compared to when q was high. This trend was similar to those observed in Section 3.2.2.1 for the variation of δ, where a high value of δ implied a higher Kalman gain and vice-versa. This can also be explained physically, by noting that, since the channel used was stationary as seen in Table 3.6, $\mathbf{Q}(k-1) = \mathbf{0}$ will model the channel perfectly as it was argued earlier. Therefore as q decreased, the **RKCE** modelled the actual CIR more closely.

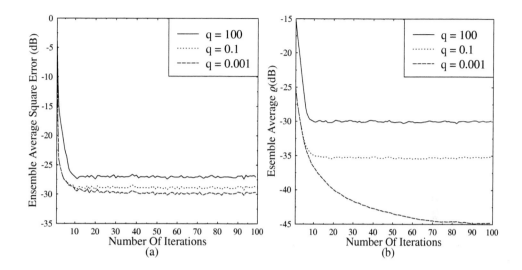

Figure 3.7: Convergence analysis of the **RKCE** parameterized with q of Equation 3.62, when the system error covariance matrix $\mathbf{Q}(k)$ of Equation 3.45 was varied using the simulation parameters listed in Table 3.6. (a) Ensemble average square error $E|e(k)|^2$ versus number of iterations. (b) Ensemble average CIR estimation error of Equation 3.54, ϱ versus number of iterations.

This, in turn, provided a better CIR estimate and consequently yielded a lower final MSE and CIR estimation error.

From our previous arguments we can see that the matrix $\mathbf{Q}(k-1)$ introduced a degree of sub-optimality in a stationary environment at the expense of stability. In the last few sections we have discussed and quantified the effects of certain Kalman variables on the convergence of the **RKCE** algorithm. In the next section, we will determine the optimum values of these Kalman variables under the constraint of a limited number of iterations or, equivalently, for a fixed training sequence length.

3.2.2.4   Recursive Kalman Channel Estimator Parameter Settings

In this section we will determine the desirable settings of the Kalman variables, $\mathbf{R}(k)$, $\mathbf{Q}(k)$ as well as δ and therefore create an optimum **RKCE** for a stationary environment. The performance measure used to determine these settings is the CIR estimation error ϱ. In these experiments, the Kalman variables are varied one at a time and the associated results are shown in Figure 3.8. The simulation parameters are once again listed in Table 3.6 and the training sequence length L_T, was 20 symbols.

Upon observing Figure 3.8(a), the variable δ had no effect on the CIR estimation error, irrespective of the channel SNR, where for $SNR = 30$dB, ϱ was approximately -35dB. Previously, in Section 3.2.2.1, when opting for $\delta = 1$, the number of iterations needed for the

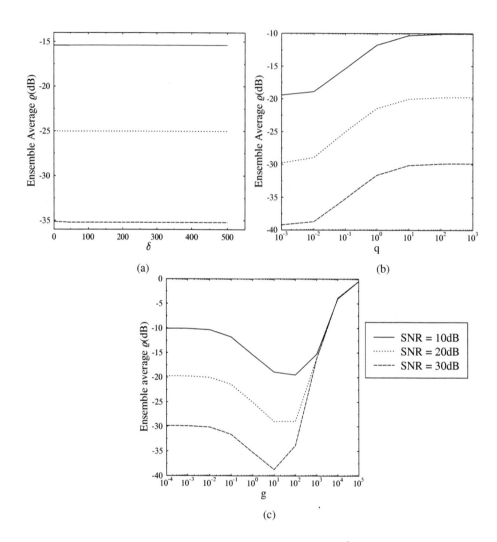

Figure 3.8: CIR estimation error analysis of the **RKCE** for channel SNRs of 10, 20 and 30dB, when
(a) δ was varied (b) $\mathbf{Q}(k) = q\mathbf{I}$ was varied (c) $\mathbf{R}(k) = g\mathbf{I}$ was varied. The simulation
parameters are listed in Table 3.6 and the training sequence length was $L_T = 20$ symbols.

algorithm to converge was approximately 20 and the CIR estimation error was constant after convergence, as it was shown in Figure 3.5(b). This implied that the algorithm converged within the training sequence length of $L_T = 20$ symbols and consequently varying δ had no effect on the CIR estimation error. We shall now consider the optimum setting for the matrix $\mathbf{Q}(k)$.

The simulation results related to the variation of $\mathbf{Q}(k)$ are shown in Figure 3.8(b). These performance curves corroborated the previous results shown in Figure 3.7, where the CIR estimation error degraded, when q in Equation 3.62 was increased. Specifically there was a marked degradation of approximately 8dB, when q was increased from 0.01 to 1, as seen in Figure 3.8(b). From this result, we can estimate the optimum setting for $\mathbf{Q}(k)$ as :

$$\mathbf{Q}(k) \approx 0.01\mathbf{I} \qquad\qquad (3.63)$$

For the investigated channel SNRs of 10, 20 and 30dB, the optimum setting for $\mathbf{R}(k)$ can be found by referring to Figure 3.8(c). For $g > 10$, the CIR estimation error degraded significantly at all the investigated channel SNRs. For example, for SNR = 20dB, the CIR estimation error was degraded by approximately 13dB, when $\mathbf{R}(k)$ was increased from 10$\mathbf{I}$ to 1000$\mathbf{I}$. Therefore, we can assign $\mathbf{R}(k)$ as :

$$\mathbf{R}(k) = g\mathbf{I} \quad \text{where} \quad g \leq 10 \qquad\qquad (3.64)$$

By considering the channel model and system parameters of Table 3.6, we have now defined the desirable settings for the **RKCE** with respect to the CIR estimation error. This implied that the settings were chosen such that the CIR estimates approximated the actual CIR as closely as possible. In the next section we will highlight the application of the Kalman algorithm directly in the context of the DFE, rather than using it as a channel estimator.

3.2.3   Recursive Kalman Decision Feedback Equalizer

The application of the Kalman algorithm to equalization is similar to that of the **RKCE** portrayed in Figure 3.4. The schematic of the adaptive Recursive Kalman Decision Feedback Equalizer (RKDFE) is shown in Figure 3.9, where the adaptive **RKDFE** consists of a DFE and a Kalman process.

The Kalman process provides the estimate of the DFE coefficient vector $\hat{\mathbf{c}}(k)$ and it continuously updates these coefficient estimates based on the error signal $\mathbf{e}(k)$ and the received training sequence, $\mathbf{R_x}(k)$, during the entire length of the training sequence.

Similarly to the **RKCE**, we can now proceed to rewrite the state model equations for the **RKDFE** as :

$$\mathbf{c}(k) = \mathbf{c}(k-1) + \mathbf{W}(k-1) \qquad\qquad (3.65)$$

$$\mathbf{T_x}(k) = \mathbf{R_x}(k)\mathbf{c}(k-1) + \mathbf{V}(k) \qquad\qquad (3.66)$$

where these equations correspond to Equations 3.48 and 3.49 in the context of **RKCE** of Figure 3.4. Physically Equation 3.65 implies that the previous and current DFE coefficients $\mathbf{c}(k-1)$ and $\mathbf{c}(k)$, differ from each other by a random vector $\mathbf{W}(k-1)$. Similarly, the transmitted training sequence $\mathbf{T_x}(k)$ and the convolution of the received one $\mathbf{R_x}(k)$, with

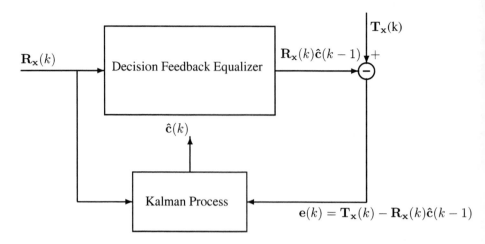

Figure 3.9: Schematic of the adaptive recursive Kalman DFE.

the DFE coefficients $c(k - 1)$ differ also by a random vector, namely $\mathbf{V}(k)$ in Equation 3.66. We can mirror the procedure defined in Section 3.2.1 for the **RKCE** and rewrite the recursive Kalman equations listed in Table 3.3 for the **RKDFE** as :

$$\hat{\mathbf{c}}(k) = \hat{\mathbf{c}}(k - 1) + \mathbf{K}(k)[\mathbf{T_x}(k) - \mathbf{R_x}(k)\hat{\mathbf{c}}(k - 1)], \qquad (3.67)$$

$$\mathbf{K}(k) = \mathbf{P}(k, k - 1)\mathbf{R_x}(k)^*[\mathbf{R_x}(k)\mathbf{P}(k, k - 1)\mathbf{R_x}(k)^{*\mathbf{T}} + \mathbf{R}(k)]^{-1}, \qquad (3.68)$$

$$\mathbf{P}(k, k - 1) = \mathbf{P}(k - 1) + \mathbf{Q}(k - 1), \qquad (3.69)$$

$$\mathbf{P}(k) = \mathbf{P}(k, k - 1) - \mathbf{K}(k)\mathbf{R_x}(k)\mathbf{P}(k, k - 1). \qquad (3.70)$$

We note that the Kalman process in this application treated the forward and backward co-efficients of the DFE of Figure 3.9 identically. The generic recursive Kalman process was depicted in Figure 3.2. In the context of the **RKDFE**, the estimated parameter is the DFE coefficient vector $\hat{\mathbf{c}}(k)$, instead of $\hat{\mathbf{x}}(k)$. As before, we will now investigate the convergence capability of the **RKDFE** with the aim of establishing a desirable parameter setting for the adaptive **RKDFE**.

3.2.4   Convergence Analysis of the Recursive Kalman Decision Feedback Equalizer

Three different performance measures are used below for investigating the convergence performance of the **RKDFE**. As before, the classic learning curve analysis was used as one of the measures. The other measures used are the Equalizer Coefficient Estimation Error (ECEE) ξ and the Average Bit Error Rate (ABER).

	δ Variation	$\mathbf{R}(k)$ Variation	$\mathbf{Q}(k)$ Variation
Modulation	BPSK	BPSK	BPSK
Number of tr. bursts	100	100	100
Length of tr. bursts	500 symbols	500 symbols	500 symbols
Non-fading CIR	$0.707 + 0.707z^{-1}$	$0.707 + 0.707z^{-1}$	$0.707 + 0.707z^{-1}$
Channel SNR	30dB	30dB	30dB
No. of feedfwd. taps, N_f	7	7	7
No. of feedback taps, N_b	1	1	1
Decision feedback	Correct	Correct	Correct
$\mathbf{R}(k)$	$\mathbf{I}$	-	$\mathbf{I}$
$\mathbf{Q}(k)$	$0.1 \times \mathbf{I}$	$0.1 \times \mathbf{I}$	$-$
δ (see Equation 3.47)	-	300	300
$\hat{\mathbf{c}}(\mathrm{o})$	$\mathbf{0}$	$\mathbf{0}$	$\mathbf{0}$

Table 3.7: Simulation parameters used to quantify the effects of δ, $\mathbf{R}(k)$ and $\mathbf{Q}(k)$ on the convergence of the **RKDFE**.

The ECEE is defined similarly to the CIR estimation error, namely as :

$$\xi(dB) = 10log_{10}\left[\frac{\sum_{i=0}^{N_f+N_b}|\hat{c}_i - c_i|^2}{\sum_{i=0}^{N_f+N_b}|c_i|^2}\right], \tag{3.71}$$

where N_f and N_b represent the length of the feedforward filter and the feedback filter of the DFE, respectively, while c_i denotes the optimum coefficients of the DFE, when perfect channel estimation is applied.

The ABER is simply defined as :

$$\text{ABER} = \frac{\text{Total number of bit errors}}{\text{Total number of bits used in the simulation}}. \tag{3.72}$$

This ABER was calculated based on an average of 100 transmission bursts, each containing 500 symbols per iteration. The same Kalman variables are varied as in Section 3.2.1 and the convergence effects are quantified and interpreted in the next section.

3.2.4.1   Effects of Varying δ in a Recursive Kalman Equalizer

In this section, the convergence effects of varying the parameter δ defined by Equation 3.47 are investigated. The matrix size of the Kalman variables corresponds to the dimensions of $M = 1$ and $N = N_f + N_b$, when applied to Table 3.4 and the simulation parameters are defined in Table 3.7.

The simulation results shown in Figure 3.10 revealed similar characteristics to those in Figure 3.5, where the convergence rate degraded, when δ was decreased. These characteristics were justified in the context of the **RKCE** with reference to Equation 3.59 and in our elaborations following Equation 3.59. Upon observing Figure 3.10(c), for $\delta = 300$, the number of iterations needed for the algorithm to converge was approximately 20, which was about twice that of the **RKCE**. This degraded performance, when compared to the **RKCE**

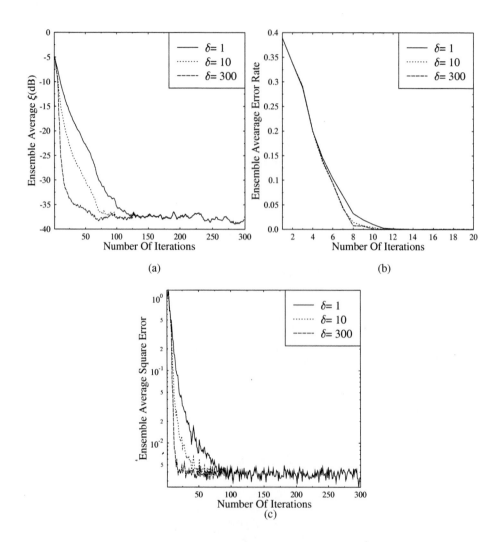

Figure 3.10: Convergence analysis of the **RKDFE**, when δ of Equation 3.47 was varied using the simulation parameters listed in Table 3.7. (a) Ensemble average equalizer coefficients estimation error ξ, versus the number of iterations. (b) Ensemble average error rate versus the number of iterations. (c) Ensemble average square error versus number of iterations.

was simply because the **RKDFE** had to estimate more taps than the **RKCE**. In general, the speed of convergence was inversely proportional to the length of the estimated vectors and hence applications having a low number of taps to be estimated converged typically faster. This was consistent with the results shown by Haykin [118] and Shukla *et al.* [125].

The degradation of the convergence time for different δ values was also more significant for the **RKDFE** than for the **RKCE**, where for $\delta = 1$ the convergence time was approximately 100 iterations in Figure 3.10. This was roughly a factor of five degradation, when compared to the convergence time shown for $\delta = 300$. The ABER exhibited minimal degradation in terms of its convergence time, where for $\delta = 300$ the convergence time was eight iterations, while for $\delta = 1$ the required number of iterations was 12.

3.2.4.2   Effects of Varying $\mathbf{R}(k)$ in a Kalman Equalizer

Below, we shall repeat the analysis completed in Section 3.2.2.2, but applied to the **RKDFE**. The parameter g of Equation 3.60 previously defined in Section 3.2.2.2 was varied and the convergence performance was quantified in Figure 3.11. Once again, the ensemble average squared error and the ECEE metric of Equation 3.71 ξ, exhibited the same trend, as that of the **RKCE** and the interpretations in Section 3.2.2.2 can be utilized here in the context of a **RKDFE**. The ABER performance was approximately the same for all values of g investigated.

3.2.4.3   Effects of Varying $\mathbf{Q}(k)$ in a Kalman Equalizer

Let us now study the effects of $\mathbf{Q}(k-1)$, which can be varied with the aid of a real constant q, according to the relationship:

$$\mathbf{Q}(k-1) = q\mathbf{I}, \tag{3.73}$$

where q is a real positive constant.

The ensemble average ECEE and the ensemble average square error shown in Figure 3.12(a) and Figure 3.12(c) exhibited similar characteristics to those of the **RKCE**. Therefore, the justifications of the trends provided in Section 3.2.2.3 can also be applied to the **RKDFE**. The ABER shown in Figure 3.12(b) converged to its minimum in eight iterations, when q was 0.1. We shall now proceed to quantify the desirable setting for the **RKDFE**.

3.2.4.4   Recursive Kalman Decision Feedback Equalizer Desirable Settings

Here, we continue our discourse by quantifying the parameter settings for the **RKDFE** in a stationary environment following the philosophy of Section 3.2.2.4. The ECEE is used as the performance measure in order to obtain the required settings. The simulation parameters are listed in Table 3.7 and the training sequence length L_T was 20 symbols.

Referring to Figure 3.13(a), the ECEE improved up to the threshold value of $\delta = 100$ and subsequently it remained constant. Thus, for $L_T = 20$ the minimum setting for δ is 100. The ECEE remained constant for q values less than 0.1 and subsequently for increased q values, the estimation error degraded significantly, as shown in Figure 3.13(b). Similar tendencies can be observed in Figure 3.13(c), where for g higher than 1 there was a significant degradation in the ECEE.

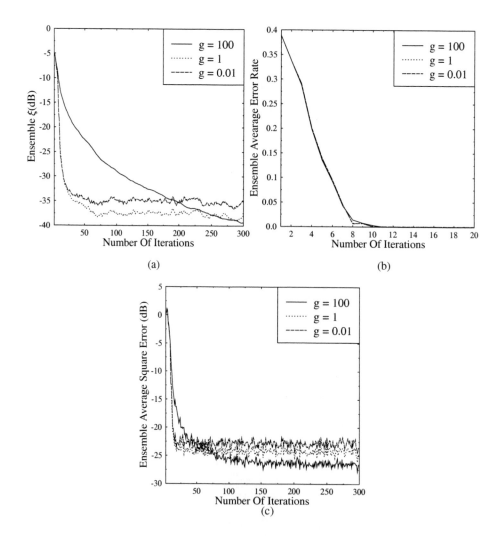

Figure 3.11: Convergence analysis of the **RKDFE**, when $\mathbf{R}(k) = g\mathbf{I}$ was varied using the simulation parameters listed in Table 3.7. (a) Ensemble average equalizer coefficient estimation error ξ, versus the number of iterations. (b) Ensemble average error rate versus the number of iterations. (c) Ensemble average square error versus the number of iterations.

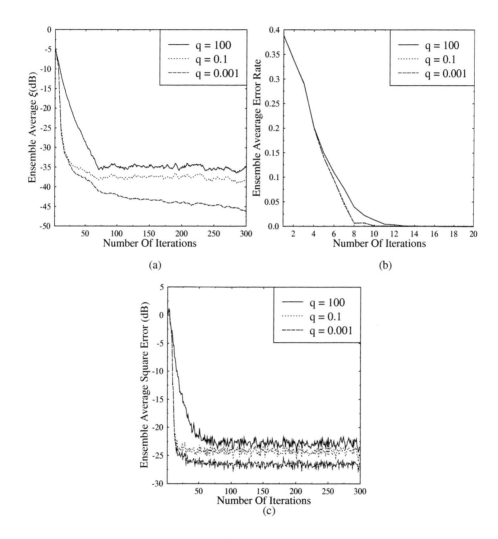

Figure 3.12: Convergence analysis of the **RKDFE**, when $\mathbf{Q}(k-1) = q\mathbf{I}$ was varied employing the simulation parameters listed in Table 3.7. (a) Ensemble average equalizer coefficient estimation error ξ, versus the number of iterations. (b) The Ensemble average error rate versus the number of iterations (c) Ensemble average square error versus the number of iterations.

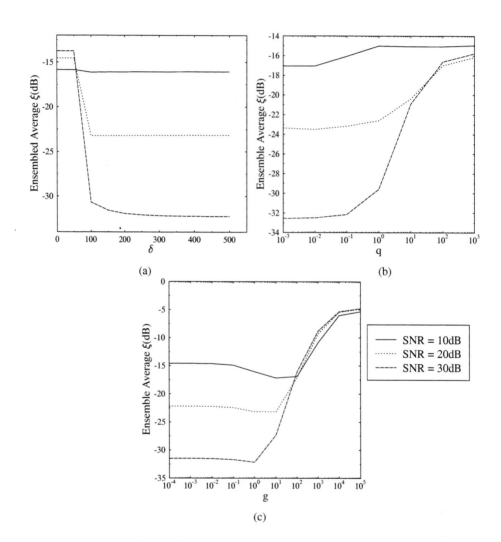

Figure 3.13: Equalizer coefficient estimation error analysis of the **RKDFE** for SNRs of 10, 20 and 30dB, when (a) δ was varied (b) $\mathbf{Q}(k) = q\mathbf{I}$ was varied (c) $\mathbf{R}(k) = g\mathbf{I}$ was varied. The simulation parameters are listed in Table 3.7 and the training sequence length was $L_T = 20$ symbols.

Since we have utilized the ECEE ξ, as our performance measure, the optimum setting for the Kalman variables investigated here can be written as :

$$\delta \geq 100. \tag{3.74}$$

$$\mathbf{Q}(k) = 0.1 \times \mathbf{I}. \tag{3.75}$$

$$\mathbf{R}(k) = g\mathbf{I} \quad \text{where} \quad g \leq 1. \tag{3.76}$$

In summary we have exemplified the determination of the desirable settings for the **RKDFE**. In the next section we will explore the complexity of the algorithm used in both the channel estimator and DFE.

3.3 Complexity Study

In this section the complexity of the **RKCE** and the **RKDFE** is quantified, which was calculated based on the number of complex operations required by the Kalman algorithm for each application. The generic complexity of the Kalman algorithm is summarized in Table 3.8, where the number of complex operations of each of the recursive Kalman equations is calculated as follows :

- The multiplication of two complex matrices of size $M \times N$ and $N \times P$ requires MNP complex additions and MNP complex multiplications.

- The addition of two complex matrices of size $M \times N$ and $M \times N$ requires MN complex additions.

The complexity of the **RKCE** and **RKDFE** can be calculated by assigning $M = 1$ in the complexity calculations listed in Table 3.8, which implied that the mean square error was calculated on a symbol by symbol basis. Physically, the dimensions M and N represented the number of measurement signals used to generate the mean square error and the number of parameters to be estimated, respectively. Consequently, based on the last line of Table 3.8 the total complexity of the **RKCE** and the **RKDFE** per iteration can be written as:

$$\text{number of complex additions or subtractions} = 5N^2 + 5N + 3, \tag{3.77}$$

$$\text{number of complex multiplications or divisions} = 3N^2 + 4N + 1. \tag{3.78}$$

In calculating the above complexity we have assumed that the inversion of the unit matrix $[\mathbf{R}(k) + \mathbf{CP}(k, k-1)\mathbf{C^{*T}}]^{-1}$, has the same complexity of the Gauss-Jordan elimination process [143], namely M^3 complex additions and M^3 complex multiplications.

Thus, for the **RKCE**, which contains five taps, i.e $N = 5$, the complexity was equivalent to 153 complex additions and 96 complex multiplications per iteration. Similarly, for the **RKDFE** having seven forward taps and four feedback taps, which resulted in $N = 11$, the complexity was 663 complex additions and 408 complex multiplications. Let us now focus our attention on the equalization performance of multilevel modems.

Kalman Recursive Operations from Table 3.3	Addition/ Subtraction	Multiplication
Kalman recursive estimator:		
$\mathbf{C}\hat{\mathbf{x}}(k-1)$	MN	MN
$\mathbf{y}(k) - \mathbf{C}\hat{\mathbf{x}}(k-1)]$	M	-
$\mathbf{K}(k)[\mathbf{y}(k) - \mathbf{C}\hat{\mathbf{x}}(k-1)]$	MN	MN
$\hat{\mathbf{x}}(k-1) + \mathbf{K}(k)[\mathbf{y}(k) - \mathbf{C}\hat{\mathbf{x}}(k-1)]$	N	-
Kalman gain:		
$\mathbf{P}(k, k-1)\mathbf{C}^{*\mathbf{T}}$	MN^2	MN^2
$\mathbf{C}\mathbf{P}(k, k-1)\mathbf{C}^{*\mathbf{T}}$	NM^2	NM^2
$\mathbf{R}(k) + \mathbf{C}\mathbf{P}(k, k-1)\mathbf{C}^{*\mathbf{T}}$	M^2	-
$[\mathbf{R}(k) + \mathbf{C}\mathbf{P}(k, k-1)\mathbf{C}^{*\mathbf{T}}]^{-1}$	M^3	M^3
$\mathbf{P}(k, k-1)\mathbf{C}^{*\mathbf{T}}[\mathbf{R}(k)$		
$+ \mathbf{C}\mathbf{P}(k, k-1)\mathbf{C}^{*\mathbf{T}}]^{-1}$	M^2N	M^2N
Predicted state error matrix:		
$\mathbf{P}(k, k-1) = \mathbf{P}(k-1) + \mathbf{Q}(k-1)$	N^2	-
State Error Matrix :		
$\mathbf{C}\mathbf{P}(k, k-1)$	MN^2	MN^2
$\mathbf{K}(k)\mathbf{C}\mathbf{P}(k, k-1)$	MN^2	MN^2
$\mathbf{P}(k)] = \mathbf{P}(k, k-1) - \mathbf{K}(k)\mathbf{C}\mathbf{P}(k, k-1)$	N^2	-
Total complexity :	$N^2(3M+2)+$ $N(2M^2 + 2M + 1)+$ $M^3 + M^2 + M$	$3MN^2+$ $N(2M^2 + 2M)+$ M^3

Table 3.8: Complexity calculation of the Kalman algorithm per iteration based on the Multi-dimensional Kalman Recursive Equations listed in Table 3.3. Note that the state transition matrix **A** has been assigned as an identity matrix **I** in order to facilitate the complexity calculation of the **RKDFE** and **RKCE**, where the channel or equalizer coefficient variation cannot be anticipated.

3.4   Adaptive Equalization in Multilevel Modems

In this section we will quantify the performance of the **RKCE** and **RKDFE** in a multi-path fading environment [88] using the modems specified in Figures 2.2 of Section 2.1, namely BPSK, QPSK, 16QAM, 64QAM. The two different receiver structures applied in the following experiments are shown in Figure 3.14. Structure 1, shown in Figure 3.14(a) incorporates the **RKCE**, the DFE coefficient computation, corresponding to Equations 2.54 and 2.55 and the DFE. The **RKCE** with the assistance of the transmitted and received training sequence, provides the CIR for the DFE coefficient computation. Subsequently, the coefficients are applied to the DFE structure in order to equalize the entire received frame. In Structure 2, the DFE coefficients are generated by the **RKDFE** of Figure 3.9 using the transmitted and received training sequence. Consequently the DFE structure utilized the estimated DFE co-

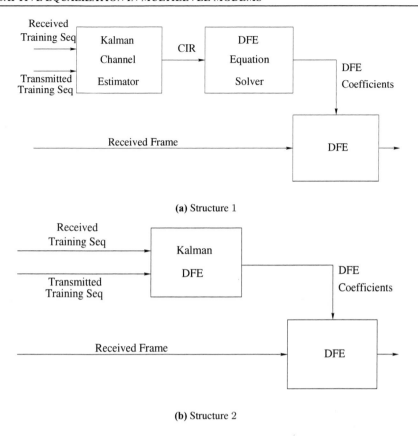

(a) Structure 1

(b) Structure 2

Figure 3.14: The receiver structures featuring (a) Structure 1, which consisted of the recursive Kalman CIR estimator and the DFE coefficient computation (b) Structure 2 using the recursive Kalman DFE.

efficients in order to produce the equalized output.

The transmitted frame structure is shown in Figure 3.15, which consists of L_d data and L_T training symbols. The training symbols are always BPSK modulated - irrespective of the data modulation - for reasons of adaptation robustness and the sequence was chosen such that it possessed a narrow, impulse-like cyclic autocorrelation function (ACF) as discussed by Milewski [140] and Steele [13]. This implied that for a certain training sequence length, a computer search was conducted on all possible binary sequences for which the ACF was evaluated. The sequence, which produced the highest ratio of the main correlation lobe power to the first side lobe power was chosen as the training sequence.

The two structures depicted in Figure 3.14 were modelled employing the simulation parameters listed in Table 3.9. A three-path equal-weight Rayleigh faded CIR was utilized, where the normalized Doppler frequency was set to 6.015×10^{-4}. Again, the simulation parameters used are listed in Table 3.9 and throughout these investigations, a slow fading channel model was used, where the CIR was kept constant for the duration of the transmitted burst, but a different CIR was used for the next burst. This model is also often referred to as

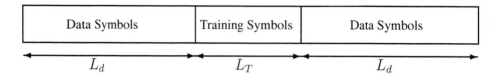

Figure 3.15: The transmitted frame structure depicting the position of the data and training symbols.

	Structure 1	Structure 2
Transmission burst format:	(See Figure 3.15)	(See Figure 3.15)
No. of data symbols, L_d	72	72
No. of training symbols, L_T	26	26
DFE parameters :		
No. of feedfwd. taps, N_f	7	7
No. of feedback taps, N_b	2	2
Decision feedback	Correct	Correct
Kalman algorithm parameters:		
Measurement error cov. matrix, $\mathbf{R}(k) = g\mathbf{I}$	$0 < g \leq 1$	$0 < g \leq 1$
System error cov. matrix, $\mathbf{Q}(k)$	$0.01\mathbf{I}$	$0.1\mathbf{I}$
δ (see Equation 3.47)	≥ 300	≥ 300
Initial channel estimate vector $\hat{\mathbf{h}}(\mathrm{o})$	$\mathbf{0}$	-
Initial DFE coeff. estimate vector $\hat{\mathbf{c}}(\mathrm{o})$	-	$\mathbf{0}$
Number of **RKCE** taps	5	-
Number of **RKDFE** taps	−	9
Channel parameters:		
Three-path Rayleigh-faded weights	$0.577 + 0.577z^{-1}$ $+0.577z^{-2}$	$0.577 + 0.577z^{-1}$ $+0.577z^{-2}$
Normalized Doppler frequency	6.015×10^{-4}	6.015×10^{-4}

Table 3.9: Simulation parameters used to quantify the performance of the two receiver structures shown in Figure 3.14, where the CIR was constant for a transmission burst.

a frame-invariant fading model.

The results of our experiments are shown in Figure 3.16. The tabulated results of the simulations are listed in Table 3.10, where for a BER of 10^{-3}, the approximate required average channel SNR was catalogued for the structures depicted in Figure 3.14. The approximate performance degradation of these adaptive structures was also compared to that of the receiver, which utilized perfect CIR estimation.

The performance of the multi-level modems in a flat Rayleigh fading channel under perfect channel estimation was also shown in Figure 3.16 in order to highlight the benefits of the inherent multi-path diversity effect, where as expected, the multi-path diversity assisted receiver outperformed the flat-fading scenario. This was evidenced, for example by Figure 3.16, where there was an approximately 10dB channel SNR gain at a BER of 10^{-3} for all modulation schemes used. This resulted from the diversity effect of the multi-path fading channels, where it was fairly unlikely for all the paths to fade simultaneously. Consequently the DFE-assisted receiver possessed a higher probability of resolving each of the multi-path contributions.

Referring to Table 3.10, the CIR-estimation induced performance degradation of Structure 1 was approximately 1dB for all modulation schemes used, when compared to that of the perfect CIR estimation scenario. Similarly, an approximate degradation of 4dB was inflicted by Structure 2 in the same context. This was, because Structure 2, which utilized the **RKDFE**, possessed a slower convergence rate, when compared to Structure 1, which employed a **RKCE** as seen in Section 3.2.4.1. Thus Structure 1 provided a superior DFE coefficient estimate due to its more accurate estimation of the CIR. The degradation for the higher-order modulation schemes was also higher due to their reduced Euclidean distance between neighbouring phasor constellation points, which were more susceptible to inaccurate CIR estimation.

The results summarized in Table 3.10 clearly showed the superiority of Structure 1, when the Kalman algorithm was adapted throughout the duration of the entire training sequence interval. In the next section we will examine and compare the complexity of both structures used in these experiments.

	Perfect estimation	Structure 1	Structure 2	Degradation of Structure 1	Degradation Structure 2
BPSK:					
2 path	14.5dB	15.5dB	19.0dB	1.0dB	4.5dB
3 path	12.5dB	13.5dB	16.5dB	1.0dB	4.0dB
QPSK:					
2 path	17.5dB	18.5dB	21.5dB	1.0dB	4.0dB
3 path	15.5dB	16.5dB	19.5dB	1.0dB	4.0dB
QAM16:					
2 path	25.0dB	26.0dB	28.0dB	1.0dB	4.0dB
3 path	23.0dB	24.0dB	26.5dB	1.0dB	4.5dB
QAM64:					
2 path	31.0dB	32.5dB	34.5dB	1.5dB	3.5dB
3 path	29.5dB	31.0dB	33.5dB	1.5dB	4.0dB

Table 3.10: Required channel SNR values for the equalizer structures shown in Figure 3.14 and for the bench-mark structure employing perfect CIR estimation as well as the channel SNR degradation of each Structure with respect to the perfect CIR estimation bench-marker. The results were tabulated from Figure 3.16 for a BER of 10^{-3} employing the parameters of Table 3.9.

3.4.1 Complexity of the Receiver Structures

The complexity of the receiver structures shown in Figure 3.14 was derived and summarized in Table 3.11. The complexity of the **RKCE** and the **RKDFE** was derived in Section 3.3 and the complexity of the DFE coefficient computation was based on Equations 2.54 and 2.55.

Referring to Table 3.11, the total complexity of Structure 1 and 2 per transmission frame

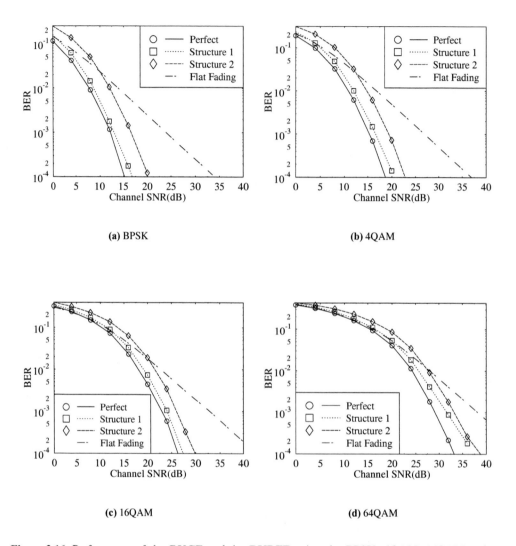

Figure 3.16: Performance of the **RKCE** and the **RKDFE** using the **BPSK**, **4QAM**, **16QAM** and **64QAM**, which employed the simulation parameters listed in Table 3.9.

for L_T Kalman iterations can be written as :

$$\text{Structure 1}_{\text{complexity}} = [L_T(5N^2 + 5N + 3) + 2N_f^2 + N_b(N_f - 1) + \frac{N_f^3}{3}]\text{additions and}$$

$$[L_T(3N^2 + 4N + 1) + 2N_f + \frac{N_f^3}{3} + 3N_f^2 + N_bN_f]$$

$$\text{multiplications,} \tag{3.79}$$

$$\text{Structure } 2_{\text{complexity}} = [L_T(5N^2 + 5N + 3)]\text{additions and}$$
$$[L_T(3N^2 + 4N + 1)]\text{multiplications.} \tag{3.80}$$

The first term in Equation 3.79 represented the required number of required complex additions, while the second term was the complexity incurred in terms of complex multiplications. The term $L_T(5N^2 + 5N + 3)$ was the product of the **RKCE** complexity per iteration according to Equation 3.77 and the number of iterations, which was represented by L_T. The terms associated with $N_b(N_f - 1)$ complex additions and $N_b N_f$ complex multiplications represented the complexity of calculating the feedback coefficients of the DFE, while the rest of the terms contributed to the complexity of calculating the feedforward coefficients of the DFE. These complexity terms are tabulated explicitly in Table 3.11.

By applying $L_T = 26$ and $N = 5$ in Equation 3.79, the complexity of Structure 1 was 4214 complex additions and 2799 complex multiplications per transmission frame for L_T iterations. The dimension N, in the context of the **RKCE** represents the number of paths the **RKCE** can estimate. It also implied that the maximum CIR length the **RKCE** can resolve was five symbols. Similarly, by applying Equation 3.80, the complexity of Structure 2, when $N = N_f + N_b = 9$ was 11778 complex additions and 7280 complex multiplications per transmission frame for L_T iterations.

By this simple complexity calculation, we deduced that the complexity of Structure 2 was higher by about a factor of three, than that of Structure 1 during the training sequence interval, even though the SNR degradation of Structure 2 was higher. Let us now summarize our findings in the next section.

3.5 Review and Discussion

In this chapter, the Recursive Kalman algorithm was presented and used as a tool in order to adapt the coefficients of the DFE in a Rayleigh fading wideband channel environment. The algorithm was derived and applied to an adaptive CIR estimator and the adaptive DFE. In the former application, the algorithm was used to estimate the CIR based on a training sequence of a given length. As for the adaptive DFE, the algorithm was utilized in order to derive the coefficients of the DFE.

For each of the applications, the convergence performance of the algorithm was quantified and investigated with respect to the variation of the system parameters, δ, $\mathbf{R}(k)$ and $\mathbf{Q}(k)$. These parameters influenced the convergence speed, the final accuracy of the estimate and the stability of the algorithm. Subsequently, the derivation of the desired values for these parameters was exemplified for the experimental conditions of Tables 3.6 and 3.7, in order to produce the final algorithm that was stable, fast and accurate in the context of CIR estimation and equalization.

The performance of the algorithm using these parameters was then evaluated over the multi-path Rayleigh fading channel of Table 3.9, using two different receiver structures, which were depicted in Figure 3.14. The first structure utilized the **RKCE**, while the other structure incorporated the **RKDFE**. In these two structures, the Kalman algorithm was used to adaptively equalize the received signal based on the training sequence located in the middle of the transmitted frame shown in Figure 3.15. These receiver structures were then applied to multi-level modems, where Structure 1 performed better by approximately 3dB in channel

	Addition/ Subtraction	Multiplication/ Division
Structure 1		
RKCE per iteration (see Equations 3.77 and 3.78) **DFE coefficient computation per frame:**	$5N^2 + 5N + 3$	$3N^2 + 4N + 1$
Cov. matrix generation from Equation 2.54: $\sum_{m=0}^{N_f-1} C_m \left[\sum_{v=0}^{l} h_v^* h_{v+m-l}\sigma^2 + N_o\delta_{m-l} \right]$ $= h_l^*\sigma_A^2; l = 0.....N_f - 1$	N_f^2	$2N_f(N_f + 1)$
Linear coefficient computation using the LU decomposition with fwd. and bwd. substitution [143] for the DFE forward coefficients	$\frac{N_f^3}{3} + N_f^2$	$\frac{N_f^3}{3} + N_f^2$
DFE feedback coefficients calculations from Equation 2.55: $b_q = \sum_{m=0}^{N_f-1} C_m h_{m+q} \ \ q = 1.....N_b$	$N_b(N_f - 1)$	$N_b N_f$
Structure 2 **Recursive Kalman DFE per iteration**	$5N^2 + 5N + 3$	$3N^2 + 4N + 1$

Table 3.11: The complexity of Structures 1 and 2 per iteration in terms of complex multiplications and additions, where N_f and N_b are the number of feedforward and feedback coefficients of the DFE, respectively. $N = N_f + N_b$, when applied to Structure 2 and similarly for Structure 2, N represents the maximum number of CIR paths the **RKCE** can detect, which in this experiment was set to five.

SNR terms for a BER of 10^{-3}, as quantified in Table 3.10. We then proceeded to calculate and compare the complexity of the above two structures for a fixed training sequence length, which was summarized in Table 3.11.

In summary, the application of the recursive Kalman algorithm was feasible with respect to the achieved performance, although extreme care had to be taken, when setting the Kalman parameters of Table 3.5. The values of these parameters influenced the state space model of Equations 3.42 and 3.43 used by the algorithm and the choice of these values had to reflect the actual characteristics of the propagation environment for the sake of achieving optimum performance. Clearly, in a strongly non-stationary propagation environment when the initial CIR estimate was unavailable, the parameters had to be chosen such that the measurement model of Equation 3.43 was dominant initially and subsequently the system model of Equation 3.42 became influential at the later stages of iterations. This was necessary, in order for the algorithm to converge during the fastest possible time, which was determined by the length of the training sequence.

On the other hand, if a reliable initial CIR estimate was available, the system model of

Equation 3.42 could be made dominant most of the time, while the measurement model of Equation 3.43 fine tuned the estimate in order to provide a more accurate CIR estimate. This scenario occurred, when a so-called decision directed mechanism was invoked, where the coefficients of the DFE were adapted based on the past detected symbols. In other words, the CIR variations throughout the entire frame and not just during the training period were tracked by the algorithm. The performance of the algorithm in this situation was quantified recently by Narayanan *et al.* [144], Harun *et al.* [124] and Shukla *et al.* [125], where the Kalman algorithm and its variants like the square root Kalman algorithm [135] provided good convergence properties. However, Harun *et al.* [124] showed that there was little difference in the performance of the Kalman algorithm and the LMS algorithm in this scenario using the assumption of perfect i.e error free decision feedback in order to adapt the equalizer coefficients.

As it was shown in Figure 3.16, the performance of the **RKCE** was superior to that of the **RKDFE**, even though the complexity of the CIR estimator was lower. However, if the CIR estimator had to adapt throughout the entire transmission frame, the complexity of the DFE coefficient computation portrayed in Figure 3.14 and Table 3.11 would grow due to its proportionality to N_f^3. This would offset the complexity advantage of the **RKCE**.

In the next chapter, we will investigate the employment of adaptive modulation in a narrow-band [145] and wideband channel environment. In the narrow-band environment, the application of power control techniques will be implemented in the context of adaptive modulation and we will also explore the utilization of equalization techniques in conjunction with adaptive modulation in a dispersive i.e wideband channel environment.

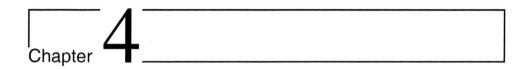

Chapter 4

Adaptive Modulation

In this chapter, the concept of Adaptive Quadrature Amplitude Modulation (AQAM) is introduced, whereby the modulation mode is adapted at the transmitter on a burst by burst basis. This adaptation is implemented based on the receiver's perceived channel quality and its main motivation is to maximise the transmission throughput at a given target BER.

In our investigations AQAM is initially applied in a narrow-band environment in conjunction with power control, where the transmitted power is only varied near the modem mode switching thresholds of the AQAM scheme. Subsequently, AQAM is investigated in a wideband channel in conjunction with a DFE-assisted receiver. In this context, a new channel quality metric is proposed in order to control the choice of AQAM modes. Let us now commence with a brief overview of the AQAM scheme.

4.1 Adaptive Modulation for Narrow-Band Fading Channels

A brief overview of the principles of AQAM in a narrow-band Rayleigh fading channel environment is given here. In a narrow-band channel, as a result of its rapid fading, the short term SNR can be severely degraded, especially if the channel exhibits a deep fade. The general philosophy of AQAM is to employ a higher-order modulation mode, when the channel quality is favourable in order to increase the transmission throughput and conversely, a more robust lower-order modulation mode is invoked, when the channel quality is low. This is achieved at a **constant symbol-rate**, regardless of the modulation mode selected and hence at a constant bandwidth requirement. Therefore the impact of AQAM mode switching on the system's design remains as low as possible.

The concept of invoking AQAM is hence to a certain extent analogous to employing power control schemes, which are typically used to combat the effects of pathloss and slow fading. However, whilst power up in order to compensate for degrading channel conditions may inflict increased cochannel interference upon other users, which in turn may require further power increments for maintaining the target quality, AQAM accommodates these channel quality fluctuations without disadvantaging other users in the system. AQAM can also be applied in order to support users, when traversing a variety of propagation environments, such as indoor or outdoor environments. In a friendly propagation environment, where the impact of ISI and co-channel interference is significantly lower, than in an outdoor environment [146], the less robust higher-order modulation modes can be activated more frequently. This results in a higher average transmission throughput, while ensuring an acceptable BER performance. By contrast, the more robust lower-order modulation modes can be utilized more frequently in a more hostile outdoor environment. These characteristics of the AQAM switching regime will be shown more explicitly in Section 4.3.4.

In adapting the modulation mode, a signalling regime has to be implemented, in order to harmonise the operation of the transmitter and receiver with regards to the adaptive modem mode parameters. In this respect, the Time Division Duplex (TDD) scheme [4, 146] is employed in order to implement an open loop channel quality signalling system. Unlike in Frequency Division Duplex (FDD), where the uplink (UL) and downlink (DL) transmission frequency bands are different, the UL and DL transmissions of the TDD scheme are time multiplexed onto the same carrier. This results in a correlation in the fading characteristics of the UL and DL propagation channels, yielding near-reciprocal channel conditions [146], provided that the channel quality is slowly varying. In a loose sense, we will often use the term reciprocity in order to indicate the similarity of the UL and DL in TDD environments. Consequently an open loop signalling system can be implemented, where the modulation mode can be adapted at the transceiver based on the information acquired during its receiving mode. This open loop system is encapsulated in Figure 4.1(a). In contrast, if near-reciprocal channel conditions are not applicable, a closed-loop based signalling system shown in Figure 4.1(b) can be implemented in a FDD based system. These signalling regimes will be further elaborated in Section 7.2.

Having discussed briefly the principle of AQAM and the possible scenarios where it can be applied, this leads us to explore the criterion and methodology of selecting the transmitter's modulation mode. The criterion used by Torrance [145] was the instantaneous received power, which was estimated by exploiting the reciprocal nature of the channel in a TDD environment. This estimate was then used to select a suitable modulation mode by comparing the channel quality estimate against a set of switching threshold levels l_n, as depicted in Figure 4.2. For example, if the estimated instantaneous received power was between the values of l_1 and l_2, according to Figure 4.2, BPSK was chosen for the next transmission burst. However, when the received near-instantaneous power was below l_1, where the channel was in a deep fade, the transmission was disabled. This was termed as the transmission blocking mode.

AQAM is not only used to combat the fading effects of a narrow-band channel, but it also attempts to maximize the transmission throughput. This is achieved, when a higher order modulation mode is used, if the short term SNR is favourable. Conversely, the scheme also attempts to optimize the mean BER by employing a more robust modulation mode, when the channel quality is degraded. As a result, there is a trade-off between the mean BER and Bits per Symbol (BPS) performance. This trade-off is governed by the values of the switching thresholds l_n. As the values decrease, the probability of employing higher-order modulation modes increases, thus yielding a better BPS performance. Conversely, if the values of l_n are increased, lower-order modulation modes are employed more frequently, resulting in an improved mean BER performance. In the next section, we will review some of the advances achieved using AQAM.

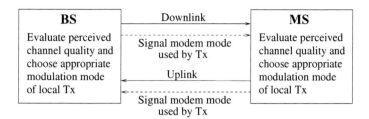

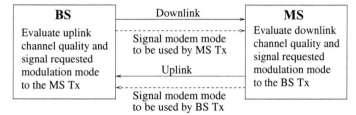

(a) Open-loop based signalling

(b) Closed-loop based signalling

Figure 4.1: Closed- and open-loop signalling regimes for the AQAM schemes, where BS represents the Base Station, MS denotes the Mobile Station, the transmitter is represented by Tx and the receiver is denoted by Rx.

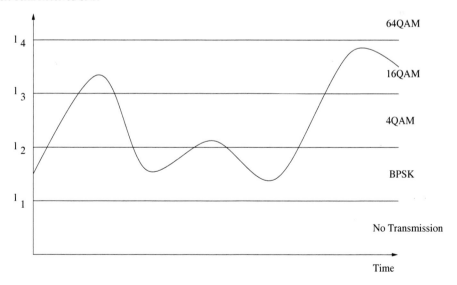

Figure 4.2: Stylised profile of the short term received SNR in a narrow-band channel, which is used to choose the next modulation mode.

4.1.1   Literature Review on Adaptive Modulation

Since we have discussed the basic principles of AQAM, we can now describe the advances achieved so far in this field. Historically, interest in techniques of adapting the modulation and transmission rate parameters began in 1968, when Hayes [147] adapted the signal amplitude according to the prevalent channel environment by utilizing a feedback channel between the transmitter and receiver that was assumed noiseless and free from latency. The adaptation of the transmission rate was then recorded by Cavers [9] in a slow Rayleigh fading environment, where the adaptation parameters were transported, again via a feedback channel.

Recent work was pioneered by Webb and Steele [1], where the modulation adaptation was analysed in a flat Rayleigh fading environment with applications in a Digital European Cordless Telecommunications (DECT)-like system. Star QAM [4] was used instead of Square QAM and the channel's reciprocity was exploited in a TDD scenario in order to adapt the modulation parameters. The metric used to quantify the channel quality was the received signal strength and the BER. Reference [1] also recorded the effects of block size, fading rate and co-channel interference on the performance. Slightly later, Sampei, Komaki and Morinaga introduced a variable-rate, variable-modulation-mode adaptive modulation scheme [148]. In this paper, an extended symbol duration of $\frac{1}{4}$ rate QPSK, $\frac{1}{2}$ rate QPSK, full-rate QPSK, 16QAM and 64QAM were used for adaptation in a narrow-band channel environment. The modulation modes and rate were switched according to the signal to co-channel interference ratio and the expected delay-spread of the channel. The signal to co-channel interference ratio was estimated perfectly and the modulation control parameters were accessed via the control channels in a TDMA scenario. The channel assumed a slowly varying statistic and the normalized delay spread was less than unity. The results were recorded in terms of spectral efficiency and BER performance for different cellular configurations and it was concluded that adaptive modulation showed promising advantages, when compared to fixed-mode modulation in terms of spectral efficiency, BER performance and robustness against channel delay-spread. In another contribution, the numerical upper bound performance of adaptive modulation in a slow Rayleigh fading channel was then evaluated by Torrance et al. [31] and subsequently, the optimisation of the switching threshold levels using Powell minimization [143] was proposed, in order to achieve a certain target performance [26].

Subsequent papers were published with more emphasis on the system aspects of adaptive modulation in a narrow-band environment. A reliable method of transmitting the modulation control parameters was proposed by Otsuki et al. [21], where the parameters were embedded in the transmission frame's mid-amble using Walsh codes. Subsequently, at the receiver the Walsh sequences were decoded using maximum likelihood detection. Another technique of estimating the required modulation mode used was proposed by Torrance et al. [37], where the modulation control symbols were represented by unequal error protection 5-PSK symbols. The adaptive modulation philosophy was then extended to the wideband multi-path environments by Kamio et al. [39] by utilizing a bi-directional DFE in a micro- and macro-cellular environment. This equalization technique employed both forward and backward oriented channel estimation based on the pre-amble and post-amble symbols in the transmitted frame. Equalizer tap gain interpolation across the transmitted frame was also utilized in order to reduce the complexity in conjunction with space diversity [39]. The authors concluded that the cell radius could be enlarged in a macro-cellular system and a higher area-spectral efficiency could be attained for micro-cellular environments by utilizing adaptive modulation.

The latency effect, which occurred when the input data rate was higher than the instantaneous transmission throughput, was studied and solutions were formulated using frequency hopping [40] and statistical multiplexing, where the number of slots allocated to a user was adaptively controlled [41].

In reference [42] symbol rate adaptive modulation was applied, where the symbol rate or the number of modulation levels was adapted by using $\frac{1}{8}$-rate 16QAM, $\frac{1}{4}$-rate 16QAM, $\frac{1}{2}$-rate 16QAM as well as full-rate 16QAM and the criterion used to adapt the modem modes was based on the instantaneous received signal to noise ratio and channel delay spread. The slowly varying channel was rendered near-reciprocal by utilizing short frame duration TDD and the maximum normalized delay spread simulated was 0.1. A variable channel coding rate was then introduced by Matsuoka *et al.* in conjunction with adaptive modulation in Reference [34], where the transmitted burst incorporated an outer Reed Solomon code and an inner convolutional code in order to achieve a higher quality data transmission. The coding rate was varied according to the prevalent channel quality using the same method, as in adaptive modulation in order to achieve a certain target BER performance. A so-called channel margin was introduced in this paper, which basically changed the switching thresholds in order to incorporate the effects of channel quality estimation errors. The utilization of channel coding in conjunction with adaptive modulation in a narrow-band environment was also recorded by Chua and Goldsmith [35]. In this contribution, trellis and lattice codes were used without channel interleaving, invoking a feedback path between the transmitter and receiver for modem mode control purposes. The effects of the delay in the feedback path on the adaptive modem's performance were studied and this scheme exhibited a higher spectral efficiency, when compared to the non-adaptive trellis coded performance.

Subsequent contributions incorporated space-diversity and power-adaptation in conjunction with adaptive modulation, for example by Suzuki *et al.* in order to combat effects of the multi-path channel environment in Reference [54] at a 10Mbits/s transmission rate. The maximum delay-spread used was one symbol duration for a target mean BER performance of 0.1%. This was achieved in a TDMA scenario, where the channel estimates were predicted based on the extrapolation of previous channel quality estimates. Variable transmitted power was then applied in combination with adaptive modulation in Reference [36], where the transmission rate and power adaptation was optimized in order to achieve an increased spectral efficiency. In this treatise, a slowly varying channel was assumed and the instantaneous received power required in order to achieve a certain upper bound performance was known prior to transmission. Power control in conjunction with a pre-distortion type nonlinear power amplifier compensator was studied in the context of adaptive modulation in Reference [55]. This method was used to mitigate the non-linearity effects associated with the power amplifier, when QAM modulators were used.

Results were also recorded concerning the performance of adaptive modulation in different multiple access schemes in a narrow-band channel environment. In a TDMA system, dynamic channel assignment was employed by Ikeda *et al.*, where in addition to assigning a different modulation mode to a different channel quality, priority was always given to the users in obtaining the time-slots, which benefitted from the best channel quality [56]. The performance was compared to fixed channel assignment systems, where gains were achieved in terms of system capacity. Furthermore, a lower call termination probability was recorded. However, the probability of intra-cell hand-off increased as a result of the dynamic channel assignment scheme, which constantly searched for a high-quality, high-throughput time-slot for the existing active users. The application of adaptive modulation in packet transmission was introduced by Ue, Sampei and Morinaga [57], where the results showed improved data throughput. Recently, the performance of adaptive modulation in an automatic repeat request (ARQ) system was published in Reference [58], where the transmitted bits were encoded using a cyclic redundant code (CRC) and a convolutional punctured code in order to increase the data throughput.

A recent treatise was published by Sampei, Morinaga and Hamaguchi [59] on laboratory test results concerning the utilization of adaptive modulation in a TDD scenario, where the modem mode switching criterion was based on the signal to noise ratio and on the normalized delay-spread. In these experimental results, the channel quality estimation errors degraded the performance and consequently a channel estimation error margin was devised, in order to mitigate this degradation. Explicitly, the channel estimation error margin was defined as the measure of how much extra protection margin must be added to the switching threshold levels, in order to minimise the effects of the channel estimation error. The delay-spread also degraded the performance due to the associated irreducible BER, which was not compensated by the receiver. However, the performance of the adaptive scheme in a delay-spread channel environment was better than that of fixed modulation scheme. Lastly, the experiment also concluded that the AQAM scheme can be operated for $f_d = 10$Hz with a normalized delay spread of 0.1 or for $f_d = 14$Hz with a normalized delay spread of 0.02, which produced a mean BER of 0.1% at a transmission rate of 1 Mbits/s. In this respect, f_d was the Doppler frequency.

With the above background on adaptive modulation, we shall now investigate the application of power control near the switching threshold levels, which is termed here as threshold-based power control.

4.2 Power Control Assisted Adaptive Modulation Over Narrow-band Rayleigh Fading Channels

In this section power control is utilized in conjunction with AQAM over a narrow-band Rayleigh fading channel, where its benefits as well as disadvantages are analysed. In this discourse, perfect power control is assumed. Furthermore, we will show that a maximum power control range of ± 2 dB around the modem mode switching thresholds is sufficient in our analysis of the effects of threshold-based power control on AQAM, an issue to be clarified during our further discourse. Threshold-based power control is only applied, when the expected received power is within a certain range of the AQAM switching thresholds. This is best explained graphically by referring to Figure 4.3. Power control is only applied, when the expected received SNR is within a certain range of the AQAM switching thresholds $l_1 - l_4$, and this range is denoted by the Power Control Zone (PCZ) in Figure 4.3. The width of this range is controlled by the power control's maximum dynamic range κ, which is also depicted in Figure 4.3. Thus, if the expected received SNR is within the PCZ of Figure 4.3, power control is applied, where the transmitted power can be increased or decreased within the maximum dynamic range κ or alternatively, the transmitted power can be left unchanged.

The main purpose of employing the threshold-based power control scheme is to optimize the system performance of AQAM, where for example, if the expected received SNR level is

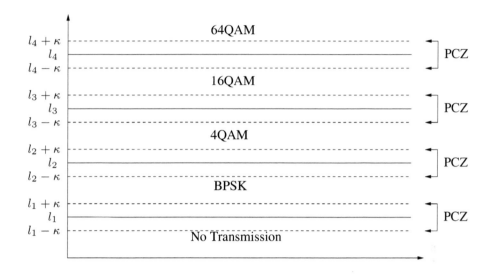

PCZ - Threshold-based Power Control Zone

$l_1 - l_4$ Adaptive Modulation Switching Thresholds

κ - Maximum dynamic range of Threshold-based power control

Figure 4.3: Schematic of the threshold-based power control scheme depicting the power control zone (PCZ), where power control is applied. The power control zones are defined by the switching thresholds, l_n, and the maximum dynamic range of the threshold-based power control scheme κ.

just below a particular adaptive switching threshold, the transmitted power can be increased to ensure that the actual received SNR level is above that particular adaptive threshold level. Consequently, a higher-order modulation mode can be used, hence increasing the throughput of the system. Alternatively, provided that the expected received SNR level is just above the adaptive switching threshold, the transmitted power can be decreased sufficiently, in order to utilize a lower-order modulation mode, thus ensuring an improved BER performance. Another possible benefit of employing the threshold-based power control scheme is to reduce the modulation mode switching frequency of the transmitter. This scheme can be utilized in order to maintain the previous modulation mode by increasing or decreasing the transmitted power, whenever the expected received SNR switching level is within the power control zone. In summary, the proposed threshold-based power control can be utilized in order to improve the AQAM performance in terms of its mean BER, mean BPS and modulation switching frequency.

The performance of this power control scheme is evaluated in terms of its mean BER, throughput expressed in average BPS, power control utilization percentage and switching frequency. Power control utilization is defined here as the average relative frequency that the transmitted power has to be changed, given a certain maximum dynamic power control range and this is used as a measure of the added complexity needed to implement the threshold-based power control scheme. Similarly, the modulation mode switching frequency is defined as the average relative frequency of the transmitter changing its modulation mode with respect to its previous modulation mode. This provided a measure of the frequency at which the modulation mode of the transmitter is changed. In the following experiments, the threshold-based power control scheme is investigated in three different scenarios in order to achieve firstly, an improved mean BER performance and secondly, a better BPS throughput. Finally, the scheme is also invoked in order to achieve a lower reduced modulation mode switching frequency.

These experiments were conducted with varying maximum dynamic power control range of ±0.5dB, ±1.0dB and ±1.5dB. Since we were mainly interested in the effects of the power control scheme, the adaptive modem mode switching thresholds were not optimized and were assigned as follows : $l_1 = -\infty$, $l_2 = 8$dB, $l_3 = 14$dB, and $l_4 = 20$dB for the non-blocking AQAM scheme, where data was constantly transmitted. Similarly, for the blocking AQAM scheme the adaptive threshold levels were $l_1 = 5$dB, $l_2 = 8$dB, $l_3 = 14$dB, and $l_4 = 20$dB, where the transmitter was disabled, when the instantaneous power was below l_1. Perfect narrow-band channel quality estimation and compensation was assumed at the receiver. Let us now concentrate on our experimental results.

4.2.1 Threshold-based Power Control Designed for an Improved Bit Error Ratio Performance

In this section the proposed threshold-based power control scheme was optimized in order to achieve an improved mean BER performance. As a result of this design criterion, whenever the expected received SNR level was above the switching threshold, but within the power control's dynamic range on Figure 4.3, the transmitted power was reduced in order to ensure that a lower-order modulation mode was employed. As a result, the BER was improved due to the employment of a more robust modulation mode.

On the basis of this criterion, an AQAM mode transition table can be formulated, as seen in Table 4.1, which can be studied with reference to Figure 4.3. The adaptive switching thresholds and the PCZ, where the power control scheme can be employed is specified in the transition table. The width of the PCZ depends on the maximum dynamic range, where the higher the range, the wider the PCZ. The allocation of the present chosen modulation mode was based on the expected received instantaneous SNR in conjunction with the threshold-based power control scheme as well as on the previous modulation mode. Again, the power control mechanism was characterized in Table 4.1 by the arrow notations ↑ and ↓, indicating the powering-up mode and powering-down mode, respectively.

The mean BER and BPS performance of our threshold-based power control assisted AQAM scheme for both non-blocking and transmission blocking scenarios is depicted in Figures 4.4(a) and 4.4(b), respectively, where its performance was quantified for different dynamic ranges κ. The performance was also compared to that of the conventional AQAM scheme without power control. As expected the BER performance of the AQAM scheme with power control improved, when compared to the conventional AQAM scheme, although the mean BPS performance was slightly degraded. This characteristic was observed for both the transmission-blocking and the non-blocking scenarios, which manifested another example of the trade-off between the mean BER and BPS performance. The mean BER performance of

Previous modulation:	No TX	BPSK	4QAM	16QAM	64QAM
SNR Level					
Below $l_1 - \kappa$	No TX	No TX	No TX	No TX	No TX
$l_1 - \kappa$ to l_1	No TX	No TX	No TX	No TX	No TX
l_1 to $l_1 + \kappa$	No TX $\downarrow$	No TX $\downarrow$	No TX $\downarrow$	No TX $\downarrow$	No TX $\downarrow$
$l_1 + \kappa$ to $l_2 - \kappa$	BPSK	BPSK	BPSK	BPSK	BPSK
$l_2 - \kappa$ to l_2	BPSK	BPSK	BPSK	BPSK	BPSK
l_2 to $l_2 + \kappa$	BPSK $\downarrow$	BPSK $\downarrow$	BPSK $\downarrow$	BPSK $\downarrow$	BPSK $\downarrow$
$l_2 + \kappa$ to $l_3 - \kappa$	4QAM	4QAM	4QAM	4QAM	4QAM
$l_3 - \kappa$ to l_3	4QAM	4QAM	4QAM	4QAM	4QAM
l_3 to $l_3 + \kappa$	4QAM $\downarrow$	4QAM $\downarrow$	4QAM $\downarrow$	4QAM $\downarrow$	4QAM $\downarrow$
$l_3 + \kappa$ to $l_4 - \kappa$	16QAM	16QAM	16QAM	16QAM	16QAM
$l_4 - \kappa$ to l_4	16QAM	16QAM	16QAM	16QAM	16QAM
l_4 to $l_4 + \kappa$	16QAM $\downarrow$	16QAM $\downarrow$	16QAM $\downarrow$	16QAM $\downarrow$	16QAM $\downarrow$
Above $l_4 + \kappa$	64QAM	64QAM	64QAM	64QAM	64QAM

Table 4.1: The AQAM transition table, which was designed in order to achieve a **low mean BER**. The previous modulation mode and the expected instantaneous SNR was stated and used in order to select the present modulation mode. The notations $\uparrow$ and $\downarrow$ indicated the powering-up and powering-down mode, respectively and κ represented the maximum dynamic range of the threshold-based power control scheme.

the blocking scheme shown in Figure 4.4(b) was lower, than that of the non-blocking scheme for channel SNRs less than 20dB. This was due to the employment of the no transmission mode (NO TX) in the blocking scheme, where the transmission was disabled until the channel quality became more favourable. The other notable characteristic was that as the power control's dynamic range was increased, the mean BER performance improved. This was consistent with our expectations, since the power control zone was wider as a result of the increasing dynamic range. Consequently, the threshold-based power control scheme could be applied over a wider range of instantaneous SNRs, thus facilitating the employment of a more robust modulation mode, which resulted in a reduced BER.

The modulation mode switching relative frequency was also analysed and the associated results are shown in Figure 4.5(a). The relative switching frequency was approximately the same as that of the conventional AQAM scheme, since the threshold-based power control assisted AQAM scheme was not optimized in order to influence the switching frequency, i.e the present modulation mode was chosen on the basis of the channel quality, irrespective of the previous modulation mode. A high switching probability was observed at an average channel SNR of between 10dB and 25dB, as evidenced by Figure 4.5(a). This corresponded to the channel SNR range, where most of the switching was expected to occur. Furthermore, the dynamic range κ, had only a slight influence on the switching performance, as depicted in Figure 4.5(a). When studying our modulation mode switching relative frequency results, it is worthwhile noting that a 100% switching utilization corresponds to the event that the modulation mode is switched for every new transmission burst. However, the power control utilization frequency increased, as the dynamic range increased. This was attributed to the wider power control zone, where the transmitted power can be increased or decreased more frequently. This is evidenced by the results shown in Figure 4.5(b).

In summary, the employment of threshold-based power control in order to improve the BER performance resulted in a slight degradation of the BPS performance, while maintaining a near constant switching frequency. In the next section the scheme was optimized in order to achieve an improved BPS performance.

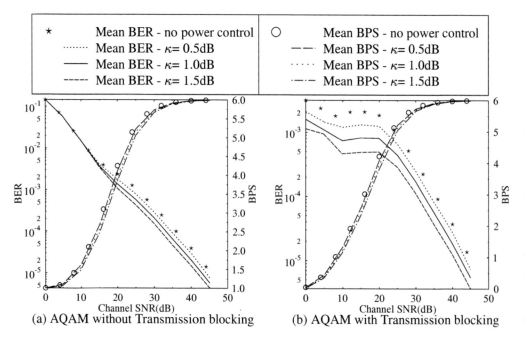

Figure 4.4: The mean BER and BPS AQAM performance employing the threshold-based power control scheme for different dynamic ranges κ, which was designed according to Table 4.1 in order to achieve a **low mean BER**. The switching threshold levels were set to $l_1 = -\infty$dB, $l_2 = 8$dB, $l_3 = 14$dB, $l_4 = 24$dB and $l_1 = 5$dB, $l_2 = 8$dB, $l_3 = 14$dB, $l_4 = 24$dB for the non-blocking and blocking schemes, respectively. Perfect channel envelope inversion was also assumed in this narrow-band channel environment at the receiver.

4.2.2 Threshold-based Power Control Designed for an Improved Bits Per Symbol Performance

In this scenario, threshold-based power control was utilized, in order to increase the mean BPS performance. This was achieved by increasing the transmitted power within the power control zone, whenever the expected received instantaneous power was below a certain threshold level. This resulted in the employment of a higher-order modulation mode, which increased the mean BPS performance at the expense of the mean BER. From this criterion,

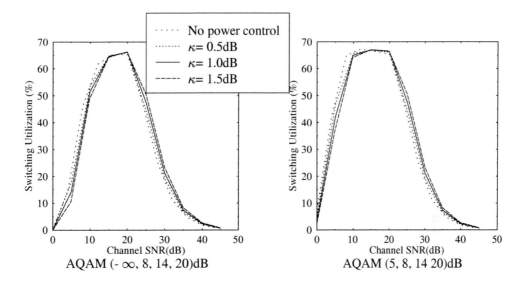

(a) Relative frequency of modulation mode switching with and without transmission blocking.

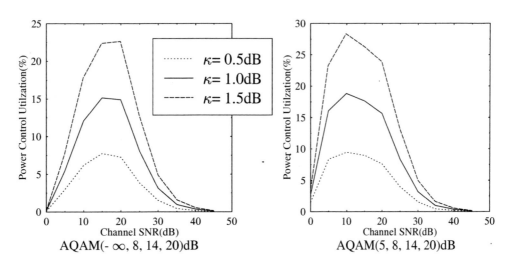

(b) Relative frequency of power control utilization with and without transmission blocking.

Figure 4.5: Relative frequency of power control and AQAM mode switching using the threshold-based power control scheme for different dynamic ranges κ, which was designed according to Table 4.1 in order to achieve a **low mean BER**. The switching threshold levels were set to $l_1 = -\infty$dB, $l_2 = 8$dB, $l_3 = 14$dB, $l_4 = 24$dB and $l_1 = 5$dB, $l_2 = 8$dB, $l_3 = 14$dB, $l_4 = 24$dB for the non-blocking and blocking schemes, respectively and perfect channel envelope inversion was assumed at the receiver.

Previous modulation:	No TX	BPSK	4QAM	16QAM	64QAM
SNR Level					
Below $l_1 - \kappa$	No TX	No TX	No TX	No TX	No TX
$l_1 - \kappa$ to l_1	BPSK ↑	BPSK ↑	BPSK ↑	BPSK ↑	BPSK ↑
l_1 to $l_1 + \kappa$	BPSK	BPSK	BPSK	BPSK	BPSK
$l_1 + \kappa$ to $l_2 - \kappa$	BPSK	BPSK	BPSK	BPSK	BPSK
$l_2 - \kappa$ to l_2	4QAM ↑	4QAM ↑	4QAM ↑	4QAM ↑	4QAM ↑
l_2 to $l_2 + \kappa$	4QAM	4QAM	4QAM	4QAM	4QAM
$l_2 + \kappa$ to $l_3 - \kappa$	4QAM	4QAM	4QAM	4QAM	4QAM
$l_3 - \kappa$ to l_3	16QAM ↑	16QAM ↑	16QAM ↑	16QAM ↑	16QAM ↑
l_3 to $l_3 + \kappa$	16QAM	16QAM	16QAM	16QAM	16QAM
$l_3 + \kappa$ to $l_4 - \kappa$	16QAM	16QAM	16QAM	16QAM	16QAM
$l_4 - \kappa$ to l_4	64QAM ↑	64QAM ↑	64QAM ↑	64QAM ↑	64QAM ↑
l_4 to $l_4 + \kappa$	64QAM	64QAM	64QAM	64QAM	64QAM
Above $l_4 + \kappa$	64QAM	64QAM	64QAM	64QAM	64QAM

Table 4.2: The AQAM transition table, which was designed to achieve a **high mean BPS**. The previous modulation mode and the expected SNR level was stated and used in order to select the present modulation mode. The notations ↑ and ↓, indicated the powering-up and powering-down mode, respectively and κ represented the maximum dynamic range of the threshold-based power control scheme.

the AQAM mode transition table can be formulated as shown in Table 4.2. The main difference between the transition tables shown in Figure 4.2 and Figure 4.1 is that the powering-up mode is used repeatedly for increasing the transmitted power, in order to utilize a higher-order modulation mode, hence improving the mean BPS performance.

The mean BER and BPS results are shown in Figures 4.6(a) and 4.6(b) for the blocking and non-blocking AQAM schemes, respectively. The effects of different dynamic ranges κ, was also highlighted in these figures. As expected, due to the BPS maximisation criterion and the resulting power control methodology, the mean BPS performance improved slightly at the expense of a higher mean BER. The increase in the power control's dynamic range κ, also improved the mean BPS performance, as depicted in Figures 4.6(a) and 4.6(b). This trend can be explained similarly to the trends in Section 4.2.1, where a wider power control zone resulted in a more frequent utilization of the threshold-based power control scheme, hence increasing the mean BPS performance. This increase in the power control utilization frequency was evidenced by the results shown in Figures 4.7(b). The relative switching frequency also displayed the same characteristics as those observed in Section 4.2.1, where the utilization frequency for different dynamic ranges κ, was approximately the same as shown in Figure 4.7(a). The arguments of Section 4.2.1 can also be applied here.

In summary, the employment of threshold-based power control in order to improve the mean BPS performance resulted in the degradation of the mean BER, while the switching frequency was more or less unchanged, when compared to the conventional AQAM scheme. In the next section, the utilization of the power control scheme is investigated in order to reduce the switching frequency and its effects on the mean BER and mean BPS performance

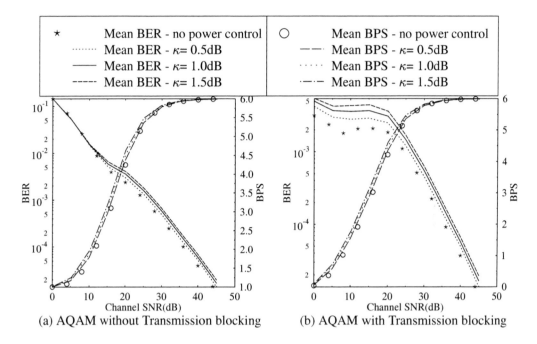

Figure 4.6: The mean BER and BPS AQAM performance employing the threshold-based power control scheme for different dynamic ranges κ. The threshold-based power control scheme was designed according to Table 4.2 in order to achieve a **high mean BPS**. The switching thresholds were set to $l_1 = -\infty$dB, $l_2 = 8$dB, $l_3 = 14$dB, $l_4 = 24$dB and $l_1 = 5$dB, $l_2 = 8$dB, $l_3 = 14$dB, $l_4 = 24$dB for the non-blocking and blocking schemes, respectively. Perfect channel envelope inversion was also assumed in this narrow-band channel environment at the receiver.

are observed.

4.2.3 Threshold-based Power Control Designed for Minimum Switching Utilization

In optimising the AQAM scheme for a lower switching frequency, the threshold-based power control scheme was designed to maintain the previous employed modulation mode if the short term SNR was within the power control zone and the previous modulation mode was a legitimate one in the power control zone. As a result of this criterion, the threshold-based power control scheme attempted to reduce the switching frequency. The corresponding modulation mode transition table was formulated, which is shown in Table 4.3, utilizing this criterion. In this table, the powering-up and powering-down mode was used appropriately to ensure that the modulation mode remained unchanged, whenever possible. This was different from Table 4.1, where the powering-down mode was used exclusively for reducing the mean BER

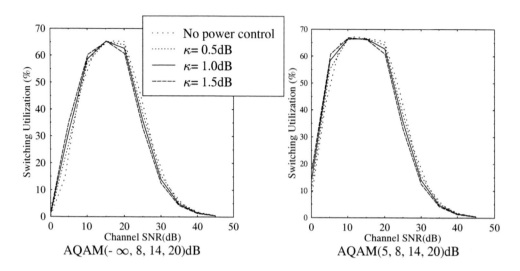

(a) Relative frequency of modulation mode switching with and without transmission blocking.

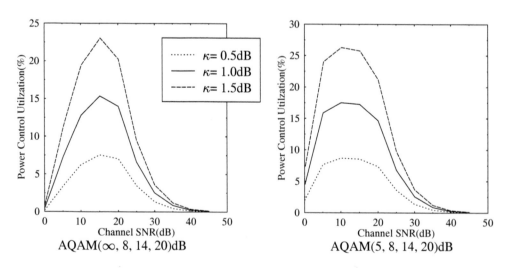

(b) Relative frequency of power control utilization with and without transmission blocking.

Figure 4.7: Relative frequency of power control and AQAM mode switching using the proposed threshold-based power control scheme for different dynamic ranges κ, which was designed according to Table 4.2 to achieve a **high mean BPS**. The switching thresholds were set to $l_1 = -\infty$dB, $l_2 = 8$dB, $l_3 = 14$dB, $l_4 = 24$dB and $l_1 = 5$dB, $l_2 = 8$dB, $l_3 = 14$dB, $l_4 = 24$dB for the non-blocking and blocking schemes, respectively and perfect channel envelope inversion was assumed at the receiver.

Previous modulation:	No TX	BPSK	4QAM	16QAM	64QAM
SNR Level					
Below $l_1 - \kappa$	No TX	No TX	No TX	No TX	No TX
$l_1 - \kappa$ to l_1	No TX	BPSK ↑	No TX	No TX	No TX
l_1 to $l_1 + \kappa$	BPSK	BPSK	BPSK	BPSK	BPSK
$l_1 + \kappa$ to $l_2 - \kappa$	BPSK	BPSK	BPSK	BPSK	BPSK
$l_2 - \kappa$ to l_2	BPSK	BPSK	4QAM ↑	BPSK	BPSK
l_2 to $l_2 + \kappa$	4QAM	BPSK ↓	4QAM	4QAM	4QAM
$l_2 + \kappa$ to $l_3 - \kappa$	4QAM	4QAM	4QAM	4QAM	4QAM
$l_3 - \kappa$ to l_3	4QAM	4QAM	4QAM	16QAM ↑	4QAM
l_3 to $l_3 + \kappa$	16QAM	16QAM	4QAM ↓	16QAM	16QAM
$l_3 + \kappa$ to $l_4 - \kappa$	16QAM	16QAM	16QAM	16QAM	16QAM
$l_4 - \kappa$ to l_4	16QAM	16QAM	16QAM	16QAM	64QAM ↑
l_4 to $l_4 + \kappa$	64QAM	64QAM	64QAM	16QAM ↓	64QAM
Above $l_4 + \kappa$	64QAM	64QAM	64QAM	64QAM	64QAM

Table 4.3: The AQAM transition table, which was designed to achieve a **low switching utilization**. The previous modulation mode and the expected SNR level was stated and used to choose the present modulation mode. The notations ↑ and ↓ indicated the powering-up and powering-down mode, respectively and κ represented the maximum dynamic range of the threshold-based power control scheme.

and from Table 4.2, where only the powering-up mode was applied in order to increase the mean BPS performance. As before, the performance of this power control scheme is analysed in terms of its associated BER, BPS, switching utilization and power control utilization frequency.

The associated mean BER and mean BPS performance results are shown in Figures 4.8(a) and 4.8(b) for the transmission blocking and non-blocking cases, respectively. The performance results were compared to the conventional AQAM scheme without power control, where there is a slight degradation in the mean BER at average channel SNRs in excess of 20dB for different dynamic ranges κ. At low average channel SNRs the utilization of the lower-order modulation modes was dominant, which enabled the power control scheme to minimise the switching utilization frequency without degrading the mean BER or mean BPS performance. However, at average channel SNRs in excess of 20 dB, the higher-order modulation modes were selected more frequently. Furthermore, the utilization of these higher-order modulation modes was maintained by the threshold-based power control regime, which was associated with a higher probability of errors. This led to the slight degradation in the mean BER at average channel SNRs in excess of 20dB, as evidenced by Figures 4.8(a) and 4.8(b). As expected, under the switching frequency minimization regime, the switching utilization decreased with increasing power control dynamic ranges, when compared to the conventional AQAM scheme. This was evident in Figure 4.9(a), where there was a switching utilization reduction of approximately 15%, when comparing the switching utilization of the conventional AQAM scheme and the power control assisted AQAM scheme, which employed a dynamic range of $\kappa = 1.5$dB. Finally, the power control utilization frequency,

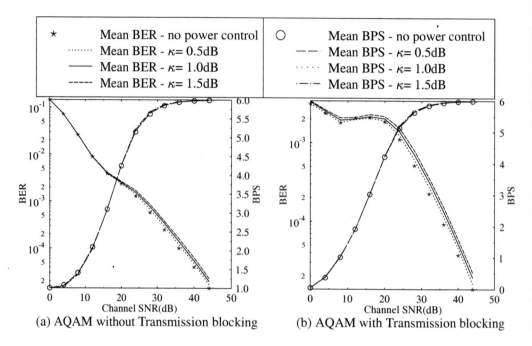

Figure 4.8: The mean BER and BPS AQAM performance employing the threshold-based power control scheme for different dynamic ranges κ. The threshold-based power control scheme was designed according to Table 4.3 in order to achieve a **low switching utilization**. The switching thresholds were set to $l_1 = -\infty$dB, $l_2 = 8$dB, $l_3 = 14$dB, $l_4 = 24$dB and $l_1 = 5$dB, $l_2 = 8$dB, $l_3 = 14$dB, $l_4 = 24$dB for the non-blocking and blocking schemes, respectively. Perfect channel envelope inversion was also assumed in this narrow-band channel environment at the receiver.

which is depicted in Figure 4.9(b), increased as the dynamic range of the power control zone was increased. This can be explained using the same arguments as in Section 4.2.1.

The employment of threshold-based power control scheme according to Table 4.3 resulted in a reduced switching utilization frequency and a slight degradation of the mean BER at high SNRs, when compared to the conventional AQAM scheme. Let us now summarize the performance aspects of the different threshold-based power control schemes that we have discussed so far. The summary of these performances is displayed in Table 4.4, where the performance was quantified by the mean BER, BPS, switching utilization frequency and power utilization frequency at a channel SNR of 20dB for the AQAM scheme with transmission blocking.

In observing Table 4.4, we note that depending on the design of the threshold-based power control scheme, the mean BER can be lowered without changing the switching threshold levels l_n. In doing so, a degradation in the mean BPS performance was incurred, once again, highlighting the mean BER and mean BPS performance trade-offs. Similarly, the mean BPS

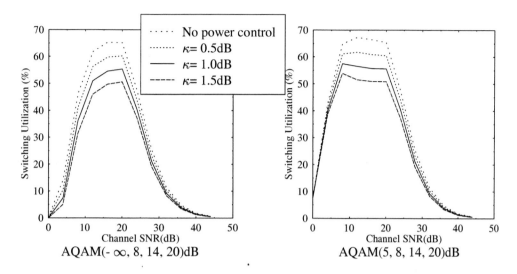

(a) Relative frequency of modulation mode switching with and without transmission blocking.

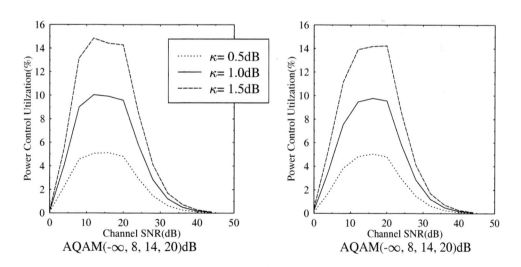

(b) Relative frequency of power control utilization performance with and without transmission blocking.

Figure 4.9: Relative frequency of power control and modulation mode switching using the threshold-based power control scheme for different dynamic ranges κ, which was designed according to Table 4.3 in order to achieve a **low switching utilization**. The switching thresholds were set to $l_1 = -\infty$dB, $l_2 = 8$dB, $l_3 = 14$dB, $l_4 = 24$dB and $l_1 = 5$dB, $l_2 = 8$dB, $l_3 = 14$dB, $l_4 = 24$dB for the non-blocking and blocking schemes, respectively and perfect channel envelope inversion was assumed at the receiver.

Threshold Level Power Control for lower Mean BER

	Mean BER	Mean BPS	Switching Util.(%)	Power Util.(%)
$\kappa = 0.5$	1.25×10^{-3}	4.07	66.1	7.6
$\kappa = 1.0$	8.06×10^{-4}	3.93	66.5	15.7
$\kappa = 1.5$	4.96×10^{-4}	3.78	66.6	23.9
Conventional	1.84×10^{-3}	4.22	65.4	-

Threshold-based Power Control for higher Mean BPS

	Mean BER	Mean BPS	Switching Util.(%)	Power Util.(%)
$\kappa = 0.5$	2.49×10^{-3}	4.35	64.1	7.4
$\kappa = 1.0$	3.09×10^{-3}	4.49	62.7	14.6
$\kappa = 1.5$	3.62×10^{-3}	4.61	60.7	21.2
Conventional	1.84×10^{-3}	4.22	65.4	-

Threshold-based Power Control for lower Switching Utilization

	Mean BER	Mean BPS	Switching Util.(%)	Power Util.(%)
$\kappa = 0.5$	1.9×10^{-3}	4.22	60.4	4.8
$\kappa = 1.0$	2.0×10^{-3}	4.23	55.6	9.6
$\kappa = 1.5$	2.1×10^{-3}	4.23	50.8	14.2
Conventional	1.84×10^{-3}	4.22	65.4	-

Table 4.4: The mean BER, mean BPS, switching utilization and power control utilization relative frequency for the three different threshold-based power control designs of Tables 4.1, 4.2 and 4.3 for different dynamic ranges κ. These performance measures were recorded at an average channel SNR of 20dB, where the switching threshold levels for this transmission blocking AQAM scheme were set to $l_1 = 5\text{dB}$, $l_2 = 8\text{dB}$, $l_3 = 14\text{dB}$, $l_4 = 24\text{dB}$. These performances were compared to that of the conventional AQAM scheme, which did not utilize any power control scheme.

performance was improved at the expense of the mean BER performance. The scheme was also utilized in order to lower the modulation switching utilization frequency, while maintaining the mean BER and BPS performance. However, these performance enhancements resulted in more frequent utilization of power control, which was an additional system overhead. The performance of the threshold-based power control scheme in all scenarios was influenced by the dynamic range κ, where, the scheme achieved its aim better for a higher value of κ, as evidenced by the results in Table 4.4. However, the disadvantage in doing this was that the power control utilization frequency increased.

In the next section, we shall concentrate on the application of AQAM in a wideband channel environment, where we can still use the principles of AQAM developed for a narrowband environment, which we have discussed in this section.

4.3   Adaptive Modulation and Equalization in a Wideband Fading Environment

In the last section the concepts of AQAM were discussed in the context of a narrow-band fading channel environment [145]. In this section we will extend those concepts in order to apply AQAM in a wideband fading channel environment, where equalization plays an important role as evidenced by the results presented in Chapters 2 and 3. Consequently, when applying AQAM in a wideband environment, the joint optimisation of the equalizer and the AQAM scheme is necessitated.

In the narrow-band channel environment the quality of the channel was determined by the short-term SNR of the received burst, which was then used as a metric in order to invoke the appropriate modulation mode at the transmitter, based on a list of switching threshold levels l_n [145]. However, in a wideband environment this metric is not applicable as an estimate of the quality of the channel, where the existence of dispersive multi-path components in the wideband channel produces not only power attenuation of the transmission burst, but also intersymbol interference, as discussed in Section 2.2. Consequently, the metric used to estimate the channel's quality has to be redefined.

The wideband channel will introduce transmission degradation in terms of signal power fluctuations and intersymbol interference as a result of its dispersive fading multi-path nature. Thus the criterion used to switch the modulation modes must incorporate these two effects of the wideband channel. Accordingly, in this wideband channel environment the channel-induced degradation is combated not only by the employment of AQAM but also by equalization. In following this line of thought, we can formulate a two-step methodology in mitigating the effects of the dispersive wideband channel. In the first step, the equalization process will eliminate most of the intersymbol interference based on a CIR estimate and consequently, the signal to noise plus residual interference ratio at the output of the equalizer is calculated based on Equations 2.58, 2.59, 2.60 and 2.61 of Section 2.4, termed as the pseudo-SNR output. This pseudo-SNR at the output of the equalizer is then used as a metric to switch the modulation modes. By utilizing this pseudo-SNR, we are ensuring that the system performance is optimized by jointly employing equalization and AQAM techniques in order to mitigate the effects of the dispersive multi-path fading channel.

In the forthcoming sections, the proposed wideband AQAM and equalization scheme is explored with the aim of characterizing its upper bound performance. The optimisation of the switching threshold levels is also analysed, in order to achieve a certain target mean BER and BPS performance. Finally, the performance of this wideband AQAM scheme and that of its individual fixed modulation modes is compared in terms of their transmission throughput. Before proceeding further, let us state the assumptions used in the employment of this scheme in a wideband channel environment.

4.3.1   Assumptions

In deriving the upper bound performance of this wideband AQAM and equalization scheme, the following assumptions are made:

1. The CIR is constant across the transmission frame, but varies from burst to burst by assuming that the channel is slowly varying, which we refer to here as frame-invariant

fading.

2. The pseudo-SNR at the output of the equalizer is estimated perfectly prior to transmission. This can only be assumed in the context of a TDD scenario, where the channel is considered to be reciprocal in the uplink and downlink and where the channel is slowly varying. Thus the transmitter is capable of utilizing its own receiver's channel quality estimate for the next transmission, given the close channel correlation between the channel quality of the transmitter and receiver slots, even though the associated latency will affect the estimation quality. However for the purpose of generating an upper bound performance, the latency is neglected. This assumption can also be applied, when there exists a reliable, low-delay feedback path between the transmitter and the channel quality estimator at the receiver [35, 36]. The impact of this assumption on the wideband AQAM schemes performance will be carefully analysed in Section 7.2.

3. At the receiver, perfect channel compensation is applied in order to achieve the upper bound performance. In Chapters 2 and 3, we have noted the performance degradation due to incorrect CIR estimation for fixed-mode modulation schemes. However, this degradation is neglected here with the aim of achieving the upper-bound performance. Nevertheless CIR estimation techniques will be invoked and studied in a wideband AQAM scheme in Chapter 7.

4. The receiver assumed perfect knowledge of the modulation mode used in its received transmission burst. In reality, some form of modem mode control signalling must be employed to convey the modulation mode used to the receiver [21, 37]. As a result of the dispersive channel, it is likely that these control symbols may become corrupted sufficiently for the receiver to make an erroneous decision on the modulation mode, which in turn might corrupt the demodulation process. However in Section 5.6 we will attempt to detect the modulation mode by exploiting the extra information provided by the channel decoder at the receiver.

5. The equalizer employed in this scheme is the DFE, where it is assumed that error propagation can be neglected. This will simplify the calculation of the associated numerical upper bound performance. Nevertheless, the impact of error propagation is included and studied at a later stage in Section 7.1 for the wideband AQAM scheme.

6. The residual ISI at the output of the DFE is assumed to be Gaussian distributed for the purpose of mean BER calculations. This assumption will be justified with the aid of an experiment at a later stage.

Above, we have outlined and justified the assumptions required, in order to achieve the upper bound performance of this wideband AQAM and equalization scheme. Let us now concentrate our attention on the methodology of this scheme.

4.3.2 Adaptive Modulation and Equalization System Overview

The system schematic of the wideband AQAM and equalization scheme is depicted in Figure 4.10. At the receiver the channel quality is estimated, which is then used to calculate the DFE coefficients via the DFE coefficient estimator block of Figure 4.10 by solving Equations 2.54

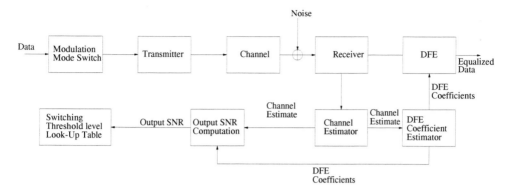

Figure 4.10: Schematic system overview of the wideband AQAM and equalization scheme.

and 2.55 of Section 2.3.4. Subsequently, the coefficients are used to equalize the corrupted received signal. In addition to that, both the CIR estimate and the DFE coefficients are utilized for computing the pseudo-SNR at the output of the DFE. The calculated pseudo-SNR is then compared against a set of optimized switching threshold levels t_n, stored in a look-up table. Consequently, a modulation mode is selected for the next transmission burst, assuming reciprocity of the uplink and downlink TDD slots, where there is a close correlation between the uplink and downlink CIR estimates. This implies that the reciprocity of the pseudo-SNR for the uplink and downlink transmission can be exploited, in order to set the next modulation mode at the transmitter. The modulation modes that are utilized in this scheme are BPSK, 4QAM, 16QAM, 64QAM and a no transmission (NO TX) mode, which were also used by Torrance [31,145] in a narrow-band channel environment. The methodology of switching the modulation modes is similar to that of Torrance [31,145], but instead of using the short term transmission SNR as a switching criterion, the pseudo-SNR at the output of the DFE γ_{DFE}, is used as follows:

$$\text{Modulation Mode} = \begin{cases} NOTX & \text{if } \gamma_{DFE} < t_1 \\ BPSK & \text{if } t_1 \leq \gamma_{DFE} < t_2 \\ 4QAM & \text{if } t_2 \leq \gamma_{DFE} < t_3 \\ 16QAM & \text{if } t_3 \leq \gamma_{DFE} < t_4 \\ 64QAM & \text{if } \gamma_{DFE} \geq t_4, \end{cases} \qquad (4.1)$$

where $t_n, n = 1...4$ are the pseudo-SNR threshold levels, which are set according to the required mean BER and BPS throughput. Let us now investigate the validity of using the pseudo-SNR at the output of the DFE γ_{DFE}, as a switching criterion in this wideband AQAM and equalization scheme.

4.3.3 The Output Pseudo Signal to Noise Ratio of the Decision Feedback Equalizer

In the last section, we introduced the concept of using the pseudo-SNR at the output of the DFE as our channel quality metric for switching the modulation modes and hence the applicability of this metric in applying AQAM is investigated in a wideband channel environment in conjunction with a DFE. Let us commence by rewriting Equations 2.58, 2.59, 2.60 and 2.61 for calculating the pseudo-SNR [105]:

$$SNR_{output} = \frac{\text{Wanted Signal Power}}{\text{Residual ISI Power + Effective Noise Power}}, \quad (4.2)$$

$$\text{Wanted Signal Power} = E\left[\left|s_k \sum_{m=0}^{N_f-1} C_m h_m\right|^2\right], \quad (4.3)$$

$$\text{Effective Noise Power} = N_o \sum_{i=0}^{N_f-1} |C_i|^2, \quad (4.4)$$

$$\text{Residual ISI Power} = \sum_{q=-(N_f-1)}^{-1} E\left[|f_q s_{k-q}|^2\right], \quad (4.5)$$

where $f_q = \sum_{m=0}^{N_f-1} C_m h_{m+q}$ and the remaining notations accrue from Figure 2.9. By substituting Equations 4.3, 4.4 and 4.5 into Equation 4.2, we can write the pseudo-SNR at the output of the DFE γ_{DFE}, as :

$$\gamma_{DFE} = \frac{E\left[\left|s_k \sum_{m=0}^{N_f-1} C_m h_m\right|^2\right]}{\sum_{q=-(N_f-1)}^{-1} E\left[|f_q s_{k-q}|^2\right] + N_o \sum_{m=0}^{N_f-1} |C_m|^2}. \quad (4.6)$$

The applicability of using this pseudo-SNR for selecting the modulation mode is based on the assumption that it can be utilized to estimate the mean BER for the constituent fixed modulation modes and consequently also for the proposed wideband AQAM scheme. This pseudo-SNR is composed of the wanted signal, the effective noise at the output of the DFE and the residual interference, as seen in Equation 4.2, hence, the term pseudo-SNR at the output of the DFE. The inclusion of the residual interference makes it difficult to estimate the actual output SNR of the DFE, since the actual distribution of the residual interference shown in Figure 4.11 is difficult to characterize in terms of a known analytically describable distribution. We attempted to characterize this distribution by using the Kolmogrov-Smirnov test [143], where the Cumulative Distribution Function (CDF) of the residual interference was compared to that of the Gaussian CDF. A low confidence level was produced, which implied that the distribution of the residual interference signal was statistically different from the Gaussian distribution.

However, in obtaining a mean BER performance estimate, we can continue our investigations by approximating the distribution of the residual interference as Gaussian and hence we tentatively view the residual interference as additional noise at the output of the DFE. This approximation was also applied by Monsen [149] in his calculation of the theoretical performance of the DFE in the fading multi-path environment. However, in order to ensure that this assumption was valid, when applied to the calculation of the mean BER, several experiments were conducted using fixed modulation schemes in a dispersive fading multi-path environment.

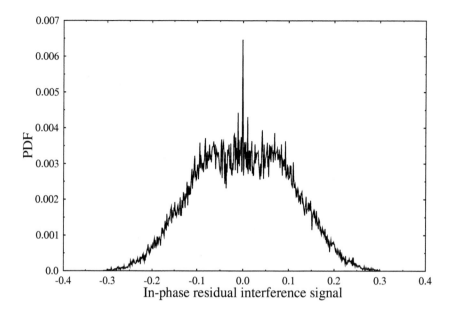

Figure 4.11: Discretised PDF of the in-phase residual interference signal for the two-path Rayleigh
fading channel of Table 3.9 for a channel SNR of 0dB, assuming error-free feedback of
the DFE and perfect channel estimation.

The fading channel models used in this experiment were a two-path CIR and the COST207
Typical Urban (TU) channel. The discretised COST207 Typical Urban (TU) channel impulse
response [150] is shown in Figure 4.12 and tabulated in Table 4.5, while the two-path channel is described in Table 3.9. The Rayleigh fading statistics assumed a carrier frequency
of 1.9GHz, a vehicular speed of 30mph and a transmission rate of 2.6MSymbols/s [151],
yielding a normalized Doppler frequency of 3.25×10^{-5}. Variations due to path-loss and
shadowing were assumed to be eliminated by power control.

The mean BER performance of the constituent fixed modulation modes in these fading
multi-path channels was numerically evaluated following the previously outlined methodol-

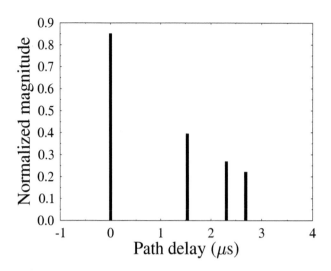

Figure 4.12: The impulse response of a COST 207 [150] Typical Urban (TU) channel specified by the parameters listed in Table 4.5.

Typical Urban	
Position (μs)	Relative Power (dB)
0.00	-1.405
1.54	-8.086
2.31	-11.427
2.69	-13.097

Table 4.5: The power and delay of each path in the COST207 [150] Typical Urban channel of Figure 4.12.

ogy of Equation 2.15 for narrow-band channels :

$$P^b_{\mathrm{BPSK_{MR}}}(\bar{\gamma}) \quad = \quad \int_0^\infty P^b_{\mathrm{BPSK}}(\gamma_{DFE})\, p(\gamma_{DFE}) d\gamma_{DFE}, \qquad (4.7)$$

$$P^b_{\mathrm{4QAM_{MR}}}(\bar{\gamma}) \quad = \quad \int_0^\infty P^b_{\mathrm{4QAM}}(\gamma_{DFE})\, p(\gamma_{DFE}) d\gamma_{DFE}, \qquad (4.8)$$

$$P^b_{\mathrm{16QAM_{MR}}}(\bar{\gamma}) \quad = \quad \int_0^\infty P^b_{\mathrm{av16QAM}}(\gamma_{DFE})\, p(\gamma_{DFE}) d\gamma_{DFE}, \qquad (4.9)$$

$$P^b_{\mathrm{64QAM_{MR}}}(\bar{\gamma}) \quad = \quad \int_0^\infty P^b_{\mathrm{av64QAM}}(\gamma_{DFE})\, p(\gamma_{DFE}) d\gamma_{DFE}, \qquad (4.10)$$

where $\bar{\gamma}$ is the average channel SNR and $p(\gamma_{DFE})$ is the PDF of γ_{DFE} at the given average channel SNR. The instantaneous values of γ_{DFE} were calculated using Equation 4.6

throughout this experiment. Subsequently, the discretised PDF of the pseudo SNR was constructed with an accuracy of 0.001dB. The PDF was then inserted into Equations 4.7, 4.8, 4.9 and 4.10 in order to calculate the numerical mean BER performance of the associated fixed modulation modes. The integration was approximated by the trapezoidal rule [143] with the integration limits set to the lowest and highest values of the instantaneous pseudo-SNR encountered at the output of the DFE, respectively. The notations P^b_{BPSK}, P^b_{4QAM}, $P^b_{\text{av}_{16\text{QAM}}}$ and $P^b_{\text{av}_{64\text{QAM}}}$ indicate the individual theoretical BER performances of each modulation mode in a Gaussian channel, which were derived in Section 2.1.2.

The fixed modulation modes were simulated, employing the simulation parameters listed in Table 4.6 and the experimental results are shown in Figures 4.14 and 4.15. The results displayed close correspondence between the numerical and simulated performance [33]. Consequently, the utilization of the output pseudo-SNR of the DFE γ_{DFE}, in the calculation of the mean BER was justified. This also implied that the quality of the wideband channel could be estimated using this criterion, which was then used to select the appropriate modulation mode in the wideband AQAM and DFE scheme. This criterion will be used to derive the numerical upper bound performance of this wideband AQAM scheme in the next section.

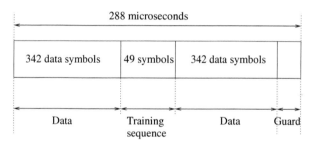

non-spread data burst

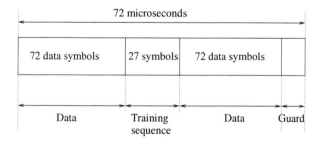

non-spread speech burst

Figure 4.13: Transmission burst structure of the FMA1 non-spread data and non-spread speech modes, as specified in the FRAMES proposal [151].

	Two-path Rayleigh fading channel	TU Rayleigh fading channel
Transmission Burst Format	non-spread speech (See Figure 4.13)	non-spread speech (See Figure 4.13)
DFE Parameters:		
No. of feedforward taps, N_f	15	35
No. of backward taps, N_b	2	7
Decision Feedback	Correct	Correct
Channel Parameters:		
Channel Estimation	Perfect	Perfect
Normalized Doppler Frequency:	3.25×10^{-5}	3.25×10^{-5}
Channel Weights	$0.707 + 0.707z^{-1}$	See Table 4.5

Table 4.6: Simulation parameters used to quantify the performance of the different fixed modulation modes, where the CIR is constant for a transmission burst. The burst format was constructed based on the Future Radio Wideband Multiple Access System (FRAMES) framework [151], which is shown in Figure 4.13.

4.3.4 Numerical Average Upper Bound Performance of the Adaptive Modulation and Decision Feedback Equalization in a Wideband Channel Environment

In the last section, we justified the utilization of the pseudo-SNR at the output of the DFE as a metric used to switch the modulation modes. By utilizing this metric and invoking the concepts and principles of narrow-band AQAM, the numerical upper bound performance for this wideband AQAM scheme can be evaluated.

In reference [31], Torrance quantified the numerical upper bound performance of AQAM in a narrow-band and interference-free scenario. This numerical method can be modified for employment in our AQAM and DFE scheme in a wideband channel environment by utilizing the pseudo-SNR at the output of the DFE, in order to arrive at the wideband numerical mean BER upper bound performance as follows [32, 33]:

$$
P_a(\gamma) = B^{-1} \cdot \left[\begin{array}{l} 1 \cdot \int_{t_1}^{t_2} P_{\mathrm{BPSK}}^b(\gamma_{DFE})\, p(\gamma_{DFE}) d\gamma_{DFE} \\ + 2 \cdot \int_{t_2}^{t_3} P_{\mathrm{4QAM}}^b(\gamma_{DFE})\, p(\gamma_{DFE}) d\gamma_{DFE} \\ + 4 \cdot \int_{t_3}^{t_4} P_{av_{16QAM}}^b(\gamma_{DFE})\, p(\gamma_{DFE}) d\gamma_{DFE} \\ + 6 \cdot \int_{t_4}^{\infty} P_{av_{64QAM}}^b(\gamma_{DFE})\, p(\gamma_{DFE}) d\gamma_{DFE} \end{array} \right], \tag{4.11}
$$

where B is the mean number of bits per symbol (BPS), which can be written as [32, 33]:

$$
B = \left[\begin{array}{l} 1 \cdot \int_{t_1}^{t_2} p(\gamma_{DFE}) d\gamma_{DFE} \\ + 2 \cdot \int_{t_2}^{t_3} p(\gamma_{DFE}) d\gamma_{DFE} \\ + 4 \cdot \int_{t_3}^{t_4} p(\gamma_{DFE}) d\gamma_{DFE} \\ + 6 \cdot \int_{t_4}^{\infty} p(\gamma_{DFE}) d\gamma_{DFE} \end{array} \right]. \tag{4.12}
$$

The wideband AQAM and equalization numerical upper bound performance was calculated similarly to that of the fixed modulation modes' mean BER numerical performance, as

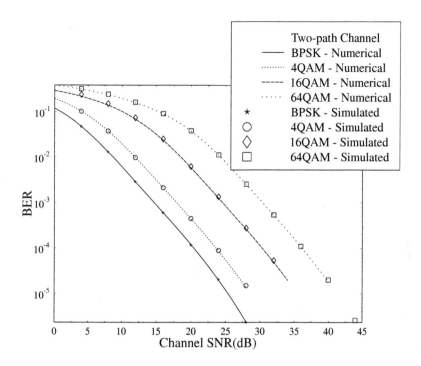

Figure 4.14: Numerical and simulated performance of BPSK, 4QAM, 16QAM and 64QAM in a **two-path Rayleigh fading channel** using the simulation parameters listed in Table 4.6. The numerical solution was calculated using Equations 4.7, 4.8, 4.9 and 4.10 and assuming that the residual interference at the output of the DFE was Gaussian distributed.

discussed in Section 4.3.3. Specifically, the numerical performance was calculated for two different AQAM schemes. In the first scheme, using the two-path Rayleigh fading channel, the pseudo-SNR switching thresholds were set for each of the modulation modes such that a BER of approximately 1% was maintained according to the fixed modulation modes' BER performance of Figure 4.14. Consequently, the levels were set as follows : $t_1 = -\infty$dB, $t_2 = 12$dB, $t_3 = 18$dB and $t_4 = 24$dB, where the transmitter constantly transmitted data. By contrast, in the second scheme, where transmission blocking was incorporated, the thresholds were $t_1 = 6$dB, $t_2 = 12$dB, $t_3 = 18$dB and $t_4 = 24$dB, where the transmission was disabled, whenever the instantaneous pseudo-SNR at the output of the DFE dipped below the threshold t_1. These experiments were then repeated for the TU Rayleigh fading channel, where the switching thresholds were $t_1 = -\infty$dB, $t_2 = 12$dB, $t_3 = 19$dB and $t_4 = 25$dB for the non-blocking transmission scheme as well as opting for $t_1 = 6$dB, $t_2 = 12$dB, $t_3 = 19$dB and $t_4 = 25$dB for the blocking-assisted scheme. Again, these levels were set within the channel SNR range of achieving a mean BER of approximately 1% for the fixed modulation modes' performance curves shown in Figure 4.15. For this experiment, the switching thresholds were set on a rather heuristic basis only for the purpose of studying the effects of

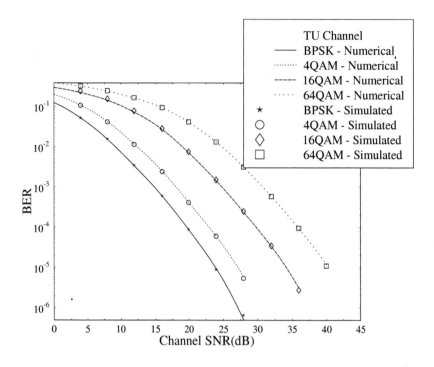

Figure 4.15: Numerical and simulated performance of BPSK, 4QAM, 16QAM and 64QAM over the **TU Rayleigh fading channel** of Figure 4.12 using the simulation parameters listed in Table 4.6. The numerical solution was calculated using Equations 4.7, 4.8, 4.9 and 4.10 and assuming that the residual interference at the output of the DFE was Gaussian distributed.

this wideband AQAM scheme. However, in Section 4.3.5 it will be shown that the switching threshold levels can be optimized for a given target performance.

The performance of the wideband AQAM and equalization scheme was investigated over the two-path and TU channels, which were described in Section 4.3.3, and the results are displayed in Figures 4.16 and 4.17, respectively. The simulated results were also compared to the numerical upper bound performance obtained using Equations 4.11 and 4.12. It was observed that a good correspondence was achieved between the numerical and simulation performance for the two different wideband channels, further justifying our numerical model for the wideband AQAM scheme.

Let us now analyse in detail the results shown in Figure 4.16 for the two-path channel. In the AQAM scheme, which did not incorporate transmission blocking (indicated by the $-\infty$ lowest threshold in the legends), the adaptive performance was better or equal to the performance using BPSK in terms of the mean BER and BPS for the channel SNR range between 0dB and 24dB. At a channel SNR of 24dB the mean BER performance was similar for the adaptive scheme and BPSK, but the mean BPS of the adaptive scheme was better by approxi-

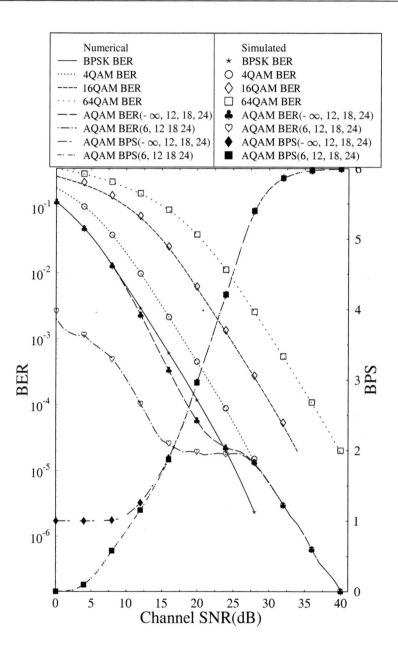

Figure 4.16: Numerical and simulated performance of the wideband AQAM and DFE scheme as well as BPSK, 4QAM, 16QAM and 64QAM over the **two-path Rayleigh fading channel** using the simulation parameters listed in Table 4.6 and the assumptions in Section 4.3.1. The numerical solutions were calculated using Equations 4.7, 4.8, 4.9 and 4.10 for the fixed modulation modes and Equations 4.11 and 4.12 were used to calculate the numerical solution for the wideband AQAM and DFE scheme. The threshold values in the legend are expressed in dB.

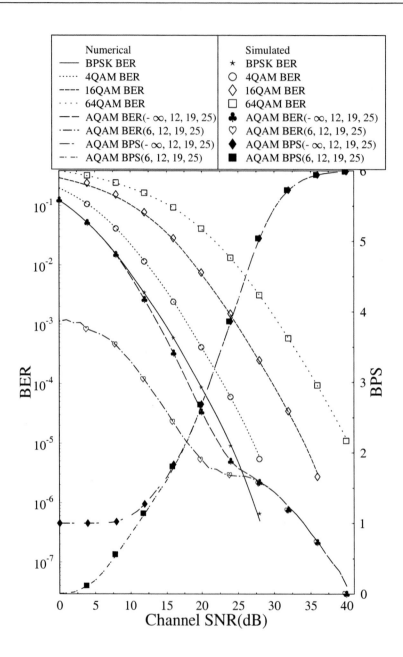

Figure 4.17: Numerical and simulated performance of the wideband AQAM and DFE scheme as well as BPSK, 4QAM, 16QAM and 64QAM over the **TU Rayleigh fading channel** using the simulation parameters listed in Table 4.6 and the assumptions in Section 4.3.1. The numerical solutions were calculated using Equations 4.7, 4.8, 4.9 and 4.10 for the fixed modulation schemes and Equations 4.11 and 4.12 were used to calculate the numerical solution for the wideband AQAM and DFE scheme. The threshold values in the legend are expressed in dB.

mately a factor of four, resulting in a mean throughput of approximately four bits per symbol. The adaptive mean BER performance was even better, than the BPSK performance in the channel SNR range of 8dB to 24dB, which is justified below. This phenomenon was also observed in the narrow-band AQAM scheme of reference [145]. In this range of $8 - 24$dB, the instantaneous pseudo-SNR at the output of the DFE increased, yielding a better mean BPS performance than that of BPSK. Consequently the adaptive mean BER was lowered, when compared to the BPSK scheme due to the averaging over a higher number of bits. In the channel SNR range of $24 - 28$dB the adaptive scheme outperformed 4QAM, where at a channel SNR of 28dB, the mean BER values for both schemes were equal again, but the mean throughput of the adaptive scheme was approximately 5.5 bits per symbol, resulting in an improvement factor of 2.5, when compared to 4QAM. These BPS throughput gains are recorded in Table 4.7. At higher channel SNRs the mean BER and BPS performance converged to the 64QAM performance, since the probability of encountering a low instantaneous pseudo-SNR at the output of the DFE was low due to the high average channel SNR encountered. Consequently, 64QAM became the dominant modulation mode with infrequent switching to the lower-order modulation modes. The same characteristics were also valid for the wideband AQAM and DFE scheme over the TU Rayleigh fading channel.

Channel SNR(dB)	Fixed Modulation	AQAM
Channel SNR(dB)	BPS throughput	BPS throughput
24	1 (BPSK)	4
28	2 (4QAM)	5.5

Table 4.7: BPS throughput comparison between the fixed modulation modes and AQAM for similar BER over the **two-path Rayleigh fading channel**. The values were tabulated from Figure 4.16.

The probability of each modulation mode being employed in the adaptive scheme is shown in Figure 4.18(a) and Figure 4.18(b) over the two-path channel and the TU Rayleigh fading channel, respectively. This accumulated probability at each particular average channel SNR was equal to unity and the effect of the switching thresholds upon the transmitted modulation mode can be observed. At low average channel SNRs the lower-order modulation modes of BPSK and 4QAM were dominant, while at higher average channel SNRs, the higher-order modulation modes became dominant. At each average channel SNR, the contribution of different modulation modes was observed, highlighting the switching mechanism of the AQAM scheme. Consequently, the pseudo-SNR - which was changing instantaneously - determined the modulation mode in order to maximise both the mean BER and BPS performance of this wideband AQAM scheme. A more explicit representation of the wideband AQAM regime is shown in Figure 4.19, which displays the variation of the modulation mode with respect to the pseudo SNR. In these figures, it can be seen explicitly that the lower-order modulation modes were chosen, when the pseudo SNR was low. In contrast, when the pseudo SNR was high, the higher-order modulation modes were selected in order to increase the transmission throughput. These figures can also be used to exemplify the application of wideband AQAM in an indoor and outdoor environment. In this respect, Figure 4.19(a) can be used to characterize a hostile outdoor environment, where the perceived channel quality was low. This resulted in the utilization of predominantly more robust modulation modes, such

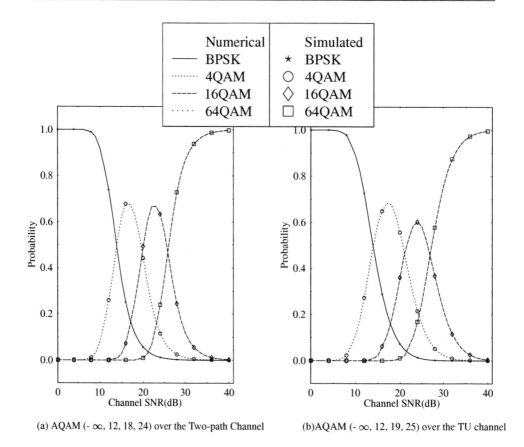

(a) AQAM (- ∞, 12, 18, 24) over the Two-path Channel   (b)AQAM (- ∞, 12, 19, 25) over the TU channel

Figure 4.18: Numerical and simulated probabilities of each modulation mode utilized by the wideband
AQAM scheme without transmission blocking over the (a) **two-path Rayleigh Fading
channel** and (b)the **TU Rayleigh fading channel** using the simulation parameters listed
in Table 4.6 and the assumptions in Section 4.3.1.

as BPSK and 4QAM. Conversely, a less hostile indoor environment is exemplified by Figure
4.19(b), where the perceived channel quality was high. As a result, the wideband AQAM
regime can adapt suitably by invoking higher-order modulation modes, as evidenced by Figure 4.19(b). In this simple example, wideband AQAM can be utilized in order to provide
a seamless transition between an indoor and outdoor environment. In the wideband AQAM
scheme utilizing transmission blocking, which was characterized in Figure 4.16 (indicated
by the lowest threshold of 6dB in the legends) over the two-path Rayleigh fading channel, at
low channel SNRs the mean BER performance was below 0.1%. This low mean BER was
achieved as a result of the employment of transmission blocking, when the pseudo-SNR was
below 6dB. Conversely, the mean BPS performance degraded, when compared to the non-
blocking adaptive scheme for average channel SNRs below 10dB. However at higher average
channel SNRs, the performance of the two different schemes converged, as the probability of

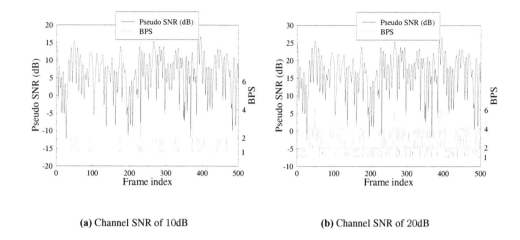

(a) Channel SNR of 10dB   **(b)** Channel SNR of 20dB

Figure 4.19: Modulation mode variation with respect to the pseudo SNR defined by Equation 4.6 over the **TU Rayleigh fading channel**. The BPS throughputs of 1, 2, 4 and 6 represent BPSK, 4QAM, 16QAM and 64QAM, respectively. The simulation parameters listed in Table 4.6 and the assumptions of Section 4.3.1 were utilized in this experiment.

the transmission blocking reduced. This is evidenced by the transmission blocking probability curve shown in Figure 4.20 over the TU Rayleigh fading channel, where the probability decreased, as the average channel SNR increased.

In summary, the wideband AQAM and DFE scheme possessed certain advantages, when compared to any of the fixed modulation modes in terms of mean BER and BPS performance. The BPS throughput comparison between fixed modulation modes and AQAM is shown in Table 4.16. This comparison will be discussed in more depth at a later stage. The other advantage of using the adaptive scheme in conjunction with transmission blocking was that the mean BER and BPS performance could be tailored or targeted to a certain required performance. However, the utilization of transmission blocking resulted in a transmission latency due to the associated buffering. Finally, we have also demonstrated the good correspondence between the numerical performance characterized by Equations 4.11 and 4.12 and the simulated performance. In the next section, the switching levels are numerically optimized, in order to achieve a certain target mean BER and BPS performance.

4.3.5   Switching Level Optimisation

In the last section we observed that transmission blocking can be used in the wideband AQAM and DFE scheme in order to target a certain required BER performance. Consequently, we can now concentrate on setting the switching levels appropriately in order to achieve a transmission quality of 1% and 4.5 in terms of mean BER and BPS, respectively, termed as the **High-BER** regime. By contrast, a higher transmission quality was targeted in our **Low-BER** regime, where a mean BER of 0.01% and a mean BPS value of 3 was expected.

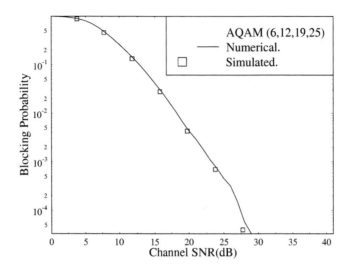

Figure 4.20: Numerical and simulated transmission blocking probabilities of the wideband AQAM scheme over the (a) **two-path Rayleigh Fading channel** and (b) the **TU Rayleigh fading channel** using the simulation parameters listed in Table 4.6 and the assumptions in Section 4.3.1.

In utilizing the pseudo-SNR at the output of the DFE as the metric to switch the modulation modes, some of the concepts of narrow-band AQAM was reused. In optimising the switching levels the Powell Multi-dimensions Line Minimization technique [143] was utilized, which was identical to the technique used in narrow-band AQAM [145]. In this technique, the inputs required were a set of initial switching level estimates hosted by the vector **P**, a cost function $T(\mathbf{P})$, which was dependent on **P**, and a vector direction **n**, which provided the minimization technique with a direction in order to reduce the cost function $T(\mathbf{P})$. Thus given these inputs, the technique sought a scalar value λ, that minimized $T(\mathbf{P} + \lambda \mathbf{n})$ and subsequently replaced the initial value of the switching level vector estimate **P** by $(\mathbf{P} + \lambda \mathbf{n})$. This was then repeated with a new direction vector **n**, which was changed based on the previous and present switching level vector, **P** and the scalar value λ. The cost function $T()$, which was similar to the function defined by Torrance [145] can be written as :

$$\text{Total Cost} = \sum_{i=\gamma_{DFE_{min}}}^{\gamma_{DFE_{max}}} \text{BER Cost}(i) + \text{BPS Cost}(i), \qquad (4.13)$$

where $\gamma_{DFE_{min}}$ and $\gamma_{DFE_{max}}$ were set to 0dB and 40dB, respectively, quantized in intervals of 1dB in order to evaluate the cost function. The BER Cost and BPS Cost were defined as [145]:

$$\text{BER Cost}(i) = \begin{cases} 10 \cdot \left[log_{10}\left(\frac{\text{BER}_m(i)}{\text{BER}_d(i)} \right) \right] & \text{if BER}_m > \text{BER}_d \\ 0 & \text{otherwise,} \end{cases} \qquad (4.14)$$

$$\text{BPS Cost}(i) = \begin{cases} \text{BPS}_d(i) - \text{BPS}_m(i) & \text{if BPS}_d > \text{BPS}_m \\ 0 & \text{otherwise.} \end{cases} \qquad (4.15)$$

The measured quantities, $\text{BER}_m(i)$ and $\text{BPS}_m(i)$ were numerically evaluated using Equations 4.11 and 4.12, respectively and $\text{BER}_d(i)$ and $\text{BPS}_d(i)$ constituted the desired performance at the average channel SNR of i. The cost function was minimised based on Equations 4.14 and 4.15, where the minimization technique attempted to ensure that the measured BER and BPS performance was within the targeted range. The BER Cost function of Equation 4.14 incorporated the logarithm function, in order to increase the significance of the small differences between the measured and targeted BER and a weighting factor of 10 was applied, in order to ensure that the minimization procedure was biased towards achieving the required target BER in preference over the targeted BPS.

The threshold optimisation was conducted for both the two-path and the TU Rayleigh fading channels for both the **High-BER** and **Low-BER** transmission systems. The initial switching level estimates were set to $t_1 = 6\text{dB}$, $t_2 = 12\text{dB}$, $t_3 = 18\text{dB}$, $t_4 = 24\text{dB}$ and $t_1 = 6\text{dB}$, $t_2 = 12\text{dB}$, $t_3 = 19\text{dB}$, $t_4 = 25\text{dB}$ for the two-path channel and the TU Rayleigh fading channel, respectively. The initial cost function minimization direction, **n**, was set to a unit vector for each of the switching levels. The results of the optimisation process are listed in Table 4.8 [32].

	t_1(dB)	t_2(dB)	t_3(dB)	t_4(dB)
High-BER (Two-path channel)	3.68026	6.3488	11.7181	17.8342
Low-BER (Two-path channel)	8.30459	10.4541	16.8846	23.051
High-BER (TU channel)	3.63628	6.2258	11.6450	17.6846
Low-BER (TU channel)	8.24582	10.4579	16.7980	23.7589

Table 4.8: The optimized switching levels t_n of the wideband AQAM and DFE scheme for **High-BER** and **Low-BER** transmission over the two-path and the TU Rayleigh fading channel. The targeted mean BER and BPS performance for **High-BER** transmission were 1% and 4.5, respectively, while for the **Low-BER** regime, the target requirements were 0.01% and 3 in terms of mean BER and BPS characteristics.

The mean BER and BPS performances were numerically calculated utilizing Equations 4.11 and 4.12 and the switching levels listed in Table 4.8 for the **High-BER** and **Low-BER** transmission systems, respectively. The results are shown in Figures 4.21(a) and 4.21(b) for the two-path and the TU Rayleigh fading channels, respectively. The targeted mean BERs of the **High-BER** and **Low-BER** regime of 1% and 0.01% over the two channels were achieved for all average channel SNRs investigated. At average channel SNRs below 20dB, the lower-order modulation modes were dominant, producing a robust system in order to achieve the targeted BER. Similarly, at high average channel SNRs the higher-order modulation mode of 64QAM dominated the transmission regime, yielding a lower mean BER than the target,

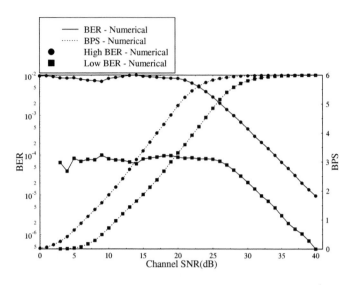

(a) AQAM performance over the two-path Rayleigh fading channel

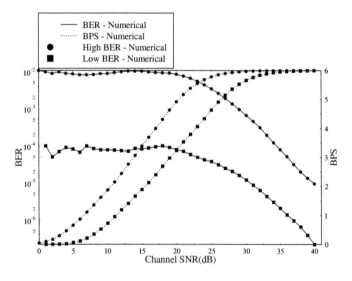

(b) AQAM Performance over the TU Rayleigh fading channel

Figure 4.21: Numerical mean BER and BPS performance of the wideband equalized AQAM scheme for the **High-BER** and **Low-BER** regime using the simulation parameters listed in Table 4.6 and the assumptions in Section 4.3.1. The switching levels used in these experiments are listed in Table 4.8.

since no higher-order modulation mode could be legitimately invoked. This is evidenced by the modulation mode probability results shown in Figures 4.22 and 4.23 for the two-path and TU Rayleigh fading channels, respectively. The targeted mean BPS values for the **High-BER** and **Low-BER** regime of 4.5 and 3 were achieved at approximately 18dB and 19dB channel SNRs for the two-path and TU Rayleigh fading channels, respectively. However, at average channel SNRs of below 3dB transmission blocking was dominant in the **Low-BER** transmission system and thus the mean BER performance was not recorded for that range of average channel SNRs.

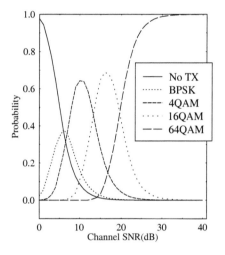 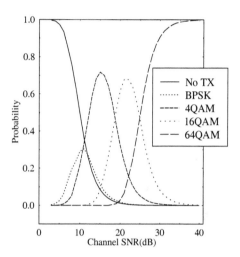

(a) **High-BER** transmission regime over the two-path Rayleigh fading channel

(b) **Low-BER** transmission regime over the two-path Rayleigh fading channel

Figure 4.22: Numerical probabilities of each modulation mode utilized for the wideband AQAM and DFE scheme over the **two-path Rayleigh Fading channel** for the **(a) High-BER Transmission** regime and **(b) Low-BER Transmission** regime using the simulation parameters listed in Table 4.6 and the assumptions in Section 4.3.1. The switching levels were optimized and set according to Table 4.8.

In summary, the Powell Multi-dimensions Line Minimization technique was applied, in order to optimize the switching threshold levels of the wideband AQAM and equalization scheme for the desired **High-BER** and **Low-BER** transmission performance. In the next section, the performance of this wideband AQAM scheme and the individual fixed modulation mode performances are compared and subsequently the advantages and disadvantages of the wideband AQAM scheme are discussed.

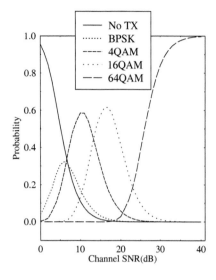
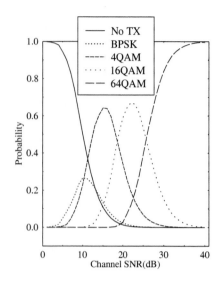

(a) High-BER transmission regime over the TU
Rayleigh fading channel

(b) Low-BER transmission regime over the TU
Rayleigh fading channel

Figure 4.23: Numerical probabilities of each modulation mode utilized for the wideband AQAM and
DFE scheme over the **TU Rayleigh Fading channel** for the **(a) Low-BER Transmission**
regime and **(b) Low-BER Transmission** regime using the simulation parameters listed in
Table 4.6 and the assumptions in Section 4.3.1. The switching levels were optimized and
set according to Table 4.8.

4.3.6   The Throughput Performance of the Fixed Modulation Modes and the Wideband Adaptive Modulation and Decision Feedback Equalization Scheme

The transmission throughput quantified in terms of the number of bits per symbol is used here
as a measure of comparing the performances of the fixed modulation modes and the wideband
AQAM and DFE scheme. The transmission throughput was evaluated for the individual fixed
modulation modes at the targeted mean BER of 1% and 0.01% by utilizing the results shown
in Figures 4.14 and 4.15 for the two-path and TU Rayleigh fading channels, respectively.
Similarly, the transmission throughput of the wideband AQAM scheme was evaluated by
using the results displayed in Figures 4.21(a) and 4.21(b).

The transmission throughput achieved for the **High-BER** and **Low-BER** transmission
regimes over both channels is shown in Figures 4.24 and 4.25. The transmission throughput
for the **High-BER** transmission regime was higher than that of the **Low-BER** transmission
regime for the same transmitted signal energy due to the more relaxed BER requirement of

the **High-BER** transmission regime, as evidenced by Figures 4.24 and 4.25. The achieved transmission throughput of the wideband AQAM scheme was higher than that of the BPSK, 4QAM and 16QAM schemes for the same average channel SNR. However, at higher average channel SNRs the throughput performance of both schemes converged, since the 64QAM became the dominant modulation mode for the wideband AQAM scheme. Improvements of $1-3$dB and $8-10$dB were observed for the **High-BER** and **Low-BER** transmission regimes, respectively for the wideband AQAM scheme over the two-path Rayleigh fading channel. Similarly, for the TU channel, $1-3$dB and $7-9$dB gains were recorded for the **High-BER** and **Low-BER** transmission schemes, respectively. These gains were considerably lower than those associated with narrow-band AQAM, where 5 - 7dB and 10 - 18dB of gains were reported for the **High-BER** and **Low-BER** transmission scheme, respectively [145]. This was expected, since in the narrow-band environment the fluctuation of the instantaneous SNR was more severe, resulting in increased utilization of the modulation switching mechanism. Consequently, the instantaneous transmission throughput increased, whenever the fluctuations yielded a high received instantaneous SNR. Conversely, in a wideband channel environment the channel quality fluctuations perceived by the DFE were less severe due to the associated multi-path diversity, which was exploited by the equalizer.

In this section, we have compared the performances of a range of fixed modulation modes and the wideband AQAM and DFE scheme in terms of their transmission throughput and recorded the SNR gains of the adaptive scheme. However, this comparison is incomplete, since there are many further issues to consider, when comparing these two schemes, which will be further explored in Chapter 7. In the next section, we will conclude this chapter by highlighting the underlying results and their corresponding implications.

4.4 Review and Discussion

In this chapter, we have focused our attention on the application of AQAM, which was discussed in the context of both narrow-band and wideband channel environments. In the narrow-band environment, threshold-based power control was incorporated into the AQAM scheme, where power control was utilized in SNR regions near the switching thresholds l_n. This was implemented firstly in order to lower the mean BER, secondly in order to increase the mean BPS performance and thirdly to decrease the modulation mode switching frequency. The power control scheme was designed according to the modulation mode transition matrix shown in Table 4.1 in order to achieve a better mean BER performance, as shown in Figures 4.4(a) and 4.4(b). Subsequently the threshold-based power control scheme was designed to increase the mean BPS performance and to reduce the modulation mode switching frequency according to Tables 4.2 and 4.3, respectively. The mean BPS and switching utilization performance of these designs was shown in Figures 4.6(a), 4.6(b) and 4.9(a), respectively. The results of these three different designs were tabulated for a channel SNR of 20dB in Table 4.4 in terms of their mean BER, BPS, modulation mode switching utilization frequency and power control utilization frequency for different dynamic ranges κ. As expected, the threshold-based power control scheme managed to slightly reduce the mean BER, but at the expense of the mean BPS performance and increased complexity in terms of the power control utilization frequency. This was also observed for the design, which increased the mean BPS performance, where the mean BER performance was slightly degraded. In both these

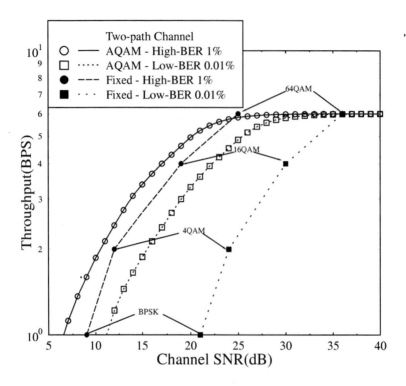

Figure 4.24: Transmission throughput of the wideband AQAM and DFE scheme and fixed modulation
modes over the **two-path Rayleigh Fading channel** for both the **High-BER** and **Low-
BER** transmission regimes using the simulation parameters listed in Table 4.6 and the
assumptions in Section 4.3.1. The switching levels used in these experiments are listed in
Table 4.8.

cases, the modulation mode switching relative frequency remained approximately the same.
However, in the third case, the threshold-based power control design managed to reduce the
modulation mode switching relative frequency with a concomitant minimal degradation to the
mean BER and BPS performance, but at the cost of increased power control utilization relative frequency. In all these cases, the higher the dynamic range κ, the higher the performance
gain as evidenced in Table 4.4, although the power control utilization relative frequency increased. In conclusion, the utilization of the threshold-based power control scheme improved
the system performance of the AQAM, but at the cost of increased complexity in handling the
power control scheme. Moreover, in these experiments perfect power control was assumed
and consequently the performance improvements shown here constituted an upper bound
case.

In Section 4.3, the AQAM scheme was applied in conjunction with the DFE in a wide-

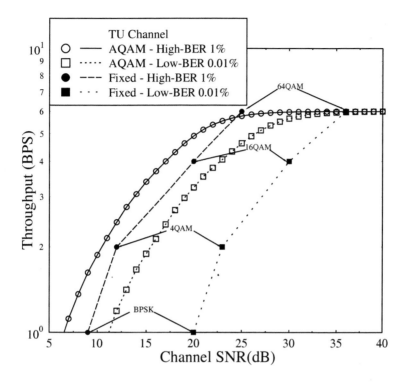

Figure 4.25: Transmission throughput of the wideband AQAM and DFE scheme and fixed modulation modes over the **TU Rayleigh Fading channel** for both the **High-BER** and **Low-BER** transmission regimes using the simulation parameters listed in Table 4.6 and the assumptions in Section 4.3.1. The switching levels used in these experiments are listed in Table 4.8.

band channel environment. In doing so, we have jointly optimized the performance of the DFE and the AQAM scheme in order to mitigate the effects of the dispersive channel. In utilizing the assumptions listed in Section 4.3.1, the pseudo-SNR at the output of the DFE was characterized by Equation 4.6, and it was used as the metric controlling the modulation modes. This metric gave a joint measure of the attenuation and ISI inflicted by the channel after equalization. As such, the utilization of this metric incorporated the contribution of the DFE and the AQAM scheme in mitigating the effects of the dispersive channel. The validity of this metric in estimating the mean BER of the system was tested in Section 4.3.3 and the results exhibited good correspondence between the numerical and simulated mean BER performance of the fixed modulation modes, as evidenced by Figures 4.14 and 4.15.

The wideband AQAM and DFE scheme was then invoked over the two-path and TU Rayleigh fading channels and its performance was then compared to the numerical upper bound performance, which was governed by Equations 4.11 and 4.12 for the mean BER and mean BPS, respectively. The results are shown in Figures 4.16 and 4.17, where there was a good correspondence between the numerical and simulated performance for both the transmission-blocking based and non-blocking scheme. The results also showed the performance improvement in terms of mean BER and BPS, when compared to the fixed modulation modes.

The optimisation of the modulation mode switching threshold levels was then implemented using the Powell Multi-dimensions Line Minimization technique for both the **High-BER** and **Low-BER** transmission regimes [26, 143]. The performance targets for the **High-BER** transmission scheme comprised of a mean BER of 1% and a mean BPS value of 3, and for the **Low-BER** transmission, a mean BER of 0.01% and BPS of 4.5 was targeted. The optimisation technique optimized the cost function, which was characterized by Equations 4.13, 4.14 and 4.15, yielding the switching threshold levels listed in Table 4.8. These optimized switching levels were then applied to the wideband AQAM scheme and the mean BER and BPS were numerically calculated and shown in Figures 4.21(a) and 4.21(b). The results indicated that the optimisation process could be utilized in order to configure the wideband AQAM scheme for different target performances.

The comparison between the wideband AQAM scheme's performance and the fixed modulation modes in terms of the transmission throughput was investigated in Section 4.3.6. The transmission throughput gain of the wideband AQAM scheme over the fixed modulation modes for the **High-BER** and **Low-BER** transmission regimes was displayed in Figures 4.24 and 4.25. Channel SNR gains of approximately $1 - 3dB$ and $8 - 10dB$ were recorded for the **High-BER** and **Low-BER** transmission regimes, respectively, over the two-path Rayleigh fading channel. Similarly, for the TU channel improvements of $1-3dB$ and $7-9dB$ were observed for **High-BER** and **Low-BER** transmission, respectively. These gains were achieved using the assumptions listed in Section 4.3.1 and hence represented an upper bound performance gain. However, in Chapter 7 these assumptions are discarded, in order to create a more practical wideband AQAM and equalization scheme. In the next chapter, adaptive channel coding techniques are invoked in conjunction with AQAM in order to increase the transmission throughput on a burst by burst basis.

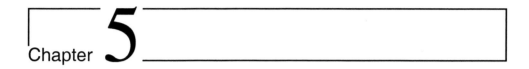

Chapter **5**

Turbo-Coded and Turbo-Equalised Wideband Adaptive Modulation

In the previous chapter, we introduced the joint Adaptive Quadrature Amplitude Modulation (AQAM) and equalization scheme, where the pseudo-SNR at the output of the DFE was used as the modulation mode switching metric in order to mitigate the effects of a wideband fading channel. In this chapter, the wideband AQAM scheme is extended to incorporate the benefits of channel coding. The general motivation for using channel coding is to exploit the error correction and the error detection capability of the channel codes in order to improve the BER and throughput performance of the wideband AQAM scheme.

As we have shown in Chapter 4, the wideband AQAM scheme was capable of yielding an improved BER and BPS performance, when compared to each individual fixed modulation mode. Since the wideband AQAM scheme improves the BER performance, high coding rate channel codes can be utilized in our coded AQAM scheme. The utilization of these high coding rate channel codes is essential to produce a better coded throughput performance, when compared to the uncoded wideband AQAM scheme, which was discussed in the previous chapter.

Since the wideband AQAM scheme always attempts to invoke the appropriate modulation mode in order to combat the wideband channel effects, the probability of encountering a received transmitted burst with a high instantaneous BER is low, when compared to the constituent fixed modulation modes. This characteristic is advantageous, since due to the less bursty error distribution, the coded wideband AQAM scheme can be implemented without the utilization of high-delay channel interleavers. Consequently we can exploit the error detection capability of the channel codes almost instantaneously at the receiver for every received transmission burst. This is essential, since the error detection capability of the channel codes can provide the receiver with extra intelligence, in order to detect the modulation mode that was utilized. The channel codecs' error detection capability can also be exploited in order to gauge the short term BER of each individual transmitted burst. Hence the short term BER can be used as a modulation mode switching metric, since it can quantify the impact of virtually all channel-induced impairments, such as signal strength variation, ISI, etc. For example, to a certain extent, this metric can incorporate the impact of co-channel interference. In our subsequent discussions the short term BER metric is not exploited, hence the interested reader is referred to the contributions by Yee and Hanzo [43, 44] for more details.

In Section 5.1 turbo coding [152] is invoked in conjunction with AQAM and its performance is compared to that of the fixed modulation modes as well as to that of the uncoded AQAM scheme presented in Section 4.3.5. Furthermore, in Section 5.6 channel coding is also exploited for detecting the modulation modes at the receiver. In Section 5.7 it is shown that employing adaptive-rate turbo channel coding in conjunction with adaptive modulation results in a higher effective throughput, than fixed-rate channel coding. Our wideband AQAM scheme is then invoked in the context of turbo equalization in Section 5.10, where channel equalization [153] and channel decoding is implemented jointly and iteratively. The chapter is concluded in Section 5.11 with a system design example cast in the context of a number of powerful wideband joint coding and modulation schemes, namely Trellis Coded Modulation (TCM), Turbo Trellis Coded Modulation (TTCM) and Bit Interleaved Coded Modulation (BICM).

Recent work on combining conventional channel coding with adaptive modulation has been conducted for example by Matsuoka *et al.* [34], where punctured convolutional coding with and without an outer Reed Solomon (RS) code was invoked in a TDD environment. Convolutional coding was also used in conjunction with adaptive modulation by Lau in reference [52], where results were presented in a Frequency Division Multiple Access (FDMA) and Time Division Multiple Access (TDMA) environment, when assuming the presence of a channel feedback path between the receiver and transmitter. Finally, Goldsmith *et al.* [154] demonstrated that in adaptive coded modulation the simulation and theoretical results confirmed a 3dB coding gain at a BER of 10^{-6} for a 4-state trellis code and a coding gain of 4dB was achieved by an 8-state trellis code over Rayleigh-fading channels, while a 128-state code performed within 5dB of the Shannonian capacity limit. Let us now briefly review the concept of turbo coding.

5.1 Turbo Coding

Turbo coding is a form of iterative channel decoding that produces excellent results as demonstrated by Berrou *et al.* [152, 155] in 1993. The concept of turbo coding can be best explained by referring to its encoder and decoder structures. The schematic of the turbo encoder is shown in Figure 5.1. Explicitly, two component encoders are utilized, in order to produce the turbo code, where a so-called random turbo interleaver [152, 156] is placed before the second encoder. The general aim of the turbo encoder is to generate two independent component codes, which encode the same information bits. The role of the turbo interleaver is to ensure that the two encoded bit streams are independent from each other, due to the scrambling of the information bits by the interleaver. The component codes used in the encoder can be either block or convolutional codes. An example of a binary block code, which is amenable to turbo coding is the family of Bose-Chaudhuri-Hocquenghem (BCH) codes [157] that possess multiple error detection and correction capabilities. Explicitly, each BCH code is represented by the notation BCH (n, k, d_{min}), where n, k and d_{min} denote the number of the encoded

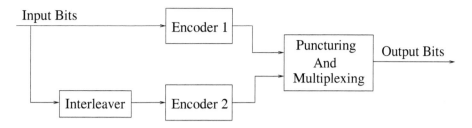

Figure 5.1: Turbo encoder schematic.

bits, the number of information bits and the minimum Hamming distance, respectively. The number of parity bits is equal to $n - k$ and the coding rate is $\frac{k}{n}$. The BCH encoder accepts k information bits and by using a specific polynomial code generator [157], the parity bits are added in order to produce n coded bits. Thus, according to the encoding rules, only certain encoded sequences are legitimate. It is this distinction that enables the decoder to recognize and correct corrupted or illegitimate codewords. We will refrain from discussing the code generation and decoding mechanism, referring the reader to references [13, 157–159]. By referring to Figure 5.1, the generated codewords are punctured and multiplexed, in order to produce the turbo code. However, puncturing of the parity bits is not applied to turbo BCH codes as proposed by Hagenauer [160] and Pyndiah [161]. Consequently, for example, a component code of BCH (31, 26, 3) will yield a turbo block code of BCH (36, 26), where the additional five parity bits of the second encoder are included in the output turbo block code, while the systematic information bits produced by the second encoder are discarded.

The other family of constituent turbo encoders that can be utilized is Recursive Systematic Convolutional (RSC) codes, which is shown in Figure 5.2. Here, the constraint length is set to $K = 3$, and the generator polynomials are set in octal terms, to 7 and 5 [157]. Referring to Figure 5.2, a stream of systematic bits, which represents the original information sequence is generated along with the corresponding parity sequence. In forming the convolutional-based turbo code, the systematic bits of the second convolutional encoder are discarded and the two sets of parity sequences are punctured accordingly. The puncturing pattern can be varied in order to produce different code rates.

The iterative decoding structure of the turbo decoder is shown in Figure 5.3. The component decoders require soft inputs and produce soft outputs. Consequently, special decoding algorithms such as the Maximum A Posteriori (MAP) [162] and the Log-MAP [163] algorithms can be invoked, which were proposed by Bahl and Robertson, respectively. These algorithms are highlighted in Appendix A.1 [164]. Essentially, the soft output generated by either decoder determines whether the decoded bit is a binary 1 or 0 as well as the reliability of the output bit decision. Let us now analyse in detail the decoder structure shown in Figure 5.3, where the notations L_{a1} and L_{a2} represent the so-called *a priori* information produced by the first and second decoder, respectively. Similarly, L_p^{D1} and L_p^{D2} denote the so-called *a posteriori* information of the first and second decoders, respectively. Finally, the so-called extrinsic information of the first and second decoders is labelled as L_e^{D1} and L_e^{D2}, respectively.

At the receiver the soft channel outputs are generated, which consist of the systematic

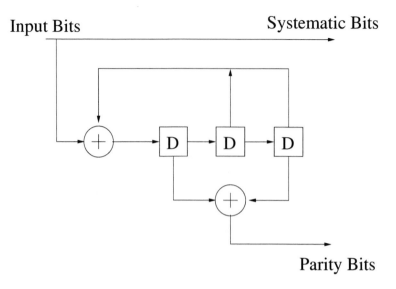

Figure 5.2: Recursive Systematic Convolutional (RSC) encoder.

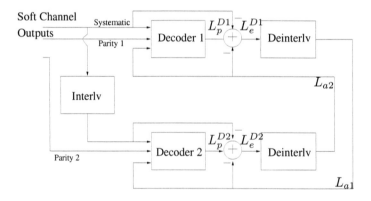

Figure 5.3: Schematic of the turbo decoder.

and the parity bits, as illustrated in Figure 5.3. The two parity sequences generated by the turbo encoder are utilized by the corresponding decoders. In this respect, the punctured parity bits are replaced by zeros during the decoding process. In the first iteration, the first component decoder accepts the soft channel outputs and by utilizing the Log-MAP algorithm of Appendix A.1 [164], the decoder produces the a *posteriori* log likelihood ratio (LLR) L_p^{D1}, which is defined as $L(u_n|r)$ in Appendix A.1 [164]. Essentially this LLR represents the log-domain probability that the bit was decoded error freely. The polarity of this LLR can also be used in order to determine whether the decoded bit is a binary 1 or 0. Subsequently, the extrinsic information L_e^{D1}, is generated by subtracting the contribution of the channel outputs, as shown in Figure 5.3. This justifies the terminology 'extrinsic', since it represents the information related to a certain bit carried by sources other than the channel output itself related to this specific bit. Hence the extrinsic information is only influenced by the first decoder, which is then interleaved in order to generate the *a priori* LLR information L_{a1}.

For the sake of presenting the information to the second decoder in the right order, the systematic bits are interleaved in order to form the soft channel outputs as depicted in Figure 5.3. Subsequently, the second decoder utilizes not only the soft channel outputs but also the independent *a priori* LLR values L_{a1}, from the first decoder in order to produce the *a posteriori* LLR L_p^{D2}. This *a posteriori* LLR value is improved at this stage, since it was influenced by the estimates of both decoders. As before, the extrinsic information of the second decoder L_e^{D2}, is generated by subtracting the channel information and the *a priori* information of the first decoder. This essentially removes any contribution generated by the first decoder, when producing the *a priori* information L_{a2}, of the second decoder, which is used in the first decoder for the subsequent iteration. This subtraction process allows us to maintain the independence of the decoding process, which is important for the sake of attaining independent estimates from the two separate decoders for each decoded bit. This process constitutes one turbo decoding iteration and it is repeated, in order to achieve better consecutive estimates of the decoded bits.

After each iteration, the output *a posteriori* LLR is improved, since the decoder can exploit the independent *a priori* information generated by the other decoder. Consequently, as the number of iterations increases, the estimation of the decoded bit improves. The performance of the turbo decoder will vary depending on the size of the turbo interleaver, where a larger interleaver will provide a higher degree of independence of the *a priori* information that is being passed from one decoder to another. This high degree of independence is exploited by both decoders in order to yield an improved decoding performance. The number of iterations also plays an important role, where a higher number of iterations will generally result in a better performance, although at the expense of a higher complexity. However, the gain achieved by each iteration reduces with increasing numbers of iterations, which will be exemplified by Figure 5.4. This is because the two decoders' information becomes more dependent on each other, diminishing the benefits of acquiring two 'opinions' concerning a given received bit. In the next section, the implementation of turbo coding in a wideband AQAM scheme is highlighted.

5.2 System Parameters

The system parameters that were used throughout our associated investigations are listed in Table 5.1. The channel coder parameters, which include the turbo interleaver size and code rate will be varied according to the different system requirements as it will be demonstrated at a later stage.

The generic setup of the turbo coded AQAM scheme consists of the modulation switching mechanism, the turbo coding parameters and the switching thresholds. The modulation mode switching mechanism is identical to that discussed in Section 4.3.2 with the exception that the coding rate and the size of the turbo interleaver is varied according to the modulation mode

Channel Type	COST207 TU(see Figure 4.12)
Normalized Doppler Frequency	3.25×10^{-5}
Data Modulation	AQAM (NOTX, BPSK, 4QAM, 16QAM, 64QAM) with perfect channel estimation
Receiver Type	Decision Feedback Equalizer Number of Forward Taps = 35 Number of Backward Taps = 7 Correct Feedback
Turbo Coding Parameters: Number of Iterations Decoding Algorithm	6 Log-MAP

Table 5.1: Generic system parameters of the turbo coded AQAM scheme.

selected. The modulation mode switching mechanism can be summarized as follows:

$$
\text{Modulation Mode} = \begin{cases} NOTX & \text{if } \gamma_{DFE} \leq t_1^c \\ BPSK, I_0, R_0 & \text{if } t_1^c < \gamma_{DFE} \leq t_2^c \\ 4QAM, I_1, R_1 & \text{if } t_2^c < \gamma_{DFE} \leq t_3^c \\ 16QAM, I_2, R_2 & \text{if } t_3^c < \gamma_{DFE} \leq t_4^c \\ 64QAM, I_3, R_3 & \text{if } \gamma_{DFE} > t_4^c, \end{cases} \tag{5.1}
$$

where I_n represents the random turbo interleaver size in terms of the number of bits. The coding rate is denoted by R_n and t_n^c represents the coded switching thresholds.

The switching thresholds for the coded AQAM scheme are difficult to numerically optimize in order to achieve a certain target BER due to the non-linear BER versus SNR characteristics of the scheme. However, the switching thresholds for the different turbo coded AQAM schemes are intuitively optimised, in order to achieve target BERs of below 1% and 0.01%, which are termed as the **High-BER** and **Low-BER** schemes, respectively. These coded schemes will be compared to the uncoded AQAM scheme, where the uncoded switching thresholds are set according to Table 4.8 for target BERs of 1% and 0.01%. The burst structures used for the **High-** and **Low-BER** schemes are the non-spread speech and the non-spread data bursts, respectively, which were shown in Figure 4.13.

5.3 Turbo Block Coding Performance of the Fixed QAM Modes

Before we attempt to characterize the Turbo Block Coded AQAM (TBCH-AQAM) scheme, let us study the performance of turbo coding, when applied to the constituent fixed modulation modes. In our experiments the component turbo channel code used was the BCH(31, 26, 3) scheme and a random turbo interleaver [165] of size 9984 bits was chosen. The block channel interleaver size was set to 13824 bits, which corresponded to the channel-coded block-length of the turbo interleaver. The turbo coding performance of the BPSK modulation mode is shown in Figure 5.4, which displayed the BER performance for different number of

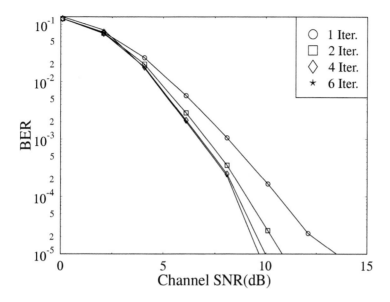

Figure 5.4: Turbo block coded performance of BPSK for different number of iterations with a component code of BCH(31, 26, 3). The system parameters of Table 5.1 and the non-spread speech burst of Figure 4.13 was utilized. The channel interleaver size was set to 13824 bits.

turbo iterations. The performance improved, as the number of turbo iterations was increased, which illustrated the improvement in estimating the decoded bit as a result of the iterative decoding regime. The iteration gain was approximately 2.1dB after six iterations at a BER of 0.01%. Explicitly, the iteration gain measured the difference between the average channel SNR required in order to achieve a particular BER and the corresponding average channel SNR required after n iterations for the same BER. However, the improvements achieved upon each iteration decreased, as the number of iterations increased, as evidenced by Figure 5.4.

The turbo block coded BER performance of the BPSK, 4QAM, 16QAM and 64QAM modes is shown in Figure 5.5 after six iterations using different channel block interleavers, where the uncoded performance is also displayed for comparison. Based on the turbo interleaver size of 9984 bits, the size of the channel interleaver was set to 13824 bits, which corresponded to the channel-coded block-length of the turbo block encoder. In order to assess the impact of the channel interleaver size, a larger channel interleaver of size 4×13824 was also utilized. As expected, in Figure 5.5 the BER performance using the larger channel interleaver was superior, when compared to that using the smaller channel interleaver, although at the cost of an associated higher transmission delay.

Referring again to Figure 5.5, substantial SNR gains were achieved, when comparing

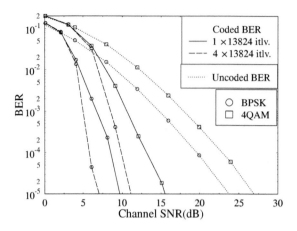

(a) Turbo block coded performance of BPSK and 4QAM.

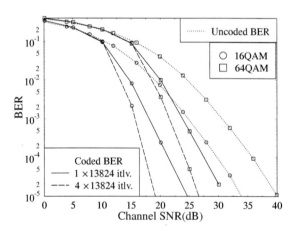

(b) Turbo block coded performance of 16QAM and 64QAM.

Figure 5.5: Turbo block coded performance of each individual modulation modes after six turbo iterations utilizing the system parameters of Table 5.1 and the non-spread speech burst of Figure 4.13. A component code BCH(31, 26, 3) was utilized in conjunction with channel interleavers of size 13824 bits and 4×13824 bits.

the coded and uncoded performance. However the gains achieved at a BER of 0.01% was higher than those achieved at a BER of 1%, as evidenced by Figure 5.5. This observation is important in the context of turbo block coded AQAM scheme, which will be presented in the next section.

5.4 Fixed Coding Rate and Fixed Interleaver Size Turbo Block Coded Adaptive Modulation

In this Fixed Coding Rate and Fixed Interleaver Size Turbo BCH Coded AQAM (**FCFI-TBCH-AQAM**) scheme, we utilized a random turbo interleaver of fixed size and a fixed coding rate for all modulation modes [166]. The turbo interleaver size was set to 9984 and a coding rate of 0.7222 was utilized, which corresponds to a component code of BCH (31, 26, 3). The switching mechanism described by Equation 5.1 was utilized in conjunction with the coded switching thresholds shown in Table 5.2 for the target BERs of 1%, 0.01% and for a near-error-free system.

Target BER	Coded Switching Thresholds (dB)				Burst Type see Figure 4.13
	t_1^c	t_2^c	t_3^c	t_4^c	
$\leq 1\%$	0.1363	2.7258	8.1450	14.1846	Speech
$\leq 0.01\%$	1.2458	3.4579	9.7980	16.7589	Data
Near-Error-Free	2.2458	4.4579	10.7980	17.7589	Data

Table 5.2: The coded switching thresholds, which were experimentally set in order to achieve the target BERs of below 1%, 0.01% and near-error-free for the **FCFI-TBCH-AQAM** scheme described in Section 5.4. The corresponding transmission burst types utilized are shown in Figure 4.13 and the switching mechanism was characterized by Equation 5.1.

The BER and BPS performance of the **FCFI-TBCH-AQAM** scheme is shown in Figure 5.6 for the target BERs of 1% and 0.01%. The uncoded **FCFI-TBCH-AQAM** performance is also depicted for comparison. As expected, the coded BER performance improved significantly, when compared to the uncoded performance, and the target BERs of 1% and 0.01% were achieved. Conversely, the coded BPS was reduced by a factor equal to the coding rate.

The BPS performance of the turbo block coded AQAM scheme was also compared to that of the fixed modulation modes for different channel interleaver sizes, as illustrated by Figure 5.7, where the throughput values were extracted from Figures 5.5 and 5.6. Referring to Figure 5.7(a), where the channel interleaver was set to 13824 bits, the wideband AQAM scheme displayed throughput SNR gains of approximately 1.0dB and 5.0dB for target BERs of 1% and 0.01%, respectively, when considering the corresponding BPS curves. However, by referring to Figure 5.7(b), when the larger channel interleaver size was utilized for the fixed modulation modes, the BPS/SNR gain was minimal for a target BER of 1% while a BPS/SNR gain of approximately 1.5dB was observed for a target BER of 0.01%. The reduction in the throughput SNR gain achieved by the wideband turbo block coded AQAM scheme was due to the superior performance of the larger channel interleaved fixed modulation modes. However, it is important to note that an associated high transmission delay was incurred.

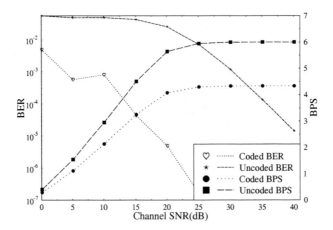

(a) Turbo block coded performance for a target BER of below 1%.

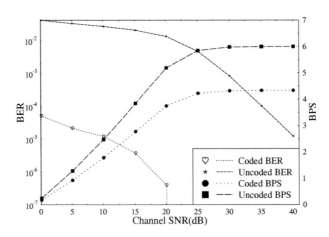

(b) Turbo block coded performance for a target BER of below 0.01%.

Figure 5.6: Turbo block coded and uncoded performance of the **FCFI-TBCH-AQAM** scheme described in Section 5.4, where the generic system parameters of Table 5.1 were utilized. The coded switching regime was characterized by Equation 5.1 with the coding rate and turbo interleaver size set to 0.7222 and 9984 bits, respectively. The coded switching thresholds and transmission burst type were set according to Table 5.2.

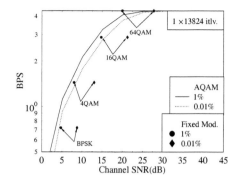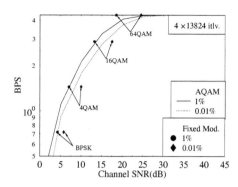

(a) Channel Interleaver of size 1 × 13824 bits   (b) Channel Interleaver of size 4 × 13824 bits

Figure 5.7: Throughput comparison between the **FCFI-TBCH-AQAM** scheme and the constituent fixed modulation modes for target BERs of 1% and 0.01%, which were evaluated from Figures 5.5 and 5.6. Different sized channel interleavers were used for the fixed modulation modes whereas the **FCFI-TBCH-AQAM** scheme employed no channel interleavers.

5.4.1   Comparisons with the Uncoded Adaptive Modulation Scheme

The performance comparison of the **FCFI-TBCH-AQAM** scheme and the uncoded AQAM scheme for the same target BER is presented here. In Section 4.3.5 the uncoded AQAM performance was optimized using the switching thresholds of Table 4.8, in order to achieve target BERs of 1% and 0.01 as evidenced by Figure 4.21(b). These uncoded results were compared to the **FCFI-TBCH-AQAM** scheme in terms of BPS and BER performance. This comparison is exemplified in Figure 5.8.

For the **High-BER** scheme, the coded BER was lower than that for the uncoded case, where a high average channel SNR gain of about 20dB was observed across the BER range of 10^{-5} and 10^{-3}. Similarly, in the channel SNR range of 0 to 15dB, the coded BPS performance was better than that of the uncoded AQAM scheme with a maximum SNR gain of 3dB at a channel SNR of 0dB, as evidenced by Figure 5.8(a). However, the BPS performance of the **FCFI-TBCH-AQAM** scheme deteriorated at high average channel SNRs, since its throughput was limited by its coding rate, which converged to a throughput of approximately 4.33 bits per symbol.

For the **Low-BER** scheme of Figure 5.8(b) the same characteristics were observed. However, the BPS gain was higher than that of the **High-BER** scheme. The coded BPS performance was higher than that of the uncoded scheme for the channel SNR range of 0 to 23dB with a maximum SNR gain of 7dB at a channel SNR of 0dB, as evidenced by the BPS curves of Figure 5.8(b). These SNR gains attained by the **Low-BER** scheme were higher than those of the **High-BER** scheme due to the higher coding gain achieved at a lower target BER. This characteristic was observed also for the fixed modulation modes of Section 5.3, where higher coding gains were recorded for lower BERs due to the steeper decay of the coded

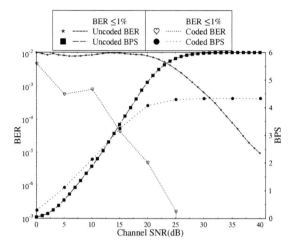

(a) Turbo block coded performance for a target BER of below 1% using
the non-spread speech burst of Figure 4.13.

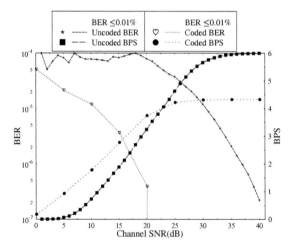

(b) Turbo block coded performance for a target BER of below 0.01% using
the non-spread data burst of Figure 4.13.

Figure 5.8: Turbo block coded performance of the **FCFI-TBCH-AQAM** scheme described in Section
5.4, where the generic system parameters of Table 5.1 were utilized. The coded switching
regime was characterized by Equation 5.1 with the coding rate and turbo interleaver size set
to 0.7222 and 9984 bits, respectively. The coded and uncoded AQAM switching thresholds
were set according to Table 5.2 and 4.8, respectively.

BER versus SNR curves. Consequently, for the **Low-BER** scheme, the switching threshold values were lowered by a margin of approximately 7dB, when compared to the **Low-BER** uncoded AQAM scheme. This is evident, when the coded switching thresholds of Table 5.2 are compared to those of the uncoded switching thresholds of Table 4.8. The lowering of the coded switching thresholds resulted in the more frequent utilization of higher-order modulation modes at lower average channel SNRs. Consequently the BPS performance improved, when compared to the uncoded AQAM scheme. By contrast, for the **High-BER** scheme the switching threshold reduction margin was only 3.5dB. The effect of the higher margin for the **Low-BER** scheme was an improved BPS performance, when compared to the **Low-BER** uncoded AQAM scheme.

The switching thresholds for the **FCFI-TBCH-AQAM** scheme were also experimentally determined, which are shown in Table 5.2 in order to achieve a near-error-free communications system. The BER and BPS performance of this near-error-free scheme is shown in Figure 5.9, where the corresponding curves of the **Low-BER** uncoded AQAM scheme were also plotted for comparison. The results characterized a near-error-free system, where the throughput was higher than that of the uncoded AQAM scheme for the channel SNR range of 0 to 22dB. The maximum average channel SNR gain of 6dB was recorded, when considering the associated throughput performance at a channel SNR of 0dB, as evidenced by Figure 5.9.

In summary, we have quantified the average channel SNR gains achieved by the **FCFI-TBCH-AQAM** scheme, when compared to the uncoded AQAM scheme, which was targeted at achieving the same BER performance. We have also noted the associated throughput degradation at high average channel SNRs as a result of the coding rate limitation imposed by the scheme. Subsequently, we revised the coded switching thresholds in order to create a near-error-free **FCFI-TBCH-AQAM** scheme, which also exhibited substantial SNR gains, when compared to the uncoded AQAM scheme. In the next section we shall introduce a range of coded AQAM schemes, which utilizes different interleaver sizes depending on the modulation mode selected. In order to remove the BPS limitation of the rate 0.7222 coded AQAM scheme and to increase its flexibility, it is feasible to introduce a range of further transmission code rates, which will be discussed in Section 5.7.

5.5 Fixed Coding Rate and Variable Interleaver Size Turbo Block Coded Adaptive Modulation

The main motivation in implementing a coded AQAM scheme in conjunction with a variable turbo interleaver size for each modulation mode is to provide the receiver with an error detection capability for each received AQAM data burst without any delay, as well as to vary the coding rate for each modulation mode. In doing so, an intelligent receiver will be capable of blindly detecting the modulation mode without explicit signalling, which will be discussed at a later stage. In order to provide an instantaneous error detection capability at the receiver, the turbo interleaver size must be equal or less than the number of transmitted bits for the transmission burst. This ensures that the received burst can be demodulated and decoded immediately on a burst by burst basis.

This Fixed Coding Rate and Variable Interleaver size Turbo Block Coded AQAM (**FCVI-TBCH-AQAM**) scheme is implemented in conjunction with a fixed coding rate of 0.7222,

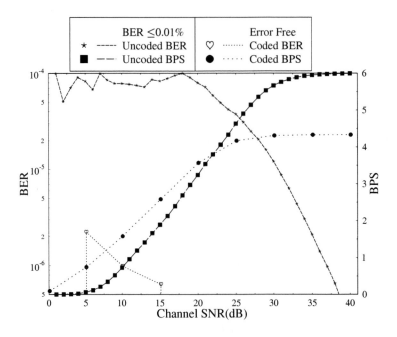

Figure 5.9: Performance of the near-**error-free** of the **FCFI-TBCH-AQAM** scheme of Section 5.4,
where the generic system parameters of Table 5.1 were utilized. The coded switching
regime was characterized by Equation 5.1 with the coding rate and turbo interleaver size
set to 0.7222 and 9984 bits, respectively. The coded switching thresholds and transmission
burst type were set according to Table 5.2. The performance was compared to the uncoded
AQAM scheme, which was optimized for a target BER of 0.01% according to Table 4.8.

corresponding to the component code of BCH (31, 26, 3) for all modulation modes [167].
The turbo interleaver size is varied according to the modulation mode selected as well as the
size of the transmission burst. The general switching regime is summarized in Equation 5.1,
where the turbo interleaver size and the switching threshold are listed in Table 5.3 and 5.4,
respectively, for the **High-BER**, **Low-BER** and for the near-error-free system. The remaining
experimental parameters are listed in Table 5.1.

   The BER and BPS performance of the **High-** and **Low-BER FCVI-TBCH-AQAM** scheme is shown in Figure 5.10. The corresponding **High-** and **Low-BER** uncoded AQAM performance curves are also depicted in Figure 5.10 for comparison. The characteristics of the
results were similar to those shown in Figure 5.8 of Section 5.4.1 and can be explained similarly. For the **High-BER FCVI-TBCH-AQAM** scheme the throughput was higher than that
of the uncoded scheme for the channel SNR range of 0 to 11dB, with a maximum SNR gain
of approximately 2.3dB at a channel SNR of 0dB. Similarly, an average channel SNR gain of
8dB was achieved, when comparing the BER performance of the **Low-BER FCVI-TBCH-
AQAM** scheme and the uncoded AQAM scheme at an average channel SNR of 20dB.

   The throughput performance of **FCVI-TBCH-AQAM** was also compared to that of the

Target	Turbo Interleaver Size (Bits)				Burst Type
BER	I_0	I_1	I_2	I_3	see Figure 4.13
$\leq 1\%$	104	208	416	624	Speech
$\leq 0.01\%$	494	988	1976	2964	Data
Near-Error-Free	494	988	1976	2964	Data

Table 5.3: The turbo interleaver size associated with each modulation mode characterized by Equation 5.1 for the **FCVI-TBCH-AQAM** scheme described in Section 5.5. The target BERs were set to be below 1%, 0.01% and near-error-free, where the corresponding transmission burst types utilized are shown in Figure 4.13.

Target	Coded Switching Thresholds (dB)				Burst Type
BER	t_1^c	t_2^c	t_3^c	t_4^c	see Figure 4.13
$\leq 1\%$	0.6363	3.2258	8.6450	14.6846	Speech
$\leq 0.01\%$	1.9958	4.2079	10.5480	17.5089	Data
Near-Error-Free	3.2458	5.4579	11.7980	18.7589	Data

Table 5.4: The coded switching thresholds, which were experimentally determined in order to achieve the target BERs of below 1%, 0.01% and near-error-free for the **FCVI-TBCH-AQAM** scheme described in Section 5.5. The corresponding transmission burst types utilized are shown in Figure 4.13 and the switching mechanism was characterized by Equation 5.1.

fixed modulation modes shown in Figure 5.5 for target BERs of 1% and 0.01%. For the **Low-BER FCVI-TBCH-AQAM** scheme, a BPS/SNR gain of approximately 1.5dB was achieved, when compared to the fixed modulation modes utilizing the large channel interleavers, as evidenced by Figure 5.11(b). However, by referring to Figure 5.11(a) a more substantial gain of approximately 5.0dB was achieved, when compared to the fixed modulation modes utilizing the smaller channel interleavers. For the **High-BER FCVI-TBCH-AQAM** scheme, minimal gains were achieved, when compared to both the large- and small-channel interleaved fixed modulation modes. It is important to note here that these low gains were achieved despite the larger turbo interleaver and channel interleaver utilized by the fixed modulation modes. This resulted in a high transmission delay for the fixed modulation modes, whereas the **FCVI-TBCH-AQAM** scheme employed low-latency instantaneous burst-by-burst decoding.

In the **Low-BER FCVI-TBCH-AQAM** scheme the SNR gains achieved in the context of the associated BER and BPS performance curves were higher than those of the **High-BER FCVI-TBCH-AQAM** scheme. Explicitly, a higher throughput performance was observed across the average channel SNR range of 0 to 22dB, with the maximum SNR gain of 6dB at an average channel SNR of 0dB. Similarly, a SNR gain of 16dB was achieved at an average channel SNR of 20dB, when the BER performances were compared, as evidenced by Figure 5.8. The higher gains achieved by the **Low-BER** scheme were due to the lower BER requirement, which was justified in Section 5.4.1. The other contributing factor was due to the higher turbo interleaver size that was utilized for the **Low-BER** scheme, which possessed a longer transmission burst structure. Consequently, the turbo block coded bits were more decorrelated, which provided a higher coding gain, as it was argued in Section 5.1.

Lastly, the **FCVI-TBCH-AQAM** scheme was optimized in order to yield a near-error-

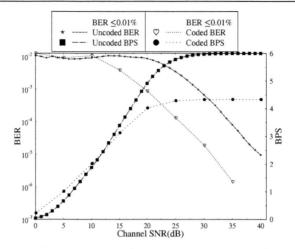

(a) Turbo block coded performance for a target BER of below 1% using the non-spread speech burst of Figure 4.13.

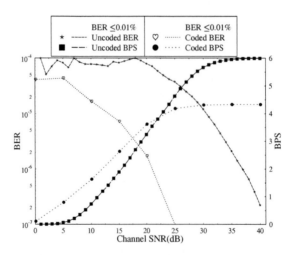

(b) Turbo block coded performance for a target BER of below 0.01% using the non-spread data burst of Figure 4.13.

Figure 5.10: Turbo block coded performance of the **FCVI-TBCH-AQAM** scheme described in Section 5.5, where the generic system parameters of Table 5.1 were utilized. The coded switching regime was characterized by Equation 5.1, where the coding rate was 0.7222 and variable turbo interleaver sizes were listed in Table 5.3, respectively. The coded and uncoded AQAM switching thresholds were set according to Table 5.4 and 4.8, respectively.

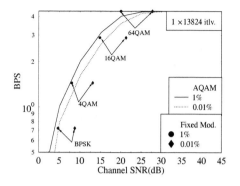 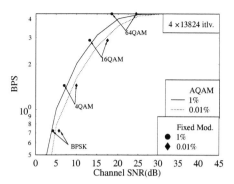

(a) Channel Interleaver of size 1×13824 bits   **(b)** Channel Interleaver of size 4×13824 bits

Figure 5.11: Throughput comparison between the **FCVI-TBCH-AQAM** scheme and the constituent fixed modulation modes for target BERs of 1% and 0.01%, which were evaluated from Figures 5.5 and 5.6. Different sized channel interleavers were used for the fixed modulation modes whereas the **FCFI-TBCH-AQAM** scheme employed no channel interleavers.

free communication system with the turbo coding parameters and the switching thresholds shown in Tables 5.3 and 5.4, respectively. The corresponding BER and BPS performance is shown in Figure 5.12, where the system was near-error-free. The throughput performance was also better for the average channel SNR range between 0 to 20dB, when compared to that of the uncoded AQAM scheme optimized for a target BER of 0.01%, as evidenced by Figure 5.12.

However, the SNR gains recorded for this variable-sized turbo interleaver scheme were lower than those of the fixed turbo interleaver scheme of Section 5.4 for both target BERs. This gain degradation was due to the reduced turbo interleaver size utilized in the **FCVI-TBCH-AQAM** scheme. Nevertheless, the **FCVI-TBCH-AQAM** scheme can provide a burst by burst error detection capability, which we will exploit in the next section.

5.6 Blind Modulation Detection

In Section 4.3.1 the receiver assumed that the modulation mode of the received packet was known. In reality, some form of signalling is needed in order to convey this information from the transmitter to the receiver [21] [37]. Recently, a blind modulation detection algorithm was proposed by Keller *et al.* in an adaptive OFDM scheme [168]. In this scheme, the mean square phasor error - which is defined as the Euclidean distance between the received equalized data symbols and the nearest legitimate constellation point for a particular AQAM mode - was evaluated. This was repeated for all valid modulation modes utilized in the wideband AQAM scheme. Subsequently, the modulation mode that produced the minimum mean square phasor error was selected. This is an example of a blind detection algorithm,

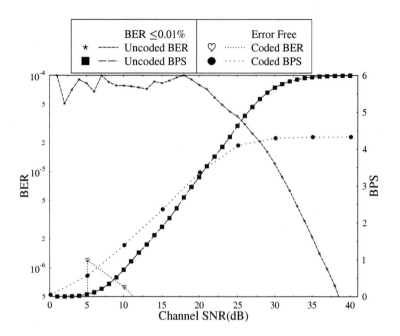

Figure 5.12: The near-**error-free** turbo block coded performance of the **FCVI-TBCH-AQAM** scheme
described in Section 5.5. The coded switching regime was characterized by Equation 5.1,
where the coding rate was set to 0.7222 and the turbo interleaver sizes were set according
to Table 5.3, respectively. The coded switching thresholds and transmission burst type
were set according to Table 5.4 and the other generic system parameters were listed Table
5.1. The performance was compared to that of the uncoded AQAM scheme, which was
optimized for a target BER of 0.01% according to Table 4.8.

where the receiver is capable of detecting the modulation mode used without any signalling
information from the transmitter. The primary motivation for the blind modulation detection
algorithm is to reduce the amount of signalling between the receiver and the transmitter,
consequently yielding an improved information throughput. This blind MSE modulation
detection algorithm can be summarized as follows upon evaluating the accumulated MSE of
a transmission burst for all legitimate modem modes :

$$
\begin{aligned}
e_m &= \frac{\sum_n \mid (R_{n,m}^{eq} - \hat{R}_{n,m}) \mid^2}{n} \\
mod_c &= min(e_m) \quad \text{for } m = \text{BPSK, 4QAM, 16QAM, 64QAM,} \quad (5.2)
\end{aligned}
$$

where m is the number of possible modulation modes and mod_c is the selected modulation
mode based on the minimum average square error of the Euclidean distance e, for all the valid
modulation modes. The function $min(e_m)$ is the selection function that selects the minimum
of all e_m values, while $R_{n,m}^{eq}$ and $\hat{R}_{n,m}$ is the nth equalized symbol and the corresponding
legitimate demapped constellation point of modulation mode m, respectively.

In exploring the performance of this blind MSE modulation detection algorithm, the PDF of all possible mean square phasor errors e_m for the four valid modulation modes is plotted and shown in Figure 5.13. In each of the sub-figures, the actual modulation mode utilized was stated in the respective captions and the PDF of the other valid modulation modes was also displayed. The common trend shown in Figure 5.13 was that the higher-order modulation modes of 64QAM and 16QAM constantly yielded the lowest mean square phasor error, independently of the actual modulation mode that was utilized, which was detrimental as regards to the performance of the blind modulation detection scheme. This characteristic can be explained by noting that the higher-order modulation modes of 64QAM and 16QAM possessed a higher number of legitimate constellation points. Consequently, the probability that the received equalized data symbol situated near a valid constellation point increased, which yielded a lower mean square phasor error. However, when BPSK was utilized, there was sufficient separation between the PDF of the BPSK and 4QAM modes, as evidenced by Figure 5.13(a). Thus this algorithm was capable of detecting the BPSK mode, if BPSK and 4QAM were the only possible valid modulation modes.

Since we have observed the deficiencies in the blind MSE-based algorithm, we will investigate the utilization of channel coding in order to blindly detect the modulation modes in the TBCH-AQAM scheme.

5.6.1 Blind Soft Decision Ratio Modulation Detection Scheme

Before elaborating further on this blind Soft Decision Ratio (SD) based modulation detection algorithm, we will address the concept of transmission blocking in AQAM. Practically, whenever the transmission is disabled, a transmission burst of a known sequence is transmitted, which is used to estimate the channel quality and hence to aid the selection of the next modulation mode. This burst is always BPSK modulated, in order to provide maximum error protection. However, this known sequence must be unique and easily identifiable by the receiver, in order to aid its NOTX mode detection. Consequently, we propose to use binary maximal-length shift register sequences $C^{(a)}$, commonly known as m-sequences, that have the following correlation properties [169]:

$$\begin{aligned} \theta_a(0) &= Q, \\ \theta_a(r) &= -1 \quad \cdot \text{for } \dot{r} \neq 0, \end{aligned} \quad (5.3)$$

where $\theta_a(r) = \sum_{i=0}^{Q-1} C_i^{(a)} C_{r+i}^{(a)}$ and Q is the length of the known m-sequence. Explicitly, at the transmitter, if the NOTX mode is selected, the same m-sequences are concatenated in order to form the transmission burst. Consequently, at the receiver the demodulated burst is correlated with the locally stored known m-sequence and if a maximum amplitude of Q is detected periodically corresponding to the correlation time-shift of zero, then the burst is deemed to be a NOTX mode burst. Having proposed a sequence for the NOTX mode and a technique for detecting it, we will now focus our attention on the detection of the BPSK, 4QAM, 16QAM and 64QAM modes.

Since the variable interleaver-based turbo block coded AQAM scheme employed burst by burst decoding at the receiver, we can exploit the error correction capability of the turbo codec. Consequently, we can utilize the information provided by the channel decoder in terms of its input bit probability and the corresponding output bit probability. In this so-called blind

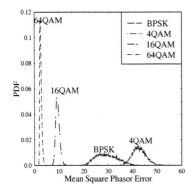

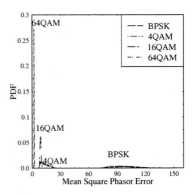

(a) The actual modulation mode was BPSK and the channel SNR was set to 8dB.

(b) The actual modulation mode was 4QAM and the channel SNR was set to 12dB.

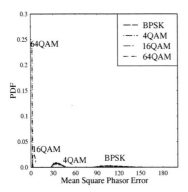

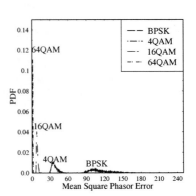

(c) The actual modulation mode was 16QAM and the channel SNR was set to 16dB.

(d) The actual modulation mode was 64QAM and the channel SNR was set to 20dB.

Figure 5.13: The PDF of the mean square phasor error defined in Equation 5.2 for each individual modulation mode and for various channel SNRs.

Soft Decision Ratio (SD) modulation detection scheme, each input bit's probability upon entering the channel decoder is compared against its corresponding output bit probability for each possible modulation mode. The results are then classified into two categories, where one category consists of the number of times the input bit probability is less than the output bit probability and vice-versa for the other category. These two categories are then used to update a Soft Decision counter SD_{ratio}, as follows:

$$SD_{ratio}^{n,m} = \begin{cases} SD_{ratio}^{n,m} & \text{if } p_{ipbit}^{n,m} \leq p_{opbit}^{n,m} \\ SD_{ratio}^{n,m} + 1 & \text{if } p_{ipbit}^{n,m} > p_{opbit}^{n,m}, \end{cases} \qquad (5.4)$$
$$\text{for } m = \text{BPSK, 4QAM, 16QAM, 64QAM,}$$

where $p_{ipbit}^{n,m}$ represents the nth input bit probability, which is demodulated using the modulation mode m. Similarly, $p_{opbit}^{n,m}$ denotes the output bit probability of the channel decoder. Subsequently, the average soft decision ratio is calculated for all possible valid modulation modes and the final modulation mode is chosen as follows:

$$\begin{aligned} \text{Average } SD_{ratio}^m &= \frac{\sum_n SD_{ratio}^{n,m}}{N}, \\ mod_c &= min(\text{Average } SD_{ratio}^m), \\ \text{for } m &= \text{BPSK, 4QAM, 16QAM, 64QAM,} \end{aligned} \qquad (5.5)$$

where mod_c denotes the chosen modulation mode and $min(a^m)$ is the selection function that selects the minimum of all a^m values, while N represents the number of coded bits in a transmission burst.

The PDF of the average SD_{ratio} of all the possible modulation modes is shown in Figure 5.14. In each of the sub-figures the actual modulation mode used was stated in the respective captions. Referring to Figure 5.14, there was a clear PDF separation between the actual modulation mode and the other modulation modes, where the SD_{ratio} of the actual modulation mode was centred at the minimum end of the average SD_{ratio} scale. It was this PDF separation that supported the feasibility of the proposed blind SD modulation detection scheme.

This blind modulation detection algorithm was implemented using the simulation parameters of Table 5.1. A conventional binary BCH(31 ,26 ,3) was utilized without channel interleavers for simplicity, although this algorithm can be applied to turbo encoding, since its component code was identical to the above BCH code. The speech-type burst of Figure 4.13 was used and the m-sequence length $Q = 31$. The performance of this algorithm in terms of its modulation Detection Error Rate (DER) is depicted in Figure 5.15(a). The detection algorithm yielded a DER below 10^{-4} at a channel SNR of approximately 24dB. However, a severe DER degradation was observed for channel SNRs between 10 - 20dB. In order to investigate this degradation, the individual Wrong Modulation Error Percentage (WME) was plotted in Figure 5.15(b). This measure recorded the relative frequency of the modulation mode detected by the algorithm, when the detection scheme was in error. As it can be observed in Figure 5.15, whenever the detection algorithm failed, the BPSK mode was frequently chosen compared to the other modulation modes. Referring to Figures 5.14(b) - 5.14(d), we observed that the SD_{ratio} PDF of the BPSK mode had the greatest overlapping region with the PDF of the actual modulation mode at low SD_{ratio} values. This implied that the receiver had a higher probability of selecting BPSK, even though it was the wrong modulation mode.

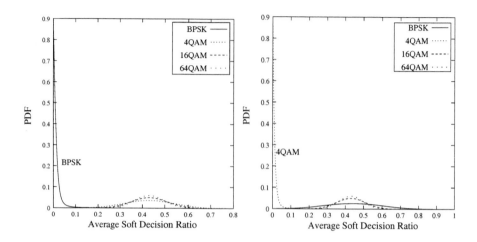

(a) The actual modulation mode was BPSK and the channel SNR was set to 8dB.

(b) The actual modulation mode was 4QAM and the channel SNR was set to 12dB.

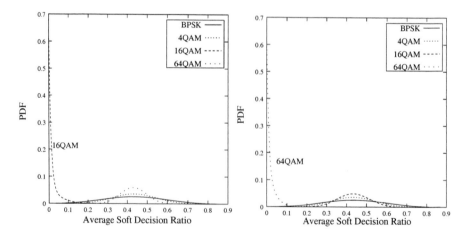

(c) The actual modulation mode was 16QAM and the channel SNR was set to 16dB.

(d) The actual modulation mode was 64QAM and the channel SNR was set to 20dB.

Figure 5.14: The PDF of the average soft decision ratio defined in Equation 5.5 for each individual modulation mode and for various channel SNRs, using a conventional binary BCH(31 ,26 ,3) coding scheme.

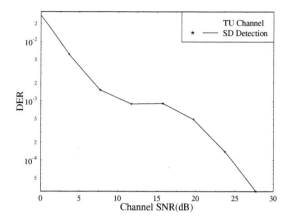

(a) DER performance.

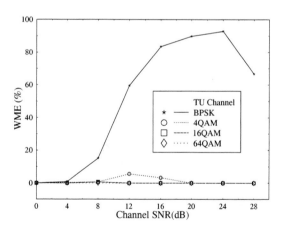

(b) WME performance.

Figure 5.15: The DER and WME performance of the SD algorithm characterized by Equation 5.5. The system parameters of Table 5.1 were utilized and the AQAM switching thresholds were set according to Table 4.8 for the target BER of 1%. The DER and WME measures were defined in Section 5.6.1 and a conventional binary BCH(31, 26, 3) coding scheme was utilized.

In order to improve the DER performance, the detection of the BPSK mode has to be more robust. Consequently, here we propose to utilize a hybrid Soft Decision Mean Square Error (SD-MSE) based blind modulation detection algorithm for the coded AQAM scheme.

5.6.2 Hybrid Soft Decision Mean Square Error Modulation Detection Algorithm

In Section 5.6 the concept of utilizing the mean square phasor error at the receiver in order to blindly detect the modulation mode was presented. In Figures 5.13 a - 5.13 d we have observed that this measure was not sufficiently reliable in order to detect the modulation modes. However, when the BPSK mode was actually utilized, there was a sufficient PDF separation between the mean square phasor error PDF of the BPSK and 4QAM modes, as evidenced by Figure 5.13(a). Consequently we exploited this property in order to detect the BPSK mode. In this hybrid algorithm the BPSK mode is detected by comparing the average square error of the BPSK and 4QAM modes. The other modulation modes - namely 4QAM, 16QAM and 64QAM - were then detected using the SD algorithm of Section 5.6.1. This SD-MSE algorithm can be summarized as follows [167, 170]:

$$
mod_c = \begin{cases} \text{BPSK} & \text{if } e_{BPSK} \leq e_{4QAM} \\ min(\text{Average } \text{SD}^m_{ratio}) & \text{if } e_{BPSK} > e_{4QAM}, \end{cases} \tag{5.6}
$$
$$
\text{for } m = 4QAM, 16QAM, 64QAM,
$$

where Average SD^m_{ratio} and e_q were defined in Equations 5.5 and 5.2, respectively. The DER performance of this hybrid algorithm is presented in Figure 5.16, where the experimental parameters were identical to those used by the SD algorithm of Section 5.6.1. In Figure 5.16 the performance of the SD detection algorithm is shown as a comparison to that of the SD-MSE algorithm. The hybrid SD-MSE algorithm achieved a DER of 10^{-4} at a channel SNR of approximately 15dB [167, 170]. The improvement of the SD-MSE algorithm was clearly seen in Figure 5.16 where the associated performance was superior to that of the SD-based technique, in the channel SNR range of between 10 - 20dB. Furthermore, the complexity of this SD-MSE algorithm was reduced, since the channel decoder was only used to detect three modes instead of the four modes of the SD algorithm.

In this section we have demonstrated that channel coding can be utilized for detecting the modulation mode at the receiver in a coded AQAM scheme. We have presented three different blind detection algorithms, where the MSE algorithm was deemed unreliable for detecting the four modes. The higher complexity SD and hybrid SD-MSE algorithms were then proposed, where the latter exhibited a better performance in terms of DER.

5.7 Variable Coding Rate Turbo Block Coded Adaptive Modulation

In Sections 5.4 and 5.5 we have characterized a range of turbo block coded AQAM schemes having fixed coding rates for all modulation modes, where a throughput degradation was observed at high channel SNRs, when compared to the uncoded AQAM schemes for similar target BERs. However, with the aim of improving the throughput of the turbo block coded

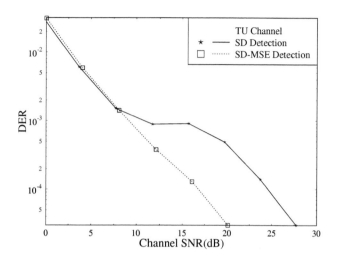

Figure 5.16: The DER performance of the SD-MSE algorithm characterized by Equation 5.6. The system parameters of Table 5.1 were utilized and the AQAM switching thresholds were set according to Table 4.8 for the target BER of 1%. The DER measure was defined in Section 5.6.1 and a conventional binary BCH(31, 26, 3) coding scheme was utilized.

AQAM scheme at high average channel SNRs, here we will introduce the concept of variable rate turbo coding AQAM schemes.

Explicitly, we will implement two types of variable code rate schemes. In the first scheme we invoke a switching mechanism that is capable of disabling and enabling the channel encoder for a chosen modulation mode. This scheme will be described in detail in the next section. For the second variable-rate scheme the coding rate is varied by utilizing different BCH component codes for the different modulation modes. Let us now describe the first variable-rate turbo block coded AQAM scheme.

5.7.1 Partial Turbo Block Coded Adaptive Modulation Scheme

In this Partial Turbo Block Coded Adaptive Modulation (**P-TBCH-AQAM**) scheme the option to disable or enable the channel encoder for each individual modulation mode is made available to the transmitter. In order to ensure that the transmitted bits are in their original sequence irrespective of the coding rate, the turbo interleaver size is varied according to the modulation mode selected, as discussed in Section 5.5. The corresponding switching mechanism for this scheme can be summarized as follows:

$$
\text{Modulation Mode} = \begin{cases}
NOTX & \text{if } \gamma_{DFE} < t_1^c \\
BPSK, I_0, R_0 & \text{if } t_1^c < \gamma_{DFE} \le t_2^c \\
BPSK, \text{Coding Disabled} & \text{if } t_2^c < \gamma_{DFE} \le t_3^c \\
4QAM, I_1, R_1 & \text{if } t_3^c < \gamma_{DFE} \le t_4^c \\
4QAM, \text{Coding Disabled} & \text{if } t_4^c < \gamma_{DFE} \le t_5^c \\
16QAM, I_2, R_2 & \text{if } t_5^c < \gamma_{DFE} \le t_6^c \\
16QAM, \text{Coding Disabled} & \text{if } t_6^c < \gamma_{DFE} \le t_7^c \\
64QAM, I_3, R_3 & \text{if } t_7^c < \gamma_{DFE} \le t_8 \\
64QAM, \text{Coding Disabled} & \text{if } \gamma_{DFE} > t_8^c,
\end{cases} \tag{5.7}
$$

where the notations are identical to those in Equation 5.1.

This scheme was simulated with the coding rate set to 0.7222, which corresponded to a turbo component code of BCH (31, 26, 3). The coded switching thresholds were chosen in order to achieve target BERs of 1% and 0.01% as shown in Table 5.6 with the corresponding turbo interleaver size for each modulation mode shown in Table 5.5. The **P-TBCH-AQAM** switching thresholds were set by combining the coded thresholds set in Table 5.4 and the uncoded switching thresholds of Table 4.8. The resulting switching thresholds are shown in Table 5.6, where if any two different switching thresholds associated with their modulation/coding mode exhibited identical values, this implied that the corresponding modulation/coding mode that is selected by these two switching thresholds is discarded. Consequently, in the **High-BER** scheme the un-coded BPSK mode was disabled, whereas for the **Low-BER** scheme the un-coded BPSK, non-coded 4QAM and non-coded 16QAM modes were disabled.

Target	Turbo Interleaver Size (Bits)				Burst Type
BER	I_0	I_1	I_2	I_3	see Figure 4.13
$\le 1\%$	104	208	416	624	Speech
$\le 0.01\%$	494	988	1976	2964	Data
Near-Error-Free	494	988	1976	2964	Data

Table 5.5: The turbo interleaver size associated with each modulation mode characterized by Equation 5.7 for the **P-TBCH-AQAM** scheme described in Section 5.7.1. The target BERs were set to be below 1%, 0.01% and near-error-free, where the corresponding transmission burst types utilized are shown in Figure 4.13.

The BER and BPS performance of the **Low-** and **High-BER P-TBCH-AQAM** scheme is shown in Figure 5.17, where the uncoded AQAM performance optimized for similar target BERs is depicted for comparison. For the **High-BER** scheme the BER performance of the **P-TBCH-AQAM** and uncoded AQAM schemes was similar and the target BER of 1% was maintained. In terms of BPS performance, at low to medium channel SNRs the coded scheme performed better, but at higher SNRs, their BPS performances converged to that of 64QAM, since the uncoded 64QAM mode was the dominant transmission mode chosen at high average channel SNRs. The same characteristics can be observed for the **Low-BER P-TBCH-AQAM** scheme, where the channel coding was only disabled, when the 64QAM

Target	Coded Switching Thresholds (dB)								Burst Type in
BER	t_1^c	t_2^c	t_3^c	t_4^c	t_5^c	t_6^c	t_7^c	t_8^c	Figure 4.13
$\leq 1\%$	0.64	3.23	3.23	6.23	8.65	11.65	14.68	17.83	Speech
$\leq 0.01\%$	1.99	4.21	4.21	10.55	10.55	17.51	17.51	23.76	Data

Table 5.6: The coded switching thresholds, which were intuitively optimized in order to achieve the target BERs of below 1% and 0.01% for the **P-TBCH-AQAM** scheme described in Section 5.7.1. The corresponding transmission burst types utilized are shown in Figure 4.13 and the switching mechanism was characterized by Equation 5.7.

mode was selected. The BER performance of the **Low-BER P-TBCH-AQAM** scheme improved at low to medium channel SNRs due to the channel codec's contribution associated with the BPSK, 4QAM and 16QAM modes. However, at channel SNRs of above 20dB the uncoded 64QAM mode became dominant, degrading slightly the BER and converging to the uncoded 64QAM performance. Nevertheless, the target BER of 0.01% was still maintained. The BPS performance of the **Low-BER P-TBCH-AQAM** scheme was similar or superior to that of the **Low-BER** uncoded AQAM scheme, where a maximum SNR gain of approximately 6dB was recorded at an average channel SNR of 0dB, as evidenced by Figure 5.17.

From these results we concluded that - as expected - the **P-TBCH-AQAM** scheme improved the throughput of the system, especially at high channel SNR values, when the channel coding was disabled. However in doing so, the BER performance slightly degraded, although it was still within the target BER limits for which it was optimised. Furthermore, the number of transmission modes was also increased, which increased the amount of signalling between the transmitter and receiver. In the next section, we will introduce another variable rate turbo block coded AQAM scheme, where the coding rate was varied in conjunction with each modulation mode by using different BCH component codes.

5.7.2 Variable Rate Turbo Block Coded Adaptive Modulation Scheme

In this Variable Rate Turbo Block Coded Adaptive Modulation (**VR-TBCH-AQAM**) scheme, a specific BCH code is assigned to each individual modulation mode [166, 170]. The higher-order modulation modes are assigned a higher code rate, in order to improve the effective data throughput at medium to high average channel SNRs and conversely, the lower-order modulation modes will be accompanied by lower code rates, in order to ensure maximum error protection at low average channel SNRs, where these modes have a high selection probability.

The modulation mode switching regime is identical to that of Equation 5.1, where the turbo interleaver size, switching levels and coding rates for all modulation modes are listed in Tables 5.7, 5.8 and 5.9, respectively. The remaining system parameters are listed in Table 5.1.

The turbo interleaver sizes were chosen with the objective of ensuring burst-by-burst turbo decoding at the receiver. Consequently the decoded bits are in the right sequence, irrespective of the different component codes used. However, due to the longer codes used by the 16QAM and 64QAM modes, dummy bits were also included in order to ensure that the number of turbo encoded bits was equal to the transmission burst size. These dummy bits could be used for conveying control or signalling information. Alternatively, these dummy

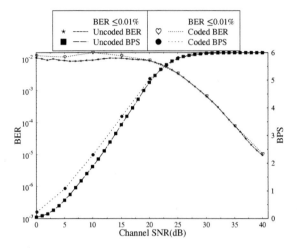

(a) Turbo block coded performance for a target BER of below 1% using the non-spread speech burst of Figure 4.13.

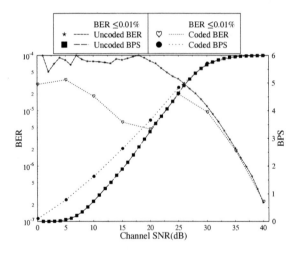

(b) Turbo block coded performance for a target BER of below 0.01% using the non-spread data burst of Figure 4.13.

Figure 5.17: Turbo block coded performance of the **P-TBCH-AQAM** scheme, which was described in Section 5.7.1, where the generic system parameters of Table 5.1 were utilized. The coded switching regime was characterized by Equation 5.7, where the coding rate was 0.7222 and the turbo interleaver sizes were listed in Table 5.5, respectively. The coded and uncoded AQAM switching thresholds were set according to Table 5.6 and 4.8, respectively.

Target BER	Turbo Interleaver Size (Bits)				Burst Type see Figure 4.13
	I_0	I_1	I_2	I_3	
$\leq 1\%$	104	208	456	720	Speech
$\leq 0.01\%$	494	988	2223	3600	Data
Near-Error-Free	494	988	2223	3600	Data

Table 5.7: The turbo interleaver size associated with each modulation mode characterized by Equation 5.1 for the **VR-TBCH-AQAM** scheme described in Section 5.7.2. The target BERs were set to be below 1%, 0.01% and near-error-free, where the corresponding transmission burst types utilized are shown in Figure 4.13.

Target BER	Coded Switching Thresholds (dB)				Burst Type see Figure 4.13
	t_1^c	t_2^c	t_3^c	t_4^c	
$\leq 1\%$	0.6363	3.2258	9.6450	15.6846	Speech
$\leq 0.01\%$	1.9958	4.2079	11.5480	18.5089	Data
Near-Error-Free	3.2458	5.4579	12.7980	19.7589	Data

Table 5.8: The coded switching thresholds, which were experimentally determined in order to achieve target BERs of below 1%, 0.01% and near-error-free for the **VR-TBCH-AQAM** scheme described in Section 5.7.2. The corresponding transmission burst types utilized are shown in Figure 4.13 and the switching mechanism was characterized by Equation 5.1.

	R_0	R_1	R_2	R_3
Turbo Code Rate	0.722	0.722	0.826	0.896
BCH (n, k, d_{min})	$(31, 26, 3)$	$(31, 26, 3)$	$(63, 57, 3)$	$(127, 120, 3)$

Table 5.9: The coding rate and the corresponding BCH component code associated with each modulation mode characterized by Equation 5.1 for the **VR-TBCH-AQAM** scheme described in Section 5.7.2.

bits could remain uncoded. In our subsequent discussions concerning this scheme, these dummy bits were not utilized for information transmission.

The corresponding BER and BPS performances are depicted in Figure 5.18. For the **High-BER VR-TBCH-AQAM** scheme, which was targeted at a BER of 1%, the coded BER performance was similar to that of the **High-BER** uncoded AQAM scheme, where a slight SNR gain was observed at average channel SNRs above 25dB. The BPS performance improved for channel SNRs between 0 to 10dB. However, the coded BPS performance degraded at high channel SNRs, when compared to the uncoded AQAM case as a result of the throughput reduction caused by the channel coding scheme. These low SNR gains observed in terms of both the BER and BPS curves were due to the smaller turbo interleaver sizes with respect to the code length as well as due to the higher code rate imposed on the higher-order modulation modes.

The BER performance of the **Low-BER VR-TBCH-AQAM** scheme was similar to that of the **Low-BER** uncoded AQAM scheme for channel SNRs below 15dB. However, at higher average channel SNRs the coded BER performance was superior, where a channel SNR gain of approximately 10dB was recorded across a wide range of BERs. The BPS performance of the **Low-BER VR-TBCH-AQAM** scheme improved for channel SNRs between 0 to 28dB, when compared to the **Low-BER** uncoded AQAM scheme. However, at higher average channel SNRs, the coded throughput was limited by the coding rate and consequently converged to a throughput of approximately 5.3 bits per symbol.

The coded switching thresholds were also re-adjusted experimentally, in order to create a near-error-free system, where the values of the coded switching thresholds are listed in Table 5.8. The BER and BPS performance is shown in Figure 5.19, where the **VR-TBCH-AQAM** scheme was near-error-free. The coded BPS performance, when compared to the **Low-BER** uncoded AQAM, exhibited an SNR gain for channel SNRs between 0 to 25dB. However, the coded BPS curve converged to a throughput of 5.3 bits per symbol at high average channel SNRs due to the limitation imposed by the coding rate of the scheme.

In conjunction with this scheme we have noted a substantial SNR gain for the **Low-BER** and near-error-free **VR-TBCH-AQAM** scheme of Figures 5.18(b) and 5.19. However only a slight SNR gain was observed for the **High-BER** scheme as evidenced by Figure 5.18(a). In the next section, we will analyse the four different turbo block coded AQAM schemes that we have introduced in this treatise and discuss their relative merits and disadvantages.

5.8   Comparisons of the Turbo Block Coded AQAM Schemes

In this section, the relative merits and disadvantages of the various turbo block coded AQAM schemes designed for target BERs of 1%, 0.01% and for near-error-free communication systems are summarized. We compared and contrasted each of these schemes in terms of their BER and BPS performance, considering also their relative complexity and their error detection capabilities. Their coded BER and BPS performances were compared to the uncoded AQAM performance. Comparisons were carried out firstly for similar target BERs in terms of the associated maximum SNR gain observed from the BPS performance curves, secondly, the maximum achievable BPS throughput and thirdly, the gain observed from the BER performance curves were recorded. These measures were termed as the BPS/SNR gain, the maximum BPS and BER/SNR gain, respectively. The BPS/SNR and BER/SNR gain was measured against the corresponding curves of the uncoded AQAM scheme for similar target BERs, where the optimized switching thresholds are listed in Table 4.8. An additional throughput-related measure was the range of channel SNRs, where the coded BPS was higher than that of the uncoded AQAM scheme for similar target BERs. This measure was termed as the effective BPS gain range.

The relative complexity of the scheme was approximated by each individual channel decoder's complexity. The complexity was measured in terms of the number of states generated by the trellis decoding algorithm in order to decode the received bits. The number of trellis states needed for each scheme provided an indication of the amount of floating-point computation needed. The complexity was calculated based on the complexity of the BCH decoder, instead of the total turbo decoding complexity, since the number of turbo decoding iterations was identical for each turbo block coded AQAM scheme. The decoder complexity in terms

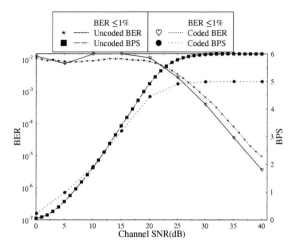

(a) Turbo block coded performance for a target BER of below 1% using the non-spread speech burst of Figure 4.13.

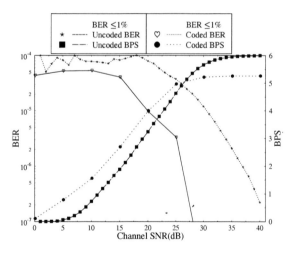

(b) Turbo block coded performance for a target BER of below 0.01% using the non-spread data burst of Figure 4.13.

Figure 5.18: Turbo block coded performance of the **VR-TBCH-AQAM** scheme, which was described in Section 5.7.2, where the generic system parameters of Table 5.1 were utilized. The coded switching regime was characterized by Equation 5.1, where the coding rates and turbo interleaver sizes were listed in Tables 5.9 and 5.7, respectively. The coded and uncoded AQAM switching thresholds were set according to Table 5.8 and 4.8, respectively.

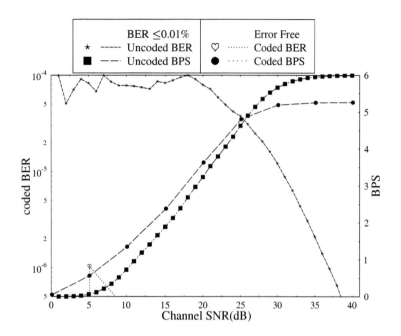

Figure 5.19: The near-**error-free** performance of the **VR-TBCH-AQAM** scheme described in Section 5.7.2. The coded switching regime was characterized by Equation 5.1, where the coding rates and turbo interleaver sizes were set according to Tables 5.9 and 5.7, respectively. The coded switching thresholds and transmission burst types were set according to Table 5.8 while the other generic system parameters were listed in Table 5.1. The performance was also compared to that of the uncoded AQAM scheme, which was optimized for a target BER of 0.01% according to Table 4.8.

of the trellis states was approximated as follows:

$$
\begin{aligned}
\text{comp}_m &\approx (2k_m - n_m + 3)(2^{n_m - k_m}), \\
average\ comp &\approx \frac{\sum_m \text{comp}_m}{4}, \\
\text{for}\ m &= \text{BPSK, 4QAM, 16QAM, 64QAM},
\end{aligned}
\tag{5.8}
$$

where comp_m is the complexity of the decoder associated with modulation mode m. Furthermore n_m and k_m denotes the number of coded bits and uncoded information bits of a certain BCH code associated with the modulation mode m. The term $2^{n_m - k_m}$ represents the total number of trellis states for a particular time instant, although the actual number of states visited in a codeword varies for different time instants, which is quantified by the term $2k_m - n_m + 3$. By assuming that the modulation modes have an equal probability of being selected, the *average comp* was the average complexity of the decoder after taking into account the complexity related to each of the four different modes.

The other complexity consideration with regards to these schemes was the number of coded transmission modes that was utilized by each scheme, which incorporated the modulation and coding parameters. The number of modes affected the amount of signalling or modulation detection complexity, where a higher number of modes required a more complex modulation detection scheme. There are four turbo block coded AQAM schemes to be compared, which were described in Sections 5.4, 5.5, 5.7.1 and 5.7.2. Explicitly, their system characteristics and their relative complexity measures are shown in Table 5.10. We will explore their complexity, BPS/SNR and BER/SNR gain comparisons for the **Low-BER** turbo block coded AQAM scheme in the next section.

Turbo Block Coded AQAM Scheme	Interleaver Size	Coding Rate	Total modes	*average comp* see Equation 5.8
FCFI-TBCH-AQAM	Fixed	Fixed	5	768
FCVI-TBCH-AQAM	Varied	Fixed	5	768
P-TBCH-AQAM	Varied	Varied	9	768
VR-TBCH-AQAM	Varied	Varied	5	4960

Table 5.10: Complexity comparisons of the **FCFI-TBCH-AQAM, FCVI-TBCH-AQAM, P-TBCH-AQAM** and **VR-TBCH-AQAM** schemes, where their characteristics were described in Sections 5.4, 5.5, 5.7.1 and 5.7.2. The channel decoder's complexity was calculated using Equation 5.8.

5.8.1   Comparison of Low-BER Turbo Block Coded AQAM Schemes

In these **Low-BER** Turbo Block Coded AQAM schemes the data burst of Figure 4.13 was utilized and their performance was compared to that of the **Low-BER** uncoded AQAM scheme, which utilized the switching thresholds of Table 4.8. The gain comparisons discussed in Section 5.8 for the different turbo block coded AQAM schemes are tabulated in Table 5.11 and depicted in Figure 5.20.

Turbo Block Coded AQAM Scheme	BPS/SNR gain (dB)	Maximum BPS	Effective BPS gain range(dB)	BER/SNR gain(dB)
FCFI-TBCH-AQAM	7.0	4.3	$0 - 23$	21.0
FCVI-TBCH-AQAM	6.0	4.3	$0 - 22$	17.5
P-TBCH-AQAM	6.0	6.0	$0 - 40$	≈ 0
VR-TBCH-AQAM	6.0	5.3	$0 - 26$	10.0

Table 5.11: Performance comparisons of the **Low-BER FCFI-TBCH-AQAM, FCVI-TBCH-AQAM, P-TBCH-AQAM** and **VR-TBCH-AQAM** schemes for a target BER of below 0.01%, where their system characteristics were described in Sections 5.4, 5.5, 5.7.1 and 5.7.2. Their performances were compared to the uncoded AQAM performance optimized for a target BER of 0.01% according to Table 4.8. The performance gains of each scheme were extracted from Figure 5.20.

The 0.01% target BER i.e **Low-BER-FCFI-TBCH-AQAM** scheme provided a high

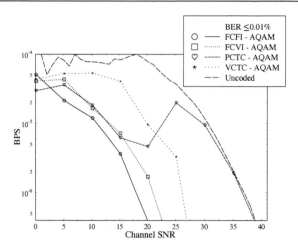

(a) Turbo block coded BER performance for a target BER of below 0.01%
using the non-spread data burst of Figure 4.13.

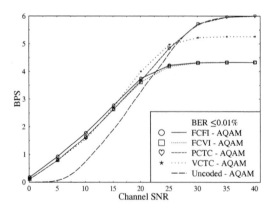

(b) Turbo block coded BPS performance for a target BER of below 0.01%
using the non-spread data burst of Figure 4.13.

Figure 5.20: Performance comparisons of the **Low-BER FCFI-TBCH-AQAM, FCVI-TBCH-AQAM, P-TBCH-AQAM** and **VR-TBCH-AQAM** schemes, where the system characteristics were described in Sections 5.4, 5.5, 5.7.1 and 5.7.2. Their performances were compared to that of the uncoded AQAM scheme optimized for a target BER of 0.01% according to Table 4.8. The performance gains of each scheme were tabulated in Table 5.11.

SNR gain in terms of its BER performance, although a limited throughput of 4.3 bits per symbol was exhibited as evidenced by Table 5.11. The BER/SNR gain of 21dB, which was measured at a channel SNR of 20dB was achieved due to the large turbo interleaver size of 9984 bits and a relatively low code rate, when compared to the other schemes. However, the maximum throughput was limited to 4.3 bits per symbol due to the low coding rate. This scheme also exhibited a better BPS performance, than the uncoded AQAM for the channel SNR range of 0 to 23dB with a maximum SNR gain of 7dB in term of its BPS performance. Furthermore, as a result of the large turbo interleaver size, the burst by burst error detection capability of the receiver in detecting the modulation modes had to be sacrificed.

The 0.01% target BER i.e **Low-BER-FCVI-TBCH-AQAM** scheme yielded a BER/SNR performance gain of 17.5 dB at a channel SNR of 20dB, which was lower than that of the corresponding **FCFI-TBCH-AQAM** scheme of Table 5.11. This was mainly due to the smaller turbo interleaver size used for each modulation mode in the **FCVI-TBCH-AQAM** scheme, which degraded the BER performance of the turbo codec. Consequently, - as seen in Table 5.4 - a set of more conservative coded switching thresholds was invoked for ensuring that the target BER was achieved. This degraded the BPS performance slightly, when compared to the **FCFI-TBCH-AQAM** scheme as it is evidenced by the 6dB maximum BPS/SNR gain and the effective BPS gain range shown in Table 5.11. However, the utilization of the variable turbo interleaver provided the burst-by-burst error detection capability of the receiver. With the exception of the turbo interleavers, both the **FCFI-TBCH-AQAM** and **FCVI-TBCH-AQAM** schemes have the same decoder complexity associated with 768 trellis states and an identical number of switching modes of 5, as evidenced by Table 5.10.

In the **P-TBCH-AQAM** scheme a high coded throughput of 6 bits per symbol was achieved but the BER/SNR gain was approximately zero in Table 5.11, although the target BER was achieved. This highest throughput was achieved as a result of the utilization of un-coded transmission modes, as shown in Equation 5.7. However, when this mode was invoked, the BER performance degraded, resulting in the minimal BER/SNR gain. Nevertheless, the BPS performance improved, when compared to the uncoded AQAM scheme, where an effective BPS gain range was observed for the entire channel SNR range, with a maximum BPS gain of 6dB. The utilization of the variable-sized turbo interleaver was essential, in order to preserve the ordering of bit sequence, when the coding rate was varied. However, due to the un-coded modes, the error detection capability was sacrificed. Furthermore, with the inclusion of the un-coded modes, the number of transmission modes increased to a maximum of 9 modes. Consequently, the adaptive switching regime and the signalling protocol between the transmitter and receiver was more complex.

In the last scheme, the **VR-TBCH-AQAM** scheme provided an average BER.SNR gain of 10dB at an average channel SNR of 20dB and a high average throughput of 5.3 bits per symbol, as a result of the higher coding rate used for this scheme. The utilization of higher coding rates for the higher-order modulation modes degraded the BER performance, when compared to the **FCVI-TBCH-AQAM** scheme, where a constant coding rate was used. However, the throughput performance improved, when compared to the **FCVI-TBCH-AQAM** scheme, where a positive BPS gain was observed for the channel SNR range between 0 to 26dB. The utilization of a variable interleaver size in the **VR-TBCH-AQAM** scheme provided the desirable burst-by-burst error detection capability in order to assist in blind modem mode detection. However, the relative decoder complexity of this scheme increased compared to the other schemes as shown in Table 5.10. This was due to the longer and more complex BCH codes that were used in order to increase the code rate for the higher-order modulation modes.

5.8.2   Comparison of High-BER Turbo Block Coded AQAM Schemes

The speech transmission burst shown in Figure 4.13 was utilized in these turbo block coded AQAM schemes. The associated performances were compared to that of the uncoded **High-BER** AQAM scheme, where the switching thresholds were set according to Table 4.8. The gain comparisons, which were defined in Section 5.8 and were extracted from Figure 5.21 are shown in Table 5.12.

Turbo Block Coded AQAM Scheme	BPS/SNR gain (dB)	Maximum BPS	Effective BPS gain(dB)	BER/SNR gain(dB)
FCFI-TBCH-AQAM	2.5	4.3	$0 - 13$	18.0
FCVI-TBCH-AQAM	2.3	4.3	$0 - 11$	8.5
P-TBCH-AQAM	2.3	6.0	$0 - 40$	≈ 0
VR-TBCH-AQAM	2.3	5.0	$0 - 11$	≈ 0

Table 5.12: Performance comparisons of the **High-BER FCFI-TBCH-AQAM, FCVI-TBCH-AQAM, P-TBCH-AQAM** and **VR-TBCH-AQAM** schemes for a target BER of below **1%**, where their characteristics were described in Sections 5.4, 5.5, 5.7.1 and 5.7.2. Their performances were compared to that of the uncoded AQAM scheme optimized for a target BER of 1% according to Table 4.8. The performance gains of each scheme were extracted from Figure 5.21.

Similar analysis to that discussed in Section 5.8.1 can be applied here, where all the **High-BER** turbo block coded AQAM schemes exhibited the same trends as those of the **Low-BER** turbo block coded AQAM schemes. However, the BER and BPS gain was significantly lower than that of the **Low-BER** schemes, as evidenced by Tables 5.12 and 5.11. The reduction in gain was caused by the higher target BER, which yielded a lower coding gain, since the higher steepness of the turbo-coded BER versus SNR curves became more effective for lower target BER schemes. This was also observed for the fixed modulation modes, where a higher coding gain was observed at lower BERs. Consequently, the **High-BER** coded switching thresholds did not reduce significantly, when compared to the switching thresholds of the **High-BER** uncoded AQAM scheme. This meant that a lower BPS/SNR gain was observed for the **High-BER** turbo block coded AQAM schemes, as evidenced by Table 5.12. Furthermore, the shorter speech transmission burst resulted in a smaller interleaver size for the variable interleaver-size turbo block coded AQAM scheme. This further degraded the coded BER and BPS performance, although the target BER of 1% was still achieved.

5.8.3   Near-Error-Free Turbo Block Coded AQAM Schemes

In these schemes, the coded switching thresholds were experimentally determined in order to achieve a near-error-free communication system. The data burst of Figure 4.13 was utilized and the gain comparisons are listed in Table 5.13, which were extracted from Figure 5.22.

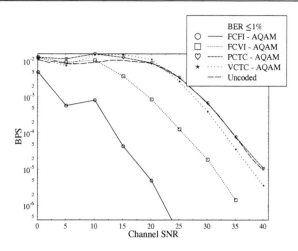

(a) Turbo block coded BER performance for a target BER of below 1% using the non-spread speech burst of Figure 4.13.

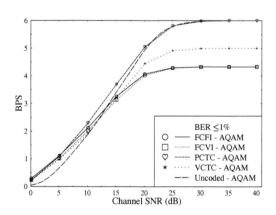

(b) Turbo block coded BER performance for a target BER of below 1% using the non-spread speech burst of Figure 4.13.

Figure 5.21: Performance comparisons of the **High-BER FCFI-TBCH-AQAM, FCVI-TBCH-AQAM, P-TBCH-AQAM** and **VR-TBCH-AQAM** schemes where the system characteristics were described in Sections 5.4, 5.5, 5.7.1 and 5.7.2. Their performances were compared to that of the uncoded AQAM scheme optimized for a target BER of 1% according to Table 4.8. The performance gains of each scheme were tabulated in Table 5.12.

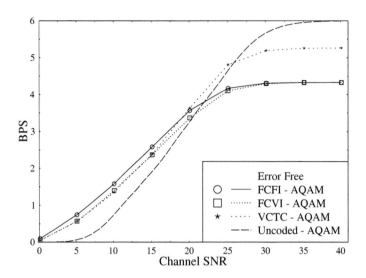

Figure 5.22: Performance comparison of the near-**error-free** FCFI-TBCH-AQAM, FCVI-TBCH-
AQAM and **VR-TBCH-AQAM** schemes, where the system characteristics were described in Sections 5.4, 5.5 and 5.7.2. Their performances were compared to that of
the uncoded AQAM scheme optimized for a target BER of 0.01% according to Table 4.8.
The performance gains of each scheme are tabulated in Table 5.12.

No near-error-free **P-TBCH-AQAM** scheme was implemented due to the inclusion of the
un-coded modes.

Turbo Block Coded AQAM Scheme	BPS/SNR gain(dB)	Maximum BPS	Effective BPS gain(dB)
FCFI-TBCH-AQAM	6.0	4.3	0 − 22.5
FCVI-TBCH-AQAM	5.0	4.3	0 − 20.0
VR-TBCH-AQAM	5.0	5.0	0 − 26.0

Table 5.13: Performance comparisons of the **FCFI-TBCH-AQAM, FCVI-TBCH-AQAM** and **VR-
TBCH-AQAM** schemes for a near-**error-free** communication system which were described in Sections 5.4, 5.5 and 5.7.2. Their performances were compared to that of the
uncoded AQAM scheme optimized for a target BER of 0.01% according to Table 4.8. The
performance gains of each scheme were extracted from Figure 5.22.

The near-error-free turbo block coded AQAM gains for the various schemes exhibited similar trends to those of the **Low-BER** turbo block coded AQAM schemes. Consequently, the analysis presented in Section 5.8.1 can be applied here. In the next section, convolutional codes are utilized as the turbo component codes, in order to reduce the complexity of the turbo block coded AQAM scheme.

5.9 Turbo Convolutional Coded AQAM Schemes

In this section, convolutional codes are utilized as the component code in the turbo codec. Explicitly, a half-rate Recursive Systematic Convolutional (RSC) encoder - which was shown in Figure 5.2 - is used, where $n = 2$, $k = 1$ and the constraint length is set to $K = 3$. This is denoted by CC $(2, 1, 3)$, where the octally represented generator polynomials were set to seven (for the feedback path) and five. The decoding algorithm used in the turbo convolutional scheme was the Log-MAP algorithm described in Appendix A.1 [164]. Let us now present and analyse the performance of the turbo convolutional scheme in the context of fixed modulation modes.

5.9.1 Turbo Convolutional Coded Fixed Modulation Mode Performance

In this section the performance of the turbo convolutional scheme is compared against that of the turbo block coded scheme for different fixed modulation modes and for similar coding rates. The simulation parameters of the turbo block coded scheme are identical to those set in Section 5.3. Similarly, in the turbo convolutional coded scheme, the code rate was set to 0.75 by applying a random puncturing pattern [171]. The sizes of the turbo interleaver and channel interleaver were chosen to be 9990 bits and 13320 bits, respectively, in order to closely match to the parameter set used for the turbo block coded scheme. The remaining simulation parameters are listed in Table 5.1.

The results are shown in Figure 5.23, where the turbo block coded performance is also displayed for comparison. A BER/SNR performance degradation of approximately $1 - 2$dB was observed at a BER of 1×10^{-4} for the turbo convolutional coded scheme, when compared to that of the turbo block coded scheme [170]. In terms of its complexity, the number of states in the block decoder trellis can be approximated upon following the philosophy of Equation 5.9 :

$$\text{No. of States for Block Codes} = \frac{\text{Encoder Input Block Length}}{k} \times (2k - n + 3) \times 2^{n-k},$$
(5.9)

where the encoder's input block length is equal to the turbo interleaver size.

Hence for an encoder input block length of 9984 bits and upon using a component code of BCH (31,26,1), the number of states produced by the block decoder is 294912. Similarly, the total number of states in a convolutional decoder trellis can be approximated by:

$$\text{Number of States for Convolutional Codes} = \text{Encoder Input Block Length} \times 2^{K-1},$$
(5.10)

where K in the constraint length of the encoder.

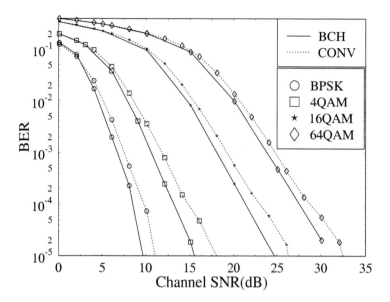

Figure 5.23: The turbo convolutional coded performance of the BPSK, 4QAM, 16QAM and 64QAM modulation modes utilizing the RSC component code CC (2, 1, 3) and the non-spread data burst of Figure 4.13. The other simulation parameters are listed in Table 5.1 and the equivalent turbo block coded performance is also shown for comparison using the BCH(31, 26, 3) component code. The turbo interleaver was of 9990 bits in depth, while the channel interleaver was of size 13320 bits.

Consequently, by applying an encoder input block length of 9990 bits and a RSC component code of CC(2,1,3), the total number of states generated by the convolutional decoder is 39960. This was approximately a factor of seven lower in terms of its complexity, when compared to the block decoder.

5.9.2   Turbo Convolutional Coded AQAM Scheme

In this section, the performance of the turbo convolutional coded scheme is evaluated in the context of a wideband AQAM scheme and compared against the performance of the turbo block coded AQAM schemes. The system parameters are identical to those of the turbo block coded AQAM schemes, which were described in Section 5.3. Essentially, the switching regime of the AQAM scheme is governed by Equation 5.1, where each modulation mode was associated with a certain code rate and turbo interleaver size. In the subsequent experiments the target BER was set to 0.01% and the non-spread data burst of Figure 4.13 was utilized as the transmission burst format. The other simulation parameters are listed in Table 5.1.

As we have seen in conjunction with the turbo block coded AQAM schemes, the performance of the turbo convolutional coded AQAM schemes is analysed by utilizing different turbo interleaver sizes and different code rates for each modulation mode. Explicitly, three different types of the turbo convolutional coded AQAM schemes are studied here:

1. Fixed Coding Rate and Variable Turbo Interleaver Turbo Convolutional Coded AQAM (**FCVI-TCONV-AQAM**) : In this scheme the convolutional coding rate was set to 0.75 and the turbo interleaver size was varied according to Table 5.14 for the different modulation modes, which ensured burst by burst decoding. This scheme was comprehensively described in Section 5.5 in the context of turbo block coded AQAM schemes. In order to provide a fair and pertinent comparison with the turbo block coded AQAM schemes, the switching thresholds were set according to Table 5.15 for achieving a target BER of 0.01%.

2. Partial Turbo Convolutional Coded AQAM (**P-TCONV-AQAM**): This scheme was identical to that described in Section 5.7.1, where the channel encoder could be disabled or enabled for each individual modulation mode. The switching regime was shown in Equation 5.7, where the coding rate was set to 0.75 and the corresponding turbo interleaver size is shown in Table 5.14. Similarly the switching thresholds - which are listed in Table 5.15 - were experimentally chosen, in order to achieve a target BER of 0.01%.

3. Variable Rate Turbo Convolutional Coded AQAM (**VR-TCONV-AQAM**): In this scheme, a specific coding rate was chosen for each modulation mode, which was discussed in Section 5.7.2 in the context of turbo block coded AQAM schemes. The coding rate was varied by utilizing different random puncturing patterns and the resulting code rates are listed in Table 5.14 for each corresponding modulation mode. The codes rates were chosen to be similar to those used for the turbo block coded scheme for comparison purposes. Finally, the switching thresholds for the AQAM scheme are shown in Table 5.15, which were experimentally set, in order to achieve a target BER of 0.01%.

AQAM	Turbo Interleaver (Bits)				Code Rates			
Scheme	I_0	I_1	I_2	I_3	R_0	R_1	R_2	R_3
FCVI-TCONV-AQAM	513	1026	2052	3078	0.75	0.75	0.75	0.75
P-TCONV-AQAM	513	1026	2052	3078	0.75	0.75	0.75	0.75
VR-TCONV-AQAM	513	1026	2280	3694	0.75	0.75	0.83	0.90

Table 5.14: The turbo interleaver size and the corresponding code rates for each modulation mode utilized in the **FCVI-TCONV-AQAM, P-TCONV-AQAM** and **VR-TCONV-AQAM** turbo convolutional coded schemes. A RSC code of CC(2, 1, 3) was used and the notations shown accrued from Equations 5.1 and 5.7.

The performance results of these schemes are shown in Figure 5.24, where the equivalent turbo block coded AQAM performance is displayed for comparison. The BER performance of the turbo convolutional coded AQAM schemes and the turbo block coded AQAM schemes were similar and in both cases the target BER of 0.01% was achieved. The characteristics of the results are similar to those of the turbo block coded AQAM schemes, which

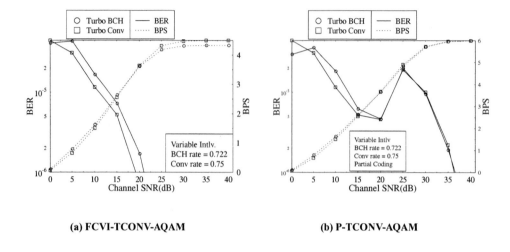

(a) FCVI-TCONV-AQAM   (b) P-TCONV-AQAM

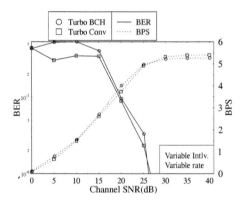

(c) VR-TCONV-AQAM

Figure 5.24: Performance of various turbo convolutional coded AQAM schemes, which utilized the RSC code CC (2, 1, 3) and the non-spread data burst of Figure 4.13. The turbo interleaver size, code rates and switching thresholds are listed in Tables 5.14 and 5.15. The generic simulation parameters are shown in Table 5.1 and the turbo block coded performance with similar parameters is also shown for comparison.

AQAM	Coded Switching Thresholds (dB)							
Scheme	t_1^c	t_2^c	t_3^c	t_4^c	t_5^c	t_6^c	t_7^c	t_8^c
FCVI-TCONV-AQAM	2.99	5.01	11.55	18.01	–	–	–	–
P-TCONV-AQAM	2.99	2.99	5.01	5.01	11.55	11.55	18.01	23.76
VR-TCONV-AQAM	2.99	5.01	12.05	20.01	–	–	–	–

Table 5.15: The switching thresholds for each modulation mode utilized in the **FCVI-TCONV-AQAM**, **P-TCONV-AQAM** and **VR-TCONV-AQAM** turbo convolutional coded schemes. A RSC code of CC(2, 1, 3) was used and the notations shown accrued from Equations 5.1 and 5.7.

were discussed in Sections 5.5, 5.7.1 and 5.7.2 and hence can be interpreted similarly. At low to medium average channel SNRs, a slight SNR gain was achieved by the turbo block coded AQAM schemes in terms of the associated BPS performances, when compared to the BPS curve of the turbo convolutional coded AQAM schemes, as evidenced by Figure 5.24. This was consistent with our expectations, since the switching thresholds of the turbo convolutional coded AQAM schemes were higher than that of the turbo block coded AQAM schemes. Consequently, the higher-order modulation modes were utilized more often at low average channel SNRs in the turbo block coded AQAM schemes. However at high average channel SNRs the maximum throughput of the turbo convolutional coded AQAM schemes were higher due to their slightly higher code rate, when compared to the turbo block coded AQAM schemes. Their associated BER versus channel SNR curves were fairly similar in all these sub-figures of Figure 5.24, although the turbo convolutional code had typically a slightly better BER.

In this section, we have applied a RSC code as the component code in our turbo codec. Subsequently, the turbo codec was applied in the context of fixed modulation modes, where the BER performance degraded by approximately $1 - 2$dB, when compared to the turbo block coded schemes. However, the complexity of the turbo convolutional coded scheme was significantly lower than that of the turbo block coded schemes, as discussed in Section 5.9. Consequently, we applied the turbo convolutional coded scheme in a wideband AQAM scheme, where the performance achieved was comparable to that of the turbo block coded AQAM schemes. In the next section we will explore the recently developed family of iterative equalization and channel decoding techniques, a scheme which is termed as turbo equalization.

5.10   Turbo Equalization

The concept of turbo equalizers is based on a joint iterative channel equalization and decoding technique [153], whereby the channel decoder is utilized in order to improve the performance of the equalization process and vice-versa in an iterative regime. Turbo equalization was pioneered by Douillard, Picart, Jézéquel, Didier, Berrou and Glavieux in 1995 [153]. In this contribution, the implementation of the turbo equalizer was derived by utilizing the previous iterative turbo decoding techniques, which were appropriately modified and incorporated in a so-called serially concatenated system, as shown in Figure 5.25. The detailed schematic of the turbo equalizer is shown in Figure 5.26, which consists of a Soft In/Soft Out (SISO) equalizer and a SISO convolutional decoder. These components are implemented based on the Log-MAP algorithm, which utilizes soft inputs and produces soft outputs. The implementation of the Log-MAP algorithm is similar to that of the turbo decoding scheme, which is described in Appendix A.1 [164]. Furthermore, these components are separated by a channel interleaver and deinterleaver, as shown in Figure 5.26. In our subsequent discussions the notations L^E and L^D represent the output Log Likelihood Ratio (LLR) of the SISO equalizer and that of the SISO decoder, respectively. The subscripts a, p and e denote the *a priori*, *a posteriori* and extrinsic values, respectively.

Figure 5.25: Schematic of the serially concatenated convolutional coded BPSK system, which performed the equalization, demodulation and channel decoding iteratively, as proposed by Douillard *et al.* [153].

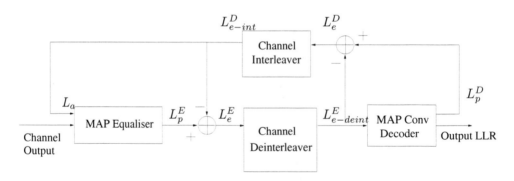

Figure 5.26: Schematic of the turbo equalizer [153] portraying the iterative structure of the equalizer and channel decoder.

Referring to Figure 5.26, the *a posteriori* information L_p^E, for the coded bits was produced by the SISO equalizer upon utilizing the channel outputs and the *a priori* information L_a, which was derived by the SISO decoder in the previous iteration. The extrinsic information of the equalizer L_e^E, was calculated by removing the contribution of the *a priori* information L_a, from the *a posteriori* information L_p^E of the equalizer, as shown in Figure 5.26. This essentially removed the contribution of the SISO decoder due to the previous iteration. The extrinsic information L_e^E, of the equalizer was subsequently deinterleaved, yielding $L_{e-deint}$ and utilized as the input to the SISO channel decoder. Consequently the SISO decoder produced the *a posteriori* information L_p^D, of the coded bits. The extrinsic information L_e^D, produced by the SISO decoder was derived by removing the deinterleaved extrinsic information $L_{e-deint}$, of the SISO equalizer from the SISO decoder's output L_p^D as shown in Figure 5.26. Consequently, the *a priori* information L_a for the next iteration, which was produced after interleaving the extrinsic information L_p^D, of the SISO decoder was independent of any

Transmission Burst type:	Non-Spread Speech Burst of Figure 4.13.
SISO Equalizer: Algorithm :	Log-MAP
SISO Decoder: Algorithm : Code Type :	Log-MAP Recursive Systematic Convolutional Code
Rate : Constraint Length, K : Octal Generator Polynomials : Channel Interleaver size :	$\frac{1}{2}$ 5 $G0 = 35, G1 = 23$ 4032
Channel Parameters: Three Equal Rayleigh-faded Weights Normalized Doppler Frequency:	$0.5773 + 0.5773z^{-1} + 0.5773z^{-2}$ 3.25×10^{-5}

Table 5.16: Generic simulation parameters used in our turbo equalization experiments.

contribution by the SISO equalizer. This process was repeated in an iterative regime, in order to produce a better estimate of the coded bits and consequently a better BER performance.

It is important to note that, unlike in the turbo decoding process, the output of the SISO convolutional decoder consists of both the source and parity LLR values. The technique used to calculate these parity LLR values was derived as an extension of the Log-MAP algorithm, which is described in Appendix A.1.3 [164]. Let us now consider the performance of our turbo-equalized fixed mode modems.

5.10.1 Fixed Modulation Performance With Perfect Channel Estimation

In this section the turbo equalizer is implemented in the context of fixed modulation modes of BPSK, 4QAM and 16QAM. The results are shown in Figure 5.27, where the experimental parameters of Table 5.16 were utilized. In these results the performance of the turbo equalizer in conjunction with one to four iterations was shown for comparison. Referring to Figure 5.27, the BER performance improved upon increasing the number of iterations for all modulation modes. This illustrated the improvement of the estimation of the coded bits as a result of the iterative decoding and equalization process. The maximum iteration gains achieved after 4 iterations were 0.7dB, 1.0dB and 2.0dB for the modulation modes of BPSK, 4QAM and 16QAM, respectively. The iteration gain was defined as the difference between the channel SNR required in order to achieve a certain BER after one iteration and the corresponding channel SNR required after n number of iterations. In Figure 5.27 the law of diminishing returns was observed on the iteration gain, where the gain of subsequent iterations decreased.

The iteration gain was higher for the higher-order modulation mode of 16QAM, when compared to that of the BPSK and 4QAM modulation modes. In this respect, the Euclidean distance between two neighbouring points in the 16QAM constellation was smaller and hence

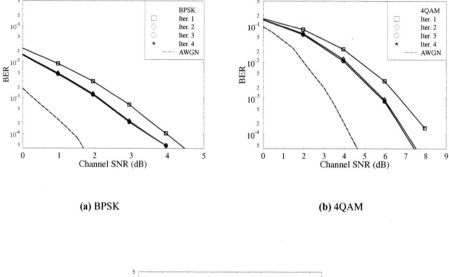

(a) BPSK　　　　　　　　　　　　　　　　(b) 4QAM

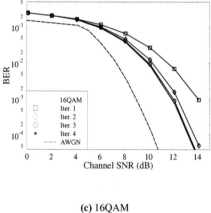

(c) 16QAM

Figure 5.27: Performance of the turbo equalizer for one to four iterations in conjunction with different modulation modes and using perfect CIR estimation. The generic simulation parameters are listed in Table 5.16.

it was more gravely affected by ISI and noise. Consequently the BER performance after one iteration incurred higher degradation, when compared to the more robust lower-order modulation modes. However, the impact of ISI was reduced significantly for the subsequent iterations, which resulted in a higher iteration gain for the 16QAM mode. Lastly, a $2-3$dB extra channel SNR was required for maintaining a BER similar to that over the non-dispersive AWGN channel.

5.10.2 Fixed Modulation Performance With Iterative Channel Estimation

In our previous experiments, perfect CIR estimation was utilized, which produced an upper-bound performance for the turbo equalizer. However, in this section, the estimation of the fading CIR is implemented iteratively.

In order to exploit the iterative regime of the turbo equalizer, the CIR was estimated after each iteration, in order to produce a more accurate CIR estimation. In the first iteration, the CIR was estimated by utilizing the mid-amble sequence shown in Figure 4.13 and subsequently the CIR was utilized in the equalization process. However, for the subsequent iterations, the CIR was re-estimated by utilizing the entire transmission frame's symbols derived from the *a posteriori* coded bits of the SISO channel decoder. The *a posteriori* information was transformed from the log domain to modulated symbols using the approach employed by Glavieux *et al.* in Reference [172].

Figure 5.28: The Gray mapping of the 16QAM mode depicting the in-phase or quadrature-phase components and the corresponding bits assignments.

In order to highlight the philosophy of the soft mapper approach, let us consider an example using the in-phase component of the 16QAM modulation mode, where the constellation points employed Gray mapping, as shown in Figure 5.28. In this constellation mapping, the nth 16QAM symbol $d_n = a_n + jb_n$, was associated with four coded data bits represented by $C_{n,i}$, where $i = 1, 2, 3, 4$. Consequently, the first two bits $C_{n,1}$ and $C_{n,2}$, determined the in-phase component of the 16QAM symbol a_n, and similarly, the last two bits $C_{n,3}$ and $C_{n,4}$ determined the quadrature-phase component of the 16QAM symbol b_n. By utilizing the LLR values of each individual bit at the output of the SISO decoder, the average soft in-phase and quadrature-phase component of the symbol, denoted by $\bar{a}_n$ and $\bar{b}_n$ can be calculated as follows [172]:

$$
\begin{aligned}
\bar{a}_n &= 3 \cdot P\{C_{n,1} = 1, C_{n,2} = 1\} + 1 \cdot P\{C_{n,1} = 1, C_{n,2} = 0\} \\
&\quad -1 \cdot P\{C_{n,1} = 0, C_{n,2} = 0\} - 3 \cdot P\{C_{n,1} = 0, C_{n,2} = 1\}, \quad (5.11)
\end{aligned}
$$

$$\bar{b}_n = 3 \cdot P\{C_{n,3} = 1, C_{n,4} = 1\} + 1 \cdot P\{C_{n,3} = 1, C_{n,4} = 0\}$$
$$-1 \cdot P\{C_{n,3} = 0, C_{n,4} = 0\} - 3 \cdot P\{C_{n,3} = 0, C_{n,4} = 1\}, \qquad (5.12)$$

where $P\{x = 1, y = 1\}$ represented the joint probability that the variable x was a logical one and y was a logical one. The constant factors of 3 and 1 in Equations 5.11 and 5.12 denoted the amplitude imposed by the mapping constellation shown in Figure 5.28. The output LLR $L\{C_{n,i}\}$, of the coded bits was defined as the log of the ratio of the probabilities of the bit taking its two possible values :

$$L\{C_{n,i}\} = ln\left\{ \frac{P(C_{n,i} = 1)}{P(C_{n,i} = 0)} \right\}, \qquad for\ i = 1, 2, 3, 4. \qquad (5.13)$$

By exploiting the relationship that $P(C_{n,i} = 1) = 1 - P(C_{n,i} = 0)$ and $P(C_{n,i} = 0) = 1 - P(C_{n,i} = 1)$, we can rewrite Equation 5.11 and 5.12 as :

$$\bar{a}_n = \frac{e^{L(C_{n,1})}\left(3.e^{L(C_{n,2})} + 1\right) - 1 - 3.e^{L(C_{n,2})}}{(1 + e^{L(C_{n,1})})(1 + e^{L(C_{n,2})})}, \qquad (5.14)$$

$$\bar{b}_n = \frac{e^{L(C_{n,3})}\left(3.e^{L(C_{n,4})} + 1\right) - 1 - 3.e^{L(C_{n,4})}}{(1 + e^{L(C_{n,3})})(1 + e^{L(C_{n,4})})}. \qquad (5.15)$$

Hence, for every iteration $\bar{a}_n$ and $\bar{b}_n$ were calculated, in order to represent the estimated transmitted symbols of the entire frame, which was subsequently used in the CIR estimator. Due to the iterative structure of the CIR estimator, the simple Least Mean Square (LMS) adaptive algorithm [118] was implemented, which obeyed the following equation:

$$\hat{\mathbf{h}}(k + 1) = \hat{\mathbf{h}}(k) + \mu \mathbf{u}(k)[r^*(k) - \mathbf{u}^{*T}(k)\hat{\mathbf{h}}(k)], \qquad (5.16)$$

where $\hat{\mathbf{h}}(k)$ and $\mathbf{u}(k)$ represented the estimated CIR vector and the training sequence vector at time n, respectively. The channel's output symbol was denoted by $r(k)$ and μ was termed as the step-size of the LMS algorithm. Since this algorithm is widely known and utilized [118], the detailed mechanism of this algorithm is not explored here any further.

The step-size for the initial CIR estimation in the first iteration was set to 0.05 and for subsequent iterations the step-size was reduced to 0.01. In the first iteration the training of the CIR estimator was restricted to the length of the mid-amble sequence. Hence, in this situation, a higher step-size was chosen in order to facilitate fast convergence at the expense of the accuracy of the CIR estimates [118]. However for subsequent iterations the training length was extended to the entire transmitted frame. Consequently the step-size was reduced, in order to improve the accuracy of the CIR estimates [118]. The specific step size values were chosen in order to satisfy the convergence limits specified by Haykin [118].

The performance of the iterative CIR estimator in a turbo equalizer is shown in Figure 5.29 for the modulation modes of BPSK, 4QAM and 16QAM [173]. The simulation parameters used in this experiment are listed in Table 5.16. In each of these figures, the performance of the system employing perfect CIR estimation after four iterations was also depicted for comparison. Referring to Figure 5.29, the performance upon utilizing the iterative CIR estimator approached that of the perfect estimation case. This was a result of the iterative nature

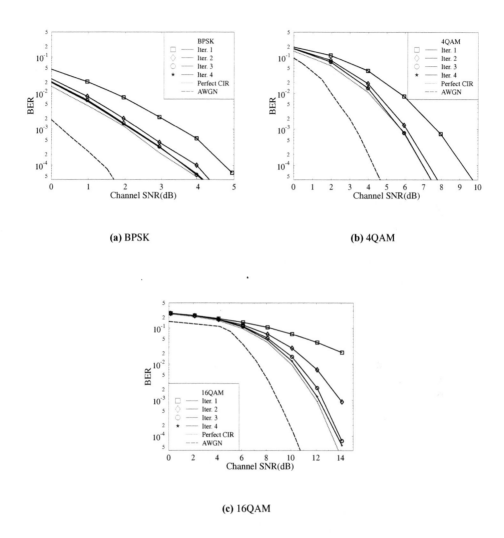

Figure 5.29: Performance of the turbo equalizer for one to four iterations in conjunction with different modulation modes. The iterative LMS CIR estimator described in Section 5.10.2 was utilized and the other simulation parameters are listed in Table 5.16. The performance with perfect CIR estimation is also shown for comparison.

of the turbo equalizer, where the reliability of the decoded information increased for every subsequent iteration. Consequently, the iterative CIR estimator exploited the increased reliability of the decoded symbols, in order to yield an improved channel estimate for the SISO equalizer. This then yielded a more reliable output from the SISO equalizer. Following the above rudimentary introduction to turbo equalization, let us now quantify its performance in a wideband AQAM scenario.

5.10.3   Turbo Equalization in Wideband Adaptive Modulation

In our previous experiments involving the wideband AQAM scheme, the output SNR of the DFE was used as a measure of the channel quality and subsequently used as a switching metric. However, in implementing the turbo equalizer in the context of wideband AQAM scheme, the SISO equalizer did not provide an SNR estimate in order to ascertain the channel quality on a burst by burst basis. Consequently, we propose to utilize the output SNR of the DFE - which was defined in Equation 4.6 - as the switching metric in an amalgamated wideband AQAM and turbo equalization scheme. In order to justify its utilization, the channel quality, which was quantified in terms of the output SNR of the DFE also had to indicate the performance of the SISO equalizer in terms of the BER. Consequently, we evaluated the correlation between the number of erroneous decisions produced by the SISO equalizer and the output SNR of the DFE on a burst-by-burst basis. This is shown in Figure 5.30, where a 4QAM mode was utilized at a channel SNR of 8dB over a symbol-spaced, equal-weight three-path fading channel. Referring to this figure, the number of error events for the SISO equalizer exhibited a good correlation with the output SNR of the DFE, where the number of error events increased, whenever the output SNR of the DFE decreased and vice versa. Consequently, we can justify the utilization of the output SNR of the DFE as a switching metric in the SISO equalizer [173]. Due to the iterative structure of the turbo equalizer, the switching thresholds were set experimentally, in order to achieve a target BER of approximately 0.01%. Hence the switching thresholds were set as follows : $t_1 = -1.5$dB, $t_2 = 2.5$dB, $t_3 = 6.5$dB and $t_4 = \infty$dB. The iterative LMS-based CIR estimator of Equation 5.16 was also utilized and the simulation parameters were set according to Table 5.16. The BER performance of the joint turbo equalization and AQAM scheme is depicted in Figure 5.31, where the wideband AQAM upper-bound performance employing perfect CIR estimation after four turbo equalization iterations was also shown for comparison [173]. Referring to Figure 5.31, the approximate target BER of 0.01% was achieved and maintained after four iterations. Furthermore, the performance of the wideband AQAM scheme with iterative CIR estimation approached that of the perfect CIR-assisted upper-bound scenario. The BPS performance of the wideband AQAM scheme is shown in Figure 5.32, where the associated performance of the fixed modulation modes of BPSK and 4QAM at a target BER of 0.01% are also depicted for comparison. For a BPS of 0.5, which was equivalent to the throughput of a half-rate coded BPSK mode, a channel SNR gain of 1.7dB was achieved by the wideband AQAM scheme. Similarly an SNR gain of 1.5dB was recorded by the wideband AQAM scheme at a target throughput of one bit per symbol, which was the throughput achieved by a half-rate coded 4QAM mode.

In implementing the turbo equalizer, the complexity incurred in terms of the number of states for the SISO equalizer was equal to m^L, where m was the number of constellation points of the modulation mode and L was the CIR memory length. Hence for higher-order modulation modes such as 64QAM, the complexity incurred was impractical even with a CIR memory length of two symbol-durations. Consequently, throughout our discussions on the turbo equalizer, the 64QAM mode was not utilized. Furthermore, we have restricted the length of the channel memory to two symbol-durations in order to reduce the complexity. Clearly, the complexity issues of this system may render it impractical especially in high delay-spread environments. Let us now review the findings of this chapter.

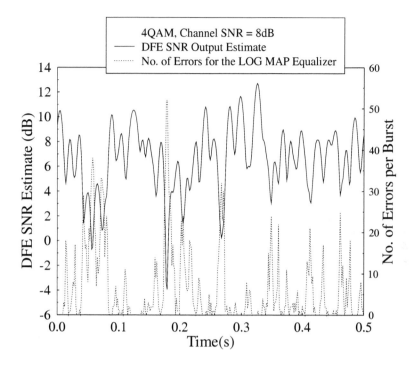

Figure 5.30: Variation of the number of errors per transmission burst produced by the Log-MAP equalizer and the corresponding output SNR estimate of the DFE using the 4QAM mode at a channel SNR of 8dB. The simulation parameters are listed in Table 5.16.

5.11 Burst-by-Burst Adaptive Wideband Coded Modulation

S. X. Ng, C. H. Wong and L. Hanzo

5.11.1 Introduction

Trellis Coded Modulation (TCM) [174], which is based on combining the functions of coding and modulation, is a bandwidth efficient scheme that has been widely recognized as an excellent error control technique suitable for applications in mobile communications [175, 176]. Turbo Trellis Coded Modulation (TTCM) [177] is a more recent channel coding scheme that has a structure similar to that of the family of power efficient binary turbo codes [155], but employs TCM codes as component codes. Rate 2/3 TTCM was shown in [177] to be 0.5 dB better in Signal-to-Noise Ratio (SNR) terms, than binary turbo codes over AWGN channels using 8-level Phase Shift Keying (8PSK). TTCM was also shown to outperform a similar-complexity TCM scheme in the context of Orthogonal Frequency Division Multiplexing (OFDM) transmission over various dispersive channels [178]. In this latter context,

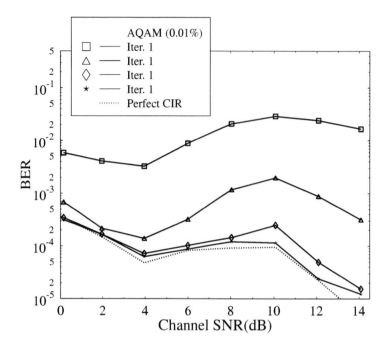

Figure 5.31: Performance of the turbo equalizer for four iterations in conjunction with the wideband
AQAM scheme. The iterative LMS CIR estimator described in Section 5.10.2 was utilized
and the other simulation parameters are listed in Table 5.16. The performance with perfect
CIR estimation is also shown for comparison.

the individual OFDM subcarriers experienced effectively narrowband fading and the TCM as
well as TTCM complexity were rendered similar by adjusting the number of turbo iterations
and code constraint length. However, the above fixed mode transceiver failed to exploit the
time varying nature of the mobile radio channel.

By contrast, in BbB adaptive schemes [1, 33, 34, 36, 60, 154, 167] a higher order modulation mode is employed, when the instantaneous estimated channel quality is high in order to increase the number of Bits Per Symbol (BPS) transmitted and conversely, a more
robust lower order modulation mode is employed, when the instantaneous channel quality
is low, in order to improve the mean Bit Error Rate (BER) performance. Uncoded adaptive schemes [1, 33, 36, 60] and coded adaptive schemes [34, 154] have been investigated for
narrowband fading channels. Finally, a turbo coded wideband adaptive scheme assisted by
Decision Feedback Equalizer (DFE) was investigated in [167].

In our practical approach the transmitter A obtains the channel quality estimate generated
by receiver B upon receiving the transmission of transmitter B. In other words, the modem
mode required by receiver B is superimposed on the transmission burst of transmitter B.
Hence a delay of one transmission burst duration is incurred. In the literature, adaptive coding

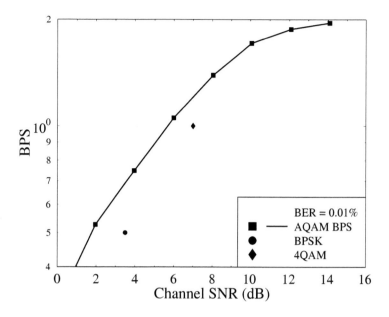

Figure 5.32: Throughput performance of the turbo equalizer for four iterations in conjunction with the wideband AQAM scheme. The iterative LMS CIR estimator described in Section 5.10.2 was utilized and the other simulation parameters are listed in Table 5.16. The throughput of the half-rate coded BPSK and 4QAM modes was also depicted for comparison.

for time-varying channels using outdated fading estimates has been investigated in [49].

Over wideband fading channels the DFE employed will eliminate most of the intersymbol interference (ISI). Consequently, the mean-squared error (mse) at the output of the DFE can be calculated and used as the metric invoked to switch the modulation modes [33]. This ensures that the performance is optimised by employing equalization and BbB adaptive TCM/TTCM jointly, in order to combat the signal power fluctuations and the ISI of the wideband channel.

This section is organized as follows. In Section 5.11.2, the system is outlined. In Section 5.11.3, the performance of fixed-mode TCM and TTCM schemes is evaluated. Section 5.11.4 contains the detailed characterization of the BbB adaptive TCM/TTCM schemes in **System I** and **System II**. In Section 5.11.5, we compare the proposed schemes with other adaptive coded modulation schemes such as Bit-Interleaved Coded Modulation [179]. Finally, we will conclude in Section 5.11.6.

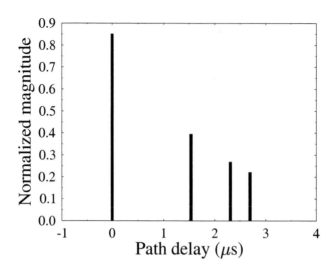

Figure 5.33: The impulse response of a COST 207 Typical Urban (TU) channel [150].

5.11.2   System Overview

The multi-path channel model is characterized by its discretised symbol-spaced COST207 Typical Urban (TU) channel impulse response [150], as shown in Figure 5.33. Each path is faded independently according to a Rayleigh distribution and the corresponding normalized Doppler frequency is 3.25×10^{-5}, the system Baud rate is $2.6 \ MBd$, the carrier frequency is $1.9 \ GHz$ and the vehicular speed is $30 \ mph$. The DFE incorporated 35 feed-forward taps and 7 feedback taps and the transmission burst structure used is shown in Figure 5.34. When considering a Time Division Multiple Access (TDMA)/Time Division Duplex (TDD) system of 16 slots per $4.615 \ ms$ TDMA frame, the transmission burst duration is $288 \ \mu s$, as specified in the Pan-European FRAMES proposal [151].

The following assumptions are stipulated. Firstly, we assume that the equalizer is capable of estimating the Channel Impulse Response (CIR) perfectly from the equaliser training sequence of Figure 5.34. Secondly, the CIR is time-invariant for the duration of a transmission burst, but varies from burst to burst according to the Doppler frequency, which corresponds to assuming that the CIR is slowly varying. The error propagation of the DFE will degrade the estimated performance, but the effect of error propagation is left for further study.

At the receiver, the CIR is estimated, which is then used to calculate the DFE coefficients [4]. Subsequently, the DFE is used to equalize the ISI-corrupted received signal. In addition, both the CIR estimate and the DFE feed-forward coefficients are utilized to compute the SNR at the output of the DFE. More specifically, by assuming that the residual ISI is near-Gaussian distributed and that the probability of decision feedback errors is negligible, the SNR at the

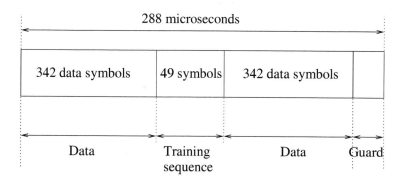

Figure 5.34: Transmission burst structure of the FMA1 non-spread data as specified in the FRAMES proposal [151].

output of the DFE, γ_{dfe}, is calculated as [33]:

$$\gamma_{dfe} = \frac{\text{Wanted Signal Power}}{\text{Residual ISI Power} + \text{Effective Noise Power}}.$$

$$= \frac{E\left[|s_k \sum_{m=0}^{N_f} C_m h_m|^2\right]}{\sum_{q=-(N_f-1)}^{-1} E\left[|\sum_{m=0}^{N_f-1} C_m h_{m+q} s_{k-q}|^2\right] + N_o \sum_{m=0}^{N_f} |C_m|^2}, \quad (5.17)$$

where C_m and h_m denotes the DFE's feed-forward coefficients and the CIR, respectively. The transmitted signal is represented by s_k and N_o denotes the noise spectral density. Lastly, the number of DFE feed-forward coefficients is denoted by N_f.

The equalizer's SNR, γ_{dfe}, in Equation 5.17, is then compared against a set of adaptive modem mode switching thresholds f_n, and subsequently the appropriate modulation mode is selected [33]. The modem mode required by receiver B is then fed back to transmitter A. The modulation modes that are utilized in this scheme are 4QAM, 8PSK, 16QAM and 64QAM [4].

The simplified block diagram of the BbB adaptive TCM/TTCM **System I** is shown in Figure 5.35, where no channel interleaving is used. Transmitter A extracts the modulation mode required by receiver B from the reverse-link transmission burst in order to adjust the adaptive TCM/TTCM mode suitable for the channel. This incurs one TDMA/TDD frame delay between estimating the actual channel condition at receiver B and the selected modulation mode of transmitter A. Better channel quality prediction can be achieved using the techniques proposed in [14]. We invoke four encoders, each adding one parity bit to each information symbol, yielding the coding rate of $1/2$ in conjunction with the TCM/TTCM mode of 4QAM, $2/3$ for 8PSK, $3/4$ for 16QAM and $5/6$ for 64QAM.

The design of TCM schemes for fading channels relies on the time and space diversity provided by the associated coder [175, 180]. Diversity may be achieved by repetition coding (which reduces the effective data rate), spaced-time coded multiple transmitter/receiver structures [181] (which increases cost and complexity) or by simple interleaving (which induces

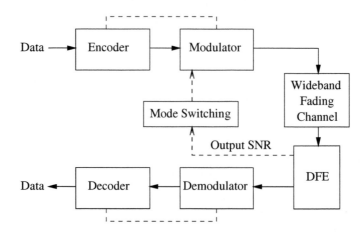

Figure 5.35: System I without channel interleaver. The equalizer's output SNR is used to select a suitable modulation mode, which is fed back to the transmitter on a burst-by-burst basis.

latency). In [182] adaptive TCM schemes were designed for narrowband fading channels utilizing repetition-based transmissions during deep fades along with ideal channel interleavers and assuming zero delay for the feedback of the channel quality information.

Figure 5.36 shows the block diagram of **System II**, where symbol-based channel interleaving over four transmission bursts is utilized, in order to disperse the bursty symbol errors. Hence, the coded modulation module assembles four bursts using an identical modulation mode, so that they could be interleaved using the symbol-by-symbol random channel interleaver without the need of adding dummy bits. Then, these four-burst TCM/TTCM packets are transmitted to the receiver. Once the receiver has received the 4th burst, the equalizer's output SNR for this most recent burst is used to choose a suitable modulation mode. The selected modulation mode is fed back to the transmitter on the reverse link burst. Upon receiving the modulation mode required by receiver B (after one TDMA frame delay), the coded modulation module assembles four bursts of data from the input buffer for coding and interleaving, which are then stored in the output buffer ready for the next four bursts transmission. Thus the first transmission burst exhibits one TDMA/TDD frame delay and the fourth transmission burst exhibits four frame delay which is the worst-case scenario.

Soft decision trellis decoding utilizing the Log-Maximum A Posteriori (Log-MAP) algorithm [163] was invoked for TCM/TTCM decoding. The Log-MAP algorithm is a numerically stable version of the MAP algorithm operating in the log-domain, in order to reduce its complexity and to mitigate the numerical problems associated with the MAP algorithm [162]. The TCM scheme invokes Ungerboeck's codes [174], while the TTCM scheme invokes Robertson's codes [177]. A component TCM code memory of 3 was used for the TTCM scheme. The number of turbo iterations for TTCM was fixed to 4 and hence it exhibited a similar decoding complexity to the TCM code of memory 6.

In the next section we present simulation results for our fixed-mode transmissions.

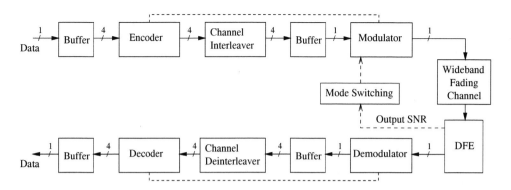

Figure 5.36: System II with channel interleaver length of 4 TDMA/TDD bursts. Data is entered into the input buffer on a burst-by-burst basis and the modulator modulates coded data from the output buffer for transmission on a burst-by-burst basis. The encoder and channel interleaver as well as the decoder and channel deinterleaver operate on a 4-burst basis. The equalizer's output SNR of the 4th burst is used to select a suitable modulation mode and fed back to the transmitter on the reverse link burst.

5.11.3 Performance of the Fixed Modem Modes

Before characterising the proposed wideband BbB adaptive scheme, the BER performance of the fixed modem modes of 4QAM, 8PSK, 16QAM and 64QAM are studied both with and without channel interleavers. These results are shown in Figure 5.37 for TCM, while in Figure 5.38 for TTCM. The random TTCM symbol-interleaver memory was set to 684 symbols, corresponding to the number of data symbols in the transmission burst structure of Figure 5.34, where the corresponding number of bits was the number of data bits per symbol $(BPS) \times 684$. A channel interleaver of 4×684 symbols was utilized, where the number of bits was $(BPS + 1) \times 4 \times 684$ bits, since one parity bit was added to each TCM/TTCM symbol.

As expected, in Figures 5.37 and 5.38 the BER performance of the channel-interleaved scenario was superior compared to that without channel interleaver, although at the cost of an associated higher transmission delay. The SNR-gain difference between the channel interleaved and non-interleaved scenarios was about $5\ dB$ in the TTCM/4QAM mode, but this difference reduced for higher-order modulation modes. Again, this gain was obtained at the cost of a four-burst channel interleaving delay. This SNR-gain difference shows the importance of time diversity in coded modulation schemes.

TTCM has been shown to be more efficient than TCM for transmissions over AWGN channels and narrowband fading channels [177, 178]. Here, we illustrate the advantage of TTCM in comparison to TCM over the dispersive or wideband Gaussian CIR of Figure 5.33 as seen in Figure 5.39. In conclusion, TTCM is superior to TCM in a variety of channels.

Let us now compare the performance of the BbB adaptive TCM/TTCM **system I** and **II**.

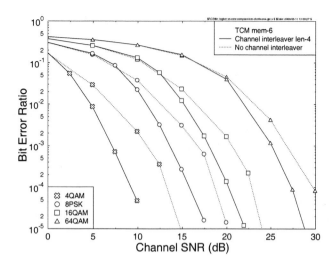

Figure 5.37: TCM performance of each individual modulation mode over the Rayleigh fading
COST207 TU channel of Figure 5.33. A TCM code memory of 6 was used, since it had a
similar decoding complexity to TTCM in conjunction with 4 iterations using a component
TCM code memory of 3.

5.11.4   Performance of System I and System II

The modem mode switching mechanism of the adaptive schemes is characterized by a set of
switching thresholds, the corresponding random TTCM symbol-interleavers and the component codes, as follows:

$$
\text{Modulation Mode} = \begin{cases}
4QAM, I_0 = 684, R_0 = 1/2 & \text{if } \gamma_{DFE} \leq f_1 \\
8PSK, I_1 = 1368, R_1 = 2/3 & \text{if } f_1 < \gamma_{DFE} \leq f_2 \\
16QAM, I_2 = 2052, R_2 = 3/4 & \text{if } f_2 < \gamma_{DFE} \leq f_3 \\
64QAM, I_3 = 3420, R_3 = 5/6 & \text{if } \gamma_{DFE} > f_3,
\end{cases}
\tag{5.18}
$$

where $f_n, n = 1...3$ are the equalizer's output SNR thresholds, while I_n represents the random TTCM symbol-interleaver size in terms of the number of bits, which is not used for the
TCM schemes. The switching thresholds f_n were chosen experimentally, in order to maintain
a BER of below 0.01% and these thresholds are listed in Table 5.17.

Let us consider the adaptive TTCM scheme in order to investigate the performance of
System I and **System II**. The BER and BPS performances of both adaptive TTCM systems
using 4 iterations are shown in Figure 5.40, where we observed that the throughput of **System II** was superior to that of **System I**. Furthermore, the overall BER of **System II** was
lower than that of **System I**. In order to investigate the switching dynamics of both systems,
the mode switching together with the equalizer's output SNR was plotted versus time at an
average channel SNR of 25 dB in Figures 5.41 and 5.42. Observe in Table 5.17 that the
switching thresholds, f_n of **System II** are lower than those of **System I**, since the fixed mode
based results of **System II** in Figure 5.38 were better. Hence higher-order modulation modes

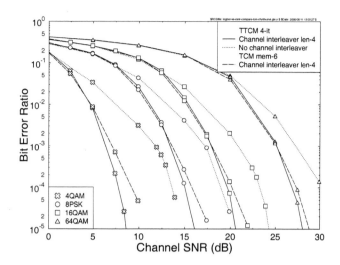

Figure 5.38: TTCM performance of each individual modulation mode over the Rayleigh fading COST207 TU channel of Figure 5.33. A component TCM code memory of 3 was used and the number of turbo iterations was 4. The performance of the TCM code with memory 6 utilizing a channel interleaver was also plotted for comparison.

BER < 0.01 %		Switching Thresholds		
Adaptive System Type		f_1	f_2	f_3
TCM, Memory 3	System I	19.56	23.91	30.52
	System II	17.17	21.91	29.61
TCM, Memory 6	System I	19.56	23.88	30.07
	System II	17.14	21.45	29.52
TTCM, 4 iterations	System I	19.69	23.45	30.29
	System II	16.66	21.40	28.47
BICM, Memory 3	System I	19.94	24.06	31.39
BICM-ID, 8 iterations	System II	16.74	21.45	28.97

Table 5.17: The switching thresholds were set experimentally in order to achieve a target BER of below 0.01%. **System I** does not utilize a channel interleaver, while **System II** uses a channel interleaver length of 4 TDMA/TDD bursts.

were chosen more frequently than in **System I**, giving a better BPS throughput. From Figure 5.41 and 5.42, it is clear that **System I** was more flexible in terms of mode switching, while **System II** benefitted from higher diversity gains due to the 4-burst channel interleaver. This diversity gain compensated for the loss of switching flexibility, ultimately providing a better performance in terms of BER and BPS, as seen in Figure 5.40.

In our next endeavour, the adaptive TCM and TTCM schemes of **System I** and **System II** are compared. Figure 5.43 shows the BER and BPS performance of **System I** for adaptive

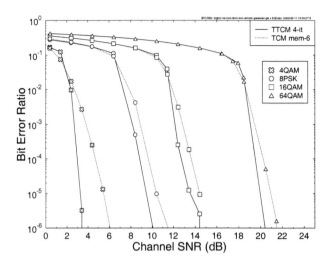

Figure 5.39: TTCM and TCM performance of each individual modulation mode over the unfaded COST207 TU channel of Figure 5.33. The TTCM scheme used component TCM codes of memory 3 and the number of turbo iterations was 4. The performance of the TCM scheme with memory 6 was plotted for comparison with the similar-complexity TTCM scheme.

TTCM using 4 iterations, adaptive TCM of memory 3 (which was the component code of our TTCM scheme) and adaptive TCM of memory 6 (which had a similar decoding complexity to our TTCM scheme). As it can be seen from the fixed mode results of Figures 5.37 and 5.38 in the previous section, TCM and TTCM performed similarly in terms of their BER, when no channel interleaver was used for this slow fading wideband channel. Hence, they exhibited a similar performance in the adaptive schemes of System I, as shown in Figure 5.43. Even the TCM scheme of memory 3 associated with a lower complexity could give a similar BER and BPS performance. This shows that the equalizer plays a dominant role in **System I**, where the coded modulation schemes could not benefit from sufficient diversity due to the lack of interleaving.

When the channel interleaver is introduced in **System II**, the bursty symbol errors are dispersed. Figure 5.44 illustrates the BER and BPS performance of **System II** for adaptive TTCM using 4 iterations, adaptive TCM of memory 3 and adaptive TCM of memory 6. The performance of all these schemes improved in the context of **System II**, as compared to the corresponding schemes in **System I**. The TCM scheme of memory 6 had a lower BER, than TCM of memory 3, and also exhibited a small BPS improvement. As expected, TTCM had the lowest BER and also the highest BPS throughput compared to the other coded modulation schemes.

In summary, we have observed BER and BPS gains for the channel interleaved adaptive coded schemes of **System II** in comparison to the schemes without channel interleaver in **System I**. Adaptive TTCM exhibited a superior performance in comparison to adaptive TCM in **Systems II**.

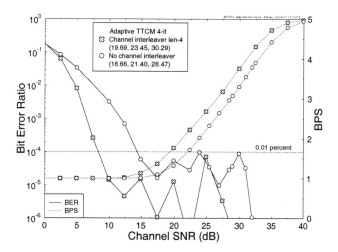

Figure 5.40: BER and BPS performance of adaptive TTCM with 4 turbo iterations in **System I** (without channel interleaver) and in **System II** (with a channel interleaver length of 4 bursts) for a target BER of less than 0.01 %. The legends indicate the associated switching thresholds expressed in dB, as seen in the round brackets.

5.11.5 Performance of Bit-Interleaved Coded Modulation

The above adaptive TCM and TTCM schemes invoked Set-Partitioning (SP) based signal labelling, in order to achieve a higher Euclidean distance between the unprotected bits of the constellation, so that parallel trellis transitions can be associated with the unprotected data bits. This reduced the decoding complexity. In TCM and TTCM random symbol interleavers were utilized for both the turbo interleaver and the channel interleaver.

Another powerful coded modulation scheme utilizing bit-based channel interleaving in conjunction with Gray signal labelling is referred to as Bit-Interleaved Coded Modulation (BICM) was proposed in [179, 183]. It combines conventional convolutional codes with several independent bit interleavers, in order to increase the associated diversity order. With bit interleavers, the code diversity order can be increased to the binary Hamming distance of a code, and the number of parallel bit-interleavers equals the number of coded bits in a symbol [183]. The performance of BICM is better than that of TCM over uncorrelated (or fully interleaved) fading channels but worse than that of TCM in Gaussian channels due to the reduced Euclidean distance imposed by the associated "random modulation" inherent in a bit-interleaved scheme [183].

Recently, iteratively decoded BICM with SP signal labelling, referred to as BICM-ID has also been proposed [184–187]. The philosophy of BICM-ID is to increase the Euclidean distance of the BICM code and to exploit the full advantage of bit interleaving by a simple iterative decoding technique. BICM-ID was shown to be better than TCM and BICM in both AWGN and uncorrelated Rayleigh fading channels in the references. Rate $2/3$ BICM-ID was shown in [185] to be only about $0.5\ dB$ away from TTCM over AWGN channels using 8PSK.

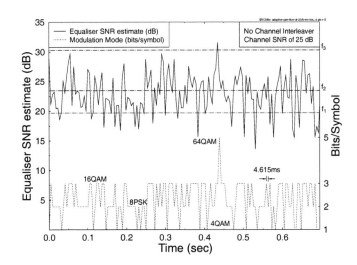

Figure 5.41: Equaliser output SNR estimate and Bits/Symbol versus time plot for adaptive TTCM with
4 turbo iterations in **System I** at an average channel SNR of 25 dB, where the modulation mode switching is based upon the equalizer's output SNR, which is compared to the
switching thresholds f_n defined in Table 5.17. The duration of one TDMA/TDD frame is
4.615 ms. The TTCM mode can be switched after one frame duration.

In order to further benchmark the performance of our BbB adaptive TCM/TTCM system,
an adaptive BICM scheme was constructed using Paaske's convolutional codes [158] for rate
$1/2, 2/3, 3/4$ and $5/6$, which provides the largest free Hamming distance. The rate $5/6$ code
was constructed using a rate $1/2$ code and puncturing, following the approach of [188]. An
adaptive BICM-ID scheme was also constructed with soft-decision feedback based iterative
decoding method [187].

Figure 5.45 shows the fixed modem modes performance for TCM, TTCM, BICM and
BICM-ID in the context of **System II**. For the sake of a fair comparison of the decoding
complexity, we used a TCM code memory of 6, TTCM code memory of 3 with 4 turbo iterations, BICM code memory of 6 and a BICM-ID code memory of 3 with 8 decoding iterations.
However, BICM-ID had a slightly higher decoding complexity, since the demodulator was
invoked in each BICM-ID iteration, whereas in the BICM, TCM and TTCM schemes the
demodulator was only visited once in each decoding process. As illustrated in the figure, the
BICM scheme performed marginally better than the TCM scheme at a BER below 0.01 %,
except in the 64QAM mode. Hence, adaptive BICM is also expected to be better than adaptive TCM in the context of **System II**, when a target BER of less than 0.01 % is desired. This
is because when the channel interleaver depth is sufficiently high, the diversity gain of the
BICM's bit-interleaver is higher than that of the TCM's symbol-interleaver [179, 183].

Figure 5.46 compares the adaptive BICM and TCM schemes in the context of **System
I**, i.e. without channel interleaving, although the BICM scheme invoked an internal bit-
interleaver of one burst memory. As it can be seen from the figure, adaptive TCM exhibited
a better BPS throughput and BER performance than BICM, due to insufficient channel inter-

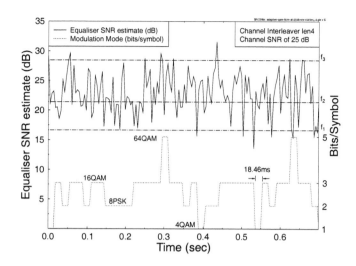

Figure 5.42: Equaliser output SNR estimate and Bits/Symbol versus time plot for adaptive TTCM with 4 turbo iterations in **System II** at an average channel SNR of 25 dB, where the modulation mode switching is based upon the equalizer's output SNR which is compared to the switching thresholds f_n defined in Table 5.17. The duration of one TDMA/TDD frame is 4.615 ms. The TTCM mode is maintained for four frame durations, i.e. for 18.46 ms.

leaving depth for BICM scheme in our slow fading wideband channels.

As observe in Figure 5.45, we noticed that BICM-ID had the worst performance at low SNRs in each modulation mode compared to other coded modulation schemes. However, it exhibited a steep slope and therefore at high SNRs it approached the performance of TTCM scheme. The adaptive BICM-ID and TTCM schemes in the context of **System II** were compared in Figure 5.46. The adaptive TTCM exibited a better BPS throughput than adaptive BICM-ID since TTCM had a better performance in fixed modem modes at BER of 0.01 %. However, adaptive BICM-ID exibited a lower BER performance than adaptive TTCM due to the high steepness of BICM-ID in fixed modem modes.

5.11.6 Summary and Conclusions

In this section BbB adaptive TCM and TTCM were proposed for wideband fading channels both with and without channel interleaving and they were characterized in performance terms over the COST 207 TU fading channel. When observing the associated BPS curves, adaptive TTCM exhibited up to 2.5 dB SNR-gain for a channel interleaver length of 4 bursts in comparison to the non-interleaved scenario, as evidenced in Figure 5.40. Upon comparing the BPS curves, adaptive TTCM also exhibited up to 0.7 dB SNR-gain compared to adaptive TCM of the same complexity in the context of **System II** for a target BER of less than 0.01 %, as shown in Figure 5.44. Lastly, adaptive TCM performed better than the adaptive BICM benchmarker in **System I** and the adaptive BICM-ID was marginally worse, than adaptive TTCM in **System II** as discussed in Section 5.11.5. Our future work will consider

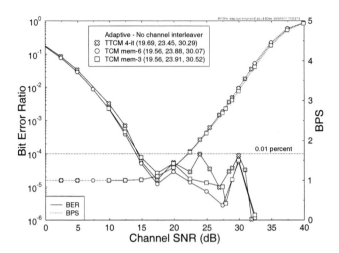

Figure 5.43: BER and BPS performance of adaptive TCM and TTCM without channel interleaving in **System I**, over the Rayleigh fading COST207 TU channel of Figure 5.33. The switching mechanism was characterized by Equation 5.18. The switching thresholds were set experimentally, in order to achieve a BER of below 0.01%, as shown in Table 5.17.

space-time coded BbB adaptive schemes.

5.12   Review and Discussion

In this chapter, we have demonstrated the benefits of turbo channel coding in conjunction with AQAM and channel equalization. In our turbo-coded AQAM schemes we have exploited the error correction capability of turbo coding in order to improve the coded BPS and BER performance, where the gains were recorded in Tables 5.11, 5.12 and 5.13 for different turbo block coded AQAM schemes, which are summarized as follows:

1. **FCFI-TBCH-AQAM** : In this scheme, a fixed turbo interleaver of size 9984 bits and a fixed coding rate of 0.7222 was implemented in conjunction with the switching regime of Equation 5.1.

2. **FCVI-TBCH-AQAM** : The code rate was fixed to 0.7222 and the turbo interleaver was varied according to Table 5.3 for each modulation mode, in order to ensure burst by burst decoding at the receiver.

3. **P-TBCH-AQAM** : Un-coded modes were added to the switching regime described by Equation 5.7, in order to increase the system's throughput. The code rate was set to 0.7222 and the turbo interleaver size was varied according to the size of the transmission burst, as shown in Table 5.5.

4. **VR-TBCH-AQAM** : In this final scheme, the code rate and turbo interleaver size were varied according to Table 5.9 and 5.7, respectively, for each modulation mode.

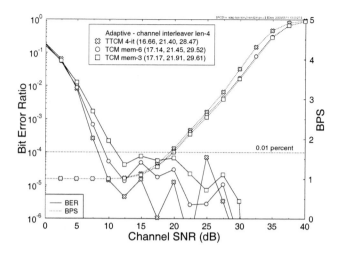

Figure 5.44: BER and BPS performance of adaptive TCM and TTCM using a channel interleaver
length of 4 bursts, in **System II** over the Rayleigh fading COST207 TU channel of Figure 5.33. The switching mechanism was characterized by Equation 5.18. The switching
thresholds were set experimentally, in order to achieve a BER of below 0.01%, as shown
in Table 5.17.

The **FCVI-TBCH-AQAM** scheme demonstrated significant throughput gains, when compared to the individual fixed modulation modes, as evidenced by Figure 5.11. In these comparisons there were minimal SNR gains for the **High-BER** schemes, while the **Low-BER**
FCVI-TBCH-AQAM scheme provided SNR gains of 5.0dB and 1.0dB, when compared to
the small and large channel interleaved based fixed modulation modes, respectively. As a
result of the employment of the channel interleavers, a high transmission delay was incurred
by the constituent fixed modulation modes. By contrast, the **FCVI-TBCH-AQAM** scheme
employed low-latency burst-by-burst decoding, which might benefit real-time applications,
such as video transmission [189, 190].

These four turbo block coded AQAM schemes were also compared and contrasted in
terms of their BER/SNR gains, BPS/SNR gains and the relative complexities of the schemes,
when compared to the uncoded AQAM scheme having target BERs of 1% and 0.01%. These
complexity comparisons were listed in Table 5.10 for all the coded schemes, where the **VR-
TBCH-AQAM** scheme exhibited the highest channel decoder complexity as a result of the
utilization of more complex BCH codes for this scheme. The complexity of the **P-TBCH-
AQAM** scheme also increased due to its higher number of transmission modes. The gains
achieved by the **Low-** and **High-BER** coded schemes are listed in Table 5.11 and 5.12, respectively. The gains achieved when the schemes were experimentally optimized for the sake
of an creating a near-error-free system was tabulated in Table 5.13. The variation and trends
of these gains were linked to the size of the turbo interleaver, the coding rate, the transmission
frame size and the targeted BER, as explained in Section 5.8.

In terms of the error correction aspects of these turbo block coded schemes, we have observed the trade-offs involving the BER, BPS throughput, and the complexity of each scheme.

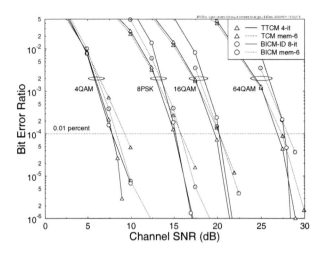

Figure 5.45: BER performance of the fixed modem modes of 4QAM, 8PSK, 16QAM and 64QAM
utilizing TCM, TTCM, BICM and BICM-ID schemes in the context of **System II**. For the
sake of maintaining a similar decoding complexity, we used a TCM code memory of 6,
TTCM code memory of 3 with 4 turbo iterations, BICM code memory of 6 and a BICM-
ID code memory of 3 with 8 decoding iterations. However, BICM-ID had a slightly higher
complexity than the other systems, since the demodulator module was invoked 8 times as
compared to only once for its counterparts during each decoding process.

With the exception of the **P-TBCH-AQAM** scheme, the BER/SNR and BPS/SNR gains were
evident at low to medium channel SNRs. However, the BPS throughput gain degraded at
high channel SNRs due to the associated coding rate limitations. In order to increase the
BPS throughput, the **P-TBCH-AQAM** and **VR-TBCH-AQAM** schemes were designed with
added complexity. In comparing these two schemes, the **VR-TBCH-AQAM** provided a better BER/SNR versus BPS/SNR gain trade-off in conjunction with a more complex channel
decoder. However, it is important to note that all these schemes achieved the targeted BER.

The size of the turbo interleaver was also crucial to the BER and BPS performance, where
in the variable turbo interleaver size assisted schemes the gains were significantly lower than
those of the **FCFI-TBCH-AQAM** scheme, when a large turbo interleaver size was used.
Since the variable turbo interleaver sizes were chosen for supporting burst-by-burst channel
decoding, a smaller speech frame shown in Figure 4.13 resulted in a smaller turbo interleaver.
The implementation of burst-by-burst decoding also provided near-instantaneous error detection capability at the receiver, which was exploited in blindly detecting the modulation modes.

For the turbo block coded AQAM scheme of Section 5.4, which utilized large fixed-sized
turbo interleavers, we have observed that the scheme, which targeted a lower BER displayed
a better BER/SNR and BPS/SNR gain, as evidenced by the gains recorded in Tables 5.11
and 5.12. This was as a result of the superior coding gain at lower BERs, which was also
observed for the fixed modulation modes of Section 5.3. As a result of the higher coding gain,
the coded switching thresholds of the turbo coded AQAM schemes were lowered resulting in
higher SNR gains, when compared to the uncoded AQAM schemes for similar target BERs.

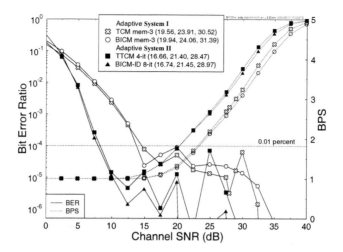

Figure 5.46: BER and BPS performance of the adaptive TCM/BICM **System I**, using memory 3 codes and that of the adaptive TTCM/BICM-ID **System II**, over the Rayleigh fading COST207 TU channel of Figure 5.33. The switching mechanism was characterized by Equation 5.18. The switching thresholds were set experimentally, in order to achieve a BER of below 0.01%, as shown in Table 5.17.

The channel codes were also exploited in order to detect the modulation modes, where the so-called hybrid SD-MSE modulation detection algorithm of Section 5.6.2 was implemented and its performance was shown in Figure 5.16. This hybrid algorithm, which was characterized by Equation 5.6 utilized the MSE algorithm in order to detect the BPSK mode, while the other modes were detected using the SD algorithm of Section 5.6.1. In this respect, a specific transmission frame structure was introduced for the NOTX mode, where a known m-sequence was transmitted in order to estimate the channel quality. Furthermore the unique correlation properties of the m-sequence - described by Equation 5.3 - supported a detection scheme at the receiver as discussed in Section 5.6.1. However, this modulation detection scheme incurred a high complexity as a result of the utilization of the channel decoders for detecting the possible modulation modes.

Convolutional codes were then utilized as the component codes for the turbo codec in Section 5.9. In comparing its performance to that of the turbo block coded schemes, a slight SNR degradation of $1 - 2$dB was observed for the fixed modulation modes, as evidenced by Figure 5.23. However, the computational complexity incurred by the turbo block coded schemes was higher by a factor of approximately seven when compared to the turbo convolutional schemes. In implementing the turbo convolutional coded AQAM schemes, the performance was similar to that of the turbo block coded AQAM schemes at a reduced complexity.

The concept of iterative channel equalization and decoding, termed as turbo equalization was then introduced in Section 5.10, where gains of approximately $0.7 - 2.0$dB were observed for the modulation modes of BPSK, 4QAM and 16QAM. Subsequently, an iterative LMS-based CIR estimation technique was proposed, which exploited the iterative nature of this scheme. The CIR estimation based performance approached that of the perfect CIR estimation based performance, as evidenced by Figure 5.29. The turbo equalizer was then implemented in an AQAM scheme, which yielded a gain of approximately $1.0 - 2.0$dB, when compared to the fixed modulation modes.

The application of turbo coding in a wideband AQAM scheme resulted in substantial performance gains, when compared to the fixed modulation modes and to the uncoded wideband AQAM scheme at a certain targeted BER. Furthermore, with the implementation of burst-by-burst decoding at the receiver the error detection capability of the channel codec was exploited, in order to detect the modulation modes as well as to potentially provide channel quality estimates. However, the complexity incurred in these turbo coded AQAM schemes was high due to the iterative regime of the channel decoding process. In the context of turbo equalization the complexity of this scheme increased exponentially, when higher-order modulation modes were used or, when channels exhibiting a long memory were encountered. This severely hindered the implementation of such AQAM schemes.

In this chapter, the performance of the wideband AQAM scheme based on the assumptions listed in Section 4.3.1 was investigated. However, in order to invoke a more practical wideband AQAM scheme, these assumptions are discarded and the resulting performance is analysed in the next chapter. Furthermore, the impact of co-channel interference is also considered, which will form the core of our next chapter.

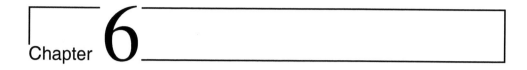

Chapter 6

Adaptive Modulation Mode Switching Optimization

B.J. Choi, L. Hanzo

6.1 Introduction

Mobile communications channels typically exhibit time-variant channel quality fluctuations [13] and hence conventional fixed-mode modems suffer from bursts of transmission errors, even if the system was designed to provide a high link margin. As argued throughout this monograph, an efficient approach of mitigating these detrimental effects is to adaptively adjust the transmission format based on the near-instantaneous channel quality information perceived by the receiver, which is fed back to the transmitter with the aid of a feedback channel [15]. This scheme requires a reliable feedback link from the receiver to the transmitter and the channel quality variation should be sufficiently slow for the transmitter to be able to adapt. *Hayes* [15] proposed transmission power adaptation, while *Cavers* [9] suggested invoking a variable symbol duration scheme in response to the perceived channel quality at the expense of a variable bandwidth requirement. Since a variable-power scheme increases both the average transmitted power requirements and the level of co-channel interference [17] imposed on other users of the system, instead variable-rate Adaptive Quadrature Amplitude Modulation (AQAM) was proposed by *Steele* and *Webb* as an alternative, employing various star-QAM constellations [16, 17]. With the advent of Pilot Symbol Assisted Modulation (PSAM) [18–20], *Otsuki et al.* [21] employed square constellations instead of star constellations in the context of AQAM, as a practical fading counter measure. Analyzing the channel capacity of Rayleigh fading channels [22–24], *Goldsmith et al.* showed that variable-power, variable-rate adaptive schemes are optimum, approaching the capacity of the channel and characterized the throughput performance of variable-power AQAM [23] . However, they also found that the extra throughput achieved by the additional variable-power assisted adap-

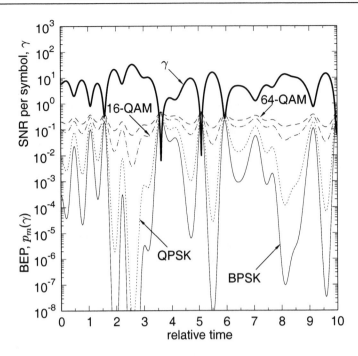

Figure 6.1: Instantaneous SNR per transmitted symbol, γ, in a flat Rayleigh fading scenario and the associated instantaneous bit error probability, $p_m(\gamma)$, of a fixed-mode QAM. The average SNR is $\bar{\gamma} = 10$dB. The fading magnitude plot is based on a normalized Doppler frequency of $f_N = 10^{-4}$ and for the duration of $100ms$, corresponding to a mobile terminal travelling at the speed of $54km/h$ and operating at $f_c = 2GHz$ frequency band at the sampling rate of $1MHz$.

tation over the constant-power, variable-rate scheme is marginal for most types of fading channels [23, 25].

6.2 Increasing the Average Transmit Power as a Fading Counter-Measure

The radio frequency (RF) signal radiated from the transmitter's antenna takes different routes, experiencing defraction, scattering and reflections, before it arrives at the receiver. Each multi-path component arriving at the receiver simultaneously adds constructively or destructively, resulting in fading of the combined signal. When there is no line-of-sight component amongst these signals, the combined signal is characterized by Rayleigh fading. The instantaneous SNR (iSNR), γ, per transmitted symbol[1] is depicted in Figure 6.1 for a typical Rayleigh fading using the thick line. The Probability Density Function (PDF) of γ is given

---

[1]When no diversity is employed at the receiver, the SNR per symbol, γ, is the same as the channel SNR, γ_c. In this case, we will use the term "SNR" without any adjective.

as [87]:

$$f_{\bar{\gamma}}(\gamma) = \frac{1}{\bar{\gamma}} e^{\gamma/\bar{\gamma}} , \qquad (6.1)$$

where $\bar{\gamma}$ is the average SNR and $\bar{\gamma} = 10$dB was used in Figure 6.1.

The instantaneous Bit Error Probability (iBEP), $p_m(\gamma)$, of BPSK, QPSK, 16-QAM and 64-QAM is also shown in Figure 6.1 with the aid of four different thin lines. These probabilities are obtained from the corresponding bit error probability over AWGN channel conditioned on the iSNR, γ, which are given as [4]:

$$p_m(\gamma) = \sum_i A_i Q(\sqrt{a_i \gamma}) , \qquad (6.2)$$

where $Q(x)$ is the Gaussian Q-function defined as $Q(x) \triangleq \frac{1}{\sqrt{2\pi}} \int_x^\infty e^{-t^2/2} dt$ and $\{A_i, a_i\}$ is a set of modulation mode dependent constants. For the Gray-mapped square QAM modulation modes associated with $m = 2, 4, 16, 64$ and 256, the sets $\{A_i, a_i\}$ are given as [4, 191]:

$$
\begin{array}{lll}
m = 2, & \text{BPSK} & \{(1,2)\} \\
m = 4, & \text{QPSK} & \{(1,1)\} \\
m = 16, & \text{16-QAM} & \left\{ \left(\frac{3}{4}, \frac{1^2}{5}\right), \left(\frac{2}{4}, \frac{3^2}{5}\right), \left(-\frac{1}{4}, \frac{5^2}{5}\right) \right\} \\
m = 64, & \text{64-QAM} & \left\{ \left(\frac{7}{12}, \frac{1^2}{21}\right), \left(\frac{6}{12}, \frac{3^2}{21}\right), \left(-\frac{1}{12}, \frac{5^2}{21}\right), \left(\frac{1}{12}, \frac{9^2}{21}\right), \left(-\frac{1}{12}, \frac{13^2}{21}\right) \right\} \\
m = 256, & \text{256-QAM} & \left\{ \left(\frac{15}{32}, \frac{1^2}{85}\right), \left(\frac{14}{32}, \frac{3^2}{85}\right), \left(\frac{5}{32}, \frac{5^2}{85}\right), \left(-\frac{6}{32}, \frac{7^2}{85}\right), \left(-\frac{7}{32}, \frac{9^2}{85}\right), \right. \\
& & \left. \left(\frac{6}{32}, \frac{11^2}{85}\right), \left(\frac{9}{32}, \frac{13^2}{85}\right), \left(\frac{8}{32}, \frac{15^2}{85}\right), \left(-\frac{7}{32}, \frac{17^2}{85}\right), \left(-\frac{6}{32}, \frac{19^2}{85}\right), \right. \\
& & \left. \left(-\frac{1}{32}, \frac{21^2}{85}\right), \left(\frac{2}{32}, \frac{23^2}{85}\right), \left(\frac{3}{32}, \frac{25^2}{85}\right), \left(-\frac{2}{32}, \frac{27^2}{85}\right), \left(-\frac{1}{32}, \frac{29^2}{85}\right) \right\} .
\end{array}
$$
$$(6.3)$$

As we can observe in Figure 6.1, $p_m(\gamma)$ exhibits high values during the deep channel envelope fades, where even the most robust modulation mode, namely BPSK, exhibits a bit error probability $p_2(\gamma) > 10^{-1}$. By contrast even the error probability of the high-throughput 16-QAM mode, namely $p_{16}(\gamma)$, is below 10^{-2}, when the iSNR γ exhibits a high peak. This wide variation of the communication link's quality is a fundamental problem in wireless radio communication systems. Hence, numerous techniques have been developed for combating this problem, such as increasing the average transmit power, invoking diversity, channel inversion, channel coding and/or adaptive modulation techniques. In this section we will investigate the efficiency of employing an increased average transmit power.

As we observed in Figure 6.1, the instantaneous Bit Error Probability (BEP) becomes excessive for sustaining an adequate service quality during instances, when the signal experiences a deep channel envelope fade. Let us define the cut-off BEP p_c, below which the Quality Of Service (QOS) becomes unacceptable. Then the outage probability P_{out} can be defined as:

$$P_{out}(\bar{\gamma}, p_c) \triangleq \Pr[p_m(\gamma) > p_c] , \qquad (6.4)$$

where $\bar{\gamma}$ is the average channel SNR dependent on the transmit power, p_c is the cut-off BEP and $p_m(\gamma)$ is the instantaneous BEP, conditioned on γ, for an m-ary modulation mode, given

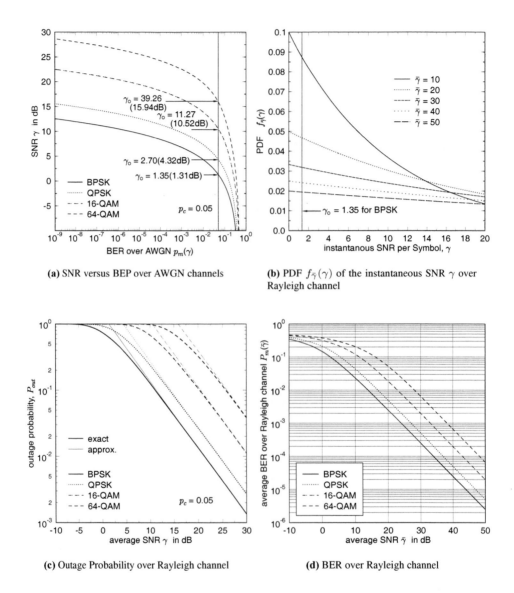

(a) SNR versus BEP over AWGN channels

(b) PDF $f_{\bar{\gamma}}(\gamma)$ of the instantaneous SNR γ over Rayleigh channel

(c) Outage Probability over Rayleigh channel

(d) BER over Rayleigh channel

Figure 6.2: The effects of an increased average transmit power. (a) The cut-off SNR γ_o versus the cut-off BEP p_c for BPSK, QPSK, 16-QAM and 64-QAM. (b) PDF of the iSNR γ over Rayleigh channel, where the outage probability is given by the area under the PDF curve surrounded by the two lines given by $\gamma = 0$ and $\gamma = \gamma_o$. An increased transmit power increases the average SNR $\bar{\gamma}$ and hence reduces the area under the PDF proportionately to $\bar{\gamma}$. (c) The exact outage probability versus the average SNR $\bar{\gamma}$ for BPSK, QPSK, 16-QAM and 64-QAM evaluated from (6.7) confirms this observation. (d) The average BEP is also inversely proportional to the transmit power for BPSK, QPSK, 16-QAM and 64-QAM.

for example by (6.2). We can reduce the outage probability of (6.4) by increasing the transmit power, and hence increasing the average channel SNR $\bar{\gamma}$. Let us briefly investigate the efficiency of this scheme.

Figure 6.2(a) depicts the instantaneous BEP as a function of the instantaneous channel SNR. Once the cut-off BEP p_c is determined as a QOS-related design parameter, the corresponding cut-off SNR γ_o can be determined, as shown for example in Figure 6.2(a) for $p_c = 0.05$. Then, the outage probability of (6.4) can be calculated as:

$$P_{out} = \Pr[\gamma < \gamma_o] \,, \tag{6.5}$$

and in physically tangible terms its value is equal to the area under the PDF curve of Figure 6.2(b) surrounded by the left y-axis and $\gamma = \gamma_o$ vertical line. Upon taking into account that for high SNRs the PDFs of Figure 6.2(b) are near-linear, this area can be approximated by $\gamma_o/\bar{\gamma}$, considering that $f_{\bar{\gamma}}(0) = 1/\bar{\gamma}$. Hence, the outage probability is inversely proportional to the transmit power, requiring an approximately 10-fold increased transmit power for reducing the outage probability by an order of magnitude, as seen in Figure 6.2(c). The exact value of the outage probability is given by:

$$P_{out} = \int_0^{\gamma_o} f_{\bar{\gamma}}(\gamma) \, d\gamma \tag{6.6}$$

$$= 1 - e^{-\gamma_o/\bar{\gamma}} \,, \tag{6.7}$$

where we used the PDF $f_{\bar{\gamma}}(\gamma)$ given in (6.1). Again, Figure 6.2(c) shows the exact outage probabilities together with their linearly approximated values for several QAM modems recorded for the cut-off BEP of $p_c = 0.05$, where we can confirm the validity of the linearly approximated outage probability[2], when we have $P_{out} < 0.1$.

The average BEP $P_m(\bar{\gamma})$ of an m-ary Gray-mapped QAM modem is given by [4,87,192]:

$$P_m(\bar{\gamma}) = \int_0^{\infty} p_m(\gamma) f_{\bar{\gamma}}(\gamma) \, d\gamma \tag{6.8}$$

$$= \frac{1}{2} \sum_i A_i \{1 - \mu(\bar{\gamma}, a_i)\} \,, \tag{6.9}$$

where a set of constants $\{A_i, a_i\}$ is given in (6.3) and $\mu(\bar{\gamma}, a_i)$ is defined as:

$$\mu(\bar{\gamma}, a_i) \triangleq \sqrt{\frac{a_i \bar{\gamma}}{1 + a_i \bar{\gamma}}} \,. \tag{6.10}$$

In physical terms (6.8) implies weighting the BEP $p_m(\gamma)$ experienced at an iSNR γ by the probability of occurrence of this particular value of γ - which is quantified by its PDF $f_{\bar{\gamma}}(\gamma)$ - and then averaging, *i.e.* integrating, this weighted BEP over the entire range of γ. Figure 6.2(d) displays the average BER evaluated from (6.9) for the average SNR rage of -10dB $\geq \bar{\gamma} \geq 50$dB. We can observe that the average BEP is also inversely proportional to the transmit power.

---

[2]The same approximate outage probability can be derived by taking the first term of the Taylor series of e^x of (6.7).

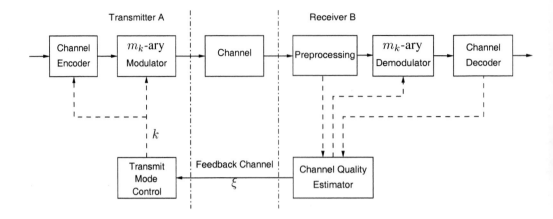

Figure 6.3: Stylised model of near-instantaneous adaptive modulation scheme.

In conclusion, we studied the efficiency of increasing the average transmit power as a fading counter-measure and found that the outage probability as well as the average bit error probability are inversely proportional to the average transmit power. Since the maximum radiated powers of modems are regulated in order to reduce the co-channel interference and transmit power, the acceptable transmit power increase may be limited and hence employing this technique may not be sufficiently effective for achieving the desired link performance. We will show that the AQAM philosophy of the next section is a more attractive solution to the problem of channel quality fluctuation experienced in wireless systems.

6.3   System Description

A stylised model of our adaptive modulation scheme is illustrated in Figure 6.3, which can be invoked in conjunction with any power control scheme. In our adaptive modulation scheme, the modulation mode used is adapted on a near-instantaneous basis for the sake of counteracting the effects of fading. Let us describe the detailed operation of the adaptive modem scheme of Figure 6.3. Firstly, the channel quality ξ is estimated by the remote receiver B. This channel quality measure ξ can be the instantaneous channel SNR, the Radio Signal Strength Indicator (RSSI) output of the receiver [17], the decoded BER [17], the Signal to Interference-and-Noise Ratio (SINR) estimated at the output of the channel equalizer [33], or the SINR at the output of a CDMA joint detector [193]. The estimated channel quality perceived by receiver B is fed back to transmitter A with the aid of a feedback channel, as seen in Figure 6.3. Then, the transmit mode control block of transmitter A selects the highest-throughput modulation mode k capable of maintaining the target BEP based on the channel quality measure ξ and the specific set of adaptive mode switching levels s. Once k is selected, m_k-ary modulation is performed at transmitter A in order to generate the transmitted signal $s(t)$, and the signal $s(t)$ is transmitted through the channel.

The general model and the set of important parameters specifying our constant-power adaptive modulation scheme are described in the next subsection in order to develop the underlying general theory. Then, in Subsection 6.3.2 several application examples are introduced.

6.3.1 General Model

A K-mode adaptive modulation scheme adjusts its transmit mode k, where $k \in \{0, 1 \cdots K-1\}$, by employing m_k-ary modulation according to the near-instantaneous channel quality ξ perceived by receiver B of Figure 6.3. The mode selection rule is given by:

$$\text{Choose mode } k \text{ when } s_k \leq \xi < s_{k+1} , \qquad (6.11)$$

where a switching level s_k belongs to the set $\mathbf{s} = \{s_k \mid k = 0, 1, \cdots, K\}$. The Bits Per Symbol (BPS) throughput b_k of a specific modulation mode k is given by $b_k = \log_2(m_k)$ if $m_k \neq 0$, otherwise $b_k = 0$. It is convenient to define the incremental BPS c_k as $c_k = b_k - b_{k-1}$, when $k > 0$ and $c_0 = b_0$, which quantifies the achievable BPS increase, when switching from the lower-throughput mode $k-1$ to mode k.

6.3.2 Examples

6.3.2.1 Five-Mode AQAM

A five-mode AQAM system has been studied extensively by many researchers, which was motivated by the high performance of the Gray-mapped constituent modulation modes used. The parameters of this five-mode AQAM system are summarised in Table 6.1. In our inves-

k	0	1	2	3	4
m_k	0	2	4	16	64
b_k	0	1	2	4	6
c_k	0	1	1	2	2
modem	No Tx	BPSK	QPSK	16-QAM	64-QAM

Table 6.1: The parameters of five-mode AQAM system.

tigation, the near-instantaneous channel quality ξ is defined as instantaneous channel SNR γ. The boundary switching levels are given as $s_0 = 0$ and $s_5 = \infty$. Figure 6.4 illustrates operation of the five-mode AQAM scheme over a typical narrow-band Rayleigh fading channel scenario. Transmitter A of Figure 6.3 keeps track of the channel SNR γ perceived by receiver B with the aid of a low-BER, low-delay feedback channel - which can be created for example by superimposing the values of ξ on the reverse direction transmitted messages of transmitter B - and determines the highest-BPS modulation mode maintaining the target BEP depending on which region γ falls into. The channel-quality related SNR regions are divided by the modulation mode switching levels s_k. More explicitly, the set of AQAM switching levels $\{s_k\}$ is determined such that the average BPS throughput is maximised, while satisfying the average target BEP requirement, P_{target}. We assumed a target BEP of $P_{target} = 10^{-2}$ in Figure 6.4. The associated instantaneous BPS throughput b is also depicted using the thick stepped line at the bottom of Figure 6.4. We can observe that the throughput varied from

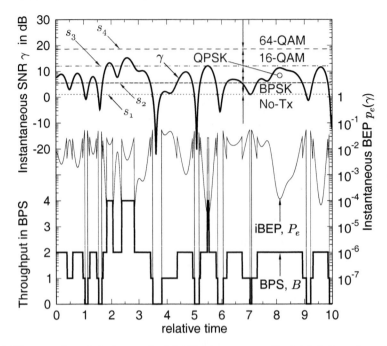

Figure 6.4: The operation of the five-mode AQAM scheme over a Rayleigh fading channel. The instantaneous channel SNR γ is represented as a thick line at the top part of the graph, the associated instantaneous BEP $P_e(\gamma)$ as a thin line at the middle, and the instantaneous BPS throughput $b(\gamma)$ as a thick line at the bottom. The average SNR is $\bar{\gamma} = 10\text{dB}$, while the target BEP is $P_{target} = 10^{-2}$.

0 BPS, when the no transmission (No-Tx) QAM mode was chosen, to 4 BPS, when the 16-QAM mode was activated. During the depicted observation window the 64-QAM mode was not activated. The instantaneous BEP, depicted as a thin line using the middle trace of Figure 6.4, is concentrated around the target BER of $P_{target} = 10^{-2}$.

6.3.2.2   Seven-Mode Adaptive Star-QAM

Webb and Steele revived the research community's interest on adaptive modulation, although a similar concept was initially suggested by Hayes [15] in the 1960s. Webb and Steele reported the performance of adaptive star-QAM systems [17]. The parameters of their system are summarised in Table 6.2.

6.3.2.3   Five-Mode APSK

Our five-mode Adaptive Phase-Shift-Keying (APSK) system employs m-ary PSK constituent modulation modes. The magnitude of all the constituent constellations remained constant, where adaptive modem parameters are summarised in Table 6.3.

k	0	1	2	3	4	5	6
m_k	0	2	4	8	16	32	64
b_k	0	1	2	3	4	5	6
c_k	0	1	1	1	1	1	1
modem	No Tx	BPSK	QPSK	8-QAM	16-QAM	32-QAM	64-QAM

Table 6.2: The parameters of a seven-mode adaptive star-QAM system [17], where 8-QAM and 16-QAM employed four and eight constellation points allocated to two concentric rings, respectively, while 32-QAM and 64-QAM employed eight and 16 constellation points over four concentric rings, respectively.

k	0	1	2	3	4
m_k	0	2	4	8	16
b_k	0	1	2	3	4
c_k	0	1	1	1	1
modem	No Tx	BPSK	QPSK	8-PSK	16-PSK

Table 6.3: The parameters of the five-mode APSK system.

6.3.2.4   Ten-Mode AQAM

Hole, Holm and Øien [50] studied a trellis coded adaptive modulation scheme based on eight-mode square- and cross-QAM schemes. Upon adding the No-Tx and BPSK modes, we arrive at a ten-mode AQAM scheme. The associated parameters are summarised in Table 6.4.

k	0	1	2	3	4	5	6	7	8	9
m_k	0	2	4	8	16	32	64	128	256	512
b_k	0	1	2	3	4	5	6	7	8	9
c_k	0	1	1	1	1	1	1	1	1	1
modem	No Tx	BPSK	QPSK	8-Q	16-Q	32-C	64-Q	128-C	256-Q	512-C

Table 6.4: The parameters of the ten-mode adaptive QAM scheme based on [50], where m-Q stands for m-ary square QAM and m-C for m-ary cross QAM.

6.3.3   Characteristic Parameters

In this section, we introduce several parameters in order to characterize our adaptive modulation scheme. The constituent mode selection probability (MSP) $\mathcal{M}_k$ is defined as the probability of selecting the k-th mode from the set of K possible modulation modes, which can be calculated as a function of the channel quality metric ξ, regardless of the specific metric used, as:

$$\mathcal{M}_k = \Pr[s_k \leq \xi < s_{k+1}] \tag{6.12}$$

$$= \int_{s_k}^{s_{k+1}} f(\xi) \, d\xi \,, \tag{6.13}$$

where s_k denotes the mode switching levels and $f(\xi)$ is the probability density function (PDF) of ξ. Then, the average throughput B expressed in terms of BPS can be described as:

$$B = \sum_{k=0}^{K-1} b_k \int_{s_k}^{s_{k+1}} f(\xi) \, d\xi \tag{6.14}$$

$$= \sum_{k=0}^{K-1} b_k \, \mathcal{M}_k \,, \tag{6.15}$$

which in simple verbal terms can be formulated as the weighted sum of the throughput b_k of the individual constituent modes, where the weighting takes into account the probability $\mathcal{M}_k$ of activating the various constituent modes. When $s_K = \infty$, the average throughput B can also be formulated as:

$$B = \sum_{k=0}^{K-1} b_k \int_{s_k}^{s_{k+1}} f(\xi) \, d\xi \tag{6.16}$$

$$= \sum_{k=0}^{K-1} c_k \int_{s_k}^{\infty} f(\xi) \, d\xi \tag{6.17}$$

$$= \sum_{k=0}^{K-1} c_k \, F_c(s_k), \tag{6.18}$$

where $F_c(\xi)$ is the complementary Cumulative Distribution Function (CDF) defined as:

$$F_c(\xi) \triangleq \int_{\xi}^{\infty} f(x) \, dx \,. \tag{6.19}$$

Let us now assume that we use the instantaneous SNR γ as the channel quality measure ξ, which implies that no co-channel interference is present. By contrast, when operating in a co-channel interference limited environment, we can use the instantaneous SINR as the channel quality measure ξ, provided that the co-channel interference has a near-Gaussian distribution. In such scenario, the mode-specific average BEP P_k can be written as:

$$P_k = \int_{s_k}^{s_{k+1}} p_{m_k}(\gamma) f(\gamma) \, d\gamma \,, \tag{6.20}$$

where $p_{m_k}(\gamma)$ is the BEP of the m_k-ary constituent modulation mode over the AWGN channel and we used γ instead of ξ in order to explicitly indicate the employment of γ as the channel quality measure. Then, the average BEP P_{avg} of our adaptive modulation scheme can be represented as the sum of the BEPs of the specific constituent modes divided by the average adaptive modem throughput B, formulated as [31]:

$$P_{avg} = \frac{1}{B} \sum_{k=0}^{K-1} b_k \, P_k \,, \qquad (6.21)$$

where b_k is the BPS throughput of the k-th modulation mode, P_k is the mode-specific average BEP given in (6.20) and B is the average adaptive modem throughput given in (6.15) or in (6.18).

The aim of our adaptive system is to transmit as high a number of bits per symbol as possible, while providing the required Quality of Service (QOS). More specifically, we are aiming for maximizing the average BPS throughput B of (6.14), while satisfying the average BEP requirement of $P_{avg} \le P_{target}$. Hence, we have to satisfy the constraint of meeting P_{target}, while optimizing the design parameter of s, which is the set of modulation-mode switching levels. The determination of optimum switching levels will be investigated in Section 6.4. Since the calculation of the optimum switching levels typically requires the numerical computation of the parameters introduced in this section, it is advantageous to express the parameters in a closed form, which is the objective of the next section.

6.3.3.1 Closed Form Expressions for Transmission over Nakagami Fading Channels

Fading channels often are modelled as Nakagami fading channels [194]. The PDF of the instantaneous channel SNR γ over a Nakagami fading channel is given as [194]:

$$f(\gamma) = \left(\frac{m}{\bar{\gamma}} \right)^m \frac{\gamma^{m-1}}{\Gamma(m)} \, e^{-m\gamma/\bar{\gamma}} \,, \quad \gamma \ge 0 \,, \qquad (6.22)$$

where the parameter m governs the severity of fading and $\Gamma(m)$ is the Gamma function [90]. When $m = 1$, the PDF of (6.22) is reduced to the PDF of γ over Rayleigh fading channel, which is given in (6.1). As m increases, the fading behaves like Rician fading, and it becomes the AWGN channel, when m tends to ∞. Here we restrict the value of m to be a positive integer. In this case, the Nakagami fading model of (6.22), having a mean of $\bar{\gamma}_s = m\,\bar{\gamma}$, will be used to describe the PDF of the SNR per symbol γ_s in an m-antenna based diversity assisted system employing Maximal Ratio Combining (MRC).

When the instantaneous channel SNR γ is used as the channel quality measure ξ in our adaptive modulation scheme transmitting over a Nakagami channel, the parameters defined in Section 6.3.3 can be expressed in a closed form. Specifically, the mode selection probability $\mathcal{M}_k$ can be expressed as:

$$\mathcal{M}_k = \int_{s_k}^{s_{k+1}} f(\gamma) \, d\gamma \qquad (6.23)$$

$$= F_c(s_k) - F_c(s_{k+1}) \,, \qquad (6.24)$$

where the complementary CDF $F_c(\gamma)$ is given by:

$$F_c(\gamma) = \int_{\gamma}^{\infty} f(x)\, dx \tag{6.25}$$

$$= \int_{\gamma}^{\infty} \left(\frac{m}{\bar{\gamma}}\right)^m \frac{x^{m-1}}{\Gamma(m)} e^{-mx/\bar{\gamma}}\, dx \tag{6.26}$$

$$= e^{-m\gamma/\bar{\gamma}} \sum_{i=0}^{m-1} \frac{(m\gamma/\bar{\gamma})^i}{\Gamma(i+1)} \,. \tag{6.27}$$

In deriving (6.27) we used the result of the indefinite integral of [195]:

$$\int x^n e^{-ax}\, dx = -(e^{-ax}/a) \sum_{i=0}^{n} x^{n-i}/a^i\, n!/(n-i)!) \,. \tag{6.28}$$

In a Rayleigh fading scenario, *i.e.* when $m = 1$, the mode selection probability $\mathcal{M}_k$ of (6.24) can be expressed as:

$$\mathcal{M}_k = e^{-s_k/\bar{\gamma}} - e^{-s_{k+1}/\bar{\gamma}} \,. \tag{6.29}$$

The average throughput B of our adaptive modulation scheme transmitting over a Nakagami channel is given by substituting (6.27) into (6.18), yielding:

$$B = \sum_{k=0}^{K-1} c_k\, e^{-ms_k/\bar{\gamma}} \left\{ \sum_{i=0}^{m-1} \frac{(ms_k/\bar{\gamma})^i}{\Gamma(i+1)} \right\} \,. \tag{6.30}$$

Let us now derive the closed form expressions for the mode specific average BEP P_k defined in (6.20) for the various modulation modes when communicating over a Nakagami channel. The BER of a Gray-coded square QAM constellation for transmission over AWGN channels was given in (6.2) and it is repeated here for convenience:

$$p_{m_k,QAM}(\gamma) = \sum_i A_i\, Q(\sqrt{a_i\gamma}) \,, \tag{6.31}$$

where the values of the constants A_i and a_i were given in (6.3). Then, the mode specific average BEP $P_{k,QAM}$ of m_k-ary QAM over a Nakagami channel can be expressed in Appendix A.6 as:

$$P_{k,QAM} = \int_{s_k}^{s_{k+1}} p_{m_k,QAM}(\gamma)\, f(\gamma)\, d\gamma \tag{6.32}$$

$$= \sum_i A_i \int_{s_k}^{s_{k+1}} Q(\sqrt{a_i\gamma}) \left(\frac{m}{\bar{\gamma}}\right)^m \frac{\gamma^{m-1}}{\Gamma(m)} e^{-m\gamma/\bar{\gamma}}\, d\gamma \tag{6.33}$$

$$= \sum_i A_i \left\{ -e^{-m\gamma/\bar{\gamma}} Q(\sqrt{a_i\gamma}) \sum_{j=0}^{m-1} \frac{(m\gamma/\bar{\gamma})^j}{\Gamma(j+1)} \right]_{s_k}^{s_{k+1}} + \sum_{j=0}^{m-1} X_j(\gamma, a_i) \right]_{s_k}^{s_{k+1}} \right\} , \tag{6.34}$$

where $g(\gamma)]_{s_k}^{s_{k+1}} \triangleq g(s_{k+1}) - g(s_k)$ and $X_j(\gamma, a_i)$ is given by:

$$X_j(\gamma, a_i) = \frac{\mu^2}{\sqrt{2a_i\pi}} \left(\frac{m}{\bar{\gamma}}\right)^j \frac{\Gamma(j + \frac{1}{2})}{\Gamma(j + 1)} \sum_{k=1}^j \left(\frac{2\mu^2}{a_i}\right)^{j-k} \frac{\gamma^{k-\frac{1}{2}}}{\Gamma(k + \frac{1}{2})} e^{-a_i\gamma/(2\mu^2)}$$

$$+ \left(\frac{2\mu^2 m}{a_i\bar{\gamma}}\right)^j \frac{1}{\sqrt{\pi}} \frac{\Gamma(j + \frac{1}{2})}{\Gamma(j + 1)} \mu Q\left(\sqrt{a_i\gamma}/\mu\right), \qquad (6.35)$$

where, again, $\mu \triangleq \sqrt{\frac{a_i\bar{\gamma}}{2 + a_i\bar{\gamma}}}$ and $\Gamma(x)$ is the Gamma function.

On the other hand, the high-accuracy approximated BEP formula of a Gray-coded m_k-ary PSK scheme ($k \geq 3$) transmitting over an AWGN channel is given as [196]:

$$p_{m_k, PSK} \simeq \frac{2}{k} \left\{ Q\left(\sqrt{2\gamma}\sin(\pi/2^k)\right) + Q\left(\sqrt{2\gamma}\sin(3\pi/2^k)\right) \right\} \qquad (6.36)$$

$$= \sum_i A_i Q(\sqrt{a_i\gamma}), \qquad (6.37)$$

where the set of constants $\{(A_i, a_i)\}$ is given by $\{(2/k, 2\sin^2(\pi/m_k)), (2/k, 2\sin^2(3\pi/m_k))\}$. Hence, the mode-specific average BEP $P_{k, PSK}$ can be represented using the same equation, namely (6.34), as for $P_{k, QAM}$.

6.4 Optimum Switching Levels

In this section we restrict our interest to adaptive modulation schemes employing the SNR per symbol γ as the channel quality measure ξ. We then derive the optimum switching levels as a function of the target BEP and illustrate the operation of the adaptive modulation scheme. The corresponding performance results of the adaptive modulation schemes communicating over a flat-fading Rayleigh channel are presented in order to demonstrate the effectiveness of the schemes.

6.4.1 Limiting the Peak Instantaneous BEP

The first attempt of finding the optimum switching levels that are capable of satisfying various transmission integrity requirements was made by Webb and Steele [17]. They used the BEP curves of each constituent modulation mode, obtained from simulations over an AWGN channel, in order to find the Signal-to-Noise Ratio (SNR) values, where each modulation mode satisfies the target BEP requirement [4]. This intuitive concept of determining the switching levels has been widely used by researchers [21,25] since then. The regime proposed by Webb and Steele can be used for ensuring that the instantaneous BEP always remains below a certain threshold BEP P_{th}. In order to satisfy this constraint, the first modulation mode should be "no transmission". In this case, the set of switching levels s is given by:

$$\mathbf{s} = \{ s_0 = 0, \; s_k \mid p_{m_k}(s_k) = P_{th} \; k \geq 1 \}. \qquad (6.38)$$

Figure 6.5 illustrates how this scheme operates over a Rayleigh channel, using the example of the five-mode AQAM scheme described in Section 6.3.2.1. The average SNR was $\bar{\gamma} = 10$dB.

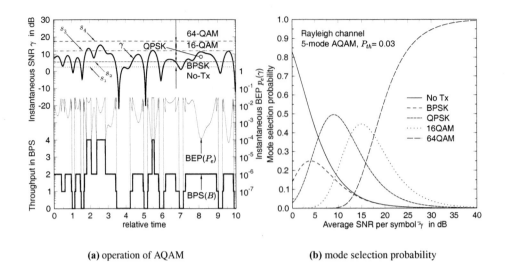

(a) operation of AQAM   **(b)** mode selection probability

Figure 6.5: Various characteristics of the five-mode AQAM scheme communicating over a Rayleigh fading channel employing the specific set of switching levels designed for limiting the peak instantaneous BEP to $P_{th} = 3 \times 10^{-2}$. (a) The evolution of the instantaneous channel SNR γ is represented by the thick line at the top of the graph, the associated instantaneous BEP $p_e(\gamma)$ by the thin line in the middle and the instantaneous BPS throughput $b(\gamma)$ by the thick line at the bottom. The average SNR is $\bar{\gamma} = 10$dB. (b) As the average SNR increases, the higher-order AQAM modes are selected more often.

and the instantaneous target BEP was $P_{th} = 3 \times 10^{-2}$. Using the expression given in (6.2) for p_{m_k}, the set of switching levels can be calculated for the instantaneous target BEP, which is given by $s_1 = 1.769$, $s_2 = 3.537$, $s_3 = 15.325$ and $s_4 = 55.874$. We can observe that the instantaneous BEP represented as a thin line by the middle of trace of Figure 6.5(a) was limited to values below $P_{th} = 3 \times 10^{-2}$.

At this particular average SNR predominantly the QPSK modulation mode was invoked. However, when the instantaneous channel quality is high, 16-QAM was invoked in order to increase the BPS throughput. The mode selection probability $\mathcal{M}_k$ of (6.24) is shown in Figure 6.5(b). Again, when the average SNR is $\bar{\gamma} = 10$dB, the QPSK mode is selected most often, namely with the probability of about 0.5. The 16-QAM, No-Tx and BPSK modes have had the mode selection probabilities of 0.15 to 0.2, while 64-QAM is not likely to be selected in this situation. When the average SNR increases, the next higher order modulation mode becomes the dominant modulation scheme one by one and eventually the highest order of 64-QAM mode of the five-mode AQAM scheme prevails.

The effects of the number of modulation modes used in our AQAM scheme on the performance are depicted in Figure 6.6. The average BEP performance portrayed in Figure 6.6(a) shows that the AQAM schemes maintain an average BEP lower than the peak instantaneous BEP of $P_{th} = 3 \times 10^{-2}$ even in the low SNR region, at the cost of a reduced average throughput, which can be observed in Figure 6.6(b). As the number of the constituent modulation modes employed of the AQAM increases, the SNR regions, where the average BEP is near

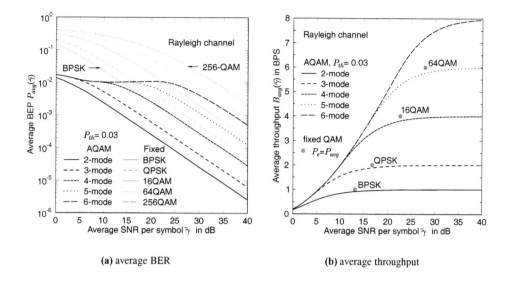

(a) average BER   (b) average throughput

Figure 6.6: The performance of AQAM employing the specific switching levels defined for limiting
the peak instantaneous BEP to $P_{th} = 0.03$. (a) As the number of constituent modulation
modes increases, the SNR region where the average BEP remains around $P_{avg} = 10^{-2}$
widens. (b) The SNR gains of AQAM over the fixed-mode QAM scheme required for
achieving the same BPS throughput at the same average BEP of P_{avg} are in the range of
5dB to 8dB.

constant around $P_{avg} = 10^{-2}$ expands to higher average SNR values. We can observe that
the AQAM scheme maintains a constant SNR gain over the highest-order constituent fixed
QAM mode, as the average SNR increases, at the cost of a negligible BPS throughput degradation. This is because the AQAM activates the low-order modulation modes or disables
transmissions completely, when the channel envelope is in a deep fade, in order to avoid
inflicting bursts of bit errors.

Figure 6.6(b) compares the average BPS throughput of the AQAM scheme employing
various numbers of AQAM modes and those of the fixed QAM constituent modes achieving
the same average BER. When we want to achieve the target throughput of $B_{avg} = 1$ BPS using the AQAM scheme, Figure 6.6(b) suggest that 3-mode AQAM employing No-Tx, BPSK
and QPSK is as good as four-mode AQAM, or in fact any other AQAM schemes employing
more than four modes. In this case, the SNR gain achievable by AQAM is 7.7dB at the average BEP of $P_{avg} = 1.154 \times 10^{-2}$. For the average throughputs of $B_{avg} = 2$, 4 and 6, the
SNR gains of the 6-mode AQAM schemes over the fixed QAM schemes are 6.65dB, 5.82dB
and 5.12dB, respectively.

Figure 6.7 shows the performance of the six-mode AQAM scheme, which is an extended
version of the five-mode AQAM of Section 6.3.2.1, for the peak instantaneous BEP values
of $P_{th} = 10^{-1}, 10^{-2}, 10^{-3}, 10^{-4}$ and 10^{-5}. We can observe in Figure 6.7(a) that the corresponding average BER P_{avg} decreases as P_{th} decreases. The average throughput curves seen
in Figure 6.7(b) indicate that as anticipated the increased average SNR facilitates attaining
an increased throughput by the AQAM scheme and there is a clear design trade-off between

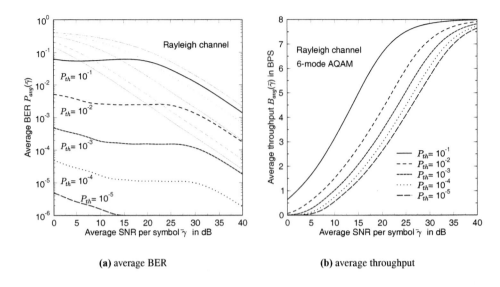

(a) average BER   (b) average throughput

Figure 6.7: The performance of the six-mode AQAM employing the switching levels of (6.38) designed for limiting the peak instantaneous BEP.

the achievable average throughput and the peak instantaneous BEP. This is because predominantly lower-throughput, but more error-resilient AQAM modes have to be activated, when the target BER is low. By contrast, higher-throughput but more error-sensitive AQAM modes are favoured, when the tolerable BEP is increased.

In conclusion, we introduced an adaptive modulation scheme, where the objective is to limit the peak instantaneous BEP. A set of switching levels designed for meeting this objective was given in (6.38), which is independent of the underlying fading channel and the average SNR. The corresponding average BEP and throughput formulae were derived in Section 6.3.3.1 and some performance characteristics of a range of AQAM schemes for transmitting over a flat Rayleigh channel were presented in order to demonstrate the effectiveness of the adaptive modulation scheme using the analysis technique developed in Section 6.3.3.1. The main advantage of this adaptive modulation scheme is in its simplicity regarding the design of the AQAM switching levels, while its drawback is that there is no direct relationship between the peak instantaneous BEP and the average BEP, which was used as our performance measure. In the next section a different switching-level optimization philosophy is introduced and contrasted with the approach of designing the switching levels for maintaining a given peak instantaneous BEP.

6.4.2 Torrance's Switching Levels

Torrance and Hanzo [26] proposed the employment of the following cost function and applied Powell's optimization method [29] for generating the optimum switching levels:

$$\Omega_T(\mathbf{s}) = \sum_{\bar{\gamma}=0\mathrm{dB}}^{40\mathrm{dB}} \left[10 \log_{10}(\max\{P_{avg}(\bar{\gamma};\mathbf{s})/P_{th}, 1\}) + B_{max} - B_{avg}(\bar{\gamma};\mathbf{s}) \right], \qquad (6.39)$$

where the average BEP P_{avg} is given in (6.21), $\bar{\gamma}$ is the average SNR per symbol, s is the set of switching levels, P_{th} is the target average BER, B_{max} is the BPS throughput of the highest order constituent modulation mode and the average throughput B_{avg} is given in (6.14). The idea behind employing the cost function Ω_T is that of maximizing the average throughput B_{avg}, while endeavouring to maintain the target average BEP P_{th}. Following the philosophy of Section 6.4.1, the minimization of the cost function of (6.39) produces a set of constant switching levels across the entire SNR range. However, since the calculation of P_{avg} and B_{avg} requires the knowledge of the PDF of the instantaneous SNR γ per symbol, in reality the set of switching levels s required for maintaining a constant P_{avg} is dependent on the channel encountered and the receiver structure used.

Figure 6.8 illustrates the operation of a five-mode AQAM scheme employing *Torrance's* SNR-independent switching levels designed for maintaining the target average BEP of $P_{th} = 10^{-2}$ over a flat Rayleigh channel. The average SNR was $\bar{\gamma} = 10$dB and the target average BEP was $P_{th} = 10^{-2}$. *Powell's* minimization [29] involved in the context of (6.39) provides the set of optimised switching levels, given by $s_1 = 2.367$, $s_2 = 4.055$, $s_3 = 15.050$ and $s_4 = 56.522$. Upon comparing Figure 6.8(a) to Figure 6.5(a) we find that the two schemes are nearly identical in terms of activating the various AQAM modes according to the channel envelope trace, while the peak instantaneous BEP associated with Torrance's switching scheme is not constant. This is in contrast to the constant peak instantaneous BEP values seen in Figure 6.5(a). The mode selection probabilities depicted in Figure 6.8(b) are similar to those seen in Figure 6.5(b).

The average BEP curves, depicted in Figure 6.9(a) show that *Torrance's* switching levels support the AQAM scheme in successfully maintaining the target average BEP of $P_{th} = 10^{-2}$ over the average SNR range of 0dB to 20dB, when five or six modem modes are employed by the AQAM scheme. Most of the AQAM studies found in the literature have applied *Torrance's* switching levels owing to the above mentioned good agreement between the design target P_{th} and the actual BEP performance P_{avg} [197].

Figure 6.9(b) compares the average throughputs of a range of AQAM schemes employing various numbers of AQAM modes to the average BPS throughput of fixed-mode QAM arrangements achieving the same average BEP, *i.e.* $P_e = P_{avg}$, which is not necessarily identical to the target BEP of $P_e = P_{th}$. Specifically, the SNR values required by the fixed mode scheme in order to achieve $P_e = P_{avg}$ are represented by the markers '⊗', while the SNRs, where the target average BEP of $P_e = P_{th}$ is achieved, is denoted by the markers '⊙'. Compared to the fixed QAM schemes achieving $P_e = P_{avg}$, the SNR gains of the AQAM scheme were 9.06dB, 7.02dB, 5.81dB and 8.74dB for the BPS throughput values of 1, 2, 4 and 6, respectively. By contrast, the corresponding SNR gains compared to the fixed QAM schemes achieving $P_e = P_{th}$ were 7.55dB, 6.26dB, 5.83dB and 1.45dB. We can observe that the SNR gain of the AQAM arrangement over the 64-QAM scheme achieving a BEP of $P_e = P_{th}$ is small compared to the SNR gains attained in comparison to the lower-throughput

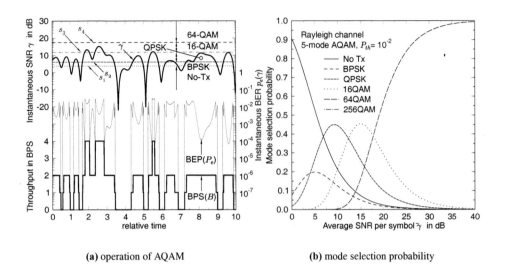

(a) operation of AQAM   (b) mode selection probability

Figure 6.8: Performance of the five-mode AQAM scheme over a flat Rayleigh fading channel employing the set of switching levels derived by Torrance and Hanzo [26] for achieving the target average BEP of $P_{th} = 10^{-2}$. (a) The instantaneous channel SNR γ is represented as a thick line at the top part of the graph, the associated instantaneous BEP $p_e(\gamma)$ as a thin line at the middle, and the instantaneous BPS throughput $b(\gamma)$ as a thick line at the bottom. The average SNR is $\bar{\gamma} = 10$dB. (b) As the SNR increases, the higher-order AQAM modes are selected more often.

fixed-mode modems. This is due to the fact that the AQAM scheme employing *Torrance*'s switching levels allows the target BEP to drop at a high average SNR due to its sub-optimum thresholds, which prevents the scheme from increasing the average throughput steadily to the maximum achievable BPS throughput. This phenomenon is more visible for low target average BERs, as it can be observed in Figure 6.10.

In conclusion, we reviewed an adaptive modulation scheme employing Torrance's switching levels [26], where the objective was to maximize the average BPS throughput, while maintaining the target average BEP. Torrance's switching levels are constant across the entire SNR range and the average BEP P_{avg} of the AQAM scheme employing these switching levels shows good agreement with the target average BEP P_{th}. However, the range of average SNR values, where $P_{avg} \simeq P_{th}$ was limited up to 25dB.

6.4.3   Cost Function Optimization as a Function of the Average SNR

In the previous section, we investigated *Torrance*'s switching levels [26] designed for achieving a certain target average BEP. However, the actual average BEP of the AQAM system was not constant across the SNR range, implying that the average throughput could potentially be further increased. Hence here we propose a modified cost function $\Omega(\mathbf{s}; \bar{\gamma})$, putting more emphasis on achieving a higher throughput and optimise the switching levels for a given SNR,

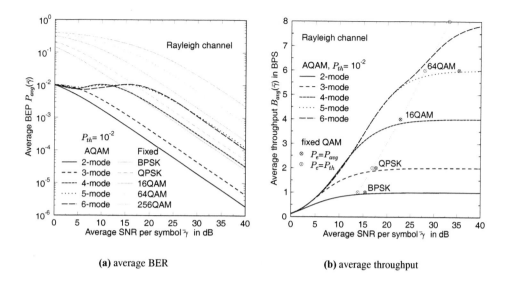

(a) average BER   **(b)** average throughput

Figure 6.9: The performance of various AQAM systems employing *Torrance*'s switching levels [26] designed for the target average BEP of $P_{th} = 10^{-2}$. (a) The actual average BEP P_{avg} is close to the target BEP of $P_{th} = 10^{-2}$ over an average SNR range which becomes wider, as the number of modulation modes increases. However, the five-mode and six-mode AQAM schemes have a similar performance across much of the SNR range. (b) The SNR gains of the AQAM scheme over the fixed-mode QAM arrangements, while achieving the same throughput at the same average BEP, *i.e.* $P_e = P_{avg}$, range from 6dB to 9dB, which corresponds to a 1dB improvement compared to the SNR gains observed in Figure 6.6(b). However, the SNR gains over the fixed mode QAM arrangement achieving the target BEP of $P_e = P_{avg}$ are reduced, especially at high average SNR values, namely for $\bar{\gamma} > 25$dB.

rather than for the whole SNR range [28]:

$$\Omega(\mathbf{s}; \bar{\gamma}) = 10 \log_{10}(\max\{P_{avg}(\bar{\gamma}; \mathbf{s})/P_{th}, 1\}) + \rho \log_{10}(B_{max}/B_{avg}(\bar{\gamma}; \mathbf{s})), \quad (6.40)$$

where $\mathbf{s}$ is a set of switching levels, $\bar{\gamma}$ is the average SNR per symbol, P_{avg} is the average BEP of the adaptive modulation scheme given in (6.21), P_{th} is the target average BEP of the adaptive modulation scheme, B_{max} is the BPS throughput of the highest order constituent modulation mode. Furthermore, the average throughput B_{avg} is given in (6.14) and ρ is a weighting factor, facilitating the above-mentioned BPS throughput enhancement. The first term at the right hand side of (6.40) corresponds to a cost function, which accounts for the difference, in the logarithmic domain, between the average BEP P_{avg} of the AQAM scheme and the target BEP P_{th}. This term becomes zero, when $P_{avg} \leq P_{th}$, contributing no cost to the overall cost function Ω. On the other hand, the second term of (6.40) accounts for the logarithmic distance between the maximum achievable BPS throughput B_{max} and the average BPS throughput B_{avg} of the AQAM scheme, which decreases, as B_{avg} approaches B_{max}. Applying Powell's minimization [29] to this cost function under the constraint of $s_{k-1} \leq s_k$, the optimum set of switching levels $\mathbf{s}_{opt}(\bar{\gamma})$ can be obtained, resulting in the highest average BPS throughput, while maintaining the target average BEP.

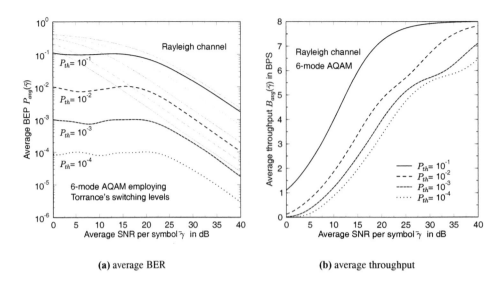

(a) average BER   **(b)** average throughput

Figure 6.10: The performance of the six-mode AQAM scheme employing Torrance's switching levels [26] for various target average BERs. When the average SNR is over 25dB and the
target average BEP is low, the average BEP of the AQAM scheme begins to decrease,
preventing the scheme from increasing the average BPS throughput steadily.

Figure 6.11 depicts the switching levels versus the average SNR per symbol optimised in
this manner for a five-mode AQAM scheme achieving the target average BEP of $P_{th} = 10^{-2}$
and 10^{-3}. Since the switching levels are optimised for each specific average SNR value, they
are not constant across the entire SNR range. As the average SNR $\bar{\gamma}$ increases, the switching
levels decrease in order to activate the higher-order mode modulation modes more often in
an effort to increase the BPS throughput. The low-order modulation modes are abandoned
one by one, as $\bar{\gamma}$ increases, activating always the highest-order modulation mode, namely
64-QAM, when the average BEP of the fixed-mode 64-QAM scheme becomes lower, than
the target average BEP P_{th}. Let us define the *avalanche SNR* $\bar{\gamma}_\alpha$ of a K-mode adaptive
modulation scheme as the lowest SNR, where the target BEP is achieved, which can be
formulated as:

$$P_{e,m_K}(\bar{\gamma}_\alpha) = P_{th} , \qquad (6.41)$$

where m_K is the highest order modulation mode, P_{e,m_K} is the average BEP of the fixed-
mode m_K-ary modem activated at the average SNR of $\bar{\gamma}$ and P_{th} is the target average BEP
of the adaptive modulation scheme. We can observe in Figure 6.11 that when the average
channel SNR is higher than the avalanche SNR, *i.e.* $\bar{\gamma} \geq \bar{\gamma}_\alpha$, the switching levels are reduced
to zero. Some of the optimised switching level versus SNR curves exhibit glitches, indicating
that the multi-dimensional optimization might result in local optima in some cases.

The corresponding average BEP P_{avg} and the average throughput B_{avg} of the two to six-
mode AQAM schemes designed for the target average BEP of $P_{th} = 10^{-2}$ are depicted in
Figure 6.12. We can observe in Figure 6.12(a) that now the actual average BEP P_{avg} of the
AQAM scheme is exactly the same as the target BEP of $P_{th} = 10^{-2}$, when the average SNR $\bar{\gamma}$

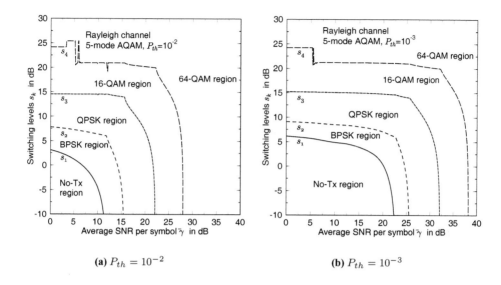

(a) $P_{th} = 10^{-2}$ **(b)** $P_{th} = 10^{-3}$

Figure 6.11: The switching levels optimised at each average SNR value in order to achieve the target average BEP of (a) $P_{th} = 10^{-2}$ and (b) $P_{th} = 10^{-3}$. As the average SNR $\bar{\gamma}$ increases, the switching levels decrease in order to activate the higher-order mode modulation modes more often in an effort to increase the BPS throughput. The low-order modulation modes are abandoned one by one as $\bar{\gamma}$ increases, activating the highest-order modulation mode, namely 64-QAM, all the time when the average BEP of the fixed-mode 64-QAM scheme becomes lower than the target average BEP P_{th}.

is less than or equal to the avalanche SNR $\bar{\gamma}_\alpha$. As the number of AQAM modulation modes K increases, the range of average SNRs where the design target of $P_{avg} = P_{th}$ is met extends to a higher SNR, namely to the avalanche SNR. In Figure 6.12(b), the average BPS throughputs of the AQAM modems employing the 'per-SNR optimised' switching levels introduced in this section are represented in thick lines, while the BPS throughput of the six-mode AQAM arrangement employing Torrance's switching levels [26] is represented using a solid thin line. The average SNR values required by the fixed-mode QAM scheme for achieving the target average BEP of $P_{e,m_K} = P_{th}$ are represented by the markers '⊙'. As we can observe in Figure 6.12(b) the new per-SNR optimised scheme produces a higher BPS throughput, than the scheme using Torrance's switching regime, when the average SNR $\bar{\gamma} > 20$dB. However, for the range of 8dB $< \bar{\gamma} < 20$dB, the BPS throughput of the new scheme is lower than that of *Torrance*'s scheme, indicating that the multi-dimensional optimization technique might reach local minima for some SNR values.

Figure 6.13(a) shows that the six-mode AQAM scheme employing 'per-SNR optimised' switching levels satisfies the target average BEP values of $P_{th} = 10^{-1}$ to 10^{-4}. However, the corresponding average throughput performance shown in Figure 6.13(b) also indicates that the thresholds generated by the multi-dimensional optimization were not satisfactory. The BPS throughput achieved was heavily dependent on the value of the weighting factor ρ in (6.40). The glitches seen in the BPS throughput curves in Figure 6.13(b) also suggest that the optimization process might result in some local minima.

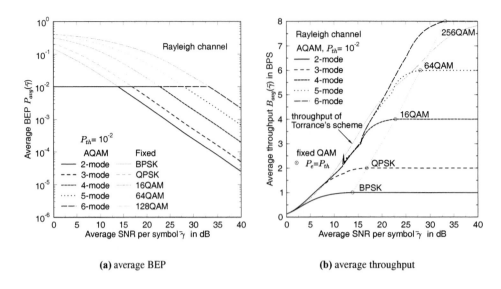

(a) average BEP   (b) average throughput

Figure 6.12: The performance of K-mode AQAM schemes for $K = 2, 3, 4, 5$ and 6, employing the switching levels optimised for each SNR value designed for the target average BEP of $P_{th} = 10^{-2}$. (a) The actual average BEP P_{avg} is exactly the same as the target BER of $P_{th} = 10^{-2}$, when the average SNR $\bar{\gamma}$ is less than or equal to the so-called avalanche SNR $\bar{\gamma}_\alpha$, where the average BEP of the highest-order fixed-modulation mode is equal to the target average BEP. (b) The average throughputs of the AQAM modems employing the 'per-SNR optimised' switching levels are represented in the thick lines, while that of the six-mode AQAM scheme employing Torrance's switching levels [26] is represented by a solid thin line.

We conclude that due to these problems it is hard to achieve a satisfactory BPS throughput for adaptive modulation schemes employing the switching levels optimised for each SNR value based on the heuristic cost function of (6.40), while the corresponding average BEP exhibits a perfect agreement with the target average BEP.

6.4.4 Lagrangian Method

As argued in the previous section, Powell's minimization [29] of the cost function often leads to a local minimum, rather than to the global minimum. Hence, here we adopt an analytical approach to finding the globally optimised switching levels. Our aim is to optimise the set of switching levels, **s**, so that the average BPS throughput $B(\bar{\gamma}; \mathbf{s})$ can be maximized under the constraint of $P_{avg}(\bar{\gamma}; \mathbf{s}) = P_{th}$. Let us define P_R for a K-mode adaptive modulation scheme as the sum of the mode-specific average BEP weighted by the BPS throughput of the individual constituent mode:

$$P_R(\bar{\gamma}; \mathbf{s}) \triangleq \sum_{k=0}^{K-1} b_k\, P_k\,, \tag{6.42}$$

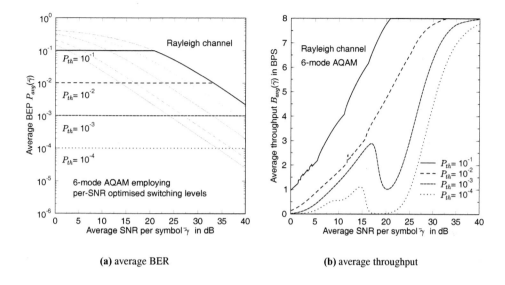

(a) average BER

(b) average throughput

Figure 6.13: The performance of six-mode AQAM employing 'per-SNR optimised' switching levels for various values of the target average BEP. (a) The average BEP P_{avg} remains constant until the average SNR $\bar{\gamma}$ reaches the avalanche SNR, then follows the average BEP curve of the highest-order fixed-mode QAM scheme, *i.e.* that of 256-QAM. (b) For some SNR values the BPS throughput performance of the six-mode AQAM scheme is not satisfactory due to the fact that the multi-dimensional optimization algorithm becomes trapped in local minima and hence fails to reach the global minimum.

where $\bar{\gamma}$ is the average SNR per symbol, s is the set of switching levels, K is the number of constituent modulation modes, b_k is the BPS throughput of the k-th constituent mode and the mode-specific average BEP P_k is given in (6.20) as:

$$P_k = \int_{s_k}^{s_{k+1}} p_{m_k}(\gamma) \, f(\gamma) \, d\gamma \,, \tag{6.43}$$

where again, $p_{m_k}(\gamma)$ is the BEP of the m_k-ary modulation scheme over the AWGN channel and $f(\gamma)$ is the PDF of the SNR per symbol γ. Explicitly, (6.43) implies weighting the BEP $p_{m_k}(\gamma)$ by its probability of occurrence quantified in terms of its PDF and then averaging, *i.e.* integrating it over the range spanning from s_k to s_{k+1}. Then, with the aid of (6.21), the average BEP constraint can also be written as:

$$P_{avg}(\bar{\gamma}; \mathbf{s}) = P_{th} \iff P_R(\bar{\gamma}; \mathbf{s}) = P_{th} \, B(\bar{\gamma}; \mathbf{s}) \,. \tag{6.44}$$

Another rational constraint regarding the switching levels can be expressed as:

$$s_k \leq s_{k+1} \,. \tag{6.45}$$

As we discussed before, our optimization goal is to maximize the objective function $B(\bar{\gamma}; \mathbf{s})$ under the constraint of (6.44). The set of switching levels s has $K + 1$ levels in it. However, considering that we have $s_0 = 0$ and $s_K = \infty$ in many adaptive modulation schemes, we have $K - 1$ independent variables in s. Hence, the optimization task is a $K - 1$ dimensional optimization under a constraint [198]. It is a standard practice to introduce a modified object function using a Lagrangian multiplier and convert the problem into a set of one-dimensional optimization problems. The modified object function Λ can be formulated employing a Lagrangian multiplier λ [198] as:

$$\Lambda(\mathbf{s}; \bar{\gamma}) = B(\bar{\gamma}; \mathbf{s}) + \lambda \left\{ P_R(\bar{\gamma}; \mathbf{s}) - P_{th} \, B(\bar{\gamma}; \mathbf{s}) \right\} \tag{6.46}$$

$$= (1 - \lambda P_{th}) \, B(\bar{\gamma}; \mathbf{s}) + \lambda P_R(\bar{\gamma}; \mathbf{s}) \, . \tag{6.47}$$

The optimum set of switching levels should satisfy:

$$\frac{\partial \Lambda}{\partial \mathbf{s}} = \frac{\partial}{\partial \mathbf{s}} \left(B(\bar{\gamma}; \mathbf{s}) + \lambda \left\{ P_R(\bar{\gamma}; \mathbf{s}) - P_{th} \, B(\bar{\gamma}; \mathbf{s}) \right\} \right) = 0 \quad \text{and} \tag{6.48}$$

$$P_R(\bar{\gamma}; \mathbf{s}) - P_t \, B(\bar{\gamma}; \mathbf{s}) = 0 \, . \tag{6.49}$$

The following results are helpful in evaluating the partial differentiations in (6.48) :

$$\frac{\partial}{\partial s_k} P_{k-1} = \frac{\partial}{\partial s_k} \int_{s_{k-1}}^{s_k} p_{m_{k-1}}(\gamma) \, f(\gamma) \, d\gamma = p_{m_{k-1}}(s_k) \, f(s_k) \tag{6.50}$$

$$\frac{\partial}{\partial s_k} P_k = \frac{\partial}{\partial s_k} \int_{s_k}^{s_{k+1}} p_{m_k}(\gamma) \, f(\gamma) \, d\gamma = -p_{m_k}(s_k) \, f(s_k) \tag{6.51}$$

$$\frac{\partial}{\partial s_k} F_c(s_k) = \frac{\partial}{\partial s_k} \int_{s_k}^{\infty} f(\gamma) \, d\gamma = -f(s_k) \, . \tag{6.52}$$

Using (6.50) and (6.51), the partial differentiation of P_R defined in (6.42) with respect to s_k can be written as:

$$\frac{\partial P_R}{\partial s_k} = b_{k-1} \, p_{m_{k-1}}(s_k) \, f(s_k) - b_k \, p_{m_k}(s_k) \, f(s_k) \, , \tag{6.53}$$

where b_k is the BPS throughput of an m_k-ary modem. Since the average throughput is given by $B = \sum_{k=0}^{K-1} c_k \, F_c(s_k)$ in (6.18), the partial differentiation of B with respect to s_k can be written as, using (6.52) :

$$\frac{\partial B}{\partial s_k} = -c_k \, f(s_k) \, , \tag{6.54}$$

where c_k was defined as $c_k \triangleq b_k - b_{k-1}$ in Section 6.3.1. Hence (6.48) can be evaluated as:

$$\left[-c_k(1 - \lambda \, P_{th}) + \lambda \left\{ b_{k-1} \, p_{m_{k-1}}(s_k) - b_k p_{m_k}(s_k) \right\} \right] \, f(s_k) = 0 \quad \text{for } k = 1, 2, \cdots, K - 1 \, . \tag{6.55}$$

A trivial solution of (6.55) is $f(s_k) = 0$. Certainly, $\{s_k = \infty, \, k = 1, 2, \cdots, K - 1\}$ satisfies this condition. Again, the lowest throughput modulation mode is 'No-Tx' in our model, which corresponds to no transmission. When the PDF of γ satisfies $f(0) = 0$, $\{s_k = 0, \, k = 1, 2, \cdots, K - 1\}$ can also be a solution, which corresponds to the fixed-mode m_{K-1}-ary modem. The corresponding avalanche SNR $\bar{\gamma}_\alpha$ can obtained by substituting $\{s_k = 0, \, k = 1, 2, \cdots, K - 1\}$ into (6.49), which satisfies:

$$p_{m_{K-1}}(\bar{\gamma}_\alpha) - P_{th} = 0 \, . \tag{6.56}$$

When $f(s_k) \neq 0$, Equation (6.55) can be simplified upon dividing both sides by $f(s_k)$, yielding:

$$-c_k(1 - \lambda P_{th}) + \lambda \left\{ b_{k-1} p_{m_{k-1}}(s_k) - b_k p_{m_k}(s_k) \right\} = 0 \text{ for } k = 1, 2, \cdots, K - 1 \,.$$
(6.57)

Rearranging (6.57) for $k = 1$ and assuming $c_1 \neq 0$, we have:

$$1 - \lambda P_{th} = \frac{\lambda}{c_1} \left\{ b_0 \, p_{m_0}(s_1) - b_1 p_{m_1}(s_1) \right\} \,.$$
(6.58)

Substituting (6.58) into (6.57) and assuming $c_k \neq 0$ for $k \neq 0$, we have:

$$\frac{\lambda}{c_k} \left\{ b_{k-1} \, p_{m_{k-1}}(s_k) - b_k p_{m_k}(s_k) \right\} = \frac{\lambda}{c_1} \left\{ b_0 \, p_{m_0}(s_1) - b_1 p_{m_1}(s_1) \right\} \,.$$
(6.59)

In this context we note that the Lagrangian multiplier λ is not zero because substitution of $\lambda = 0$ in (6.57) leads to $-c_k = 0$, which is not true. Hence, we can eliminate the Lagrangian multiplier dividing both sides of (6.59) by λ. Then we have:

$$y_k(s_k) = y_1(s_1) \text{ for } k = 2, 3, \cdots K - 1 \,,$$
(6.60)

where the function $y_k(s_k)$ is defined as:

$$y_k(s_k) \triangleq \frac{1}{c_k} \left\{ b_k p_{m_k}(s_k) - b_{k-1} \, p_{m_{k-1}}(s_k) \right\} \,, \quad k = 2, 3, \cdots K - 1 \,,$$
(6.61)

which does not contain the Lagrangian multiplier λ and hence it will be referred to as the 'Lagrangian-free function'. This function can be physically interpreted as the normalized BEP difference between the adjacent AQAM modes. For example, $y_1(s_1) = p_2(s_1)$ quantifies the BEP increase, when switching from the No-Tx mode to the BPSK mode, while $y_2(s_2) = 2 p_4(s_2) - p_2(s_2)$ indicates the BEP difference between the QPSK and BPSK modes. These curve will be more explicitly discussed in the context of Figure 6.14. The significance of (6.60) is that the relationship between the optimum switching levels s_k, where $k = 2, 3, \cdots K - 1$, and the lowest optimum switching level s_1 is independent of the underlying propagation scenario. Only the constituent modulation mode related parameters, such as b_k, c_k and $p_{m_k}(\gamma)$, govern this relationship.

Let us now investigate some properties of the Lagrangian-free function $y_k(s_k)$ given in (6.61). Considering that $b_k > b_{k-1}$ and $p_{m_k}(s_k) > p_{m_{k-1}}(s_k)$, it is readily seen that the value of $y_k(s_k)$ is always positive. When $s_k = 0$, $y_k(s_k)$ becomes:

$$y_k(0) \triangleq \frac{1}{c_k} \left\{ b_k p_{m_k}(0) - b_{k-1} p_{m_{k-1}}(0) \right\} = \frac{1}{c_k} \left\{ \frac{b_k}{2} - \frac{b_{k-1}}{2} \right\} = \frac{1}{2} \,.$$
(6.62)

The solution of $y_k(s_k) = 1/2$ can be either $s_k = 0$ or $b_k p_{m_k}(s_k) = b_{k-1} p_{m_{k-1}}(s_k)$. When $s_k = 0$, $y_k(s_k)$ becomes $y_k(\infty) = 0$. We also conjecture that

$$\frac{d \, s_k}{d \, s_1} = \frac{y_1'(s_1)}{y_k'(s_k)} > 0 \text{ when } y_k(s_k) = y_1(s_1),$$
(6.63)

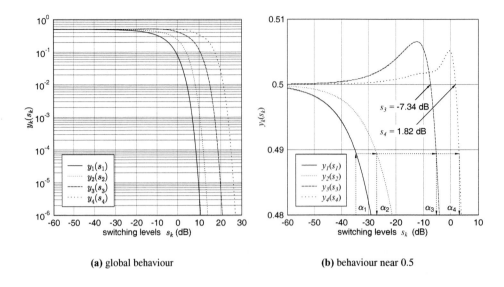

(a) global behaviour **(b)** behaviour near 0.5

Figure 6.14: The Lagrangian-free functions $y_k(s_k)$ of (6.64) through (6.67) for Gray-mapped square-shaped QAM constellations. As s_k becomes lower $y_k(s_k)$ asymptotically approaches 0.5. Observe that while $y_1(s_1)$ and $y_2(s_2)$ are monotonic functions, $y_3(s_3)$ and $y_4(s_4)$ cross the $y = 0.5$ line.

which states that the k-th optimum switching level s_k always increases, whenever the lowest optimum switching level s_1 increases. Our numerical evaluations suggest that this conjecture appears to be true.

As an example, let us consider the five-mode AQAM scheme introduced in Section 6.3.2.1. The parameters of the five-mode AQAM scheme are summarised in Table 6.1. Substituting these parameters into (6.60) and (6.61), we have the following set of equations.

$$y_1(s_1) = p_2(s_1) \tag{6.64}$$

$$y_2(s_2) = 2\,p_4(s_2) - p_2(s_2) \tag{6.65}$$

$$y_3(s_3) = 2\,p_{16}(s_3) - p_4(s_3) \tag{6.66}$$

$$y_4(s_4) = 3\,p_{64}(s_4) - 2\,p_{16}(s_4) \tag{6.67}$$

The Lagrangian-free functions of (6.64) through (6.67) are depicted in Figure 6.14 for Gray-mapped square-shaped QAM. As these functions are basically linear combinations of BEP curves associated with AWGN channels, they exhibit waterfall-like shapes and asymptotically approach 0.5, as the switching levels s_k approach zero (or $-\infty$ expressed in dB). While $y_1(s_1)$ and $y_2(s_2)$ are monotonic functions, $y_3(s_3)$ and $y_4(s_4)$ cross the $y = 0.5$ line at $s_3 = -7.34$ dB and $s_4 = 1.82$ dB respectively, as it can be observed in Figure 6.14(b). One should also notice that the trivial solutions of (6.60) are $y_k = 0.5$ at $s_k = 0$, $k = 1, 2, 3, 4$, as we have discussed before.

For a given value of s_1, the other switching levels can be determined as $s_2 = y_2^{-1}(y_1(s_1))$, $s_3 = y_3^{-1}(y_1(s_1))$ and $s_4 = y_4^{-1}(y_1(s_1))$. Since deriving the analytical inverse function of y_k is an arduous task, we can rely on a graphical or a numerical method. Figure 6.14(b) illus-

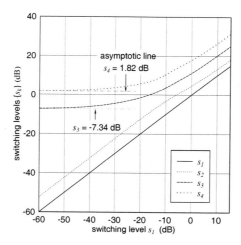

Figure 6.15: Optimum switching levels as functions of s_1, where the linear relationship of s_1 versus s_1 was also plotted for completeness. Observe that while the optimum value of s_2 shows a linear relationship with respect to s_1, those of s_3 and s_4 asymptotically approach constant values as s_1 is reduced.

trates an example of the graphical method. Specifically, when $s_1 = \alpha_1$, we first find the point on the curve y_1 directly above the abscissa value of α_1 and then draw a horizontal line across the corresponding point. From the crossover points found on the curves of y_2, y_3 and y_4 with the aid of the horizontal line, we can find the corresponding values of the other switching levels, namely those of α_2, α_3 and α_4. In a numerical sense, this solution corresponds to a one-dimensional (1-D) root finding problem [29, Ch 9]. Furthermore, the $y_k(s_k)$ values are monotonic, provided that we have $y_k(s_k) < 0.5$ and this implies that the roots found are unique. The numerical results shown in Figure 6.15 represent the direct relationship between the optimum switching level s_1 and the other optimum switching levels, namely s_2, s_3 and s_4. While the optimum value of s_2 shows a near-linear relationship with respect to s_1, those of s_3 and s_4 asymptotically approach two different constants, as s_1 becomes smaller. This corroborates the trends observed in Figure 6.14(b), where $y_3(s_3)$ and $y_4(s_4)$ cross the $y = 0.5$ line at $s_3 = -7.34$ dB and $s_4 = 1.82$ dB, respectively. Since the low-order modulation modes are abandoned at high average channel SNRs in order to increase the average throughput, the high values of s_1 on the horizontal axis of Figure 6.15 indicate encountering a low channel SNR, while low values of s_1 suggest that high channel SNRs are experienced, as it transpires for example from Figure 6.11.

Since we can relate the other switching levels to s_1, we have to determine the optimum value of s_1 for the given target BEP, P_{th}, and the PDF of the instantaneous channel SNR, $f(\gamma)$, by solving the constraint equation given in (6.49). This problem also constitutes a 1-D root finding problem, rather than a multi-dimensional optimization problem, which was the case in Sections 6.4.2 and 6.4.3. Let us define the constraint function $Y(\bar{\gamma}; \mathbf{s}(s_1))$ using (6.49) as:

$$Y(\bar{\gamma}; \mathbf{s}(s_1)) \triangleq P_R(\bar{\gamma}; \mathbf{s}(s_1)) - P_{th} B(\bar{\gamma}; \mathbf{s}(s_1)), \tag{6.68}$$

where we represented the set of switching levels as a vector, which is the function of s_1, in order to emphasise that s_k satisfies the relationships given by (6.60) and (6.61).

More explicitly, $Y(\bar{\gamma}; \mathbf{s}(s_1))$ of (6.68) can be physically interpreted as the difference between $P_R(\bar{\gamma}; \mathbf{s}(s_1))$, namely the sum of the mode-specific average BEPs weighted by the BPS throughput of the individual AQAM modes, as defined in (6.42) and the average BPS throughput $B(\bar{\gamma}; \mathbf{s}(s_1))$ weighted by the target BEP P_{th}. Considering the equivalence relationship given in (6.44), (6.68) reflects just another way of expressing the difference between the average BEP P_{avg} of the adaptive scheme and the target BEP P_{th}.

Even though the relationships implied in $\mathbf{s}(s_1)$ are independent of the propagation conditions and the signalling power, the constraint function $Y(\bar{\gamma}; \mathbf{s}(s_1))$ of (6.68) and hence the actual values of the optimum switching levels are dependent on propagation conditions through the PDF $f(\gamma)$ of the SNR per symbol and on the average SNR per symbol $\bar{\gamma}$.

Let us find the initial value of $Y(\bar{\gamma}; \mathbf{s}(s_1))$ defined in (6.68), when $s_1 = 0$. An obvious solution for s_k when $s_1 = 0$ is $s_k = 0$ for $k = 1, 2, \cdots, K - 1$. In this case, $Y(\bar{\gamma}; \mathbf{s}(s_1))$ becomes:

$$Y(\bar{\gamma}; 0) = b_{K-1} \left(P_{m_{K-1}}(\bar{\gamma}) - P_{th} \right), \tag{6.69}$$

where b_{K-1} is the BPS throughput of the highest-order constituent modulation mode, while $P_{m_{K-1}}(\bar{\gamma})$ is the average BEP of the highest-order constituent modulation mode for transmission over the underlying channel scenario and P_{th} is the target average BEP. The value of $Y(\bar{\gamma}; 0)$ could be positive or negative, depending on the average SNR $\bar{\gamma}$ and on the target average BEP P_{th}. Another solution exists for s_k when $s_1 = 0$, if $b_k\, p_{m_k}(s_k) = b_{k-1}\, p_{m_{k-1}}(s_k)$. The value of $Y(\bar{\gamma}; 0^+)$ using this alternative solution turns out to be close to $Y(\bar{\gamma}; 0)$. However, in the actual numerical evaluation of the initial value of Y, we should use $Y(\bar{\gamma}; 0^+)$ for ensuring the continuity of the function Y at $s_1 = 0$.

In order to find the minima and the maxima of Y, we have to evaluate the derivative of $Y(\bar{\gamma}; \mathbf{s}(s_1))$ with respect to s_1. With the aid of (6.50) to (6.54), we have:

$$\begin{aligned}
\frac{dY}{ds_1} &= \sum_{k=1}^{K-1} \frac{\partial Y}{\partial s_k} \frac{ds_k}{ds_1} \\
&= \sum_{k=1}^{K-1} \frac{\partial}{\partial s_k} \{P_R - P_{th}\, B\} \frac{ds_k}{ds_1} \\
&= \sum_{k=1}^{K-1} \left\{ b_{k-1}\, p_{m_{k-1}}(s_k) - b_k\, p_{m_k}(s_k) + P_{th}\, c_k \right\} f(s_k) \frac{ds_k}{ds_1} \\
&= \sum_{k=1}^{K-1} \left[\frac{c_k}{c_1} \left\{ b_0\, p_{m_0}(s_1) - b_1\, p_{m_1}(s_1) \right\} + P_{th}\, c_k \right] f(s_k) \frac{ds_k}{ds_1} \\
&= \frac{1}{c_1} \left\{ b_0\, p_{m_0}(s_1) - b_1\, p_{m_1}(s_1) + P_{th} \right\} \sum_{k=1}^{K-1} c_k\, f(s_k) \frac{ds_k}{ds_1}.
\end{aligned} \tag{6.70}$$

Considering $f(s_k) \geq 0$ and using our conjecture that $\frac{ds_k}{ds_1} > 0$ given in (6.63), we can conclude from (6.70) that $\frac{dY}{ds_1} = 0$ has roots, when $f(s_k) = 0$ for all k or when $b_1\, p_{m_1}(s_1) - b_0\, p_{m_0}(s_1) = P_{th}$. The former condition corresponds to either $s_i = 0$ for some PDF $f(\gamma)$ or to $s_k = \infty$ for all PDFs. By contrast, when the condition of $b_1\, p_{m_1}(s_1) - b_0\, p_{m_0}(s_1) = P_{th}$ is met, $dY/ds_1 = 0$ has a unique solution. Investigating the sign of the first derivative between these zeros, we can conclude that $Y(\bar\gamma; s_1)$ has a global minimum of Y_{min} at $s_1 = \zeta$ such that $b_1\, p_{m_1}(\zeta) - b_0\, p_{m_0}(\zeta) = P_{th}$ and a maximum of Y_{max} at $s_1 = 0$ and another maximum value at $s_1 = \infty$.

Since $Y(\bar\gamma; s_1)$ has a maximum value at $s_1 = \infty$, let us find the corresponding maximum value. Let us first consider $\lim_{s_1 \to \infty} P_{avg}(\bar\gamma; s(s_1))$, where upon exploiting (6.21) and (6.42) we have:

$$\lim_{s_1 \to \infty} P_{avg}(\bar\gamma; s_k) = \frac{\lim_{s_1 \to \infty} P_R}{\lim_{s_1 \to \infty} B} \tag{6.71}$$

$$= \frac{0}{0}. \tag{6.72}$$

When applying l'Hopital's rule and using Equations (6.50) through (6.54), we have:

$$\frac{\lim_{s_1 \to \infty} P_R}{\lim_{s_1 \to \infty} B} = \frac{\lim_{s_1 \to \infty} \frac{d}{ds_1} P_R}{\lim_{s_1 \to \infty} \frac{d}{ds_1} B} \tag{6.73}$$

$$= \lim_{s_1 \to \infty} \frac{1}{c_1} b_1\, p_{m_1}(s_1) - b_0\, p_{m_0}(s_1) \tag{6.74}$$

$$= 0^+, \tag{6.75}$$

implying that $P_{avg}(\bar\gamma; s_k)$ approaches zero from positive values, when s_1 tends to ∞. Since according to (6.21), (6.42) and (6.68) the function $Y(\bar\gamma; s(s_1))$ can be written as $B(P_{avg} - P_{th})$, we have:

$$\lim_{s_1 \to \infty} Y(\bar\gamma; s_1) = \lim_{s_1 \to \infty} B(P_{avg} - P_{th}) \tag{6.76}$$

$$= \lim_{s_1 \to \infty} B(0^+ - P_{th}) \tag{6.77}$$

$$= 0^-, \tag{6.78}$$

Hence $Y(\bar\gamma; s(s_1))$ asymptotically approaches zero from negative values, as s_1 tends to ∞. From the analysis of the minimum and the maxima, we can conclude that the constraint function $Y(\bar\gamma; s(s_1))$ defined in (6.68) has a unique zero only if $Y(\bar\gamma; 0^+) > 0$ at a switching value of $0 < s_1 < \zeta$, where ζ satisfies $b_1\, p_{m_1}(\zeta) - b_0\, p_{m_0}(\zeta) = P_{th}$. By contrast, when $Y(\bar\gamma; 0^+) < 0$, the optimum switching levels are all zero and the adaptive modulation scheme always employs the highest-order constituent modulation mode.

As an example, let us evaluate the constraint function $Y(\bar\gamma; s_1)$ for our five-mode AQAM scheme operating over a flat Rayleigh fading channel. Figure 6.16 depicts the values of $Y(s_1)$ for several values of the target average BEP P_{th}, when the average channel SNR is 30dB. We can observe that $Y(s_1) = 0$ may have a root, depending on the target BEP P_{th}. When $s_k = 0$ for $k < 5$, according to (6.21), (6.42) and (6.68) $Y(s_1)$ is reduced to

$$Y(\bar\gamma; 0) = 6(P_{64}(\bar\gamma) - P_{th}), \tag{6.79}$$

where $P_{64}(\bar\gamma)$ is the average BEP of 64-QAM over a flat Rayleigh channel. The value of $Y(\bar\gamma; 0)$ in (6.79) can be negative or positive, depending on the target BEP P_{th}.

We can observe in Figure 6.16 that the solution of $Y(\bar\gamma; s(s_1)) = 0$ is unique, when it exists. The locus of the minimum $Y(s_1)$, i.e. the trace curve of points $(Y_{min}(s_{1,min}), s_{1,min})$,

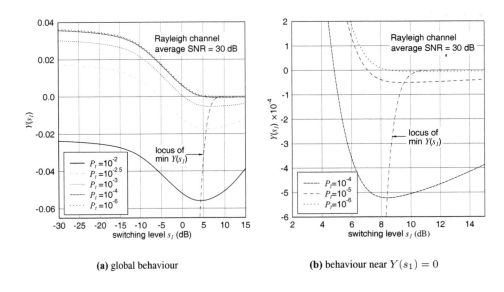

(a) global behaviour   (b) behaviour near $Y(s_1) = 0$

Figure 6.16: The constraint function $Y(\bar{\gamma}; \mathbf{s}(s_1))$ defined in (6.68) for our five-mode AQAM scheme employing Gray-mapped square-constellation QAM operating over a flat Rayleigh fading channel. The average SNR was $\bar{\gamma} = 30$ dB and it is seen that Y has a single minimum value, while approaching 0^-, as s_1 increases. The solution of $Y(\bar{\gamma}; \mathbf{s}(s_1)) = 0$ exists, when $Y(\bar{\gamma}; 0) = 6\{p_{64}(\bar{\gamma}) - P_{th}\} > 0$ and is unique.

where Y has the minimum value, is also depicted in Figure 6.16. The locus is always below the horizontal line of $Y(s_1) = 0$ and asymptotically approaches this line, as the target BEP P_{th} becomes smaller.

Figure 6.17 depicts the switching levels optimised in this manner for our five-mode AQAM scheme maintaining the target average BEPs of $P_{th} = 10^{-2}$ and 10^{-3}. The switching levels obtained using Powell's optimization method in Section 6.4.3 are represented as the thin grey lines in Figure 6.17 for comparison. In this case all the modulation modes may be activated with a certain probability, until the average SNR reaches the avalanche SNR value, while the scheme derived using Powell's optimization technique abandons the lower throughput modulation modes one by one, as the average SNR increases.

Figure 6.18 depicts the average throughput B expressed in BPS of the AQAM scheme employing the switching levels optimised using the Lagrangian method. In Figure 6.18(a), the average throughput of our six-mode AQAM arrangement using Torrance's scheme discussed in Section 6.4.2 is represented as a thin grey line. The Lagrangian multiplier based scheme showed SNR gains of 0.6dB, 0.5dB, 0.2dB and 3.9dB for a BPS throughput of 1, 2, 4 and 6, respectively, compared to Torrance's scheme. The average throughput of our six-mode AQAM scheme is depicted in Figure 6.18(b) for the several values of P_{th}, where the corresponding BPS throughput of the AQAM scheme employing per-SNR optimised thresholds determined using Powell's method are also represented as thin lines for $P_{th} = 10^{-1}$, 10^{-2} and 10^{-3}. Comparing the BPS throughput curves, we can conclude that the per-SNR optimised Powell method of Section 6.4.3 resulted in imperfect optimization for some values of the average SNR.

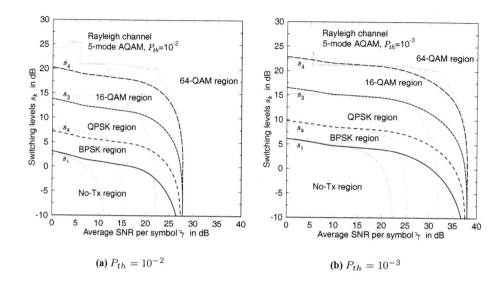

(a) $P_{th} = 10^{-2}$ **(b)** $P_{th} = 10^{-3}$

Figure 6.17: The switching levels for our five-mode AQAM scheme optimised at each average SNR
value in order to achieve the target average BEP of (a) $P_{th} = 10^{-2}$ and (b) $P_{th} = 10^{-3}$
using the Lagrangian multiplier based method of Section 6.4.4. The switching levels
based on Powell's optimization are represented in thin grey lines for comparison.

In conclusion, we derived an optimum mode-switching regime for a general AQAM
scheme using the Lagrangian multiplier method and presented our numerical results for various AQAM arrangements. Since the results showed that the Lagrangian optimization based
scheme is superior in comparison to the other methods investigated, we will employ these
switching levels in order to further investigate the performance of various adaptive modulation schemes.

6.5 Results and Discussions

The average throughput performance of adaptive modulation schemes employing the globally optimised mode-switching levels of Section 6.4.4 is presented in this section. The mobile
channel is modelled as a Nakagami-m fading channel. The performance results and discussions include the effects of the fading parameter m, that of the number of modulation modes,
the influence of the various diversity schemes used and the range of Square QAM, Star QAM
and MPSK signalling constellations.

6.5.1 Narrow-band Nakagami-m Fading Channel

The PDF of the instantaneous channel SNR γ of a system transmitting over the Nakagami
fading channel is given in (6.22). The parameters characterising the operation of the adaptive
modulation scheme were summarised in Section 6.3.3.1.

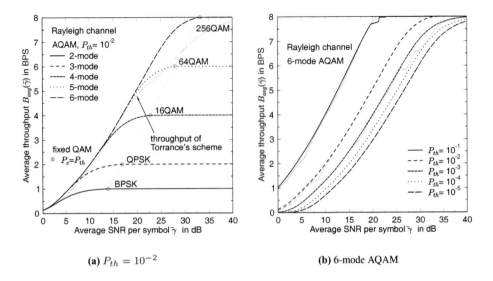

(a) $P_{th} = 10^{-2}$ **(b)** 6-mode AQAM

Figure 6.18: The average BPS throughput of various AQAM schemes employing the switching levels
optimised using the Lagrangian multiplier method (a) for $P_{th} = 10^{-2}$ employing two
to six-modes and (b) for $P_{th} = 10^{-2}$ to $P_{th} = 10^{-5}$ using six-modes. The average
throughput of the six-mode AQAM scheme using Torrance's switching levels [26] is represented for comparison as the thin grey line in figure (a). The average throughput of the
six-mode AQAM scheme employing per-SNR optimised thresholds using Powell's optimization method are represented by the thin lines in figure (b) for the target average BEP
of $P_{th} = 10^{-1}$, 10^{-2} and 10^{-3}.

6.5.1.1 Adaptive PSK Modulation Schemes

Phase Shift Keying (PSK) has the advantage of exhibiting a constant envelope power, since
all the constellation points are located on a circle. Let us first consider the BEP of fixed-mode
PSK schemes as a reference, so that we can compare the performance of adaptive PSK and
fixed-mode PSK schemes. The BEP of Gray-coded coherent M-ary PSK (MPSK), where
$M = 2^k$, for transmission over the AWGN channel can be closely approximated by [196]:

$$p_{MPSK}(\gamma) \simeq \sum_{i=1}^{2} A_i Q(\sqrt{a_i \gamma}) , \tag{6.80}$$

where $M \geq 8$ and the associated constants are given by [196]:

$$A_1 = A_2 = 2/k \tag{6.81}$$

$$a_1 = 2\sin^2(\pi/M) \tag{6.82}$$

$$a_2 = 2\sin^2(3\pi/M) . \tag{6.83}$$

Figure 6.19(a) shows the BEP of BPSK, QPSK, 8PSK, 16PSK, 32PSK and 64PSK for transmission over the AWGN channel. The differences of the required SNR per symbol, in order

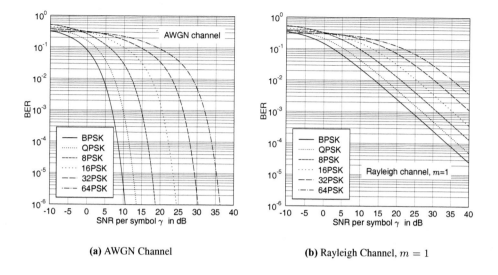

(a) AWGN Channel **(b)** Rayleigh Channel, $m = 1$

Figure 6.19: The average BEP of various MPSK modulation schemes.

to achieve the BER of $p_{MPSK}(\gamma) = 10^{-6}$ for the modulation modes having a throughput difference of 1 BPS are around 6dB, except between BPSK and QPSK, where a 3dB difference is observed.

The average BEP of MPSK schemes over a flat Nakagami-m fading channel is given as:

$$P_{MPSK}(\bar{\gamma}) = \int_0^{\infty} p_{MPSK}(\gamma) \, f(\gamma) \, d\gamma \,, \qquad (6.84)$$

where the BEP $p_{MPSK}(\gamma)$ for a transmission over the AWGN channel is given by (6.80) and the PDF $f(\gamma)$ is given by (6.22). A closed form solution of (6.84) can be readily obtained for an integer m using the results given in [87, (14-4-15)], which can be expressed as:

$$P_{MPSK}(\bar{\gamma}) = \sum_{i=1}^{2} A_i \left[\tfrac{1}{2}(1 - \mu_i) \right]^m \sum_{j=0}^{m-1} \binom{m-1+j}{j} \left[\tfrac{1}{2}(1 + \mu_i) \right]^j \,, \qquad (6.85)$$

where μ_i is defined as:

$$\mu_i \triangleq \sqrt{\frac{a_i \bar{\gamma}}{2m + a_i \bar{\gamma}}} \,. \qquad (6.86)$$

Figure 6.19(b) shows the average BEP of the various MPSK schemes for transmission over a flat Rayleigh channel, where $m = 1$. The BEP of MPSK over the AWGN channel given in (6.80) and that over a Nakagami channel given in (6.85) will be used in comparing the performance of adaptive PSK schemes.

The parameters of our nine-mode adaptive PSK scheme are summarised in Table 6.5 following the definitions of our generic model used for the adaptive modulation schemes developed in Section 6.3.1. The models of other adaptive PSK schemes employing a different

k	0	1	2	3	4	5	6	7	8
m_k	0	2	4	8	16	32	64	128	256
b_k	0	1	2	3	4	5	6	7	8
c_k	0	1	1	1	1	1	1	1	1
mode	No Tx	BPSK	QPSK	8PSK	16PSK	32PSK	64PSK	128PSK	256PSK

Table 6.5: Parameters of a nine-mode adaptive PSK scheme following the definitions of the generic adaptive modulation model developed in Section 6.3.1.

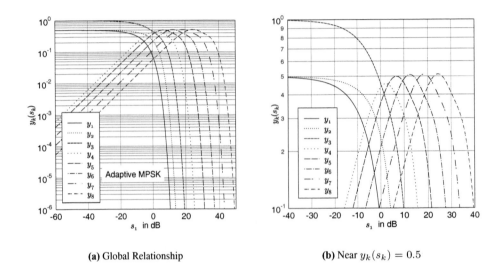

(a) Global Relationship

(b) Near $y_k(s_k) = 0.5$

Figure 6.20: 'Lagrangian-free' functions of (6.61) for a nine-mode adaptive PSK scheme. For a given value of s_1, there exist two solutions for s_k satisfying $y_k(s_k) = y_1(s_1)$. However, only the higher value of s_k satisfies the constraint of $s_{k-1} \leq s_k$, $\forall\, k$.

number of modes can be readily obtained by increasing or reducing the number of columns in Table 6.5. Since the number of modes is $K = 9$, we have $K + 1 = 10$ mode-switching levels, which are hosted by the vector $\mathbf{s} = \{s_k \mid k = 0, 1, 2, \cdots, 9\}$. Let us assume $s_0 = 0$ and $s_9 = \infty$. In order to evaluate the performance of the nine-mode adaptive PSK scheme, we have to obtain the optimum switching levels first. Let us evaluate the 'Lagrangian-free' functions defined in (6.61), using the parameters given in Table 6.5 and the BEP expressions given in (6.80). The 'Lagrangian-free' functions of our nine-mode adaptive PSK scheme are depicted in Figure 6.20. We can observe that there exist two solutions for s_k satisfying $y_k(s_k) = y_1(s_1)$ for a given value of s_1, which are given by the crossover points over the horizontal lines at the various coordinate values scaled on the vertical axis. However, only the higher value of s_k satisfies the constraint of $s_{k-1} \leq s_k$, $\forall\, k$. The enlarged view near $y_k(s_k) = 0.5$ seen in Figure 6.20(b) reveals that $y_4(s_4)$ may have no solution of $y_4(s_4) = y_1(s_1)$, when $y_1(s_1) > 0.45$. One option is to use a constant value of $s_4 = 2.37$dB, where $y_4(s_4)$ reaches its peak value. The other option is to set $s_4 = s_3$, effectively eliminating

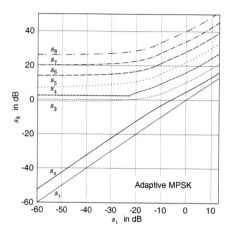

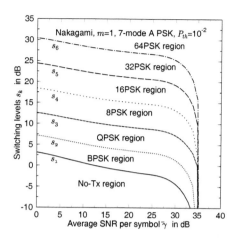

(a) Relationship between the optimum switching levels of a nine-mode PSK scheme

(b) Optimum switching levels of a seven-mode adaptive PSK scheme

Figure 6.21: Optimum switching levels. (a) Relationships between s_k and s_1 in a nine-mode adaptive PSK scheme. (b) Optimum switching levels for 7-mode adaptive PSK scheme operating over a Rayleigh channel at the target BEP of $P_{th} = 10^{-2}$.

16PSK from the set of possible modulation modes. It was found that both policies result in the same performance up to four effective decimal digits in terms of the average BPS throughput.

Upon solving $y_k(s_k) = y_1(s_1)$, we arrive at the relationships sought between the first optimum switching level s_1 and the remaining optimum switching levels s_k. Figure 6.21(a) depicts these relationships. All the optimum switching levels, except for s_1 and s_2, approach their asymptotic limit monotonically, as s_1 decreases. A decreased value of s_1 corresponds to an increased value of the average SNR. Figure 6.21(b) illustrates the optimum switching levels of a seven-mode adaptive PSK scheme operating over a Rayleigh channel associated with $m = 1$ at the target BEP of $P_{th} = 10^{-2}$. These switching levels were obtained by solving (6.68). The optimum switching levels show a steady decrease in their values as the average SNR increases, until it reaches the avalanche SNR value of $\bar{\gamma} = 35$dB, beyond which always the highest-order PSK modulation mode, namely 64PSK, is activated.

Having highlighted the evaluation of the optimum switching levels for an adaptive PSK scheme, let us now consider the associated performance results. We are reminded that the average BEP of our optimised adaptive scheme remains constant at $P_{avg} = P_{th}$, provided that the average SNR is less than the avalanche SNR. Hence, the average BPS throughput and the relative SNR gain of our APSK scheme in comparison to the corresponding fixed-mode modem are our concern.

Let us now consider Figure 6.22, where the average BPS throughput of the various adaptive PSK schemes operating over a Rayleigh channel associated with $m = 1$ are plotted, which were designed for the target BEP of $P_{th} = 10^{-2}$ and $P_{th} = 10^{-3}$. The markers '$\otimes$' and '$\odot$' represent the required SNR of the various fixed-mode PSK schemes, while achiev-

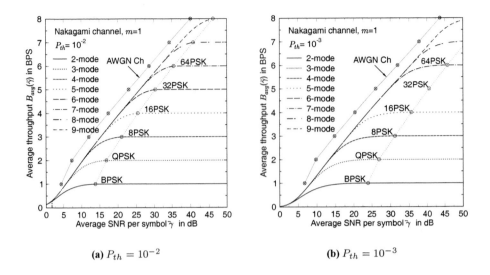

(a) $P_{th} = 10^{-2}$ **(b)** $P_{th} = 10^{-3}$

Figure 6.22: The average BPS throughput of various adaptive PSK schemes operating over a Rayleigh channel ($m = 1$) at the target BEP of (a) $P_{th} = 10^{-2}$ and (b) $P_{th} = 10^{-3}$. The markers '⊗' and '⊙' represent the required SNR of the corresponding fixed-mode PSK scheme, while achieving the same target BEP as the adaptive schemes, operating over an AWGN channel and a Rayleigh channel, respectively.

ing the same target BER as the adaptive schemes, operating over an AWGN channel and a Rayleigh channel, respectively. It can be observed that introducing an additional constituent mode into an adaptive PSK scheme does not make any impact on the average BPS throughput, when the average SNR is relatively low. For example, when the average SNR $\bar{\gamma}$ is less than 10dB in Figure 6.22(a), employing more than four APSK modes for the adaptive scheme does not improve the average BPS throughput. In comparison to the various fixed-mode PSK modems, the adaptive modem achieved the SNR gains between 4dB and 8dB for the target BEP of $P_{th} = 10^{-2}$ and 10dB to 16dB for the target BEP of $P_{th} = 10^{-3}$ over a Rayleigh channel. Since no adaptive scheme operating over a fading channel can outperform the corresponding fixed-mode counterpart operating over an AWGN channel, it is interesting to investigate the performance differences between these two schemes. Figure 6.22 suggests that the required SNR of our adaptive PSK modem achieving 1BPS for transmission over a Rayleigh channel is approximately 1dB higher, than that of fixed-mode BPSK operating over an AWGN channel. Furthermore, this impressive performance can be achieved by employing only three modes, namely No-Tx, BPSK and QPSK for the adaptive PSK modem. For other BPS throughput values, the corresponding SNR differences are in the range of 2dB to 3dB, while maintaining the BEP of $P_{th} = 10^{-2}$ and 4dB for the BEP of $P_{th} = 10^{-3}$.

We observed in Figure 6.22 that the average BPS throughput of the various adaptive PSK schemes is dependent on the target BEP. Hence, let us investigate the BPS performances of the adaptive modems for the various values of target BEPs using the results depicted in Figure 6.23. The average BPS throughputs of a nine-mode adaptive PSK scheme are represented as various types of lines without markers depending on the target average BERs, while those

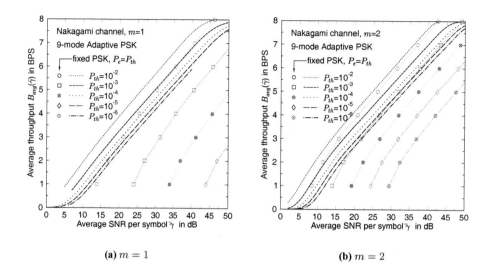

Figure 6.23: The average BPS throughput of a nine-mode adaptive PSK scheme operating over a Nakagami fading channel (a) $m = 1$ and (b) $m = 2$. The markers represent the SNR required for achieving the same BPS throughput and the same average BEP as the adaptive schemes.

of the corresponding fixed PSK schemes are represented as various types of lines with markers according to the key legend shown in Figure 6.23. We can observe that the difference between the required SNRs of the adaptive schemes and fixed schemes increases, as the target BEP decreases. It is interesting to note that the average BPS curves of the adaptive PSK schemes seem to converge to a set of densely packed curves, as the target BEP decreases to values around $10^{-4} - 10^{-6}$. In other words, the incremental SNR required for achieving the next target BEP, which is an order of magnitude lower, decreases as the target BEP decreases. On the other hand, the incremental SNR for the same scenario of fixed modems seems to remain nearly constant at 10dB. Comparing Figure 6.23(a) and Figure 6.23(b), we find that this seemingly constant incremental SNR of the fixed-mode modems is reduced to about 5dB, as the fading becomes less severe, *i.e.* when the fading parameter becomes $m = 2$.

Let us now investigate the effects of the Nakagami fading parameter m on the average BPS throughput performance of various adaptive PSK schemes by observing Figure 6.24. The BPS throughput of the various fixed PSK schemes for transmission over an AWGN channel is depicted in Figure 6.24 as the ultimate performance limit achievable by the adaptive schemes operating over Nakagami fading channels. For example, when the channel exhibits Rayleigh fading, *i.e.* when the fading parameter becomes $m = 1$, the adaptive PSK schemes show 3dB to 4dB SNR penalty compared to their fixed-mode counterparts operating over the AWGN channel. Compared to fixed-mode BPSK, the adaptive scheme required only a 1dB higher SNR. As the fading becomes less severe, the average BPS throughput of the adaptive PSK schemes approaches that of fixed-mode PSK operating over the AWGN channel. For the target BEP of $P_{th} = 10^{-3}$, the SNR gap between the BPS throughput curves becomes higher. The adaptive PSK scheme operating over the Rayleigh channel required 4dB to 5dB

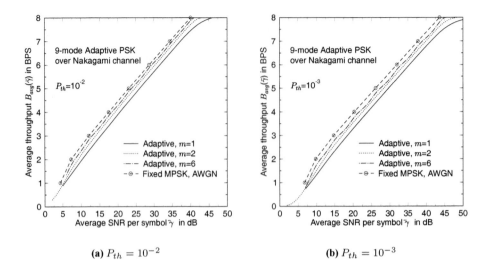

(a) $P_{th} = 10^{-2}$ (b) $P_{th} = 10^{-3}$

Figure 6.24: The effects of the Nakagami fading parameter m on the average BPS throughput of a nine-mode adaptive PSK scheme designed for the target BEP of (a) $P_{th} = 10^{-2}$ and (b) $P_{th} = 10^{-3}$. As m increases, the average throughput of the adaptive modem approaches the throughput of fixed PSK modems operating over an AWGN channel.

higher SNR for achieving the same throughput compared to the fixed PSK schemes operating over the AWGN channel.

Figure 6.25 summarises the relative SNR gains of our adaptive PSK schemes over the corresponding fixed PSK schemes. For the target BEP of $P_{th} = 10^{-3}$ the relative SNR gain of the nine-mode adaptive scheme compared to BPSK changes from 15.5dB to 1.3dB, as the Nakagami fading parameter changes from 1 to 6. Observing Figure 6.25(a) and Figure 6.25(b) we conclude that the advantages of employing adaptive PSK schemes are more pronounced when

1. the fading is more severe,

2. the target BEP is lower, and

3. the average BPS throughput is lower.

Having studied the range of APSK schemes, let us in the next section consider the family of adaptive coherently detected Star-QAM schemes.

6.5.1.2   Adaptive Coherent Star QAM Schemes

In this section, we study the performance of adaptive coherent QAM schemes employing Type-I Star constellations [4]. Even though non-coherent Star QAM (SQAM) schemes are more popular owing to their robustness to fading without requiring pilot symbol assisted channel estimation and Automatic Gain Control (AGC) at the receiver, the results provided

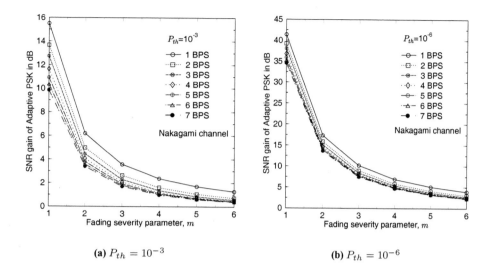

(a) $P_{th} = 10^{-3}$ **(b)** $P_{th} = 10^{-6}$

Figure 6.25: The SNR gain of adaptive PSK schemes in comparison to the corresponding fixed-mode PSK schemes yielding the same BPS throughput for the target BEP of (a) $P_{th} = 10^{-3}$ and (b) $P_{th} = 10^{-6}$. The performance advantage of employing adaptive PSK schemes decreases, as the fading becomes less severe.

in this section can serve as benchmark results for non-coherent Star QAM schemes and the coherent Square QAM schemes.

The BEP of coherent Star QAM over an AWGN channel is derived in Appendix A.4. It is shown that their BEP can be expressed as:

$$p_{SQAM}(\gamma) \simeq \sum_i A_i \, Q(\sqrt{a_i \gamma}), \tag{6.87}$$

where A_i and a_i are given in Appendix A.4 for 8-Star, 16-Star, 32-Star and 64-Star QAM. The SNR-dependent optimum ring ratios were also derived in Appendix A.4 for these Star QAM modems. Figure 6.26(a) shows the BEP of BPSK, QPSK, 8-Star QAM, 16-Star QAM, 32-Star QAM and 64-Star QAM employing the optimum ring ratios over the AWGN channel. Comparing Figure 6.19(a) and Figure 6.26(a), we can observe that 16-Star QAM, 32-Star QAM and 64-Star QAM are more power-efficient than 16 PSK, 32 PSK and 64 PSK, respectively. However, the envelope power of the Star QAM signals is not constant, unlike that of the PSK signals. Following an approach similar to that used in (6.84) and (6.85), the average BEP of the various SQAM schemes over a flat Nakagami-m fading channel can be expressed as:

$$P_{SQAM}(\bar{\gamma}) = \sum_i A_i \left[\tfrac{1}{2}(1 - \mu_i)\right]^m \sum_{j=0}^{m-1} \binom{m-1+j}{j} \left[\tfrac{1}{2}(1 + \mu_i)\right]^j, \tag{6.88}$$

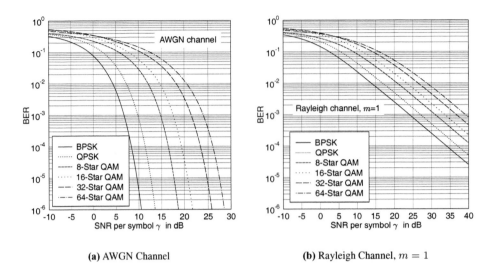

(a) AWGN Channel   (b) Rayleigh Channel, $m = 1$

Figure 6.26: The average BEP of various SQAM modulation schemes.

where μ_i is defined as:

$$\mu_i \triangleq \sqrt{\frac{a_i \bar{\gamma}}{2m + a_i \bar{\gamma}}} \ . \tag{6.89}$$

Figure 6.26(b) shows the average BEP of various SQAM schemes for transmission over a flat Rayleigh channel, where $m = 1$. It can be observed that the 16-Star, 32-Star and 64-Star QAM schemes exhibit SNR advantages of around 3.5dB, 4dB, and 7dB compared to 16-PSK, 32-PSK and 64-PSK schemes at a BEP of 10^{-2}. The BEP of SQAM for transmission over the AWGN channel given in (6.87) and that over a Nakagami channel given in (6.88) will be used in comparing the performance of the various adaptive SQAM schemes.

k	0	1	2	3	4	5	6
m_k	0	2	4	8	16	32	64
b_k	0	1	2	3	4	5	6
c_k	0	1	1	1	1	1	1
mode	No Tx	BPSK	QPSK	8-Star	16-Star	32-Star	64-Star

Table 6.6: Parameters of a seven-mode adaptive Star QAM scheme following the definitions developed in Section 6.3.1 for the generic adaptive modulation model.

The parameters of a seven-mode adaptive Star QAM scheme are summarised in Table 6.6 following the definitions of the generic model developed in Section 6.3.1 for adaptive modulation schemes. Since the number of modes is $K = 7$, we have $K + 1 = 8$ mode-switching levels hosted by the vector $\mathbf{s} = \{s_k \mid k = 0, 1, 2, \cdots, 7\}$. Let us assume that $s_0 = 0$ and $s_7 = \infty$. Then, we have to determine the optimum values for the remaining six switching

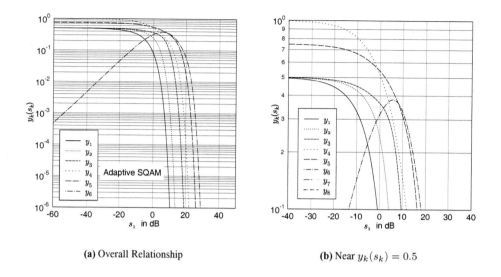

(a) Overall Relationship   **(b)** Near $y_k(s_k) = 0.5$

Figure 6.27: 'Lagrangian-free' functions of (6.61) for a seven-mode adaptive Star QAM scheme.

levels using the technique developed in Section 6.4.4. The 'Lagrangian-free' functions corresponding to a seven-mode Star QAM scheme are depicted in Figure 6.27 and the relationships obtained for the switching levels are displayed in Figure 6.28(a). We can observe that as seen for APSK in Figure 6.20 there exist two solutions for s_6 satisfying $y_6(s_6) = y_1(s_1)$ for a given value of s_1, when $y_1 \leq 0.382$. However, only the higher value of s_k satisfies the constraint of $s_6 \geq s_5$. When $s_1 \leq 7.9$dB, the optimum value of s_6 should be set to s_5, in order to guarantee $s_6 \geq s_5$. Figure 6.28(b) illustrates the optimum switching levels of a seven-mode adaptive Star QAM scheme operating over a Rayleigh channel at the target BEP of $P_{th} = 10^{-2}$. These switching levels were obtained by solving (6.68). The optimum switching levels show a steady decrease in their values, as the average SNR increases, until they reach the avalanche SNR value of $\bar{\gamma} = 28.5$dB, beyond which always the highest-order modulation mode, namely 64-Star QAM, is activated.

Let us now investigate the associated performance results. We are reminded that the average BEP of our optimised adaptive scheme remains constant at $P_{avg} = P_{th}$, provided that the average SNR is less than the avalanche SNR. Hence, the average BPS throughput and the SNR gain of our adaptive modem in comparison to the corresponding fixed-mode modems are our concern.

Let us first consider Figure 6.29, where the average BPS throughput of the various adaptive Star QAM schemes operating over a Rayleigh channel associated with $m = 1$ is shown at the target BEP of $P_{th} = 10^{-2}$ and $P_{th} = 10^{-3}$. The markers '⊗' and '⊙' represent the required SNR of the corresponding fixed-mode Star QAM schemes, while achieving the same target BEP as the adaptive schemes, operating over an AWGN channel and a Rayleigh channel, respectively. Comparing Figure 6.22(a) and Figure 6.29(a), we find that the tangent of the average BPS curves of the adaptive Star QAM schemes is higher than that of adaptive PSK schemes. Explicitly, the tangent of the Star QAM schemes is around 0.3BPS/dB, whereas that of the APSK schemes was 0.18BPS/dB. This is due to the more power-efficient constel-

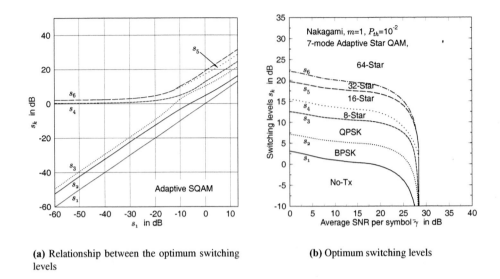

(a) Relationship between the optimum switching levels

(b) Optimum switching levels

Figure 6.28: Optimum switching levels of a seven-mode Adaptive Star QAM scheme. (a) Relationships between s_k and s_1. (b) Optimum switching levels of a seven-mode adaptive Star QAM scheme operating over a Rayleigh channel at the target BEP of $P_{th} = 10^{-2}$.

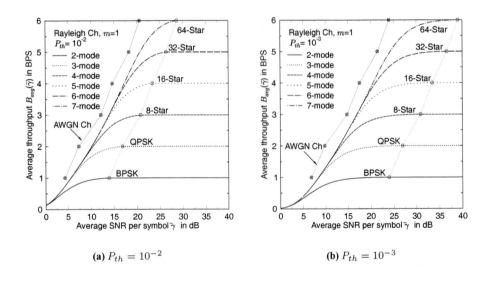

(a) $P_{th} = 10^{-2}$

(b) $P_{th} = 10^{-3}$

Figure 6.29: The average BPS throughput of the various adaptive Star QAM schemes operating over a Rayleigh fading channel associated with $m = 1$ at the target BEP of (a) $P_{th} = 10^{-2}$ and (b) $P_{th} = 10^{-3}$. The markers '$\otimes$' and '$\odot$' represent the required SNR of the corresponding fixed-mode Star QAM schemes, while achieving the same target BEP as the adaptive schemes, operating over an AWGN channel and a Rayleigh channel, respectively.

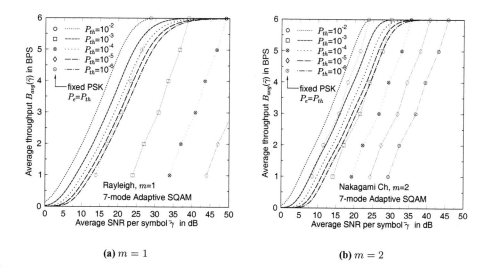

(a) $m = 1$ **(b)** $m = 2$

Figure 6.30: The average BPS throughput of a seven-mode adaptive Star QAM scheme operating over a Nakagami fading channel (a) $m = 1$ and (b) $m = 2$. The markers represent the SNR required by the fixed-mode schemes for achieving the same BPS throughput and the same average BER as the adaptive schemes.

lation arrangement of Star QAM in comparison to the single-ring constellations of the PSK modulations schemes. In comparison to the corresponding fixed-mode Star QAM modems, the adaptive modem achieved an SNR gain of 6dB to 8dB for the target BEP of $P_{th} = 10^{-2}$ and 12dB to 16dB for the target BEP of $P_{th} = 10^{-3}$ over a Rayleigh channel. Compared to the fixed-mode Star QAM schemes operating over an AWGN channel, our adaptive schemes approached their performance within about 3dB in terms of the required SNR value, while achieving the same target BEP of $P_{th} = 10^{-2}$ and $P_{th} = 10^{-3}$.

Since Figure 6.29 suggests that the relative SNR gain of the adaptive schemes is dependent on the target BER, let us investigate the effects of the target BEP in more detail. Figure 6.30 shows the BPS throughput of the various adaptive schemes at the target BEP of $P_{th} = 10^{-2}$ to $P_{th} = 10^{-6}$. The average BPS throughput of a seven-mode adaptive Star QAM scheme is represented with the aid of the various line types without markers, depending on the target average BERs, while those of the corresponding fixed-mode Star QAM schemes are represented as various types of lines having markers according to the legends shown in Figure 6.30. We can observe that the difference between the SNRs required for the adaptive schemes and fixed schemes increases, as the target BEP decreases. The fixed-mode Star QAM schemes require additional SNRs of 10dB and 6dB in order to achieve an order of magnitude lower BEP for the Nakagami fading parameters of $m = 1$ and $m = 2$, respectively. However, our adaptive schemes require additional SNRs of only 1dB to 3dB for achieving the same goal.

Let us now investigate the effects of the Nakagami fading parameter m on the average BPS throughput performance of the various adaptive Star QAM schemes by observing Figure 6.31. The BPS throughput of the fixed-mode Star QAM schemes for the transmission over

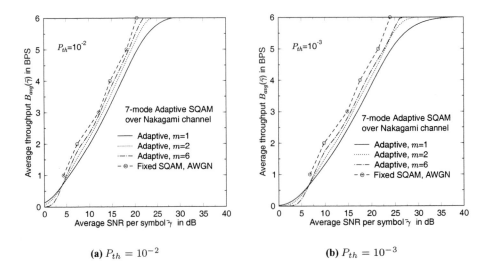

(a) $P_{th} = 10^{-2}$ (b) $P_{th} = 10^{-3}$

Figure 6.31: The effects of the Nakagami fading parameter m on the average BPS throughput of a seven-mode adaptive Star QAM scheme at the target BEP of (a) $P_{th} = 10^{-2}$ and (b) $P_{th} = 10^{-3}$. As m increases, the average throughput of the adaptive modem approaches the throughput of the fixed-mode Star QAM modems operating over an AWGN channel.

an AWGN channel is depicted in Figure 6.31 as the ultimate performance limit achievable by the adaptive schemes operating over Nakagami fading channels. As the Nakagami fading parameter m increases from 1 to 2 and to 6, the SNR gap between the adaptive schemes operating over a Nakagami fading channel and the fixed-mode schemes decreases. When the average SNR is less than $\bar{\gamma} \leq 6$dB, the average BPS throughput of our adaptive schemes decreases, when the fading parameter m increases. The rationale of this phenomenon is that as the channel becomes more and more like an AWGN channel, the probability of activating the BPSK mode is reduced, resulting in more frequent activation of the No-Tx mode and hence the corresponding average BPS throughput inevitably decreases.

The effects of the Nakagami fading factor m on the SNR gain of our adaptive Star QAM scheme can be observed in Figure 6.32. As expected, the relative SNR gain of the adaptive schemes at a throughput of 1 BPS is the highest among the BPS throughputs considered. However, the order observed in terms of the SNR gain of the adaptive schemes does not strictly follow the increasing BPS order at the target BEP of $P_{th} = 10^{-3}$ and $P_{th} = 10^{-6}$, as it did for the adaptive PSK schemes of Section 6.5.1.1. Even though the adaptive Star QAM schemes exhibit a higher throughput, than the adaptive PSK schemes, the SNR gains compared to their fixed-mode counterparts are more or less the same, showing typically less than 1dB difference, except for the 5 BPS throughput scenario, where the adaptive QAM scheme gained up to 1.3dB more in terms of the required SNR than the adaptive PSK scheme.

Having studied the performance of a range of adaptive Star QAM schemes, in the next section we consider adaptive modulation schemes employing the family of square-shaped QAM constellations.

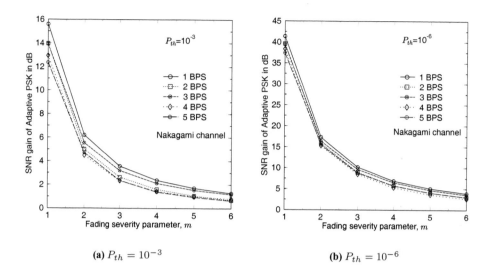

(a) $P_{th} = 10^{-3}$ **(b)** $P_{th} = 10^{-6}$

Figure 6.32: The SNR gain of the various adaptive Star QAM schemes in comparison to the fixed-mode Star QAM schemes yielding the same BPS throughput at the target BEP of (a) $P_{th} = 10^{-3}$ and (b) $P_{th} = 10^{-6}$. The advantage of the adaptive Star QAM schemes decreases, as the fading becomes less severe.

6.5.1.3 Adaptive Coherent Square QAM Modulation Schemes

Since coherent Square M-ary QAM (MQAM) is the most power-efficient M-ary modulation scheme [4] and the accurate channel estimation becomes possible with the advent of Pilot Symbol Assisted Modulation (PSAM) techniques [18–20], *Otsuki*, *Sampei* and *Morinaga* proposed to employ coherent square QAM as the constituent modulation modes for an adaptive modulation scheme [21] instead of non-coherent Star QAM modulation [17]. In this section, we study the various aspects of this adaptive square QAM scheme employing the optimum switching levels of Section 6.4.4. The closed form BEP expressions of square QAM over an AWGN channel can be found in (6.2) and that over a Nakagami channel can be expressed using a similar form given in (6.88). The optimum switching levels of adaptive Square QAM were studied in Section 6.4.4 as an example.

The average BEP of our six-mode adaptive Square QAM scheme operating over a flat Rayleigh fading channel is depicted in Figure 6.33(a), which shows that the modem maintains the required constant target BER, until it reaches the BER curve of the specific fixed-mode modulation scheme employing the highest-order modulation mode, namely 256-QAM, and then it follows the BEP curve of the 256-QAM mode. The various grey lines in the figure represent the BEP of the fixed constituent modulation modes for transmission over a flat Rayleigh fading channel. An arbitrarily low target BEP could be maintained at the expense of a reduced throughput.

The average throughput is shown in Figure 6.33(b) together with the estimated channel capacity of the narrow-band Rayleigh channel [22, 23] and with the throughput of several variable-power, variable-rate modems reported in [25]. Specifically, Goldsmith and

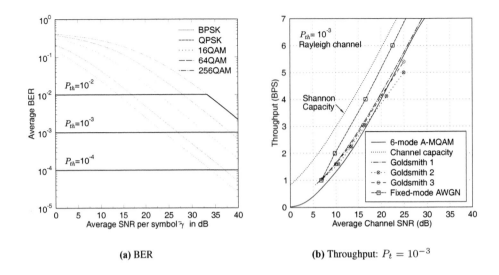

(a) BER

(b) Throughput: $P_t = 10^{-3}$

Figure 6.33: The average BEP and average throughput performance of a six-mode adaptive Square QAM scheme operating over a flat Rayleigh channel ($m = 1$). (a) The constant target average BEP is maintained over the entire range of the average SNR values up to the avalanche SNR. (b) The average BPS throughput of the equivalent constant-power adaptive scheme is compared to *Goldsmith*'s schemes [25]. The 'Goldsmith 1' and 'Goldsmith 2' schemes represent a variable-power adaptive scheme employing hypothetical continuously variable-BPS QAM modulation modes and Square QAM modes, respectively. The 'Goldsmith 3' scheme represents the simulation results associated with a constant-power adaptive Square QAM reported in [25].

Chua [25] studied the performance of their variable-power variable-rate adaptive modems based on a BER bound of m-ary Square QAM, rather than using an exact BER expression. Since our adaptive Square QAM schemes do not vary the transmission power, our scheme can be regarded as a sub-optimal policy viewed for their respective [25]. However, the throughput performance of Figure 6.33(b) shows that the SNR degradation is within 2dB in the low-SNR region and within half a dB in the high-SNR region, in comparison to the ideal continuously variable-power adaptive QAM scheme employing a range of hypothetical continuously variable-BPS QAM modes [25], represented as the 'Goldsmith 1' scheme in the figure. *Goldsmith* and *Chua* [25] also reported the performance of a variable-power discrete-rate and a constant-power discrete-rate scheme, which we represented as the 'Goldsmith 2' and 'Goldsmith 3' scenarios in Figure 6.33(b), respectively. Since their results are based on approximate BER formulas, the average BPS throughput performance of the 'Goldsmith 3' scheme is optimistic, when the average SNR γ is less than 17dB. Considering that our scheme achieves the maximum possible throughput the given average SNR value with the aid of the globally optimised switching levels, the average throughput of the 'Goldsmith 3' scheme is expected to be lower, than that of our scheme, as is the case when the average SNR γ is higher than 17dB.

Figure 6.34(a) depicts the average BPS throughput of our various adaptive Square QAM

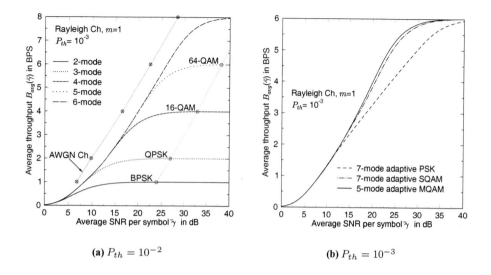

(a) $P_{th} = 10^{-2}$ **(b)** $P_{th} = 10^{-3}$

Figure 6.34: The average BPS throughput of various adaptive Square QAM schemes operating over a
Rayleigh channel ($m = 1$) at the target BEP of $P_{th} = 10^{-3}$. (a) The markers '$\otimes$' and
'$\odot$' represent the required SNR of the corresponding fixed-mode Square QAM schemes
achieving the same target BEP as the adaptive schemes, operating over an AWGN channel
and a Rayleigh channel, respectively. (b) The comparison of the various adaptive schemes
employing PSK, Star QAM and Square QAM as the constituent modulation modes.

schemes operating over a Rayleigh channel associated with $m = 1$ at the target BEP of
$P_{th} = 10^{-3}$. Figure 6.34(a) shows that even though the constituent modulation modes of our
adaptive schemes do not include 3, 5 and 7-BPS constellations, the average BPS throughput
steadily increases without undulations. Compared to the fixed-mode Square QAM schemes
operating over an AWGN channel, our adaptive schemes require additional SNRs of less
than 3.5dB, when the throughput is below 6.5 BPS. The comparison of the average BPS
throughputs of the adaptive schemes employing PSK, Star QAM and Square QAM modems,
as depicted in Figure 6.34(b), confirms the superiority of Square QAM over the other two
schemes in terms of the required average SNR for achieving the same throughput and the
same target average BEP. Since all these three schemes employ BPSK, QPSK as the second
and the third constituent modulation modes, their throughput performance shows virtually no
difference, when the average throughput is less than or equal to $B_{avg} = 2$ BPS.

Let us now investigate the effects of the Nakagami fading parameter m on the average
BPS throughput performance of the adaptive Square QAM schemes observing Figure 6.35.
The BPS throughput of the fixed-mode Square QAM schemes over an AWGN channel is
depicted in Figure 6.35 as the ultimate performance limit achievable by the adaptive schemes
operating over Nakagami fading channels. Similar observations can be made for the adaptive
Square QAM scheme, like for the adaptive Star QAM arrangement characterized in Figure 6.31. A specific difference is, however, that the average BPS throughput recorded for
the fading parameter of $m = 6$ exhibits an undulating curve. For example, an increased m
value results in a limited improvement of the corresponding average BPS throughput near

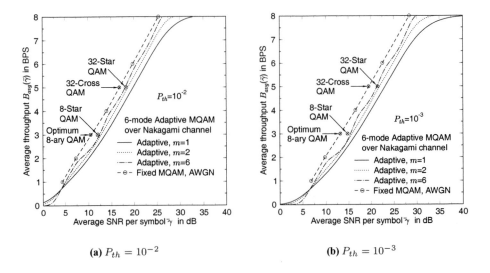

(a) $P_{th} = 10^{-2}$ **(b)** $P_{th} = 10^{-3}$

Figure 6.35: The effects of the Nakagami fading parameter m on the average BPS throughput of a seven-mode adaptive Square QAM scheme at the target BEP of (a) $P_{th} = 10^{-2}$ and (b) $P_{th} = 10^{-3}$. As m increases, the average throughput of the adaptive modem approaches the throughput of the corresponding fixed Square QAM modems operating over an AWGN channel.

the throughput values of 2.5, 4.5 and 6.5 BPS. This is because our adaptive Square QAM schemes do not use 3-, 5- and 7-BPS constituent modems, unlike the adaptive PSK and adaptive Star QAM schemes. Figure 6.36 depicts the corresponding optimum mode-switching levels for the six-mode adaptive Square QAM scheme. The black lines represent the switching levels, when the Nakagami fading parameter is $m = 6$ and the grey lines when $m = 1$. In general, the lower the switching levels, the higher the average BPS throughput of the adaptive modems. When the Nakagami fading parameter is $m = 1$, the switching levels decrease monotonically, as the average SNR increases. However, when the fading severity parameter is $m = 6$, the switching levels fluctuate, exhibiting several local minima around 8dB, 15dB and 21dB. In the extreme case of $m \to \infty$, *i.e.* when operating over an AWGN-like channel, the switching levels would be $s_1 = s_2 = 0$ and $s_k = \infty$ for other k values in the SNR range of 7.3dB $< \bar{\gamma} < 14$dB, $s_1 = s_2 = s_3 = 0$ and $s_4 = s_5 = \infty$ when we have 14dB $< \bar{\gamma} < 20$dB, $s_k = 0$ except for $s_5 = \infty$ when the SNR is in the range of 20dB $< \bar{\gamma} < 25$dB and finally, all $s_k = 0$ for $\forall k$, when $\bar{\gamma} > 25$dB, when considering the fixed-mode Square QAM performance achieved over an AWGN channel represented by markers '⊙' in Figure 6.35. Observing Figure 6.37, we find that our adaptive schemes become highly 'selective', when the Nakagami fading parameter becomes $m = 6$, exhibiting narrow triangular shapes. As m increases, the shapes will eventually converge to Kronecker delta functions.

A possible approach to reducing the undulating behaviour of the average BPS throughput curve is the introduction of a 3-BPS and a 5-BPS mode as additional constituent modem modes. The power-efficiency of 8-Star QAM and 32-Star QAM is insufficient for maintaining a linear growth of the average BPS throughput, as we can observe in Figure 6.35. Instead, the

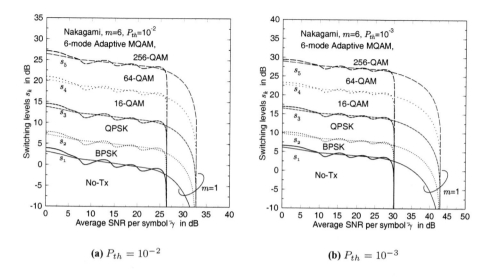

Figure 6.36: The switching levels of the six-mode adaptive Square QAM scheme operating over Nakagami fading channels at the target BER of (a) $P_{th} = 10^{-2}$ and (b) $P_{th} = 10^{-3}$. The bold lines are used for the fading parameter of $m = 6$ and the grey lines are for $m = 1$.

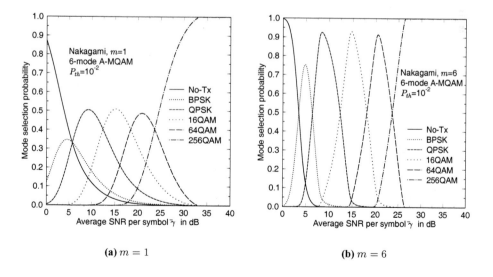

Figure 6.37: The mode selection probability of a six-mode adaptive Square QAM scheme operating over Nakagami fading channels at the target BEP of $P_{th} = 10^{-2}$. When the fading becomes less severe, the mode selection scheme becomes more 'selective' in comparison to that for $m = 1$.

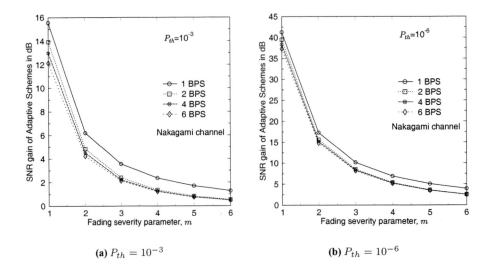

(a) $P_{th} = 10^{-3}$ **(b)** $P_{th} = 10^{-6}$

Figure 6.38: The SNR gain of the six-mode adaptive Square QAM scheme in comparison to the various fixed-mode Square QAM schemes yielding the same BPS throughput at the target BEP of (a) $P_{th} = 10^{-3}$ and (b) $P_{th} = 10^{-6}$. The performance advantage of the adaptive Square QAM schemes decreases, as the fading becomes less severe.

most power-efficient 8-ary QAM scheme [87, pp 279] and 32-ary Cross QAM scheme [4, pp 236] have a potential of reducing these undulation effects. However, since we observed in Section 6.5.1.1 and Section 6.5.1.2 that the relative SNR advantage of employing adaptive Square QAM rapidly reduces, when the Nakagami fading parameter increases, even though the additional 3-BPS and 5-BPS modes are also used, there seems to be no significant benefit in employing non-square shaped additional constellations.

Again, we can observe in Figure 6.35 that when the average SNR is less than $\bar{\gamma} \leq 6$dB, the average BPS throughput of our adaptive Square QAM scheme decreases, as the Nakagami fading parameter m increases. As we discussed in Section 6.5.1.2, this is due to the less frequent activation of the BPSK mode in comparison to the 'No-Tx' mode, as the channel variation is reduced.

The effects of the Nakagami fading factor m on the relative SNR gain of our adaptive Square QAM scheme can be observed in Figure 6.38. The less severe the fading, the smaller the relative SNR advantage of employing adaptive Square QAM in comparison to its fixed-mode counterparts. Except for the 1-BPS mode, the SNR gains become less than 0.5dB, when m is increased to 6 at the target BEP of $P_{th} = 10^{-3}$. The trend observed is the same at the target BEP of $P_{th} = 10^{-6}$, showing relatively higher gains in comparison to the $P_{th} = 10^{-3}$ scenario.

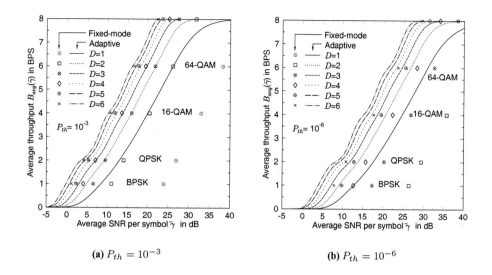

(a) $P_{th} = 10^{-3}$ **(b)** $P_{th} = 10^{-6}$

Figure 6.39: The average BPS throughput of the MRC-aided antenna-diversity assisted adaptive Square QAM scheme operating over independent Rayleigh fading channels at the target average BEP of (a) $P_{th} = 10^{-3}$ and (b) $P_{th} = 10^{-6}$. The markers represent the corresponding fixed-mode Square QAM performances.

6.5.2 Performance over Narrow-band Rayleigh Channels Using Antenna Diversity

In the last section, we observed that the adaptive modulation schemes employing Square QAM modes exhibit the highest BPS throughput among the schemes investigated, when operating over Nakagami fading channels. Hence, in this section we study the performance of the adaptive Square QAM schemes employing antenna diversity operating over independent Rayleigh fading channels. The BEP expression of the fixed-mode coherent BPSK scheme can be found in [87, pp 781] and those of coherent Square QAM can be readily extended using the equations in (6.2) and (6.3). Furthermore, the antenna diversity scheme operating over independent narrow-band Rayleigh fading channels can be viewed as a special case of the two-dimensional (2D) Rake receiver analysed in Appendices A.5 and A.6. The performance of antenna-diversity assisted adaptive Square QAM schemes can be readily analysed using the technique developed in Section 6.4.4.

Figure 6.39 depicts the average BPS throughput performance of our adaptive schemes employing Maximal Ratio Combining (MRC) aided antenna diversity [199, Ch 5, 6] operating over independent Rayleigh fading channels at the target average BEP of $P_{th} = 10^{-3}$ and $P_{th} = 10^{-6}$. The markers represent the performance of the corresponding fixed-mode Square QAM modems in the same scenario. The average SNRs required achieving the target BEP of the fixed-mode schemes and that of the adaptive schemes decrease, as the antenna diversity order increases. However, the differences between the required SNRs of the adaptive schemes and their fixed-mode counterparts also decrease, as the antenna diversity order increases. The SNRs of both schemes required for achieving the target BEPs of $P_{th} = 10^{-3}$.

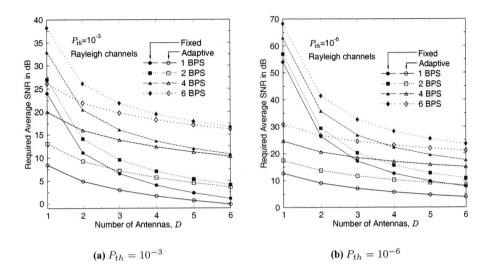

(a) $P_{th} = 10^{-3}$ **(b)** $P_{th} = 10^{-6}$

Figure 6.40: The SNR required for the MRC-aided antenna-diversity assisted adaptive Square QAM schemes and the corresponding fixed-mode modems operating over independent Rayleigh fading channels at the target average BEP of (a) $P_{th} = 10^{-3}$ and (b) $P_{th} = 10^{-6}$.

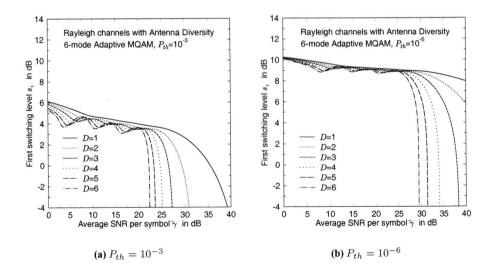

(a) $P_{th} = 10^{-3}$ **(b)** $P_{th} = 10^{-6}$

Figure 6.41: The first switching level s_1 of the MRC-aided antenna-diversity assisted adaptive Square QAM scheme operating over independent Rayleigh fading channels at the target average BEP of (a) $P_{th} = 10^{-3}$ and (b) $P_{th} = 10^{-6}$.

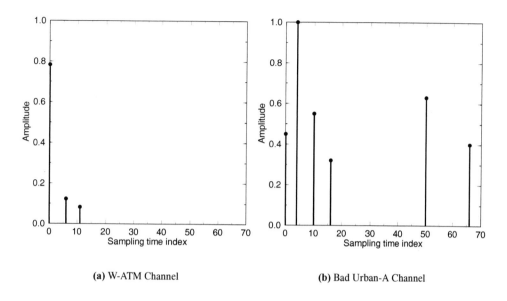

(a) W-ATM Channel **(b)** Bad Urban-A Channel

Figure 6.42: Multi-path Intensity Profiles (MIPs) of the Wireless Asynchronous Transfer Mode (W-ATM) indoor channel [4, Ch 20] and that of the Bad-Urban Reduced-model A (BU-RA) channel [200].

and $P_{th} = 10^{-6}$ are displayed in Figure 6.40, where we can observe that dual antenna diversity is sufficient for the fixed-mode schemes in order to obtain half of the achievable SNR gain of the six-antenna aided diversity scheme, whereas triple-antenna diversity is required for the adaptive schemes operating in the same scenario. The corresponding first switching levels s_1 are depicted in Figure 6.41 for different orders of antenna diversity up to an order of six. As the antenna diversity order increases, the avalanche SNR becomes lower and the switching-threshold undulation effects begin to appear. The required values of the first switching level s_1 are within a range of about 1dB and 0.5dB for the target BEPs of $P_{th} = 10^{-3}$ and $P_{th} = 10^{-6}$, respectively, before the avalanche SNR is reached. This suggests that the optimum mode-switching levels are more dependent on the target BEP, than on the number of diversity antennas.

6.5.3 Performance over Wideband Rayleigh Channels using Antenna Diversity

Wideband fading channels are characterized by their multi-path intensity profiles (MIP). In order to study the performance of the various adaptive modulation schemes, we employ two different MIP models in this section, namely a shortened Wireless Asynchronous Transfer Mode (W-ATM) channel [4, Ch 20] for an indoor scenario and a Bad-Urban Reduced-model A (BU-RA) channel [200] for a hilly urban outdoor scenario. Their MIPs are depicted in Figure 6.42. The W-ATM channel exhibits short-range, low-delay multi-path components, while the BU-RA channel exhibits six higher-delay multi-path components. Again, let us

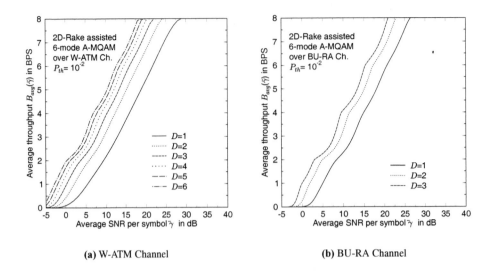

(a) W-ATM Channel   **(b)** BU-RA Channel

Figure 6.43: The effects of the number of diversity antennas D on the average BPS throughput of the 2D-Rake assisted six-mode adaptive Square QAM scheme operating over the wideband independent Rayleigh fading channels characterized in Figure 6.42 at the target BEP of $P_{th} = 10^{-2}$.

assume that our receivers are equipped with MRC Rake receivers [201], employing a sufficiently higher number of Rake fingers, in order to capture all the multi-path components generated by our channel models. Furthermore, we employ antenna diversity [199, Ch 5] at the receivers. This combined diversity scheme is often referred to as a two-dimensional (2D) Rake receiver [202, pp 263]. The BEP of the 2D Rake receiver transmission over wide-band independent Rayleigh fading channels is analysed in Appendix A.5. A closed-form expression for the mode-specific average BEP of a 2D-Rake assisted adaptive Square QAM scheme is also given in Appendix A.6. Hence, the performance of our 2D-Rake assisted adaptive modulation scheme employing the optimum switching levels can be readily obtained.

The average BPS throughputs of the 2D-Rake assisted adaptive schemes operating over the two different types of wideband channel scenarios are presented in Figure 6.43 at the target BEP of $P_{th} = 10^{-2}$. The throughput performance depicted corresponds to the upper-bound performance of Direct-Sequence Code Division Multiple Access (DS-CDMA) or Multi-Carrier CDMA employing Rake receivers and the MRC-aided diversity assisted scheme in the absence of Multiple Access Interference (MAI). We can observe that the BPS throughput curves undulate, when the number of antennas D increases. This effect is more pronounced for transmission over the BU-RA channel, since the BU-RA channel exhibits six multi-path components, increasing the available diversity potential of the system approximately by a factor of two in comparison to that of the W-ATM channel. The performance of our adaptive scheme employing more than three antennas for transmission over the BU-RA channel could not be obtained owing to numerical instability, since the associated curves become similar to a series of step-functions, which is not analytic in mathematical terms. A similar observation can be made in the context of Figure 6.44, where the target BEP is $P_{th} = 10^{-3}$. Comparing

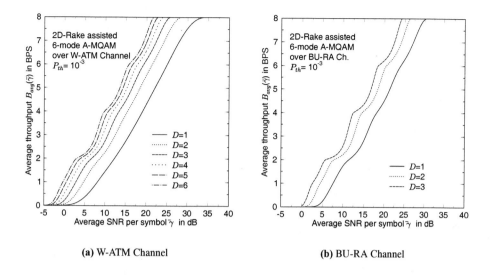

(a) W-ATM Channel   (b) BU-RA Channel

Figure 6.44: The effects of the number of diversity antennas D on the average BPS throughput of the 2D-Rake assisted six-mode adaptive Square QAM scheme operating over the wideband independently Rayleigh fading channels characterized in Figure 6.42 at the target BEP of $P_{th} = 10^{-3}$.

Figure 6.43 and Figure 6.44, we observe that the BPS throughput curves corresponding to $P_{th} = 10^{-3}$ are similar to shifted versions of those corresponding to $P_{th} = 10^{-2}$, which are shifted in the direction of increasing SNRs. On the other hand, the BPS throughput curves corresponding to $P_{th} = 10^{-3}$ undulate more dramatically. When the number of antennas is $D = 3$, the BPS throughput curves of the BU-RA channel exhibit a stair-case like shape. The corresponding mode switching levels and mode selection probabilities are shown in Figure 6.45. Again, the switching levels heavily undulate. The mode-selection probability curve of BPSK has a triangular shape, increasing linearly, as the average SNR $\bar{\gamma}$ increases to 2.5dB and decreasing linearly again as $\bar{\gamma}$ increases from 2.5dB. On the other hand, the mode-selection probability curve of QPSK increases linearly and decreases exponentially, since no 3-BPS mode is used. This explains, why the BPS throughput curves increase in a near-linear fashion in the SNR range of 0 to 5dB and in a stair-case fashion beyond that point. We can conclude that the stair-case like shape in the upper SNR range of SNR is a consequence of the absence of the 3-BPS, 5-BPS and 7-BPS modulation modes in the set of constituent modulation modes employed. As we discussed in Section 6.5.1.3, this problem may be mitigated by introducing power-efficient 8 QAM, 32 QAM and 128 QAM modes.

The average SNRs required achieving the target BEP of $P_{th} = 10^{-3}$ by the 2D-Rake assisted adaptive scheme and the fixed-mode schemes operating over wide-band fading channels are depicted in Figure 6.46. Since the fixed-mode schemes employing Rake receivers are already enjoying the diversity benefit of multi-path fading channels, the SNR advantages of our adaptive schemes are less than 8dB and 2.6dB over the W-ATM channel and over the BU-RA channel, respectively, even when a single antenna is employed. This relatively small SNR gain in comparison to those observed over narrow-band fading channels in Figure 6.40 erodes

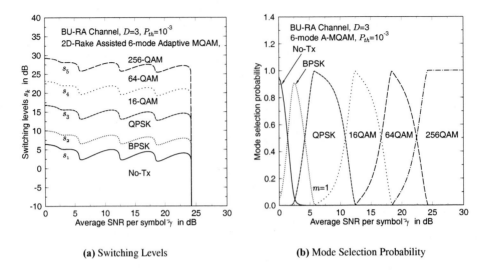

(a) Switching Levels   **(b)** Mode Selection Probability

Figure 6.45: The mode switching levels and mode selection probability of the 2D-Rake assisted six-mode adaptive Square QAM scheme using $D = 3$ antennas operating over the BU-RA channel characterized in Figure 6.42(b) at the target BEP of $P_{th} = 10^{-3}$.

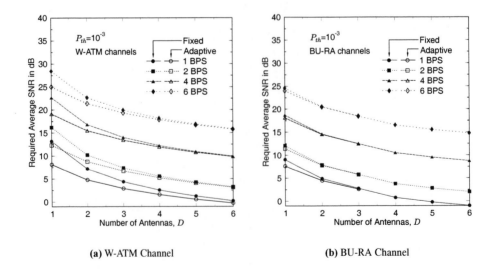

(a) W-ATM Channel   **(b)** BU-RA Channel

Figure 6.46: The average SNRs required for achieving the target BEP of $P_{th} = 10^{-3}$ by the 2D-Rake assisted adaptive schemes and by the fixed-mode schemes operating over (a) the W-ATM channel and (b) the BU-RA channel.

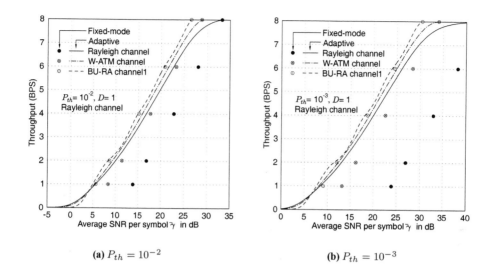

(a) $P_{th} = 10^{-2}$ **(b)** $P_{th} = 10^{-3}$

Figure 6.47: The average BPS throughput of the adaptive schemes and fixed-mod schemes transmission over a narrow-band Rayleigh channel, the W-ATM channel and BU-RA channel of Figure 6.42 at the target BEP of (a) $P_{th} = 10^{-2}$ and (b) $P_{th} = 10^{-3}$.

as the number of antennas increases. For example, when the number of antennas is $D = 6$, the SNR gains of the adaptive schemes operating over the W-ATM channel of Figure 6.42(a) become virtually zero, where the combined channel becomes an AWGN-like channel. On the other hand, $D = 3$ number of antennas is sufficient for the BU-RA channel for exhibiting such a behaviour, since the underlying multi-path diversity provided by the six-path BU-RA channel is higher than that of the tree-path W-ATM channel.

6.5.4 Uncoded Adaptive Multi-Carrier Schemes

The performance of the various adaptive Square QAM schemes has been studied also in the context of multi-carrier systems [4, 203, 204] . The family of Orthogonal Frequency Division Multiplex (OFDM) [205] systems converts frequency selective Rayleigh channels into frequency non-selective or flat Rayleigh channels for each sub-carrier, provided that the number of sub-carriers is sufficiently high. The power and bit allocation strategy of adaptive OFDM has attracted substantial research interests [4]. OFDM is particularly suitable for combined time-frequency domain processing [204]. Since each sub-carrier of an OFDM system experiences a flat Rayleigh channel, we can apply adaptive modulation for each sub-carrier independently from other sub-carriers. Although a practical scheme would group the sub-carriers into similar-quality sub-bands for the sake of reducing the associated modem mode signalling requirements. The performance of this AQAM assisted OFDM (A-OFDM) scheme is identical to that of the adaptive scheme operating over flat Rayleigh fading channels, characterized in Section 6.5.2.

MC-CDMA [206, 207] receiver can be regarded as a frequency domain Rake-receiver, where the multiple carriers play a similar role to that of the time-domain Rake fingers. Our simulation results showed that the single-user BEP performance of MC-CDMA employing multiple antennas is essentially identical to that of the time-domain Rake receiver using antenna diversity, provided that the spreading factor is higher than the number of resolvable multi-path components in the channel. Hence, the throughput of the Rake-receiver over the three-path W-ATM channel [4] and the six-path BU-RA channel [200] studied in Section 6.5.3 can be used for investigating the upper-bound performance of adaptive MC-CDMA schemes over these channels. Figure 6.47 compares the average BPS throughput performances of these schemes, where the throughput curves of the various adaptive schemes are represented as three different types of lines, depending on the underlying channel scenarios, while the fixed-mode schemes are represented as three different types of markers. The solid line corresponds to the performance of A-OFDM and the marker '•' corresponds to that of the fixed-mode OFDM. On the other hand, the dotted lines correspond to the BPS throughput performance of adaptive MC-CDMA operating over wide-band channels and the markers '⊙' and '⊗' to those of the fixed-mode MC-CDMA schemes.

It can be observed that fixed-mode MC-CDMA has a potential to outperform A-OFDM, when the underlying channel provides sufficient diversity due to the high number of resolvable multi-path components. For example, the performance of fixed-mode MC-CDMA operating over the W-ATM channel of Figure 6.42(a) is slightly lower than that of A-OFDM for the BPS range of less than or equal to 6 BPS, owing to the insufficient diversity potential of the wide-band channel. On the other hand, fixed-mode MC-CDMA outperforms A-OFDM, when the channel is characterized by the BU-RA model of Figure 6.42(b). We have to consider several factors, in order to answer, whether fixed-mode MC-CDMA is better than A-OFDM. Firstly, fully loaded MC-CDMA, which can transmit the same number of symbols as OFDM, suffers from multi-code interference and our simulation results showed that the SNR degradation is about 2-4dB at the BEP of 10^{-3}, when the Minimum Mean Square Error Block Decision Feedback Equalizer (MMSE-BDFE) [208] based joint detector is used at the receiver. Considering these SNR degradations, the throughput of fixed-mode MC-CDMA using the MMSE-BDFE joint detection receiver falls just below that of the A-OFDM scheme, when the channel is characterized by the BU-RA model. On the other hand, the adaptive schemes may suffer from inaccurate channel estimation/prediction and modem mode signalling feedback delay [25]. Hence, the preference order of the various schemes may depend on the channel scenario encountered, on the interference effects and other practical issues, such as the aforementioned channel estimation accuracy, feedback delays, etc.

6.5.5   Concatenated Space-Time Block Coded and Turbo Coded Symbol-by-Symbol Adaptive OFDM and Multi-Carrier CDMA[3]

In the previous sections we studied the performance of uncoded adaptive schemes. Since a Forward Error Correction (FEC) code reduces the SNR required for achieving a given target BEP at the expense of a reduced BPS throughput, it is interesting to investigate the performance of adaptive schemes employing FEC techniques. These investigations will allow us to gauge, whether channel coding is capable of increasing the system's effective throughput, when aiming for a specific target BER. Another important question to be answered is whether there are any further potential performance advantages, when we combine adaptive modula-

---

[3]This section was based on collaborative research with the contents of [209].

tion with space-time coding. We note in advance that our related investigations are included here with a view to draw the reader's attention to the associated system design trade-offs, rather than to provide an indepth comparative study of adaptive modulation and space-time coding. Hence here we will be unable to elaborate on the philosophy of space-time coding, we will simply refer to the associated literature for background reading. However, the topic of space-time coding will be revisited in significantly more depth in Chapter 14.

A variety of FEC techniques has been used in the context of adaptive modulation schemes. In their pioneering work on adaptive modulation, Webb and Steele [17] used a set of binary BCH codes. Vucetic [210] employed various punctured convolutional codes in response to the time-variant channel status. On the other hand, various Trellis Coded Modulation (TCM) [174,211] schemes were used in the context of adaptive modulation by Alamouti and Kallel [182], Goldsmith and Chua [212], as well as Hole, Holm and Øien [50].

Keller, Liew and Hanzo studied the performance of Redundant Residue Number System (RRNS) codes in the context of adaptive multi-carrier modulation [213,214]. Various turbo coded adaptive modulation schemes have been investigated also by Liew, Wong, Yee and Hanzo [44,215,216]. With the advent of space-time (ST) coding techniques [7,217,218], various concatenated coding schemes combining ST coding and FEC coding can be applied in adaptive modulation schemes. In this section, we investigate the performance of various concatenated space-time block-coded and turbo-coded adaptive OFDM and MC-CDMA schemes.

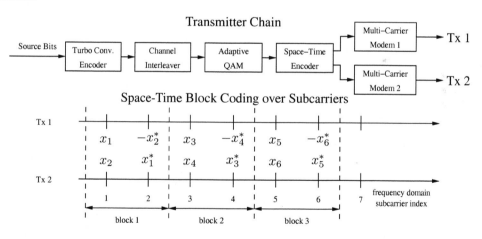

Figure 6.48: Transmitter structure and space-time block encoding scheme

Figure 6.48 portrays the stylised transmitter structure of our system. The source bits are channel coded by a half-rate turbo convolutional encoder [152] using a constraint length of $K = 3$ as well as an interleaver having a memory of $L = 3072$ bits and interleaved by a random block interleaver. Then, the AQAM block selects a modulation mode from the set of no transmission, BPSK, QPSK, 16-QAM and 64-QAM depending on the instantaneous channel quality perceived by the receiver, according to the SNR-dependent optimum switching levels derived in Section 6.4.4. It is assumed that the perfectly estimated channel quality experienced by receiver A is fed back to transmitter B superimposed on the next burst transmitted to receiver B. The modulation mode switching levels of our AQAM scheme determine the

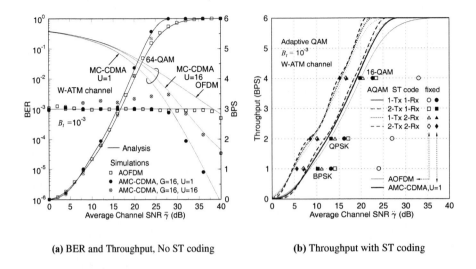

(a) BER and Throughput, No ST coding   (b) Throughput with ST coding

Figure 6.49: Performance of uncoded five-mode AOFDM and AMC-CDMA. The target BEP is $B_t = 10^{-3}$ when transmitting over the W-ATM channel [4, pp.474]. (a) The constant average BEP is maintained for AOFDM and single user AMC-CDMA, while 'full-user' AMC-CDMA exhibits a slightly higher average BEP due to the residual MUI. (b) The SNR gain of the adaptive modems decreases, as ST coding increases the diversity order. The BPS curves appear in pairs, corresponding to AOFDM and AMC-CDMA - indicated by the thin and thick lines, respectively - for each of the four different ST code configurations. The markers represent the SNRs required by the fixed-mode OFDM and MC-CDMA schemes for maintaining the target BER of 10^{-3} in conjunction with the four ST-coded schemes considered.

average BEP as well as the average throughput.

The modulated symbol is now space-time encoded. As seen at the bottom of Figure 6.48, Alamouti's space-time block code [7] is applied across the frequency domain. A pair of the adjacent sub-carriers belonging to the same space-time encoding block is assumed to have the same channel quality. We employed a Wireless Asynchronous Transfer Mode (W-ATM) channel model [4, pp.474] transmitting at a carrier frequency of 60GHz, at a sampling rate of 225MHz and employing 512 sub-carriers. Specifically, we used a three-path fading channel model, where the average SNR of each path is given by $\bar{\gamma}_1 = 0.79192\bar{\gamma}$, $\bar{\gamma}_2 = 0.12424\bar{\gamma}$ and $\bar{\gamma}_3 = 0.08384\bar{\gamma}$. The Multi-path Intensity Profile (MIP) of the W-ATM channel is illustrated in Figure 6.42(a) in Section 6.5.3. Each channel associated with a different antenna is assumed to exhibit independent fading.

The simulation results related to our uncoded adaptive modems are presented in Figure 6.49. Since we employed the optimum switching levels derived in Section 6.4.4, both our adaptive OFDM (AOFDM) and the adaptive single-user MC-CDMA (AMC-CDMA) modems maintain the constant target BER of 10^{-3} up to the 'avalanche' SNR value, and then follow the BER curve of the 64-QAM mode. However, 'full-user' AMC-CDMA, which is defined as an AMC-CDMA system supporting $U = 16$ users with the aid of a spreading factor of $G = 16$ and employing the MMSE-BDFE Joint Detection (JD) receiver [219], exhibits a slightly higher average BER, than the target of $B_t = 10^{-3}$ due to the residual Multi-User Interference (MUI) of the imperfect joint detector. Since in Section 6.4.4 we derived the optimum switching levels based on a single-user system, the levels are no longer optimum, when residual MUI is present. The average throughputs of the various schemes expressed in terms of BPS steadily increase and at high SNRs reach the throughput of 64-QAM, namely 6 BPS. The throughput degradation of 'full-user' MC-CDMA imposed by the imperfect JD was within a fraction of a dB. Observe in Figure 6.49(a) that the analytical and simulation results are in good agreement, which we denoted by the lines and distinct symbols, respectively.

The effects of ST coding on the average BPS throughput are displayed in Figure 6.49(b). Specifically, the thick lines represent the average BPS throughput of our AMC-CDMA scheme, while the thin lines represent those of our AOFDM modem. The four pairs of hollow and filled markers associated with the four different ST-coded AOFDM and AMC-CDMA scenarios considered represent the BPS throughput versus SNR values associated with fixed-mode OFDM and fixed-mode MMSE-BDFE JD assisted MC-CDMA schemes. Specifically, observe for each of the 1, 2 and 4 BPS fixed-mode schemes that the right most markers, namely the circles, correspond to the 1-Tx / 1-Rx scenario, the squares to the 2-Tx / 1-Rx scheme, the triangles to the 1-Tx / 2-Rx arrangement and the diamonds to the 2-Tx / 2-Rx scenarios. First of all, we can observe that the BPS throughput curves of OFDM and single-user MC-CDMA are close to each other, namely within 1 dB for most of the SNR range. This is surprising, considering that the fixed-mode MMSE-BDFE JD assisted MC-CDMA scheme was reported to exhibit around 10dB SNR gain at a BEP of 10^{-3} and 30dB gain at a BEP of 10^{-6} over OFDM [220]. This is confirmed in Figure 6.49(b) by observing that the SNR difference between the ○ and ● markers is around 10dB, regardless whether the 4, 2 or 1 BPS scenario is concerned.

Let us now compare the SNR gains of the adaptive modems over the fixed modems. The SNR difference between the BPS curve of AOFDM and the fixed-mode OFDM represented by the symbol ○ at the same throughput is around 15dB. The corresponding SNR difference between the adaptive and fixed-mode 4, 2 or 1 BPS MC-CDMA modem is around 5dB. More explicitly, since in the context of the W-ATM channel model [4, pp.474] fixed-mode MC-CDMA appears to exhibit a 10dB SNR gain over fixed-mode OFDM, the additional 5dB SNR gain of AMC-CDMA over its fixed-mode counterpart results in a total SNR gain of 15dB over fixed-mode OFDM. Hence ultimately the performance of AOFDM and AMC-CDMA becomes similar.

Let us now examine the effect of ST block coding. The SNR gain of the fixed-mode schemes due to the introduction of a 2-Tx / 1-Rx ST block code is represented as the SNR difference between the two right most markers, namely circles and squares. These gains are nearly 10dB for fixed-mode OFDM, while they are only 3dB for fixed-mode MC-CDMA modems. However, the corresponding gains are less than 1dB for both adaptive modems, namely for AOFDM and AMC-CDMA. Since the transmitter power is halved due to using two Tx antennas in the ST codec, a 3dB channel SNR penalty was already applied to the curves in Figure 6.49(b). The introduction of a second receive antenna instead of a second transmit antenna eliminates this 3dB penalty, which results in a better performance for the 1-Tx/2-Rx scheme than for the 2-Tx/1-Rx arrangement. Finally, the 2-Tx / 2-Rx system gives around 3-4dB SNR gain in the context of fixed-mode OFDM and a 2-3dB SNR gain for fixed-mode MC-CDMA, in both cases over the 1-Tx / 2-Rx system. By contrast, the SNR

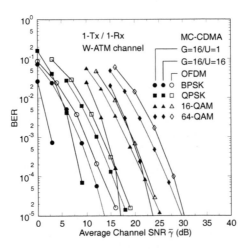

Figure 6.50: Performance of turbo convolutional coded fixed-mode OFDM and MC-CDMA for transmission over the W-ATM channel of [4, pp.474], indicating that JD MC-CDMA still outperforms OFDM. However, the SNR gain of JD MC-CDMA over OFDM is reduced to
1-2dB at a BEP of 10^{-4}.

gain of the 2-Tx / 2-Rx scheme over the 1-Tx / 2-Rx based adaptive modems was, again, less
than 1dB in Figure 6.49(b). More importantly, for the 2-Tx / 2-Rx scenario the advantage of
employing adaptive modulation erodes, since the fixed-mode MC-CDMA modem performs
as well as the AMC-CDMA modem in this scenario. Moreover, the fixed-mode MC-CDMA
modem still outperforms the fixed-mode OFDM modem by about 2dB. We conclude that
since the diversity-order increases with the introduction of ST block codes, the channel quality variation becomes sufficiently small for the performance advantage of adaptive modems to
erode. This is achieved at the price of a higher complexity due to employing two transmitters
and two receivers in the ST coded system.

When channel coding is employed in the fixed-mode multi-carrier systems, it is expected that OFDM benefits more substantially from the frequency domain diversity than
MC-CDMA, which benefited more than OFDM without channel coding. The simulation
results depicted in Figure 6.50 show that the various turbo-coded fixed-mode MC-CDMA
systems consistently outperform OFDM. However, the SNR differences between the turbo-
coded BER curves of OFDM and MC-CDMA are reduced considerably.

The performance of the concatenated ST block coded and turbo convolutional coded
adaptive modems is depicted in Figure 6.51. We applied the optimum set of switching levels
designed in Section 6.4.4 for achieving an uncoded BEP of 3×10^{-2}. This uncoded target
BEP was stipulated after observing that it is reduced by half-rate, $K = 3$ turbo convolutional
coding to a BEP below 10^{-7}, when transmitting over AWGN channels. However, our simulation results yielded zero bit errors, when transmitting 10^9 bits, except for some SNRs, when
employing only a single antenna.

Figure 6.51(a) shows the BEP of our turbo coded adaptive modems, when a single antenna
is used. We observe in the figure that the BEP reaches its highest value around the 'avalanche'
SNR point, where the adaptive modulation scheme consistently activates 64-QAM. The sys-

tem is most vulnerable around this point. In order to interpret this phenomenon, let us briefly consider the associated interleaving aspects. For practical reasons we have used a fixed interleaver length of $L = 3072$ bits. When the instantaneous channel quality was high, the $L = 3072$ bits were spanning a shorter time-duration during their passage over the fading channel, since the effective BPS throughput was high. Hence the channel errors appeared more bursty, than in the lower-throughput AQAM modes, which conveyed the $L = 3072$ bits over a longer time duration, hence dispersing the error bursts over a longer duration of time. The uniform dispersion of erroneous bits versus time enhances the error correction power of the turbo code. On the other hand, in the SNR region beyond the 'avalanche' SNR point seen in Figure 6.51(a) the system exhibited a lower uncoded BER, reducing the coded BEP even further. This observation suggests that further research ought to determine the set of switching thresholds directly for a coded adaptive system, rather than by simply estimating the uncoded BER, which is expected to result in near-error-free transmission.

We can also observe that the turbo coded BEP of AOFDM is higher than that of AMC-CDMA in the SNR range of 10-20dB, even though the uncoded BER is the same. This appears to be the effect of the limited exploitation of frequency domain diversity of coded OFDM, compared to MC-CDMA, which leads to a more bursty uncoded error distribution, hence degrading the turbo coded performance. The fact that ST block coding aided multiple antenna systems show virtually error free performance corroborates our argument.

Figure 6.51(b) compares the throughputs of the coded adaptive modems and the uncoded adaptive modems exhibiting a comparable average BER. The SNR gains due to channel coding were in the range of 0dB to 8dB, depending on the SNR region and on the scenarios employed. Each bundle of throughput curves corresponds to the scenarios of 1-Tx/1-Rx OFDM, 1-Tx/1-Rx MC-CDMA, 2-Tx/1-Rx OFDM, 2-Tx/1-Rx MC-CDMA, 1-Tx/2-Rx OFDM, 1-Tx/2-Rx MC-CDMA, 2-Tx/2-Rx OFDM and 2-Tx/2-Rx MC-CDMA starting from the far right curve, when viewed for throughput values higher than 0.5 BPS. The SNR difference between the throughput curves of the ST and turbo coded AOFDM and those of the corresponding AMC-CDMA schemes was reduced compared to the uncoded performance curves of Figure 6.49(b). The SNR gain owing to ST block coding assisted transmit diversity in the context of AOFDM and AMC-CDMA was within 1dB due to the halved transmitter power. Therefore, again, ST block coding appears to be less effective in conjunction with adaptive modems.

In conclusion, the performance of ST block coded constant-power adaptive multi-carrier modems employing optimum SNR-dependent modem mode switching levels were investigated in this section. The adaptive modems maintained the constant target BEP stipulated, whilst maximizing the average throughput. As expected, it was found that ST block coding reduces the relative performance advantage of adaptive modulation, since it increases the diversity order and eventually reduces the channel quality variations. When turbo convolutional coding was concatenated to the ST block codes, near-error-free transmission was achieved at the expense of halving the average throughput. Compared to the uncoded system, the turbo coded system was capable of achieving a higher throughput in the low SNR region at the cost of a higher complexity. The study of the relationship between the uncoded BEP and the corresponding coded BEP showed that adaptive modems obtain higher coding gains, than that of fixed modems. This was due to the fact that the adaptive modem avoids burst errors even in deep channel fades by reducing the number of bits per modulated symbol eventually to zero.

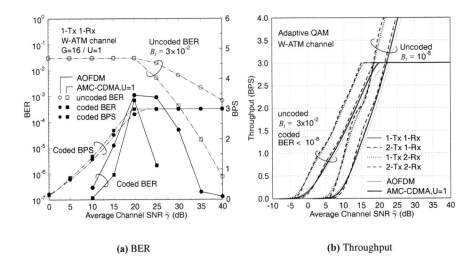

(a) BER   (b) Throughput

Figure 6.51: Performance of the concatenated ST block coded and turbo convolutional coded adaptive OFDM and MC-CDMA systems over W-ATM channel of [4, pp.474]. The uncoded target BEP is 3×10^{-2}. The coded BEP was less than 10^{-8} for most of the SNR range, resulting in virtually error free transmission. (a) The coded BEP becomes higher near the 'avalanche' SNR point, when a single antenna was used. (b) The coded adaptive modems have SNR gains up to 7dB compared to their uncoded counterparts achieving a comparable average BER.

6.6 Review and Discussion

Following a brief introduction to several fading counter-measures, a general model was used to describe several adaptive modulation schemes employing various constituent modulation modes, such as PSK, Star QAM and Square QAM, as one of the attractive fading counter-measures. In Section 6.3.3.1, the closed form expressions were derived for the average BER, the average BPS throughput and the mode selection probability of the adaptive modulation schemes, which were shown to be dependent on the mode-switching levels as well as on the average SNR. After reviewing in Section 6.4.1, 6.4.2 and 6.4.3 the existing techniques devised for determining the mode-switching levels, in Section 6.4.4 the optimum switching levels achieving the highest possible BPS throughput while maintaining the average target BEP were developed based on the Lagrangian optimization method.

Then, in Section 6.5.1 the performance of uncoded adaptive PSK, Star QAM and Square QAM was characterized, when the underlying channel was a Nakagami fading channel. It was found that an adaptive scheme employing a k-BPS fixed-mode as the highest throughput constituent modulation mode was sufficient for attaining all the benefits of adaptive modulation, while achieving an average throughput of up to $(k-1)$ BPS. For example, a three-mode adaptive PSK scheme employing No-Tx, 1-BPS BPSK and 2-BPS QPSK modes attained the maximum possible average BPS throughput of 1 BPS and hence adding higher-throughput modes, such as 3-BPS 8-PSK to the three-mode adaptive PSK scheme resulting in a four-mode adaptive PSK scheme did not achieve a better performance across the 1 BPS through-put range. Instead, this four-mode adaptive PSK scheme extended the maximal achievable BPS throughput by any adaptive PSK scheme to 2 BPS, while asymptotically achieving a throughput of 3 BPS as the average SNR increases.

On the other hand, the relative SNR advantage of adaptive schemes in comparison to fixed-mode schemes increased as the target average BER became lower and decreased as the fading became less severe. More explicitly, less severe fading corresponds to an increased Nakagami fading parameter m, to an increased number of diversity antennas, or to an increased number of multi-path components encountered in wide-band fading channels. As the fading becomes less severe, the average BPS throughput curves of our adaptive Square QAM schemes exhibit undulations owing to the absence of 3-BPS, 5-BPS and 7-BPS square QAM modes.

The comparisons between fixed-mode MC-CDMA and adaptive OFDM (AOFDM) were made based on different channel models. In Section 6.5.4 it was found that fixed-mode MC-CDMA might outperform adaptive OFDM, when the underlying channel provides sufficient diversity. However, a definite conclusion could not be drawn since in practice MC-CDMA might suffer from MUI and AOFDM might suffer from imperfect channel quality estimation and feedback delays.

Concatenated space-time block coded and turbo convolutional-coded adaptive multi-carrier systems were investigated in Section 6.5.5. The coded schemes reduced the required average SNR by about 6dB-7dB at throughput of 1 BPS achieving near error-free transmission. It was also observed in Section 6.5.5 that increasing the number of transmit antennas in adaptive schemes was not very effective, achieving less than 1dB SNR gain, due to the fact that the transmit power per antenna had to be reduced in order to limit the total transmit power for the sake of fair comparison.

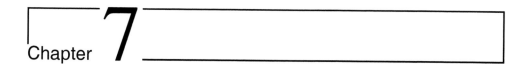

Practical Considerations of Wideband AQAM

In deriving the upper bound performance of wideband AQAM portrayed in Figure 4.21(b), various assumptions were made and stated in Section 4.3.1. However, in order to provide a more accurate comparison between AQAM and its constituent fixed modulation modes, those assumptions must be justified and their effects have to be investigated. Specifically, perfect, i.e. error-free feedback was assumed for the DFE, while in practice erroneous decision can be fed back, which results in error propagation. Consequently the impact of error propagation is studied in the context of both fixed and adaptive QAM schemes. Furthermore, as stated in Section 4.3.1, perfect modulation mode selection was assumed, whereby the output SNR of the DFE was estimated perfectly prior to transmission. However, in stipulating this assumption, the delay incurred between channel quality estimation and the actual utilization of the estimate was neglected in the wideband AQAM scheme.

In this chapter the impact of co-channel interference on the wideband AQAM scheme is also investigated. In this respect, interference compensation techniques are invoked in order to reduce the degradation resulting from the co-channel interference. Let us now commence our investigations by studying the error propagation phenomenon in the DFE.

7.1 Impact of Error Propagation

Error propagation is a phenomenon that occurs, whenever an erroneous decision is fed back into the feedback filter of the DFE. When a wrong decision is fed back, the feedback filter produces an output estimate which is erroneous. The incorrect estimate precipitates further errors at the output of the equalizer. This leads to another erroneous decision being fed back into the feedback filter. Consequently, this recursive phenomenon degrades the BER performance of the DFE. Intuitively, the effects of this error will last throughout the memory span of the feedback filter. This causes an error propagation throughout the feedback filter, until the memory of the feedback filter is cleared of any erroneous feedback inputs.

The performance of the fixed modulation modes of our AQAM scheme in conjunction

Transmission Burst type:	Non-Spread Speech Burst of Figure 4.13.
DFE Parameters: No. of feedforward taps, N_f No. of feedback taps,N_b Decision Feedback	35 7 Past Decision
Recursive Kalman Channel Estimator Parameters: Measurement error covariance Matrix, $\mathbf{R}(k) = g\mathbf{I}$ System error covariance Matrix, $\mathbf{Q}(k)$ δ (see Equation 3.47) Initial Channel Estimate Vector $\hat{\mathbf{h}}(o)$ Number of RKCE taps	$0 < g \le 1$ $0.01\mathbf{I}$ ≥ 300 **0** 8
Channel Parameters: Typical Urban Rayleigh-faded Weights Normalized Doppler Frequency:	See Figure 4.12 and Table 4.5 3.25×10^{-5}

Table 7.1: Generic simulation parameters that were utilized in our experiments.

with error propagation is depicted in Figure 7.1, where the corresponding curve of the error-free feedback scenario is also displayed for comparison. Perfect channel compensation was applied at the receiver and the other simulation parameters are listed in Table 7.1. There was only a slight degradation in the BER performance of the BPSK and 4QAM modes, as evidenced by Figure 7.1. However, for the higher-order modulation modes of 16QAM and 64QAM, a more severe degradation of approximately 1.5 and 3.0dB was recorded, respectively. These results were expected, since the higher-order modulation modes were more susceptible to feedback errors due to the smaller Euclidean distance of their constellation points.

The impact of error propagation on the wideband AQAM scheme over a TU Rayleigh fading channel was also investigated and the results are shown in Figure 7.2. The corresponding curve of the wideband AQAM scheme with error-free decision feedback was also shown for comparison and the switching thresholds of the wideband AQAM scheme were set according to Table 4.8 for target BERs of 1% and 0.01%. At low to medium average channel SNRs the BER performance of the wideband AQAM scheme exposed to error propagation was similar to that of the AQAM scheme with error-free decision feedback. However, at higher average channel SNRs, as a result of error propagation, a BER/SNR degradation of approximately 3dB was observed. These results were consistent with the results shown for the fixed modulation modes of Figure 7.1. At low to medium average channel SNRs, the impact of error propagation was negligible due to two factors. Firstly, at those channel SNRs the lower-order modulation modes, which were more robust against error propagation were utilized more frequently. Secondly, the higher-order modulation modes were only utilized, when the channel quality was favourable, which resulted in low instantaneous BERs. Consequently, less erroneous decisions were made, which reduced the impact of error propagation.

However, at higher average channel SNRs, the probability of modulation mode switching

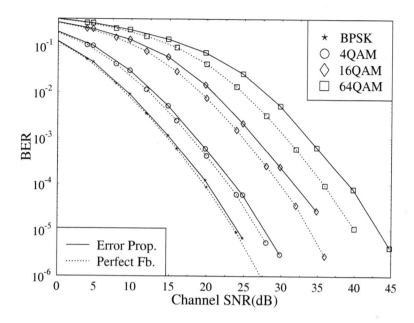

Figure 7.1: Impact of error propagation on the modulation modes of BPSK, 4QAM, 16QAM and 64QAM over the TU Rayleigh fading channel of Figure 4.12. Perfect channel compensation was applied and the simulation parameters are listed in Table 7.1.

was low, where the 64QAM mode was frequently chosen. Consequently, the impact of error propagation was more apparent, as it was observed the case of the fixed modulation mode of 64QAM in Figure 7.1. Nevertheless, the target performance of 1% and 0.01% was achieved even in the presence of erroneous decision feedback.

7.2 Channel Quality Estimation Latency

The estimation of the channel quality prior to transmission is vital in the implementation of the wideband AQAM scheme, since it is used in the selection of the appropriate modulation mode for the next transmission burst. In generating the upper bound performance curves depicted in Figure 4.21(b), we assumed that the required modulation mode was selected perfectly prior to transmission, as stated in Section 4.3.1. However, in a realistic and practical wideband AQAM scheme this assumption must be discarded as a result of the inherent channel quality estimation delay incurred by the scheme. Nevertheless, it must be stressed that

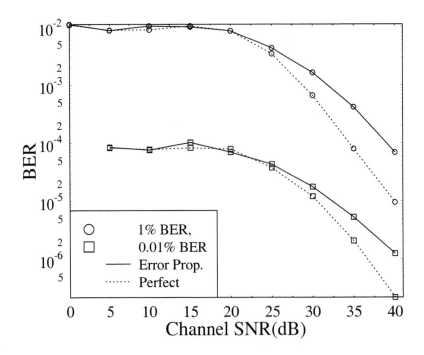

Figure 7.2: Impact of error propagation on the wideband AQAM scheme over a TU Rayleigh fading
channel, where the switching thresholds were set according to Table 4.8 for target BERs of
1% and 0.01%. Perfect channel compensation was applied and the simulation parameters
are listed in Table 7.1.

the assumption was essential in order to record the upper bound performance of the AQAM
scheme.

The channel quality estimation latency is defined as the delay incurred between the event
of estimating the channel quality to the actual moment of transmission using the modem mode
deemed optimum at the instant of the channel quality estimation. During this delay, the fading channel quality varies according to the Doppler frequency and consequently, the channel
quality estimates perceived prior to transmission may become obsolete. Consequently, the
chosen modulation mode is not optimum with regards to the actual channel quality and this
degrades the BER performance of the wideband AQAM scheme. This degradation is dependent on the amount of delay incurred and the rate at which the fading channel quality
fluctuates, as quantified by its Doppler frequency. Before we proceed to investigate the performance degradation as a result of the channel quality estimation latency, let us present two
possible time-frame structures, where wideband AQAM can be implemented. This will provide us with a clearer understanding concerning the amount of delay incurred by the scheme.

7.2.1 Sub-frame Based Time Division Duplex/Time Division Multiple Access System

In this sub-frame based Time Division Duplex/Time Division Multiple Access (TDD / TDMA) system, the uplink and downlink time-slots are separated equally into two halves of the TDMA frame, as shown in Figure 7.3. In this respect the time-slot is defined as the window in time, in which the transmission burst is received or transmitted. By utilizing the time-frame configuration shown in Figure 7.3, we will explain the operation of the wideband AQAM scheme and the corresponding channel quality estimation latency that is incurred. In the up-link transmission, shown in Figure 7.3, the channel quality was estimated at the Base Station (BS) and subsequently an appropriate modulation mode was selected for its next downlink transmission. This was achieved by exploiting the channel's reciprocity during the uplink and downlink transmissions, since the transmission frequencies for both links were identical in a TDD system. Having selected the modulation mode, a delay of half a TDMA frame was incurred at the BS before the downlink transmission was activated as shown in Figure 7.3. We refer to this regime as open-loop controlled AQAM. Let us now in the next section consider closed-loop control.

7.2.2 Closed-Loop Time Division Multiple Access System

The corresponding closed-loop TDMA construction was similar to that of the sub-frame TDD/TDMA with the exception that the uplink and downlink transmission frequencies were different. Hence this was a Frequency Division Duplex (FDD) system. Consequently, the assumed channel reciprocity - which was invoked in the sub-frame based TDD/TDMA sys-tem - was less applicable. Hence a closed-loop signalling system was required in order to implement the wideband AQAM scheme, which is shown in Figure 7.4. In the uplink trans-mission, the channel quality was estimated at the BS, in order to select the next uplink modu-lation mode. Subsequently, the selected uplink modulation mode was conveyed to the Mobile Station (MS) with the aid of control symbols during the next downlink transmission. Conse-quently, the selected modulation mode was utilized by the MS in its next uplink transmission. As a result of the closed-loop signalling regime, the delay incurred by the system was equal to the duration of one TDMA time-frame. Consequently, the open-loop system described in Section 7.2.1 was more applicable to AQAM transmission as a result of its lower delay, when compared to the close-loop system. This latency can be substantially reduced using slot-by-slot TDD/TDMA, where the uplink and downlink slots are adjacent, which is also supported by the third-generation Universal Mobile Telecommunication System (UMTS) [221].

7.2.3 Impact of Channel Quality Estimation Latency

Regardless of the type of wideband AQAM scheme that was implemented, we investigated the maximum delay that could be tolerated by the AQAM scheme by assuming that the per-formance degradation in the uplink and downlink transmission was identical. In our experi-ments, the delay was measured in terms of a time-slot duration of $72\mu s$, as proposed in the Pan European FRAMES framework [151]. Mid-amble associated CIR estimation based on the Kalman algorithm - which was discussed in Chapter 3 - was implemented, in order to es-timate the channel quality. The normalized Doppler frequency was set to 3.25×10^{-5}, which

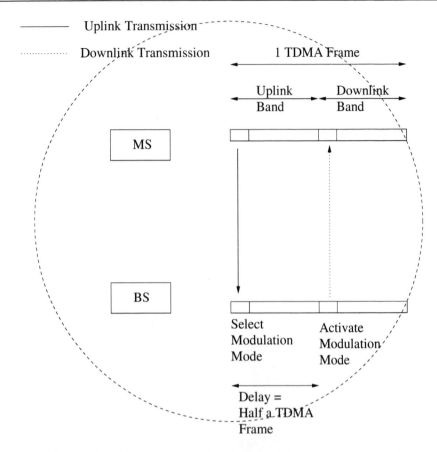

Figure 7.3: Sub-frame based TDD/TDMA system for the uplink and downlink transmission, as described in Section 7.2.1. Channel reciprocity was exploited in this system and the channel
quality estimation latency was equivalent to half a TDMA frame.

was equivalent to a TDMA system using a 1.9GHz in carrier frequency, transmission rate
of 2.6 MSymbols/s and a vehicular speed of 13.33m/s. The specific simulation parameters
used in our subsequent experiments are listed in Table 7.1. The AQAM switching thresholds
were set according to Table 4.8, which were optimised for maintaining target BERs of 1%
and 0.01%.

The results of our investigations are shown in Figures 7.5(a) and 7.5(b) for target BERs
of 1% and 0.01%, respectively. In these figures the wideband AQAM scheme was subjected
to a delay of 8, 16 and 32 time-slots and the performance was compared to that of the zero-
delay upper bound performance. For the target BER of 1% we can observe that the BER
performance degradation increased, as delay was increased as evidenced by Figure 7.5(a). At
high average channel SNRs, the BER degradation was minimal as a result of the reduction
of modulation mode switching frequency, where the 64QAM mode was frequently selected.
The BER degradation was more evident for the AQAM scheme designed for a low target BER
of 0.01% as a result of its increased sensitivity to errors. By referring to Figure 7.5(b), at a

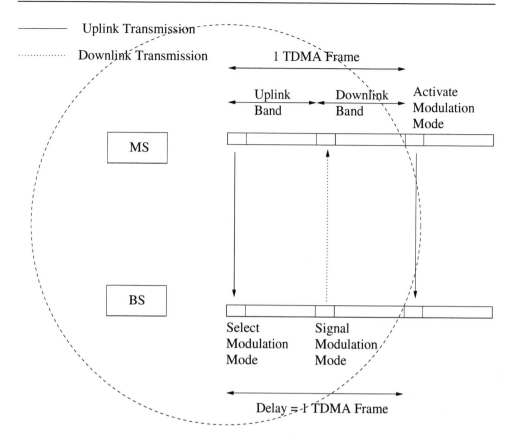

Figure 7.4: Closed-loop FDD/TDMA system for the uplink and downlink transmission, as described in Section 7.2.2. Channel reciprocity was not assumed in this system in favour of a closed-loop signalling regime and the channel quality estimation latency was equivalent to the duration of one TDMA frame.

channel SNR of 20dB and at a delay of 32 time-slots, the BER performance was degraded by approximately two orders of magnitude in comparison to the upper bound performance. In these experiments, the modulation mode selection regime was affected by the delay incurred by the system. The impact was especially significant, when the channel quality was low and a less robust higher-order modulation mode was utilized erroneously. The BPS performance in Figures 7.5(a) and 7.5(b) remained unchanged for different delays. This can be readily explained by observing that on average the throughput was the same even if the modulation mode selected was erroneous.

As discussed previously, the performance of the wideband AQAM scheme depended on the channel quality estimation delay incurred, as well as on the Doppler frequency of the fading channel. In order to investigate the system's performance dependency on the Doppler frequency, a slower fading channel having a normalized Doppler frequency of 2.17×10^{-6} was utilized. This corresponded to a carrier frequency of 1.9GHz, transmission rate of 2.6 Msymbols/s and a pedestrian speed of 0.89m/s in the Pan European FRAMES Proposal [151].

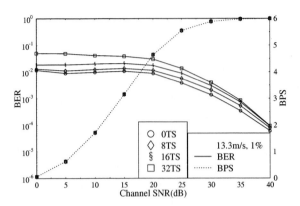

(a) Performance at a target BER of 1% at channel quality estimation delays of 8, 16 and 32 time-slots, where each time-slot is of $72\mu s$ duration.

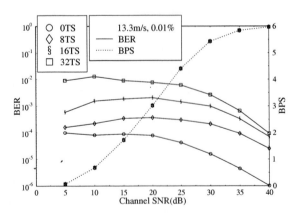

(b) Performance at a target BER of 0.01% at channel quality estimation delays of 8, 16 and 32 time-slots, where each time-slot is of $72\mu s$ duration.

Figure 7.5: Impact of channel quality estimation latency upon the wideband AQAM scheme, where the modem mode switching thresholds were set according to Table 4.8. The normalized Doppler frequency was set to 3.25×10^{-5} and the other simulation parameters are listed in Table 7.1.

The other simulation parameters were set according to Table 7.1. The BER and BPS performances of the AQAM scheme over this slower fading channel are shown in Figures 7.6(a) and 7.6(b) for a target BER of 1% and 0.01%, respectively. In these figures, the characteristics observed in Figures 7.5(a) and 7.5(b) were also evident and hence the associated trends can be explained similarly. However, in order to investigate the impact of the Doppler frequency, the BER performance at an average channel SNR over the two fading channels exhibiting different Doppler frequencies were recorded against different delays in Figures 7.7(a) and 7.7(b). For a target BER of 1% a higher BER degradation was experienced by the higher Doppler frequency scheme, where at a BER of 2×10^{-2} the lower Doppler frequency scheme can tolerate an additional delay of 7 time-slots, as evidenced by Figure 7.7(a). Similarly, at a BER of 1×10^{-3} for the scheme having a target BER of 0.01%, an additional 5 time-slots delay can be tolerated by the scheme with the lower Doppler frequency.

From the above experiments, we can conclude that as the channel quality estimation delay and Doppler frequency increased, the performance degradation of the wideband AQAM scheme was higher. Furthermore, the impact of channel quality estimation latency was more evident at low target BERs due to its increased error sensitivity. In order to improve the robustness of the AQAM scheme against channel quality estimation delay, in the next section we will invoke a simple channel quality prediction method and experimentally optimise the modem mode switching thresholds.

7.2.4   Linear Prediction of Channel Quality

In order to mitigate the effects of channel quality estimation delay on the wideband AQAM scheme, the next channel quality estimate can be predicted using linear prediction. This simple technique utilizes the previous channel estimates for linear prediction, in order to predict the next channel quality estimate. Subsequently, if the prediction is accurate, the modulation mode selection errors will decrease, yielding a more delay-robust wideband AQAM scheme. This linear prediction technique was applied to the wideband AQAM scheme in conjunction with two different Doppler frequencies and various time delays for target BERs of 1% and 0.01%. The results are depicted in Figures 7.8(a) and 7.8(b) for an average channel SNR of 20dB, where the performance without linear prediction is also shown for comparison. In these figures, the linearly predictive scheme exhibited a higher tolerance against channel quality estimation delay. The maximum delays that can be tolerated for a target BER of 1% and 0.01% are tabulated in Table 7.2 for the schemes with and without linear prediction. From this table, channel quality estimation delay gains of approximately 8 time-slots can be achieved using the above linear predictive techniques for the lower Doppler frequency scheme. Similarly, delay gains of 6 time-slots were recorded for the higher Doppler frequency scheme.

In these experiments we have highlighted that a simple channel quality prediction technique can substantially improve the robustness of the wideband AQAM scheme against channel quality estimation delay. However, it must be stressed that the AQAM scheme performed better in a slowly varying environment, which also facilitated a better channel prediction performance.

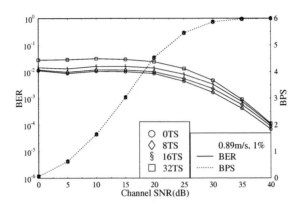

(a) Performance at a target BER of 1% at channel quality estimation delays of 8, 16 and 32 time-slots, where each time-slot is of 72μs duration.

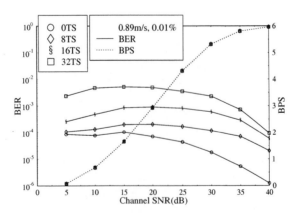

(b) Performance at a target BER of 0.01% at channel quality estimation delays of 8, 16 and 32 time-slots, where each time-slot is of 72μs duration.

Figure 7.6: Impact of channel quality estimation latency upon the wideband AQAM scheme, where the modem mode switching thresholds were set according to Table 4.8. The normalized Doppler frequency was set to 2.17×10^{-6} and the other simulation parameters are listed in Table 7.1.

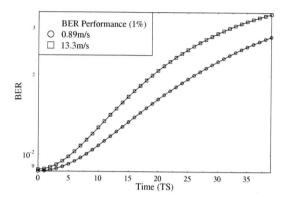

(a) Performance at a target BER of 1% for different channel quality estimation delays in terms of time-slots (TS), where each time-slot is of $72\mu s$ duration.

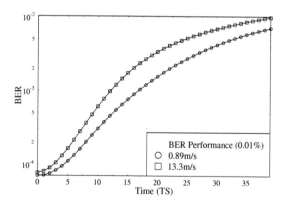

(b) Performance at a target BER of 0.01% for different channel quality estimation delays in terms of time-slots (TS), where each time-slot is of $72\mu s$ duration.

Figure 7.7: Impact of channel quality estimation latency upon the wideband AQAM scheme for two different normalized Doppler frequencies of 3.25×10^{-5} (at 13.3m/s) and 2.17×10^{-6} (at 0.89m/s), where the modem mode switching thresholds were set according to Table 4.8. The average channel SNR was set to 20dB and the other simulation parameters are listed in Table 7.1.

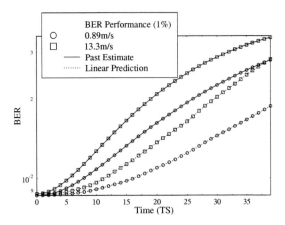

(a) Performance at a target BER of 1% for different channel quality estimation delays in terms of time-slots (TS), where each time-slot is of 72μs
duration.

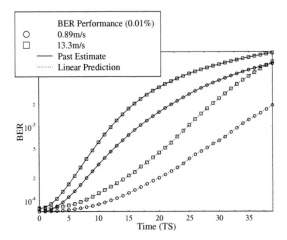

(b) Performance at a target BER of 0.01% for different channel quality estimation delays in terms of time-slots (TS), where each time-slot is of 72μs
duration.

Figure 7.8: Impact of channel quality estimation latency upon the wideband AQAM scheme for two
different normalized Doppler frequencies of 3.25×10^{-5} (at 13.3m/s) and 2.17×10^{-6}(at
0.89m/s), where the switching thresholds were set according to Table 4.8. The performance
utilizing the linear prediction technique (denoted by Linear Prediction) was compared to
the conventional non-predicted technique (denoted by Past Estimate). The average channel
SNR was set to 20dB and the other simulation parameters are listed in Table 7.1.

Speed(m/s)	1%		0.01%	
	Linear(TS)	Past(TS)	Linear(TS)	Past(TS)
0.89	16	8	12	4
13.33	11	5	8	3

Table 7.2: The channel quality estimation delays in an AQAM wideband scheme in order to achieve target BERs of 1% and 0.01%, which were extracted from Figure 7.8. The delays were measured for different normalized Doppler frequencies of 3.25×10^{-5} (at a vehicular speed of 13.3m/s) and 2.17×10^{-6} (at a vehicular speed of 0.89m/s). Further comparisons were made between the performance achieved by utilizing the linear prediction technique of Section 7.2.4 (denoted by Linear) and the conventional non-predicted scheme (denoted by Past). The delays were measured in terms of time-slots (TS), where each time-slot was of 72μs duration.

7.2.5 Sub-frame TDD/TDMA Wideband AQAM Performance

Having considered the implications of channel quality estimation latency, we will now investigate the performance of wideband AQAM in a sub-frame based TDD/TDMA scheme, which was discussed in Section 7.2.1. The channel quality estimation latency incurred was equivalent to half of a TDMA frame, which was set to 4.615ms according to the Pan European FRAMES proposal [151]. Hence the channel quality estimation latency incurred was 2.3075ms or 32 time-slots, where each time-slot was of 72μs duration. The linear prediction technique of Section 7.2.4 was invoked, in order to predict the next channel quality. The modem mode switching thresholds - which are shown in Table 7.3 - were experimentally determined in order to achieve the target BERs of 1% and 0.01%, since the impact of delay prohibited the utilization of the Powell optimization technique discussed in Section 4.3.5. The normalized Doppler frequency was set to 2.17×10^{-6} in order to create a slowly varying propagation environment and the other simulation parameters were set according to Table 7.1. The associated wideband AQAM performances for target BERs of 1% and 0.01% are shown in Figures 7.9 and 7.10, respectively. In both of these figures, the corresponding upper bound performance was also included for benchmarking.

	$t_1(dB)$	$t_2(dB)$	$t_3(dB)$	$t_4(dB)$
1%	5.64	8.00	13.65	18.68
0.01%	11.25	13.56	18.80	25.00

Table 7.3: The switching thresholds that were manually optimised in order to achieve target BERs of 1% and 0.01%. The wideband AQAM regime was implemented in a sub-frame based TDD/TDMA system having a channel quality estimation latency of 32 time-slots or 2.3075ms as shown in Figures 7.9 and 7.10.

Referring to Figures 7.9 and 7.10, the target BERs of 1% and 0.01% were achieved with slight degradation in terms of its throughput performance, when compared to the upper bound performance. Explicitly, a BPS/SNR degradation of 0.9dB and 1.8dB was observed for target BERs of 1% and 0.01%, respectively. The BPS throughput performance of the latency-impaired wideband AQAM scheme was also compared to that of the fixed modulation modes

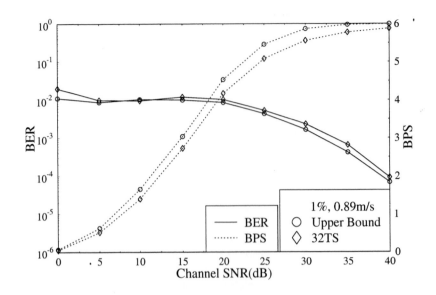

Figure 7.9: The performance of a sub-frame TDD/TDMA based wideband AQAM scheme having a channel quality estimation latency of 32 time-slots or 2.3075ms, as described in Section 7.2.1. The switching thresholds are set according to Table 7.3 for a target **BER of 1%** and the simulation parameters are listed in Table 7.1. The upper-bound performance without channel quality estimation delay was also displayed for comparison.

of Figure 7.1 for target BERs of 1% and 0.01%. The results are shown in Figure 7.11 exhibited the same characteristics as Figure 4.25 and hence can be justified similarly. The gains achieved by the latency-impaired wideband AQAM are tabulated in Table 7.4 for a throughput of 1, 2 and 4 bits per symbol, corresponding to the throughput of BPSK, 4QAM and 16QAM modes, respectively.

BPS	1%			0.01%		
	Fixed(dB)	AQAM(dB)	Gain(dB)	Fixed(dB)	AQAM(dB)	Gain(dB)
1	9.70	11.10	1.40	20.30	26.70	6.40
2	13.00	13.70	0.70	23.60	29.00	5.40
4	21.00	22.40	1.40	32.00	25.70	5.70

Table 7.4: The channel SNR gain achieved by the sub-frame based TDD/TDMA wideband AQAM scheme with a channel quality estimation latency of 32 time-slots or 2.3075ms, when compared to the fixed modulation modes, at throughputs of 1, 2 and 4 bits per symbol (BPS). The values were extracted from Figure 7.11.

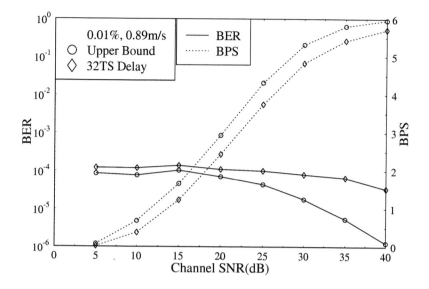

Figure 7.10: The performance of a sub-frame TDD/TDMA based wideband AQAM scheme with a channel quality estimation latency of 32 time-slots or 2.3075ms, as described in Section 7.2.1. The switching thresholds are set according to Table 7.3 for a target **BER of 0.01%** and the simulation parameters are listed in Table 7.1. The upper-bound performance without channel quality estimation delay was also displayed for comparison.

In this section, we have analysed and recorded the impact of channel quality estimation latency over channels having different Doppler frequencies upon a wideband AQAM scheme. We invoked a simple linear prediction technique, in order to predict the next channel quality estimate, which allowed the wideband AQAM scheme to be more robust against channel quality estimation delay. Subsequently, the maximum channel quality estimation delay that can be tolerated by a wideband AQAM scheme was recorded in Table 7.2 for target BERs of 1% and 0.01%. Finally, we characterized a realistic and practical sub-frame TDD/TDMA based AQAM system, which was robust up to delays of 2.3ms and still achieved substantial BER/SNR gains over fixed modulation modes, as evidenced by Figure 7.11 and Table 7.4.

7.3   Effect of Co-channel Interference on AQAM

In all our previous experiments our work has been restricted to a noise limited environment. However, in a cellular mobile environment the impact of interference - in particular co-channel interference - has to be considered in a wideband AQAM scheme. In order to increase the capacity of a cellular mobile environment, tight frequency reuse techniques are frequently utilized [222]. This is a technique, whereby a particular radio channel of a cell can

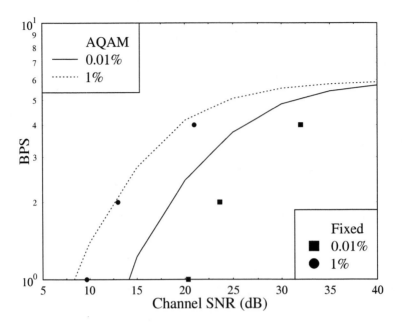

Figure 7.11: The BPS throughput of the sub-frame TDD/TDMA based wideband AQAM scheme with a channel quality estimation latency of 32 time-slots or 2.3075ms and that of the individual fixed modulation modes of BPSK, 4QAM, 16QAM. The BPS throughput values were extracted from Figures 7.10 and 7.2 and the simulation parameters are listed in Table 7.1.

be reused in another cell, which is separated by a certain distance. As a result of the utilization of a common radio channel in these two cells, transmission in one cell can propagate to and distort the co-channel transmissions in the other cell. Hence the interference is termed as Co-Channel Interference (CCI).

In our subsequent experiments the interferer was assumed to be temporally synchronous, in other words the signals transmitted by the interferer and the reference user were perfectly synchronous at the receiver. This approach was also adopted by - amongst others - Torrance and Webb [61, 223]. The signal of the interferer was also assumed to be phase non-coherent with the reference signal at the receiver. In this respect, we have assumed that the independent fading nature of each user's channel resulted in a phase non-coherent scenario. The channel model of the interferer and desired user was assumed identical, as described by Table 4.5 and Figure 4.12. The Signal to Interference Ratio (SIR) is a parameter which characterizes an interference-limited environment and is defined as follows [222, 224] :

$$\text{SIR} = \frac{S}{\sum_{k=1}^{K_1} P_k^{intf}}, \tag{7.1}$$

where K_1 is the number of interferers, S is the signal power of the reference user and P_k^{intf} is the power transmitted by the kth interferer.

7.3.1 Impact of Co-Channel Interference on Channel Quality Estimation

In a wideband AQAM scheme the presence of CCI can potentially degrade the accuracy of the demodulation process and the channel quality estimation, which is needed for AQAM mode selection. The issues associated with the impact of CCI upon the demodulation process is discussed in Section 7.3.2 in more depth, while here we focus our attention on the performance degradation inflicted by the channel quality estimation in this section.

In Section 7.2 we have discussed the importance of channel quality estimation, in order to ensure that the selected modulation mode was optimum. However, in an interference-limited environment the performance of a wideband AQAM scheme is degraded due to two factors:

- The presence of CCI degrades the ability of the receiver to accurately estimate the channel quality on a burst-by-burst basis. Consequently the modulation mode selection errors increase, yielding a degraded BER performance.

- In a TDD/TDMA AQAM scheme, the channel's reciprocity is exploited in the uplink and downlink transmission, in order to estimate the channel quality, as highlighted in Section 7.2. However, this reciprocity is not applicable in estimating the CCI, since the uplink and downlink CCI possess different propagation paths and different transmitted powers. Consequently the modulation mode selection regime of the receiver - which is subjected to uncorrelated uplink and downlink CCI - may not be optimum.

By assuming that statistically speaking the impact of CCI on the receiver is identical in the uplink and downlink transmission, we can focus our investigations on the downlink performance for simplicity. In order to isolate and study the impact of CCI on the modem mode switching regime, we assumed that the CCI was only present during the uplink transmission, but not during the downlink transmission. Again, this was a hypothetical situation, but it was necessary to stipulate these conditions, in order to analyse the impact of CCI on the switching regime. With this assumption in place, the reception at the BS was contaminated with CCI while the MS experienced a interference-free demodulation conditions.

In our subsequent discussion, we considered only a single-interferer scenario and the modulation mode of the interferer in the wideband AQAM scheme was chosen randomly from the set of permissible modes of the AQAM regime. The uplink channel quality was estimated using the CIR estimator based on the Kalman algorithm of Section 3.2.1 with the aid of the mid-amble sequence of the FRAMES non-spread speech burst of Figure 4.13. The switching thresholds were set according to Table 4.8 for target BERs of 1% and 0.01%. The other simulation parameters are listed in Table 7.1

An informative insight into the impact of CCI on the AQAM switching regime can be obtained by observing the average channel quality estimation errors for different average

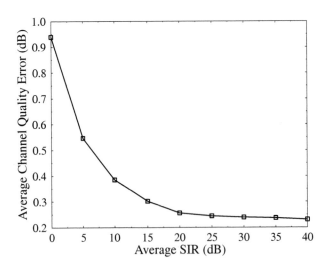

Figure 7.12: The downlink average channel quality error performance defined by Equation 7.2 for a wideband AQAM scheme at an average SNR of 20dB. An interference-free scenario was assumed at the MS and the switching thresholds were set according to Table 4.8 for a target BER of 0.01%. The performance was averaged over 10000 transmission bursts and the simulation parameters were set according to Table 7.1.

SIRs, which is shown in Figure 7.12. The average channel quality estimation error is defined as follows :

$$\text{Average Channel Quality Estimation Error} = \frac{\sum_{i=1}^{N}(\gamma_{dfe}^{acc} - \gamma_{dfe}^{est})}{N}\text{dB}, \qquad (7.2)$$

where γ_{dfe}^{acc} is the accurate SNR output of the DFE without CCI and with perfect channel estimation, while γ_{dfe}^{est} is the estimated SNR output of the DFE in the presence of uplink CCI. The average value was obtained over $N = 10000$ transmission bursts.

In Figure 7.12 the average difference between the actual and estimated SNR output of the DFE was recorded for different SIRs over the COST207 CIR of Table 4.5 and Figure 4.12, measured at an average channel SNR of 20dB. As expected, at low average SIRs the average channel quality estimation error was high as a result of inaccurate channel estimation of the reference user. Conversely, at higher average SIRs, the magnitude of the CCI was lower, resulting in better channel estimation of the reference user. Consequently the average channel quality estimation error converged to an average minimum of 0.25dB, as evidenced by Figure 7.12.

The BER and BPS performances for target BERs of 1% and 0.01% for average uplink SIRs of 0, 10, 20 and 30dB are shown in Figures 7.13 and 7.14, respectively. In terms of BER performance, as the average SIR increased, the CCI induced BER and BPS degrada-

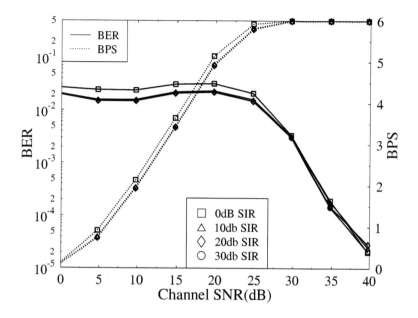

Figure 7.13: The **downlink performance** of a wideband AQAM scheme contaminated with co-channel interference at the BS and that of an interference-free scenario at the MS. This hypothetical situation was necessary in order to study the impact of CCI on the channel quality estimation process, as explained in Section 7.3.1. The modem mode switching thresholds were set according to Table 4.8 for a target BER of 1% and the simulation parameters are listed in Table 7.1. Midamble channel estimation was applied at both the BS and MS.

tion decreased, as evidenced by Figures 7.13 and 7.14. The degradation was more evident in the context of the wideband AQAM scheme, which was optimised for a low BER of 0.01% due to its increased sensitivity to modulation mode selection errors. At a low average SIR of 0dB, the estimation of the reference user's channel quality degraded and hence the effect of modulation selection errors increased. Consequently, both schemes encountered severe degradation in terms of their BER performance. However, at higher average SIRs - above 10dB - the BER performance approached the target BER, for which it was optimised. This was achieved as a result of sufficiently accurate channel quality estimates. The BPS performance did not change significantly for different average SIRs. However, at an average SIR of 0dB, the BPS throughput increased, which was consistent with the corresponding BER degradation. In this respect, as a result of channel quality estimation errors, less robust modulation modes were erroneously selected, yielding a degraded BER performance and an increased BPS throughput.

By referring to Figures 7.13 and 7.14, the wideband AQAM scheme was sufficiently robust against CCI in terms of its modem mode switching regime performance for average

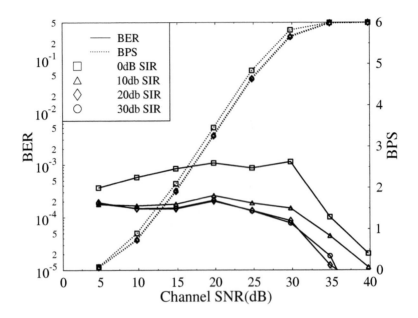

Figure 7.14: The **downlink performance** of a wideband AQAM scheme contaminated with co-channel
interference at the BS and an interference-free scenario at the MS. This hypothetical situation was necessary in order to study the impact of CCI on the channel quality estimation
process, as explained in Section 7.3.1. The modem mode switching thresholds were set
according to Table 4.8 for a target BER of 0.01% and the simulation parameters are listed
in Table 7.1. Midamble channel estimation was applied at both the BS and MS.

SIRs above 10dB. Hence, the subsequent experiments incorporating CCI will only consider
average SIRs equal to or in excess of 10dB.

7.3.2 Impact of Co-Channel Interference on the Demodulation Process

In the last section we have quantified and investigated the impact of CCI on the modem mode
switching regime of the AQAM scheme. Consequently, in this section we will investigate
the impact of CCI on the AQAM demodulation process in the presence of CCI. In order to
isolate and study the impact of CCI on the demodulation process, we have considered an
interference-free scenario at the BS and a CCI-impaired receiver at the MS in a downlink
transmission scenario. This is justified upon assuming that the average SIRs considered here
are in excess of 10dB, which was shown to be sufficient for an AQAM scheme in terms of
reliable channel quality estimation.

In order to mitigate the impact of CCI on the demodulation process, two different approaches are presented here. In the first approach we will utilize Joint Detection (JD) tech-

niques [208], in order to jointly mitigate the impact of CCI, ISI and noise. Alternatively, in the second approach, we will exploit the modem mode switching regime of the AQAM scheme in reducing the impact of CCI on the demodulation process. Before we invoke these two approaches, let us first quantify the impact of CCI on both fixed and adaptive modulation modes without the aid of CCI compensation techniques.

In the following fixed modulation mode based experiments, the modulation mode of the interferer and reference user was identical. We also assumed perfect CIR quality estimation for the reference user, although no information regarding the interferer was utilized in the demodulation process at the reference receiver. The other simulation parameters are listed in Table 7.1.

The associated results for the BPSK, 4QAM, 16QAM and 64QAM modulation modes are shown in Figures 7.15(a), 7.15(b), 7.15(c) and 7.15(d), respectively with their individual single-user performances depicted for comparison purposes. The single-user performance was quantified, when there was no CCI in the system. In each of these figures we can observe different characteristics at low and high average channel SNRs. At low average channel SIRs, where noise was dominant, the performance of all modulation modes approached that of the single-user scenario. However, at high average channel SNRs, the impact of CCI became dominant, resulting in an error floor for all modulation modes. The value of the error floors decreased with increasing average SIRs for the simple reason that the CCI became less dominant. Consequently, as the average SIR increased, the BER performance for all modulation modes approached that of the single-user scenario. The other notable characteristic was that the impact of CCI was more severe in the higher-order modulation modes of 16QAM and 64QAM, which resulted in higher error floors. This was consistent with our expectations, since the higher-order modulation modes, with their relatively small constellation Euclidean distances, were more susceptible to CCI.

Having studied the CCI-impairment on the performance of the individual modem modes in Figure 7.15, the impact of CCI on the wideband AQAM scheme is quantified in Figures 7.16 and 7.17 for target BERs of 1% and 0.01%, respectively for average SIRs of 10, 20 and 30dB. The switching thresholds were set according to Table 4.8 for the respective target BERs and the modulation mode of the interferer was selected randomly. As evidenced by Figures 7.16 and 7.17, the BER degradation increased as the average SIR decreased. The observed BER degradation was several orders of magnitude for the wideband AQAM scheme designed for a target BER of 0.01%. This was significantly more severe, when compared to the wideband AQAM scheme with a target BER of 1%. This highlighted again the increased error sensitivity of the wideband AQAM scheme designed for a lower target BER. The BER deterioration for a certain average SIR was higher as the average channel SNR increased. Again, this was consistent with our expectations, since the higher-order modulation modes which were less robust to CCI - as evidenced by Figure 7.15 - were utilized more frequently, as the average channel SNR increased. The BPS performance was unchanged, since a CCI-free scenario was imposed at the BS, which resulted in optimum modulation mode selection, with regards to the channel quality estimation. In order to mitigate the impact of CCI, joint detection techniques [225] are invoked in the next section, which are well known in the field of CDMA-based systems.

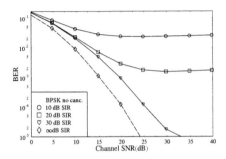

(a) BPSK performance with a BPSK interferer

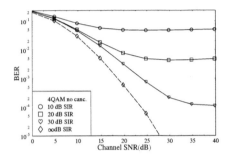

(b) 4QAM performance with a 4QAM interferer

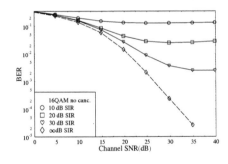

(c) 16QAM performance with a 16QAM interferer

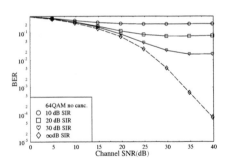

(d) 64QAM performance with a 64QAM interferer

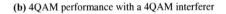

Figure 7.15: BER performance of the fixed modulation modes in the presence of CCI **without the utilization of CCI compensation schemes**. Perfect CIR and channel quality estimation was used for the reference user and the other simulation parameters of the reference user are listed in Table 7.1.

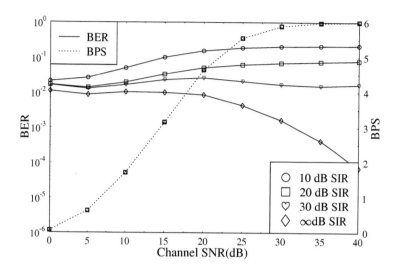

Figure 7.16: Downlink BER and BPS performance of the wideband AQAM scheme in the presence of a co-channel interferer with associated random modulation modes at the MS, **without the utilization of CCI compensation schemes**. Perfect CIR and channel quality estimation of the reference user was implemented and the other simulation parameters of the reference user are listed in Table 7.1. The performance was targeted at a BER of 1% using the switching thresholds of Table 4.8 and an CCI-free scenario was assumed at the BS, which was justified in Section 7.3.2.

7.3.3 Joint Detection Based CCI Compensation Scheme

The Joint Detection (JD) receivers are derivatives of the single-user equalizers described in Chapter 2, which are used to equalize signals that have been distorted by inter-symbol interference (ISI) due to multi-path channels. The problem of CCI is very similar to that of multi-path propagation-induced ISI. Each user in a K-user system suffers from CCI due to the other $(K - 1)$ users. This CCI can also be viewed as a single-user signal perturbed by ISI from $(K - 1)$ paths in a multi-path channel. Therefore, classic equalization techniques [87, 118] used to mitigate the effects of ISI were modified for multiuser detection and these types of multiuser detectors were classified as joint detection receivers. The concept of joint detection for the uplink was proposed by Klein and Baier [226] for synchronous burst transmission, where the performance of a zero-forcing block linear equalizer (ZF-BLE) was investigated for frequency-selective channels. Other joint detection schemes for uplink situations were also proposed by Jung, Blanz, Nasshan, Steil, Baier and Klein, such as

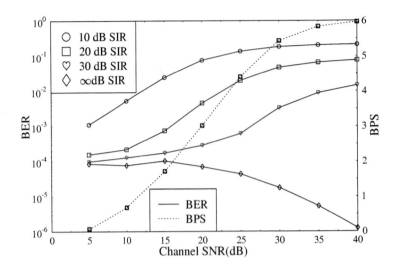

Figure 7.17: Downlink BER and BPS performance of the wideband AQAM scheme in the presence of
a co-channel interferer associated with random modulation modes at the MS, **without the
utilization of CCI compensation schemes.** Perfect CIR and channel quality estimation
of the reference user was implemented and the other simulation parameters of the reference user are listed in Table 7.1. The performance was targeted at a BER of 0.01% using
the switching thresholds of Table 4.8 and a CCI-free scenario was assumed at the BS,
which was justified in Section 7.3.2.

the minimum mean-square error block linear equalizer (MMSE-BLE) [208, 219, 227, 228],
the zero-forcing block decision feedback equalizer (ZF-BDFE) [219, 228] and the minimum
mean-square error block decision feedback equalizer (MMSE-BDFE) [219, 228]. The utilization of the multiuser detection concept in a TDMA environment for ISI and co-channel
interference cancellation was implemented by amongst others Yoshino *et al.* [229], Valenti *et
al.* [230] and Joung *et al.* [231].

The main motivation in these JD techniques was to jointly mitigate the effects of ISI, noise
and Multiple Access Interference (MAI), which was dominant in a CDMA multiuser system.
This was somewhat analogous to the conventional MMSE-type equalization schemes, where
noise and ISI were jointly optimised in order to reduce the effective mean square error at
the input of the detector. In a TDMA environment, we can exploit the JD techniques by
considering the co-channel interference as multiple-access interference and assuming that
the spreading sequence length was restricted to a single-chip, which conformed to symbol-
based TDMA transmission. Since the DFE with its MMSE criterion was mainly featured

in our work, we have opted for invoking the so-called Minimum Mean Square Error Block Decision Feedback Equalizer (JD-MMSE-BDFE) [208] in order to mitigate the impact of CCI. We will now present a rudimentary introduction to the operation of a JD-MMSE-BDFE receiver, when applied in a TDMA system. For a more detailed exposure, the interested reader is referred to [232].

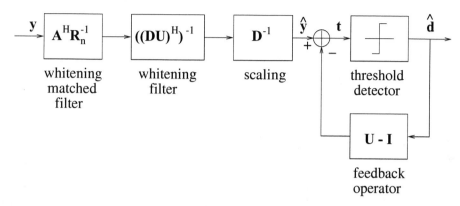

Figure 7.18: The schematic of the JD-MMSE-BDFE receiver structure featuring the different filter functionalities and their corresponding coefficients as explained in Section 7.3.3.1.

7.3.3.1   Theory of the JD-MMSE-BDFE

The basic structure of the JD-MMSE-BDFE is shown in Figure 7.18, where the feedforward filter of the structure consists of a whitening matched filter, a whitening filter and a scaling filter, while the feedback filter is similar to that of the DFE. Let us now analyse in detail the structure shown in Figure 7.18. The received signal vector $\mathbf{y}$ is represented as follows:

$$\mathbf{y} = \mathbf{A}\mathbf{d} + \mathbf{n}, \tag{7.3}$$

where $\mathbf{d}$ is the concatenated data vector of the reference user and the interferers, while $\mathbf{n}$ is the corresponding noise sample vector. The vector $\mathbf{d}$ represents the data frame defined by the data symbols of the reference and interfering users as follows:

$$\mathbf{d} = d_1^1, d_2^1, d_3^1 \ldots d_1^2, d_2^2, d_3^2 \ldots d_N^k, \tag{7.4}$$

where d_n^k represents the nth symbol of the kth user for $n = 1, 2 \ldots N$ and $k = 1, 2 \ldots K$.

The system matrix, $\mathbf{A}$, which contains the CIR estimates of the reference user and interferers, is of size KN columns and $(N + W - 1)$ rows, where W is the length of the CIR. Physically, this matrix describes the ISI and co-channel interference inflicted on the reference user. The matrix is constructed from a set of column vectors $\mathbf{a}_j$ for $j = 1, 2, \ldots KN$, which can be written as :

$$\mathbf{a_j} = \begin{pmatrix} \mathbf{0}_{(n-1)} \\ \mathbf{h}_{(W)}^{(k)} \\ \mathbf{0}_{(N-n)} \end{pmatrix}, \tag{7.5}$$

where $j = (k-1)N + n$ for $n = 1, 2...N$ and $k = 1, 2...K$. The vector $\mathbf{0}_i$ is a column vector of zeroes having a corresponding vector length of i. Furthermore, $\mathbf{h}_{\mathbf{W}}^{(\mathbf{k})}$ is the column vector containing the symbol rate based CIR estimates of the kth user and its vector length is equal to W.

By referring to Figure 7.18, the received signal $\mathbf{y}$ is first processed by a whitening matched filter. This filter is characterized by the conjugate transpose of the system matrix $\mathbf{A}^H$ and the noise covariance matrix, $\mathbf{R}_n = E[\mathbf{nn}^H]$, where H represents conjugate transpose and $E(.)$ is the expected value of $(.)$. The output of the whitening matched filter is subsequently processed by a whitening filter and a scaling filter, as shown in Figure 7.18. These filters are defined by a real-valued diagonal matrix $\mathbf{D}$ and an upper triangular matrix $\mathbf{U}$ can be obtained by the Cholesky Decomposition [233] of the following matrices:

$$(\mathbf{DU})^H \mathbf{DU} = \mathbf{A}^H \mathbf{R}_n^{-1} \mathbf{A} + \mathbf{R}_d^{-1}, \tag{7.6}$$

where $\mathbf{R}_d = E[\mathbf{dd}^H]$ is the covariance matrix of the data. The feedback filter is characterized by the matrix $\mathbf{U}$ and an identity matrix $\mathbf{I}$, as depicted in Figure 7.18.

The operation of the feedforward and feedback filters of the JD-MMSE-BDFE is similar to that of the DFE. The feedforward filter removes not just the ISI, but it also attempts to mitigate the CCI and similarly, the feedback filter removes the ISI and CCI contributions of the interferers. We will now investigate the performance of the above JD-MMSE-BDFE scheme, when applied to fixed modulation modes in the presence of CCI.

7.3.3.2 Performance of the JD-MMSE-BDFE

In our following experiments the SIR was varied from 0 to 30dB and the CIRs of the desired user and the interferer were estimated perfectly at the receiver. The BER performances of the BPSK, 4QAM, 16QAM and 64QAM modulation modes are shown in Figure 7.19. In all these figures the performance without CCI compensation techniques is also shown for comparison. The results reflected similar characteristics to those seen in Figure 7.15, where at low average channel SNRs, the performance approached that of the single-user scenario. However, at high average channel SNRs, the performance was limited by the CCI, although resulting in a lower error floor, when compared to the non-CCI compensated results. The gains recorded by the JD-based receivers were dependent on the modulation modes, where an order of magnitude of reduction in the error floor was observed for the BPSK mode at an average SIR of 10dB. Conversely, the gain was minimal for the higher-order modulation modes of 16QAM and 64QAM. There are two main contributing factors to this minimal performance gain exhibited by the JD-MMSE-BDFE. In CDMA systems, typically a long spreading sequence is utilized, in order to distinguish and separate the signals of different users. This was extremely useful in a CDMA-based joint detector, where a separation between the reference user and the interferer aided the recovery of the signal of the reference user. However, in our TDMA system a single-chip spreading sequence was utilized, which prohibited user-separation by the joint detector. Consequently the gains recorded in Figure 7.19 were modest. The second contributing factor was the impact of error propagation in the feedback filter of the JD-MMSE-BDFE. As highlighted in Section 7.1, error propagation is a phenomenon, whereby any detection errors of data estimates resulted in further data estimation errors. The impact of error propagation was even more severe in the JD-MMSE-BDFE

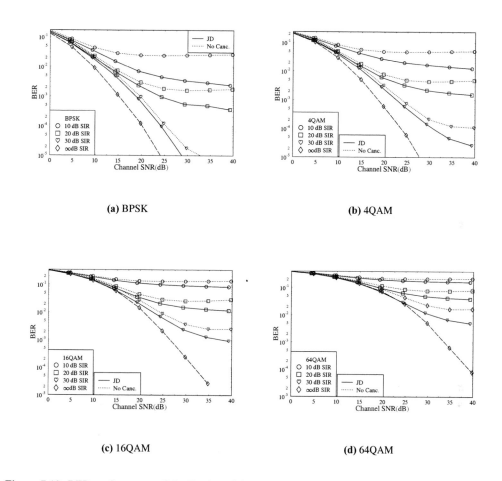

(a) BPSK

(b) 4QAM

(c) 16QAM

(d) 64QAM

Figure 7.19: BER performance of the fixed modulation modes in the presence of an identically modulated interferer **with the assistance of the JD-MMSE-BDFE receiver** shown in Figure 7.18. Perfect CIR estimation of the reference user and interferer was implemented and the other simulation parameters of the reference user are listed in Table 7.1.

receiver, since any errors corrupted the data estimation of both the reference user and the interferer. This was detrimental, since the data estimation of the reference user was dependent on the data estimates of the interferer. Consequently, this inter-dependency severely degraded the performance of the JD-MMSE-BDFE receiver. The higher-order modulation modes were also more susceptible to these error propagation effects due to the smaller Euclidean distances of their constellation points, as explained in Section 7.1. Hence, the performance gains were smaller for the higher-order modulation modes, as evidenced by Figure 7.19.

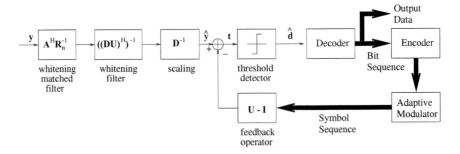

Figure 7.20: The schematic of the embedded convolutional decoding assisted JD-MMSE-BDFE receiver structure featuring the different filter functionalities and the sequence feedback mechanism for the reference user as explained in Section 7.3.3.3.

7.3.3.3   Embedded Convolutionally-coded JD-MMSE-BDFE

In order to reduce the impact of error propagation, convolutional coding was invoked in an embedded structure, as shown in Figure 7.20 [234]. In this convolutional decoding assisted structure the output of the convolutional decoder was utilized in the feedback filter of the JD-MMSE-BDFE for the reference user. However, instead of utilizing a symbol by symbol feedback, a sequence-based feedback was implemented, where the feedback filter inputs were replaced by a sequence of symbols produced by the decoder. This was also proposed by Cheung [105] and Mak [235] in the context of a Continuous Phase Modulation (CPM) and Trellis Coded Modulation (TCM) based scheme, respectively in a noise limited environment. Consequently, the input symbols of the feedback filter were replaced by more reliable information from the convolutional decoder on a burst-by-burst basis. Hence the impact of error propagation in the feedback filter of the JD-MMSE-BDFE was reduced.

The performance of the embedded structure of Figure 7.20 is depicted in Figure 7.21 for the fixed modulation modes of BPSK, 4QAM and 16QAM in the presence of a BPSK interferer [234]. The utilization of an interferer using BPSK represented an upper bound performance of this embedded structure, since the impact of error propagation was less severe, as evidenced by Figure 7.19. The least robust modulation mode of 64QAM was not utilized, since the impact of error propagation was too severe. In our subsequent experiments, the half-rate RSC encoder shown in Figure 5.2 having a constraint length of $K = 5$ was utilized in conjunction with a Viterbi decoder. The performance of a conventional separately decoded DFE structure - where the convolutional decoder was not embedded into the JD-MMSE-BDFE structure - was also shown in Figure 7.21 for comparison.

A better BER performance was produced by the convolutional decoding assisted structure, when compared to the conventional scheme for all modulation modes. In this respect, the error floor was reduced by an order of magnitude, as evidenced by Figure 7.21. This was attributed to the reduction of error propagation due to the sequence-based feedback of more reliable convolutionally decoded symbols into the feedback filter.

The embedded structure of Figure 7.20 was also implemented in the context of a wideband AQAM scheme. In our subsequent experiments, the switching thresholds were experimentally adjusted in order to achieve target BERs of 1% and 0.01%, which are shown in

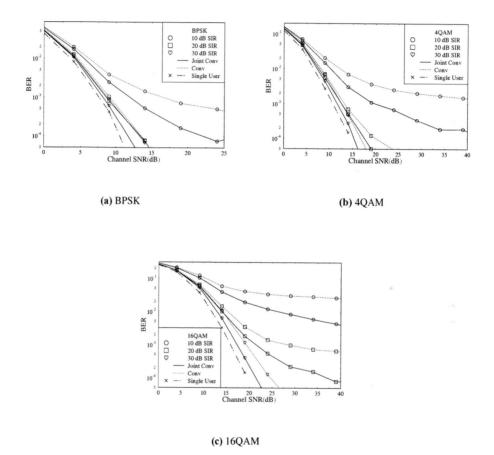

(a) BPSK

(b) 4QAM

(c) 16QAM

Figure 7.21: Performance of the fixed modulation modes in the presence of an identically modulated
interferer **using the embedded convolutional decoding assisted JD-MMSE-BDFE receiver** shown in Figure 7.20 [234]. Perfect CIR estimation of the reference user and the
interferer was implemented and the other simulation parameters of the reference user are
listed in Table 7.1. The performance utilizing a conventional - separately decoded - JD-
MMSE-BDFE receiver structure was also shown for comparison (denoted by Conv in the
figure).

SIR(dB)	1%			0.01%		
	t_1(dB)	t_2(dB)	t_3(dB)	t_1(dB)	t_2(dB)	t_3(dB)
10	0.0	2.0	12.0	-	-	-
20	−1.0	2.0	6.0	2.0	5.0	12.0
30	−1.0	2.0	6.0	2.0	5.0	10.0

Table 7.5: The experimentally determined switching thresholds of the wideband AQAM scheme **utilizing the embedded convolutional decoding assisted JD-MMSE-BDFE receiver** for target BERs of 1% and 0.01%. These thresholds were utilized in Figure 7.22.

SIR(dB)	1%		0.01%	
	Fixed(dB)	AQAM(dB)	Fixed(dB)	AQAM(dB)
10	5.50	1.30	-	-
20	4.50	2.50	14.10	5.00
30	4.50	2.50	11.80	5.00

Table 7.6: The average channel SNRs required in order to achieve a BPS throughput of 0.5 for the fixed modulation mode of BPSK and for the wideband AQAM scheme for target BERs of 1% and 0.01%. The required average channel SNR was extracted from Figures 7.21(a) and 7.22, which utilized the embedded convolutional decoding assisted JD-MMSE-BDFE receiver structure of Figure 7.20.

Table 7.5. The results over 10, 20 and 30dB average SIR channels are depicted in Figure 7.22 for target BERs of 1% and 0.01% [234]. Referring to Figure 7.22(a), the target BER was achieved at all average SIRs. However, at an average SIR of 10dB the BPS throughput degraded, when compared to the performance achieved at average SIRs of 20dB and 30dB. This reflected the conservative approach of the wideband AQAM mode switching regime, where the switching thresholds were increased, in order to reduce the impact of CCI at the expense of a reduced throughput. The wideband AQAM scheme targeted at a BER of 0.01% exhibited similar characteristics, as evidenced by Figure 7.22(b). However, the target BER was not achieved for an average SIR of 10dB and hence it was not shown.

The BPS throughput of the fixed modulation modes and that of the wideband AQAM scheme along with the corresponding average channel SNRs required were recorded from Figures 7.21 and 7.22. These values are compared in Tables 7.6 and 7.7 for a BPS performance of 0.5 and 1, respectively [234], which are equivalent to the half-rate coded throughput of fixed mode BPSK and 4QAM. For the target BER of 1%, BPS/SNR gains of approximately $2 − 4$dB were recorded for AQAM and similarly, approximately $7 − 9$dB of BPS/SNR gain was observed for the AQAM scheme with a target BER of 0.01%.

7.3.3.4   Segmented Wideband AQAM

In the last section we utilized the JD-based receivers, in order to mitigate the impact of CCI in an AQAM wideband scheme. However, in this section we will present a more simple approach of minimizing the impact of CCI by invoking a more stringent modem mode switching regime for the wideband AQAM scheme.

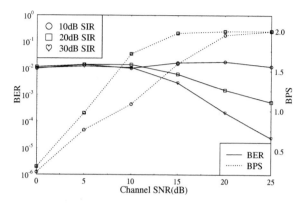

(a) Target BER of 1%.

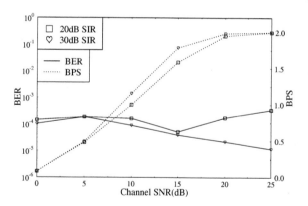

(b) Target BER of 0.01%.

Figure 7.22: Downlink BER and BPS performance of the wideband AQAM scheme in the presence of a co-channel interferer with associated random modulation modes at the MS, **using the embedded convolutional decoding assisted JD-MMSE-BDFE receiver** shown in Figure 7.20. Perfect CIR of the reference user and interferer was implemented and the other simulation parameters of the reference user are listed in Table 7.1. The switching thresholds were set according to Table 7.5 and a CCI-free scenario was assumed at the BS, which was justified in Section 7.3.2.

	1%		0.01%	
SIR(dB)	Fixed(dB)	AQAM(dB)	Fixed(dB)	AQAM(dB)
10	10.50	8.70	-	-
20	7.50	5.00	17.50	10.00
30	7.00	5.00	15.50	8.80

Table 7.7: The average channel SNRs required in order to achieve a BPS throughput of 0.5 for the fixed modulation mode of 4QAM and for the wideband AQAM scheme for target BERs of 1% and 0.01%. The required average channel SNR was extracted from Figures 7.21(b) and 7.22, which utilized the embedded convolutional decoding assisted JD-MMSE-BDFE receiver structure of Figure 7.20.

In Figure 7.15 we have noted the fact that the performance of the fixed modulation modes exhibited different characteristics at low and high average channel SNRs. The BER performance was limited by noise at low channel SNRs and conversely, at high average channel SNRs CCI was the limiting factor, which resulted in an error floor. Furthermore, the higher-order modulation modes suffered a higher performance degradation, when compared to that of the lower-order modulation modes. These two characteristics can be exploited in a wideband AQAM scheme, where it was possible to segment the switching thresholds based on the average channel SNR. Consequently the switching thresholds were optimised for a noise- and CCI-limited environment at low and high average channel SNRs, respectively. The obvious method of achieving this was to optimise the switching thresholds against noise at low average channel SNRs and similarly, against CCI at high average channel SNRs. However, this required a different set of switching thresholds for a given average channel SNR and average SIR.

A more practical approach was to create a set of switching thresholds, which could be invoked for any given average channel SNR and average SIR. Consequently, a switching regime termed as *segmented switching*, which was characterized by an inner and outer switching threshold was introduced. The functionality of the inner and outer switching thresholds was different, where the former provided protection against noise impairment and similarly, the latter was developed in order to mitigate the effects of CCI. The inner switching thresholds, t_n, which were based on the instantaneous SNR output of the DFE γ_{dfe}, were used to select the appropriate modulation mode. This was similar to the conventional AQAM modem mode switching regime of Equation 4.1 with the exception that the number of legitimate AQAM modem modes was dependent on the outer switching thresholds. The outer switching thresholds m_n, which were based on the instantaneous SIR γ_{cci}, determined the number of modulation modes that were utilized in the AQAM modem mode switching regime. This regime is illustrated by Table 7.8, which characterized the methodology, in which the modulation mode was selected based on the inner switching thresholds, t_n and the outer switching thresholds, m_n. The number of modes that was used by the switching regime on a burst-by-burst basis was determined by the instantaneous SIR, γ_{cci} and the corresponding outer switching thresholds, m_n for $n = 1, 2, 3, 4$. When the instantaneous SIR was high and above the outer switching threshold, m_4, the switching regime illustrated by column two of Table 7.8 was implemented. In this regime all the possible AQAM modem modes were utilized and the modulation mode was selected based on the instantaneous SNR output of the DFE,

γ_{dfe} and the corresponding inner switching thresholds, t_n for $n = 1, 2, 3, 4$. However, if the instantaneous SIR was low, for example $m_2 \leq \gamma_{cci} < m_3$, the switching regime illustrated by column four of Table 7.8 was implemented, where the 16QAM and 64QAM modulation modes were disabled. The modulation mode was subsequently chosen according to the instantaneous SNR output of the DFE and according to the corresponding reduced number of inner switching thresholds, t_n for $n = 1, 2, 3$.

Mode	$\gamma_{cci} \geq m_4$	$m_3 \leq \gamma_{cci} < m_4$	$m_2 \leq \gamma_{cci} < m_3$	$m_1 \leq \gamma_{cci} < m_2$	$\gamma_{cci} < m_1$
NOTX	$\gamma_{dfe} \leq t_1$	$\gamma_{dfe} \leq t_1$	$\gamma_{dfe} \leq t_1$	$\gamma_{dfe} \leq t_1$	$\gamma_{dfe} \leq \infty$
BPSK	$t_1 \leq \gamma_{dfe} < t_2$	$t_1 \leq \gamma_{dfe} < t_2$	$t_1 \leq \gamma_{dfe} < t_2$	$\gamma dfe \geq t_1$	-
4QAM	$t_2 \leq \gamma_{dfe} < t_3$	$t_2 \leq \gamma_{dfe} < t_3$	$\gamma_{dfe} \geq t_3$	-	-
16QAM	$t_3 \leq \gamma_{dfe} < t_4$	$\gamma_{dfe} \geq t_4$	-	-	-
64QAM	$\gamma_{dfe} \geq t_4$	-	-	-	-

Table 7.8: The switching regime of the segmented switching based wideband AQAM scheme as discussed in Section 7.3.3.4. The regime was characterized by the inner switching thresholds, t_n and the outer switching thresholds, m_n for $n = 1, 2, 3, 4$.

The general philosophy of this segmented switching regime was to exploit the switching structure of the wideband AQAM scheme, in order to mitigate the impact of noise and CCI. The inner switching thresholds were developed in order to reduce the impact of noise regardless of the average channel SNR. However, in the presence of CCI a more conservative switching regime was needed, which led to the development of the outer switching thresholds. The main functionality of the outer switching threshold was to restrict the number of modes that could be used by the switching mechanism. Consequently, the higher-order modulation modes, which were susceptible to CCI-induced degradation were only used, when the instantaneous SIR was high.

During the initial uplink transmission, the BS receiver estimated the output SNR of the DFE γ_{dfe} as well as the instantaneous uplink SIR, γ_{cci}. The output SNR of the DFE was estimated based on the mid-amble assisted CIR estimation of the reference user in the presence of noise and CCI. Consequently, by utilizing the CIR estimates, the reference user's data was demodulated. The received burst of the reference user was reconstructed by corrupting the estimated data of the reference user by the wideband fading CIR estimates, in order to measure the reference user's signal power. The reconstructed signal was then subtracted from the composite received burst of the reference user, the interferer and the noise, in order to estimate the total noise plus interference level. According to this procedure all impairments were attributed to the CCI, which is a good approximation in interference-limited environments. Finally, the SIR was estimated by calculating the ratio of the reference user's signal power to the interference plus noise power.

Subsequently, the AQAM modulation mode was selected utilizing the output SNR of the DFE by exploiting the channel's reciprocity. Furthermore, by referring to Table 7.8, the BS determined the number of legitimate modes for the next uplink transmission. This information was then relayed to the mobile station superimposed on the subsequent downlink transmission with the aid of control information in the transmission burst. At the mobile station's receiver, the instantaneous output SNR of the DFE and the previous instantaneous uplink SIR were then used for selecting the appropriate uplink modulation mode based on AQAM switching regime illustrated by Table 7.8. In this transmission scenario we have exploited the channel reciprocity of the reference user. However, the uplink and downlink instantaneous SIR were estimated and signalled in a closed-loop system without exploiting the concept of reciprocity.

In order to devise an upper bound wideband AQAM performance utilizing the segmented switching regime, we have assumed that the instantaneous burst-by-burst SIR was estimated perfectly prior to transmission. However, in our subsequent experiments this assumption was discarded in order to create a more practical wideband AQAM scheme. Unlike in Section 4.3.5, where Powell optimization was utilized in order to set the switching thresholds, the presence of CCI and the proposed twin layered switching mechanism resulted in an intractable optimization problem. Consequently, the inner switching thresholds, t_n were set according to Table 4.8 and the outer switching thresholds, m_n were adjusted experimentally in order to achieve target BERs of 1% and 0.01%, as shown in Table 7.9. The associated

	m_1	m_2	m_3	m_4
Upper Bound, 1%	3.64	6.22	13.64	20.98
Upper Bound, 0.01%	8.24	11.46	18.79	25.76
Latency, 1%	3.64	7.22	15.64	23.98
Latency, 0.01%	9.24	13.46	21.79	28.76

Table 7.9: The outer switching thresholds, which were experimentally adjusted for target BERs of 1% and 0.01% for the segmented switching based wideband AQAM scheme. The thresholds were set for attaining an upper bound performance with no delay in estimating the interference level and also for a system, which possessed a 2.3075ms delay. The corresponding performances are shown in Figures 7.23, and 7.24.

upper bound performance results are shown in Figure 7.23 for target BERs of 1% and 0.01%. The average SIR was varied in the range of $0 - 30$dB average SIR and the performance in a noise limited environment was also shown for comparison. The target BERs were achieved for all investigated average SIRs and the BPS performance improved, as the average SIR increased. For a fixed average SIR, as the average channel SNR increased, the higher-order modulation modes were used more frequently, resulting in an increased BPS performance. However, at high average channel SNRs, the BPS performance varied for different average SIRs. This demonstrated the effects of the segmented switching regime of the wideband AQAM scheme, where at a' low average SIR of 10dB the higher-order modulation modes were frequently disabled resulting in a lower BPS performance. Conversely, at a high average SIR of 30dB - due to the favourable channel and interference conditions - the higher-order modulation modes were frequently utilized, yielding an improved BPS performance.

The impact of channel quality estimate latency was severe in a practical AQAM scheme, which was demonstrated in Section 7.2. In the proposed segmented wideband AQAM scheme, the delay incurred by the closed-loop estimation of the interference levels was twice that of the channel quality estimation, as discussed in Section 7.2.2. These delays were potentially detrimental to the upper bound performance of the segmented wideband AQAM scheme. Subsequently a channel quality estimation delay of 16 time-slots was used, where each time-slot was of duration 72μs. The delay of 16 time-slots was selected, as it was the maximum delay that could be incurred by the wideband AQAM scheme for a target BER of 1% in an open-loop system, as shown by Table 7.2. Consequently, the delay in estimating the interference level in a closed-loop signalling regime was set to 32 time-slots which corresponded

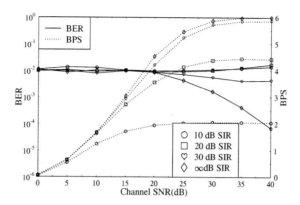

(a) Target BER of 1%.

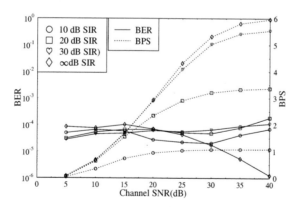

(b) Target BER of 0.01%.

Figure 7.23: The downlink channel quality estimation latency-free upper bound performance of the segmented AQAM scheme defined by Table 7.8. The corresponding inner- and outer-switching thresholds were set according to Tables 4.8 and 7.9, respectively. Perfect CIR estimation was applied for the reference user and the interference level was measured using the Kalman-filter based mid-amble assisted CIR estimation scheme of Table 7.1. The interferer used a random modulation mode in the downlink transmission and an interference-free scenario was assumed at the BS.

to twice the channel quality estimation delay. A mid-amble based channel quality estimation was performed using the Kalman filtering based CIR estimator. Subsequently, the instantaneous burst-by-burst SIR was estimated following the methodology described previously in this section. The outer switching thresholds m_n, for this latency-affected wideband AQAM scheme are shown in Table 7.9 for target BERs of 1% and 0.01% and the inner switching thresholds, t_n are listed in Table 4.8.

The performance of the latency-affected segmented wideband AQAM scheme is shown in Figure 7.24 for target BERs of 1% and 0.01%, respectively. The target BERs were achieved and the BPS performance exhibited similar characteristics to those of the upper bound curves shown in Figure 7.23. However, the BPS throughput degraded, when compared to the upper bound performance. This was consistent with our expectations, since the switching thresholds of the latency-affected segmented wideband AQAM scheme were set more conservatively, in order to reduce the impact of latency and to achieve the target BER.

The throughput comparison between the segmented wideband AQAM scheme and the fixed modulation modes is shown in Figure 7.25 for target BERs of 1% and 0.01%, respectively. The throughput and the corresponding average channel SNR required was obtained from Figure 7.15 for the fixed modulation modes. The throughput of the upper bound segmented wideband AQAM scheme is also displayed for comparison. In some cases, the target BER was not achieved by the fixed modulation modes and hence the corresponding points were not shown. For the target BER of 1%, an approximately $3-5$dB BPS/SNR was observed for average SIRs of 20 and 30dB. Similarly, for the target BER of 0.01%, at an average SIR of 30dB, a BPS/SNR gain of approximately 11.6dB was achieved at an average throughput of 1 bit per symbol. Furthermore, at an average BPS of 2, the BPS/SNR gain achieved was approximately 20dB.

In comparing the fixed modulation modes and the segmented wideband AQAM scheme, CCI compensation techniques were not applied. Consequently, the degradation incurred by the fixed modulation modes was severe. However, in the segmented wideband AQAM scheme, we exploited the switching regime of Table 7.8 for the wideband AQAM scheme, in order to reduce the impact of CCI at the expense of a lower throughput. Nevertheless, even exposed to a CCI estimate latency equivalent to 32 time-slots, the segmented wideband AQAM scheme having instantaneous SIR estimates managed to outperform the fixed modulation modes.

7.4   Review and Discussion

In this chapter we explored and quantified the performance of the proposed wideband AQAM scheme, which was exposed to error propagation, channel quality estimation latency and CCI. This was implemented in order to discard the idealistic assumptions stipulated in Section 4.3.1 and hence to create a practical and realistic wideband AQAM scheme.

The performance in conjunction with error propagation was recorded in Figure 7.2, where at low to medium average channel SNRs, the BER degradation was minimal. However, at higher average channel SNRs an approximate SNR degradation of 3dB was observed and justified in Section 7.1. Nevertheless, the target BERs of 1% and 0.01% were achieved without any degradation in terms of the BPS throughput performance.

The impact of channel quality estimation latency - where the delay incurred by the scheme

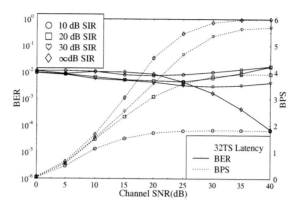

(a) Target BER of 1%.

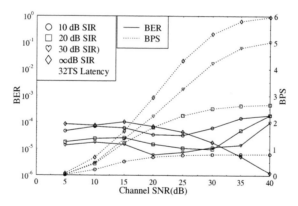

(b) Target BER of 0.01%.

Figure 7.24: The downlink performance of the segmented AQAM scheme defined by the switching regime of Table 7.8, which was exposed to a delay of 2.3075ms in estimating the interference level. The corresponding inner- and outer- switching thresholds were set according to Tables 4.8 and 7.9, respectively. Perfect CIR estimation was applied for the reference user and the interference level was measured using the Kalman-filter based mid-amble assisted CIR estimation scheme of Table 7.1. The interferer had a random modulation mode in the downlink transmission and an interference-free scenario was assumed at the BS.

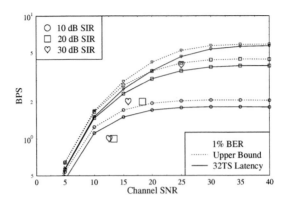

(a) Target BER of 1%.

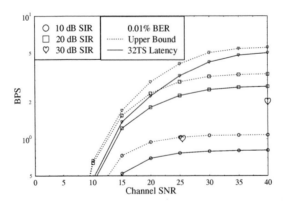

(b) Target BER of 0.01%.

Figure 7.25: The BPS throughput of the segmented AQAM scheme and the fixed modulation modes, where the throughput values were extracted from Figures 7.15, 7.23. This included the upper bound, the latency-free segmented AQAM performance and the practical segmented AQAM performance with a delay of 2.3075ms in estimating the interference levels. The target BERs were not achieved by some fixed modulation modes for certain average SIR and thus the corresponding points were not shown.

resulted in a non-optimum modulation mode selection - was studied in Section 7.2.3. Consequently, the BER performance degradation was aggravated as the channel quality estimation delay increased, as evidenced by Figures 7.5(a) and 7.5(b). The amount of delay was dependent on the time frame structure and the signalling protocol, as discussed in Section 7.2.1 and 7.2.2. Furthermore the BER performance was influenced by the fluctuation rate - i.e. the Doppler frequency - of the time varying channel, where a channel exhibiting a low normalized Doppler frequency was more favourable for the implementation of a wideband AQAM scheme, as shown in Figures 7.7(a) and 7.7(b). Subsequently, in order to reduce the impact of channel quality estimation latency, the channel quality estimates were predicted using a simple linear prediction algorithm and the associated results were compared in Table 7.2. Finally, we presented a sub-frame based TDD/TDMA wideband AQAM scheme exhibiting a channel quality estimation latency of 2.3075ms and the corresponding results are shown in Figures 7.9 and 7.10. The BPS throughput comparison between this practical sub-frame based TDD/TDMA wideband AQAM scheme and the fixed modulation modes was recorded in Table 7.4, where maximum gains of approximately 1.4dB and 6.4dB were achieved for target BERs of 1% and 0.01%. These gains were lower than the upper bound performance gains as a result of the channel quality estimation latency. In this respect, we can underline the importance of the channel quality estimates in a wideband AQAM scheme. It was also noted that the wideband AQAM scheme performed more reliably in a slowly varying channel environment, which facilitated accurate channel quality estimation. Furthermore, the estimation of the channel quality can be improved by utilizing more sophisticated prediction algorithms, than the linear prediction algorithm of Section 7.2.4.

The performance of the wideband AQAM scheme was also investigated in the presence of CCI with regards to the channel quality estimation and the demodulation process of the receiver. The channel quality estimation degraded as a result of the CCI, where in Figure 7.12 the average channel quality estimation error of Equation 7.2 increased, as the average SIR decreased. Consequently, the modulation mode switching regime was impaired, which resulted in modulation mode selection errors. The impact of these errors on the BER and BPS performance was shown in Figures 7.13 and 7.14. However, the degradation was minimal for an average SIR in excess of 10dB, which was then utilized as the minimum average SIR threshold for the employment of a wideband AQAM scheme.

The impact of CCI on the demodulation process of the receiver was studied and quantified in Figure 7.15 for fixed modulation modes. From these results, we observed the different characteristics shown at low- and high-average channel SNRs. Furthermore, the BER degradation was more severe in the context of the higher-order modulation modes of 16QAM and 64QAM. In the wideband AQAM scheme the BER degradation increased, as the average SIR decreased, as evidenced by Figures 7.16 and 7.17. The degradation was higher at high average channel SNRs due to the utilization of the less robust higher-order modulation modes.

In order to mitigate the impact of CCI, a JD-MMSE-BDFE receiver was invoked and the corresponding performance was shown in Figure 7.19 for fixed modulation modes. In these experiments, only modest gains were observed, when compared to the non-compensated fixed modulation modes, which was due to the absence of a spreading sequence and due to the impact of error propagation in the JD-MMSE-BDFE. Subsequently, convolutional coding was invoked in an embedded JD-MMSE-BDFE structure, as shown in Figure 7.20, in order to reduce the impact of error propagation. In this embedded structure the more reliable symbols from the convolutional decoder were utilized in the feedback filter via a sequence feedback regime, as explained in Section 7.3.3.3. Consequently, the impact of error propagation was reduced, which yielded an improved BER performance, as evidenced by Figure 7.21 for the fixed modulation modes and similarly, by Figure 7.22 for the wideband AQAM scheme. The wideband AQAM scheme produced BPS/SNR gains of approximately $2-4$dB and $7-9$dB at target BERs of 1% and 0.01%, respectively, when compared to the fixed modulation schemes.

The main disadvantage of the JD-MMSE-BDFE receiver was that it required accurate CIR estimates of the reference user and the interferer. It was feasible to obtain sufficiently accurate CIR estimates for the reference user based on the mid-amble sequence in the presence of CCI, as evidenced by Figure 7.12. However, the estimation of the interferer's channel was usually degraded due to insufficient information regarding the interferer. Consequently, the performance of the JD-MMSE-BDFE receiver was severely affected, since the data estimation of the reference user was dependent upon the data estimation of the interferer. Hence the results - which were generated utilizing the JD-MMSE-BDFE - represented an upper bound performance.

In our second approach in mitigating the impact of CCI upon the demodulation process of the receiver, the concept of segmented wideband AQAM was introduced. In this scheme, inner and outer switching thresholds were implemented for any given average channel SNR and average SIR. The inner switching thresholds were developed based on a noise limited environment, while the outer switching thresholds were constructed based on an interference-limited environment, following the regime outlined in Section 7.3.3.4 and Table 7.8. The outer switching thresholds restricted the utilization of higher-order modulation modes, when the instantaneous SIR was low, in order to reduce the impact of CCI and hence reduced the BPS throughput. A practical, wideband finite estimation-latency AQAM scheme was implemented, where a delay of 32 time-slots was incurred in estimating the instantaneous SIR. The corresponding results were displayed in Figure 7.24, where the target BERs were achieved at the expense of a reduced BPS throughput. Finally, substantial BPS/SNR gains were achieved and recorded for this segmented AQAM scheme as shown in Figure 7.25.

The main advantage of the segmented AQAM approach was that information regarding the interferer was not needed in order to reduce the impact of CCI. However, this approach required estimation of the instantaneous SIR and the quality of this estimation was dependent on the estimation delay incurred by the scheme. Nevertheless, given a delay of 32 time-slots in estimating the instantaneous SIR, the target BERs were achieved with significant BPS/SNR gains over the fixed modulation modes. Having quantified the performance benefits of the proposed wideband AQAM scheme, we can now conclude our findings and provide suggestions for future work in the next chapter.

Part II

Near-instantaneously Adaptive Modulation and Neural Network Based Equalisation

Chapter **8**

Neural Network Based Equalization

In this chapter, we will give an overview of neural network based equalization. Channel equalization can be viewed as a classification problem. The optimal solution to this classification problem is inherently nonlinear. Hence we will discuss, how the nonlinear structure of the artificial neural network can enhance the performance of conventional channel equalizers and examine various neural network designs amenable to channel equalization, such as the so-called multilayer perceptron network [236–240], polynomial perceptron network [241–244] and radial basis function network [85, 245–247]. We will examine a neural network structure referred to as the Radial Basis Function (RBF) network in detail in the context of equalization. As further reading, the contribution by Mulgrew [248] provides an insightful briefing on applying RBF network for both channel equalization and interference rejection problems. Originally RBF networks were developed for the generic problem of data interpolation in a multi-dimensional space [249, 250]. We will describe the RBF network in general and motivate its application. Before we proceed, our forthcoming section will describe the discrete time channel model inflicting intersymbol interference that will be used throughout this thesis.

8.1 Discrete Time Model for Channels Exhibiting Intersymbol Interference

A band-limited channel that results in intersymbol interference (ISI) can be represented by a discrete-time transversal filter having a transfer function of:

$$F(z) = \sum_{n=0}^{L} f_n z^{-n}, \tag{8.1}$$

where f_n is the nth impulse response tap of the channel and $L + 1$ is the length of the channel impulse response (CIR). In this context, the channel represents the convolution of

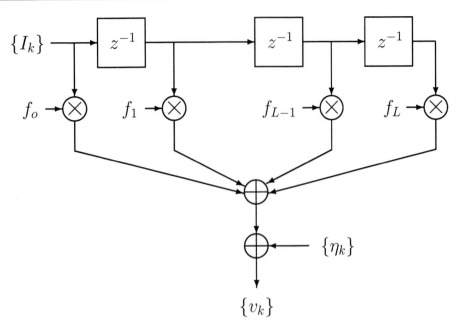

Figure 8.1: Equivalent discrete-time model of a channel exhibiting intersymbol interference and experiencing additive white Gaussian noise.

the impulse responses of the transmitter filter, the transmission medium and the receiver filter.
In our discrete-time model discrete symbols I_k are transmitted to the receiver at a rate of $\frac{1}{T}$
symbols per second and the output v_k at the receiver is also sampled at a rate of $\frac{1}{T}$ per second.
Consequently, as depicted in Figure 8.1, the passage of the input sequence $\{I_k\}$ through the
channel results in the channel output sequence $\{v_k\}$ that can be expressed as

$$v_k = \sum_{n=0}^{L} f_n I_{k-n} + \eta_k \qquad -\infty \leq k \leq \infty, \tag{8.2}$$

where $\{\eta_k\}$ is a white Gaussian noise sequence with zero mean and variance σ_η^2. The number
of interfering symbols contributing to the ISI is L. In general, the sequences $\{v_k\}$, $\{I_k\}$,
$\{\eta_k\}$ and $\{f_n\}$ are complex-valued. Again, Figure 8.1 illustrates the model of the equivalent
discrete-time system corrupted by Additive White Gaussian Noise (AWGN).

8.2   Equalization as a Classification Problem

In this section we will show that the characteristics of the transmitted sequence can be exploited by capitalising on the finite state nature of the channel and by considering the equalization problem as a geometric classification problem. This approach was first expounded
by Gibson, Siu and Cowan [237], who investigated utilizing nonlinear structures offered by
Neural Networks (NN) as channel equalisers.

We assume that the transmitted sequence is binary with equal probability of logical ones
and zeros in order to simplify the analysis. Referring to Equation 8.2 and using the notation

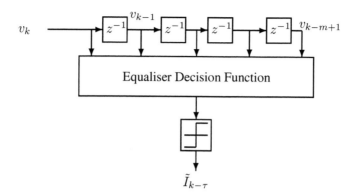

Figure 8.2: Linear m-tap equalizer schematic.

of Section 8.1, the symbol-spaced channel output is defined by

$$
\begin{aligned}
v_k &= \sum_{n=0}^{L} f_n I_{k-n} + \eta_k \\
&= \tilde{v}_k + \eta_k \qquad -\infty \leq k \leq \infty,
\end{aligned} \tag{8.3}
$$

where $\{\eta_k\}$ is the additive Gaussian noise sequence, $\{f_n\}$, $n = 0, 1, \dots, L$ is the CIR, $\{I_k\}$ is the channel input sequence and $\{\tilde{v}_k\}$ is the noise-free channel output.

The mth order equaliser, as illustrated in Figure 8.2, has m taps as well as a delay of τ, and it produces an estimate $\tilde{I}_{k-\tau}$ of the transmitted signal $I_{k-\tau}$. The delay τ is due to the precursor section of the CIR, since it is necessary to facilitate the causal operation of the equalizer by supplying the past and future received samples, when generating the delayed detected symbol $I_{k-\tau}$. Hence the required length of the decision delay is typically the length of the CIR's precursor section, since outside this interval the CIR is zero and therefore the equaliser does not have to take into account any other received symbols. The channel output observed by the linear mth order equaliser can be written in vectorial form as

$$
\mathbf{v}_k = \begin{bmatrix} v_k & v_{k-1} & \cdots & v_{k-m+1} \end{bmatrix}^T, \tag{8.4}
$$

and hence we can say that the equalizer has an m-dimensional channel output observation space. For a CIR of length $L + 1$, there are hence $n_s = 2^{L+m}$ possible combinations of the binary channel input sequence

$$
\mathbf{I}_k = \begin{bmatrix} I_k & I_{k-1} & \cdots & I_{k-m-L+1} \end{bmatrix}^T \tag{8.5}
$$

that produce $n_s = 2^{L+m}$ different possible noise-free channel output vectors

$$
\tilde{\mathbf{v}}_k = \begin{bmatrix} \tilde{v}_k & \tilde{v}_{k-1} & \cdots & \tilde{v}_{k-m+1} \end{bmatrix}^T. \tag{8.6}
$$

The possible noise-free channel output vectors $\tilde{v}_k$ or particular points in the observation space will be referred to as the desired channel states. Expounding further, we denote each of the $n_s = 2^{L+m}$ possible combinations of the channel input sequence $\mathbf{I}_k$ of length $L + m$ symbols as $\mathbf{s}_i, 1 \leq i \leq n_s = 2^{L+m}$, where the channel input state $\mathbf{s}_i$ determines the desired channel output state $\mathbf{r}_i, i = 1, 2, \ldots, n_s = 2^{L+m}$. This is formulated as:

$$\tilde{\mathbf{v}}_{\mathbf{k}} = \mathbf{r}_i \qquad \text{if } \mathbf{I}_k = \mathbf{s}_i, \qquad i = 1, 2, \ldots, n_s.$$

The desired channel output states can be partitioned into two classes according to the binary value of the transmitted symbol $I_{k-\tau}$, as seen below:

$$\begin{aligned}
V_{m,\tau}^+ &= \{\tilde{\mathbf{v}}_k | I_{k-\tau} = +1\}, \\
V_{m,\tau}^- &= \{\tilde{\mathbf{v}}_k | I_{k-\tau} = -1\},
\end{aligned} \tag{8.7}$$

and

$$V_{m,\tau} = V_{m,\tau}^+ \bigcup V_{m,\tau}^-. \tag{8.8}$$

We can denote the desired channel output states according to these two classes as follows:

$$\begin{aligned}
\mathbf{r}_i^+ &\in V_{m,\tau}^+ \qquad i = 1, 2, \ldots, n_s^+, \\
\mathbf{r}_j^- &\in V_{m,\tau}^- \qquad j = 1, 2, \ldots, n_s^-,
\end{aligned} \tag{8.9}$$

where the quantities n_s^+ and n_s^- represent the number of channel states $\mathbf{r}_i^+$ and $\mathbf{r}_j^-$ in the set $V_{m,\tau}^+$ and $V_{m,\tau}^-$, respectively.

The relationship between the transmitted symbol I_k and the channel output v_k can also be written in a compact form as:

$$\begin{aligned}
\mathbf{v}_k &= \mathbf{F}\mathbf{I}_k + \boldsymbol{\eta}_k \\
&= \tilde{\mathbf{v}}_k + \boldsymbol{\eta}_k,
\end{aligned} \tag{8.10}$$

where $\boldsymbol{\eta}_k$ is an m-component vector that represents the AWGN sequence, $\tilde{\mathbf{v}}_k$ is the noise-free channel output vector and $\mathbf{F}$ is an $m \times (m + L)$ CIR-related matrix in the form of:

$$\mathbf{F} = \begin{bmatrix} f_0 & f_1 & \cdots & f_L & \cdots & 0 \\ 0 & f_0 & \cdots & f_{L-1} & \cdots & 0 \\ \vdots & \vdots & & & & \vdots \\ 0 & 0 & f_0 & \cdots & f_{L-1} & f_L \end{bmatrix}, \tag{8.11}$$

with $f_j, j = 0, \ldots, L$ being the CIR taps.

Below we demonstrate the concept of finite channel states in a two-dimensional output observation space ($m = 2$) using a simple two-coefficient channel ($L = 1$), assuming the CIR of:

$$F(z) = 1 + 0.5z^{-1}. \tag{8.12}$$

Thus, $\mathbf{F} = \begin{bmatrix} 1 & 0.5 & 0 \\ 0 & 1 & 0.5 \end{bmatrix}$, $\tilde{\mathbf{v}}_k = \begin{bmatrix} \tilde{v}_k & \tilde{v}_{k-1} \end{bmatrix}^T$ and $\mathbf{I}_k = \begin{bmatrix} I_k & I_{k-1} & I_{k-2} \end{bmatrix}^T$.
All the possible combinations of the transmitted binary symbol I_k and the noiseless channel outputs $\tilde{v}_k, \tilde{v}_{k-1}$, are listed in Table 8.1.

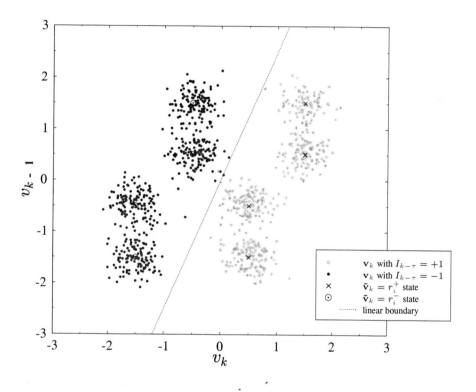

Figure 8.3: The noiseless BPSK-related channel states $\tilde{\mathbf{v}}_k = \mathbf{r}_i$ and the noisy channel outputs $\mathbf{v}_k$ of a Gaussian channel having a CIR of $F(z) = 1 + 0.5z^{-1}$ in a two-dimensional observation space. The noise variance $\sigma_\eta^2 = 0.05$, the number of noisy received $\mathbf{v}_k$ samples output by the channel and input to the equalizer is 2000 and the decision delay is $\tau = 0$. The linear decision boundary separates the noisy received $\mathbf{v}_k$ clusters that correspond to $I_{k-\tau} = +1$ from those that correspond to $I_{k-\tau} = -1$.

I_k	I_{k-1}	I_{k-2}	$\tilde{v}_k$	$\tilde{v}_{k-1}$
-1	-1	-1	-1.5	-1.5
-1	-1	+1	-1.5	-0.5
-1	+1	-1	-0.5	+0.5
-1	+1	+1	-0.5	+1.5
+1	-1	-1	+0.5	-1.5
+1	-1	+1	+0.5	-0.5
+1	+1	-1	+1.5	+0.5
+1	+1	+1	+1.5	+1.5

Table 8.1: Transmitted signal and noiseless channel states for the CIR of $F(z) = 1 + 0.5z^{-1}$ and an equalizer order of $m = 2$.

Figure 8.3 shows the 8 possible noiseless channel states $\tilde{v}_k$ for a BPSK modem and the noisy channel output v_k in the presence of zero mean AWGN with variance $\sigma_\eta^2 = 0.05$. It is seen that the observation vector v_k forms clusters and the centroids of these clusters are the noiseless channel states r_i. The equalization problem hence involves identifying the regions within the observation space spanned by the noisy channel output v_k that correspond to the transmitted symbol of either $I_k = +1$ or $I_k = -1$.

A linear equalizer performs the classification in conjunction with a decision device, which is often a simple sign function. The decision boundary, as seen in Figure 8.3, is constituted by the locus of all values of v_k, where the output of the linear equalizer is zero as it is demonstrated below. For example, for a two tap linear equalizer having tap coefficients c_1 and c_2, at the decision boundary we have:

$$v_k c_1 + v_{k-1} c_2 = 0 \tag{8.13}$$

and

$$v_{k-1} = -(\frac{c_1}{c_2}) v_k \tag{8.14}$$

gives a straight line decision boundary as shown in Figure 8.3, which divides the observation space into two regions corresponding to $I_k = +1$ and $I_k = -1$. In general, the linear equalizer can only implement a hyperplane decision boundary, which in our two-dimensional example was constituted by a line. This is clearly a non-optimum classification strategy, as our forthcoming geometric visualization will highlight. For example, we can see in Figure 8.3 that the point $\tilde{v} = [\ 0.5 \quad -0.5\ ]$ associated with the $I_k = +1$ decision is closer to the decision boundary than the point $\tilde{v} = [\ -1.5 \quad -0.5\ ]$ associated with the $I_k = -1$ decision. Therefore, in the presence of noise, there is a higher probability of the channel output centred at point $\tilde{v} = [\ 0.5 \quad -0.5\ ]$ to be wrongly detected as $I_k = -1$, than that of the channel output centred around $\tilde{v} = [\ -1.5 \quad -0.5\ ]$ being incorrectly detected as $I_k = +1$. Gibson et al. [237] have shown examples of linearly non-separable channels, when the decision delay is zero and the channel is of non-minimum phase nature. The linear separability of the channel depends on the equalizer order, m, on the delay τ and in situations where the channel characteristics are time varying, it may not be possible to specify values of m and τ, which will guarantee linear separability.

According to Chen, Gibson and Cowan [241], the above shortcomings of the linear equalizer are circumvented by a Bayesian approach [251] to obtaining an optimal equalization solution. In this spirit, for an observed channel output vector $\mathbf{v}_k$, if the probability that it was caused by $I_{k-\tau} = +1$ exceeds the probability that it was caused by $I_{k-\tau} = -1$, then we should decide in favour of $+1$ and vice versa. Thus, the optimal Bayesian equalizer solution is defined as [241]:

$$\tilde{I}_{k-\tau} = sgn(f_{Bayes}(\mathbf{v}_k)) = \begin{cases} +1 & \text{if } f_{Bayes}(\mathbf{v}_k) \geq 0 \\ -1 & \text{if } f_{Bayes}(\mathbf{v}_k) < 0, \end{cases} \qquad (8.15)$$

where the optimal Bayesian decision function $f_{Bayes}(\cdot)$, based on the difference of the associated conditional density functions is given by [85]:

$$\begin{aligned} f_{Bayes}(\mathbf{v}_k) &= P(\mathbf{v}_k|I_{k-\tau} = +1) - P(\mathbf{v}_k|I_{k-\tau} = -1) \\ &= \sum_{i=1}^{n_s^+} p_i^+ p(\mathbf{v}_k - \mathbf{r}_i^+) - \sum_{j=1}^{n_s^-} p_j^- p(\mathbf{v}_k - \mathbf{r}_j^-), \end{aligned} \qquad (8.16)$$

where p_i^+ and p_i^- is the *a priori* probability of appearance of each desired state $\mathbf{r}_i^+ \in V_{m,\tau}^+$ and $\mathbf{r}_i^- \in V_{m,\tau}^-$, respectively and $p(\cdot)$ denotes the associated probability density function. The quantities n_s^+ and n_s^- represent the number of desired channel states in $V_{m,\tau}^+$ and $V_{m,\tau}^-$, respectively, which are defined implicitly in Figure 8.3. If the noise distribution is Gaussian, Equation 8.16 can be rewritten as:

$$\begin{aligned} f_{Bayes}(\mathbf{v}_k) &= \sum_{i=1}^{n_s^+} p_i^+ (2\pi\sigma_\eta^2)^{-m/2} exp(-\|\mathbf{v}_k - \mathbf{r}_i^+\|^2/2\sigma_\eta^2) \\ &\quad - \sum_{j=1}^{n_s^-} p_j^- (2\pi\sigma_\eta^2)^{-m/2} exp(-\|\mathbf{v}_k - \mathbf{r}_j^-\|^2/2\sigma_\eta^2). \end{aligned} \qquad (8.17)$$

Again, the optimal decision boundary is the locus of all values of $\mathbf{v}_k$, where the probability $I_{k-\tau} = +1$ given a value $\mathbf{v}_k$ is equal to the probability $I_{k-\tau} = -1$ for the same $\mathbf{v}_k$.

In general, the optimal Bayesian decision boundary is a hyper-surface, rather than just a hyper-plane in the m-dimensional observation space and the realization of this nonlinear boundary requires a nonlinear decision capability. Neural networks provide this capability and the following section will discuss the various neural network structures that have been investigated in the context of channel equalization, while also highlighting the learning algorithms used.

8.3 Introduction to Neural Networks

8.3.1 Biological and Artificial Neurons

The human brain consists of a dense interconnection of simple computational elements referred to as neurons. Figure 8.4(a) shows a network of biological neurons. As seen in the

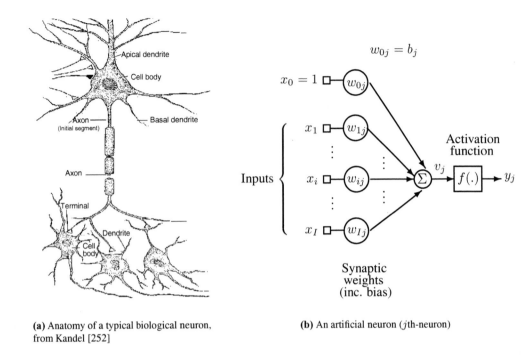

(a) Anatomy of a typical biological neuron,
from Kandel [252]

(b) An artificial neuron (jth-neuron)

Figure 8.4: Comparison between biological and artificial neurons.

figure, the neuron consists of a cell body – which provides the information-processing functions – and of the so-called axon with its terminal fibres. The dendrites seen in the figure are
the neuron's 'inputs', receiving signals from other neurons. These input signals may cause
the neuron to *fire*, i.e. to produce a rapid, short-term change in the potential difference across
the cell's membrane. Input signals to the cell may be excitatory, increasing the chances of
neuron firing, or inhibitory, decreasing these chances. The axon is the neuron's transmission
line that conducts the potential difference away from the cell body towards the terminal fibres. This process produces the so-called *synapses*, which form either excitatory or inhibitory
connections to the dendrites of other neurons, thereby forming a neural network. Synapses
mediate the interactions between neurons and enable the nervous system to adapt and react
to its surrounding environment.

In Artificial Neural Networks (ANN), which mimic the operation of biological neural
networks, the processing elements are artificial neurons and their signal processing properties
are loosely based on those of biological neurons. Referring to Figure 8.4(b), the jth-neuron
has a set of I synapses or connection links. Each link is characterized by a synaptic weight
$w_{ij}, i = 1, 2, \ldots, I$. The weight w_{ij} is positive, if the associated synapse is excitatory and it
is negative, if the synapse is inhibitory. Thus, signal x_i at the input of synapse i, connected
to neuron j, is multiplied by the synaptic weight w_{ij}. These synaptic weights that store
'knowledge' and provide connectivity, are adapted during the learning process.

The weighted input signals of the neuron are summed up by an adder. If this summation exceeds a so-called firing threshold θ_j, then the neuron fires and issues an output. Otherwise it remains inactive. In Figure 8.4(b) the effect of the firing threshold θ_j is represented by a bias, arising from an input which is always 'on', corresponding to $x_0 = 1$, and weighted by $w_{0,j} = -\theta_j = b_j$. The importance of this is that the bias can be treated as just another weight. Hence, if we have a training algorithm for finding an appropriate set of weights for a network of neurons, designed to perform a certain function, we do not need to consider the biases separately.

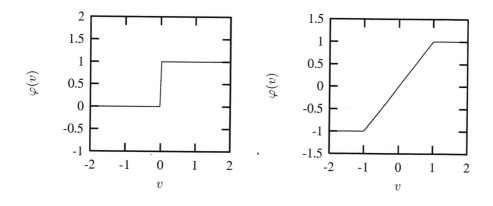

(a) Threshold activation function   **(b)** Piecewise-linear activation function

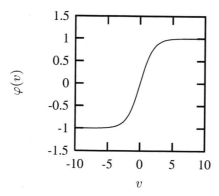

(c) Sigmoid activation function

Figure 8.5: Various neural activation functions $f(v)$.

The activation function $f(\cdot)$ of Figure 8.5 limits the amplitude of the neuron's output to some permissible range and provides nonlinearities. Haykin [253] identifies three basic types of activation functions:

1. *Threshold Function.* For the threshold function shown in Figure 8.5(a), we have

$$f(v) = \begin{cases} 1 & \text{if } v \geq 0 \\ 0 & \text{if } v < 0 \end{cases}. \tag{8.18}$$

Neurons using this activation function are referred to in the literature as the *McCulloch-Pitts model* [253]. In this model, the output of the neuron gives the value of 1 if the total internal activity level of that neuron is nonnegative and 0 otherwise.

2. *Piecewise-Linear Function.* This neural activation function, portrayed in Figure 8.5(b), is represented mathematically by:

$$f(v) = \begin{cases} 1, & v \geq 1 \\ v, & -1 > v > 1 \\ -1, & v \leq -1 \end{cases}, \tag{8.19}$$

where the amplification factor inside the linear region is assumed to be unity. This activation function approximates a nonlinear amplifier.

3. *Sigmoid Function.* A commonly used neural activation function in the construction of artificial neural networks is the sigmoid activation function. It is defined as a strictly increasing function that exhibits smoothness and asymptotic properties, as seen in Figure 8.5(c). An example of the sigmoid function is the hyperbolic tangent function, which is shown in Figure 8.5(c) and it is defined by [253]:

$$f(v) = \frac{1 - exp(-v)}{1 + exp(-v)}. \tag{8.20}$$

This activation function is differentiable, which is an important feature in neural network theory [253].

The model of the jth artificial neuron, shown in Figure 8.4(b) can be described in mathematical terms by the following pair of equations:

$$y_j = f(v_j), \tag{8.21}$$

where:

$$v_j = \sum_{i=0}^{I} w_{ij} x_i. \tag{8.22}$$

Having introduced the basic elements of neural networks, we will focus next on the associated network structures or architectures. The different neural network structures yield different functionalities and capabilities. The basic structures will be described in the following section.

8.3.2 Neural Network Architectures

The network's architecture defines the neurons' arrangement in the network. Various neural network architectures have been investigated for different applications, including for example

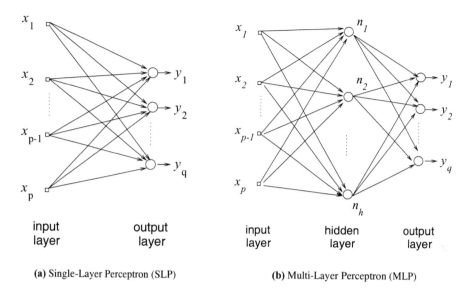

(a) Single-Layer Perceptron (SLP) **(b)** Multi-Layer Perceptron (MLP)

Figure 8.6: Layered feedforward networks.

channel equalization. Distinguishing the different structures can assist us in their design, analysis and implementation.We can identify three different classes of network architectures, which are the subjects of our forthcoming deliberations.

The so-called *layered feedforward networks* of Figure 8.6 exhibit a layered structure, where all connection paths are directed from the input to the output, with no feedback. This implies that these networks are unconditionally stable. Typically, the neurons in each layer of the network have only the output signals of the preceding layer as their inputs.

Two types of layered feedforward networks are often invoked, in order to introduce neural networks, namely the

- *Single-Layer Perceptrons* (SLP) which have a single layer of neurons.

- *Multi-Layer Perceptrons* (MLP) which have multiple layers of neurons.

Again, these structures are shown in Figure 8.6. The MLP distinguishes itself from the SLP by the presence of one or more *hidden layers* of neurons. Figure 8.6(b) illustrates the layout of a MLP having a single hidden layer. It is referred to as a p-h-q network, since it has p source nodes, h hidden neurons and q neurons in the output layer. Similarly, a layered feedforward network having p source nodes, h_1 neurons in the first hidden layer, h_2 neurons in the second hidden layer, h_3 neurons in the third layer and q neurons in the output layer is referred to as a p-h_1-h_2-h_3-q network. If the SLP has a differentiable activation function, such as the sigmoid function given in Equation 8.20, the network can learn by optimizing its weights using a variety of gradient-based optimization algorithms, such as the *gradient descent* method, described briefly in Appendix A.2. The interested reader can refer to the monograph by Bishop [254] for further gradient-based optimization algorithms used to train neural networks.

Input layer

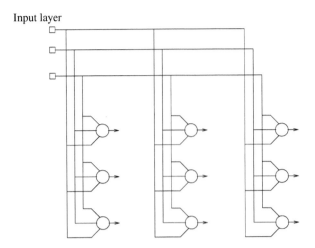

Figure 8.7: Two-dimensional lattice of 3-by-3 neurons.

The addition of hidden layers of nonlinear nodes in MLP networks enables them to extract or learn nonlinear relationships or dependencies from the data, thus overcoming the restriction that SLP networks can only act as linear discriminators. Note that the capabilities of MLPs stem from the nonlinearities used within neurons. If the neurons of the MLP were linear elements, then a SLP network with appropriately chosen weights could carry out exactly the same calculations, as those performed by any MLP network. The downside of employing MLPs however, is that their complex connectivity renders them more implementationally complex and they need nonlinear training algorithms. The so-called *error back propagation* algorithm popularized in the contribution by Rumelhart *et al.* [255, 256] is regarded as the standard algorithm for training MLP networks, against which other learning algorithms are often benchmarked [253].

Having considered the family of layered feedforward networks we note that a so-called *recurrent neural network* [253] distinguishes itself from a layered feedforward network by having at least one *feedback* loop.

Lastly, lattice structured neural networks [253] consist of networks of a one-dimensional, two-dimensional or higher-dimensional array of neurons. The lattice network can be viewed as a feedforward network with the output neurons arranged in rows and columns. For example, Figure 8.7 shows a two-dimensional lattice of 3-by-3 neurons fed from a layer of 3 source nodes.

Neural network models are specified by the nodes' characteristics, by the network topology, and by their training or learning rules, which set and adapt the network weights appropriately, in order to improve performance. Both the associated design procedures and training rules are the topic of much current research [257]. The above rudimentary notes only give a brief and basic introduction to neural network models. For a deeper introduction to other neural network topologies and learning algorithms, please refer for example to the review by Lippmann [258]. Let us now provide a rudimentary overview of the associated equalization concepts in the following section.

8.4 Equalization Using Neural Networks

A few of the neural network architectures that have been investigated in the context of channel equalization are the so-called Multilayer Perceptron (MLP) advocated by Gibson, Siu and Cowan [236–240], as well as the Polynomial-Perceptron (PP) studied by Chen, Gibson, Cowan, Chang, Wei, Xiang, Bi, L.-Ngoc *et al.* [241–244]. Furthermore, the RBF was investigated by Chen, McLaughlin, Mulgrew, Gibson, Cowan, Grant *et al.* [85, 245–247], the recurrent network [259] was proposed by Sueiro, Rodriguez and Vidal, the Functional Link (FL) technique was introduced by Gan, Hussain, Soraghan and Durrani [260–262] and the Self-Organizing Map (SOM) was proposed by Kohonen *et al.* [263].

Various neural network based equalisers have also been implemented and investigated for transmission over satellite mobile channels [264–266]. The following section will present and summarise some of the neural network based equalisers found in literature. We will investigate the RBF structure in the context of equalization in more detail during our later discourse in the next few sections.

8.5 Multilayer Perceptron Based Equaliser

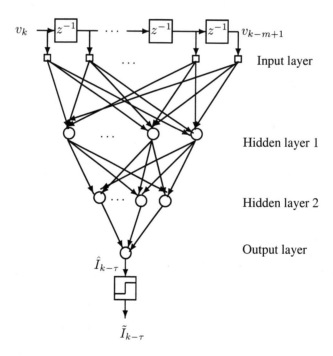

Figure 8.8: Multilayer perceptron model of the m-tap equalizer of Figure 8.2.

Multilayer perceptrons (MLPs), which have three layers of neurons, i.e. two hidden layers and one output layer, are capable of forming any desired decision region for example in the context of modems, which was noted by Gibson and Cowan [267]. This property renders them attractive as nonlinear equalisers. The structure of a MLP network has been described in Section 8.3.2 as a layered feedforward network. As an equaliser, the input of the MLP network is the sequence of the received signal samples $\{v_k\}$ and the network has a single output, which gives the estimated transmitted symbol $\tilde{I}_{k-\tau}$, as shown in Figure 8.8. Figure 8.8 shows the $m - h_1 - h_2 - 1$ MLP network as an equaliser. Referring to Figure 8.9, the jth neuron $(j = 1, \ldots, h_l)$ in the lth layer $(l = 0, 1, 2, 3$, where the 0th layer is the input layer and the third layer is the output layer) accepts inputs $\mathbf{v}^{(l-1)} = [v_1^{(l-1)} \ldots v_{h_{l-1}}^{(l-1)}]^T$ from the $(l-1)$th layer and returns a scalar $v_j^{(l)}$ given by

$$v_j^{(l)} = f(\sum_{i=1}^{h_{l-1}} w_{ij}^{(l)} v_i^{(l-1)}) \qquad j = 1, \ldots, h_l, \qquad l = 0, 1, 2, 3, \tag{8.23}$$

where $h_0 = m$ is the number of nodes at the input layer, which is equivalent to the equalizer order and h_3 is the number of neurons at the output layer, which is one according to Figure 8.8. The output value $v_j^{(l)}$ serves as an input to the $(l+1)$th layer. Since the transmitted binary symbol taken from the set $\{+1,-1\}$ has a bipolar nature, the sigmoid type activation function $f(\cdot)$ of Equation 8.20 is chosen to provide an output in the range of $[-1,+1]$, as shown in Figure 8.5(c). The MLP equalizer can be trained adaptively by the so-called error back propagation algorithm described for example by Rumelhart, Hinton and Williams [255].

The major difficulty associated with the MLP is that training or determining the required weights is essentially a nonlinear optimization problem. The mean squared error surface corresponding to the optimization criterion is multi-modal, implying that the mean squared error surface has local minima as well as a global minimum. Hence it is extremely difficult to design gradient type algorithms, which guarantee finding the global error minimum corresponding to the optimum equalizer coefficients under all input signal conditions. The error back propagation algorithm to be introduced during our further discourse does not guarantee convergence, since the gradient descent might be trapped in a local minimum of the error surface. Furthermore, due to the MLP's typically complicated error surface, the MLP equaliser using the error back propagation algorithm has a slower convergence rate than the conventional adaptive equalizer using the Least Mean Square (LMS) algorithm described in Appendix A.2. This was illustrated for example by Siu *et al.* [240] using experimental results. The introduction of the so-called momentum term was suggested by Rumelhart *et al.* [256] for the adaptive algorithm to improve the convergence rate. The idea is based on sustaining the weight change moving in the same direction with a 'momentum' to assist the back propagation algorithm in moving out of a local minimum. Nevertheless, it is still possible that the adaptive algorithm may become trapped at local minima. Furthermore, the above-mentioned

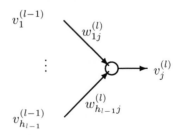

Figure 8.9: The jth neuron in the mth layer of the MLP.

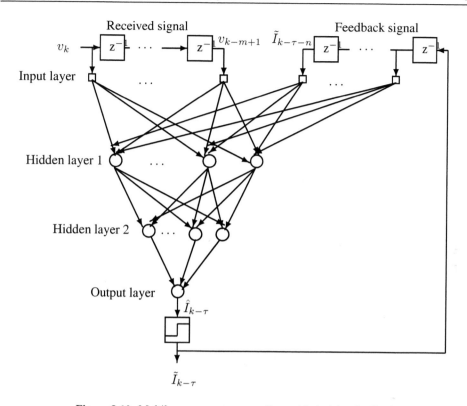

Figure 8.10: Multilayer perceptron equalizer with decision feedback.

momentum term may cause oscillatory behaviour close to a local or global minimum. Interested readers may wish to refer to the excellent monograph by Haykin [253] that discusses the virtues and limitations of the error back propagation algorithm invoked to train the MLP network, highlighting also various methods for improving its performance. Another disadvantage of the MLP equalizer with respect to conventional equalizer schemes is that the MLP design incorporates a three-layer perceptron structure, which is considerably more complex.

Siu *et al.* [240] incorporated decision feedback into the MLP structure, as shown in Figure 8.10 with a feedforward order of m and a feedback order of n. The authors provided simulation results for binary modulation over a dispersive Gaussian channel, having an impulse response of $F(z) = 0.3482 + 0.8704z^{-1} + 0.3482z^{-2}$. Their simulations show that the MLP DFE structure offers superior performance in comparison to the LMS DFE structure. They also provided a comparative study between the MLP equalizer with and without feedback. The performance of the MLP equalizer was improved by about 5dB at a BER of 10^{-4} relative to the MLP without decision feedback and having the same number of input nodes. Siu, Gibson and Cowan also demonstrated that the performance degradation due to decision errors is less dramatic for the MLP based DFE, when compared to the conventional LMS DFE, especially at poor signal-to-noise ratio (SNR) conditions. Their simulations showed that the MLP DFE structure is less sensitive to learning gain variation and it is capable of converging to a lower mean square error value. Despite providing considerable performance improvements, MLP equalisers are still problematic in terms of their convergence performance and due to their more complex structure relative to conventional equalisers.

8.6   Polynomial Perceptron Based Equaliser

The so-called PP or Volterra series structure was proposed for channel equalization by Chen, Gibson and Cowan [241]. The PP equaliser has a simpler structure and a lower computational complexity, than the MLP structure, which makes it more attractive for equalization. A perceptron structure is employed, combined with polynomial approximation techniques, in order to approximate the optimal nonlinear equalization solution. The design is justified by the so-called *Stone-Weierstrass theorem* [268], which states that any continuous function can be approximated within an arbitrary accuracy by a polynomial of a sufficiently high order. The model of the PP was investigated in detail by Xiang *et al.* [244]. The nonlinear equalizer is constructed according to [241]:

$$
\begin{aligned}
f_p(\mathbf{v}_k) &= \sum_{i_1=0}^{m-1} c_{i_1} v_{k-i_1} + \sum_{i_1=0}^{m-1}\sum_{i_2=i_1}^{m-1} c_{i_1 i_2} v_{k-i_1} v_{k-i_2} + \cdots \\
&\quad + \sum_{i_1=0}^{m-1} \cdots \sum_{i_l=i_{l-1}}^{m-1} c_{i_1 \ldots i_l} v_{k-i_1} \cdots v_{k-i_l}, \\
&= \sum_{i=0}^{n} w_i x_{i,k}, & (8.24) \\
f_{PP}(\mathbf{v}_k) &= f(f_p(\mathbf{v}_k)), & (8.25) \\
\tilde{I}_{k-\tau} &= sgn[f_{PP}(\mathbf{v}_k)], & (8.26)
\end{aligned}
$$

where l is the polynomial order, m is the equalizer order, $x_{i,k}$ are the so-called monomials (polynomial with a single power term) corresponding to the power terms of the equalizer inputs from v_{k-i_1} to $v_{k-i_1} \cdots v_{k-i_l}$, w_i are the corresponding polynomial coefficients c_{i_1} to $c_{i_1 \ldots i_l}$ and n is the number of terms in the polynomial. Here, the term w_i and $x_{i,k}$ of Equation 8.24 correspond to the synaptic weights and inputs of the perceptron/neuron described in Figure 8.4(b), respectively.

The function $f_p(\mathbf{v}_k)$ in Equation 8.25 is the polynomial that approximates the Bayesian decision function $f_{Bayes}(\mathbf{v}_k)$ of Equation 8.16 and the function $f_{PP}(\mathbf{v}_k)$ in Equation 8.25 is the PP decision function. The activation function of the perceptron $f(\cdot)$ is the sigmoid function given by Equation 8.20. The reasons for applying the sigmoidal function were highlighted by Chen, Gibson and Cowan [241], which are briefly highlighted below. In theory the number of terms in Equation 8.24 can be infinite. However, in practice only a finite number of terms can be implemented, which has to be sufficiently high to achieve a low received signal mis-classification probability, i.e. a low decision error probability. The introduction of the sigmoidal activation function $f(x)$ is necessary, since it allows a moderate polynomial degree to be used, while having an acceptable level of mis-classification of the equalizer input vector corresponding to the transmitted symbols. This was demonstrated by Chen *et al.* [241] using a simple classifier example. Chen *et al.* [241] reported that a polynomial degree of $l = 3$ or

5 was sufficient with the introduction of the sigmoidal activation function judging from their simulation results for the experimental circumstances stipulated.

From a conceptual point of view, the PP structure expands the input space of the equaliser, which is defined by the dimensionality of $\{\mathbf{v}_k\}$, into an extended nonlinear space and then employs a neuron element in this space. Consider a simple polynomial perceptron based

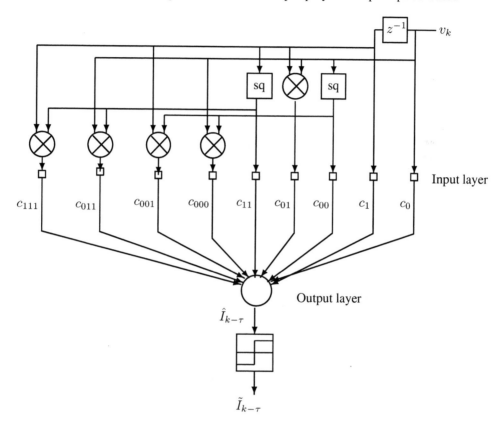

Figure 8.11: Polynomial perceptron equalizer using an equalizer order of $m = 2$ and polynomial order of $l = 3$.

equaliser, where the equaliser order is $m = 2$ and the polynomial order is $l = 3$. Then the polynomial decision function is given by:

$$
\begin{aligned}
f_{PP}(\mathbf{v}_k) = \ & f(c_0 v_k + c_1 v_{k-1} + c_{00} v_k^2 + c_{01} v_k v_{k-1} + c_{11} v_{k-1}^2 + \\
& c_{000} v_k^3 + c_{001} v_k^2 v_{k-1} + c_{011} v_k v_{k-1}^2 + c_{111} v_{k-1}^3).
\end{aligned}
\tag{8.27}
$$

The structure of the equalizer defined by Equation 8.27 is illustrated in Figure 8.11. The simulation results of Chen *et al.* [241] using binary modulation show close agreement with the bit error rate performance of the MLP equaliser. However, the training of the PP equaliser is much easier compared to the MLP equaliser, since only a single-layer perceptron is involved in the PP equaliser. The nonlinearity of the sigmoidal activation function introduces local minima to the error surface of the otherwise linear perceptron structure. Thus, the stochastic gradient algorithm [255,256] assisted by the previously mentioned momentum term [256] can be invoked in their scheme in order to adaptively train the equaliser. The decision feedback structure of Figure 8.10 can be incorporated into Chen's design [241] in order to further improve the performance of the equaliser.

The PP equalizer is attractive, since it has a simpler structure than that of the MLP. The PP equalizer also has a multi-modal error surface – exhibiting a number of local minima and a global minimum – and thus still retains some problems associated with its convergence performance, although not as grave as the MLP structure. Another drawback is that the number of terms in the polynomial of Equation 8.24 increases exponentially with the polynomial order l and with the equaliser order m, resulting in an exponential increase of the associated computational complexity.

8.7   Radial Basis Function Networks

8.7.1   Introduction

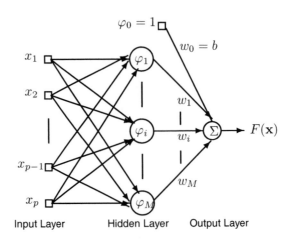

Figure 8.12: Architecture of a radial basis function network.

In this section, we will introduce the concept of the so-called *Radial Basis Function* (RBF) networks and highlight their architecture. The RBF network [253] consists of three different layers, as shown in Figure 8.12. The input layer is constituted by p source nodes. A set of M nonlinear activation functions $\varphi_i, i = 1, \ldots, M$, constitutes the hidden second layer. The output of the network is provided by the third layer, which is comprised of output nodes. Figure 8.12 shows only one output node, in order to simplify our analysis. This construction is based on the basic neural network design. As suggested by the terminology, the activation functions in the hidden layer take the form of radial basis functions [253]. Radial functions are characterized by their responses that decrease or increase monotonically with distance from a central point, $\mathbf{c}$, i.e. as the Euclidean norm $\|\mathbf{x} - \mathbf{c}\|$ is increased, where $\mathbf{x} = [x_1 \ x_2 \ \ldots \ x_p]^T$ is the input vector of the RBF network. The central points in the vector

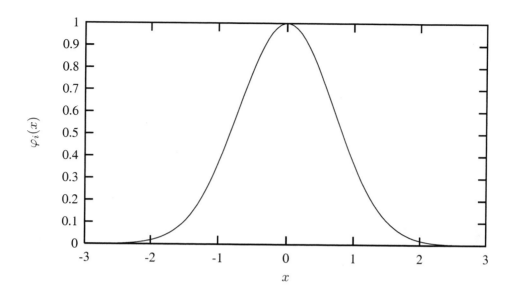

Figure 8.13: Gaussian radial basis function described by Equation 8.29 with centre $c_i = 0$ and spread of $2\sigma_i^2 = 1$.

c are often referred to as the RBF centres. Therefore, the radial basis functions take the form

$$\varphi_i(\mathbf{x}) = \varphi(\|\mathbf{x} - \mathbf{c}_i\|), \qquad i = 0, \dots, M, \tag{8.28}$$

where M is the number of independent basis functions in the RBF network. This justifies the 'radial' terminology. A typical radial function is the Gaussian function which assumes the form:

$$\varphi_i(\mathbf{x}) = exp\left(-\frac{\|\mathbf{x} - \mathbf{c}_i\|^2}{2\sigma_i^2}\right), \qquad i = 0, \dots, M, \tag{8.29}$$

where $2\sigma_i^2$ is representative of the 'spread' of the Gaussian function that controls the radius of influence of each basis function. Figure 8.13 illustrates a Gaussian RBF, in the case of a scalar input, having a scalar centre of $c = 0$ and a spread or width of $2\sigma_i^2 = 1$. Gaussian-like RBFs are localized, i.e. they give a significant response only in the vicinity of the centre and $\varphi(x) \to 0$ as $x \to \infty$. As well as being localized, Gaussian basis functions have a number of useful analytical properties, which will be highlighted in our following discourse.

Referring to Figure 8.12, the RBF network can be represented mathematically as follows:

$$F(\mathbf{x}) = \sum_{i=0}^{M} w_i \varphi_i(\mathbf{x}). \tag{8.30}$$

The bias b in Figure 8.12 is absorbed into the summation as w_0 by including an extra basis function φ_0, whose activation function is set to 1. Bishop [254] gave an insight into the role of the bias w_0 when the network is trained by minimizing the sum-of-squared error between the

RBF network output vector and the desired output vector. The bias is found to compensate for the difference between the mean of the RBF network output vector and the corresponding mean of the target data evaluated over the training data set.

Note that the relationship between the RBF network and the Bayesian equalization solution expressed in Equation 8.17, can be given explicitly. The RBF network's bias is set to $b = w_0 = 0$. The RBF centres $c_i, i = 1, \ldots, M$, are in fact the noise-free dispersion-induced channel output vectors $r_i, i = 1, \ldots, n_s$ indicated by circles and crosses, respectively, in Figure 8.3 and the number of hidden nodes M of Figure 8.12 corresponds to the number of desired channel output vectors, n_s, i.e. $M = n_s$. The RBF weights $w_i, i = 1, \ldots, M$, are all known from Equation 8.17 and they correspond to the scaling factors of the conditional probability density functions in Equation 8.17. Section 8.9.1 will provide further exposure to these issues.

Having described briefly the RBF network architecture, the next few sections will present its design in detail and also motivate its employment from the point of view of classification problems, interpolation theory and regularization. The design of the hidden layer of the RBF is justified by Cover's Theorem [269] which will be described in Section 8.7.2. In Section 8.7.3, we consider the so-called interpolation problem in the context of RBF networks. Then, we discuss the implications of sparse and noisy training data in Section 8.7.4. The solution to the problem of using regularization theory is also presented there. Lastly, in Section 8.7.5, the generalized RBF network is described, which concludes this section.

8.7.2 Cover's Theorem

The design of the radial basis function network is based on a curve-fitting (*approximation*) problem in a high-dimensional space, a concept, which was augmented for example by Haykin [253]. Specifically, the RBF network solves a complex pattern-classification problem, such as the one described in Section 8.2 in the context of Figure 8.3 for equalization, by first transforming the problem into a high-dimensional space in a nonlinear manner and then by finding a surface in this multi-dimensional space that best fits the training data, as it will be explained below. The underlying justification for doing so is provided by *Cover's theorem* on the *separability of patterns*, which states that [269]:

> a complex pattern-classification problem non-linearly cast in a high-dimensional space is more likely to become linearly separable, than in a low-dimensional space.

We commence our discourse by highlighting the pattern-classification problem. Consider a surface that separates the space of the noisy channel outputs of Figure 8.3 into two regions or classes. Let X denote a set of N patterns or points $x_1, x_2, \ldots, x_N$, each of which is assigned to one of two classes, namely X^+ and X^-. This dichotomy or binary partition of the points with respect to a surface becomes successful, if the surface separates the points belonging to the class X^+ from those in the class X^-. Thus, to solve the pattern-classification problem, we need to provide this *separating surface* that gives the decision boundary, as shown in Figure 8.14.

We will now non-linearly cast the problem of separating the channel outputs into a high-dimensional space by introducing a vector constituted by a set of real-valued functions $\varphi_i(x)$,

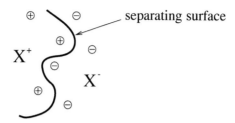

Figure 8.14: Pattern-classification into two dimensions, where the patterns are linearly non-separable, since a line cannot separate all the X^+ and X^- values, but the non-linear separating surface can – hence the term nonlinearly separable.

where $i = 1, 2, \ldots, M$, for each input pattern $\mathbf{x} \in X$, as follows:

$$\varphi(\mathbf{x}) = [\varphi_1(\mathbf{x})\ \varphi_2(\mathbf{x})\ \ldots\ \varphi_M(\mathbf{x})]^T, \qquad (8.31)$$

where pattern $\mathbf{x}$ is a vector in a p-dimensional space and M is the number of real-valued functions. Recall that in our approach M is the number of possible channel output vectors for Bayesian equalization solution. The vector $\varphi(\mathbf{x})$ maps points of $\mathbf{x}$ from the p-dimensional input space into corresponding points in a new space of dimension M, where $p < M$. The function $\varphi_i(\mathbf{x})$ of Figure 8.12 is referred to as a *hidden function*, which plays a role similar to a hidden unit in a feedforward neural network, such as that in Figure 8.6(b). A dichotomy X^+, X^- of X is said to be φ-*separable*, if there exists an M-dimensional vector $\mathbf{w}$, such that for the scalar product $\mathbf{w}^T \varphi(\mathbf{x})$ we may write

$$\mathbf{w}^T \varphi(\mathbf{x}) \geq 0, \qquad \text{if } \mathbf{x} \in X^+ \qquad (8.32)$$

and

$$\mathbf{w}^T \varphi(\mathbf{x}) < 0, \qquad \text{if } \mathbf{x} \in X^-. \qquad (8.33)$$

The hypersurface defined by the equation

$$\mathbf{w}^T \varphi(\mathbf{x}) = 0 \qquad (8.34)$$

describes the separating surface in the φ space. The inverse image of this hypersurface is

$$\{\mathbf{x} : \mathbf{w}^T \varphi(\mathbf{x}) = 0\}, \qquad (8.35)$$

which defines the separating surface in the input space.

Below we give a simple example in order to visualise the concept of Cover's theorem in the context of the separability of patterns. Let us consider the XOR problem of Table 8.2, which is not linearly separable since the XOR = 0 and XOR = 1 points of Figure 8.15(a) cannot be separated by a line. The XOR problem is transformed into a linearly separable problem by casting it from a two-dimensional input space into a three-dimensional space by the function $\varphi(\mathbf{x})$, where $\mathbf{x} = \begin{bmatrix} x_1 & x_2 \end{bmatrix}^T$ and $\varphi = \begin{bmatrix} \varphi_1 & \varphi_2 & \varphi_3 \end{bmatrix}^T$. The hidden functions of Figure 8.12 are given in our example by:

$$\varphi_1(\mathbf{x}) = x_1, \qquad (8.36)$$
$$\varphi_2(\mathbf{x}) = x_2, \qquad (8.37)$$
$$\varphi_3(\mathbf{x}) = x_1 x_2. \qquad (8.38)$$

x_1	x_2	XOR
0	0	0
0	1	1
1	0	1
1	1	0

Table 8.2: XOR truth table.

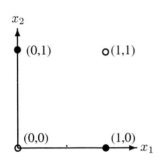

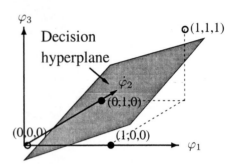

(a) XOR problem, which is not linearly separable.

(b) XOR problem mapped to the three-dimensional space by the function $\varphi(\mathbf{x})$. The mapped XOR problem is linearly separable.

Figure 8.15: The XOR problem solved by $\varphi(\mathbf{x})$ mapping. Bold dots represent XOR = 1, while hollow dots correspond to XOR = 0.

The higher-dimensional φ-inputs and the desired XOR output are shown in Table 8.3.

φ_1	φ_2	φ_3	XOR
0	0	0	0
0	1	0	1
1	0	0	1
1	1	1	0

Table 8.3: XOR truth table with inputs of φ_1, φ_2 and φ_3.

Figure 8.15(b) illustrates, how the higher-dimensional XOR problem can be solved with the aid of a linear separating surface. Note that $\varphi_i, i = 1, 2, 3$ given in the above example are not of the radial basis function type described in Equation 8.28. They are invoked as a simple example to demonstrate the general concept of Cover's theorem.

Generally, we can find a non-linear mapping $\varphi(\mathbf{x})$ of sufficiently high dimension M, such that we have linear separability in the φ-space. It should be stressed, however that in some cases the use of nonlinear mapping may be sufficient to produce linear separability without having to increase the dimensionality of the hidden unit space [253].

8.7.3 Interpolation Theory

From the previous section, we note that the RBF network can be used to solve a nonlinearly separable classification problem. In this section, we highlight the use of the RBF network for performing *exact interpolation* of a set of data points in a multi-dimensional space. The exact interpolation problem requires every input vector to be mapped exactly onto the corresponding target vector, and forms a convenient starting point for our discussion of RBF networks. In the context of channel equalization we could view the problem as attempting to map the channel output vector of Equation 8.4 to the corresponding transmitted symbol.

Consider a feedforward network with an input layer having p inputs, a single hidden layer and an output layer with a single output node. The network of Figure 8.12 performs a nonlinear mapping from the input space to the hidden space, followed by a linear mapping from the hidden space to the output space. Overall, the network represents a mapping from the p-dimensional input space to the one-dimensional output space, written as

$$s : \mathbb{R}^p \rightarrow \mathbb{R}^1, \tag{8.39}$$

where the mapping s is described by a continuous hypersurface $\Gamma \subset \mathbb{R}^{p+1}$. The continuous surface Γ is a multi-dimensional plot of the output as a function of the input. Figure 8.16 illustrates the mapping $F(x)$ from a single-dimensional input space x to a single-dimensional output space and the surface Γ. Again, in the case of an equaliser, the mapping surface Γ maps the channel output to the transmitted symbol.

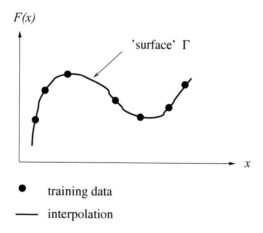

Figure 8.16: Stylised exact interpolation between the known input-output pairs by the continuous surface Γ.

In practical situations, the continuous surface Γ is unknown and the training data might be contaminated by noise. The network undergoes a so-called *learning process*, in order to find the specific surface in the multi-dimensional space that provides the best fit to the training data d_i where $i = 1, 2, \ldots, N$. The 'best fit' surface is then used to interpolate the test data or for the specific case of an equaliser, the estimated transmitted symbol. Formally, the learning process can be categorized into two phases, the training phase and the generalization phase. During the training phase, the fitting procedure for the surface Γ is optimised based on N known data points presented to the neural network in the form of input-output pairs $[\mathbf{x}_i, d_i], i = 1, 2, \ldots N$. The generalization phase constitutes the interpolation between the data points, where the interpolation is performed along the constrained surface generated by the fitting procedure, as the optimum approximation to the true surface Γ.

Thus, we are led to the theory of multivariable interpolation in high-dimensional spaces. Assuming a single-dimensional output space, the interpolation problem can be stated as follows:

> Given a set of N different points $\mathbf{x}_i \in \mathbb{R}^p, i = 1, 2, \ldots, N$, in the p-dimensional input space and a corresponding set of N real numbers $d_i \in \mathbb{R}^1, i = 1, 2, \ldots, N$, in the one-dimensional output space, find a function $F : \mathbb{R}^p \to \mathbb{R}^1$ that satisfies the interpolation condition:

$$F(\mathbf{x}_i) = d_i, \qquad i = 1, 2, \ldots, N, \tag{8.40}$$

implying that for $i = 1, 2, \ldots, N$ the function $F(\mathbf{x})$ interpolates between the values d_i. Note that for exact interpolation, the interpolating surface is constrained to pass through all the training data points $\mathbf{x}_i$. The RBF technique is constituted by choosing a function $F(x)$ that obeys the following form:

$$F(\mathbf{x}) = \sum_{i=1}^{N} w_i \varphi(\|\mathbf{x} - \mathbf{x}_i\|), \tag{8.41}$$

where $\varphi_i(\mathbf{x}) = \varphi(\|\mathbf{x} - \mathbf{x}_i\|), i = 1, 2, \ldots, N$, is a set of N nonlinear functions, known as the radial basis function, and $\|.\|$ denotes the distance *norm* that is usually taken to be Euclidean. The known training data points $\mathbf{x}_i \in \mathbb{R}^p, i = 1, 2, \ldots, N$ constitute the centroids of the radial basis functions. The unknown coefficients w_i represent the weights of the RBF network of Figure 8.12. In order to link Equation 8.41 with Equation 8.30 we note that the number of radial basis functions M is now set to the number of training data points N and the RBF centres $\mathbf{c}_i$ of Equation 8.28 are equivalent to the training data points $\mathbf{x}_i$, i.e., $\mathbf{c}_i = \mathbf{x}_i, i = 1, 2, \ldots N$. The term associated with $i = 0$ was not included in Equation 8.41, since we argued above that the RBF bias was $w_0 = 0$.

Upon inserting the interpolation conditions of Equation 8.40 in Equation 8.41, we obtain the following set of simultaneous linear equations for the unknown weights w_i:

$$\begin{bmatrix} \varphi_{11} & \varphi_{12} & \cdots & \varphi_{1N} \\ \varphi_{21} & \varphi_{22} & \cdots & \varphi_{2N} \\ \vdots & \vdots & \vdots & \vdots \\ \varphi_{N1} & \varphi_{N2} & \cdots & \varphi_{NN} \end{bmatrix} \begin{bmatrix} w_1 \\ w_2 \\ \vdots \\ w_N \end{bmatrix} = \begin{bmatrix} d_1 \\ d_2 \\ \vdots \\ d_N \end{bmatrix}, \tag{8.42}$$

where

$$\varphi_{ji} = \varphi(\|\mathbf{x}_j - \mathbf{x}_i\|), \qquad j, i = 1, 2, \ldots, N. \tag{8.43}$$

Let

$$\mathbf{d} = [d_1, d_2, \ldots, d_N]^T \tag{8.44}$$

$$\mathbf{w} = [w_1, w_2, \ldots, w_N]^T, \tag{8.45}$$

where the N-by-1 vectors $\mathbf{d}$ and $\mathbf{w}$ represent the equaliser's desired response vector and the linear weight vector, respectively. Let Φ denote an N-by-N matrix with elements of $\varphi_{ji}, j, i = 1, 2, \ldots, N$, which we refer to as the *interpolation matrix*, since it generates the interpolation $F(\mathbf{x}_i) = d_i$ through Equation 8.40 and Equation 8.41 using the weights w_i. Then Equation 8.42 can be written in the compact form of:

$$\Phi\mathbf{w} = \mathbf{d}. \tag{8.46}$$

We note that if the data points d_i are all distinct and the interpolation matrix Φ is positive definite, implying that all of its elements are positive and hence Φ is invertible, then we can solve Equation 8.46 to obtain the weight vector $\mathbf{w}$, which is formulated as:

$$\mathbf{w} = \Phi^{-1}\mathbf{d}, \tag{8.47}$$

where Φ^{-1} is the inverse of the interpolation matrix Φ.

From *Light's theorem* [270], there exists a class of radial basis functions that generates an interpolation matrix, which is positive definite. Specifically, Light's theorem applies to a range of functions, which include the *Gaussian functions* [270] of:

$$\varphi(r) = \exp\left(-\frac{r^2}{2\sigma^2}\right), \tag{8.48}$$

$$\varphi_{ji} = \exp(\frac{-\|\mathbf{x}_j - \mathbf{x}_i\|^2}{2\sigma^2}), \qquad j, i = 1, 2, \ldots, N, \tag{8.49}$$

where σ^2 is the variance of the Gaussian function. Hence the elements φ_{ji} of Φ can be determined from Equation 8.49. Since Φ is invertible, it is always possible to generate the weight vector $\mathbf{w}$ for the RBF network from Equation 8.47, in order to provide the interpolation through the training data.

In an equalization context, exact interpolation can be problematic. The training data are sparse and are contaminated by noise. This problem will be addressed in the next section.

8.7.4 Regularization Theory

The partitioning hyper-surface and the interpolation hyper-surface mentioned in the previous sections were reconstructed or approximated from a given set of data points that may be sparse or noisy during learning. Therefore, the learning process used to reconstruct or approximate the classification hyper-surface can be seen as belonging to a generic class of problems referred to as *inverse problems* [253].

An inverse problem may be 'well-posed' or 'ill-posed'. In order to explain the term 'well-posed', assume that we have a domain X and a range Y taken to be spaces obeying the properties of metrics and they are related to each other by a fixed but unknown mapping $Y = F(X)$. The problem of reconstructing the mapping F is said to be *well-posed*, if the following conditions are satisfied [271]:

1. *Existence:* For every input vector $\mathbf{x} \in X$, there exists an output $y = F(\mathbf{x})$, where $y \in Y$, as seen in Figure 8.17.

2. *Uniqueness:* For any pair of input vectors $\mathbf{x}, \mathbf{t} \in X$, we have $F(\mathbf{x}) = F(\mathbf{t})$ if, and only if, $\mathbf{x} = \mathbf{t}$.

3. *Continuity:* The mapping is continuous.

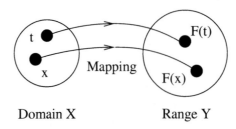

Domain X   Range Y

Figure 8.17: The mapping of the input domain X onto the output range Y.

If these conditions are not satisfied, the inverse problem of identifying x giving rise to y is said to be ill-posed.

Learning, where the partitioning or interpolation hyper-surface is approximated, is in general an ill-posed inverse problem . This is because the uniqueness criterion may be violated, since there may be insufficient information in the training data to reconstruct the input-output mapping uniquely. Furthermore, the presence of noise or other impairments in the input data adds uncertainty to the reconstructed input-output mapping. This is the case in the context of the equalization problem.

Tikhonov [272] proposed a method referred to as *regularization for solving ill-posed problems*. The basic idea of regularization is to *stabilize* the solution by means of some auxiliary non-negative function that imposes prior restrictions such as, smoothness or correlation constraints on the input-output mapping and thereby converting an ill-posed problem into a well-posed problem. This approach was treated in depth by Poggio and Girosi [273].

According to Tikhonov's regularization theory [272], the previously introduced function F is determined by minimising a *cost function* $\mathcal{E}(F)$, defined by

$$\mathcal{E}(F) = \mathcal{E}_s(F) + \lambda \mathcal{E}_c(F), \tag{8.50}$$

where λ is a positive real number referred to as the *regularization parameter* and the two terms involved are [272]:

1. *Standard Error Term:* This term, denoted by $\mathcal{E}_s(F)$, quantifies the standard error between the desired response d_i and the actual response y_i for training samples $i = 1, 2, \ldots, N$. It is defined by

$$
\begin{aligned}
\mathcal{E}_s(F) &= \frac{1}{2} \sum_{i=1}^{N} (d_i - y_i)^2 \\
&= \frac{1}{2} \sum_{i=1}^{N} [d_i - F(\mathbf{x}_i)]^2.
\end{aligned}
\tag{8.51}
$$

2. *Regularizing Term:* This term, denoted by $\mathcal{E}_c(F)$, depends on the geometric properties of the approximation function $F(\mathbf{x})$. It provides the so-called *a priori* smoothness

constraint and it is defined by

$$\mathcal{E}_c(F) = \frac{1}{2}\|\mathcal{P}F\|^2 \tag{8.52}$$

where $\mathcal{P}$ is a linear (pseudo) differential operator, referred to as a *stabilizer* [253], which stabilizes the solution F, rendering it smooth and therefore continuous.

The regularization parameter λ indicates, whether the given training data set is sufficiently extensive in order to specify the solution $F(\mathbf{x})$. The limiting case $\lambda \to 0$ implies that the problem is unconstrained. Here, the solution $F(\mathbf{x})$ is completely determined from the given data set. The other limiting case, $\lambda \to \infty$, implies that the *a priori* smoothness constraint is sufficient to specify the solution $F(\mathbf{x})$. In other words, the training data set is unreliable. In practical applications the regularization parameter λ is assigned a value between the two limiting conditions, so that both the sample data and the *a priori* information contribute to the solution $F(\mathbf{x})$.

The minimization of the cost function $\mathcal{E}(F)$ by evaluating the derivative of $\mathcal{E}(F)$ in Equation 8.50 provides the following solution to $F(\mathbf{x})$ [253]:

$$
\begin{aligned}
F(\mathbf{x}) &= \frac{1}{\lambda} \sum_{i=1}^{N} [d_i - F(\mathbf{x}_i)] G(\mathbf{x}; \mathbf{x}_i) \\
&= \sum_{i=1}^{N} w_i G(\mathbf{x}; \mathbf{x}_i),
\end{aligned}
\tag{8.53}
$$

where $G(\mathbf{x}; \mathbf{x}_i)$ denotes the so-called Green function centred at $\mathbf{x}_i$ and $w_i = \frac{1}{\lambda}[d_i - F(\mathbf{x}_i)]$. Equation 8.53 states that the solution $F(\mathbf{x})$ to the regularization problem is a linear superposition of N number of Green functions centred at the training data points $x_i, i = 1, 2, \ldots, N$. The weights w_i are the *coefficients of the expansion* of $F(\mathbf{x})$ in terms of $G(\mathbf{x}; \mathbf{x}_i)$ and x_i are the *centres of the expansion* for $i = 1, 2, \ldots, N$. The centres $\mathbf{x}_i$ of the Green functions used in the expansion are the given data points used in the training process.

We now have to determine the unknown expansion cofficients w_i denoted by

$$w_i = \frac{1}{\lambda}[d_i - F(\mathbf{x}_i)], \qquad i = 1, 2, \ldots, N. \tag{8.54}$$

Let

$$
\begin{aligned}
\mathbf{F} &= [F(\mathbf{x}_1), F(\mathbf{x}_2), \ldots, F(\mathbf{x}_N)]^T, \tag{8.55} \\
\mathbf{d} &= [d_1, d_2, \ldots, d_N]^T, \tag{8.56} \\
G &= \begin{bmatrix}
G(\mathbf{x}_1; \mathbf{x}_1) & G(\mathbf{x}_1; \mathbf{x}_2) & \ldots & G(\mathbf{x}_1; \mathbf{x}_N) \\
G(\mathbf{x}_2; \mathbf{x}_1) & G(\mathbf{x}_2; \mathbf{x}_2) & \ldots & G(\mathbf{x}_2; \mathbf{x}_N) \\
\vdots & \vdots & & \vdots \\
G(\mathbf{x}_N; \mathbf{x}_1) & G(\mathbf{x}_N; \mathbf{x}_2) & \ldots & G(\mathbf{x}_N; \mathbf{x}_N)
\end{bmatrix}, \tag{8.57} \\
\mathbf{w} &= [w_1, w_2, \ldots, w_N]^T. \tag{8.58}
\end{aligned}
$$

Rewriting Equation 8.54 and Equation 8.53 in matrix form, we obtain respectively:

$$\mathbf{w} = \frac{1}{\lambda}(\mathbf{d} - \mathbf{F}) \tag{8.59}$$

and

$$\mathbf{F} = \mathbf{G}\mathbf{w}. \tag{8.60}$$

Upon substituting Equation 8.60 into Equation 8.59, we get

$$(\mathbf{G} + \lambda\mathbf{I})\mathbf{w} = \mathbf{d}, \tag{8.61}$$

where $\mathbf{I}$ is the N-by-N identity matrix.

Invoking Light's Theorem [270] from Section 8.7.3, we may state that the matrix $\mathbf{G}$ is positive definite for certain classes of Green functions, provided that the data points $\mathbf{x}_1, \mathbf{x}_2, \dots$, $\mathbf{x}_N$ are distinct. The classes of Green functions covered by Light's theorem include the so-called multi-quadrics and Gaussian functions [253]. In practice, λ is chosen to be sufficiently large to ensure that $\mathbf{G} + \lambda\mathbf{I}$ is positive definite and therefore, invertible. Hence, the linear Equation 8.61 will have a unique solution given by

$$\mathbf{w} = (\mathbf{G} + \lambda\mathbf{I})^{-1}\mathbf{d}. \tag{8.62}$$

The set of Green functions used is characterized by the specific form adopted for the stabilizer $\mathcal{P}$ and the associated boundary conditions [253]. By definition, if the stabilizer $\mathcal{P}$ is translationally invariant, then the Green function $G(\mathbf{x}; \mathbf{x}_i)$ centred at $\mathbf{x}_i$ will depend only on the difference between the argument $\mathbf{x}$ and $\mathbf{x}_i$, i.e.:

$$G(\mathbf{x}; \mathbf{x}_i) = G(\mathbf{x} - \mathbf{x}_i). \tag{8.63}$$

If the stabilizer $\mathcal{P}$ is to be both *translationally and rotationally invariant*, then the Green function $G(\mathbf{x}; \mathbf{x}_i)$ will depend only on the *Euclidean norm* of the difference vector $\mathbf{x} - \mathbf{x}_i$, formulated as:

$$G(\mathbf{x}; \mathbf{x}_i) = G(\|\mathbf{x} - \mathbf{x}_i\|). \tag{8.64}$$

Under these conditions, the Green function must be a *radial basis function*. Therefore, the regularized solution of Equation 8.53 takes on the form:

$$F(\mathbf{x}) = \sum_{i=1}^{N} w_i G(\|\mathbf{x} - \mathbf{x}_i\|). \tag{8.65}$$

An example of a Green function, whose form is characterized by the differential operator $\mathcal{P}$ that is both translationally and rotationally invariant is the *multivariate Gaussian function* that obeys the following form

$$G(\mathbf{x}; \mathbf{x}_i) = exp\left(-\frac{1}{2\sigma_i^2}\|\mathbf{x} - \mathbf{x}_i\|^2\right), \qquad i = 1, \dots, N. \tag{8.66}$$

Equation 8.66 is characterized by a *mean vector* $\mathbf{x}_i$ and common *variance* σ_i^2.

It is important to realize that the solution described by Equation 8.65 differs from that of Equation 8.41. The solution of Equation 8.65 is *regularized* by the definition given in Equation 8.62 for the weight vector $\mathbf{w}$. The two solutions are the same only if the regularization parameter λ is equal to zero. The regularization parameter λ provides the smoothing effect in constructing the partition or interpolation hyper-surface during the learning process.

Typically, the number of training data symbols is higher than the number of basis functions required for the RBF network to give an acceptable approximation to the interpolation solution. The generalized RBF network is introduced to address this problem and its structure is discussed in the following section.

8.7.5   Generalized Radial Basis Function Networks

The one-to-one correspondence between the training input data $\mathbf{x}_i$ and the Green function $G(\mathbf{x}; \mathbf{x}_i)$ for $i = 1, 2, \ldots, N$ is prohibitively expensive to implement in computational terms for large N values. Especially the computation of the linear weights w_i is computationally demanding, which requires the inversion of an N-by-N matrix according to Equation 8.62. In order to overcome these computational difficulties, the complexity of the RBF network would have to be reduced and this requires an approximation to the regularized solution.

The approach followed here involves seeking a suboptimal solution in a lower-dimensional space that approximates the regularized solution described by Equation 8.53. This can be achieved using *Galerkin's method* [253]. According to this technique, the approximated solution $F^*(\mathbf{x})$ is expanded using a reduced $M \leq N$ number of basis functions, as follows:

$$F^*(\mathbf{x}) = \sum_{i=1}^{M} w_i \varphi_i(\mathbf{x}), \tag{8.67}$$

where $\varphi_i(\mathbf{x}), i = 1, 2, \ldots, M$, is a new set of basis functions. The number M of the basis functions M is typically less than the number of data points N and the coefficients w_i constitute a new set of weights. Using radial basis functions, we set

$$\varphi_i(\mathbf{x}) = G(\|\mathbf{x} - \mathbf{c}_i\|), \qquad i = 1, 2, \ldots, M, \tag{8.68}$$

where $\mathbf{c}_i, i = 1, 2, \ldots, M$, is the set of RBF centres to be determined. Thus, with the aid of Equation 8.67 and Equation 8.68 we have

$$
\begin{aligned}
F^*(\mathbf{x}) &= \sum_{i=1}^{M} w_i G(\mathbf{x}; \mathbf{c}_i) \\
&= \sum_{i=1}^{M} w_i G(\|\mathbf{x} - \mathbf{c}_i\|).
\end{aligned}
\tag{8.69}
$$

Now the problem we have to address is the determination of the new set of weights $w_i, i = 1, 2, \ldots, M$, based on a reduced number of $M \leq N$ basis functions so as to minimize the new cost function $\xi(F^*)$ according to Tikhonov's cost function of Equation 8.50. This new cost function is defined by

$$\xi(F^*) = \sum_{i=1}^{N} \left(d_i - \sum_{j=1}^{M} w_j G(\|\mathbf{x}_i - \mathbf{c}_j\|) \right)^2 + \lambda \|\mathcal{P}F^*\|^2. \tag{8.70}$$

Minimizing Equation 8.70 with respect to the weight vector $\mathbf{w}$ yields [253]:

$$(\mathbf{G}^T\mathbf{G} + \lambda\mathbf{G}_0)\mathbf{w} = \mathbf{G}^T\mathbf{d}, \tag{8.71}$$

where

$$\mathbf{d} = [d_1, d_2, \ldots, d_N]^T, \tag{8.72}$$

$$\mathbf{G} = \begin{bmatrix} G(\mathbf{x}_1; \mathbf{c}_1) & G(\mathbf{x}_1; \mathbf{c}_2) & \ldots & G(\mathbf{x}_1; \mathbf{c}_M) \\ G(\mathbf{x}_2; \mathbf{c}_1) & G(\mathbf{x}_2; \mathbf{c}_2) & \ldots & G(\mathbf{x}_2; \mathbf{c}_M) \\ \vdots & \vdots & & \vdots \\ G(\mathbf{x}_N; \mathbf{c}_1) & G(\mathbf{x}_N; \mathbf{c}_2) & \ldots & G(\mathbf{x}_N; \mathbf{c}_M) \end{bmatrix}, \tag{8.73}$$

$$\mathbf{w} = [w_1, w_2, \ldots, w_M]^T, \tag{8.74}$$

$$\mathbf{G}_0 = \begin{bmatrix} G(\mathbf{c}_1; \mathbf{c}_1) & G(\mathbf{c}_1; \mathbf{c}_2) & \ldots & G(\mathbf{c}_1; \mathbf{c}_M) \\ G(\mathbf{c}_2; \mathbf{c}_1) & G(\mathbf{c}_2; \mathbf{c}_2) & \ldots & G(\mathbf{c}_2; \mathbf{c}_M) \\ \vdots & \vdots & & \vdots \\ G(\mathbf{c}_M; \mathbf{c}_1) & G(\mathbf{c}_M; \mathbf{c}_2) & \ldots & G(\mathbf{c}_M; \mathbf{c}_M) \end{bmatrix}. \tag{8.75}$$

Here, the matrix $\mathbf{G}$ is a non-symmetric N-by-M matrix and the matrix $\mathbf{G}_0$ is a symmetric M-by-M matrix. Thus, upon solving Equation 8.71 to obtain the weights $\mathbf{w}$, we get:

$$\mathbf{w} = (\mathbf{G}^T\mathbf{G} + \lambda\mathbf{G}_0)^{-1}\mathbf{G}^T\mathbf{d}. \tag{8.76}$$

Observe that the solution in Equation 8.76 is different from Tikhonov's solution in Equation 8.62. Specifically, in Equation 8.57 the matrix $\mathbf{G}$ is a symmetric N-by-N matrix, while in Equation 8.73 it is a non-symmetric N-by-M matrix.

By introducing a number of modifications to the exact interpolation procedure presented in Section 8.7.3 we obtain the generalized radial basis function network model that provides a smooth interpolating function, in which the number of basis functions is determined by the affordable complexity of the mapping to be represented, rather than by the size of the data set. The modifications which are required are as follows:

1. The number of basis functions, M, need not be equal to the number of training data points, N.

2. In contrast to Equation 8.41, the centres of the basis functions are no longer constrained to be given by N training input data points $\mathbf{x}_i$. Thus, the position of the centres of the radial basis functions $\mathbf{c}_i, i = 1, 2, \ldots, M$, in Equation 8.69 are the unknown parameters that have to be 'learned' together with the weights of the output layer $w_i, i = 1, 2, \ldots, M$. A few methods of obtaining the RBF centres are as follows: random selection from the training data, the so-called Orthogonal Least Squares (OLS) learning algorithm of Chen, Cowan, Grant et al. [274,275] and the well-known K-means clustering algorithm [85]. We opted for using the K-means clustering algorithm in order to learn the RBF centres in our equalization problem and this algorithm will be described in more detail in Section 8.8.

3. Instead of having a common RBF spread or width parameter $2\sigma^2$, as described in Equation 8.48, each basis function is given its own width $2\sigma_i^2$, as in Equation 8.66. The value of the spread or width is determined during training. Bishop [254] noted that based on noisy interpolation theory, it is a useful rule of thumb when designing the RBF network with good generalization properties to set the width $2\sigma_i^2$ of the RBF large in relation to the spacing of the RBF input data.

Here, the new set of RBF network parameters, c_i, σ_i^2, and w_i, where $1 \leq i \leq M \leq N$, can be learnt in a sequential fashion. For example, a clustering algorithm can be used to estimate the RBF centres, c_i. Then, an estimate of the variance of the input vector with respect to each centre provides the width parameter, σ_i^2. Finally, we can calculate the RBF weights w_i using Equation 8.76 or adaptively using the LMS algorithm [253].

Note that apart from regularization, an alternative way of reducing the number of basis functions required and thus reduce the associated complexity is to use the OLS learning procedure proposed by Chen, Cowan and Grant [274]. This method is based on viewing the RBF network as a linear regression model, where the selection of RBF centres is regarded as a problem of subset selection. The OLS method, employed as a forward regression procedure, selects a suitable set of RBF centres, which are referred to as the regressors, from a large set of candidates for the training data, yielding $M < N$. As a further advance, Chen, Chng and Alkadhimi [275] proposed a regularised OLS learning algorithm for RBFs that combines the advantages of both the OLS and the regularization method. Indeed, it was OLS training that was used in the initial application of RBF networks to the channel equalization problem [247]. Instead of using the regularised interpolation method, we opted for invoking detection theory, in order to solve the equalization problem with the aid of RBF networks. This will be expounded further in Section 8.9.

Having described and justified the design of the RBF network of Figure 8.12 that was previously introduced in Section 8.7.1, in the next section the K-means clustering algorithm used to learn the RBF centres and to partition the RBF network input data into K subgroups or clusters is described briefly.

8.8 K-means Clustering Algorithm

In general, the task of the K-means algorithm [276] is to partition the domain of arbitrary vectors into K regions and then to find a centroid-like reference vector, $c_i, i = 1, \ldots, K$, that best represents the set of vectors in each region or partition. In the RBF network based equalizer design the vectors to be clustered are the noisy channel state vectors $v_k, k = -\infty, \ldots, \infty$ observed by the equalizer using the current tap vectors, such as those seen in Figure 8.3, where the centroid-like reference vectors are constituted by the optimal channel states $r_i, i = 1, \ldots, n_s$, as described in the previous sections. Suppose that a set of input patterns x of the algorithm is contained in a domain $\mathbb{P}$. The K-means clustering problem is formulated as finding a partition of $\mathbb{P}$, $\mathbf{P} = [\mathbb{P}_1, \ldots, \mathbb{P}_K]$, and a set of reference vectors $\mathbf{C} = \{c_1, \ldots, c_K\}$ that minimize the cluster MSE cost function defined as follows:

$$\text{MSE}(\mathbf{P}, \mathbf{C}) = \sum_{i=1}^{K} \int_{\mathbb{P}_i} p(\mathbf{x}) \cdot \|\mathbf{x} - c_i\|^2 d\mathbf{x}, \tag{8.77}$$

where $\| \, \|$ denotes the l_2 norm and $p(\mathbf{x})$ denotes the probability density function of $\mathbf{x}$.

Upon presenting a new training vector to the K-means algorithm, it repetitively updates both the reference vectors or centroids c_i and the partition $\mathbf{P}$. We define $c_{i,k}$ and x_k as the ith reference vector and the current input pattern presented to the algorithm at time k. The adaptive K-means clustering algorithm computes the new reference vector $c_{i,k+1}$ as

$$c_{i,k+1} = c_{i,k} + M_i(\mathbf{x}_k)\{\mu(\mathbf{x}_k - c_{i,k})\}, \tag{8.78}$$

where μ is the learning rate governing the speed and accuracy of the adaptation and $M_i(\mathbf{x}_k)$ is the so-called membership indicator that specifies, whether the input pattern $\mathbf{x}_k$ belongs to region $\mathbb{P}_i$ and also, whether the ith neuron is active. In the traditional adaptive K-means algorithm the learning rate μ is typically a constant and the membership indicator $M_i(\mathbf{x})$ is defined as:

$$M_i(\mathbf{x}) = \begin{cases} 1 & \text{if } \|\mathbf{x} - \mathfrak{c}_i\|^2 \leq \|\mathbf{x} - \mathfrak{c}_j\|^2 \text{ for each } i \neq j \\ 0 & \text{otherwise.} \end{cases} \tag{8.79}$$

A serious problem associated with most K-means algorithm implementations is that the clustering process may not converge to an optimal or near-optimal configuration. The algorithm can only assure local optimality, which depends on the initial locations of the representative vectors. Some initial reference vectors get 'entrenched' in regions of the algorithm's input vector domain with few or no input patterns and may not move to where they are needed. To deal with this problem, Rumelhart and Zipser [277] employed leaky learning, where in addition to adjusting the closest reference vector, other reference vectors are also adjusted, but in conjunction with smaller learning rates. Another approach, proposed by DeSieno and is referred to as the conscience algorithm [278] keeps track of how many times each reference vector has been updated in response to the algorithm's input vectors and if a reference vector gets updated or 'wins' too often, it will 'feel guilty' and therefore pulls itself out of the competition. Thus, the average rates of 'winning' for each region is equalized and no reference vectors can get 'entrenched' in that region. However, these two methods yield partitions that are not optimal with respect to the MSE cost function of Equation 8.77.

The performance of the adaptive K-means algorithm depends on the learning rate μ in Equation 8.78. There is a tradeoff between the *dynamic performance* (rate of convergence) and the *steady-state performance* (residual deviation from the optimal solution or excess MSE). When using a fixed learning rate, it must be sufficiently small for the adaptation to converge. The excess MSE is smaller at a lower learning rate. However, a smaller learning rate also results in a slower convergence rate. Because of this problem, adaptive K-means algorithms having variable learning rates have been investigated [279]. The traditional adaptive K-means algorithm can be improved by incorporating two mechanisms: by biasing the clustering towards an optimal partition and by adjusting the learning rate dynamically. The justification and explanation concerning how the two mechanisms are implemented are described in more detail by Chinrungrueng *et al.* [279].

Having described the K-means clustering algorithm, which can be used as the RBF network's learning algorithm, we proceed to further explore the RBF network structure in the context of an equalizer in the following Section.

8.9   Radial Basis Function Network Based Equalisers

8.9.1   Introduction

The RBF network is ideal for channel equalization applications, since it has an equivalent structure to the so-called optimal Bayesian equalization solution of Equation 8.17 [85]. Therefore, RBF equalisers can be derived directly from theoretical considerations related to optimal detection and all our prior knowledge concerning detection problems [251] can

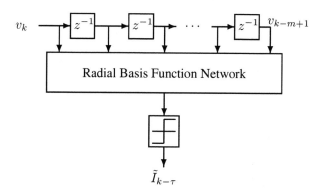

Figure 8.18: Radial Basis Function equalizer for BPSK.

be exploited. The neural network equalizer based on the MLP of Section 8.5, the polynomial perceptrons of Section 8.6 and on the so-called self-organizing map [263] constitutes a model-free classifier, thus requiring a long training period and large networks. The schematic of the RBF equalizer is depicted in Figure 8.18. The overall response of the RBF network of Figure 8.12, again, can be formulated as:

$$f_{RBF}(\mathbf{v}_k) = \sum_{i=1}^{M} w_i \varphi(\|\mathbf{v}_k - \mathbf{c}_i\|),$$

$$\varphi(x) = exp(-x^2/\rho), \tag{8.80}$$

where $\mathbf{c}_i, i = 1, \dots, M$ represents the RBF centres, which have the same dimensionality as the input vector $\mathbf{v}_k$, $\| \cdot \|$ denotes the Euclidean norm, $\varphi(\cdot)$ is the radial basis function introduced in Section 8.7, ρ are positive constants defined as the spread or width of the RBF in Section 8.7 (each of the RBFs has the same width, i.e., $2\sigma_i^2 = \rho$, since the received signal is corrupted by the same Gaussian noise source) and M is the number of hidden nodes of the RBF network. Note that the number of input nodes of the RBF network in Figure 8.12, p, is now equivalent to the order m of the equaliser, i.e. $p = m$, and the bias is set to $b = 0$. The detected symbol is given by:

$$\tilde{I}_{k-\tau} = sgn(f_{RBF}(\mathbf{v}_k)), \tag{8.81}$$

where the decision delay τ is introduced to facilitate causality in the equalizer and to provide the 'past' and the 'future' received samples, with respect to the 'delayed' detected symbol, for equalization.

The relationship between the RBF network and the Bayesian equalization solution expressed in Equation 8.17 can be established explicitly. The RBF centres $\mathbf{c}_i, i = 1, \dots, M$ are in fact constituted by the noise-free channel output vectors $\mathbf{r}_i$ indicated by the circles and crosses in Figure 8.3, while the number of hidden nodes M in Figure 8.12 corresponds to the number of desired channel output vectors, n_s, i.e., $M = n_s$. The weights w_i correspond to the scaling factors of the conditional probability density functions in Equation 8.17 given by:

$$w_i = \begin{cases} p_i(2\pi\sigma_\eta^2)^{-m/2} & \text{if } \mathbf{r}_i \in V_{m,\tau}^+, \\ -p_i(2\pi\sigma_\eta^2)^{-m/2} & \text{if } \mathbf{r}_i \in V_{m,\tau}^-, \end{cases} \tag{8.82}$$

where p_i is the *a priori* probability of occurence for the noise-free channel output vector $\mathbf{r}_i$ and σ_η^2 is the noise variance of the Gaussian channel. For equiprobable transmitted binary symbols the *a priori* probability of each state is identical. Therefore, the network can be simplified considerably in the context of binary signalling by fixing the RBF weights to $w_i = +1$, if the RBF centroids $\mathbf{c}_i$ correspond to a positive channel state $\mathbf{v}_i^+$ and to $w_i = -1$, if the centroids $\mathbf{c}_i$ correspond to a negative channel state $\mathbf{v}_i^-$. The widths ρ in Equation 8.80 are controlled by the noise variance and are usually set to $\rho = 2\sigma_\eta^2$, while $\varphi(\cdot)$ is the noise probability density function, which is usually Gaussian. When these conditions are met, the RBF network realizes precisely the Bayesian equalization solution [85], a fact, which is augmented further below.

Specifically, in order to realize the optimal Bayesian solution using the RBF network, we have to identify the RBF centres or the noise-free channel output vectors. Chen *et al.* [85] achieved this using two alternative schemes. The first method identifies the channel model using standard linear adaptive CIR estimation algorithms such as for example Kalman filtering [280] and then calculates the corresponding CIR-specific noise-free vectors. The second method estimates these vectors or centres directly using so-called supervised learning – where training data are provided – and a decision-directed clustering algorithm [85,246], which will be described in detail in Section 8.9.3.

The ultimate link between the RBF network and the Bayesian equaliser renders the RBF design an attractive solution to equalization problems. The performance of the RBF equalizer is superior to that of the MLP and PP equalisers of Sections 8.5 and 8.6 and it needs a significantly shorter training period, than these nonlinear equalisers [85]. Furthermore, Equation 8.80 shows that RBF networks are linear in terms of the weight parameter w_i, while the non-linear RBFs $\varphi(x)$ are assigned to the hidden layer of Figure 8.12. The RBF network can be configured to have a so-called uni-modal error surface where f_{RBF} in Equation 8.80 exhibits only one minimum, namely the global minimum, with respect to its weights w_i, while also having a guaranteed convergence performance. The RBF equalizer is capable of equalising nonlinear channels, can be also adapted to non-Gaussian noise distributions. Furthermore, in a recursive form, referred to as the *recurrent RBF equaliser* [259], the equalizer can provide optimal decisions based on all the previous received samples, $v_{k-i}, i = 0, \ldots, \infty$, instead of only those previous received samples, $v_{k-i}, i = 0, \ldots, v_{k-m+1}$ which are within the equaliser's memory. The RBF equaliser can be used to compute the so-called *a posteriori* probabilities of the transmitted symbols, which are constituted by their correct detection probabilities. The advantages of using the *a posteriori* symbol probabilities for blind equalization and tracking in time-variant environments have been discussed in several contributions [259,281]. Furthermore, the *a posteriori* probabilities generated can be used to directly estimate the associated BER without any reference signal. The BER estimate can be used by the receiver as a measure of reliability of the data transmission process or even to control the transmission rate in variable rate digital modems or to invoke a specific modulation in adaptive QAM systems.

The drawback of RBF networks is, however, that their complexity, i.e. the number of neurons n_s in the hidden layer of Figure 8.12 grows dramatically, when the channel memory L and the equalizer order m increase, since $n_s = 2^{L+m}$. The vector subtraction $\mathbf{v}_k - \mathbf{c}_i$ in Equation 8.80 involves m subtraction operations, while the computation of the norm $\| \cdot \|^2$ of an m-element vector involves m multiplications and $m - 1$ additions. Thus, the term $w_i\varphi(\|\mathbf{v}_k - \mathbf{c}_i\|)$ in Equation 8.80 requires $2m - 1$ additions/subtractions, $m + 1$ multipli-

Number of subtractions and additions	$2n_s m - 1$
Number of multiplications	$n_s(m + 1)$
Number of divisions	n_s
Number of exp()	n_s

Table 8.4: Computational complexity of a linear RBF network equalizer having m inputs and n_s hidden units per equalised output sample based on Equation 8.80. When the optimum Bayesian equalizer of Equation 8.17 is used, we have $n_s = 2^{L+m}$, while in Section 8.9.7 we will reduce the complexity of the RBF equalizer by reducing the value of n_s.

cations, one division and an $\exp(\cdot)$ operation. The summation $\sum_{i=1}^{M}$ in Equation 8.80 where $M = n_s$, involves $n_s - 1$ additions. Therefore the associated computational complexity of the RBF network equalizer based on Equation 8.80 is given in Table 8.4.

For non-stationary channels the values of the RBF centres, c_i, will vary as a function of time and each centre must be re-calculated, before applying the decision function of Equation 8.80. Since $n_s = 2^{L+m}$ can be high, the evaluation of Equation 8.80 may not be practical for real-time applications. A range of methods proposed for reducing the complexity of the RBF network equalizer and to render it more suitable for realistic channel equalization will be described in Section 8.9.7. Our simulation results will be presented in Section 8.12.

8.9.2 RBF-based Equalization in Multilevel Modems

In the previous sections, the transmitted symbols considered were binary. In this section, based on the suggestions of Chen, McLaughlin and Mulgrew [245], we shall extend the design of the RBF equaliser to complex $\mathcal{M}$-ary modems, where the information symbols are selected from the set of $\mathcal{M}$ complex values, $\mathcal{I}_i, i = 1, 2, \ldots, \mathcal{M}$. An example is, when a Quadrature Amplitude Modulation (QAM) scheme [4] is used.

Since the delayed transmitted symbols $I_{k-\tau}$ in the schematic of Figure 8.18 may assume any of the legitimate $\mathcal{M}$ complex values, the channel input sequence $\mathbf{I}_k$, defined in Equation 8.5, produces $n_s = \mathcal{M}^{L+m}$ different possible values for the noise-free channel output vector $\tilde{\mathbf{v}}_k$ of Figure 8.18 described in Equation 8.6, which were visualised for the binary case in Figure 8.3. The desired channel states can correspondingly be partitioned into $\mathcal{M}$ classes – rather than two – according to the value of the transmitted symbol $I_{k-\tau}$, which is formulated as follows:

$$
\begin{aligned}
V_{m,\tau}^i &= \{\tilde{\mathbf{v}}_k | I_{k-\tau} = \mathcal{I}_i\}, \\
&= \{\mathbf{r}_1^i, \ldots, \mathbf{r}_j^i, \ldots, \mathbf{r}_{n_s^i}^i\}, \qquad i = 1, 2, \ldots, \mathcal{M}, \quad (8.83)
\end{aligned}
$$

where $\mathbf{r}_j^i, j = 1, \ldots, n_s^i$, is the jth desired channel output state due to the $\mathcal{M}$-ary transmitted symbol $I_{k-\tau} = \mathcal{I}_i, i = 1, \ldots, \mathcal{M}$. More explicitly, the quantities n_s^i represent the number of channel states $\mathbf{r}_j^i$ in the set $V_{m,\tau}^i$. The number of channel states in any of the sets $V_{m,\tau}^i$ is identical for all the transmitted symbols $\mathcal{I}_i, i = 1, 2, \ldots, \mathcal{M}$, i.e. $n_s^i = n_s^j$ for $i \neq j$ and $i, j = 1, \ldots \mathcal{M}$. Lastly, we have $\sum_{i=1}^{M} n_s^i = n_s$.

Thus, the optimal Bayesian decision solution of Equation 8.15 defined for binary signalling based on Bayes' decision theory [241] has to be redefined for $\mathcal{M}$-ary signalling as follows, in order to achieve the minimum error-probability:

$$\tilde{I}_{k-\tau} = \mathcal{I}_i^*, \qquad \text{if } \zeta_i^*(k) = \max\{\zeta_i(k), 1 \le i \le \mathcal{M}\}, \tag{8.84}$$

where $\zeta_i(k)$ is the decision variable based on the conditional density function given by:

$$
\begin{aligned}
\zeta_i(k) &= P(\mathbf{v}_k | I_{k-\tau} = \mathcal{I}_i) \cdot P(I_{k-\tau} = \mathcal{I}_i) \\
&= \sum_{j=1}^{n_s^i} p_j^i p(\mathbf{v}_k - \mathbf{r}_j^i), \qquad 1 \le i \le \mathcal{M}.
\end{aligned}
\tag{8.85}
$$

The quantities $p_j^i, i = 1, \dots, \mathcal{M}, j = 1, \dots, n_s^i$ denote the *a priori* probability of appearance of each desired state $\mathbf{r}_j^i \in V_{m,\tau}^i$ associated with the transmitted $\mathcal{M}$-ary symbol $\mathcal{I}_i, i = 1, \dots, \mathcal{M}$ and $p(\cdot)$ is the probability density function of the additive noise of the channel.

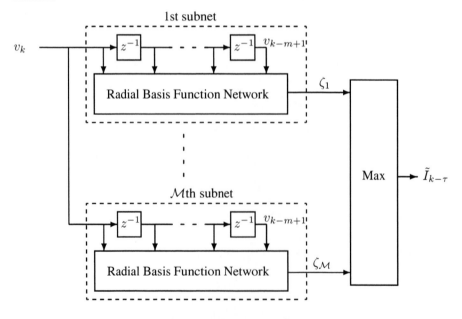

Figure 8.19: Radial Basis Function equalizer for $\mathcal{M}$-level modems.

Thus, there are $\mathcal{M}$ neural 'subnets' associated with the $\mathcal{M}$ decision variables $\zeta_i(k) = P(\mathbf{v}_k | I_{k-\tau} = \mathcal{I}_i) \cdot P(I_{k-\tau} = \mathcal{I}_i), i = 1, 2, \dots, \mathcal{M}$. The architecture of the RBF equalizer for the $\mathcal{M}$-ary multilevel modem scenario considered is shown in Figure 8.19. Note that the output of each sub-RBF network gives the corresponding conditional density function $\zeta_i(k) = P(\mathbf{v}_k | I_{k-\tau} = \mathcal{I}_i) \cdot P(I_{k-\tau} = \mathcal{I}_i)$ and this output value can be used for generating soft decision inputs in conjunction with error correction techniques. Observe that the schematic of Figure 8.19 is more explicit, than that of Figure 8.18, since for the specific case of BPSK we have $\mathcal{M} = 2$. This yields two equaliser subnets, which correspond to the transmission of a logical one as well as a logical zero, respectively.

The computational complexity of the $\mathcal{M}$-ary RBF equalizer is dependent on the order $\mathcal{M}$ of the modulation scheme, since the number of sub-RBF hidden nodes is equivalent to

$n_s^i = \mathcal{M}^{L+m}/\mathcal{M}$. Thus, its application is typically restricted to low-order $\mathcal{M}$-ary modulation schemes. The computational complexity of each subnet of the $\mathcal{M}$-ary RBF equaliser is similar to that in Table 8.4, taking into account the reduced number of hidden nodes, namely $n_s^i = n_s/\mathcal{M}$. Thus, the overall computational complexity of the $\mathcal{M}$-ary RBF equaliser described by Equation 8.84 and 8.85 is given in Table 8.5.

Number of subtractions and additions	$2n_s m - \mathcal{M}$
Number of multiplications	$n_s(m+1)$
Number of divisions	n_s
Number of exp()	n_s
Number of max operations	1

Table 8.5: Computational complexity of an mth-order RBF network equalizer per equalised output sample for $\mathcal{M}$-ary modulation based on Equation 8.84 and 8.85. The total number of hidden nodes of the RBF equalizer is n_s.

8.9.3 Adaptive RBF Equalization

The knowledge of the noise-free channel outputs is essential for the determination of the decision function associated with Equation 8.84. The channel state estimation – where the channel states were defined in Section 8.2, in particular in the context of Equation 8.7 – requires the knowledge of the CIR, but this often may not be available. Thus the channel state has to be 'learned' during the actual data transmission or inferred during the equalizer training period, when the transmitted symbols are known to the receiver. This can be achieved typically in two ways [246]:

- By invoking CIR estimation methods [245, 246, 282]

- By employing so-called clustering algorithms [85] as described in Section 8.8

These methods will be highlighted by the following two Sections.

8.9.4 Channel Estimation Using a Training Sequence

According to our approach in this section, the channel model is first estimated using algorithms such as the Least Mean Square (LMS) algorithm [280]. With the knowledge of the CIR, the channel state can then be calculated. Let us define the CIR estimate associated with the model of Figure 8.1 as:

$$\hat{\mathbf{f}}_k = \begin{bmatrix} \hat{f}_{0,k} & \cdots & \hat{f}_{L,k} \end{bmatrix}^T,$$ (8.86)

and introduce the $(L+1)$-element channel estimator input vector

$$\mathbf{I}_{f,k} = \begin{bmatrix} I_k & \cdots & I_{k-L} \end{bmatrix}^T,$$ (8.87)

where $\{I_k\}$ is the transmitted channel input sequence, which is known during the training period. Then the error between the actual channel output v_k and the estimated channel output derived using the estimated CIR $\hat{\mathbf{f}}_{k-1}$ can be expressed as:

$$\varepsilon_k = v_k - \hat{\mathbf{f}}_{k-1}^T \mathbf{I}_{f,k}. \tag{8.88}$$

The CIR estimate can then be updated following the steepest descent philosophy of Equation A.52 as follows:

$$\hat{\mathbf{f}}_k = \hat{\mathbf{f}}_{k-1} + \mu_f \varepsilon_k \mathbf{I}_{f,k}^*, \tag{8.89}$$

where μ_f is the step-size defined by the channel estimator learning rule. Note however that the LMS channel estimation technique based on the channel model described in Figure 8.1 will fail, if the channel is non-linear in its nature.

During data transmission after learning, a decision-directed and delayed version of Equation 8.88 and Equation 8.89 is used, which is formulated as:

$$\begin{aligned}
\varepsilon_{k-\tau} &= v_{k-\tau} - \hat{\mathbf{f}}_{k-\tau-1} \tilde{\mathbf{I}}_{f,k-\tau} \\
\hat{\mathbf{f}}_{k-\tau} &= \hat{\mathbf{f}}_{k-\tau-1} + \mu_f \varepsilon_{k-\tau} \tilde{\mathbf{I}}_{f,k-\tau}^*,
\end{aligned} \tag{8.90}$$

that can be employed to track time-varying channels, where

$$\tilde{\mathbf{I}}_{f,k-\tau} = \begin{bmatrix} \tilde{I}_{k-\tau} & \cdots & \tilde{I}_{k-\tau-L} \end{bmatrix}^T \tag{8.91}$$

is the channel estimator input vector associated with the CIR vector $\mathbf{f}_{k-\tau}$. Note that during data transmission, $\{\tilde{I}_{k-\tau}\}$ is the delayed symbol, detected by the equaliser. At instant $k+1$, the delayed CIR estimate $\hat{\mathbf{f}}_{k-\tau}$ is used to track the time-varying channel as though it were the most recent estimate $\hat{\mathbf{f}}_k$. The current channel model $\hat{\mathbf{f}}_{k+1}$ might have changed considerably. This tracking error owing to the inherent decision delays will degrade the performance of the channel estimator. As it will be demonstrated in Figure 8.22 at a later stage, increasing the decision delay τ first introduced in the context of Equation 8.81 improves the performance of the equalizer for a stationary channel. By contrast, this will degrade the performance of the channel estimator for a nonstationary channel environment. Thus we need to achieve a reasonable compromise and the selection of the decision delay parameter τ yielding satisfactory equalizer performance will depend on how rapidly the CIR varies.

The computational complexity of the LMS channel estimator is characterized in Table 8.6 based on Equation 8.88, which requires $L+1$ multiplication and $L+1$ addition/subtraction operations, and Equation 8.89 which involves $L+2$ multiplication and $L+1$ addition operations. On the basis of the estimated CIR $\hat{\mathbf{f}}_k$ it is straightforward to compute the estimated noise-free channel outputs $\tilde{v}_k$ using convolution and therefore to generate the channel output states $\mathbf{r}_i$. Upon substituting Equation 8.2 into the noiseless version of Equation 8.10, the channel output state $\mathbf{r}_i$ can be computed from:

$$\mathbf{r}_i = \mathbf{F} \mathbf{s}_i \tag{8.92}$$

where the elements of the CIR matrix $\mathbf{F}$ are obtained from Equation 8.89. Equation 8.92 requires $m(m+L)$ multiplication and $m(m+L-1)$ addition operations. Therefore, an additional computational load is encountered in converting the CIR estimate $\hat{\mathbf{f}}_k$ into the vector

$$2(L+1)+1 \text{ multiplications}$$
$$2(L+1) \text{ additions or subtractions}$$

Table 8.6: Computational complexity of the LMS CIR estimator for a channel having $L+1$ symbol-spaced taps per estimated CIR based on Equation 8.88 and Equation 8.89.

$$m(m+L)+2(L+1)+1 \text{ multiplications}$$
$$3L+m+1 \text{ additions or subtractions}$$

Table 8.7: Computational complexity of the m-dimensional channel output state learning algorithm using the LMS CIR estimator for a channel having $L+1$ symbol-spaced taps per channel output state based on Equation 8.88, Equation 8.89 and Equation 8.92.

$\mathbf{r}_i$ of channel output states and this has to be added to the computational complexity calculation of the CIR estimator given in Table 8.6, in order to quantify to give the total complexity for this channel state learning method, as shown in Table 8.7.

The CIR estimate can also be updated using the Recursive Least Square (RLS) algorithm [280], which has a better convergence performance compared to the LMS algorithm in most cases. However, the RLS algorithm exhibits a higher computational complexity than the LMS algorithm. For dispersive mobile radio channels the adaptive algorithm is expected to continuously operate during both the training and transmission periods in highly nonstationary environments, consequently its numerical stability is vital. Many versions of the fast RLS algorithm may not be suitable for this purpose. The CIR can also be estimated using the so-called least sum of square errors (LSSE) algorithm [283]. This algorithm is similar to the CIR estimator used in the GSM system [13] and those in [284, 285], and it exhibits a low computational complexity.

8.9.5 Channel Output State Estimation using Clustering Algorithms

Apart from training sequences, the channel states can also be estimated invoking the clustering algorithms described in Section 8.8. The computational procedures of the so-called supervised K-means clustering algorithm during the equalizer training period can be summarised as follows [85]:

$$\text{if } \mathbf{I}_k = \mathbf{s}_i, \text{ then}$$

$$\mathbf{c}_{i,k} = \mathbf{c}_{i,k-1} + \mu_c \cdot (\mathbf{v}_k - \mathbf{c}_{i,k-1}),$$

otherwise

$$\mathbf{c}_{i,k} = \mathbf{c}_{i,k-1}, \tag{8.93}$$

where μ_c is the associated learning rate, $\mathbf{s}_i, 1 \leq i \leq n_s = \mathcal{M}^{L+m}$ is the ith channel input sequence and $\mathbf{I}_k = \begin{bmatrix} I_k & \cdots & I_{k-m+1-L} \end{bmatrix}^T$ is an $(m+L)$-element transmitted symbol vector, which is known during the training phase. Explicitly, according to Equation 8.93 the clustering algorithm takes into account the most recently received m-element vector $\mathbf{v}_k$ in adapting the ith RBF centre $\mathbf{c}_{i,k}$, if the current $(L+m)$-element channel input vector $\mathbf{I}_k$ is given by the specific $(L + m)$-element vector $\mathbf{s}_i$. Initially, the RBF centres are all set to 0, i.e $c_{i,0} = 0, i = 1 \leq i \leq n_s = \mathcal{M}^{L+m}$. Equation 8.93 dictates that the previous centroid $\mathbf{c}_{i,k-1}$ has to be updated according to the 'distance' $(\mathbf{v}_k - \mathbf{c}_{i,k})$ between itself and the most recent $(L + m)$-element received vector $\mathbf{v}_k$ after scaling it by the learning rate μ_c. Otherwise the ith centre is not updated based on the information of the current received vector $\mathbf{v}_k$. Referring back to Section 8.8, the membership indicator defined by Equation 8.79 differs from that of the supervised version of the K-means clustering algorithm described by Equation 8.93. Explicitly, this modified membership indicator is defined as:

$$M_i(\mathbf{x}) = \begin{cases} 1 & \text{if } \mathbf{I}_k = \mathbf{s}_i \\ 0 & \text{otherwise.} \end{cases} \tag{8.94}$$

For time-varying channels we have to track the time-varying channel states during transmission after the training period. For tracking the channel-induced channel state variations, the following decision-directed clustering algorithm can be used to adjust the RBF centres, in order to take into account the current network input vector $\mathbf{v}_k$ in the updating of the centres as follows [85]:

$$\text{if } \tilde{\mathbf{I}}_{k-\tau} = \mathbf{s}_i, \text{then}$$

$$\mathbf{c}_{i,k} = \mathbf{c}_{i,k-1} + \mu_c \cdot (\mathbf{v}_{k-\tau} - \mathbf{c}_{i,k-1}),$$

$$\text{otherwise}$$

$$\mathbf{c}_{i,k} = \mathbf{c}_{i,k-1}, \tag{8.95}$$

where $\tilde{\mathbf{I}}_{k-\tau} = \begin{bmatrix} \tilde{I}_{k-\tau} & \cdots & \tilde{I}_{k-\tau-m+1-L} \end{bmatrix}^T$ represents the $(L + m)$ equalised demodulated symbols after decision and a delay of τ. Note that whilst in Equation 8.93 the transmitted vector $\mathbf{I}_k$ was used, in Equation 8.95 the vector $\tilde{\mathbf{I}}_{k-\tau}$ at the output of the decision device is used. The computational complexity of the clustering algorithm obeying Equation 8.93 is given in Table 8.8.

Local operation: Find $i, i = 1, \ldots, n_s$, for which $\mathbf{I}_k = \mathbf{s}_i$.
m multiplications
$2m$ additions or subtractions

Table 8.8: Computational complexity of the clustering algorithm specified by Equation 8.93 per channel output state for a RBF network having m inputs and n_s hidden nodes.

As we mentioned previously, all the RBF centres were initially set to 0. However, the centres can be initialised to the corresponding noisy channel states, in order to improve the convergence rate, since there is a higher probability that the actual channel states are nearer to the noisy channel states, than to $\mathbf{c}_{i,0} = 0, i = 1, \ldots, n_s = \mathcal{M}^{L+m}$. Thus, the algorithm described by Equation 8.93 can be adapted as follows:

$$\begin{aligned} &\textit{if } \mathbf{I}_k = \mathbf{s}_i, \text{ and } \mathbf{c}_{i,k} \text{ has not been initialised } \textit{then} \\ &\qquad \mathbf{c}_{i,k} = \mathbf{v}_k, \\ &\textit{else if } \mathbf{I}_k = \mathbf{s}_i, \text{ and } \mathbf{c}_{i,k} \text{ has been initialised } \textit{then} \\ &\qquad \mathbf{c}_{i,k} = \mathbf{c}_{i,k-1} + \mu_c \cdot (\mathbf{v}_k - \mathbf{c}_{i,k-1}). \end{aligned} \tag{8.96}$$

The achievable improvement of the convergence performance in conjunction with this algorithm will be demonstrated by our simulation results in Section 8.12.

8.9.6   Other Adaptive RBF Parameters

In the previous subsection, clustering algorithms were used for training the RBF centres. Similar procedures can be employed also for training the RBF weights as it will be outlined below. Explicitly, if our previous assumption of equiprobable symbols is violated, we have to adjust the RBF weights in order to learn the corresponding scaling factors of the conditional probability density functions in Equation 8.17 during the training period. The adaptation of the RBF weights can be achieved pursuing the approach of Chen, Mulgrew and Grant using the following supervised LMS algorithm [85]:

$$
\begin{aligned}
\epsilon_k &= I_{k-\tau} - f_{RBF}(\mathbf{v}_k) \\
\mathbf{w}_{i,k} &= \mathbf{w}_{i,k-1} + \mu_w \epsilon_k \varphi(\|\mathbf{v}_k - \mathbf{c}_i\|),
\end{aligned}
\tag{8.97}
$$

where μ_w is the learning rate for the RBF weights. Explicitly, the error $\epsilon_k = I_{k-\tau} - f_{RBF}(\mathbf{v}_k)$ between the $(L+m)$-element transmitted symbol vector $\mathbf{I}_{k-\tau}$ and the RBF's output is scaled by the RBF learning rate μ_w and this product is then used to weight $\varphi(\|\mathbf{v}_k - \mathbf{c}_i\|)$, in order to update the previous RBF weight $\mathbf{w}_{i,k-1}$, where $\varphi(\|\mathbf{v}_k - \mathbf{c}_i\|)$ is the RBF evaluated at the Euclidean norm $\|\mathbf{v}_k - \mathbf{c}_i\|$ characteristic of the 'distance' between the centroids $\mathbf{c}_i, i = 1, \ldots, n_s = \mathcal{M}^{L+m}$, and the $(L+m)$-element received vector $\mathbf{v}_k$.

Furthermore, if the exact number of RBF centres is not known precisely or if there is a deliberate attempt to use a reduced set of centres to reduce the computational complexity – as it will be described in Section 8.9.7 – it may be prudent to train the weights using the LMS algorithm of Equation 8.97, in order to make best use of the actual centres that have been provided [248]. Similarly, in noisy environments, where clustering techniques may only provide fairly crude estimates of the centres, training the RBF weights will make best use of the trained centres [248]. Another method of training the RBF weights is demonstrated in Chapter 11 where the information of the coded symbols, generated by the channel decoder is used to adapt the RBF weights.

8.9.7   Reducing the Complexity of the RBF Equaliser

In an effort to reduce the RBF equaliser's complexity, Chng et al. [286] proposed finding a RBF centre subset model in order to approximate the Bayesian decision function's response given in Equation 8.17 for the current $(L+m)$-element input vector $\mathbf{v}_k$. This implied using only the centres which are near, in Euclidean sense, to the current input vector $\mathbf{v}_k$ for the subset model. The rationale of this approach is based on the assumption that the contribution of the RBF centres to the decision function is inversely related to their distance from the input vector, as we can observe from Equation 8.17. The decision function response using only the centres within a distance of Δ from $\mathbf{v}_k$ is very similar to the full Bayesian RBF response, if the distance Δ is sufficiently large. Chng's results show that a distance of $\Delta = 4\sigma_\eta$ is sufficient and can reduce the number of centres required for the subset model to as small as 5-10% of the full model. Chng's paper [286] also provides a fast algorithm for identifying the specific centres, which are within a distance of Δ from the input vector $\mathbf{v}_k$ for the subset model .

Patra and Mulgrew [287] investigated the computational complexity aspects of RBF equalizers. They proposed an RBF equalizer using scalar centres, which can implement the Bayesian decision function of Equation 8.17, while allowing a lower computational complexity compared to previously reported RBF equalizers. This issue will be detailed in the next section, hence suffice to say here that the scalar centre c_{il} is the $(l+1)$th component of the RBF centroid vector $\mathbf{c}_i = [c_{i0} \ \cdots \ c_{il} \ \cdots \ c_{i(m-1)}]^T$, associated with the mth order equaliser, where c_{il} assumes the possible values of the noise-free channel output $\tilde{v}_k$ in order to realise the optimal Bayesian decision function of Equation 8.17. For binary transmission, there are 2^{L+1} possible noise-free channel output states, which correspond to each of the m elements of the equaliser's input vector $\tilde{\mathbf{v}}$, described in Equation 8.6, where $L+1$ is the length of the CIR. The mapping between the scalar centres and the scalar channel states will be expounded in more detail in Section 8.10.

The RBF equalizer described by the scalar centres can efficiently employ subset centre selection for computing the decision function of Equation 8.17, resulting in a substantial reduction in computational complexity. The algorithm proposed for subset centre selection by Patra [287] is more attractive compared to that suggested by Chng [286] *et al.* , since it is more efficient in terms of selecting a subset of the total set of centres in the one-dimensional space. This is because we only need to select a subset of centres from a total of 2^{L+1} possible scalar centres for Patra's method [287] compared to a total of $n_s = 2^{m+L}$ possible vector centres for Chng's method [286].

Another method of selecting a subset of significant RBF centres is to make use of past detected symbols. This idea, which incoporated decision feedback into the RBF network was proposed by Chen *et al.* [245, 246]. Section 8.11 will present this approach in more detail, together with our simulation results in Section 8.12.

In an effort to further reduce the complexity we invoke an approach often used in turbo codes [152] for complexity reduction. Specifically, we proposed generating the output of the RBF equaliser in logarithmic form by invoking the Jacobian logarithm [288, 289], in order to avoid the computation of exponentials and to reduce the number of multiplictions performed. We refer to this equaliser as the Jacobian RBF equaliser, which will be introduced in Section 10.2.

8.10 Scalar Noise-free Channel Output States

In this section, we will describe in detail the scalar noise-free channel output states and relate them to the m-element noise-free channel output state vector $\tilde{\mathbf{v}}_i$ and to the scalar RBF centres c_{il} that we have mentioned in Section 8.9.7. After defining the scalar noise-free channel output state, we will expound on how it is used to reduce the complexity of the RBF equaliser.

Referring back to Equation 8.6 and Equation 8.3, the lth element $\tilde{v}_{k-l}, l = 0, 1, \ldots, m-1$, of the $(L+1)$-element noise-free channel output vector $\tilde{\mathbf{v}}_k$ corresponds to the so-called block-convolution of a sequence of $L+1$ transmitted symbols and the $L+1$ CIR taps. In other words, the number of transmitted symbols contributing to the value of $\tilde{v}_{k-l}$ is $L+1$ and we represent these transmitted symbols by an $L+1$ element vector $\mathbf{I}_{f,k}$, as described by Equation 8.87. Let us now introduce the concept of scalar states using the channel-state example of Table 8.1, where the scalar channel output states are $r_1 = -1.5$, $r_2 = -0.5$, $r_3 = 0.5$ and $r_4 = 1.5$, while the number of scalar channel states is $n_{s,f} = 2^{L+1} = 4$ ($L =$

1). Thus, the vector channel output states can be expressed with the aid of the scalar states forming the vector as $\mathbf{r}_1 = [r_1 \ r_1]^T$, $\mathbf{r}_2 = [r_1 \ r_2]^T, \ldots$, etc. More explicitly at every instant $-\infty < k < \infty$ the noiseless scalar channel output is given by the corresponding convolution of the input bits and the CIR. In general, the number of different possible combinations of the $(L + 1)$-element transmitted symbol sequence in $\mathbf{I}_{f,k}$ is $n_{s,f} = 2^{L+1}$ for a binary modulation scheme. We represent these transmitted symbol combinations equivalently as a channel input state $\mathbf{s}_{scalar,i}$, where $i = 1, 2, \ldots, n_{s,f} = 2^{L+1}$. After convolution with the CIR, each of these channel input states $\mathbf{s}_{scalar,i}$ generates a scalar channel output state $r_i, i = 1, 2, \ldots, n_{s,f} = 2^{L+1}$. Thus, as we have seen with reference to Table 8.1 the noise-free channel output $\tilde{v}_k$ can take up any of the $n_{s,f} = 2^{L+1}$ scalar channel output states r_i, depending on $\mathbf{I}_{f,k}$, which is summarised as:

$$\tilde{v}_k = r_i \qquad \text{if } \mathbf{I}_{f,k} = \mathbf{s}_{scalar,i} \qquad i = 1, \ldots, n_{s,f}, \qquad -\infty < k < \infty. \tag{8.98}$$

Similarly to our introductory example, the scalar channel output states $r_i, i = 1, 2, \ldots, n_{s,f} = 2^{L+1}$, can be suitably combined to form the vector channel output states $\mathbf{r}_j, j = 1, 2, \ldots, n_s = 2^{m+L}$, seen in Equation 8.2.

In order to realise the optimal Bayesian decision function of Equation 8.17 , the scalar centre c_{il} – which is the $(l+1)$th component of the vector centre $\mathbf{c}_i$, where $i = 1, 2, \ldots, n_s = 2^{m+L}$ and $l = 0, 1, \ldots, m - 1$, as mentioned in Section 9.9.7 – has to assume the value of these scalar channel output states r_i. The scalar centres c_{il} can be obtained from a lookup table that provides the mapping $Q : R \rightarrow C$, where $R = \{r_1, \ldots, r_i, \ldots r_{n_{s,f}}\}$ and $C = \{c_{00}, \ldots, c_{il}, \ldots, c_{n_s(m-1)}\}$. Using again the example of Table 8.1 and letting $\mathbf{c}_i = \mathbf{r}_i, i = 1, \ldots, n_s$, the scalar centres correspond to the scalar channel output states as follows : $c_{00} = r_1, c_{01} = r_1, c_{10} = r_1, c_{11} = r_2, \ldots$, etc.

A scalar channel output state r_i is just the conditional mean of the noisy observation v_k given by $\mathbf{I}_{f,k} = \mathbf{s}_{scalar,i}$, and a clustering procedure can be used to update the scalar channel states as follows [245]:

$$\text{if } \mathbf{I}_{f,k} = \mathbf{s}_{scalar,i}, \text{then}$$
$$r_{i,k} = r_{i,k-1} + \mu_r \cdot (v_k - r_{i,k-1}),$$
$$\text{otherwise}$$
$$r_{i,k} = r_{i,k-1}, \tag{8.99}$$

where μ_r is the associated learning rate of the scalar channel states. For time-varying channels, it is necessary to continuously update r_i during data transmission. This can be achieved using the following decision-directed version of Equation 8.99:

$$\text{if } \tilde{\mathbf{I}}_{f,k-\tau} = \mathbf{s}_{scalar,i}, \text{then}$$
$$r_{i,k} = r_{i,k-1} + \mu_r \cdot (v_{k-\tau} - r_{i,k-1})$$
$$\text{otherwise}$$
$$r_{i,k} = r_{i,k-1}. \tag{8.100}$$

The computational complexity of the clustering algorithm in the context of the scalar channel states is given in Table 8.9. Note that the computational load of the clustering scheme for the scalar channel states is lower than that for the vector channel states, which becomes explicit by comparing Table 8.9 and Table 8.8 of Section 8.9.5, since $n_{s,f} < n_s$. However, some additional processing is required, in order to expand the scalar states into the vector states. This is not costly, especially, if the expansion can be done via a lookup table.

> Local operation: Find i, $i = 1, \ldots, n_{s,f}$, for which $\mathbf{I}_{f,k} = \mathbf{s}_{scalar,i}$.
> 1 multiplication
> 2 additions or subtractions

Table 8.9: Computational complexity of the clustering algorithm per scalar channel output state for $n_{s,f}$ number of scalar channel output states based on Equation 8.99.

As mentioned in Section 8.9.3, the channel states can be learnt by invoking channel estimation methods. Section 8.9.4 described a channel estimation method using the LMS algorithm. Since the number of channel taps $L + 1$ is lower than that of the scalar channel states $n_{s,f} = \mathcal{M}^{L+1}$, it becomes explicit that an adaptive scheme based on a channel estimator requires a shorter training period than the clustering approach. Thus the former is better suited for time-variant channels. However, the clustering scheme does not assume the linear channel model described by Equation 8.1 and it is immune to nonlinear distortion. When significant nonlinear distortion is inflicted for example by the system's power amplifier, the estimated channel states based on a linear model will deviate from the true states, causing a performance loss. The clustering approach does not suffer from this problem and it always converges to the set of true channel output states, regardless of whether the channel is linear or nonlinear.

The scalar channel state clustering scheme provides faster convergence compared to the vector channel state clustering scheme, since the convergence performance depends on the number of clusters or channel states and the number of scalar channel states is less than the number of vector channel states. This will be demonstrated in Section 8.12, which will provide simulation results in order to characterize the performance of the scalar channel state clustering scheme.

Upon extending the scalar channel state concept to multilevel modems, we note that the number of channel states $n_{s,f} = \mathcal{M}^{L+1}$ grows exponentially with the number $\mathcal{M}$ of symbol constellation points used in the modulation scheme. Thus, the convergence rate is dependent on the type of modulation scheme used.

8.11   Decision Feedback Assisted Radial Basis Function Network Equaliser [245, 246, 282]

In their seminal contribution Chen, Mulgrew and McLaughlin [245, 246, 282] introduced decision feedback into the RBF equalizer in order to reduce its computational complexity, as mentioned earlier in Section 8.9.7. Figure 8.20 illustrates this design for a binary modulation scheme. Observe in the figure that in contrast to conventional DFEs, where the output of the feedback section is subtracted from that of the feedforward section, here the feedback section is employed to assist in the operation of the feedforward section, as it will become explicit later in this section. The structure of a decision feedback RBF equalizer is specified by the equaliser's decision delay τ, the feedforward order m and the feedback order n.

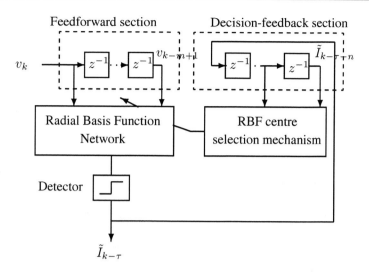

Figure 8.20: Radial basis function equalizer with decision feedback.

The n-symbol long binary feedback vector $\tilde{\mathbf{I}}_{feedback,k-\tau} = \begin{bmatrix} \tilde{I}_{k-\tau-1} & \cdots & \tilde{I}_{k-\tau-n} \end{bmatrix}^T$ is associated with $n_f = 2^n$ states. We denote the set of $n_f = 2^n$ different feedback sequences by $s_{f,j}$, $1 \leq j \leq n_f = 2^n$. The binary subset $V_{m,\tau}^+$ and $V_{m,\tau}^-$ of the channel states defined in Equation 8.7 can be further partitioned into n_f subsets, $V_{m,\tau,j}^+$ and $V_{m,\tau,j}^-$, according to the $n_f = 2^n$ possible feedback states such that the union of the $n_f = 2^n$ number of feedback states associated with the two legitimate binary transmitted symbols can be formulated as:

$$V_{m,\tau}^+ = \bigcup_{1 \leq j \leq n_f} V_{m,\tau,j}^+$$

$$V_{m,\tau}^- = \bigcup_{1 \leq j \leq n_f} V_{m,\tau,j}^-, \tag{8.101}$$

where $V_{m,\tau}^{\pm}$ is the set of possible $\tilde{\mathbf{v}}_k$ values associated with the delayed transmitted symbol $I_{k-\tau} = \pm 1$ and the feedback symbol sequence $\tilde{\mathbf{I}}_{feedback,k-\tau} = s_{f,j}$ yields the following subsets:

$$V_{m,\tau,j}^+ = \{\tilde{\mathbf{v}}_k | I_{k-\tau} = +1 \cap \tilde{\mathbf{I}}_{feedback,k-\tau} = s_{f,j}\},$$

$$V_{m,\tau,j}^- = \{\tilde{\mathbf{v}}_k | I_{k-\tau} = -1 \cap \tilde{\mathbf{I}}_{feedback,k-\tau} = s_{f,j}\},$$

$$1 \leq j \leq n_f. \tag{8.102}$$

Thus the role of the feedback symbol vector $\tilde{\mathbf{I}}_{feedback,k-\tau}$ in the decision feedback structure is to select a subset of centres for a particular decision. The proportion of channel states in the sets $V_{m,\tau,j}^+$ and $V_{m,\tau,j}^-$ is $n_{s,j}^+ = n_s^+/n_f$ and $n_{s,j}^- = n_s^-/n_f$, respectively. The total number of channel states associated with the feedback state $s_{f,j}$ is given by $n_{s,j} = n_{s,j}^+ + n_{s,j}^-$. Given the feedback vector $\tilde{\mathbf{I}}_{feedback,k-\tau} = s_{f,j}$, the Bayesian decision function of Equation 8.17

can be rewritten with a reduced number of noiseless channel states as:

$$
\begin{aligned}
f_{Bayes}(\mathbf{v}_k | \tilde{\mathbf{I}}_{feedback,k-\tau} = \mathbf{s}_{f,j}) &= \sum_{i=1}^{n_{s,j}^+} p_{j,i}^+ (2\pi\sigma_\eta^2)^{-m/2} exp(-\|\mathbf{v}_k - \mathbf{r}_{j,i}^+\|^{2\prime}/2\sigma_\eta^2) \\
&\quad - \sum_{l=1}^{n_{s,j}^-} p_{j,l}^- (2\pi\sigma_\eta^2)^{-m/2} exp(-\|\mathbf{v}_k - \mathbf{r}_{j,l}^-\|^2/2\sigma_\eta^2),
\end{aligned}
$$
$$
j = 1, \dots, n_f, \tag{8.103}
$$

where $\mathbf{r}_{j,i}^+$ and $\mathbf{r}_{j,i}^-$, $i = 1, \dots, n_{s,j}^\pm$, $j = 1, \dots, n_f$ are the ith noiseless channel states, when the feedback vector $\tilde{\mathbf{I}}_{feedback,k-\tau}$ is $\mathbf{s}_{f,j}$, while the superscripts $^+$ and $^-$ correspond to the transmitted symbols of $I_{k-\tau} = +1$ and $I_{k-\tau} = -1$, respectively. Explicitly, $\mathbf{r}_{j,i}^+ \in V_{m,\tau,j}^+$, $\mathbf{r}_{j,l}^- \in V_{m,\tau,j}^-$, while $p_{j,i}^+$ and $p_{j,i}^-$ are the *a priori* probability of occurence for each state $\mathbf{r}_{j,i}^+$ and $\mathbf{r}_{j,i}^-$, respectively. The minimum error probability decision is thus formulated as:

$$
\tilde{I}_{k-\tau} = sgn(f_{Bayes}(\mathbf{v}_k | \tilde{\mathbf{I}}_{feedback,k-\tau} = \mathbf{s}_{f,j})). \tag{8.104}
$$

The relationship between the RBF network described in Equation 8.80 and the Bayesian DFE decision function expressed in Equation 8.103 can now be given explicitly. The weights w_i in Equation 8.80 correspond to the scaling factors of the conditional probability density function given by $\pm p_{j,i}^\pm (2\pi\sigma_\eta^2)^{-m/2}$ in Equation 8.103. This was mentioned before in the context of Equation 8.82. The RBF centres $\mathbf{c}_i$ in Equation 8.80 correspond to the noise-free channel output vectors $\mathbf{r}_{j,i}^+$ and $\mathbf{r}_{j,i}^-$. That is, if the n-element feedback symbol sequence $\tilde{\mathbf{I}}_{feedback,k-\tau}$ obtained is equivalent to $\mathbf{s}_{f,j}$, we assigned the $n_{s,j}$ number of RBF centres $\mathbf{c}_i, i = 1, \dots, n_{s,j}$, to the channel output vector $\mathbf{r}_{j,i}^\pm, i = 1, \dots, n_{s,j}^\pm$. The decision feedback reduces the computational complexity of the RBF equaliser, since the number of RBF hidden nodes needed to realize the Bayesian equalization solution of Equation 8.17 is reduced from $n_s = 2^{m+L}$ to $n_{s,j} = n_s/n_f = 2^{m+L}/2^n = 2^{m+L-n}$ with the knowledge of the feedback state value. However, when the equalizer makes an incorrect decision and this decision is fed back, the wrong subset of centres is selected and this will degrade the BER performance of the RBF DFE, as it will be demonstrated in Section 8.12.

Extending the decision feedback RBF equalizer to a multilevel modem scenario is straightforward by introducing sub-RBF networks for each possible decision variable based on the conditional probability density function, as it was described in Section 8.9.2. The conditional Bayesian decision variable of Equation 8.85 can be redefined for the Bayesian DFE as:

$$
\begin{aligned}
\zeta_i(k) &= P(\mathbf{v}_k | I_{k-\tau} = \mathcal{I}_i \cap \tilde{\mathbf{I}}_{feedback,k-\tau} = \mathbf{s}_{f,l}) \\
&= \sum_{l=1}^{n_{s,j}^i} p_{j,l}^i p(\mathbf{v}_k - \mathbf{r}_{j,l}^i), \\
&\qquad 1 \leq i \leq \mathcal{M}, \\
&\qquad 1 \leq j \leq n_f, \tag{8.105}
\end{aligned}
$$

where $\mathbf{r}_{j,l}^i$ is the lth noiseless channel state, $l = 1, \dots, n_{s,j}^i$ when the feedback vector is given by $\tilde{\mathbf{I}}_{feedback,k-\tau} = \mathbf{s}_{f,j}$ and the transmitted symbol is $I_{k-\tau} = \mathcal{I}_i$, i.e., $\mathbf{r}_{j,l}^i \in V_{m,\tau,j}^i$.

Determine the feedback state	
$2n_{s,j}m - \mathcal{M}$	subtraction and addition
$n_{s,j}(m+1)$	multiplication
$n_{s,j}$	division
$n_{s,j}$	exp()
1	max evaluation

Table 8.10: Computational complexity of a decision feedback RBF network equalizer with m inputs and $n_{s,j}$ hidden units per equalised output sample based on Equation 8.103.

The computational complexity of the decision feedback assisted RBF equaliser is given in Table 8.10 based on Equation 8.103, which is similar to that without decision feedback given in Table 8.5, except for the reduced number of hidden units $n_{s,j} \leq \mathcal{M}^{L+m}$. We conclude that in general, the complexity increase of the RBF DFE is of the order of $\mathcal{M}^L$, since as $n_{s,j} = \mathcal{M}^{m+L-n}$. Hence, its application is typically restricted to low-order $\mathcal{M}$-ary modulation schemes, such as 4-QAM and to channels, where the ISI does not extend beyond four or five symbol periods [248].

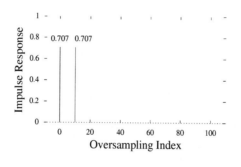

(a) Two-path channel

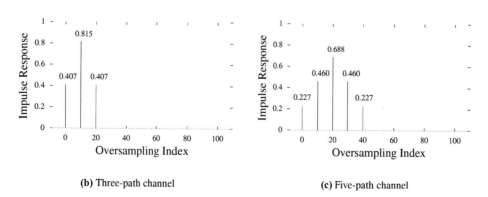

(b) Three-path channel

(c) Five-path channel

Figure 8.21: Four discrete time channel impulse responses for an oversampling ratio of 10.

The oldest symbol that influences the decision at the kth signalling instant, which produces the detected symbol $\tilde{I}_{k-\tau}$, is $I_{k-m+1-L}$, as seen in Equation 8.5. The oldest feedback
symbol is $\tilde{I}_{k-\tau-n}$. Therefore, it is sufficient to employ a feedback order of

$$n = \log_2 n_f = L + m - 1 - \tau, \tag{8.106}$$

because this will enable us to influence decisions over the memory duration L of the concatenated channel and the feedforward RBF section m. Assuming hence $n = L+m-1-\tau$, Chen,
Mulgrew and McLaughlin [246] mathematically proved that the Bayesian DFE of a feedforward order of $m = \tau + 1$ has the same conditional decision variables as those having a feedforward order of $m > \tau + 1$. The mathematical proof is given in Appendix A.3. Thus, given
the delay τ – which was defined in the context of Figure 8.18 as the total decision delay of the
feedforward shift-register of the RBF DFE – the feedforward order $m = \tau + 1$ is sufficient
for attaining the best possible BER at the lowest possible complexity [246]. This is demonstrated in Figure 8.22 over the two-path channel environment of Figure 8.21(a). Substituting
$m = \tau + 1$ in Equation 8.106 gives the corresponding feedback order of $n = \log_2 n_f = L$.
Overall, the equaliser delay τ specifies the number of channel states n_s, required for computing the decision variables and thus determines the computational complexity encountered. A
pragmatic rule is to set the equaliser's decision delay to $\tau = L$ [246]. However, note that
increasing the decision delay τ and feedforward order m will improve the performance of
the RBF equaliser, as demonstrated in Figure 8.23, at the expense of increasing the computational complexity exponentially, since the number of desired channel states $n_s = \mathcal{M}^{L+m}$
increases exponentially with m. Figure 8.24 shows the equaliser's BER performance versus
its feedforward order m. The BER performance improves almost linearly with the feedforward order, before the curves reach their SNR-dependent residual BERs. The effect of
increasing the feedforward order is more significant in BER-reduction terms at high E_b/N_0
values, as shown in Figure 8.24. For example, at an SNR of 12dB, an increase of the feedforward order from $m = 3$ to $m = 6$ improves the equaliser's BER performance by an order of
magnitude, from 10^{-5} to 10^{-6}.

In the next subsection we shall further illustrate the concept of the feedback states and the
redefined noiseless channel states using the same example as in Section 8.2.

8.11.1   Radial Basis Function Decision Feedback equalizer Example

The channel impulse response used in this example was given by Equation 8.12, which is
repeated here for convenience: $F(z) = 1 + 0.5z^{-1}$, implying that we have $L = 1$. We use
the following equaliser parameters:

- Feedforward order of $m = 2$.

- Feedback order of $n = 1$.

- Decision delay of $\tau = 1$.

Thus, in Figure 8.20 we have $\tilde{\mathbf{I}}_{feedback,k-\tau} = \left[\tilde{I}_{k-2} \right]$, since the feedforward section delays
the received signal by two sampling interval durations. Furthermore, $\tilde{\mathbf{v}}_k = \begin{bmatrix} \tilde{v}_k & \tilde{v}_{k-1} \end{bmatrix}^T$
and the delayed transmitted symbol is I_{k-1}. The number of noise-free channel output states

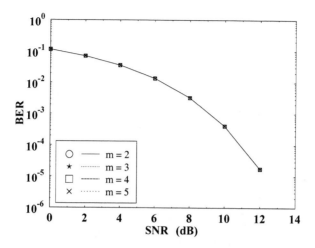

Figure 8.22: BER versus SNR performance of the RBF equalizer with correct decision feedback upon varying the feedforward order m over the dispersive two-path Gaussian channel of Figure 8.21(a). The equalizer decision delay τ was fixed to 1 symbol and the feedback order n was varied according to $n = L + m - 1 - \tau$, where $L + 1 = 2$ is the CIR length.

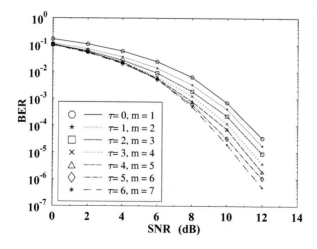

Figure 8.23: BER versus SNR performance of the RBF equalizer with correct decision feedback upon varying the decision delay τ over the dispersive two-path Gaussian channel of Figure 8.21(a). The equalizer feedforward order m was fixed to $m = \tau + 1$ and the feedback order n was varied according to $n = L + m - 1 - \tau$, where $L + 1 = 2$ is the CIR length. The equaliser's complexity increases exponentially with m, as seen in Table 8.10.

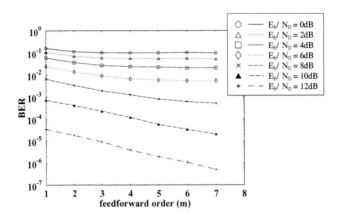

Figure 8.24: BER versus feedforward order m of the RBF equalizer with correct decision feedback for varying E_b/N_0 values over the dispersive two-path Gaussian channel of Figure 8.21(a). The equalizer decision delay τ was varied according to $\tau = m - 1$ and the equalizer complexity increases exponentially with m, as seen in Table 8.10.

is $n_s = 2^{m+L} = 2^{2+1} = 8$ in Figure 8.25, where $n_s^+ = 4$ and $n_s^- = 4$, the number of feedback states is $n_f = 2^n = 2$, while the number of subset channel states associated with the $n_f = 2^n = 2$ feedback states is $n_{s,j}^+ = 2$ and $n_{s,j}^- = 2$. We denote the feedback states $s_{f,j}$, where $j = 1, 2$ as $s_{f,1} = [-1]$ and $s_{f,2} = [+1]$. Assuming that the feedback symbols are correct, all the combinations of the transmitted binary symbols I_k, I_{k-1} and I_{k-2} as well as the noiseless channel outputs $\tilde{v}_k$, $\tilde{v}_{k-1}$, the noiseless channel output states $r_{i,j}^+$ and $r_{i,j}^-$ and the feedback states $s_{f,j}$ are listed in Table 8.11. Again, Figure 8.25 shows the noiseless channel output states observed by an equaliser having a feedforward order of $m = 2$ and decision delay of $\tau = 1$. Figure 8.26(a) and Figure 8.26(b) show the noiseless channel output states of the RBF DFE using the parameters given above, when the feedback state $s_{f,j}$ is equivalent to -1 and +1, respectively, as stated in Table 8.11. Following the spirit of Figure 8.3 in partitioning the decision space, at this stage we have to decide, what the transmitted bit I_{k-2} was. This decision can be carried out by evaluating Equation 8.105 and identifying the symbol $\mathcal{I}_i$, $i = 1, \ldots, \mathcal{M}$ associated with the highest probability.

Note that the number of channel states required, in order to estimate the transmitted symbol $I_{k-\tau}$ is now reduced from $n_s = 2^{m+L}$ to $n_{s,j} = n_s/n_f = 2^{m+L}/2^n = 2^{m+L-n}$, if we invoke the feedback state $s_{f,j}$ in order to assist in the RBF subset selection. Explicitly, in the example given above the number of channel states is reduced from 8 to 4, given the information of the feedback symbols. The computational complexity reduction factor owing to decision feedback is actually higher than n_f, since a DFE typically requires a reduced feedforward order m with respect to that, which is required without decision feedback. This is justified by the following arguments. Increasing the number of feedforward taps m extends the dimensionality of the observation space. This is necessary, in order to be able to increase the Euclidean distance between the RBF centres and thus to decrease the probability of mis-classification. It is apparent that the minimum distance amongst the constellation

I_k	I_{k-1}	I_{k-2}	$\tilde{v}_k$	$\tilde{v}_{k-1}$	$\mathbf{s}_{f,j}$	$\mathbf{r}_{j,i}$
-1	-1	-1	-1.5	-1.5	$\mathbf{s}_{f,1} = -1$	$\mathbf{r}_{1,1}^-$
+1	-1	-1	+0.5	-1.5	$\mathbf{s}_{f,1} = -1$	$\mathbf{r}_{2,1}^-$
-1	+1	-1	-0.5	+0.5	$\mathbf{s}_{f,1} = -1$	$\mathbf{r}_{1,1}^+$
+1	+1	-1	+1.5	+0.5	$\mathbf{s}_{f,1} = -1$	$\mathbf{r}_{2,1}^+$
-1	-1	+1	-1.5	-0.5	$\mathbf{s}_{f,2} = +1$	$\mathbf{r}_{1,2}^-$
+1	-1	+1	+0.5	-0.5	$\mathbf{s}_{f,2} = +1$	$\mathbf{r}_{2,2}^-$
-1	+1	+1	-0.5	+1.5	$\mathbf{s}_{f,2} = +1$	$\mathbf{r}_{1,2}^+$
+1	+1	+1	+1.5	+1.5	$\mathbf{s}_{f,2} = +1$	$\mathbf{r}_{2,2}^+$

Table 8.11: Transmitted signal I_k, I_{k-1}, I_{k-2}, noiseless channel output $\tilde{v}_k$, $\tilde{v}_{k-1}$, feedback channel states $\mathbf{s}_{f,j}$ and noiseless channel states $\mathbf{r}_{j,i}$ for the channel impulse response of $F(z) = 1 + 0.5z^{-1}$ and equaliser feedforward order of $m = 2$, feedback order of $n = 1$ and decision delay of $\tau = 1$ symbol. The coordinates $\tilde{v}_k$ and $\tilde{v}_{k-1}$ identify the points $\mathbf{r}_{j,i}$ in Figure 8.25 and 8.26. This table is the extension of Table 8.1, where the entries were rearranged appropriately, in order to separate the entries assosiated with $s_{f,1} = -1$ and $s_{f,2} = +1$.

points of the subsets $V_{m,\tau,j}^+$ and $V_{m,\tau,j}^-$ of Figure 8.26 for a particular feedback state $\mathbf{s}_{f,j}$, is larger than amongst the points of the full subsets $V_{m,\tau}^+$ and $V_{m,\tau}^-$ of Figure 8.25. Thus, with the introduction of decision feedback, the Euclidean distance between the centres is already increased and hence a smaller m is sufficient for maintaining a given equalizer performance. Again, the increased Euclidean distance can be observed by comparing the noiseless channel outputs $\tilde{v}_k$, $\tilde{v}_{k-1}$ in Figure 8.25 and those in Figure 8.26. The distance between a constellation point or state corresponding to the transmitted symbol $I_k = +1$ and the nearest point or state corresponding to the transmitted symbol $I_k = -1$ is increased, when the DFE scheme is used. Another important advantage of the decision feedback method is that the noiseless channel states $\mathbf{r}_{j,i}$ corresponding to different transmitted symbols are linearly separable, provided that the parameters of the RBF DFE are chosen to be $\tau = L$, $m = \tau + 1 = L + 1$ and $n = L + m - \tau - 1 = L$, which was proven mathematically by Chen, Mulgrew, Chng and Gibson [290] for a PAM modulation scheme. This proof can be readily extended to a QAM scheme. It should be emphasized that even though the noiseless channel states are linearly separable for the conditions stated above for the equaliser's parameters, the optimal decision boundary will generally be nonlinear. However, the linear separability is a highly desired property to have, since the equalization performance in this case is generally significantly better, than that of the nonlinearly separable case [291]. Note that the noiseless channel states $\mathbf{r}_i$ in the equaliser's observation space can be inseparable, as it will be demonstrated in Section 8.12.1.

8.11.2 Space Translation Properties of the Decision Feedback

In this section we provide a brief discourse on a technique, which can be used to reduce the number of states to be stored by the equaliser and also to eliminate the selection of the subset of states corresponding to the feedback symbol I_{k-2} in the example of Figure 8.26. In general, when $\tau > 1$, several feedback symbols influence the number of feedback states and

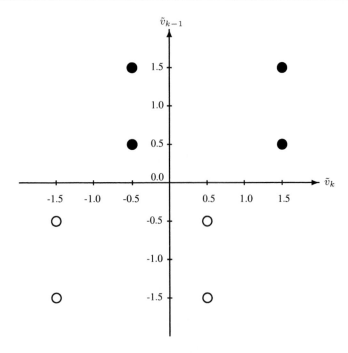

Figure 8.25: The noiseless channel states $\tilde{v}_k$, $\tilde{v}_{k-1}$ of a channel having a CIR of $F(z) = 1 + 0.5z^{-1}$ in a two-dimensional observation space. The filled circles represent the channel states in the set $V_{m,\tau}^{+1}$ corresponding to the transmitted symbol of $I_{k-\tau} = +1$ and the hollow circles represent the channel states in the set $V_{m,\tau}^{-1}$ corresponding to the transmitted symbol of $I_{k-\tau} = -1$, where the feedforward order is $m = 2$ and the decision delay of the equaliser is $\tau = 1$.

hence the associated storage and complexity reduction may be significant.

For a particular feedback state $\mathbf{s}_{f,j}$ characterized by the specific symbols $\tilde{I}_k$ in the feedback register, the subsets $V_{m,\tau,j}^+$ and $V_{m,\tau,j}^-$ are related to the subsets $V_{m,\tau,l}^+$ and $V_{m,\tau,l}^-$ having consecutive feedback states of $\mathbf{s}_{f,j}$ and $\mathbf{s}_{f,l}$, respectively, by a linear transformation. This can be shown mathematically as follows. Upon rewriting Equation 8.10, in order to take into account the decision feedback state in the expression of the noisy channel output and assuming $n \leq L$ and $m = \tau + 1$, gives:

$$\mathbf{v}_k = \mathbf{F}\mathbf{I}_k + \boldsymbol{\eta}_k, \tag{8.107}$$

where $\boldsymbol{\eta}_k = \begin{bmatrix} \eta_k & \cdots & \eta_{k-m+1} \end{bmatrix}^T$. The transmitted symbols influencing $\mathbf{v}_k$ can be divided in three classes as follows: $\mathbf{I}_k = \begin{bmatrix} \mathbf{I}_{1,k}^T & \mathbf{I}_{2,k}^T & \mathbf{I}_{3,k}^T \end{bmatrix}^T$, where $\mathbf{I}_{1,k}$ indicates those symbols, which reside in the feedforward shift register, $\mathbf{I}_{2,k}$ denotes those in the feedback register and $\mathbf{I}_{3,k}$ consists of the rest of the symbols that influence $\mathbf{v}_k$ but are left out by the

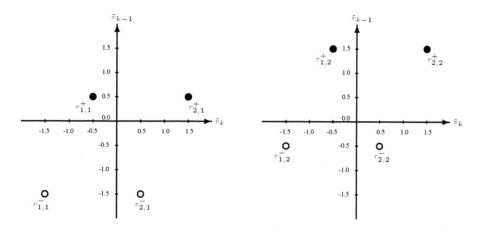

(a) The detected feedback symbol I_{k-2} is equivalent to **-1**

(b) The detected feedback symbol I_{k-2} is equivalent to **+1**

Figure 8.26: The noiseless channel states $\tilde{v}_k$, $\tilde{v}_{k-1}$ observed by the RBF DFE with feedforward order of $m = 2$, feedback order of $n = 1$ and decision delay of $\tau = 1$ symbol, assuming that the feedback symbols are correct. The channel has a CIR of $F(z) = 1 + 0.5z^{-1}$. The filled circles represent the channel states in the set $V^{+1}_{m,\tau,j}$ corresponding to the transmitted symbol of $I_{k-\tau} = +1$ and the hollow circles represent the channel states in the set $V^{-1}_{m,\tau,j}$ corresponding to the transmitted symbol of $I_{k-\tau} = -1$. Again, our final decision concerning the transmitted bit I_{k-2} is based on identifying the symbol $\mathcal{I}_i$, $i = 1, \ldots, \mathcal{M}$ associated with the highest probability.

DFE. These symbols can be written as:

$$
\left.
\begin{aligned}
\mathbf{I}_{1,k} &= \begin{bmatrix} I_k & \cdots & I_{k-\tau} \end{bmatrix}^T \\
\mathbf{I}_{2,k} &= \begin{bmatrix} I_{k-\tau-1} & \cdots & I_{k-\tau-n} \end{bmatrix}^T \\
\mathbf{I}_{3,k} &= \begin{bmatrix} I_{k-\tau-n-1} & \cdots & I_{k-m-L+1} \end{bmatrix}^T
\end{aligned}
\right\}.
\tag{8.108}
$$

Furthermore, the $m \times (m + L)$ CIR-related matrix $\mathbf{F}$ has the form

$$
\mathbf{F} = \begin{bmatrix} \mathbf{F}_1 & \mathbf{F}_2 & \mathbf{F}_3 \end{bmatrix},
\tag{8.109}
$$

with the $m \times (\tau + 1)$ matrix $\mathbf{F}_1$, $m \times n$ matrix $\mathbf{F}_2$ and $m \times (m + L - n - \tau - 1)$ matrix $\mathbf{F}_3$

defined by

$$
\mathbf{F}_1 = \begin{bmatrix} f_0 & f_1 & \cdots & f_\tau \\ 0 & f_0 & \ddots & \vdots \\ \vdots & \ddots & \ddots & f_1 \\ 0 & \cdots & 0 & f_0 \end{bmatrix}, \tag{8.110}
$$

$$
\mathbf{F}_2 = \begin{bmatrix} f_{\tau+1} & f_{\tau+2} & \cdots & f_{\tau+n} \\ f_\tau & f_{\tau+1} & \ddots & \vdots \\ \vdots & \ddots & \ddots & f_{n+1} \\ f_1 & \cdots & f_{n-1} & f_n \end{bmatrix}, \tag{8.111}
$$

$$
\mathbf{F}_3 = \begin{bmatrix} f_{\tau+n+1} & f_{\tau+n+2} & \cdots & f_{m+L-1} \\ f_{\tau+n} & f_{\tau+n+1} & \ddots & \vdots \\ \vdots & \ddots & \ddots & f_{L+1} \\ f_{n+1} & \cdots & f_{L-1} & f_L \end{bmatrix}, \tag{8.112}
$$

where $f_i = 0$ for $i < L$. Explicitly, $\mathbf{F}_1$ hosts those CIR taps, which affect the feedforward section, symbols contained by $\mathbf{I}_{1,k}$, $\mathbf{F}_2$ encompasses those, which weight the feedback symbols $\mathbf{I}_{2,k}$, while $\mathbf{F}_3$ contains the symbols not considered by the DFE. Under the assumption that the feedback symbol is correct, that is $\tilde{\mathbf{I}}_{feedback,k-\tau} = \mathbf{I}_{2,k}$ and based on Equation 8.107, the noise-free channel output vector of Equation 8.6 can be rewritten as

$$
\begin{aligned}
\tilde{\mathbf{v}}_k &= \mathbf{F}_1 \mathbf{I}_{1,k} + \mathbf{F}_2 \tilde{\mathbf{I}}_{feedback,k-\tau} + \mathbf{F}_3 \mathbf{I}_{3,k} \\
&= \tilde{\mathbf{v}}'_k + \mathbf{F}_2 \tilde{\mathbf{I}}_{feedback,k-\tau},
\end{aligned} \tag{8.113}
$$

where we introduced

$$
\tilde{\mathbf{v}}'_k = \mathbf{F}_1 \mathbf{I}_{1,k} + \mathbf{F}_3 \mathbf{I}_{3,k}. \tag{8.114}
$$

Thus the linear transformation between the consecutive noise-free channel output vectors of $\tilde{\mathbf{v}}_k$ and $\tilde{\mathbf{v}}'_k$ is provided by the term $\mathbf{F}_2 \tilde{\mathbf{I}}_{feedback,k-\tau}$ in Equation 8.113.

Using the CIR of Equation 8.12 we have $f_{\tau+1} = 0, f_1 = 0.5, f_0 = 1$ in Equations 8.110 – 8.112, yielding $\mathbf{F}_1 = \begin{bmatrix} 1 & 0.5 \\ 0 & 1 \end{bmatrix}$, $\mathbf{F}_2 = \begin{bmatrix} 0 \\ 0.5 \end{bmatrix}$ and $\mathbf{F}_3 = 0$. Assuming $I_k = +1$ and $I_{k-1} = +1$, we have $\mathbf{I}_{1,k} = \begin{bmatrix} 1 \\ 1 \end{bmatrix}$ and evaluating Equation 8.114 gives:

$$
\tilde{\mathbf{v}}'_k = \begin{bmatrix} 1 & 0.5 \\ 0 & 1 \end{bmatrix} \begin{bmatrix} 1 \\ 1 \end{bmatrix} + 0 = \begin{bmatrix} 1.5 \\ 1 \end{bmatrix}. \tag{8.115}
$$

Hence from Equation 8.113, for the specific feedback state of $\mathbf{s}_{f,1} = \tilde{\mathbf{I}}_{feedback,k-\tau} = [-1]$ the noiseless channel state is given by

$$
\mathbf{r}_{2,1}^+ = \begin{bmatrix} 1.5 \\ 1 \end{bmatrix} - \begin{bmatrix} 0 \\ -0.5 \end{bmatrix} = \begin{bmatrix} 1.5 \\ 0.5 \end{bmatrix}, \tag{8.116}
$$

while for the feedback state of $\mathbf{s}_{f,2} = \tilde{\mathbf{I}}_{feedback,k-\tau} = [+1]$ the noiseless channel state is given by

$$\mathbf{r}_{2,2}^{+} = \begin{bmatrix} 1.5 \\ 1 \end{bmatrix} - \begin{bmatrix} 0 \\ 0.5 \end{bmatrix} = \begin{bmatrix} 1.5 \\ 1.5 \end{bmatrix}, \tag{8.117}$$

as seen in Table 8.11.

We note that the linear transformation of Equation 8.113 between the consecutive noise-free channel outputs of $\tilde{v}_k$ and $\tilde{v}'_k$ depends on the feedback states $\mathbf{s}_{f,j}$ and the CIR. The geometric distance amongst the corresponding points of the set of $V_{m,\tau,j}^{+}$ and of the set $V_{m,\tau,j}^{-}$ for the same feedback state $\mathbf{s}_{f,j}$ is not altered by the transformation. Using the example in Figure 8.26, the geometric distance between the points $\mathbf{r}_{2,1}^{+}$ and $\mathbf{r}_{2,1}^{-}$ corresponding to the feedback symbol $I_{k-2} = -1$, is equivalent to the geometric distance between the points $\mathbf{r}_{2,2}^{+}$ and $\mathbf{r}_{2,2}^{-}$ corresponding to the feedback symbol $I_{k-2} = -1$. Thus it is sufficient to consider just one particular feedback state, when examining the Symbol Error Rate (SER) performance.

Previous research [290, 292] pointed out furthermore that the elements of $\tilde{v}'_k$ can be computed recursively. The ith element of $\tilde{v}'_k$, where, $i = m - 1, \ldots, 2, 1$, can be represented by its unit delayed version as follows:

$$v_{k-i} = z^{-1} v_{k+1-i} \qquad i = m - 1, \ldots, 2, 1, \tag{8.118}$$

where z^{-1} is the unit-delay operator. From Equation 8.113, the ith and $(i-1)$th elements of $\tilde{v}'_k$ can be written as

$$v_{k-i} = v'_{k-i} + \sum_{j=1}^{n} f_{m+j-1-i} I_{k-\tau-j} \tag{8.119}$$

$$v_{k-i+1} = v'_{k-i+1} + \sum_{j=1}^{n} f_{m+j-1-i+1} I_{k-\tau-j}. \tag{8.120}$$

Using Equation 8.118, 8.119 and 8.120, we have:

$$v_{k-i} = z^{-1} v_{k+1-i}$$

$$= z^{-1} \left(v'_{k-i+1} + \sum_{j=1}^{n} f_{m+j-1-i+1} I_{k-\tau-j} \right)$$

$$= z^{-1} v'_{k-i+1} + \sum_{j=1}^{n} f_{m+j-i} I_{k-1-\tau-j}$$

$$v'_{k-i} + \sum_{j=1}^{n} f_{m+j-i-1} I_{k-\tau-j} = z^{-1} v'_{k-i+1} + \sum_{j=1}^{n} f_{m+j-i} I_{k-1-\tau-j}$$

$$= z^{-1} v'_{k-i+1} + \sum_{j=2}^{n+1} f_{m+j-i-1} I_{k-\tau-j}$$

$$v'_{k-i} = z^{-1} v'_{k+1-i} - f_{m-i} I_{k-\tau-1} + f_{m-i+n} I_{k-\tau-n-1}$$
$$i = m - 1, \ldots, 2, 1. \tag{8.121}$$

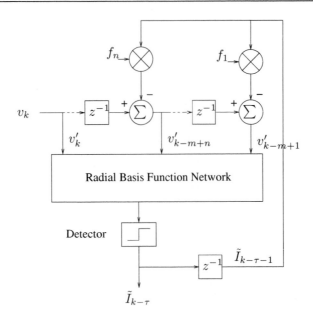

Figure 8.27: Space-translated RBF DFE.

Upon substituting $n = L$ into Equation 8.121 we arrive at:

$$v'_{k-i} = z^{-1}v'_{k+1-i} - f_{m-i}I_{k-\tau-1} \tag{8.122}$$

$$v'_k = v_k. \tag{8.123}$$

Based on this interpretation of decision feedback, an alternative DFE structure is depicted in Figure 8.27. This version of the space-translated RBF DFE realises the same optimal solution as the subset centre selection RBF DFE depicted in Figure 8.20. However, the space-translated RBF DFE of Figure 8.27 removes the requirement of different set of centres for different decision feedbacks and has hence a clear advantages in hardware implementational terms. The decision feedback 'merges' the channel states corresponding to different feedback states and hence the DFE of Figure 8.27 can be studied more conveniently in the translated $\mathbf{v}'$-space. [1]

8.12 Simulation Results

8.12.1 Performance of RBF Assisted Equalisers over Dispersive Gaussian Channels

In all our results presented in this section the transmitted symbols I_k were equiprobable binary symbols assuming values from the set $\{\pm 1\}$. Therefore the weights of the RBF network

---

[1]This property leads to the implementation of the so-called Minimum BER (MBER) DFE based on either a linear filter [290] or on the so-called support vector machine [293,294] proposed by Chen *et al.* that construct hyperplanes, which can separate the different signal classes.

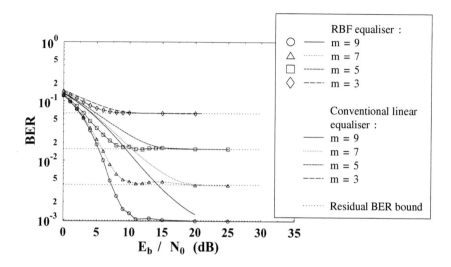

Figure 8.28: BER versus E_b/N_0 performance of the RBF equaliser using no decision feedback upon varying the number of equalizer taps m over the two-path Gaussian channel of Figure 8.21(a) using BPSK. The performace is compared to that of the linear MSE equaliser using m number of taps. The residual BER bound (= $\frac{1}{2^{m+L}}$, where $L+1$ is the CIR length) is shown for different values of m. The residual BER is due to the constellation points appearing on top of each other in Figure 8.30.

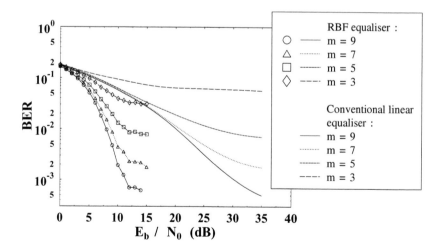

Figure 8.29: BER versus E_b/N_0 performance of the RBF equalizer using no decision feedback upon varying number of equalizer taps m over the three-path dispersive Gaussian channel of Figure 8.21(b) employing BPSK. The performace is compared to that of the linear MSE equaliser using m number of taps.

were fixed to $w_i = +1$, if the RBF centroids c_i seen in Figure 8.12 correspond to a positive channel state $\mathbf{v}_i^+$ and to $w_i = -1$, if the RBF centroids c_i correspond to a negative channel state $\mathbf{v}_i^-$, as explained in Section 8.9.1. The noise variance σ_η^2 was fixed to unity, while the power of the transmitted symbol was varied according to the SNR per bit, namely E_b/N_0. The transmitted symbol was oversampled by a factor of 10 and it was pulse-shaped. Both the transmitter and receiver had a square root Nyquist filter [4] with a roll-off factor of 0.5. The combined transfer function of these two filters produced a raised cosine filter and this design satisfies the Nyquist criterion of zero ISI at sampling instants.

Initially the centres of the RBF network were positioned at the desired channel states seen for example in Figure 8.3. The width of the RBF network ρ was set to $2\sigma_\eta^2$. We assumed that the CIR was known and the number of hidden nodes was set to $n_s = 2^{L+m}$. In practice the CIR can be estimated using channel sounding [124, 125] and using the estimated CIR would result in some performance degradation. The impulse responses of the channels used for the simulations were characterized by Figure 8.21(a) for the two-path channel, Figure 8.21(b) for the three-path channel and Figure 8.21(c) for the five-path channel.

The BER performance of the RBF network was compared with that of the linear MSE equalizer [280] (pp. 607-612). The tap weights of the linear MSE equalizer were set to obtain the best possible performance and both schemes used the same number of taps given by m. Figure 8.28 and Figure 8.29 show our BER performance comparison for the two-path channel and the three-path channel, respectively. The two-path results of Figure 8.28 show that for the same number of taps the RBF network equaliser provides superior performance in comparison to the linear MSE equaliser, before the residual BER is reached, above which the BER performance did not improve upon increasing E_b/N_0. Beyond this point the RBF network equalizer and the linear MSE equalizer have a similar BER performance.

This can be explained graphically by first observing the desired channel states in the channel observation space of Figure 8.30. For the two-path channel environment of Figure 8.21(a) and an equaliser having three taps, the desired channel states and a linear decision boundary surface provided by the linear MSE equaliser are shown in Figure 8.30. Note that the noiseless channel output $\tilde{\mathbf{v}} = \begin{bmatrix} 0 & 0 & 0 \end{bmatrix}^T$ due to the transmitted data sequence of $\{-1 + 1 - 1 + 1\}$ and $\{+1 - 1 + 1 - 1\}$ corresponds to both $I_{k-\tau} = +1$ and $I_{k-\tau} = -1$. Thus the channel states $\mathbf{r}^+$ and $\mathbf{r}^-$ are inseparable both linearly and nonlinearly at that point, even when the input dimension is increased. This provides the performance limitation manifested in terms of the residual BER for both the linear MSE equaliser and the RBF network equaliser. The value of the residual BER is dependent on the relative frequency of encountering this inseparable channel state scenario. For example, in the case of the two-path channel environment mentioned above, where there are two channel states corresponding to the noiseless channel output vector $\tilde{\mathbf{v}} = \begin{bmatrix} 0 & 0 & 0 \end{bmatrix}^T$ and both are classified as corresponding to $I_{k-\tau} = +1$ or $I_{k-\tau} = -1$, one channel state out of the total of n_s legitimate channel states will be classified wrongly, irrespective of E_b/N_0. Thus, the minimum achievable bit error rate will be $\frac{1}{n_s}$ for a particular equalizer order m. This explains the BER residual in Figure 8.28. The BER residual 'bound' of $\frac{1}{n_s}$ is also shown in Figure 8.28 using dashed lines for the various m values employed. The three-path results of Figure 8.29 also show superior performance in comparison to the linear MSE equaliser, before the residual BER is reached. Again, the residual BER 'bound' can be explained by the inseparable channel states.

The BER performance generally improves upon increasing the number of equaliser taps

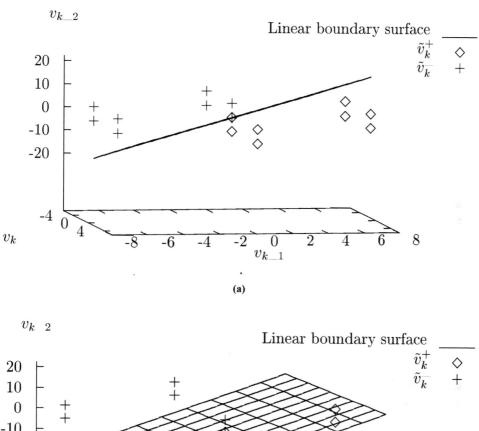

(a)

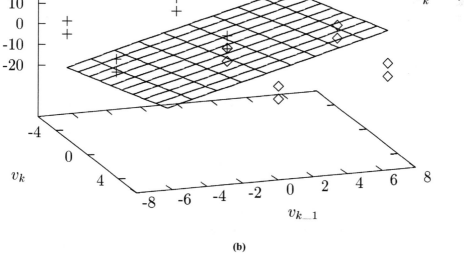

(b)

Figure 8.30: The desired channel states and a linear decision boundary surface provided by the linear MSE equalizer in a three-dimensional channel observation space corresponding to a three-tap equalizer and the CIR of $F(z) = 0.707 + 0.707z^{-1}$, viewed at different angles.

m, as does the 'bound' $\frac{1}{n_s}$. The dimension of the channel observation space that increases with increasing m has the effect of increasing the Euclidean distance between the desired channel states and therefore improves the separability between $\tilde{\mathbf{v}}^+$ and $\tilde{\mathbf{v}}^-$, but does not irradicate the ambiguity associated with $\tilde{\mathbf{v}} = \begin{bmatrix} 0 & 0 & 0 \end{bmatrix}^T$.

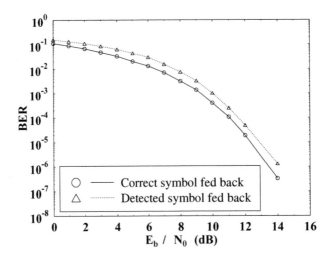

Figure 8.31: BER versus E_b/N_0 performance of the BPSK RBF equalizer with decision feedback over the dispersive **two-path** Gaussian channel of Figure 8.21(a). The equalizer has a feedforward order of $m = 2$, feedback order of $n = 1$ and decision delay of $\tau = 1$ symbol.

Figure 8.31, Figure 8.32 and Figure 8.33 show the BER performance of the RBF network equalizer in conjunction with decision feedback for the two-path channel, three-path channel and five-path channels of Figure 8.21, respectively. The equaliser feedforward order m is fixed to $\tau + 1$, while the feedback order was set to $n = L$, as described in Section 8.11. The results shows that the decision feedback structure not only decreases the computational complexity, since less taps and less hidden nodes are neccessary, it also substantially improves the BER performance. The residual BER is eliminated, since the desired states $\tilde{\mathbf{v}}^+$ and $\tilde{\mathbf{v}}^-$ that correspond to the same point in the channel observation space have now different feedback states and the set of noiseless channel states $V^+_{m,\tau,j}$ and $V^-_{m,\tau,j}$ are now separable. This confirms the findings by Chen, Mulgrew, Chng and Gibson [290] that the noiseless channel states corresponding to a different transmitted symbol are linearly separable, provided that the decision delay, feedforward section and feedback section length of the RBF DFE are chosen to be $\tau = L$, $m = \tau + 1 = L + 1$ and $n = L + m - \tau - 1 = L$.

Note furthermore that the error propagation due to erroneous decision feedback has a moderate effect on the performance of the BPSK RBF network equaliser, amounting to around 1dB performance degradation at BER $= 10^{-4}$ for all the three channels of Figure 8.21.

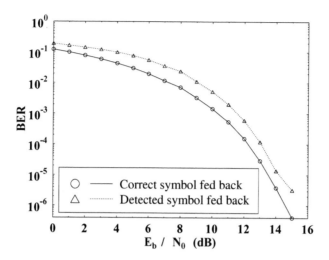

Figure 8.32: BER versus E_b/N_0 performance of the BPSK RBF equalizer with decision feedback over the dispersive **three-path** Gaussian channel of Figure 8.21(b). The equalizer has a feedforward order of $m = 3$, feedback order of $n = 2$ and decision delay of $\tau = 2$ symbols.

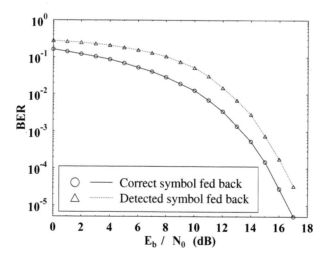

Figure 8.33: BER versus E_b/N_0 performance of the BPSK RBF equalizer with decision feedback over the dispersive **five-path** Gaussian channel of Figure 8.21(c). The equalizer has a feedforward order of $m = 5$, feedback order of $n = 4$ and decision delay of $\tau = 4$ symbols.

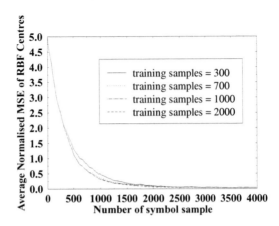

Figure 8.34: The MSE of the BPSK RBF equalizer centres versus transmitted symbol index for various numbers of training samples using the **vector centre clustering algorithm** of Section 8.9.5 over the two-path channel environment of Figure 8.21(a). The equalizer had $m = 5$ feedforward taps and a decision delay of $\tau = 2$ symbols. The centre learning rate μ_c of Equations 8.93 and 8.95 was set to 0.1 and the SNR was 10dB.

8.12.2   Performance of Adaptive RBF DFE

As our next endeavour, the adaptive performance of the RBF network equaliser employing the K-means clustering algorithm of Section 8.9.5 was investigated. Firstly, the average normalized MSE of the vector centres at signalling interval k was defined as:

$$\text{MSE}(\mathbf{c}, k) = \frac{1}{n_s \sigma_{\tilde{v}}^2} \sum_{i=1}^{n_s} \|\mathbf{c}_{i,k} - \mathbf{c}_{i,opt}\|^2, \qquad (8.124)$$

where n_s is the number of RBF centres, $\sigma_{\tilde{v}}^2$ is the variance of the noise-free received signal, $\mathbf{c}_{i,k}, i = 1, \ldots, n_s, k = 0, \ldots, \infty$ represents the ith assumed RBF centre at signalling interval k and $\mathbf{c}_{i,opt}$ is the vector associated with the ith desired or assumed 'true' RBF centre. The K-means clustering technique operates by iteratively adjusting the RBF centres upon every sampling instance according to Equation 8.93 during training mode, while Equation 8.95 is used during the decision-directed mode. The centres' MSE convergence performance is demonstrated over the two-path channel environment in Figure 8.34 using different number of training symbols, while in Figure 8.35 upon varying the learning rate μ_c of Equation 8.93 and 8.95. The results show a good convergence performance for our stationary two-path channel of Figure 8.21(a) upon invoking the decision-directed learning algorithm of Equation 8.95. However, further simulations have to be carried out, in order to investigate the effect of time-varying wideband mobile channels. As the learning rate μ_c is increased, the centres converge faster to their desired positions, but as expected, the MSE curves of the centres become more spurious, especially at low SNRs as we can see from Figure 8.35. Based on these results we recommend using a variable learning rate μ_c, where μ_c is set to a higher rate during the training mode so that the equalizer converges faster and is set to a lower rate during the decision-directed learning mode, in order to reduce the spuriosity of the centre

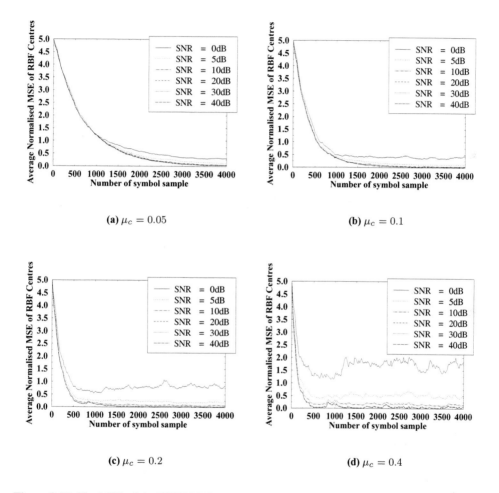

Figure 8.35: The MSE of the BPSK RBF equalizer centres versus transmitted symbol index for **various learning rates** μ_c of the centres using the **vector centre clustering algorithm** of Section 8.9.5 over the two-path channel environment of Figure 8.21(a). The equalizer had $m = 5$ feedforward taps and a decision delay of $\tau = 2$ symbols, and the number of training symbols was 700.

MSE.

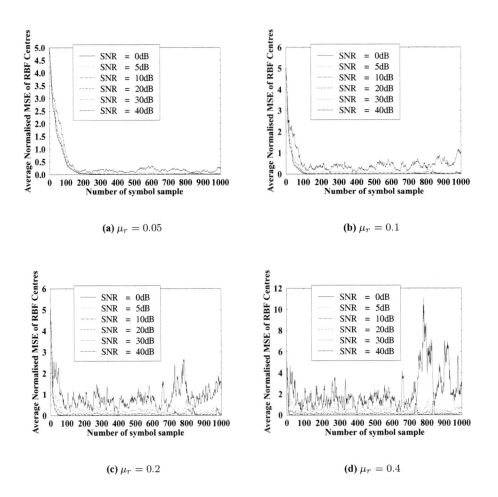

(a) $\mu_r = 0.05$ **(b)** $\mu_r = 0.1$

(c) $\mu_r = 0.2$ **(d)** $\mu_r = 0.4$

Figure 8.36: The MSE of the BPSK RBF equalizer centres versus transmitted symbol index for **various learning rates** μ_r of the centres using the **scalar centre clustering algorithm** of Section 8.10 with 700 training symbols over the two-path channel environment of Figure 8.21(a). The equalizer had $m = 5$ feedforward taps and a decision delay of $\tau = 2$ symbols.

The performance of the scalar centre clustering algorithm described in Section 8.10 is demonstrated over the same two-path Gaussian channel environment in Figure 8.36 and Figure 8.37. Comparing Figure 8.35 and Figure 8.36 using various centre learning rates μ_r, shows that the scalar centre clustering algorithm provides a significantly faster convergence rate, since the number of scalar centres $n_{s,f} = \mathcal{M}^{L+1} = 4$ ($\mathcal{M} = 2, L = 1$) is only dependent on the number of symbol constellation points, $\mathcal{M}$, and on the CIR length $L + 1$. Hence the number of scalar centres is significantly less than the number of vector centres given by $n_s = \mathcal{M}^{L+m} = 64$ ($m = 5$), which is additionally dependent on the equalizer order m as

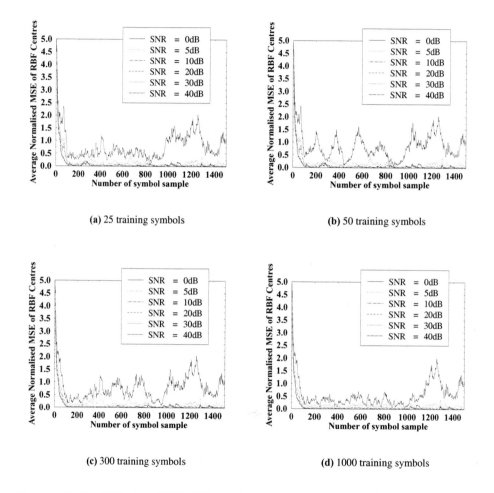

(a) 25 training symbols

(b) 50 training symbols

(c) 300 training symbols

(d) 1000 training symbols

Figure 8.37: The MSE of the BPSK RBF equalizer centres versus transmitted symbol index for **various number of training samples** using the **scalar centres clustering algorithm** of Section 8.10 over the two-path channel environment of Figure 8.21(a). The equalizer had $m = 5$ feedforward taps and a decision delay of $\tau = 2$ symbols. The centre learning rate μ_r was set to 0.1.

well. However, the MSE learning curves of the centres are more spurious in conjunction with the scalar centre clustering algorithm, since the value of a scalar centre affects the value of a few vector centres that contain that particular scalar centre and thus the estimation error of a scalar centre will be magnified, when we examine the average normalized MSE of the vector centres in Figure 8.36. Figure 8.37 shows the average normalized MSE of the RBF centres for a varying number of training symbols. Note that the algorithm still converges during the decision-directed mode, although the MSE curve behaves more spuriously during this mode compared to the learning phase, especially at low SNRs.

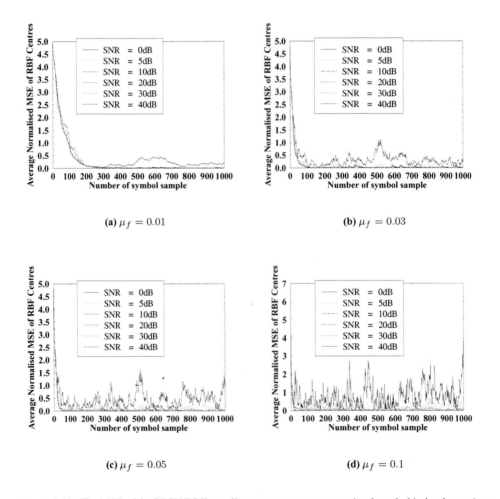

Figure 8.38: The MSE of the BPSK RBF equalizer centres versus transmitted symbol index for **various learning rates** μ_f using the **LMS channel estimator technique** of Section 8.9.4 with 300 training symbols over the two-path channel environment of Figure 8.21(a). The equalizer had $m = 5$ feedforward taps and a decision delay of $\tau = 2$ symbols.

The centres' MSE convergence performance for the channel estimation method using the LMS algorithm described in Section 8.9.4 is demonstrated in Figures 8.38 and 8.39. Compar-

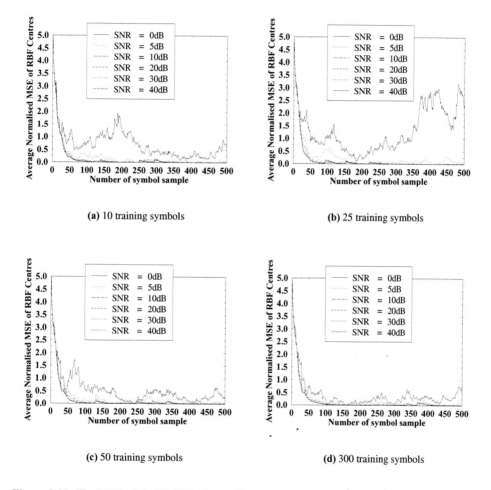

(a) 10 training symbols

(b) 25 training symbols

(c) 50 training symbols

(d) 300 training symbols

Figure 8.39: The MSE of the BPSK RBF equalizer centres versus transmitted symbol index for **various number of training samples** using the **LMS channel estimator technique** of Section 8.9.4 over the two-path channel environment of Figure 8.21(a). The equalizer had $m = 5$ feedforward taps and a decision delay of $\tau = 2$ symbols. The channel estimator learning rate μ_f was set to 0.03.

ing Figure 8.36 and 8.38 using varying learning rates shows that the LMS channel estimator technique provides faster convergence rate at a given learning rate, since the number of CIR coefficients that were adapted according to Equation 8.89 by the LMS channel estimator is less than the number of scalar centres of the scalar clustering algorithm. However, if we compare Figure 8.36(a), (b), (c) and (d) with Figure 8.38(a), (b), (c) and (d), respectively, they show rather similar convergence rates during the training mode since the number of scalar centres ($2^{L+1} = 4$) is not too high compared to the number of channel coefficients ($L + 1 = 2$) to be learnt adaptively. Figure 8.39 shows the average normalized MSE of the RBF centres for varying number of training symbols. Again, the LMS channel estimator technique of Section 8.9.4 still converges during the decision-directed mode.

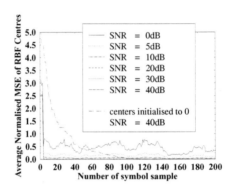

Figure 8.40: The MSE of the BPSK RBF equalizer centres versus transmitted symbol index using the **scalar centre clustering algorithm** of Section 8.10 with **centres initialised to the corresponding noisy channel states**, as described by Equation 8.96 over the two-path channel environment of Figure 8.21(a). The MSE of the BPSK RBF equalizer centres learnt using the scalar centre clustering algorithm with centres initially set to 0 for SNR = 40dB is shown for comparison. The equalizer had $m = 5$ feedforward taps and a decision delay of $\tau = 2$ symbols. The learning rate μ_r was set to 0.1 and the number of training symbols is 200.

Figure 8.40 shows the centres' MSE convergence performance, when the scalar centres clustering algorithm was initialised with the corresponding noisy channel states, as described in Section 8.10 over the same two-path Gaussian channel environment. Comparing Figure 8.36(b) and Figure 8.40 reveals that the initialization to the corresponding noisy channel states significantly increases the convergence rate of the clustering algorithm at a low additional computational cost. The convergence rate is also seen to be faster than that of the LMS channel estimator technique when we compare Figure 8.38(d) with Figure 8.40, since the number of scalar centres ($2^{L+1} = 4$) is not significantly higher, than the number of channel coefficients ($L + 1 = 2$) to be learnt adaptively. Note that since the number of scalar centres ($\mathcal{M}^{L+1}$) increases exponentially with the length L of the CIR, the LMS channel estimation technique will have a better convergence rate for high order modulation scheme and high CIR lengths, than the scalar centre clustering algorithm using the above-mentioned initialization to the noisy centres.

8.12.3 Performance of the RBF Equalizer for Square-QAM over Gaussian Channels

In this section the performance of the RBF equalizer is analysed in conjunction with multilevel modulation schemes in a Gaussian environment. We used square-shaped Quadrature Amplitude Modulation (QAM) constellations [4]. Figure 8.41 portrays the location of each constellation point in terms of their in-phase (I) and quadrature-phase (Q) components for 2-, 4-, 16- and 64-QAM. Each constellation point is assigned a bit sequence. Gray coding is applied to assign the bit sequences to their respective constellation points, ensuring that the nearest-neighbour constellation points had a Hamming distance of one. Therefore the assignment of constellation points is optimised in terms of minimising the BER. For a more in-depth understanding of QAM techniques, the interested reader is referred to [4].

We use a RBF DFE for multilevel modems as discussed in Section 8.9.2 and Section 8.11. Figure 8.42 shows the bit error rate performance for the 2-, 4-, 16- and 64-QAM schemes in conjunction with correct and detected symbol feed-back. The performance degradation due to decision errors is approximately 0.5dB for 2- and 4-QAM, 1dB for 16dB, 1.5dB for 64dB at BER = 10^{-4} and thus it has a moderate effect at low BERs. Note however the E_b/N_0 degradation increases, as the BER increases, which becomes more significant at higher order QAM.

Figure 8.43 shows the performance comparison between the conventional DFE and the RBF equalizer with decision feedback over the dispersive two-path Gaussian channel of Figure 8.21(a). The parameters of the conventional DFE were chosen such that it exhibited the best possible performance for our simulation scenario and hence a further increase of the feedforward order would not give a significant performance improvement. The conventional DFE used in our simulations had a feedforward order of $m = 7$, feedback order of $n = 1$ and decision delay of $\tau = 7$ symbols. The RBF equalizer using decision feedback was found to give a similar performance with a reduced feedforward order of 2, feedback order of 1 and decision delay of 1 symbol. The performance of the RBF assisted decision feedback equalizer can still be further improved quite significantly by increasing both the decision delay τ and the feedforward order m, as we discussed in Section 8.11 and this was demonstrated in Figure 8.23 for Binary Phase Shift Keying (BPSK).

8.12.4 Performance of the RBF Equalizer over Wideband Rayleigh Fading Channels

In this section we used $\mathcal{M}$-QAM symbols. The combined transfer function of the transmitter and receiver filters yielded a raised cosine filter with a roll-off factor of 0.5. The transmitter and receiver filters were identical and were implemented as finite-impulse-response (FIR) filters. The filter tap weights were samples of the truncated square-root-raised-cosine impulse response. The transmitted symbol was oversampled by a factor of 8 and it was pulse-shaped. The baseband time-invariant multipath fading channel was represented as follows:

$$c(t) = \sum_{i=0}^{n_c} f_i(t)\delta_{t-\tau_i(t)}, \qquad (8.125)$$

where n_c is the number of fading paths, $f_i(t)$ is the complex-valued ith CIR tap at time t, $\tau_i(t)$ is the excess delay at time t and δ_t is a delta function located at signalling instant

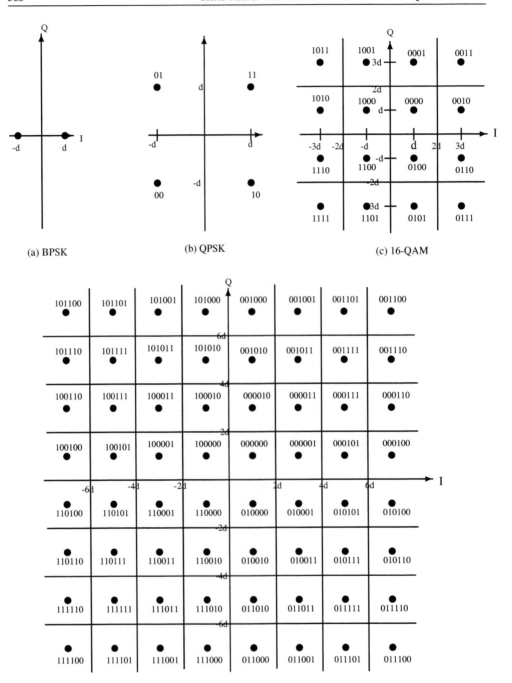

(a) BPSK

(b) QPSK

(c) 16-QAM

(d) 64-QAM

——— DECISION BOUNDARIES

Figure 8.41: QAM Phasor Constellations.

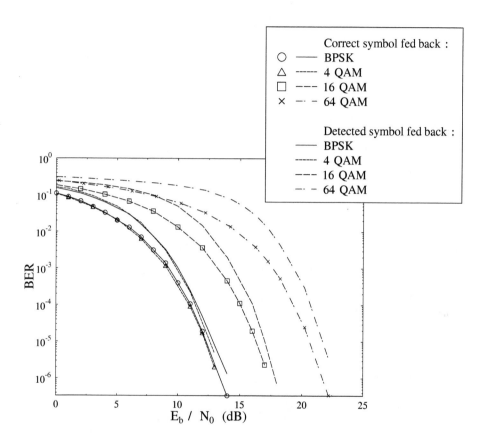

Figure 8.42: BER versus E_b/N_0 performance of the **RBF equaliser** using decision feedback over the dispersive two-path Gaussian channel for different $\mathcal{M}$-QAM schemes. The impulse response of the two-path channel is described by Figure 8.21(a). The equalizer had a feedforward order of $m = 2$, feedback order of $n = 1$ and decision delay of $\tau = 1$ symbol.

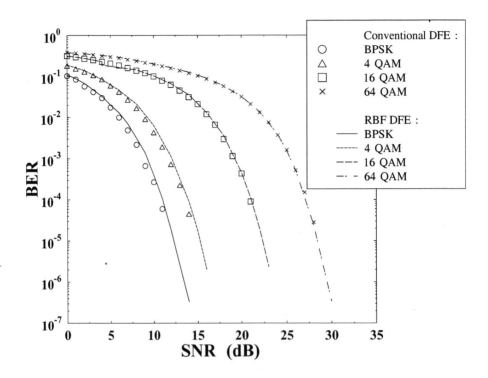

Figure 8.43: BER versus signal to noise ratio performance of the **RBF equaliser** using decision feedback and the **conventional DFE** over the dispersive two-path Gaussian channel for different $\mathcal{M}$-QAM schemes. The impulse response of the two-path channel is described by
Figure 8.21(a). The RBF equalizer had a feedforward order of $m = 2$, feedback order of
$n = 1$ and decision delay of $\tau = 1$ symbol. The conventional DFE had a feedforward
order of $m = 7$, feedback order of $n = 1$ and decision delay of $\tau = 7$ symbols. Correct
symbols were fed back.

t. The multipath components $f_i(t)$ have independent Rayleigh fading statistics, they are
uncorrelated and are scaled by their designed weights. For a more in-depth charaterization
of Rayleigh fading channels, the reader is referred to the tutorial by Sklar [295]. In our
simulations, the fading parameters of the channel are given in Table 8.12 and we employ two
symbol-spaced fading paths with the weights given by $0.707 + 0.707z^{-1}$. The structure of the
transmitted burst is given in Figure 8.44, where the training symbol sequence is implemented
as a preamble. In our simulations, the number of training symbols L_T was set to 27 and the
number of data symbols L_D was set to 144.

Figure 8.45 provides our BER performance comparison between the conventional DFE
and the RBF DFE for different $\mathcal{M}$-QAM schemes. The conventional DFE assumed perfect
channel estimation and its equalizer coefficients were optimised using the MSE criterion

Transmission Frequency	1800MHz
Transmission Rate	133kBds
Vehicular Speed	30 mph
Normalized Doppler Frequency	6×10^{-4}

Table 8.12: Simulation parameter of the Rayleigh fading channel.

Training Symbols	Data Symbols
$\longleftarrow L_T \longrightarrow$	$\longleftarrow \qquad\qquad L_D \qquad\qquad \longrightarrow$

Figure 8.44: Transmitted frame structure depicting the position of the data and training symbols. For example in the context of the COST 207 CIR of Figure 8.52 the number of training symbols L_T was 49 and the number of data symbols L_D was 122.

as described in [280] (pp. 607-612). The centres of the RBF DFE were positioned at the desired channel states. In these simulations the CIR taps were kept constant for the duration of the transmitted burst and were faded before the next burst, which we refer to as burst-invariant fading. From Figure 8.45 we note that for BPSK, the RBF DFE having a low feedforward order of $m = 2$, feedback order of $n = 1$ and decision delay of $\tau = 1$ symbol was found to give similar performance to the conventional DFE having a feedforward order of $m = 7$, feedback order of $n = 1$ and decision delay of $\tau = 7$ symbols. For 4-QAM, 16-QAM and 64-QAM, the RBF DFE having the same parameters gives inferior performance compared to the conventional DFE in the two-path Rayleigh fading channel scenario. This is dissimilar to the performance of the two-path Gaussian channel shown in Figure 8.43. The performance degradations endured by the higher order modulation schemes are higher under fading channel conditions even in conjunction with perfect channel estimation, since these schemes are more sensitive to fades due to their reduced Euclidean distance between their neighbouring constellation points. Nevertheless, the performance of the RBF DFE can be improved by increasing both the decision delay τ and the feedforward order m, as we discussed in Section 8.11, at the expense of increased computational complexity. This is demonstrated in Figure 8.46, where the performance of the RBF DFE having an increased decision delay of $\tau = 2$ and corresponding feedforward order of $m = 3$ and $n = 1$ showed an improved performance, attaining similar BER performance curves to the previously described conventional DFE for BPSK, 4-QAM and 16-QAM. The performance of 64-QAM is not shown here due to the associated high computational complexity of the simulation.

The adaptive performance of the RBF DFE was investigated over the two-path Rayleigh fading channel at a normalized Doppler frequency of 6×10^{-4} for the BPSK modulation scheme. In our adaptive RBF DFE simulations, we used a variable centre learning rate μ_r, where we had $\mu_r = 0.3$ during the training mode and $\mu_r = 0.1$ during the decision-directed learning mode. We assigned a sequence of L_T pseudo-random binary symbols as the training symbol sequence as seen in Figure 8.44. We note, however that we will have to find the symbol sequence that can give the best training performance. Figure 8.47 provides our performance comparison for the RBF DFE using perfect channel estimation and when the adaptive RBF DFE is trained with the aid of the scalar centre clustering algorithm described

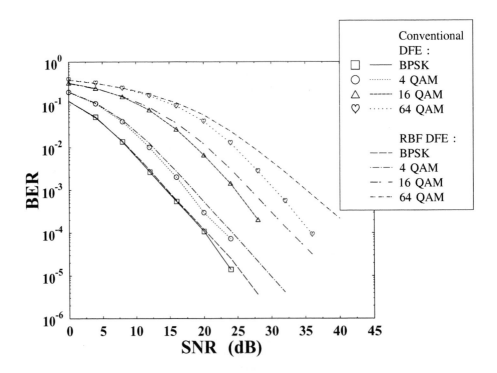

Figure 8.45: BER versus signal to noise ratio performance of the conventional DFE and the RBF equalizer with decision feedback over the two equal weight, symbol-spaced path Rayleigh fading channel of $F(z) = 0.707 + 0.707z^{-1}$ for different $\mathcal{M}$-QAM schemes. Both equalisers assume perfect CIR estimation. The conventional DFE had a feedforward order of $m = 7$, feedback order of $n = 1$ and decision delay of $\tau = 7$ symbols. The RBF DFE had a **feedforward order of** $m = 2$, feedback order of $n = 1$ and **decision delay of** $\tau = 1$ symbol. Correct symbols were fed back. The Rayleigh fading parameters are summarised in Table 8.12.

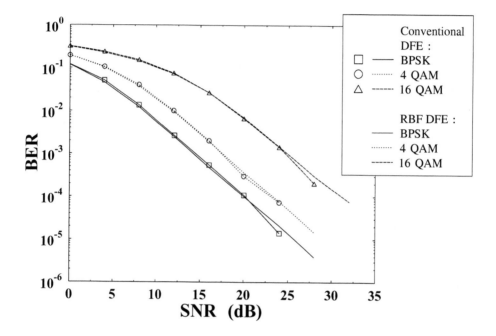

Figure 8.46: BER versus signal to noise ratio performance of the conventional DFE and the RBF equalizer with decision feedback over the two equal weight, symbol-spaced path Rayleigh fading channel of $F(z) = 0.707 + 0.707z^{-1}$ for different $\mathcal{M}$-QAM schemes. Both equalisers assume perfect CIR estimation. The conventional DFE had a feedforward order of $m = 7$, feedback order of $n = 1$ and decision delay of $\tau = 7$ symbols. The RBF DFE had a **feedforward order of** $m = 3$, feedback order of $n = 1$ and **decision delay of** $\tau = 2$ symbol. Correct symbols were fed back. The Rayleigh fading parameters are summarised in Table 8.12.

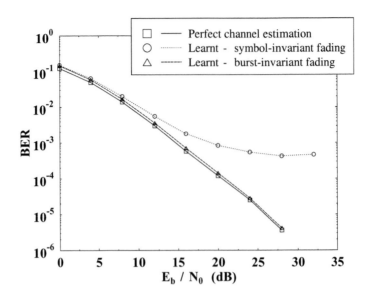

Figure 8.47: BER versus E_b/N_0 performance of the adaptive RBF DFE with correct decision fed back under burst-invariant fading and symbol-invariant fading. The RBF DFE is adapted using the scalar centre clustering algorithm described in Section 8.10. The performance of the RBF DFE using perfect CIR estimation is provided for comparison. The RBF DFE had a feedforward order of $m = 2$, feedback order of $n = 1$ and decision delay of $\tau = 1$ symbol. The Rayleigh fading parameters are summarised in Table 8.12.

in Section 8.10. Figure 8.47 shows that there is a high performance degradation due to the imperfect CIR knowledge and a residual BER is experienced in our simulations, when the wideband fading channel is symbol-invariant, as opposed to being burst-invariant, i.e. when it is kept invariant for only a symbol duration rather than for a burst duration. This phenomenon can be explained by comparing Figure 8.48 and 8.49 with Figure 8.50 and 8.51 that show the snapshots of the channel output vector $\tilde{\mathbf{v}}_k$, and that of the learnt and ideal channel states $\mathbf{r}_{j,i}$, when the feedback state is $\mathbf{s}_{f,j} = [-1]$ for the SNR of 30dB. The fades are symbol-invariant for Figure 8.48 and 8.49, and burst-invariant for Figure 8.50 and 8.51 throughout the transmission frame of $L_T + L_D = 177$ symbols. The ideal channel states were obtained from the taps of the impulse response of the channel at the start of the frame and the learnt channel states were obtained using the scalar centre clustering algorithm described in Section 8.10. We also observed from Figure 8.48 and 8.49 as well as from Figure 8.50 and 8.51 that the scalar clustering algorithm is capable of tracking the desired channel states. At high SNRs the fades dominate, rather than the Gaussian noise, resulting in error statistics, which are not Gaussian. When the fading of the CIR is symbol-invariant, the effect of the fades is evident throughout the whole transmission burst, as we can observe from Figure 8.48 and 8.49 and this degrades the BER performance and gives an increased residual BER. But if the fade is burst-invariant, the channel effects due to fades will not be evident throughout the whole transmission frame, as shown in Figure 8.50 and 8.51, and thus this effect will not manifest itself in the results. The degradation of the BER performance due to fades within the transmission burst is evident in Figure 8.48(b). Theoretically, the channel output vectors are separable at any time instance due to the appropriate setting of the equaliser's parameters. However, the clustering algorithm tracking error and the small Euclidean distance between the channel states rendered the channel output vectors inseparable for the symbol-invariant fading scenario.

The channel output vectors are separable however for burst-invariant fading, as shown in Figure 8.50(b). During our simulations, the RBF DFE produced 77 symbol errors out of 144 data symbols in the frame of Figure 8.48(b) for the symbol-invariant fading scenario, but did not give any symbol errors in the frame of Figure 8.50(b) for the burst-invariant fading scenario. The inseparable channel output vectors explain the residual BER present during our symbol-invariant fading simulations, as shown in Figure 8.47. We note that even for relatively slow fading channels, the channel states can change significantly from symbol-to-symbol within a transmission burst duration. This phenomenon was noted in our simulations. Hence, when we assume perfect channel estimation and burst-invariant fading in our simulations, the results constitute best-case estimates.

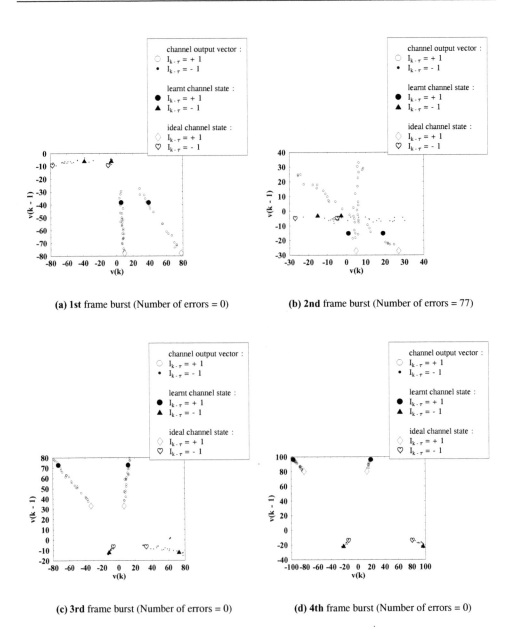

(a) 1st frame burst (Number of errors = 0)   **(b) 2nd** frame burst (Number of errors = 77)

(c) 3rd frame burst (Number of errors = 0)   **(d) 4th** frame burst (Number of errors = 0)

Figure 8.48: The channel output vectors, learnt channel output states and ideal channel output states of 1–4 transmission bursts in two-dimensional observation space, when the feedback symbol is -1 over the two-path symbol-spaced, equal-gain Rayleigh fading channel. **The fading is symbol-invariant.** The DFE had the parameters of $m = 2, n = 1$ and $\tau = 1$, while the SNR was 30dB.

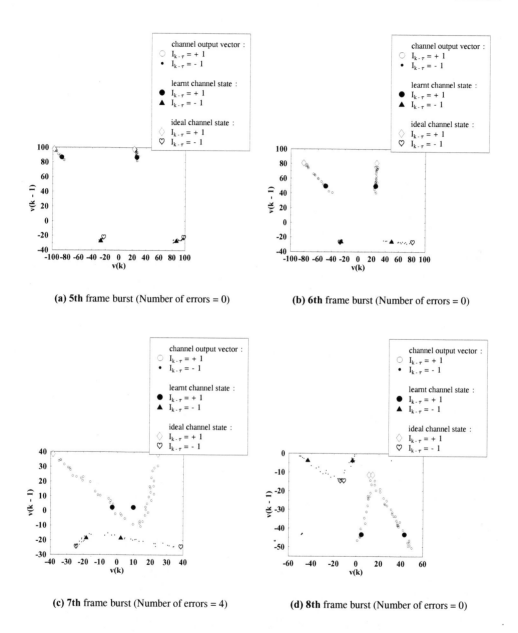

(a) 5th frame burst (Number of errors = 0)

(b) 6th frame burst (Number of errors = 0)

(c) 7th frame burst (Number of errors = 4)

(d) 8th frame burst (Number of errors = 0)

Figure 8.49: The channel output vectors, learnt channel output states and ideal channel output states of 5–8 transmission bursts in two-dimensional observation space, when the feedback symbol is -1 over the two-path symbol-spaced, equal-gain Rayleigh fading channel. **The fading is symbol-invariant**. The DFE had the parameters of $m = 2, n = 1$ and $\tau = 1$, while the SNR was 30dB.

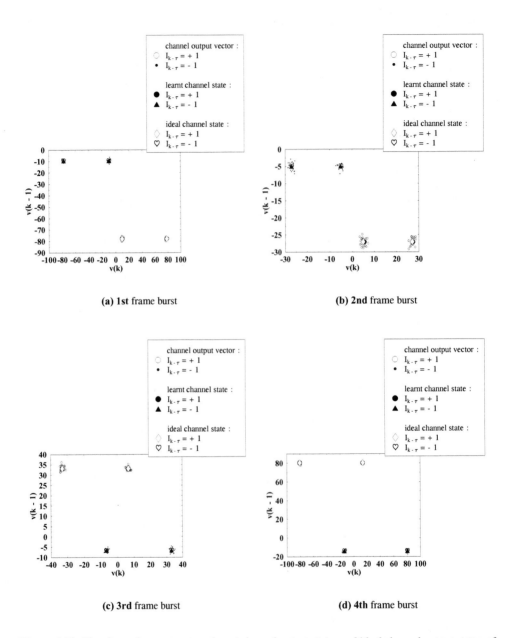

(a) 1st frame burst

(b) 2nd frame burst

(c) 3rd frame burst

(d) 4th frame burst

Figure 8.50: The channel output vectors, learnt channel output states and ideal channel output states of 1–4 transmission bursts in two-dimensional observation space when the feedback symbol is -1 over the two-path symbol-spaced, equal-gain Rayleigh fading channel. **The fading is burst-invariant.** The DFE had the parameters of $m = 2, n = 1$ and $\tau = 1$, while the SNR was 30dB.

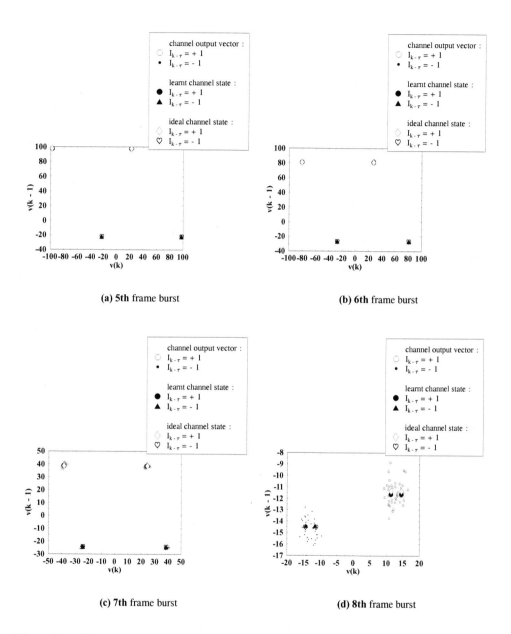

Figure 8.51: The channel output vectors, learnt channel output states and ideal channel output states in two-dimensional observation space when the feedback symbol is -1 over the two-path symbol-spaced, equal-gain Rayleigh fading channel. **The fade is burst-invariant.** The DFE had the parameters of $m = 2, n = 1$ and $\tau = 1$, and the SNR was 30dB.

8.12.5   Performance of the RBF DFE over COST 207 Channels

In this section the performance of the RBF DFE is investigated over the widely-used family of Rayleigh fading COST 207 test channels [150]. The magnitude of the impulse responses and their respective delays can be calculated by applying a set of rules, which is specified in the COST 207 report [150]. More specifically, the CIR taps may be positioned on an equispaced legitimate raster, provided the taps themselves are not equispaced. The impulse responses and the relative delays of the channels referred to as the Typical Urban (TU) and Hilly Terrain (HT) models are shown in Figure 8.52 and Table 8.13. Figure 8.53 shows the observed channel output and the learnt channel states in a two dimensional $\begin{bmatrix} v_k & v_{k-1} \end{bmatrix}$ space, when the decision delay is one symbol for the AWGN contaminated dispersive TU and HT channels without fading. Note in Figure 8.53(b) that the channel states are separable without fading. The fading parameters used in our simulations were given in Table 8.12. The structure of the transmitted burst was given in Figure 8.44. In our simulations, the number of training symbols L_T was set to 49 and the number of data symbols L_D was set to 122. The scalar centre clustering algorithm of Equation 8.99 was used in conjunction with a variable centre learning rate such that μ_r was 0.3 during the training mode and 0.1 during the decision-directed learning mode. Figure 8.54 shows the BER versus E_b/N_0 performance for the COST 207 TU and HT channels, where the RBF DFE parameters were set to be $m = 2, n = 1, \tau = 1$ for the TU channel and $m = 3, n = 2, \tau = 2$ for the HT channel, so that the decision delay covered the whole impulse response length. Thus, we assumed $L = 1$ for the TU channel and $L = 2$ for the HT channel. Figure 8.54 depicts the BER performance for both symbol-invariant and burst-invariant fading burst. From Figure 8.54 we observed again the residual BER of approximately 10^{-3} due to the desired channel states that are close together in terms of Euclidean distance, which is a consequence of the non-ideal learnt channel states and inseparable channel state clusters in the symbol-invariant scenario, as mentioned in Section 8.12.4. For the burst-invariant scenario, the residual BER was approximately 2×10^{-5} for TU channel and 5×10^{-6} for the HT channel, where again, the explanation of Section 8.12.4 applies. However, for the burst-invariant scenario, where the noiseless channel states remain the same throughout the burst duration, the performance degradation that produces the residual BER is due to the non-ideal learnt channel states, especially when the desired channel states are close together.

Hilly Terrain		Typical Urban	
Position (μs)	Relative Power (dB)	Position (μs)	Relative Power (dB)
0.00	-0.7	0.00	-0.87
0.94	-14.99	1.88	-9.03
1.88	-15.44	2.82	-13.12
15.04	-29.30	4.70	-21.28
15.98	-10.89		
17.86	-23.13		

Table 8.13: The relative power and delay of each path in the COST 207 [150] Typical Urban and Hilly Terrain channels which are depicted in Figure 8.52.

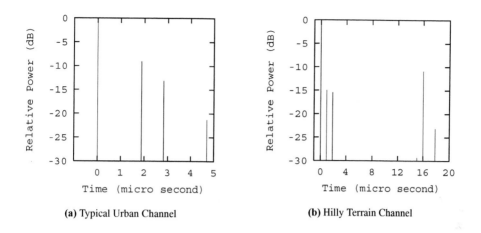

(a) Typical Urban Channel   **(b)** Hilly Terrain Channel

Figure 8.52: The impulse response of the COST 207 Typical Urban and Hilly Terrain channels depicting the relative power of each impulse and their relative delays, as shown in Table 8.13.

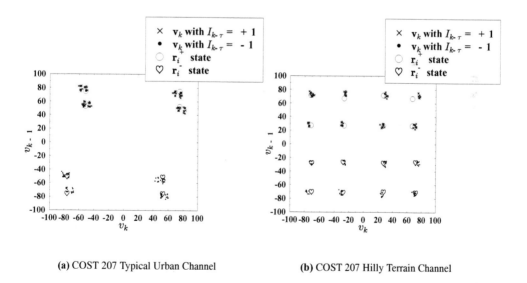

(a) COST 207 Typical Urban Channel   **(b)** COST 207 Hilly Terrain Channel

Figure 8.53: The noisy channel outputs $\mathbf{v}_k$ and the learnt channel states $\mathbf{r}_i$ of the COST 207 Typical Urban and Hilly Terrain channels with their transfer function depicted in Figure 8.52 for a BPSK modulation scheme. The SNR was 30dB, the number of samples was 171 and the decision delay was one symbol. The channel states $\mathbf{r}_i$ were learnt with the aid of the scalar clustering algorithm of Section 8.10, where the number of training symbols was 49 and the learning rate μ_r was set to 0.3 during training mode, while to 0.1 during decision-directed mode. By comparison, the corresponding quantities for the CIR of $F(z) = 0.707 + 0.707z^{-1}$ were plotted in Figures 8.50 and 8.51.

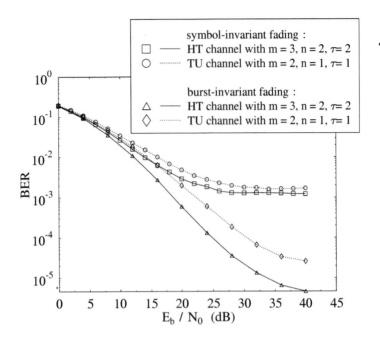

Figure 8.54: BER versus E_b/N_0 performance for the BPSK RBF DFE in conjunction with correct decision fed back over the COST 207 TU and HT channels with the impulse responses described by Table 8.13 and Figure 8.52.

8.13   Review and Discussion

In this chapter we provided a brief overview of neural networks and described, how equalization can be viewed as a classification problem. The architecture of RBF networks was presented and we described the design of the RBF equalizer based on the Bayesian equalizer solution. Our performance comparisons between the linear MSE equalizer and RBF equaliser in Figure 8.28 and 8.29 demonstrated that the RBF equaliser is capable of providing superior performance with the aid of an equivalent equalizer order at the expense of an exponential complexity increment upon increasing the equalizer order. According to Figure 8.28 and 8.29, the RBF equalizer having a feedforward order of $m = 9$ provides a performance improvement of 10dB and 20dB over the linear MSE equalizer for the two-path and three-path Gaussian channel of Figure 8.21, respectively, at a BER of 10^{-3}. We note that both the linear MSE equalizer and the RBF equaliser exhibit a residual BER characteristic, if the channel states corresponding to different transmitted symbols are inseparable in the channel observation space, as shown in Figure 8.30.

The adaptive performance of the RBF equalizer employing the vector centre clustering algorithm of Section 8.9.5, scalar centre clustering algorithm of Section 8.10 and the LMS channel estimator of Section 8.9.4 were compared. The convergence rate of the clustering algorithm depends on the number of channel coefficients to be adapted and therefore depends on the modulation scheme used and on the CIR length. However, the convergence of the LMS channel estimation technique only depends on the CIR length and therefore this technique is preferred for high-order modulation schemes and high CIR lengths. This is particularly true for the scenario, where the modulation mode of the training sequence and the data sequence differs, e.g for the adaptive QAM system described in Chapter 9, where a more robust modulation mode is used for the training sequence. The LMS channel estimation technique could only be used to obtain the correponding RBF centres, since the desired channel output differs for the training- and data sequences. However, note that the LMS channel estimation technique incurs a higher computational complexity compared to both the vector and scalar clustering algorithms, as demonstrated in Table 8.7, 8.8 and 8.9.

Decision feedback was then introduced into the RBF equaliser, in order to reduce its computational complexity. As a result, its performance improved, since the Euclidean distance between the channel states corresponding to different transmitted symbols increased, when the DFE scheme was used. Recall that the parameters of the RBF DFE were chosen to be $m = \tau + 1$ and $n = L$, where m, n, τ and $L + 1$ are the feedforward order, feedback order, delay and CIR length, which provide the best solution for a fixed equalizer delay τ. As expected, the performance degradation due to decision error propagation increased, as the BER increased, which became more significant for higher order QAM, as it was shown in Figure 8.42. For fading channel conditions, the performance degradation for higher order modulation schemes were higher, since they are more sensitive to fades due to the reduced Euclidean distance between the neighbouring channel states.

We investigated the performance of the adaptive RBF equalizer in symbol- and burst-invariant fading scenarios. We observed the effects of inseparable channel state clusters for symbol-invariant fading in Figure 8.48(b), which was due to the fast-fading effects present across the burst duration. This phenomenon, in conjunction with the non-ideal learnt channel states, explain the residual BER present in our simulations. Therefore, we have to note that even for relatively slow fading channels, the channel states value can change significantly on a symbol-by-symbol basis in a transmission burst duration.

In the next chapter, we will proceed to investigate the implementation and performance of the RBF equalizer in the context of adaptive modulation schemes.

9

RBF-Equalized Adaptive Modulation

Having considered the concepts of RBF-assisted channel equalization in Chapter 8, we are now ready to amalgamate these concepts with the AQAM philosophy detailed in Part I of the book. Although it is advantageous for the reader to consult Part I of the book, before delving into Part II dedicated to RBF-assisted arrangements, there is sufficient background information in this part of the book for the reader to be able to dispense with reading Part I. Here we will commence by providing a brief introduction to the state-of-the-art in AQAM transmissions over both narrow-band, as well as wide-band channels and then we will refer back to Part I of the book in more detail with the objective of establishing a link between the two parts.

Based on the foundations of the previous chapter, in this chapter the concept of RBF equalizers is extended to Burst-by-Burst (BbB) Adaptive QAM (AQAM) schemes. As discussed in Part I of the book, BbB AQAM schemes employ a higher-order modulation mode in transmission bursts, when the channel quality is favourable, in order to increase the throughput and conversely, a more robust but lower-order modulation mode is utilized in those transmission bursts, where the instantaneous channel quality drops. The modem mode switching regime will be detailed in more depth during our further discourse. We will show that this RBF-AQAM scheme naturally lends itself to accurate channel quality estimation. We will provide an outline of our various assumptions and the description of the simulation model, leading to our RBF-AQAM performance studies. This scheme is shown to give a significant improvement in terms of the mean BER and bits per symbol (BPS) performance compared to that of the individual fixed modulation modes. Let us now commence with a brief background on adaptive modulation in both narrow- and wide-band fading channel environments.

9.1   Background to Adaptive Modulation in a Narrowband Fading Channel

We summarise here the principles of *adaptive modulation* in a narrow-band Rayleigh fading channel environment. In a narrow-band channel, as a result of channel fading, the short-term SNR can be severely degraded. This typically degrades the short-term BER at the receiver. Again, the concept of adaptive modulation is to employ a higher modulation mode, when the channel quality is favourable, in order to increase the throughput and conversely, a more robust modulation mode is employed, in order to provide an acceptable BER, when the channel exhibits a deep fade. Thus, adaptive modulation is not only used to combat the fading effects of a narrow-band channel, but it also attempt to maximise the throughput. This idea is somewhat reminiscent of invoking a coarse power control scheme although without the detrimental effects of inflicting increased interferences upon other system users due to powering up during the intervals of low channel quality. In our work we used a variable number of modulation levels and again, we refer to this scheme as AQAM, while maintaining a constant transmitted power.

Adaptive modulation can only be invoked in the context of duplex transmissions, since some method of informing the transmitter of the quality of the link as perceived by the receiver is required unless an explicit feedback control channel is provided by the system. More explicitly, in adapting the modulation mode, a signalling regime has to be implemented in order to harmonise the operation of the transmitter and receiver with regards to the adaptive modem mode parameters. The range of signalling options is summarized in Figure 9.1 for both so-called open-loop and closed-loop signalling. For example, adaptive modulation can be applied in a time division duplex (TDD) arrangement, where the uplink and downlink transmissions are time-multiplexed onto the same carrier as depicted in Figure 9.2. If the channel quality of the uplink and downlink can be considered similar, an open-loop signalling system can be implemented, where the modulation mode can be adapted at the transmitter based on the information about the channel quality acquired during its receiving mode. This open-loop system is portrayed in Figure 9.1(a). The specific modem mode invoked has to be explicitly signalled by the transmitter to the receiver along with the reverse-direction information and it must be strongly protected against transmission errors, in order to avoid catastrophic BER degradations in case of modem mode signalling errors. By contrast, if the above channel quality predicability is not applicable – for example due to the presence of co-channel interference, etc. – the closed-loop based signalling system shown in Figure 9.1(b) can be implemented. This would be typical in a frequency division duplex (FDD) based system, where the uplink and downlink transmission frequency bands are different. Explicitly, the receiver has to instruct the remote transmitter concerning the modem mode to be used for meeting the receiver's target integrity requirements. The modem mode side-information signalling requirement is the same for both of the above signalling scenarios. For example, two bits per transmission burst are required to signal four different modem modes. However, the channel quality information will be based on a more obsolete channel quality estimate in the dissimilar uplink/downlink scenario, when the receiver instructs the remote transceiver concerning the modem mode to be used for meeting the receiver's BER target. It was shown in the context of a Kalman-filtered DFE block turbo coded AQAM scheme that it is feasible to refrain from explicitly signalling the modem modes upon invoking blind mode detection and hence increase the associated throughput [215].

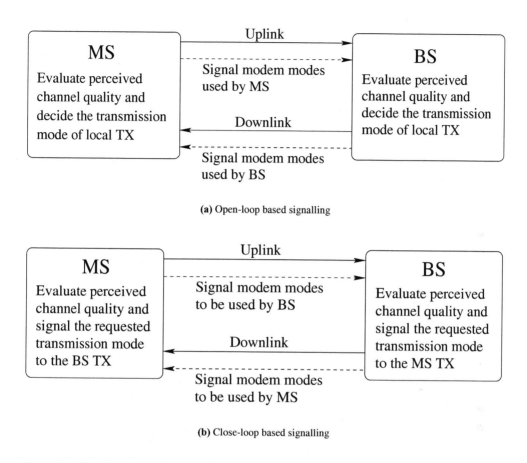

(a) Open-loop based signalling

(b) Close-loop based signalling

Figure 9.1: Closed- and open-loop signalling regimes for AQAM, where BS represents the Base Station, MS denotes the Mobile Station and the transmitter is represented by TX.

Having discussed briefly the principle of adaptive modulation and the associated scenarios, where it can be applied, we can now explore the methology used for choosing the appropriate number of modulation levels.

Torrance [145] used the instantaneous received power as the channel quality measure. The estimated instantaneous received power was used to select the suitable modulation mode by comparing the received power against a set of switching thresholds, $l_n, n = 1, \ldots, 4$, as depicted in Figure 9.3. These switching thresholds govern the tradeoff between the mean BER and the BPS performance of the system. If low switching thresholds are used, the probability of employing a high-order modulation mode increases, thus yielding a better BPS performance. Conversely, if high switching thresholds are used, a low-order modulation mode is employed more frequently, resulting in an improved mean BER performance. In his efforts to derive upper-bound performance bounds Torrance [145] assumed perfect channel quality estimation and compensation, perfect knowledge of the modulation mode at the receiver and perfect estimation of the expected received power prior to transmission.

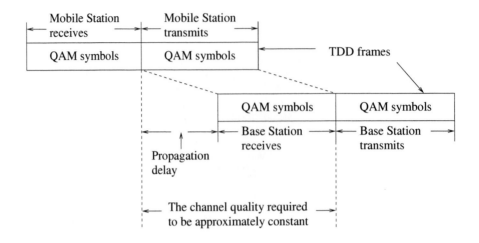

Figure 9.2: The TDD framing structure used in our AQAM system.

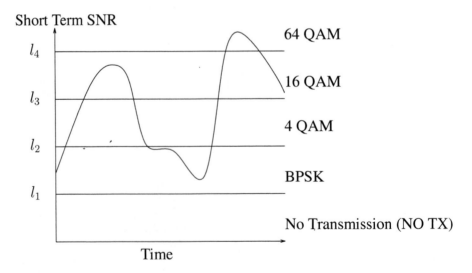

Figure 9.3: Stylised profile of the short-term received SNR, which is used to choose the next modulation mode of the transmitter in TDD mode.

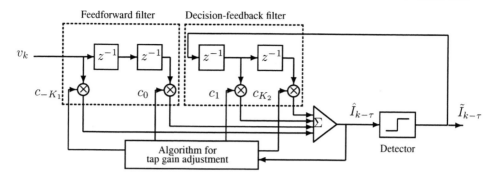

Figure 9.4: Decision-feedback equalizer schematic.

Webb and Steele [4] used the received signal strength and the BER as channel quality measures in a flat Rayleigh-fading environment. The signal to co-channel interference ratio and the expected delay spread of the channel was used by Sampei, Komaki and Morinaga [59] as the criteria to switch amongst the modulation modes and the legitimate modulation rates. They used $\frac{1}{4}$-rate QPSK, $\frac{1}{2}$-rate QPSK, QPSK, 16-QAM, 64-QAM in a narrow-band channel environment. Sampei, Morinaga and Hamaguchi utilized the signal to noise ratio and the normalized delay spread as the channel quality measure.

For a review of other work that has been conducted using adaptive modulation, the reader is referred to the previous chapters.

9.2 Background on Adaptive Modulation in a Wideband Fading Channel

In this section we will initially extend the AQAM concept to wideband fading channel environments by employing conventional channel equalization. We will briefly summarise, how the performance of the equalizer and the AQAM scheme can be jointly optimized.

As expected, the AQAM switching criteria of the narrow-band scenario mentioned in Section 9.1 has to be modified for the wideband channel environment. In Torrance's paper [145] for example, the quality of the channel was determined on the basis of the short-term SNR, which was then used as a metric in order to choose the appropriate modulation mode for the transmitter. However, in a wideband environment, the SNR metric is not reliable in quantifying the quality of the channel, where the existence of the multipath components in the wideband channel produces not only power attenuation of the transmission burst, but also intersymbol interference, as discussed in Section 8.1. Even when the channel SNR is high, QAM transmissions over wideband Rayleigh fading channels are subjected to error bursts due to ISI. Consequently, the metric required to quantify the channel quality has to be redefined, in order to incorporate the effects of the wideband channel.

Wong and Hanzo [32, 296] approached this problem by formulating a two-step methodology to mitigate the effects of the dispersive wideband channel. The first step employed a conventional Kalman-filtering based DFE, in order to eliminate most of the ISI. In the second step, the signal to noise plus residual interference ratio at the output of the equalizer was calculated based on the channel estimate. This ratio was referred to as the *pseudo SNR*, since it exhibited a Gaussian-like distribution and it was used as a metric to switch the modulation mode. Again, in [32,296], Wong used the conventional Kalman-filtering based DFE depicted in Figure 9.4. If the ISI due to past detected symbols is eliminated by the feedback filter, then the wanted signal power, the residual ISI signal power and the effective noise power can be expressed as follows [105]:

$$\text{Wanted Signal Power} \quad = \quad E\left[|q_0 I_n|^2\right], \tag{9.1}$$

$$\text{Residual ISI Signal} \quad = \quad \sum_{k=-K_1}^{-1} E\left[|q_k I_{n-k}|^2\right], \tag{9.2}$$

$$\text{Effective Noise Power} \quad = \quad N_0 \sum_{j=-K_1}^{0} |c_j|^2, \tag{9.3}$$

$$n = -\infty, \dots, \infty, \tag{9.4}$$

where $q_k = \sum_{j=-K_1}^{0} c_j f_{k-j}$, $c_j, j = -K_1, \dots, 0$ are the feedforward tap coefficients, $c_j, j = 1, \dots, K_2$ are the feedback tap coefficients, f_k is the kth impulse response tap of the channel and N_0 is the noise power. Therefore, the pseudo SNR output of the DFE, γ_{DFE}, can be calculated as follows:

$$\gamma_{\text{DFE}} = \frac{E\left[|q_0 I_n|^2\right]}{\sum_{k=-K_1}^{-1} E\left[|q_k I_{n-k}|^2\right] + N_0 \sum_{j=-K_1}^{0} |c_j|^2}. \tag{9.5}$$

The calculated pseudo SNR output of the DFE, γ_{DFE}, is then compared against a set of switching threshold levels, l_n, stored in a lookup table. The pseudo SNR output of the DFE, γ_{DFE}, is used for invoking the appropriate modem mode as follows [296]:

$$\text{Modulation Mode} = \begin{cases} \text{NO TX} & \text{if } \gamma_{DFE} < l_1 \\ \text{BPSK} & \text{if } l_1 < \gamma_{DFE} < l_2 \\ \text{4-QAM} & \text{if } l_2 < \gamma_{DFE} < l_3 \\ \text{16-QAM} & \text{if } l_3 < \gamma_{DFE} < l_4 \\ \text{64-QAM} & \text{if } \gamma_{DFE} > l_4, \end{cases} \tag{9.6}$$

where $l_n, n = 1, \dots, 4$ are the pseudo-SNR thresholds levels, and Powell's Multi-dimensional Line Minimization technique [297] was used to optimize the switching levels l_n in [32].

9.3   Brief Overview of Part I of the Book

In Part I of this monograph we commenced by analysing the performance of the DFE using multi-level modulation schemes, when communicating over static multi-path Gaussian channels as shown in Figure 2.10. These discussions were further developed in the context of a multi-path fading channel environment, where the recursive Kalman algorithm was invoked in order to track and equalize the received linearly distorted data, as evidenced by Figure 3.16. Explicitly, an adaptive CIR estimator and DFE were implemented in two different receiver structures, as shown in Figure 3.14, while their performances were compared in Figure

2.10. In this respect, Structure 1, which utilized the adaptive CIR estimator provided a better performance, when compared to that of Structure 2, which involved the adaptive DFE, as evidenced by Table 3.10. Furthermore, the complexity of Structure 2 was higher than that of Structure 1, which was studied in Section 3.4. However, these experiments were conducted in a fast start-up environment, where adaptation was restricted to the duration of the training sequence length. By contrast, if the adaptation was invoked over the entire transmission frame using a decision directed scheme, the complexity advantage of Structure 1 was eroded, as discussed in Section 3.5. The application of these fast adapting and accurate CIR estimators was crucial in a wideband AQAM scheme, where the CIR variation across the transmission frame was slow. In these experiments valuable insights were obtained with regards to the design of the equalizer and to the characteristics of the adaptive algorithm itself. This provided a firm foundation for the further investigation of the proposed wideband AQAM scheme.

Following our introductory chapters, in **Chapter 4** the concept of adaptive modulation cast in the context of a narrow-band environment was introduced in conjunction with the application of threshold-based power control. In this respect, power control was applied in the vicinity of the switching thresholds of the AQAM scheme. The associated performance was recorded in Table 4.4, where the trade-off between the BER and BPS performance was highlighted. The relative frequency of modulation mode switching was also reduced at the cost of a slight BER degradation. However, the complexity of the scheme increased due to the implementation of the power control regime. Moreover, the performance gains portrayed at this stage represented an upper-bound estimate, since perfect power control was applied. Consequently, the introduction of threshold-based power control in a narrow-band AQAM did not offer an attractive complexity versus performance gain trade-off.

The concept of AQAM was subsequently invoked in the context of a wideband channel, where the DFE was utilized in conjunction with the AQAM modem mode switching regime. Due to the dispersive multi-path characteristics of the wideband channel, a metric based on the output SNR of the DFE was proposed in order to quantify the channel's quality. This ensured that the wideband channel effects were mitigated by the employment of AQAM and equalization techniques. Subsequently a numerical model based on this criterion was established for the wideband AQAM scheme, as evidenced by Figures 4.16 and 4.17. The wideband AQAM switching thresholds were optimised for maintaining a certain target BER and BPS performance, as shown in Figure 4.3.5. The wideband AQAM BPS throughput performance was then compared to that of the constituent fixed modulation modes, where BPS/SNR gains of approximately $1 - 3dB$ and $7 - 9dB$ were observed for target BERs of 1% and 0.01%, respectively. However, as a result of the assumption made in Section 4.3.1, these gains constituted an upper bound estimate. Nevertheless, the considerable gains achieved provided further motivation for the research of wideband AQAM schemes.

The concept of wideband coded AQAM was presented in **Chapter 5**, where turbo block coding was invoked in the switching regime for different wideband AQAM schemes. The key characteristics of these four schemes, namely those of the **FCFI-TBCH-AQAM, FCVI-TBCH-AQAM, P-TBCH-AQAM** and **VR-TBCH-AQAM** arrangements, were highlighted in Table 5.10 in terms of the respective turbo interleaver size and the coding rate utilized. The general aim of using turbo block coding in conjunction with a high code rates was to increase the effective BPS transmission throughput, which was achieved, as shown in Table 5.11 for the arrangement that we referred to as the **Low-BER** scheme. In this respect, all the schemes produced gains in terms of their BER and BPS performance, when compared to the uncoded

AQAM scheme, which was optimised for a BER of 0.01%. This comparison was recorded in Table 5.11 for the four different turbo coded AQAM schemes studied.

The **FCFI-TBCH-AQAM** scheme exhibited a better throughput gain, when compared to the other schemes. This was achieved as a result of the larger turbo interleaver used in this scheme, which also incurred a higher delay. The size of the turbo interleaver was then varied, while retaining identical coding rate for each modulation mode, resulting in the **FCVI-TBCH-AQAM** scheme, where burst-by-burst decoding was achieved at the receiver. The BPS throughput performance of this scheme was also compared to that of the constituent fixed modulation modes, which utilized different channel interleaver sizes, as shown in Figure 5.11. SNR gains of approximately 1.5 and 5.0dB were achieved by the adaptive scheme for a target BERs of 0.01%, when compared to the fixed modulation modes using the small- and large-channel interleavers, respectively. By contrast, for a target BER of 1% only modest gains were achieved by the wideband AQAM scheme. These apparently low gains were the consequence of an 'unfair' comparison, since sibnificantly larger turbo interleaver and channel interleaver sizes were utilized by the fixed modulation modes. Naturally, his resulted in a high transmission delay for the fixed modulation modes. By contrast, the **FCVI-TBCH-AQAM** scheme employed low-latency instantaneous burst-by-burst decoding, which is important in real-time interactive communications.

The size of the turbo interleaver and the coding rate was then varied according to the modulation mode, in order to ensure burst-by-burst decoding at the receiver. This resulted in the **P-TBCH-AQAM** scheme, which also incorporated un-coded modes for the sake of increasing the achievable throughput. Finally, the **VR-TBCH-AQAM** scheme activated different code rates in conjunction with the different modulation modes. These schemes produced a higher maximum throughput due to the utilization of higher code rates. However, the SNR gains in terms of both the BER and BPS performance degraded, when compared to the **FCFI-TBCH-AQAM** scheme as a result of the reduced-size turbo interleaver used. Furthermore, the utilization of higher code rates for the **VR-TBCH-AQAM** arrangement resulted in a higher decoding complexity. Once again, these comparisons are recorded in Table 5.7. Similar characteristics were also observed in the context of the **High-BER** candidate scheme and in conjunction with the near-error-free schemes. However, the performance gains of the **High-BER** schemes were less than those of the **Low-BER** schemes. This was primarily due to the lower channel coding gain achieved at higher BERs and due to the smaller turbo interleaver size used.

The advantages of burst-by-burst decoding were also exploited in the context of blindly detecting the modulation modes. In this respect, the channel coding information and the mean square phasor error was utilized in the hybrid SD-MSE modulation mode detection algorithm of Section 5.6.2 characterized by Equation 5.6. Furthermore, concatenated m-sequences [169] were used in order to detect the NO TX mode while also estimating the channel's quality. The performance of this algorithm was shown in Figure 5.16, where a modulation mode detection error rate (DER) of 10^{-4} was achieved at an average channel SNR of 15dB. However, the complexity incurred by this algorithm was high due to the multiple channel decoding processes required for each individual modulation mode.

Turbo convolutional coding was then introduced and its performance using fixed modulation modes was compared to that of turbo block coding, as shown in Figure 5.23. A BER versus SNR degradation of approximately $1 - 2$dB was observed for the turbo convolutional coded performance at a BER of 10^{-4}. However, the complexity was significantly reduced, namely by a factor of seven, when compared to the previously studied turbo block coded schemes. Turbo convolutional coding was then incorporated in our wideband AQAM scheme and its performance was compared to that of the turbo block coded AQAM schemes, where the results were similar, as evidenced by Figure 5.24. Consequently the complexity versus performance gain trade-off was more attractive for our turbo convolutional coded AQAM schemes.

In our continued investigations of coded AQAM schemes, turbo equalization was invoked where BPS/SNR gains of approximately $1 - 2$dB were achieved by our AQAM scheme. In achieving this performance, iterative CIR estimation was implemented based on the LMS algorithm, which approached the perfect CIR estimation assisted AQAM performance, as shown in Figure 5.31. However, the implementation of this scheme was severely hindered by the high complexity incurred, which increased exponentially in conjunction with higher-order modulation modes and longer CIR memory.

The chapter was concluded with a system design example cast in the context of TCM, TTCM and BICM based AQAM schemes, which were studied under the constraint of a similar implementational complexity. The BbB adaptive TCM and TTCM schemes were investigated when communicating over wideband fading channels both with and without channel interleaving and they were characterised in performance terms over the COST 207 TU fading channel. When observing the associated BPS curves, adaptive TTCM exhibited up to $2.5\ dB$ SNR-gain for a channel interleaver length of four transmission bursts in comparison to the non-interleaved scenario, as it was evidenced in Figure 5.40. Upon comparing the associated BPS curves, adaptive TTCM also exhibited up to $0.7\ dB$ SNR-gain compared to adaptive TCM of the same complexity in the context of **System II**, while maintaining a target BER of less than 0.01 %, as it was shown in Figure 5.44. Finally, adaptive TCM performed better, than the adaptive BICM benchmarker in the context of **System I**, while the adaptive BICM-ID scheme performed marginally worse, than adaptive TTCM in the context of **System II**, as it was discussed in Section 5.11.5.

In **Chapter 6** following a brief introduction to several fading counter-measures, a general model was used for describing various adaptive modulation schemes employing various constituent modulation modes, such as PSK, Star QAM and Square QAM, as one of the attractive fading counter-measures. In Section 6.3.3.1, the closed form expressions were derived for the average BER, the average BPS throughput and the mode selection probability of the adaptive modulation schemes, which were shown to be dependent on the mode-switching levels as well as on the average SNR. In Sections 6.4.1, 6.4.2 and 6.4.3 we reviewed the existing techniques devised for determining the mode-switching levels. Furthermore, in Section 6.4.4 the optimum switching levels achieving the highest possible BPS throughput were studied, while maintaining the average target BER. These switching levels were developed based on the Lagrangian optimization method.

Then, in Section 6.5.1 the performance of uncoded adaptive PSK, Star QAM and Square QAM was characterised, when the underlying channel was a Nakagami fading channel. It was found that an adaptive scheme employing a k-BPS fixed-mode as the highest throughput constituent modulation mode was sufficient for attaining all the benefits of adaptive modulation, while achieving an average throughput of up to $k - 1$ BPS. For example, a three-mode adaptive PSK scheme employing No-Tx, 1-BPS BPSK and 2-BPS QPSK modes attained the maximum possible average BPS throughput of 1 BPS and hence adding higher-throughput modes, such as 3-BPS 8-PSK to the three-mode adaptive PSK scheme resulting in a four-mode adaptive PSK scheme did not achieved a better performance across the 1 BPS throughput range. Instead, this four-mode adaptive PSK scheme extended the maximum BPS throughput achievable by any adaptive PSK scheme to 2 BPS, while asymptotically achieving a throughput of 3 BPS, as the average SNR increases.

On the other hand, the relative SNR advantage of adaptive schemes in comparison to fixed-mode schemes increased as the target average BER became lower and decreased as the fading became less severe. More explicitly, less severe fading corresponds to an increased Nakagami fading parameter m, to an increased number of diversity antennas, or to an increased number of multi-path components encountered in wide-band fading channels. As the fading becomes less severe, the average BPS throughput curves of our adaptive Square QAM schemes exhibit undulations owing to the absence of 3-BPS, 5-BPS and 7-BPS square QAM modes.

The comparisons between fixed-mode MC-CDMA and adaptive OFDM (AOFDM) were made based on different channel models. In Section 6.5.4 it was found that fixed-mode MC-CDMA might outperform adaptive OFDM, when the underlying channel provides sufficient diversity. However, a definite conclusion could not be drawn since in practice MC-CDMA might suffer from MUI and AOFDM might suffer from imperfect channel quality estimation and feedback delays.

Concatenated space-time block coded and turbo convolutional-coded adaptive multi-carrier systems were investigated in Section 6.5.5. The coded schemes reduced the required average SNR by about 6dB-7dB at a throughput of 1 BPS, achieving near error-free transmission. It was also observed in Section 6.5.5 that increasing the number of transmit antennas in adaptive schemes was not very effective, achieving less than 1dB SNR gain, due the fact that the transmit power per antenna had to be reduced in order to limit the total transmit power for the sake of a fair comparison.

The practical issues regarding the implementation of the advocated wideband AQAM scheme was analysed in **Chapter 7**. The impact of error propagation in the DFE was highlighted in Figure 7.2, where the BER degradation was minimal and the target BERs were achieved without any degradation to the transmission throughput performance. The impact of channel quality estimation latency was also studied, where the sub frame based TDD/TDMA system of Section 7.2.1 was implemented. In this system, a channel quality estimation delay of 2.3075ms was imposed and the channel quality estimates were predicted using a linear prediction technique. In this practical wideband AQAM scheme, SNR gains of approximately 1.4dB and 6.4dB were achieved for target BERs of 1% and 0.01%, when compared to the performance of the constituent fixed modulation modes. This was shown graphically in Figure 7.11.

CCI was then subsequently introduced in Section 7.3, where in terms of channel quality estimation, the minimum average SIR that can be tolerated by the wideband AQAM scheme was approximately 10dB, as evidenced by Figure 7.14. In order to mitigate the impact of CCI on the demodulation process, the JD-MMSE-BDFE scheme using an embedded convolutional encoder was invoked, where the performance was shown in Figure 7.21 and 7.22 for the fixed modulation modes and for the wideband AQAM scheme, respectively. The performance gains achieved by the wideband AQAM scheme were approximately $2 - 4$dB and $7 - 9$dB for the target BERs of 1% and 0.01%, when compared to the performance of the associated fixed modulation modes. However, these gains constituted an upper bound estimate, since perfect channel estimation of the reference user and the interferer was assumed.

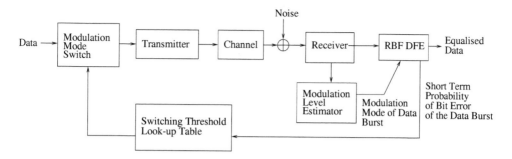

Figure 9.5: System schematic of the joint adaptive modulation and RBF equalizer scheme.

The concept of segmented wideband AQAM was then introduced, in order to reduce the impact of CCI. In this scheme, the inner and outer switching thresholds were developed based on a noise and interference limited environment, respectively, for any given average channel SNR and average SIR. A channel quality estimation delay of 2.3075ms was imposed in estimating the instantaneous SIR and the associated performance was displayed in Figure 7.24. By employing this segmented AQAM scheme, accurate estimation of the instantaneous SIR was needed, which was provided by the Kalman-filter based mid-amble assisted CIR estimation. However, information regarding the interferer was not required for reducing the impact of CCI, which was a substantial advantage.

Having studied a whole host of the associated aspects of AQAM-assisted wireless communications in Part I of the book, in the forthcoming chapters we will focus our attention on using RBF-aided equalizers as described in Section 8.9, instead of employing the conventional DFE. The joint adaptive modulation and RBF equalization scheme will be described next, followed by our simulation results.

9.4 Joint Adaptive Modulation and RBF Based Equalization

In this section, we will describe the joint AQAM and RBF network based equalization scheme and the switching metric employed. We commence by exploring the joint AQAM and RBF equalizer scheme's best-case performance. Finally, the performance of this scheme and that of the individual fixed modulation modes is compared in terms of their mean BER and BPS.

9.4.1 System Overview

The schematic of the joint AQAM and RBF network based equalization scheme is depicted in Figure 9.5. We use the RBF DFE described in Section 8.11 in this scheme. At the receiver, the RBF DFE is trained using the method described in Section 8.9.3 and then the corrupted received signal is equalized. The short-term probability of bit error or short-term BER of the transmitted burst is calculated from the output of the sub-RBF network, and is used as the switching metric. Section 9.4.2 will highlight this issue in more detail. The short-term BER is compared to a set of switching BER values corresponding to the modulation mode of the received data burst. Consequently, a modulation mode is selected for the next transmission, assuming channel quality similarity for the uplink and downlink transmissions. This implies that the similarity of the short-term BER of consecutive uplink and downlink data bursts can be exploited, in order to set the next modulation mode. The modulation modes utilized in our system are BPSK, 4-QAM, 16-QAM, 64-QAM and no transmission (NO TX), similarly to Equation 9.6. Therefore, the modulation mode is switched according to the estimated short-term BER, $P_{\text{bit, short-term}}$, as follows:

$$\text{Modulation Mode} = \begin{cases} \text{NO TX} & \text{if } P_{\text{bit, short-term}} \geq P_2^{\mathcal{M}} \\ \text{BPSK} & \text{if } P_2^{\mathcal{M}} > P_{\text{bit, short-term}} \geq P_4^{\mathcal{M}} \\ \text{4-QAM} & \text{if } P_4^{\mathcal{M}} > P_{\text{bit, short-term}} \geq P_{16}^{\mathcal{M}} \\ \text{16-QAM} & \text{if } P_{16}^{\mathcal{M}} > P_{\text{bit, short-term}} \geq P_{64}^{\mathcal{M}} \\ \text{64-QAM} & \text{if } P_{64}^{\mathcal{M}} > P_{\text{bit, short-term}}, \end{cases} \qquad (9.7)$$

where $P_i^{\mathcal{M}}, i = 2, 4, 16, 64$ are the switching BER thresholds corresponding to the various $\mathcal{M}$-QAM modes.

9.4.2 Modem Mode Switching Metric

The RBF equalizer based on the optimal Bayesian decision function of Equation 8.17, as described in Chapter 8, is capable of providing the 'on-line' estimation of the BER in the receiver without the knowledge of the transmitted symbols. This is possible, since the equalizer is capable of estimating the a *posteriori* probability of the transmitted symbols, if the CIR is known and provided that the centres of the RBF network are assigned the values of the channel states, as it was originally suggested in Section 8.9.

Referring to Section 8.9.2 and Figure 8.19, the output of the RBF networks provides the conditional probability density function of each legitimate QAM symbol, $\mathcal{I}_i, i = 1, \ldots, \mathcal{M}$ which is described by Equation 8.85. The a *posteriori* probability $\varsigma_i(k)$ of the transmitted symbols, can be evaluated from the conditional density function, $\zeta_i(k)$ as follows:

$$\begin{aligned} \varsigma_i(k) &= P(I_{k-\tau} = \mathcal{I}_i | \mathbf{v}_k) \\ &= \frac{P(\mathbf{v}_k | I_{k-\tau} = \mathcal{I}_i) \cdot P(I_{k-\tau} = \mathcal{I}_i)}{P(\mathbf{v}_k)} \\ &= \frac{\zeta_i(k)}{P(\mathbf{v}_k)}, \qquad -\infty \leq k \leq \infty. \end{aligned} \qquad (9.8)$$

The a *posteriori* probability $\tilde{\varsigma}(k)$ of the detected symbol can be obtained without the knowledge of the term $P(\mathbf{v}_k)$, if the a *posteriori* probability has unity support (i.e. the sum of the a *posteriori* probabilities of all symbols is unity):

$$\tilde{\varsigma}(k) = \frac{\zeta_i^*(k)}{\sum_{i=1}^{\mathcal{M}} \zeta_i(k)}, \qquad -\infty \leq k \leq \infty, \qquad (9.9)$$

where $\zeta_i^*(k) = \max\{\zeta_i(k), 1 \leq i \leq \mathcal{M}\}$, as defined in Equation 8.84. Therefore, the probability of a *symbol* error associated with the decision $\tilde{I}_{k-\tau} = \mathcal{I}_i^*$ is given by:

$$P_s^{'}(k) = 1 - \tilde{\varsigma}(k), \qquad -\infty \leq k \leq \infty, \qquad (9.10)$$

and the overall probability of symbol error of the detector is given by:

$$P_{\text{symbol}} = E\{P_s(k)\} \qquad -\infty \leq k \leq \infty. \tag{9.11}$$

Similarly, the probability of a *bit* error can be obtained from the *a posteriori* probability of the bits representing the QAM symbols. Below we provide an example for the 4-QAM scheme. The *a posteriori* probability of the 4 symbols, $\mathcal{I}_1, \mathcal{I}_2, \mathcal{I}_3$ and $\mathcal{I}_4$, is estimated by the RBF networks as $\varsigma_1, \varsigma_2, \varsigma_3$ and ς_4, respectively. A 4-QAM symbol is denoted by the bits $U_0 U_1$ and the symbols $\mathcal{I}_1, \mathcal{I}_2, \mathcal{I}_3$ and $\mathcal{I}_4$ correspond to $00, 01, 10\ 11$, respectively. Thus, the *a posteriori* probability of the bits is given as follows:

$$
\begin{aligned}
P(U_0 = 1) &= P(U_0 U_1 = 11 \cup U_0 U_1 = 10) = \varsigma_4 + \varsigma_3, \\
P(U_0 = 0) &= P(U_0 U_1 = 01 \cup U_0 U_1 = 00) = \varsigma_2 + \varsigma_1, \\
P(U_1 = 1) &= P(U_0 U_1 = 11 \cup U_0 U_1 = 01) = \varsigma_4 + \varsigma_2, \\
P(U_1 = 0) &= P(U_0 U_1 = 10 \cup U_0 U_1 = 00) = \varsigma_3 + \varsigma_1.
\end{aligned}
\tag{9.12}
$$

In general, the average probability of bit error for the detected symbol at signalling instant k is given by:

$$P_b(k) = \frac{\sum_{i=0}^{\text{BPS-1}} 1 - P(U_i(k) = b_i)}{\text{BPS}}, \tag{9.13}$$

where BPS denotes the number of bits per symbol and b_i is the value (either 0 or 1) of the ith bit of the symbol exhibiting the maximum *a posteriori* probability. The overall probability of bit error for the detector is given by:

$$P_{\text{bit}} = E\{P_b(k)\} \qquad -\infty \leq k \leq \infty. \tag{9.14}$$

For our joint RBF based equalization and AQAM scheme, we are unable to obtain the true probability of bit error for the detector, namely P_{bit} averaged over all data bursts, since we need to collect a large number of received samples for an accurate estimation. We can only obtain the short-term probability of bit error, $P_{\text{bit, short-term}}$, which is the average bit error probability over a data burst that was received, i.e.,

$$P_{\text{bit, short-term}} = \frac{\sum_{n=1}^{L_D} P_b(n)}{L_D}, \tag{9.15}$$

where L_D is the number of data symbols per burst. Thus, we could estimate the channel quality on a BbB basis, relying on the estimated $P_{\text{bit, short-term}}$ value. The short-term probability of bit error or BER is only an estimate of the actual P_{bit} of the system for the duration of the data burst. The accuracy of the estimation is dependent on the number of data symbols L_D in the burst. This issue will not be discussed further for now.

Having described the switching metric used by the joint AQAM and RBF equalizer scheme, we will further investigate this scheme with the aim of producing a best-case performance estimate. Before proceeding, the next section will present the assumptions used, when we employ this scheme in a wideband channel environment.

9.4.3   Best-case Performance Assumptions

In deriving the best-case performance of this joint adaptive modulation and RBF based equalization scheme, the following assumptions are made:

1. Perfect CIR estimation or channel state estimation is assumed at the receiver. The RBF's centres are assigned the values of the channel states. The associated CIR and channel state estimation techniques were presented in Section 8.9.3, 8.9.4 and 8.9.5. We note that incorrect estimation of the channel states will degrade the performance of the constituent fixed modulation modes, as it was demonstrated by our simulation results in Section 8.12. This degradation is neglected here with the aim of deriving a best-case performance estimate.

2. The CIR is time-invariant for the duration of the transmission burst, but varies from burst to burst, which corresponds to assuming that the channel is slowly varying. However, if the CIR changes during the transmission burst or if the estimation algorithm gives an inaccurate channel estimate, the effect of the channel variations can be considered by modifying the noise variance estimate, as discussed in [259, 298]. Let us briefly summarize this idea. We define the error between the noisy channel output v_k and the estimated noiseless channel state output $\hat{v}_k$ as follows:

$$
\begin{aligned}
e_k &= v_k - \hat{v}_k \\
&= v_k - \sum_{n=0}^{L} \hat{f}_n I_{k-n} \\
&= \triangle_f(I_k, \dots, I_{k-L}) + \eta_k,
\end{aligned} \tag{9.16}
$$

where $\triangle_f(\cdot)$ is an error function caused by an inaccurate estimate of the channel impulse response $\hat{f}_n, n = 0, \dots, L$. Having determined this noise term, the RBF equalizer uses the noise variance in its width parameter seen in Equation 8.80 in order to compute the conditional probability densities of each legitimate QAM symbols. Therefore, by computing the 'noise variance' as the average of e_k^2, and substituting these values in Equation 8.80 yields $\rho = 2E\left[e_k^2\right]$. Hence we translated the CIR estimation error to a noise-like term.

3. We assume furthermore that the receiver has perfect knowledge of the modulation mode used in its received transmission burst. For a practical system, control symbols must be used to convey the modulation mode employed by the transmitter to the receiver [21, 26].

4. The RBF DFE used in the system neglects error propagation by feeding the correct symbol to be used for RBF subset centre selection or space translation, as described in Section 8.11. However, at low target BERs, we will expect low performance degradation due to decision feedback error propagation, as it was demonstrated in Figure 8.42.

5. The short-term probability of error estimate, namely $P_{\text{bit, short-term}}$, is known prior to transmission for all the modulation modes used in the system. This can be assumed in

a TDD scenario, where the channel can be considered similar in the uplink and downlink transmission and when the channel is slowly varying. We also assume that given the estimated $P_{bit, short\text{-}term}$ for a particular modulation mode, the transmitter knows the corresponding short-term probability of bit error for the other modulation modes used in the system under the same channel conditions. Thus, the transmitter of a base station for example, can utilize its receiver's $P_{bit, short\text{-}term}$ estimation for its next transmission, provided that there is a high channel quality correlation between the transmitter and receiver slots. Note however that the latency between the transmitter and receiver slots can affect the quality of the estimation. This latency is mitigated, when employing slot-by-slot TDD - as in the third generation IMT-2000 and UTRA [299–301] proposals - where any TDD-slot can be configured as an uplink or downlink slot, hence reducing the latency of channel quality estimates.

During our further discourse we will gradually remove these idealistic assumptions.

Having described the assumptions stipulated, in order to derive the best-case performance of this joint adaptive modulation and RBF based equalization scheme, we now describe our simulation model.

9.4.4 Simulation Model for Best-case Performance

In our experiments, pseudo-random symbols were transmitted in a fixed-length burst for all modulation modes over the burst-invariant wideband channel to fulfil assumptions 2 and 5. The receiver received each data burst having different modulation modes and equalized each one of them independently. The estimated short-term probability of bit error or BER was obtained for each modulation mode, as described in Section 9.4.2. The highest-order modulation mode, $\mathcal{M}^*$ that provided a short-term BER $P_{bit, short\text{-}term}^{\mathcal{M}^*}$, which was below the target BER $P_{bit, target}$, when:

$$\mathcal{M}^* = \max\{\mathcal{M} = 2, 4, 16, 64, \text{ such that } P_{bit, short\text{-}term}^{\mathcal{M}} \leq P_{bit, target}\}, \qquad (9.17)$$

was chosen to be the actual modulation mode that was used by the transmitter and the received equalized burst was used for the BER estimation of the system. The notation $P_{bit, short\text{-}term}^{\mathcal{M}}$ represents the short-term BER of $\mathcal{M}$-QAM. However, if all the modulation mode could not provide the targetted BER performance, i.e. $P_{bit, short\text{-}term}^{2} > P_{bit, target}$, NO TX mode is utilized. Figure 9.6 shows the simulation schematic of the joint AQAM and RBF DFE scheme used in our best-case BER performance evaluation. The next section will present our simulation results and analysis.

9.4.5 Simulation Results

The simulation parameters are listed in Table 9.1, noting that we analysed the joint AQAM and RBF equalizer scheme over a two-path Rayleigh fading channel. The wideband fading channel was burst-invariant. The RBF DFE used in our simulations had a feedforward order of $m = 2$, feedback order of $n = 1$ and delay of $\tau = 1$.

Figure 9.7 portrays the short-term BER of the burst-invariant channel versus symbol index, as estimated by the RBF DFE. For the simulated scenario, i.e., for a Doppler frequency of

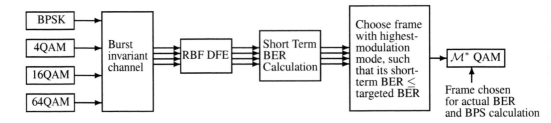

Figure 9.6: The simulation schematic of the joint AQAM and RBF DFE arrangement used for best-case
BER performance estimation.

Number of data symbols per burst, L_D	144
Number of training symbols per burst, L_T	27
Transmission Frequency	1.9GHz
Transmission Rate	2.6MBd
Vehicular Speed	30 mph
Normalized Doppler Frequency	3.3×10^{-5}
Channel weights	$0.707 + 0.707z^{-1}$
RBF DFE feedforward order, m	2
RBF DFE feedback order, n	1
RBF DFE decision delay, τ	1

Table 9.1: Simulation parameters.

3.3×10^{-5} the short-term BER is slowly varying and it is relatively predictable for a number
of consecutive data bursts. Thus, assumption 2 of Section 9.4.3 is valid for this scenario.

The probability density function (PDF) of the BER estimation error of the RBF DFE for
various channel SNRs is shown in Figure 9.8 for BPSK transmission bursts. The actual BER
is the ratio of the number of bit errors encountered in a data burst to the total number of
bits transmitted in that burst. Figure 9.8 suggests that the RBF DFE provides a good BER
estimation, especially for high channel SNRs. We note, however that the accuracy of the
actual BER evaluation is limited by the burst-length of 144 bits and its resolution is $1/144$.
Hence at high SNRs the actual number of errors registered is often 0, which portrays the BER
estimation algorithm of Equation 9.15 in a less accurate light in the PDF of Figure 9.8, than
it is in reality.

We will now analyse the best-case performance of the joint AQAM and RBF DFE scheme
in more detail, using the simulation model described in Section 9.4.4 and the assumptions
listed in Section 9.4.3. We designed two systems, a higher integrity scheme, having a target
BER of 10^{-4}, which can be rendered error-free by error correction coding and hence we refer
to this arrangement as a data transmission scheme; the lower integrity scheme was designed
for maintaining a BER of 10^{-2}, which is adequate for speech transmission especially in conjunction with FEC. The target BPS values of these schemes were 3 and 4.5 bits per symbol,
respectively, although these values can only be attained for sufficiently high SNRs.

Figure 9.9(a) and Figure 9.9(b) show the simulated best-case performance of the joint

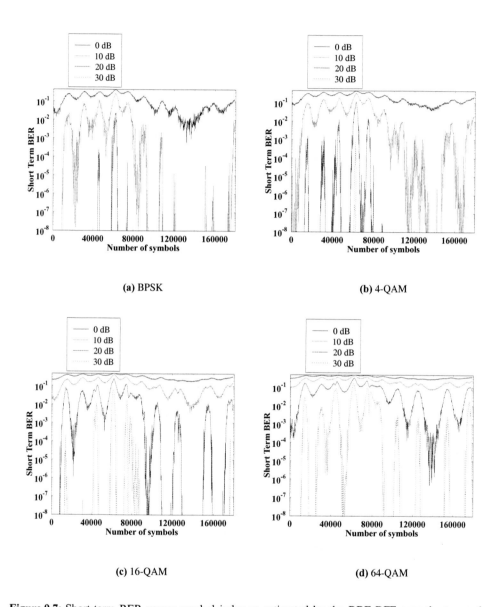

(a) BPSK

(b) 4-QAM

(c) 16-QAM

(d) 64-QAM

Figure 9.7: Short-term BER versus symbol index as estimated by the RBF DFE over the two-path equal-weight, symbol-spaced Rayleigh fading channel of Table 9.1. The RBF DFE had a feedforward order of $m = 2$, feedback order of $n = 1$ and decision delay of $\tau = 1$ symbol. Perfect channel impulse response estimation is assumed and the error propagation due to decision feedback is ignored. The transmitted burst of Figure 8.44 consists of 171 symbols (144 data symbols and 27 training symbols).

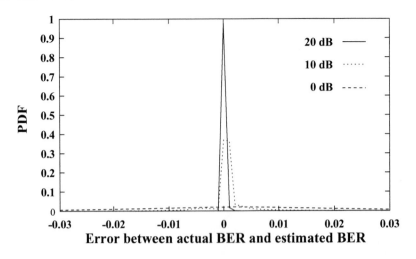

Figure 9.8: Discretised PDF of the error between the actual BER of the data burst and the BER estimated by the RBF DFE for the two-path Rayleigh fading channel of Table 9.1 using BPSK.

AQAM and RBF DFE scheme for the target BER of 10^{-2} designed for speech transmission and for the target BER of 10^{-4} created for data transmission, respectively. The BER
performance of the constituent fixed modulation modes is also depicted in both figures for
comparison. The best-case performance was evaluated for two different adaptive modulation
schemes. In the first scheme, the transmitter always transmitted data without transmission
blocking, i.e. the NO TX mode of Equation 9.7 was not invoked. By contrast, in the second
scheme, dummy data was transmitted, whenever the estimated short-term BER was higher
than the target BER, a scenario, which we referred to as transmission blocking. The transmission of dummy data during blocking allowed us to keep monitoring the BER, in order to
determine when to commence transmission and in which modem mode.

We will commence by analysing Figure 9.9(a), where the joint AQAM and RBF DFE
scheme was designed for speech transmission, i.e. for a BER of 10^{-2}. For the adaptive
scheme, which did not incorporate transmission blocking, the performance of adaptive modulation was better or equivalent to the performance of BPSK in terms of the mean BER and
mean BPS for the SNR range between 0dB and 9dB. At the channel SNR of 9dB, even though
the mean BER performance was equivalent for the adaptive scheme and the BPSK scheme,
the mean BPS for the adaptive scheme improved by a factor of 1.5, resulting in a mean BPS
of 1.5. In the SNR range of 9dB to 16dB, the adaptive scheme outperformed the 4-QAM
scheme in terms of the mean BER performance. At the channel SNR of 16dB, the mean
BERs of both schemes are equivalent, although the mean BPS of the adaptive scheme is 2.7,
resulting in a BPS improvement by a factor of 1.35, when compared to 4-QAM. At the channel SNR of 26dB, the mean BPS improvement of the adaptive scheme is by a factor of 1.3 for
an equivalent mean BER. The adaptive scheme that utilized transmission blocking achieved a
mean BER below 1%. At the channel SNR of 12dB, even though the mean BER performance
was equivalent for the BPSK scheme and the adaptive scheme with transmission blocking,
the mean BPS for the adaptive scheme improved by a factor of 2. As the SNR improved,
the performance of the adaptive schemes both with and without transmission blocking con-

verged, since the probability of encountering high short-term BERs reduced. The mean BER and mean BPS performance of both adaptive schemes converged to that of 64-QAM for high SNRs, where 64-QAM becomes the dominant modulation mode.

Similar trends were observed for data-quality transmission, i.e. for the 10^{-4} target BER scheme in Figure 9.9(b). However, we note that for the SNR range between 8dB to 20dB, the mean BER of the adaptive scheme without transmission blocking was better, than that of BPSK. This phenomenon was also observed in the narrowband adaptive modulation scheme of [145] and in the wideband joint AQAM and DFE scheme of [32, 296], which can be explained as follows. The mean BER of the system is the ratio of the total number of bit errors to the total number of bits transmitted. The mean BER will decrease with decreasing number of bit errors and with increasing number of total bits transmitted in the data burst. For a fixed number of symbols transmitted, the number of total bits transmitted in a data burst is constant for the BPSK scheme, while for the AQAM scheme the total number of bits transmitted in a data burst increased, when a higher-order AQAM mode was used. However, in this case the BER increased. If the relative bits per symbol increment upon using AQAM is higher than the relative bit error ratio increment, then the mean BER of the adaptive scheme will be improved. Consequently the adaptive mean BER can be lower than that of BPSK.

The probability of encountering each modulation mode employed in the adaptive scheme based on the estimated short-term BER switching mechanism is shown in Figure 9.10 and Figure 9.11 for the BER = 10^{-2} and BER = 10^{-4} schemes, respectively. As expected, the sum of the probabilities at each particular SNR is equal to one. At low SNRs, the lower order modulation modes (NO TX or BPSK) are dominant, producing a robust system. At higher SNRs, the higher order modulation modes become dominant, yielding a higher mean BPS and yet a reduced mean BER. From Figure 9.11(b), we observe that the transmission blocking mode was dominant in the SNR range of 0dB to 4dB and thus the mean BER performance was not recorded in that range of SNRs in Figure 9.9(b).

Comparing Figure 9.10(a) and Figure 9.11(a), the probability of transmission blocking was higher for data-quality transmission, in order to achieve a lower target BER due to the associated more stringent BER requirements of 10^{-4}. The probability of transmission blocking was close to zero, once the channel SNR increased to about 16dB and 20dB for the BER = 10^{-2} and BER = 10^{-4} schemes, respectively. These are the points, where the performance of the adaptive schemes with and without transmission blocking converged, as demonstrated in Figure 9.9. We observed that the probabilities of the 4-QAM, 16-QAM and 64-QAM modes being utilized for the adaptive scheme with and without transmission blocking were fairly similar. This is because introducing transmission blocking will predominantly affect the probability of BPSK, which will be utilized instead of no data transmission.

In summary, the AQAM RBF DFE scheme has its advantages, when compared to the individual fixed modulation modes in terms of the mean BER and mean BPS performance. Note however for the adaptive scheme without transmission blocking that the target performance of BER = 10^{-2} and BER = 10^{-4} can only be achieved, if the channel SNR is higher than 9dB and 18dB, respectively. The target mean BERs for speech transmission (BER = 10^{-2}) and data transmission (BER = 10^{-4}) were achieved for all channel SNRs, when we utilized transmission blocking. The target performance for speech (BER = 10^{-2}) and data (BER = 10^{-4}) transmission in terms of mean BPS (4.5 and 3, respectively) can only be achieved for the AQAM scheme with and without transmission blocking, if the channel SNR is in excess of about 22dB. Thus, the advantage of using an adaptive scheme with transmission blocking

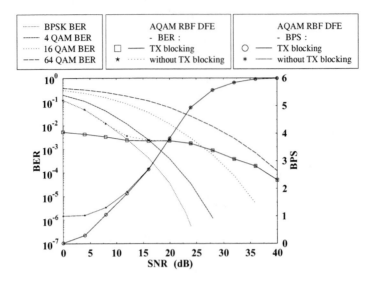

(a) Target BER is 10^{-2} (mean BER for speech transmission)

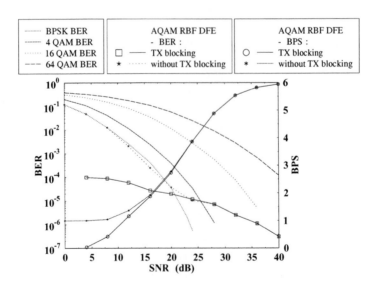

(b) Target BER is 10^{-4} (mean BER for data transmission)

Figure 9.9: The simulated best-case performance of the AQAM RBF DFE showing also the BER performance of the constituent fixed modulation schemes, namely BPSK, 4-QAM, 16-QAM and 64-QAM, over the two-path Rayleigh-fading channel of Table 9.1 and using the assumptions of Section 9.4.3.

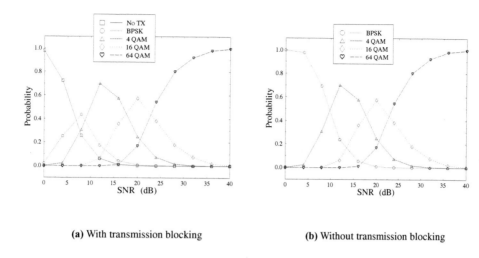

(a) With transmission blocking   **(b)** Without transmission blocking

Figure 9.10: The probability of encountering the various $\mathcal{M}$-QAM modulation modes in the joint AQAM and RBF DFE scheme for **best-case performance** during **speech-quality transmission (target BER of 0.01)** over the two-path equal-weight, symbol-spaced Rayleigh fading channel using the simulation parameters listed in Table 9.1 and the assumptions stated in Section 9.4.3.

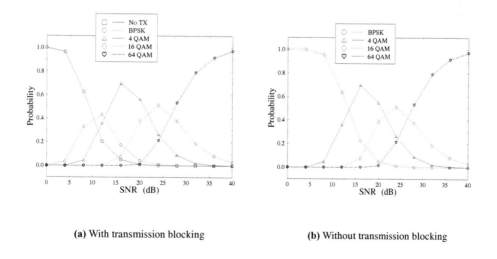

(a) With transmission blocking   **(b)** Without transmission blocking

Figure 9.11: The probability of encountering the various $\mathcal{M}$-QAM modulation modes in the joint AQAM and RBF DFE scheme for **best-case performance** during **data-quality transmission (target BER of 10^{-4})** over the two-path equal-weight, symbol-spaced Rayleigh fading channel using the simulation parameters listed in Table 9.1 and the assumptions stated in Section 9.4.3.

is that the performance of the joint AQAM and RBF DFE scheme can be 'tuned' to a certain required mean BER performance. However, the disadvantage is that the utilization of transmission blocking results in transmission latency, an issue, which was addressed for example in [40, 41]. Specifically, the interdependency of the required buffer size, doppler frequency and latency was analysed. Furthermore, frequency hopping was proposed for reducing the average duration of NO TX mode at low Doppler frequencies, where the latency and the buffer size may become excessive.

Let us now embark on a comparative analysis between the joint AQAM RBF DFE scheme and the Kalman-filtering based joint AQAM DFE scheme introduced by Wong *et al.* [32] for wideband channels. The joint AQAM DFE scheme in [32] used the *pseudo-SNR* at the output of the DFE as the switching metric, an issue discussed briefly in Section 9.2. The pseudo-SNR at the output of the DFE was compared to a set of pseudo-SNR thresholds optimized using Powell's method [297]. Table 9.2 gives the results of the optimization process invoked, in order to achieve transmission integrities of 10^{-2} and 10^{-4} over the two-path Rayleigh-fading channel of Table 9.1 [32]. The conventional DFE used in the adaptive scheme had a feedforward order of $m = 15$, feedback order of $n = 2$ and decision delay of $\tau = 15$ symbols. The parameters m, n and τ of the conventional DFE were chosen such that it exhibited the best possible performance for our simulation scenario and hence further increase of the feedforward order would not give a significant performance improvement. We note again that for our best-case performance comparisons, the switching metric used for both schemes – namely the short-term BER for the AQAM RBF DFE scheme and the pseudo SNR for the AQAM DFE scheme – was estimated perfectly prior to transmission and the appropriate AQAM mode was chosen for the data burst to be transmitted, which satisfied the target BER requirement.

	l_1(dB)	l_2(dB)	l_3(dB)	l_4(dB)
Speech	3.68026	6.3488	11.7181	17.8342
Data	8.30459	10.4541	16.8846	23.051

Table 9.2: The optimized switching levels l_n of the joint adaptive modulation and DFE scheme for speech and data transmission in the two-path Rayleigh fading channel [32]. The target mean BER and BPS performance for speech was 10^{-2} and 4.5, respectively, while for computer data, 10^{-4} and 3, respectively.

Figure 9.12 provides the BER performance comparison of the conventional DFE and the RBF DFE over the two-path Rayleigh fading channel of Table 9.1 for the constituent fixed modulation modes. The BER performance of the RBF DFE for BPSK and 4-QAM was better than that of the conventional DFE, as the SNR increased. By contrast, the BER performance of the RBF DFE was inferior compared to that of the conventional DFE for 16- and 64-QAM. The performance of the RBF DFE can be, however, improved by increasing both the decision delay τ and the feedforward order m, as argued in Section 8.11, at the expense of increased computational complexity. However, the present parameter values for the conventional DFE and RBF DFE are convenient, since they yield similar BER performances.

The performance comparison of the adaptive schemes, i.e. that of the AQAM DFE and AQAM RBF DFE, is given in Figure 9.13. For the 10^{-2} target BER system, the AQAM RBF DFE provides a better BER performance, than the Kalman-filtering based AQAM DFE in the SNR range from 0dB to 28dB at the expense of a lower BPS performance, especially for higher SNRs. As the SNR exceeds 28dB, the BER performance of the AQAM DFE scheme becomes superior to that of the AQAM RBF DFE. This is because at higher SNRs the 64-QAM modulation mode prevails and since the 64-QAM BER performance of the conventional DFE was better, than that of the RBF DFE in Figure 9.12, hence the mean BER improvement of the AQAM DFE is expected, when compared to that of the AQAM RBF DFE.

For the 10^{-4} target BER system, the BER performance of the AQAM DFE and AQAM RBF DFE is fairly similar in the SNR range from 5dB to 12dB, but the BPS performance of the AQAM RBF DFE is better, than that of the AQAM DFE in that range. In this SNR range the lower-order modulation modes dominate. Since the RBF DFE can provide a better BER performance, than that of the conventional DFE for the lower-order modulation modes, the BPS performance of the AQAM RBF DFE can be improved, while maintaining a similar BER performance to that of the AQAM DFE. As the SNR exceeds 12dB, the BER performance of the AQAM RBF DFE remains better at the expense of a lower BPS performance.

The overall results of our simulations show that the AQAM RBF DFE is capable of performing similarly to the AQAM DFE at a lower decision delay and lower feedforward and feedback order. However, the computational complexity of the RBF DFE is dependent on the modulation mode, since the number of RBF centres increases with the number of modulation levels, as discussed in Section 8.7. This is not so in the context of the conventional DFE, where the computational complexity is only dependent on the feedforward and feedback order. Table 9.3 compares the computational complexity of the RBF DFE ($m = 2$, $n = 1$, $\tau = 1$) and the conventional DFE ($m = 15$, $n = 2$) used in our simulations. The complexity analysis of the RBF DFE is based on Table 8.10. The high computational cost incurred by the RBF DFE in the high-order $\mathcal{M}$-ary modulation modes presents a drawback for the AQAM RBF DFE scheme.

Operation	RBF DFE				Conv. DFE
	BPSK	4-QAM	16-QAM	64-QAM	
subtraction and addition	15	60	1008	16320	16
multiplication	12	48	768	12288	17
division	4	16	256	4096	0
exp()	4	16	256	4096	0

Table 9.3: Computational complexity of RBF DFE and conventional DFE per equalized output sample. The RBF DFE has a feedforward order of $m = 2$, feedback order of $n = 1$ and decision delay of $\tau = 1$ symbol. The number of RBF hidden units $n_{s,j}$ is dependent on the order of the $\mathcal{M}$-QAM modes and the channel memory L where $n_{s,j} = \mathcal{M}^{m+L-n}$. The channel memory is assumed to be $L = 1$. The complexity analysis of the RBF DFE is based on Table 8.10. The conventional DFE has a feedforward order of $m = 15$, feedback order of $n = 2$ and decision delay of $\tau = 15$ symbols.

Nevertheless, we note that unlike the conventional DFE, the AQAM RBF DFE is capable of performing well over channels, which result in non-linearly separable received phasor constellations.

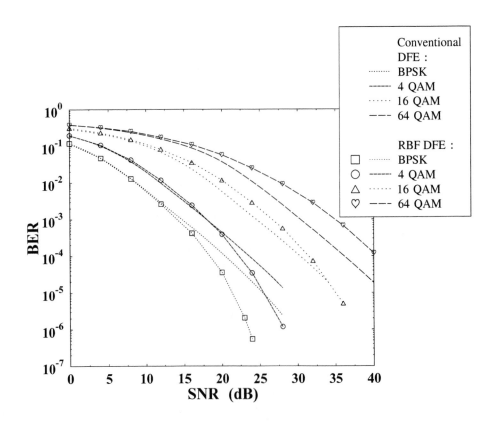

Figure 9.12: BER versus SNR performance of the conventional DFE and the RBF DFE over the two-path equal-weight symbol-spaced Rayleigh-fading channel of Table 9.1 for different $\mathcal{M}$-QAM schemes. The conventional DFE has a feedforward order of $m = 15$, feedback order of $n = 2$ and decision delay of $\tau = 15$ symbols. The RBF DFE has a feedforward order of $m = 2$, feedback order of $n = 1$ and decision delay of $\tau = 1$ symbols.

9.4.6 Discussion

In the above sections, BbB adaptive modulation was applied in conjunction with the RBF DFE of Section 8.11 in a wideband channel environment. The short-term BER of Equation 9.15 estimated by the RBF DFE was used as the modem mode switching metric in order to switch between different modulation modes. The validity of using this metric was tested in Section 9.4.5 and in Figure 9.8 it was shown that the RBF DFE gives a good BER estimate for the adaptive scheme to maintain the target mean BER performance. The simulation results also showed that there was a performance improvement in terms of the mean BER and mean BPS, when compared to the constituent fixed modulation modes. The performance of the joint AQAM RBF DFE scheme was then compared to that of the joint AQAM conventional DFE scheme investigated by Wong [32]. The AQAM RBF DFE having a lower feedforward and feedback order and a smaller decision delay, showed comparable performance to the

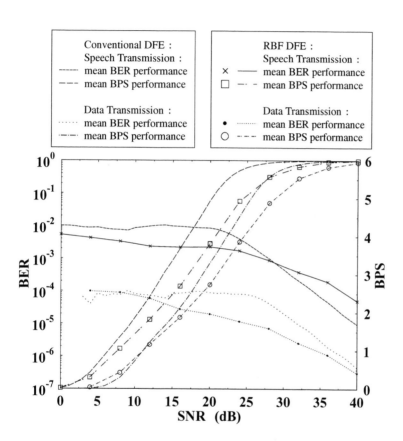

Figure 9.13: Simulated best-case performance of the AQAM RBF DFE scheme and the numerical best-case performance of the joint AQAM conventional DFE scheme for speech- and data-transmission [32], using the parameters listed in Table 9.1 and the assumptions stated in Section 9.4.3. The modem mode switching levels used for the joint AQAM conventional DFE scheme are listed in Table 9.2. The RBF DFE had a feedforward order of $m = 2$, feedback order of $n = 1$ and decision delay of $\tau = 1$ symbol and the conventional DFE had a feedforward order of $m = 15$, feedback order of $n = 2$ and decision delay of $\tau = 15$ symbols.

AQAM DFE in our simulations.

In our future work, the performance of the AQAM RBF DFE will be investigated in practical situations, where the effect of discarding the assumptions made in Section 9.4.3 is to be quantified.

9.5 Performance of the AQAM RBF DFE Scheme: Switching Metric Based on the Previous Short-term BER Estimate

In this section, we analyse the performance of the AQAM RBF DFE scheme by discarding assumption 5 of Section 9.4.3. Therefore, the estimated short-term BER of the *current* transmitted burst is used to select the modulation mode for the *next* transmission burst, as described in Equation 9.7.

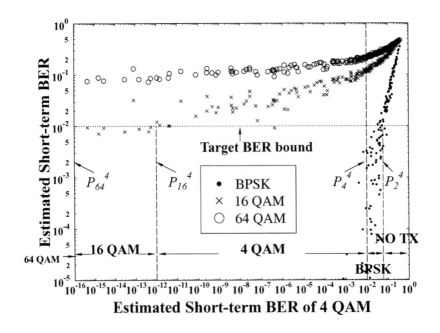

Figure 9.14: The estimated short-term BER for all the possible modulation modes that can be invoked, assuming that the current mode is 4-QAM – versus the estimated short-term BER of 4-QAM for the two-path Rayleigh fading channel of Table 9.1.

The BER switching thresholds corresponding to $\mathcal{M}$-QAM, $P_i^{\mathcal{M}}, i = 2, 4, 16, 64$, can be obtained by estimating the BER degradation/improvement, when the modulation mode is switched from $\mathcal{M}$-QAM to a higher/lower number of modulation levels. In this experiment, we obtain this BER degradation/improvement measure from the estimated short-term BER of every modulation mode used, under the same instantaneous channel conditions. Figure 9.14 shows the estimated short-term BER of all the possible modulation modes that can be invoked, assuming that the current mode is 4-QAM, versus the estimated short-term BER of 4-QAM under the same instantaneous channel conditions. The short-term BERs of the modulation modes are obtained on a burst-by-burst basis from the RBF DFE according to Equation 9.15. Each point in Figure 9.14 represents the RBF DFE's estimated short-term BER for a specific received data burst using the corresponding modulation mode. In order to maintain the target BER of 10^{-2}, Figure 9.14 demonstrates, how each switching BER threshold P_i^4 is obtained. The short-term BER of the 4-QAM transmission burst, when the corresponding BPSK, 16-QAM and 64-QAM transmission burst under the same instantaneous channel conditions has an estimated BER of 10^{-2} is approximately $6 \times 10^{-2}, 10^{-12}$ and 0, respectively. For example, if the estimated short-term BER of the received 4-QAM transmission burst is below $P_{16}^4 = 10^{-12}$, the modulation mode can be 'safely' switched to 16-QAM for the next transmission burst, since the short-term BER of this 16-QAM transmission burst is expected to be below the target BER of 10^{-2}. The 4-QAM error probability of $P_{16}^4 = 10^{-12}$ used in this example for switching to 16-QAM appears extremely conservative, but it is justified by the large uncertainty associated with the estimation of the BER due to the Rayleigh-faded impulse response taps. This manifests itself also in the rather spread nature of the BER estimates in Figure 9.14. A feasible technique for mitigating this phenomenon is employing the fade-tracking scheme of Figure 11.2 in Reference [4]. Using this method the switching BER thresholds were obtained for the target BER of 10^{-2} and 10^{-4}, as listed in Table 9.4 and Table 9.5, respectively in the context of all possible combinations of the mode transitions. Note that the extremely low values for $P_{16}^4 = 1 \times 10^{-45}$ and $P_{64}^{16} = 1 \times 10^{-50}$ in Table 9.5 were obtained by extrapolating the curves similar to Figure 9.14 but for 16-QAM and 64-QAM, respectively, in order to achieve the target BER of 10^{-4}.

	$P_2^{\mathcal{M}}$	$P_4^{\mathcal{M}}$	$P_{16}^{\mathcal{M}}$	$P_{64}^{\mathcal{M}}$
NO TX	9×10^{-3}	5×10^{-5}	0.0	0.0
BPSK	1×10^{-2}	5×10^{-5}	0.0	0.0
4-QAM	6×10^{-2}	1×10^{-2}	1×10^{-12}	0.0
16-QAM	2×10^{-1}	1×10^{-1}	1×10^{-2}	1×10^{-8}
64-QAM	3×10^{-1}	2×10^{-1}	9×10^{-2}	1×10^{-2}

Table 9.4: The switching BER thresholds $P_i^{\mathcal{M}}$ of the joint adaptive modulation and RBF DFE scheme for the target BER of 10^{-2} over the two-path Rayleigh fading channel of Table 9.1.·

Figure 9.15 shows the BER and BPS performance of the joint AQAM RBF DFE scheme designed for BER = 10^{-2} with the switching thresholds given in Table 9.4 – when using the current transmission burst's BER estimate, in order to determine the modem mode of the next transmission burst – in contrast to its best-case performance. The performance comparison shows that there is little performance degradation, when the *current* short-term BER estimate is used to control the modulation mode of the *next* transmission burst based on the switching parameters of Table 9.4 for the AQAM scheme designed for BER = 10^{-2}. Since the channel of Table 9.1 is slowly varying, the performance of the joint AQAM RBF DFE scheme based

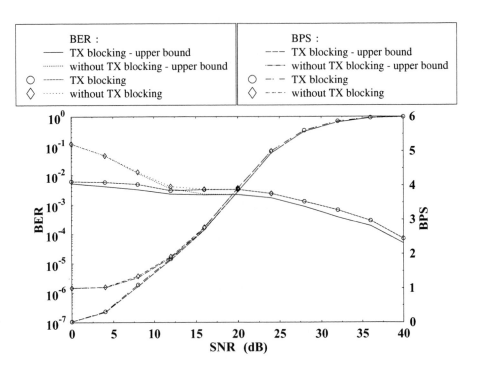

Figure 9.15: The BER and BPS performance of the joint AQAM RBF DFE scheme **using the current BER estimate in order to estimate the next burst's transmission mode**, and its best-case performance for the 10^{-2} target BER system, using the parameters listed in Table 9.1. The modem mode switching levels used for the joint AQAM RBF DFE scheme are listed in Table 9.4. The RBF DFE had a feedforward order of $m = 2$, feedback order of $n = 1$ and decision delay of $\tau = 1$ symbol.

on the switching parameters of Table 9.4 is comparable to its best-case performance.

	$P_2^{\mathcal{M}}$	$P_4^{\mathcal{M}}$	$P_{16}^{\mathcal{M}}$	$P_{64}^{\mathcal{M}}$
NO TX	9×10^{-5}	1×10^{-15}	0.0	0.0
BPSK	1×10^{-4}	1×10^{-15}	0.0	0.0
4-QAM	1.5×10^{-2}	1×10^{-4}	1×10^{-45}	0.0
16-QAM	1.2×10^{-1}	5×10^{-2}	1×10^{-4}	1×10^{-50}
64-QAM	2.2×10^{-1}	1.5×10^{-1}	3×10^{-2}	1×10^{-4}

Table 9.5: The switching BER thresholds $P_i^{\mathcal{M}}$ of the joint adaptive modulation and RBF DFE scheme for the target BER of 10^{-4} over the two-path Rayleigh fading channel of Table 9.1.

Figure 9.16 shows the BER and BPS performance of the AQAM RBF DFE scheme designed for data-transmission using the switching threshold given in Table 9.5 in comparison to its best-case performance. The degradation with respect to the best-case performance of

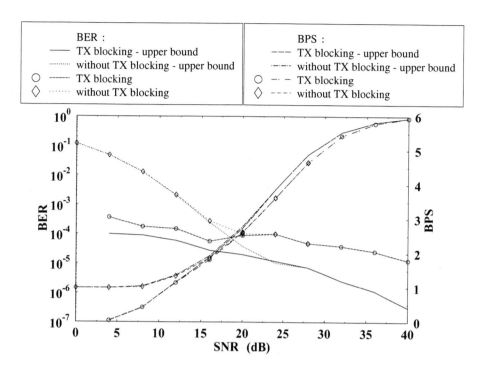

Figure 9.16: The BER and BPS performance of the joint AQAM RBF DFE scheme **using the current BER estimate in order to estimate the next burst's transmission mode**, and its best-case performance for **data-transmission**, using the parameters listed in Table 9.1. The modem mode switching levels used for the AQAM RBF DFE scheme are listed in Table 9.5. The RBF DFE had a feedforward order of $m = 2$, feedback order of $n = 1$ and decision delay of $\tau = 1$ symbol.

the AQAM RBF DFE scheme designed for data transmission at BER = 10^{-4} based on the switching threshold given in Table 9.5 is more significant compared to the adaptive scheme designed for BER = 10^{-2}, as seen in Figure 9.15 and Figure 9.16. Note that for the low BER switching thresholds of $P_{16}^4(= 1 \times 10^{-45})$ and $P_{64}^{16}(= 1 \times 10^{-50})$ in Table 9.5 – which was required by the adaptive scheme for achieving the target BER of 10^{-4} – the RBF DFE is unable to provide BER estimates of such high accuracy. As the SNR improves, the relative frequency of encountering the switching thresholds P_{16}^4 and P_{64}^{16} increases and thus the performance degradation compared to the best-case increases. The performance degradation with respect to the best-case was also contributed by the spread nature of the BER estimates due to the Rayleigh-faded CIR taps. The BER estimation spread was more evident, when the BER estimate decreased, as shown in Figure 9.14. Therefore, there is a substantial BER estimation inaccuracy associated with the switching thresholds P_{16}^4 and P_{64}^{16}.

Figures 9.17 and 9.18 compare the probability of encountering each modulation mode employed in the adaptive scheme and those employed in the best-case performance scenario for speech-quality (BER = 10^{-2}) transmission and data-quality (BER = 10^{-4}) transmission,

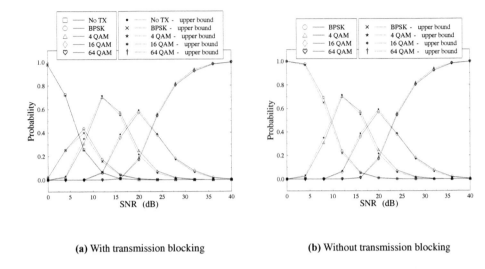

(a) With transmission blocking   (b) Without transmission blocking

Figure 9.17: The probability of encountering the various $\mathcal{M}$-QAM modulation modes in the joint AQAM and RBF DFE scheme during **speech-quality transmission (target BER of 0.01)** over the two-path equal-weight, symbol-spaced Rayleigh fading channel using the simulation parameters listed in Table 9.1. The probability of modulation mode utilization for best-case performance, as given in Figure 9.10, is provided for comparison.

respectively. Figure 9.17 shows that the switching BER thresholds of Table 9.4, determined with our suggested method and the previous short-term BER estimate is capable of providing similar modulation mode utilization for the adaptive scheme designed for speech-quality transmission compared with its best-case performance. However, for data-quality transmission, we note from Figure 9.18 that the utilization of the 64-QAM mode of the adaptive scheme is more frequent, than that of the best-case performance for high SNRs. This also explains the substantial BER degradation from its best-case performance, as the SNR improves, as demonstrate in Figure 9.16.

9.6   Review and Discussion

The RBF DFE was shown to provide a good 'on-line' BER estimation of the received data burst, which was used as the AQAM mode switching metric. Our simulation results showed that the proposed RBF DFE-assisted burst-by-burst adaptive modem outperformed the individual constituent fixed modulation modes in terms of the mean BER and BPS. Transmission blocking was utilized to maintain the target BER performance. Without transmission blocking, the target BER of 10^{-2} and 10^{-4} can only be achieved, when the channel SNR is higher than 9dB and 18dB, respectively. However, the disadvantage is that the utilization of transmission blocking results in transmission latency.

The AQAM scheme employing RBF DFE was compared to the AQAM scheme using conventional DFE in order to mitigate the effects of the dispersive wideband channel. Our

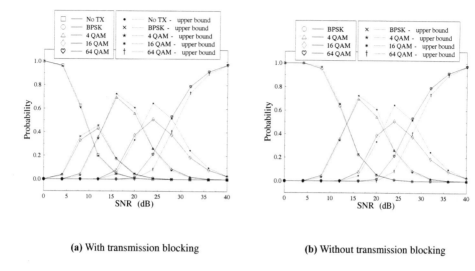

(a) With transmission blocking

(b) Without transmission blocking

Figure 9.18: The probability of encountering the various $\mathcal{M}$-QAM modulation modes in the joint AQAM and RBF DFE scheme during **data-quality transmission (target BER of 0.0001)** over the two-path equal-weight, symbol-spaced Rayleigh fading channel using the simulation parameters listed in Table 9.1. The probability of modulation mode utilization for the best-case performance, as given in Figure 9.11, is provided for comparison.

results showed that the AQAM RBF DFE scheme was capable of performing as well as the conventional AQAM DFE at a lower decision delay and lower feedforward as well as feedback order. The performance of the AQAM RBF DFE can be improved by increasing both the decision delay τ and the feedforward order m, at the expense of increased computational complexity, while the performance of the conventional AQAM DFE cannot be improved significantly by increasing its equalizer order. However, the computational complexity of the RBF DFE is dependent on the AQAM mode and increases significantly for higher-order modulation modes. This is not so in the context of the conventional DFE, where the computational complexity is only dependent on the feedforward and feedback order.

A method to obtain the switching BER thresholds of the joint AQAM RBF DFE scheme was proposed in Section 9.5 and was shown to suffer only minor performance degradation in comparison to the achievable best-case performance generated by assuming that the corresponding BER of all modulation modes was known given the estimated BER of the received burst.

Overall, we have shown that our proposed AQAM scheme improved the throughput performance compared to the constituent fixed modulation modes. The RBF DFE provided a reliable channel quality measure, which quantified all channel impairments, irrespective of their source for the AQAM scheme and at the same time it improved the BER performance. In the following chapter, we will enhance the performance of the AQAM RBF DFE by invoking turbo coding.

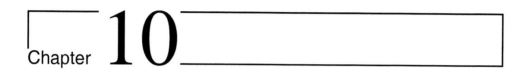

Chapter **10**

RBF Equalization Using Turbo Codes

In this chapter, the wideband AQAM scheme explored in the previous chapter is extended to incorporate the benefits of channel coding. Channel coding, with its error correction and detection capability, is capable of improving the BER and throughput performance of the wideband AQAM scheme. Since the wideband AQAM scheme always attempts to invoke the appropriate modulation mode in order to combat the wideband channel effects, the probability of encountering a received transmitted burst with a high instantaneous BER is low, when compared to the constituent fixed modulation modes. This characteristic is advantageous, since due to the less bursty error distribution, a coded wideband AQAM scheme can be implemented successfully without the utilization of long-delay channel interleavers. Therefore we can exploit the error detection capability of the channel codes near-instantaneously at the receiver for every received transmission burst.

Turbo coding [152, 155] is invoked in conjunction with the RBF assisted AQAM scheme in a wideband channel scenario in this chapter. We will first introduce the novel concept of Jacobian RBF equalizer, which is a reduced-complexity logarithmic version of the RBF equalizer. The Jacobian logarithmic RBF equalizer generates its output in the logarithmic domain and hence it can be used to provide soft outputs for the turbo decoder. We will investigate different channel quality measures – namely the short-term BER and average burst log-likelihood ratio magnitude of the bits in the received burst before and after channel decoding – for controlling the mode-switching regime of our adaptive scheme. We will now briefly review the concept of turbo coding.

10.1 Introduction to Turbo Codes

Turbo codes were introduced in 1993 by Berrou, Glavieux and Thitimajshima [152, 155]. These codes achieve a near-Shannon-limit error correction performance with relatively simple component codes and invoking large interleavers. The component codes that are usually used are either recursive systematic convolutional (RSC) codes or block codes. The general structure of the turbo encoder is shown in Figure 10.1. The information sequence is encoded twice, using an interleaver or scrambler between the two encoders, rendering the two encoded data sequences approximately statistically independent of each other. The encoders produce a so-called systematically encoded output, which is equivalent to the original information sequence, as well as a stream of parity information bits. The parity outputs of the two component codes are then often punctured in order to maintain as high a coding rate as possible, without substantially reducing the codec's performance. Finally, the bits are multiplexed before being transmitted.

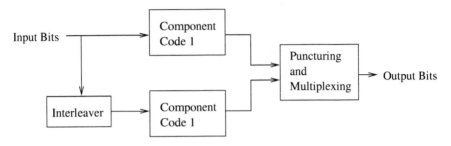

Figure 10.1: Turbo encoder schematic.

The turbo decoder consists of two decoders, linked by interleavers in a structure obeying the constraints imposed by the encoder, as seen in Figure 10.1. The turbo decoder accepts soft inputs and provides soft outputs as the decoded sequence. The soft inputs and outputs provide not only an indication of whether a particular bit was a binary 0 or a 1, but also deliver the so-called log-likelihood ratio (LLR) of the bit which constituted by the logarithm of the quotient of the probability of the bit concerned being a logical one and zero, respectively. Two often-used decoders are the Soft Output Viterbi Algorithm (SOVA) [302] and the Maximum A Posteriori (MAP) [162] algorithm.

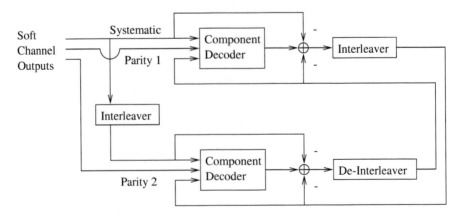

Figure 10.2: Turbo decoder schematic.

As seen in Figure 10.2, each decoder takes three types of inputs - the systematically encoded channel output bits, the parity bits transmitted from the associated component encoder and the information estimate from the other component decoder, referred to as the *a priori* information of the decoded bits. The decoder operates iteratively. In the first iteration, the first component decoder provides a soft output and the so-called extrinsic output based on the soft channnel outputs alone. The terminology 'extrinsic' implies that this information is not based on the received information directly related to the bit concerned, it is rather based on information, which is indirectly related to the bit due to the code-constraints introduced by the encoder. This extrinsic output generated by the first decoder – which constitutes the first decoder's 'opinion' as to the bit concerned – is used by the second component decoder as *a priori* information, and this information together with the channel outputs is used by the second component decoder, in order to generate its soft output and extrinsic information. Symmetrically, in the second iteration, the extrinsic information generated by the second decoder in the first iteration is used as the *a priori* information for the first decoder. Using this *a priori* information, the decoder is likely to decode more bits correctly than it did in the first iteration. This cycle continues and at each iteration the BER in the decoded sequence drops. However, the extra BER improvement obtained with each iteration diminishes, as the number of iterations increases. In order to limit the computational complexity, the number of iterations is usually fixed according to the prevalent design criteria expressed in terms of performance and complexity. When the series of iterations is curtailed, after either a fixed number of iterations or when a termination criterion is satisfied, the output of the turbo decoder is given by the de-interleaved *a posteriori* LLRs of the second component decoder. The sign of these *a posteriori* LLRs gives the hard decision output and in some applications the magnitude of these LLRs provides the confidence measure of the decoder's decision. Because of the iterative nature of the decoder, it is important not to re-use the same information more than once at each decoding step, since this would destroy the independence of the two encoded sequences which was originally imposed by the interleaver of Figure 10.2. For this reason the concept of the so-called extrinsic and intrinsic information was used in the original paper on turbo coding by Berrou *et al.* [152] to describe the iterative decoding of turbo codes.

For a more detailed exposition of the concept and algorithm used in the iterative decoding of turbo codes, the reader is referred to [152]. Other, non-iterative decoders have also been proposed [303, 304] which give optimal decoding of turbo codes, but they are rather complex and provide disproportionately low improvement in performance over iterative decoders. Therefore, the iterative scheme shown in Figure 10.2 is usually used. Continuing from our previous work, where we used an RBF equalizer to mitigate the effects of the wideband channel, we will introduce turbo coding in order to improve the BER and/or BPS performance.

In the next section, before we discuss the joint RBF equalization and turbo coding system, we will introduce the *Jacobian logarithmic RBF equalizer*, which computes the output of the RBF network in logarithmic form based on the Log-MAP algorithm [288] used in turbo codes to reduce their computational complexity.

10.2 Jacobian Logarithmic RBF Equalizer

The Bayesian-based RBF equalizer has a high computational complexity due to the evaluation of the nonlinear exponential functions in Equation 8.80 and due to the high number of additions/subtractions and multiplications/divisions required for the estimation of each symbol, as it was expounded in Section 8.9.

In this section – based on the approach often used in turbo codes – we propose generating the output of the RBF network in logarithmic form by invoking the so-called Jacobian logarithm [288, 289] , in order to avoid the computation of exponentials and to reduce the number of multiplications performed. We will refer to the RBF equalizer using the Jacobian logarithm as the *Jacobian logarithmic RBF equalizer*. Below we will present this idea in more detail.

We will first introduce the Jacobian logarithm, which is defined by the relationship [288]:

$$
\begin{aligned}
J(\lambda_1, \lambda_2) &= \ln(e^{\lambda_1} + e^{\lambda_2}) \\
&= \max(\lambda_1, \lambda_2) + \ln(1 + e^{-|\lambda_1 - \lambda_2|}) \\
&\approx \max(\lambda_1, \lambda_2) + f_c(|\lambda_1 - \lambda_2|),
\end{aligned}
\tag{10.1}
$$

where the first line of Equation 10.1 is expressed in a computationally less demanding form as $\max(\lambda_1, \lambda_2)$ plus the correction function $f_c(\cdot)$. The correction function $f_c(x) = \ln(1 + e^{-x})$ has a dynamic range of $\ln(2) \geq f_c(x) > 0$, and it is significant only for small values of x [288]. Thus, $f_c(x)$ can be tabulated in a look-up table, in order to reduce the computational complexity [288]. The correction function $f_c(\cdot)$ only depends on $|\lambda_1 - \lambda_2|$, therefore the look-up table is one dimensional and experience shows that only few values have to be stored [305]. The Jacobian logarithmic relationship in Equation 10.1 can be extended also to cope with a higher number of exponential summations, as in $\ln\left(\sum_{k=1}^{n} e^{\lambda_k}\right)$. Reference [288] showed that this can be achieved by nesting the $J(\lambda_1, \lambda_2)$ operation as follows:

$$
\ln\left(\sum_{k=1}^{n} e^{\lambda_k}\right) = J(\lambda_n, J(\lambda_{n-1}, \ldots J(\lambda_3, J(\lambda_2, \lambda_1))\ldots)).
\tag{10.2}
$$

Having presented the Jacobian logarithmic relationship, we will now describe, how this operation can be used to reduce the computational complexity of the RBF equalizer.

The overall response of the RBF network, given in Equation 8.80, is repeated here for convenience:

$$
f_{RBF}(\mathbf{v}_k) = \sum_{i=1}^{M} w_i \exp(-\|\mathbf{v}_k - \mathbf{c}_i\|^2 / \rho).
\tag{10.3}
$$

Expressing Equation 10.3 in a logarithmic form and substituting in the Jacobian logarithm, we obtain:

$$
\begin{aligned}
\ln(f_{RBF}(\mathbf{v}_k)) &= \ln(\sum_{i=1}^{M} w_i \exp(-\|\mathbf{v}_k - \mathbf{c}_i\|^2 / \rho)) \\
&= \ln(\sum_{i=1}^{M} \exp(\ln(w_i)) \exp(-\|\mathbf{v}_k - \mathbf{c}_i\|^2 / \rho)) \\
&= \ln(\sum_{i=1}^{M} \exp(w_i' + \nu_{ik})) \\
&= \ln(\sum_{i=1}^{M} \exp(\lambda_{ik})) \\
&= J(\lambda_{Mk}, J(\lambda_{(M-1)k}, \ldots J(\lambda_{2k}, \lambda_{1k})\ldots)),
\end{aligned}
\tag{10.4}
$$

where $w_i' = \ln(w_i)$, which can be considered as a transformed weight. Furthermore, we used the shorthand $\nu_{ik} = -\|\mathbf{v}_k - \mathbf{c}_i\|^2/\rho$ and $\lambda_{ik} = \nu_{ik} + w_i'$. By introducing the Jacobian logarithm, every weighted summation of two exponential operations in Equation 10.3 is substituted with an addition, a subtraction, a table look-up and a max operation according to Equation 10.1, thus reducing the computational complexity. The term $\ln(\sum_{i=1}^M \exp(w_i' + \nu_{ik}))$ requires $3M - 1$ additions/subtractions, $M - 1$ table look-up and $M - 1 \max(\cdot)$ operations. Most of the computational load arises from computing the Euclidean norm term $\|\mathbf{v}_k - \mathbf{c}_i\|^2$, and the associated total complexity will depend on the number of RBF centres and on the dimension m of both the RBF centre vector $\mathbf{c}_i$ and the channel output vector $\mathbf{v}_k$. The evaluation of the term $\nu_{ik} = -\|\mathbf{v}_k - \mathbf{c}_i\|^2/\rho$ requires $2m - 1$ additions/subtractions, m multiplications and one division operation. Therefore, the computational complexity of a RBF DFE having m inputs and $n_{s,j}$ hidden RBF nodes per equalised output sample, which was previously given in Table 8.10, is now reduced to the values seen in Table 10.1 due to employing the Jacobian algorithm.

Determine the feedback state	
$n_{s,j}(2m + 2) - 2\mathcal{M}$	subtraction and addition
$n_{s,j}m$	multiplication
$n_{s,j}$	division
$n_{s,j} - \mathcal{M} + 1$	max
$n_{s,j} - \mathcal{M}$	table look-up

Table 10.1: Computational complexity of a $\mathcal{M}$-ary Jacobian logarithmic decision feedback RBF network equalizer with m inputs and $n_{s,j}$ hidden units per equalised output sample based on Equations 8.103 and 10.4.

Exploiting the fact that the elements of the vector of noiseless channel outputs constituting the channel states $\mathbf{r}_i, i = 1, \ldots, n_s$ correspond to the convolution of a sequence of $(L + 1)$ transmitted symbols and $(L + 1)$ CIR taps – where these vector elements are referred to as the scalar channel states $r_l, l = 1, \ldots, n_{s,f}(= \mathcal{M}^{L+1})$ – we could use Patra's and Mulgrew's method [287] to reduce the computational load arising from evaluating the Euclidean norm ν_{ik} in Equation 10.4. Expanding the term ν_{ik} gives

$$
\begin{aligned}
\nu_{ik} &= \frac{-\|\mathbf{v}_k - \mathbf{c}_i\|^2}{\rho} \\
&= -\frac{(v_k - c_{i0})^2}{\rho} - \frac{(v_{k-1} - c_{i1})^2}{\rho} - \cdots \\
&\quad - \frac{(v_{k-j} - c_{ij})^2}{\rho} - \cdots - \frac{(v_{k-m+1} - c_{i(m-1)})^2}{\rho}, \\
&\quad i = 1, \ldots, M, \qquad k = -\infty, \ldots, \infty,
\end{aligned}
\tag{10.5}
$$

where v_{k-j} is the delayed received signal and c_{ij} is the jth component of the RBF centre vector $\mathbf{c}_i$, which takes the values of the scalar channel outputs $r_l, l = 1, \ldots, n_{s,f}$ as described in Section 8.10. Note from Equation 10.5 that ν_{ik} is a summation of the delayed components, $-\frac{(v_k - c_{ij})^2}{\rho}$ and the scalar centres c_{ij} take the values of the scalar channel outputs $r_l, l = 1, \ldots, n_{s,f}$. Thus, we could reduce the computational complexity of evaluating

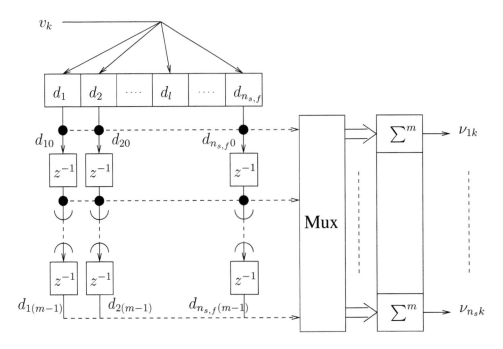

Figure 10.3: Reduced complexity computation of ν_{ik} in Equation 10.5 for substitution in Equation 10.4 based on scalar channel output.

Equation 10.5 by pre-calculating $d_l = -\frac{(v_k - r_l)^2}{\rho}, l = 1, \ldots, n_{s,f}$ for all the $n_{s,f}$ possible values of the scalar channel outputs $r_l, l = 1, \ldots, n_{s,f}$ and storing the values. From Equation 10.5 the value of ν_{ik} can be obtained by summing the corresponding delayed values of d_l, which we will define as

$$d_{lj} = -\frac{(v_{k-j} - r_l)^2}{\rho}, \qquad l = 1, \ldots, n_{s,f}, \qquad j = 0, \ldots, m-1.$$

Substituting Equation 10.2 into Equation 10.5 yields:

$$\nu_{ik} = \sum_{\substack{j=0 \\ c_{ij} = r_l}}^{m-1} d_{lj}, \qquad i = 1, \ldots, M, \qquad k = -\infty, \ldots, \infty.$$

The reduced complexity computation of ν_{ik} in Equation 10.2 for substitution in Equation 10.4 based on the scalar channel outputs r_l, can be represented as in Figure 10.3. The multiplexer (Mux) of Figure 10.3 maps d_{lj} of Equation 10.2 corresponding to the scalar centre r_l to the contribution of the vector centre's component c_{ij}.

The computation of $d_l = -\frac{(v_k - r_l)^2}{\rho}, l = 1, \ldots, n_{s,f}$ requires $n_{s,f}$ multiplication, division and subtraction operations. For every RBF centre vector $\mathbf{c}_i$, computing its corresponding ν_{ik} value according to Equation 10.2 needs $m - 1$ additions. The reduced computational complexity per equalised output sample of an $\mathcal{M}$-ary Jacobian DFE with m inputs,

$n_{s,j} = \mathcal{M}^{m+L-n}$ hidden RBF nodes derived from $n_{s,f} = \mathcal{M}^{L+1}$ scalar centres is given in Table 10.2. Comparing Table 10.1 and 10.2, we observe a substantial computational complexity reduction, especially for a high feedforward order m, since $n_{s,f} < n_{s,j}$, if $m-n < 1$. For example, for the 16-QAM mode we have $n_{s,f} = 256$ and $n_{s,j} = 256$ for the RBF DFE equalizer parameters of $m = 2$, $n = 1$ and $\tau = 1$. The total complexity reduction is by a factor of about 1.3. If we increase the RBF DFE feedforward order and use the equalizer parameters of $m = 3$, $n = 1$ and $\tau = 2$ – which gives a better BER performance – then we have $n_{s,f} = 256$ and $n_{s,j} = 4096$ – and the total complexity reduction is by a factor of about 2.1. The computational complexity can be further reduced by neglecting the RBF scalar centres situated far from the received signal v_k, since the contribution of RBF scalar centres r_l to the decision function is inversely related to their distance from the received signal v_k, as recognised by Patra [287].

Determine the feedback state	
$n_{s,j}(m+2) - 2\mathcal{M} + n_{s,f}$	subtraction and addition
$n_{s,f}$	multiplication
$n_{s,f}$	division
$n_{s,j} - \mathcal{M} + 1$	max
$n_{s,j} - \mathcal{M}$	table look-up

Table 10.2: Reduced computational complexity per equalised output sample of an $\mathcal{M}$-ary Jacobian logarithmic RBF DFE based on scalar centres. The Jacobian RBF DFE based on Equation 8.103 and 10.4 has m inputs and $n_{s,j}$ hidden RBF nodes, which are derived from the $n_{s,f}$ number of scalar centres.

Figures 10.4 and 10.5 show the BER versus SNR performance comparison of the RBF DFE and the Jacobian logarithmic RBF DFE over the two-path Gaussian channel and two-path Rayleigh fading channel of Table 9.1, respectively. For the simulation of the Jacobian logarithmic RBF DFE the correction function $f_c(\cdot)$ in Equation 10.1 was approximated by a pre-computed table having eight stored values ranging from 0 to $\ln(2)$. From these results we concluded that the Jacobian logarithmic RBF equalizer's performance was equivalent to that of the RBF equalizer, whilst having a lower computational complexity.

Having presented the proposed reduced complexity Jacobian logarithmic RBF equalizer, we will now proceed to introduce the joint RBF equalization and turbo coding system and investigate its performance in both fixed QAM and burst-by-burst (BbB) AQAM schemes.

10.3 System Overview

The structure of the joint RBF DFE and turbo decoder is portrayed in Figure 10.6. The output of the RBF DFE provides the *a posteriori* LLRs of the transmitted bits based on the *a posteriori* probability of each legitimate $\mathcal{M}$-QAM symbol. The *a posteriori* LLR of a data bit u_k is denoted by $\mathcal{L}(u_k|\mathbf{v}_k)$, which was defined as the log of the ratio of the probabilities

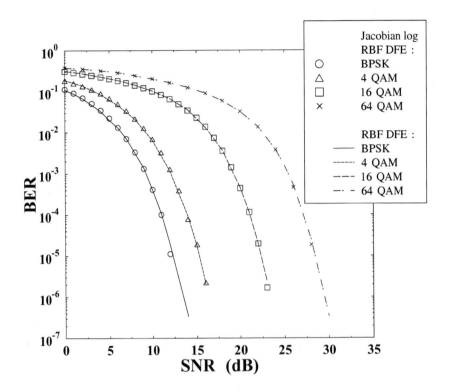

Figure 10.4: BER versus signal to noise ratio performance of the RBF DFE and the Jacobian logarithmic RBF DFE over the dispersive **two-path Gaussian channel** of Figure 8.21(a) for
different $\mathcal{M}$-QAM modes. Both equalizers have a feedforward order of $m = 2$, feedback
order of $n = 1$ and decision delay of $\tau = 1$ symbol.

of the bit being a logical 1 or a logical 0, conditioned on the received sequence $\mathbf{v}_k$:

$$
\begin{aligned}
\mathcal{L}(u_k|\mathbf{v}_k) &= \ln\left(\frac{P(u_k = +1|\mathbf{v}_k)}{P(u_k = -1|\mathbf{v}_k)}\right), \\
&= L(u_k = +1|\mathbf{v}_k) - L(u_k = -1|\mathbf{v}_k),
\end{aligned}
\tag{10.6}
$$

where the term $L(u_k = \pm1|\mathbf{v}_k) = ln(P(u_k = \pm1|\mathbf{v}_k))$ is the log-likelihood of the data bit
u_k having the value ±1 conditioned on the received sequence $\mathbf{v}_k$.

The LLR of the bits representing the QAM symbols can be obtained from the *a posteriori*
log-likelihood of the symbol. Below we provide an example for the 4-QAM mode of our
AQAM scheme. The *a posteriori* log-likelihood L_1, L_2, L_3 and L_4 of the four possible 4-
QAM symbols is given by the Jacobian RBF networks. A 4-QAM symbol is denoted by the
bits $U_0 U_1$ and the symbols $\mathcal{I}_1, \mathcal{I}_2, \mathcal{I}_3$ and $\mathcal{I}_4$ correspond to 00, 01, 10 11, respectively. Thus,

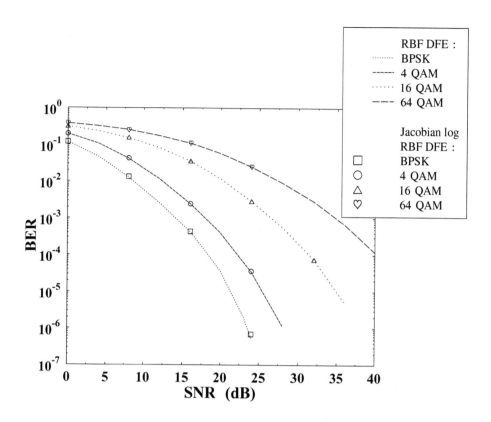

Figure 10.5: BER versus signal to noise ratio performance of the RBF DFE and the Jacobian logarithmic RBF DFE over the **two path equal weight, symbol-spaced Rayleigh fading channel**
of Table 9.1 for different $\mathcal{M}$-QAM modes. Both equalizers have a feedforward order of
$m = 2$, feedback order of $n = 1$ and decision delay of $\tau = 1$ symbol. Correct symbols
were fed back.

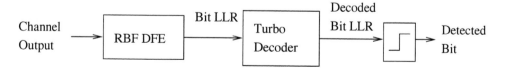

Figure 10.6: Joint RBF DFE and turbo decoder schematic.

the *a posteriori* LLRs of the bits are obtained as follows:

$$\mathcal{L}(U_0|\mathbf{v}_k) = L(U_0 = 1|\mathbf{v}_k) - L(U_0 = 0|\mathbf{v}_k),$$
$$\mathcal{L}(U_1|\mathbf{v}_k) = L(U_1 = 1|\mathbf{v}_k) - L(U_1 = 0|\mathbf{v}_k), \tag{10.7}$$

where,

$$
\begin{aligned}
L(U_0 = 1|\mathbf{v}_k) &= L(U_0U_1 = 11 \cup U_0U_1 = 10|\mathbf{v}_k) = \ln(e^{P(U_0U_1=11|\mathbf{v}_k) \cdot P(U_0U_1=10|\mathbf{v}_k)}) \\
&= J(L_4, L_3), \tag{10.8} \\
L(U_0 = 0|\mathbf{v}_k) &= L(U_0U_1 = 01 \cup U_0U_1 = 00|\mathbf{v}_k) = \ln(e^{P(U_0U_1=01|\mathbf{v}_k) \cdot P(U_0U_1=00|\mathbf{v}_k)}) \\
&= J(L_2, L_1), \tag{10.9} \\
L(U_1 = 1|\mathbf{v}_k) &= L(U_0U_1 = 11 \cup U_0U_1 = 01|\mathbf{v}_k) = \ln(e^{P(U_0U_1=11|\mathbf{v}_k) \cdot P(U_0U_1=01|\mathbf{v}_k)}) \\
&= J(L_4, L_2), \tag{10.10} \\
L(U_1 = 0|\mathbf{v}_k) &= L(U_0U_1 = 10 \cup U_0U_1 = 00|\mathbf{v}_k) = \ln(e^{P(U_0U_1=10|\mathbf{v}_k) \cdot P(U_0U_1=00|\mathbf{v}_k)}) \\
&= J(L_3, L_1), \tag{10.11}
\end{aligned}
$$

and $J(\lambda_1, \lambda_2)$ denotes the Jacobian logarithmic relationship of Equation 10.1.

Note that the Jacobian RBF equalizer will provide $\log_2(\mathcal{M})$ number of LLR values for every $\mathcal{M}$-QAM symbol. These value are fed to the turbo decoder as its soft inputs. The turbo decoder will iteratively improve the BER of the decoded bits and the detected bits will be constituted by the sign of the turbo decoder's soft output.

The probability of error for the detected bit can be estimated on the basis of the soft output of the turbo decoder. Referring to Equation 10.6 and assuming $P(u_k = +1|\mathbf{v}_k) + P(u_k = -1|\mathbf{v}_k) = 1$, the probability of error for the detected bit is given by

$$
P_{error}(u_k) = \begin{cases} 1 - P(u_k = +1|\mathbf{v}_k) = P(u_k = -1|\mathbf{v}_k), & \text{if } \mathcal{L}(u_k|\mathbf{v}_k) \geq 0 \\ 1 - p(u_k = -1|\mathbf{v}_k) = P(u_k = +1|\mathbf{v}_k), & \text{if } \mathcal{L}(u_k|\mathbf{v}_k) < 0 \end{cases}. \tag{10.12}
$$

With the aid of the definition in Equation 10.6 the probability of the bit having the value of +1 or -1 can be rewritten in terms of the *a posteriori* LLR of the bit, $\mathcal{L}(u_k|\mathbf{v}_k)$ as follows:

$$
\begin{aligned}
P(u_k = +1|\mathbf{v}_k) &= \frac{1}{1 + e^{-\mathcal{L}(u_k|\mathbf{v}_k)}}, \\
P(u_k = -1|\mathbf{v}_k) &= \frac{1}{1 + e^{\mathcal{L}(u_k|\mathbf{v}_k)}}. \tag{10.13}
\end{aligned}
$$

Upon substituting Equation 10.13 into Equation 10.12, we redefined the probability of error of a detected bit in terms of its LLR as:

$$
P_{error}(u_k) = \frac{1}{1 + e^{|\mathcal{L}(u_k|\mathbf{v}_k)|}}, \tag{10.14}
$$

where $|\mathcal{L}(u_k|\mathbf{v}_k)|$ is the magnitude of $\mathcal{L}(u_k|\mathbf{v}_k)$. Again, the average short-term probability of bit error within the decoded burst is given by:

$$
P_{\text{bit, short-term}} = \frac{\sum_{i=0}^{L_b} P_{error}(u_i)}{L_b}, \tag{10.15}
$$

where L_b is the number of decoded bits per transmitted burst and u_i is the ith decoded bit in the burst. This value, which we will refer to as the *estimated short-term BER* was found

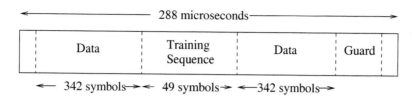

Figure 10.7: Transmission burst structure of the so-called FMA1 nonspread data mode as specified in the FRAMES proposal [307].

to give a good estimation of the actual BER of the burst, which will be demonstrated in Section 10.4. The actual BER is the ratio of the number of bit errors encountered in a data burst to the total number of bits transmitted in that burst.

In the next section we will investigate the performance of the turbo-coding assisted RBF DFE $\mathcal{M}$-QAM scheme based on our simulation results.

10.4 Turbo-coded RBF-equalized $\mathcal{M}$-QAM Performance

According to our BER versus BPS optimiztion approach high code rates in excess of $2/3$ are desirable, in order to maximise the BPS throughput of the system. Consequently, block codes were favoured as the turbo component codes in preference to the more widely used Recursive Systematic Convolutional (RSC) code based turbo-coded benchmarker scheme, since turbo block coding has been shown to perform better for coding rates in excess of $2/3$ [306]. This is demonstrated first in Figure 10.11, which will be discussed in more depth at a later stage. In our simulations, unless otherwise stated, we hence utilized the turbo coding parameters given in Table 10.3 and employed the transmission burst structure shown in Figure 10.7. The turbo encoder used two Bose-Chaudhuri-Hocquenghem BCH(31, 26) block codes in parallel. A 9984-bit random interleaver was used between the two component codes, unless otherwise stated. We used the Log-MAP decoder [288] throughout our simulations, since it offered the same performance as the optimal MAP decoder with a reduced complexity. The DFE used correct symbol feedback and we assumed perfect CIR estimation, hence the associated results indicate the system's upper-bound performance.

	BCH	RSC
Component code	BCH(31,26)	$K = 3, n = 2, k = 1$
Octal generator polynomial		$G[0] = 7_8 \; G[1] = 5_8$
Code rate, R	$0.72 = \frac{26}{36}$ [1]	0.75
Turbo interleaver type	Random	Random
Turbo interleaver size	9984-bit	9984-bit
Component decoders	Log-MAP	Log-MAP

Table 10.3: The turbo BCH and RSC coding parameters.

---

[1] The parity bits were not punctured, since block turbo codes suffer from performance loss upon puncturing.

10.4.1   Results over Dispersive Gaussian Channels

We will first investigate the performance of the joint RBF DFE $\mathcal{M}$-QAM and turbo coding scheme over the two-path Gaussian channel of Figure 8.21(a). Figure 10.8 provides our BER performance comparison between the RBF DFE scheme and the conventional DFE scheme in conjunction with the turbo BCH codec of Table 10.3. The RBF DFE has a feedforward order of 2, feedback order of 1 and decision delay of 1 symbol in Figure 10.8(a) and a feedforward order of 3, feedback order of 1 and decision delay of 1 symbol for Figure 10.8(b). The parameters of the conventional DFE were a feedforward order of 7 and feedback order of 1, which were assigned such that they gave the best possible BER performance according to our experiments and hence there was no significant BER improvement upon increasing the feedforward and feedback order. Figure 10.8 also demonstrates the effect of the number of decoding iterations used. The performance of the uncoded scheme is also provided as a comparison. Using turbo coding improves the performance by approximately 3.2dB at a BER of 10^{-2} for both the RBF DFE ($m = 2, \tau = 1$ and $m = 3, \tau = 2$) and for conventional DFE schemes. As the number of iterations used by the turbo decoder increases, both the turbo-coded RBF DFE and the turbo-coded conventional DFE scheme perform significantly better. However, the 'per-iteration' BER improvement is reduced, as the number of iterations increases. Hence, for complexity reasons, the number of decoding iterations was set to six for our forthcoming simulations.

Figure 10.8(a) indicates that the turbo-coded conventional DFE scheme performs slightly better than the turbo-coded RBF DFE ($m = 2, \tau = 1$) scheme, corresponding to approximate improvements of 0.5dB, 0.3dB and 0.1dB for one iteration, three iterations and six iterations, respectively, at a BER of 10^{-4}. However, the performance of the turbo-coded RBF DFE scheme can be further improved by increasing its feedforward order and decision delay, as demonstrated in Figure 10.8(b), unlike that of the turbo-coded conventional DFE where there is no further performance improvement upon increasing the equalizer order. The improved turbo-coded RBF DFE ($m = 3, \tau = 2$) scheme gives an SNR improvement of 0.2dB, 0.2dB and 0.5dB for one iteration, three iterations and six iterations, respectively, at a BER of 10^{-4} compared to the conventional DFE scheme. The SNR improvement at a BER of 10^{-4} compared to the uncoded conventional DFE is -0.5dB and 0.2dB for the RBF DFE using $m = 2, \tau = 1$ and $m = 3, \tau = 2$, respectively. We observed that the turbo-coded performance of the conventional DFE and RBF DFE follow the trends of their uncoded performances.

We will now extend our investigations to QAM schemes. Figure 10.9 shows the BER performance of the BCH turbo-coded RBF DFE system for various QAM modes over the two-path Gaussian channel. Introducing turbo coding into the system improves the performance by 8dB for BPSK, 4-QAM and 16-QAM and by about 9.5dB for 64-QAM at a BER of 10^{-4}. Note that turbo coding only starts to improve the uncoded performance after the uncoded BER drops below 10^{-1}, since coding could not improve the BER performance, if the number of errors in the undecoded burst exceeded a certain limit.

The Jacobian logarithmic RBF DFE introduced in Section 10.2 can be used to substitute the RBF DFE in order to reduce the computational complexity of the system. The turbo-coded performance of the Jacobian logarithmic RBF DFE is shown to be similar to that of the RBF DFE in Figure 10.10, since the Jacobian logarithmic algorithm is capable of giving a good approximation of the equalised channel output LLRs.

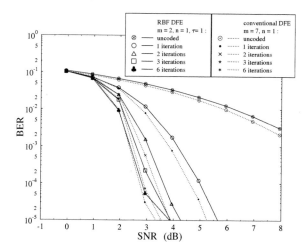

(a) RBF DFE with feedforward order of $m = 2$, feedback order of $n = 1$ and decision delay of $\tau = 1$ symbol.

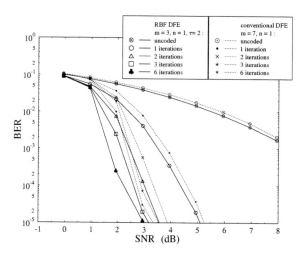

(b) RBF DFE with feedforward order of $m = 3$, feedback order of $n = 1$ and decision delay of $\tau = 2$ symbol.

Figure 10.8: BER versus SNR performance for the BPSK RBF DFE and for a conventional DFE using the turbo BCH codec of Table 10.3 with different number of iterations over the dispersive two-path Gaussian channel of Figure 8.21(a). The conventional DFE has a feedforward order of $m = 7$ and a feedback order of $n = 1$. The turbo interleaver size is 9984 bits.

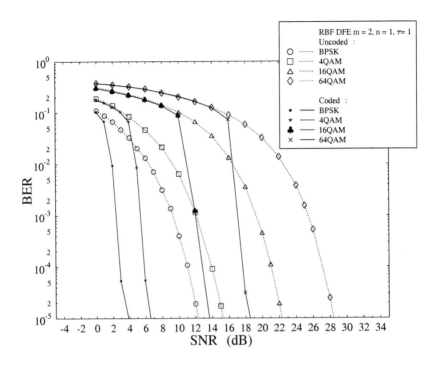

Figure 10.9: BER versus SNR performance for the RBF DFE using the turbo codec of Table 10.3 over
the dispersive two-path Gaussian channel of Figure 8.21(a) in conjunction with various
QAM modes. The RBF DFE has a feedforward order of $m = 2$, feedback order of
$n = 1$ and a decision delay of $\tau = 1$ symbol. The number of turbo BCH(31,26) decoder
iterations is six, while the random turbo interleaver size is 9984 bits.

10.4.2   Results over Dispersive Fading Channels

We will now investigate the performance of the joint RBF DFE $\mathcal{M}$-QAM and turbo coding
scheme over the wideband Rayleigh fading channel environment of Table 10.4, while the
parameters of the turbo codec are given in Table 10.3.

As noted before, Figure 10.11 shows the performance of the Jacobian RBF DFE in conjunction with both BCH and RSC based turbo coding for various QAM modes. The BCH
turbo-coded scheme improves the system performance by 5dB, 4dB, 7dB and 8dB using
BPSK, 4-QAM, 16-QAM and 64-QAM, respectively, for a BER of 10^{-4}. By contrast,
for the RSC turbo coded scheme the BER performance improves by 2dB for BPSK and
4-QAM, while 3dB for 16-QAM and 64-QAM. Similarly to the 2-path Gaussian channel,
the turbo-coded schemes only start to provide significant BER improvements with respect to
the uncoded scheme, once the uncoded BER dips below 10^{-1}. Our performance comparison
with the turbo convolutional codec of Table 10.3 given in Figure 10.11 demonstrates that the
$R = 0.72$ turbo block code provides a better BER performance than the $R = 0.75$ RSC-turbo
codec, at the cost of a higher computational complexity. As seen in Table 10.3, a half rate

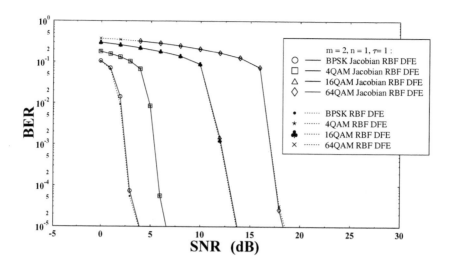

Figure 10.10: BER versus SNR performance for the RBF DFE and Jacobian logarithmic RBF DFE using the turbo codec of Table 10.3 over the dispersive two-path Gaussian channel of Figure 8.21(a) in conjunction with various QAM modes. The equalizer has a feedforward order of $m = 2$, feedback order of $n = 1$ and a decision delay of $\tau = 1$ symbol. The number of turbo BCH(31,26) decoder iterations is six, while the random turbo interleaver size is 9984 bits.

RSC encoder of constraint lenght $K = 3$ was used in the RSC turbo codec. The generator polynomials expressed in octal terms were set to seven (for the feedback path) and five. Similarly to the turbo BCH codec, the code rate was set to 0.75 by applying a random puncturing pattern in the RSC encoder. The turbo interleaver depth was also chosen to be 9984 bits.

Transmission Frequency	1.9GHz
Transmission Rate	2.6MBd
Vehicular Speed	30 mph
Normalized Doppler Frequency	3.3×10^{-5}
Channel weights	$0.707 + 0.707z^{-1}$

Table 10.4: Simulation parameters for the two-path Rayleigh fading channel.

Modulation Mode	BPSK	4-QAM	16-QAM	64-QAM
Interleaver Size	494	988	1976	2964

Table 10.5: Corresponding random interleaver sizes for each modulation mode.

10.5   Channel Quality Measure

In order to identify the potentially most reliable channel quality measure to be used in our BbB adaptive turbo-coded QAM modems to be designed during our forthcoming discourse, we will now analyse the relationship between the average burst LLR magnitude before and after channel decoding. For this reason, the random turbo interleaver size was reduced from the previously used 9984 bits and it was varied on a BbB basis, corresponding to the modulation mode used, as shown in Table 10.5, in order to enable BbB decoding so that we could obtain the average burst LLR magnitude of the coded data burst corresponding to the uncoded data burst. Explicitly, the interleaver size is set to be equivalent to the number of source bits in a data burst, in order to enable BbB decoding. Since the code rate is 0.72 and the number of coded bits is 684, 1368, 2736 and 4104 for BPSK, 4-QAM, 16-QAM and 64-QAM, respectively, for a burst length of 684 symbols, the interleaver size (= number of source bits = number of coded bits - number of parity bits) is as shown in Table 10.5. The average burst LLR magnitude is defined as follows:

$$\mathcal{L}_{\text{average}} = \frac{\sum_{i=0}^{L_b} |\mathcal{L}(u_i|\mathbf{v}_k)|}{L_b}, \tag{10.16}$$

where L_b is the number of data bits per transmitted burst and u_i is the ith data bit in the burst. Figure 10.12 shows the improvement of the average burst LLR magnitude after turbo decoding for the turbo BCH codec of Table 10.3 over the wideband Rayleigh fading channel environment of Table 10.4. As seen in the figure, the gradient of the curve is approximately unity for the average burst LLR magnitude before decoding over the range of 0 to 5 for BPSK and 4-QAM, 0 to 6 for 16-QAM and 0 to 10 for 64-QAM. Thus, there is no average LLR magnitude improvement upon introducing turbo decoding in this low reliability range. This is in harmony with our previous observations in Figures 10.10 and 10.11, namely that there is no BER improvement for BERs below 10^{-1}. Beyond this range, there is a sharp increase in the decoded LLR magnitude due to turbo decoding. Figure 10.12(a) also shows the effect of increasing the number of decoder iterations on the average burst LLR magnitude. Increasing the number of decoder iterations improves not only the BER, but also the average confidence measure of the decoder's decisions.

Figure 10.13 shows the relationship between the estimated short-term BER defined in Equation 10.15 and the average burst LLR magnitude after turbo decoding using six iterations. Note that the curves becomes more 'spread out', as the short-term BER decreases.

This is because the relationship between the probability of bit error in the decoded burst expressed in the logarithmic domain is inversely proportional to its LLR magnitude, as shown in Figure 10.14. The average of the burst LLR magnitude is dominated by the LLR values of the bits having lower probability of bit error, whereas the short-term BER of the burst is dominated by the bits with higher probability of bit error. The variance of the LLR values of the bits in the burst accounts for the 'spread' of the the estimated short-term BER versus average burst LLR magnitude curves in Figure 10.13 at low short-term BER values.

Since the average burst LLR magnitude is related to the estimated short-term BER, after accounting for the 'spread' at low short-term BERs, the average burst LLR magnitude can be used as the modem mode switching metric in our AQAM scheme, which will be discussed in Section 10.6. The average burst LLR magnitude is preferred instead of the short-term BER as the modem mode switching metric, because it can avoid the extra computational complexity of having to convert the output of the RBF DFE and the turbo decoder from the LLR values to BER values according to Equation 10.14, in order to obtain the short-term BER of the data burst.

10.6　Turbo Coding and RBF Equalizer Assisted AQAM

10.6.1　System Overview

The schematic of the joint AQAM and RBF network based equalization scheme using turbo coding is depicted in Figure 10.15. The switching thresholds can be based on the switching metric either before or after turbo decoding. In this section we will investigate the performance of the AQAM scheme using either the short-term BER or the average burst LLR magnitude as our switching metric.

For our experiments in the following sections, the simulation parameters are listed in Table 10.4, noting that we analysed the joint AQAM and RBF DFE scheme in conjunction with turbo coding over the two-path Rayleigh fading channel of Table 10.4. The wideband fading channel was burst-invariant, implying that during a transmission burst the channel impulse response was considered time-invariant. In our simulations, we used the Jacobian RBF DFE of Section 10.2, which gave a similar turbo-coded BER performance to the RBF DFE but at a lower computational complexity, as it was demonstrated in Figure 10.10. The Jacobian RBF DFE had a feedforward order of $m = 2$, feedback order of $n = 1$ and decision delay of $\tau = 1$. We used the BCH(31, 26) code of Table 10.3 as the turbo component code and the BbB random interleavers depending on the modulation mode were employed, as given in Table 10.5. The modulation modes utilized in our system are BPSK, 4-QAM, 16-QAM, 64-QAM and NO TX.

10.6.2　Performance of the AQAM Jacobian RBF DFE Scheme: Switching Metric Based on the Short-Term BER Estimate

Following from Section 9.5, where the *uncoded* AQAM RBF DFE scheme used the estimated short-term BER to switch the modem mode, we will now investigate the performance of the *turbo-coded* AQAM RBF DFE scheme based on the same switching metric. The estimated short-term BER can be obtained both before or after turbo BCH(31,26) decoding for the coded system. The estimated short-term BER before decoding can be obtained with the aid of the RBF DFE based on Equation 9.15, while that after turbo decoding can be obtained with the aid of the decoder based on Equation 10.15.

The plot of the estimated BER versus actual BER before and after turbo BCH(31,26) decoding and their corresponding PDFs of the BER estimation error for the Jacobian RBF DFE and for various channel SNRs is shown in Figures 10.16, 10.17, 10.18 and 10.19, for BPSK transmission bursts over the dispersive two-path Gaussian channel of Figure 8.21(a) and the two-path Rayleigh fading channel of Table 10.4, respectively. The actual burst-BER is the ratio of the number of bit errors encountered in a data burst to the total number of bits transmitted in that burst. The figures suggest that the Jacobian RBF DFE and the turbo BCH(31,26) decoder provide a good BER estimation, especially at higher channel SNRs. We note, however again that the accuracy of the actual BER evaluation is limited by the burst-length of 684 bits and 494 bits for the undecoded and decoded bursts, respectively. Therefore, for high SNRs the actual number of errors registered is often 0, which portrays the estimation algorithm in a less accurate light in the PDF of Figure 10.18 and 10.19 than it is in reality, since the 'resolution' of the reference BER is $1/684$ or $1/494$.

We shall refer to the AQAM scheme that utilized the switching thresholds based on the short-term BER before and after decoding, 'before decoding'-scheme and 'after decoding'-scheme, respectively. The short-term BER $P_{\text{bit, short-term}}$, obtained from either the RBF DFE or the turbo BCH(31,26) decoder is compared to a set of switching BER thresholds, P_i^M, $i = 2, 4, 16, 64$, corresponding to the various M-QAM modes, and the modulation mode is switched according to Equation 9.7.

As discussed in Section 9.5, the switching BER thresholds can be obtained by estimating the BER degradation/improvement, when the modulation mode is switched from M-QAM to a higher/lower value of M. We obtain this BER degradation/improvement measure from the estimated short-term BER of every modulation mode used under the same channel scenario.

In our experiments used to obtain the switching BER thresholds, pseudo-random symbols were transmitted in a fixed-length burst for all modulation modes across the burst-invariant wideband channel. The receiver receives each data burst having different modulation modes, equalises and turbo BCH(31,26) decodes each one of them independently. The estimated short-term BER before and after turbo BCH(31,26) decoding for all modulation modes was obtained according to Equation 9.15 and Equation 10.15, respectively. Thus, we have the estimated short-term BER of the received data burst before and after decoding for every modulation mode under the same channel conditions, which we could use to observe the BER degradation/improvement, when we switch from M-QAM to a higher/lower value of M. We could not use the BER performance versus SNR curve of Figure 10.11 generated over the dispersive two-path fading channel of Table 10.4 for the various QAM modes to estimate the BER improvement/degradation, since the BER in that figure was an average of the time-varying short-term BER of all the transmitted bursts over the faded channel. For the switching mechanism we need the 'short-term' BER measure and not the 'long-term' BER measure to configure the modem for the next transmission burst.

The switching BER thresholds for the 'before decoding'-scheme can be obtained by estimating the degradation/improvement of the short-term BER *before decoding*, when the modulation mode is switched from M-QAM to a higher/lower value of M to achieve the target BER after decoding. Figure 10.20 shows the estimated short-term BER *after decoding* for all the possible modulation modes that can be switched to versus the estimated short-term BER of 16-QAM *before decoding* under the same channel conditions. The figure shows how each switching BER threshold P_i^{16}, $i = 2, 4, 16, 64$ is obtained. For example, in order to maintain the target BER of 10^{-4}, the short-term BER of the 16-QAM transmission burst before turbo decoding has to be approximately 2.5×10^{-1}, 2×10^{-1}, 5×10^{-2} and 1×10^{-3}, when switching to BPSK, 4-QAM and 64-QAM, respectively, under the same channel conditions. Using the same method for the other modulation modes, the switching BER thresholds are obtained, as listed in Table 10.6. For the 'after decoding' switching scheme, the short-term BER thresholds P_i^M, $i = 2, 4, 16, 64$, listed in Table 10.7 were obtained. However, for NO TX bursts, where only dummy data are transmitted, turbo decoding is not necessary. Thus, for NO TX bursts we use the short-term BER *before decoding* as the switching metric.

Figure 10.21 shows the performance of the 'before decoding'-scheme and 'after decoding'-scheme using the switching thresholds given in Tables 10.6 and 10.7, respectively. Both schemes have similar BPS performances. However, the 'before decoding'-scheme performs

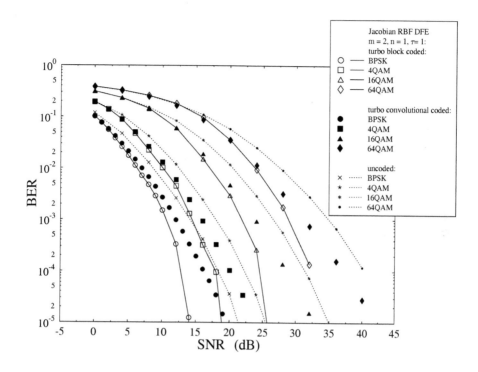

Figure 10.11: BER versus SNR performance for the Jacobian logarithmic RBF DFE using the turbo
codec of Table 10.3 over **the dispersive two-path fading channel** of Table 10.4 for
various QAM modes. The equalizer has a feedforward order of $m = 2$, feedback order
of $n = 1$ and a decision delay of $\tau = 1$ symbol. The number of **convolutional and
BCH** turbo decoder iterations is six, while the turbo interleaver size is fixed to 9984 bits.

	$P_2^{\mathcal{M}}$	$P_4^{\mathcal{M}}$	$P_{16}^{\mathcal{M}}$	$P_{64}^{\mathcal{M}}$
NO TX	2.5×10^{-2}	2×10^{-3}	1×10^{-32}	0.0
BPSK	2.5×10^{-2}	2×10^{-3}	1×10^{-32}	0.0
4-QAM	1×10^{-1}	4×10^{-2}	4×10^{-5}	0.0
16-QAM	2.5×10^{-1}	2×10^{-1}	5×10^{-2}	1×10^{-3}
64-QAM	3.2×10^{-1}	2.5×10^{-1}	1.3×10^{-1}	5×10^{-2}

Table 10.6: The switching BER thresholds $P_i^{\mathcal{M}}$ of the joint adaptive modulation and RBF DFE scheme
for the turbo-decoded target BER of 10^{-4} over the two-path Rayleigh fading channel of
Table 9.1. The switching metric is based on the estimated short-term BER obtained before
turbo decoding from the **RBF DFE**. This table explicitly indicates the uncoded modem
BER that has to be maintained by the modem modes shown at the top of the table, in order
to achieve the 10^{-4} turbo-decoded BER after switching to the various modem modes seen
in the left-most column.

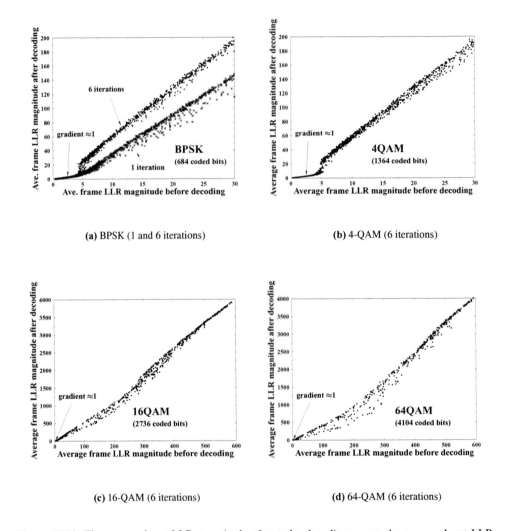

(a) BPSK (1 and 6 iterations)

(b) 4-QAM (6 iterations)

(c) 16-QAM (6 iterations)

(d) 64-QAM (6 iterations)

Figure 10.12: The average burst LLR magnitude after turbo decoding versus the average burst LLR magnitude before turbo decoding using BbB interleaving and turbo BCH decoding employing the parameters of Table 10.3 over the burst-invariant two-path fading channel of Table 10.4.

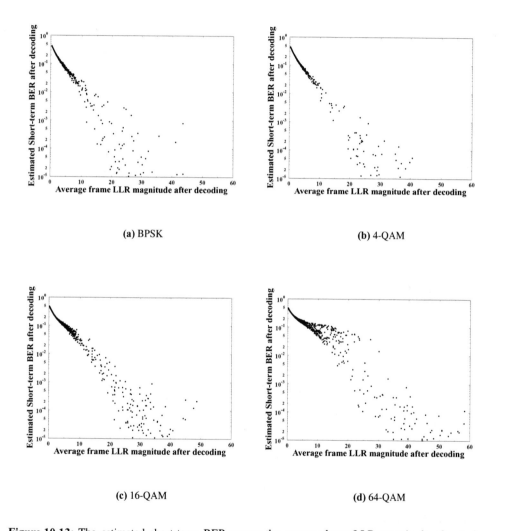

(a) BPSK

(b) 4-QAM

(c) 16-QAM

(d) 64-QAM

Figure 10.13: The estimated short-term BER versus the average burst LLR magnitude after turbo BCH(31,26) decoding using six iterations over the burst-invariant two-path fading channel of Table 10.4.

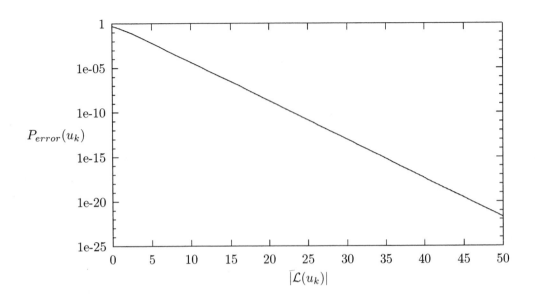

Figure 10.14: The probability of error of the detected bit versus the magnitude of its LLR.

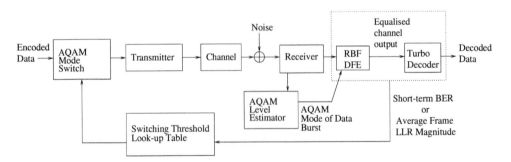

Figure 10.15: System schematic of the joint adaptive modulation and RBF equalizer scheme using turbo coding.

better, than the 'after decoding'-scheme in terms of its BER performance. Note that the 'after decoding'-scheme could only achieve the target BER of 10^{-4} beyond the SNR of 32dB. The performance degradation of the 'after decoding'-scheme can be explained by observing Figure 10.22, which shows the short-term BER fluctuation obtained before and after decoding at an SNR of 10dB for 4-QAM – the dominant modulation mode at 10dB. The BER fluctuation after decoding is more spurious and hence exhibits a higher variance than before decoding. Our modem mode switching mechanism assumes that the BER of the transmission burst is slowly varying and the estimated short-term BER of the *current* received burst is used to select the modulation mode for the *next* transmission burst. The spurious nature of the short-term BER after decoding, which is used as the switching metric, defies the BER predictability assumptions made and hence degrades the performance of the modulation mode switching

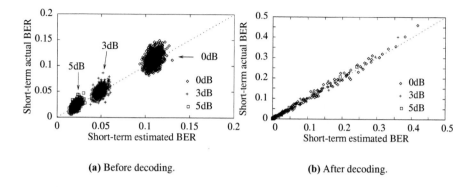

(a) Before decoding.   **(b)** After decoding.

Figure 10.16: The actual BER versus estimated BER before and after turbo BCH(31,26) decoding with the error PDF given in Figure 10.18 for the **dispersive two-path Gaussian channel** of Figure 8.21(a) using BPSK. The number of turbo BCH(31,26) decoder iterations is six while the random turbo interleaver size is 494.

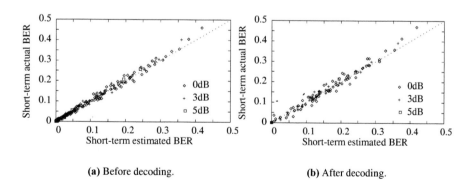

(a) Before decoding.   **(b)** After decoding.

Figure 10.17: The actual BER versus estimated BER before and after turbo BCH(31,26) decoding with error PDF given in Figure 10.19 for the **dispersive two-path Rayleigh fading channel** of Table 10.4 using BPSK. The number of turbo decoder iterations is six, while the turbo interleaver size is 494.

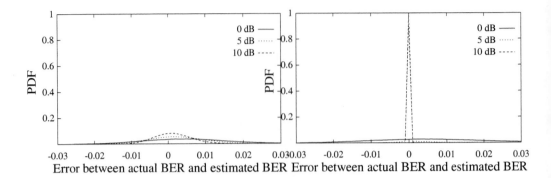

(a) Before turbo BCH(31,26) decoding.   (b) After turbo BCH(31,26) decoding.

Figure 10.18: Discretised PDF of the error between the actual BER of BPSK bursts and the BER estimated by the Jacobian RBF DFE before and after turbo BCH(31,26) decoding for the **dispersive two-path Gaussian channel** of Figure 8.21(a) using BPSK. The number of turbo BCH(31,26) decoder iterations is six, while the random turbo interleaver size is 494.

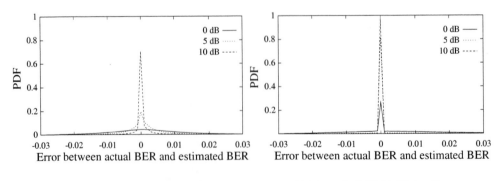

(a) Before turbo BCH(31,26) decoding.   (b) After turbo BCH(31,26) decoding.

Figure 10.19: Discretised PDF of the error between the actual BER of BPSK bursts and the BER estimated by the Jacobian RBF DFE before and after turbo BCH(31,26) decoding for the **two-path Rayleigh fading channel** of Table 10.4 using BPSK. The number of turbo decoder iterations is six, while the turbo interleaver size is 494.

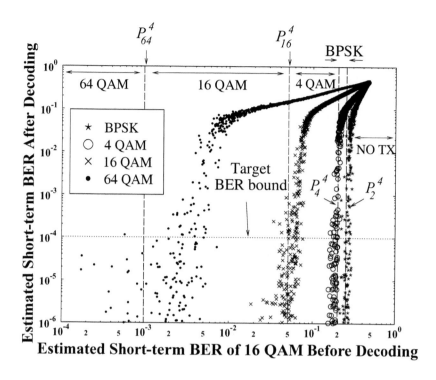

Figure 10.20: The estimated short-term BER after turbo BCH(31,26) decoding for all the possible modulation modes that can be invoked, assuming that current mode is 16-QAM – versus the estimated short-term BER of 16-QAM **before** decoding over the two-path Rayleigh fading channel of Table 9.1.

	$P_2^{\mathcal{M}}$	$P_4^{\mathcal{M}}$	$P_{16}^{\mathcal{M}}$	$P_{64}^{\mathcal{M}}$
NO TX	3×10^{-2}	2.5×10^{-3}	1×10^{-35}	0.0
BPSK	1×10^{-5}	1×10^{-30}	0.0	0.0
4-QAM	8×10^{-2}	1×10^{-5}	1×10^{-50}	0.0
16-QAM	2×10^{-1}	1.6×10^{-1}	1×10^{-5}	1×10^{-34}
64-QAM	3.2×10^{-1}	2.7×10^{-1}	1.3×10^{-1}	1×10^{-5}

Table 10.7: The switching BER thresholds $P_i^{\mathcal{M}}$ of the joint adaptive modulation and RBF DFE scheme for the turbo-decoded target BER of 10^{-4} over the two-path Rayleigh fading channel of Table 9.1. The switching metric is based on the estimated short-term BER obtained after turbo decoding from the **decoder**. This table explicitly indicates the coded modem BER that has to be maintained by the modem modes shown at the top of the table, in order to achieve the 10^{-4} turbo-decoded BER after switching to the various modem modes seen in the left-most column.

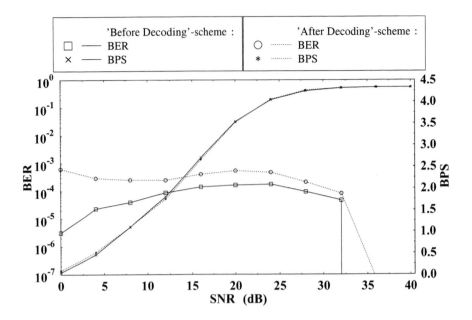

Figure 10.21: The BER and BPS performance of the turbo BCH(31,26) coded AQAM Jacobian RBF
DFE aiming for a target BER of 10^{-4} for data-transmission using the parameters listed
in Table 10.4. The 'before decoding'-scheme and 'after decoding'-scheme uses the estimated short-term BER of Equation 9.15 and 10.15 before and after decoding, respectively, as switching metric. The modem mode switching BERs used for both schemes
are listed in Table 10.6 and 10.7. The Jacobian RBF DFE had a feedforward order of
$m = 2$, feedback order of $n = 1$ and decision delay of $\tau = 1$ symbol. The turbo coding
parameters are given in Table 10.3 and the number of turbo decoder iterations is six. The
BbB turbo interleaver size was fixed according to the modulation mode used as shown
in Table 10.5.

mechanism. We also note from Table 10.7 that the thresholds required for the modem to
switch to a higher-order modulation mode are extremely low. For example, when the BER
must be lower than $P_{16}^4 = 1 \times 10^{-50}$ for the modem to switch from 4-QAM to 16-QAM. The
extremely low values of the thresholds associated with the 'after decoding'-scheme degrade
the performance of the mode switching mechanism. The 'before decoding'-scheme has a
more reasonable set of thresholds, as shown in Table 10.6 and therefore performs better.

In the following section, we will investigate the performance of the coded adaptive scheme
using the average burst LLR magnitude, defined by Equation 10.16 as an alternative switching metric.

10.6.3 Performance of the AQAM Jacobian RBF DFE Scheme:
Switching Metric Based on the Average Burst LLR Magnitude

As discussed in Section 10.3, the probability of bit error is related to the magnitude of the
bit LLR according to Equation 10.14. Thus, in addition to the BER-based switching criteria

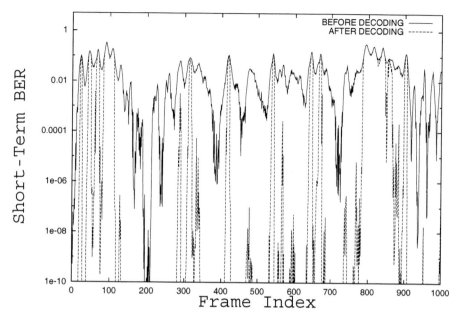

Figure 10.22: Short-term BER before and after turbo BCH(31,26) decoding versus symbol index for 4-QAM and for a channel SNR of 10dB over the two-path equal-weight, symbol-spaced Rayleigh fading channel of Table 9.1. The RBF DFE had a feedforward order of $m = 2$, feedback order of $n = 1$ and decision delay of $\tau = 1$ symbol. Perfect CIR estimation is assumed and decision fedback error propagation is ignored. These low short-term BER estimates were obtained from the average values of Equation 10.14, which was plotted in Figure 10.14.

of the previous section, the magnitude of the bit LLR can also be used as the modem mode switching metric. The turbo decoder iteratively improves the BER of the decoded bits. Since the average burst LLR magnitude before and after decoding has an approximately linear relationship, as demonstrated by Figure 10.12 in Section 10.4.2, the average probability of error for the decoded burst can be inferred from the average burst LLR magnitude provided by the RBF equalizer using Equation 10.16. Thus, this parameter can also be used as the switching metric of the turbo-coded BbB AQAM scheme.

Figures 10.23 and 10.24 portray the average burst LLR magnitude fluctuation before and after turbo decoding, respectively, over the burst-invariant channel of Table 10.4 versus the symbol index for various QAM modes, as given by the RBF DFE and the turbo decoder, which is slowly varying and predictable for a number of consecutive data bursts. Therefore in our simulated channel scenario the average burst LLR magnitude both before and after turbo decoding constitute suitable metrics for the AQAM switching mechanism.

The average burst LLR magnitude obtained from either the RBF DFE or the turbo decoder is compared to a set of switching LLR magnitudes corresponding to the modulation mode of that data burst. Consequently, a modulation mode is selected for the next transmission burst, based on the current estimated BER upon assuming slowly fading channels. More explicitly, this implies that the similarity of the average burst LLR magnitude of consecutive data bursts can be exploited, in order to set the next modulation mode. Again, the modulation modes

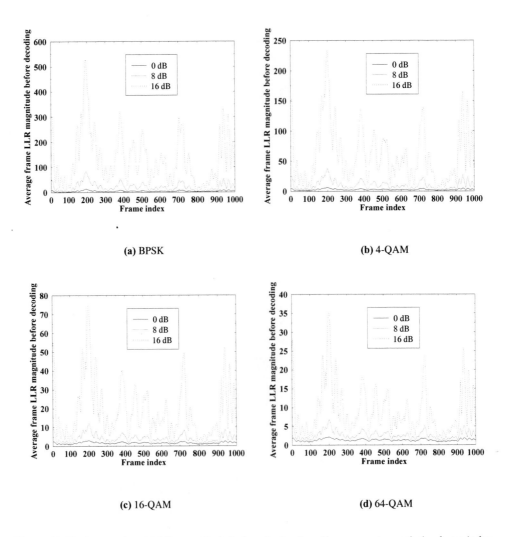

(a) BPSK

(b) 4-QAM

(c) 16-QAM

(d) 64-QAM

Figure 10.23: Average burst LLR magnitude **before turbo decoding** versus transmission burst index
for various QAM modes as given by the RBF DFE over the two-path equal-weight,
symbol-spaced Rayleigh fading channel of Table 9.1. Perfect CIR estimation is assumed
and error propagation in decision feedback is ignored.

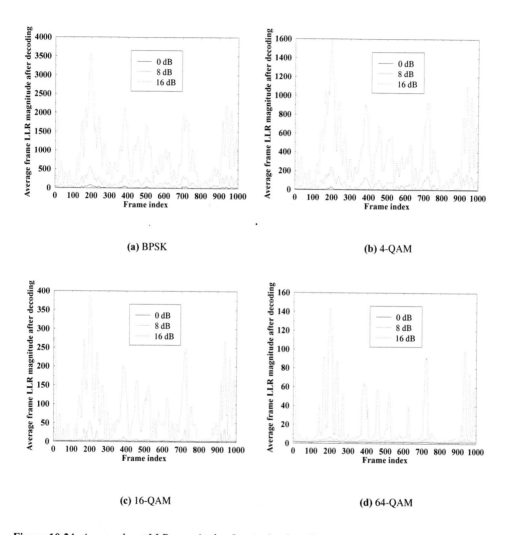

(a) BPSK

(b) 4-QAM

(c) 16-QAM

(d) 64-QAM

Figure 10.24: Average burst LLR magnitude **after turbo decoding** versus transmission burst index for various QAM modes as given by the RBF DFE over the two-path equal-weight, symbol-spaced Rayleigh fading channel of Table 9.1. Perfect CIR estimation is assumed and error propagation in decision feedback is ignored.

utilized in our system are BPSK, 4-QAM, 16-QAM, 64-QAM and no transmission (NO TX). Therefore, the modulation mode is switched according to the average burst LLR magnitude as follows:

$$\text{Modulation Mode} = \begin{cases} \text{NO TX} & \text{if } \mathcal{L}_{\text{average}} \leq \mathcal{L}_2^{\mathcal{M}} \\ \text{BPSK} & \text{if } \mathcal{L}_2^{\mathcal{M}} < \mathcal{L}_{\text{average}} \leq \mathcal{L}_4^{\mathcal{M}} \\ \text{4-QAM} & \text{if } \mathcal{L}_4^{\mathcal{M}} < \mathcal{L}_{\text{average}} \leq \mathcal{L}_{16}^{\mathcal{M}} \\ \text{16-QAM} & \text{if } \mathcal{L}_{16}^{\mathcal{M}} < \mathcal{L}_{\text{average}} \leq P_{64}^{\mathcal{M}} \\ \text{64-QAM} & \text{if } \mathcal{L}_{64}^{\mathcal{M}} < \mathcal{L}_{\text{average}}, \end{cases} \tag{10.17}$$

where $\mathcal{L}_i^{\mathcal{M}}, i = 2, 4, 16, 64$ are the switching LLR magnitude thresholds corresponding to the $\mathcal{M}$-QAM mode.

The LLR magnitude switching thresholds corresponding to $\mathcal{M}$-QAM, $\mathcal{L}_i^{\mathcal{M}}, i = 2, 4, 16, 64$, can be obtained by estimating the average burst LLR magnitude degradation/improvement, upon switching the modulation mode from $\mathcal{M}$-QAM to a higher or lower number of bits per symbol. The target BER requirement can be met by obtaining the average burst LLR magnitude of each modulation mode corresponding to the estimated channel quality and by activating the specific mode satisfying this target BER.

In our experiments, we obtained the LLR magnitude degradation/improvement upon switching from each modem mode to all other legitimate modes under the same instantaneous channel conditions. As an example for the 'before decoding'-scheme, Figure 10.25 shows the short-term BER – defined in Equation 10.15 – that would be encountered upon switching to all possible AQAM modes after BCH(31,26) turbo decoding versus the average burst LLR magnitude of 4-QAM before decoding, which was the current AQAM mode. In order to maintain the target BER of 10^{-4}, Figure 10.25 demonstrates how each switching LLR magnitude $\mathcal{L}_i^4, i = 2, 4, 16, 64$ is obtained after averaging the LLR magnitude occurances seen in the figure. More explicitly, the average burst LLR magnitudes before decoding encountered in the 4-QAM transmission burst would have to be 4.0, 7.5, 40.0 and 100.0, before switching to BPSK, 4-QAM, 16-QAM and 64-QAM AQAM bursts under the same channel conditions, leading to an estimated BER of 10^{-4} after BCH(31,26) turbo decoding. For example, if the average LLR magnitude $\mathcal{L}_{\text{average}}$ before decoding of the received 4-QAM transmission burst is in the range of $100 > \mathcal{L}_{\text{average}} \geq 40$, the modulation mode is switched from 4-QAM to 16-QAM for the next AQAM burst, since the BER of this 16-QAM transmission burst is estimated to be below the target BER of 10^{-4}. Note that due to the 'spreading' of the average burst LLR magnitude before decoding versus the short-term BER curve – especially for higher-order AQAM modes, as seen in Figure 10.25 – the threshold is estimated from the mean of this dynamic range. Using the same method for the other modulation modes, the 'before decoding' switching LLR magnitude thresholds were obtained for the turbo-decoded target BER of 10^{-4}, as listed in Table 10.8. For the 'after decoding' switching scheme, a similar method was implemented, in order to obtain the switching thresholds listed in Table 10.8. Similar to the 'after decoding' short-term BER switching metric described in Section 10.6.2, the average burst LLR magnitude *before decoding* is used as the switching metric for the NO TX bursts, since turbo decoding is not performed in this mode.

Figure 10.26 compares the performance of the adaptive schemes using the short-term BER estimate based on Equation 10.15 and the average burst LLR magnitude before and after decoding as the switching metric. Both the 'before' and 'after decoding' LLR schemes of this section have similar BER and BPS performances to the 'before decoding' short-term

	Before Decoding				After Decoding			
	$\mathcal{L}_2^M$	$\mathcal{L}_4^M$	$\mathcal{L}_{16}^M$	$\mathcal{L}_{64}^M$	$\mathcal{L}_2^M$	$\mathcal{L}_4^M$	$\mathcal{L}_{16}^M$	$\mathcal{L}_{64}^M$
NO TX	8.0	17.0	90.0	380.0	8.0	17.0	90.0	380.0
BPSK	8.0	17.0	90.0	380.0	40.0	100.0	∞	∞
4QAM	4.0	7.5	40.0	140.0	6.0	32.0	230.0	∞
16QAM	2.0	3.0	11.5	55.0	3.2	4.0	40.0	200.0
64QAM	1.7	2.2	6.2	30.0	24.0	3.0	6.5	45.0

Table 10.8: The switching LLR magnitude thresholds $\mathcal{L}_i^M$ before and after decoding of the RBF DFE BbB AQAM scheme with turbo coding for the target BER of 10^{-4} over the two-path Rayleigh fading channel of Table 10.4.

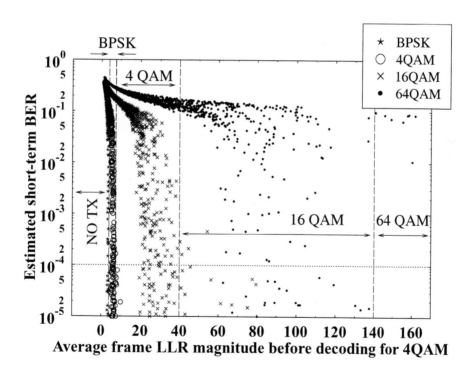

Figure 10.25: The estimated short-term BER for all the possible turbo BCH(31,26) decoded AQAM modes versus the average burst LLR magnitude of 4-QAM over the two-path Rayleigh fading channel of Table 9.1. The figure illustrates the expected spread of the short-term BER of all turbo decoded modem modes given a certain average burst LLR magnitude value in conjunction with 4-QAM as the current modem mode.

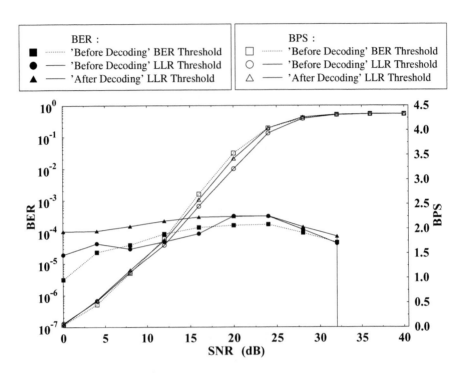

Figure 10.26: The BER and BPS performance of the turbo BCH(31,26) coded AQAM Jacobian RBF
DFE using different switching metrics for a data-transmission target BER of 10^{-4} over
the two-path Rayleigh fading channel of Table 10.4. The modem mode switching thresholds used for both scheme are listed in Tables 10.6 and 10.8, respectively. The Jacobian
RBF DFE had a feedforward order of $m = 2$, feedback order of $n = 1$ and decision
delay of $\tau = 1$ symbol. The turbo coding parameters were given in Table 10.3 and the
number of turbo decoder iterations was six. The BbB turbo interleaver size was fixed
according to the modulation mode used, as shown in Table 10.5.

BER scheme of Section 10.6.2, although the scheme using the average burst LLR magnitude
as the switching metric has a lower computational complexity. This is because the output
of the Jacobian RBF DFE is in a logarithmic form and obtaining the short-term BER values
requires us to convert the logarithmic output to the non-logarithmic domain using exponential
functions, in order to acquire the probability of bit error according to Equation 10.15.

Figure 10.27 shows our performance comparison of the AQAM Jacobian RBF DFE
scheme in conjunction with turbo BCH(31,26) coding for the target BER of 10^{-4} with the
'before decoding' LLR magnitude switching metric along with its constituent turbo-coded
fixed QAM modes. Figure 10.27 also shows the BER and BPS performance of the AQAM
RBF DFE scheme *without turbo coding*, using the *short-term BER* as the switching metric,
as described in Section 9.5 for performance comparison. The switching BER thresholds of

the AQAM RBF DFE scheme *without turbo coding* were listed in Table 9.5.

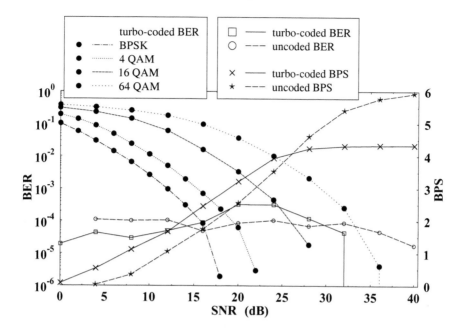

Figure 10.27: The BER and BPS performance of the uncoded and turbo BCH(31,26) coded AQAM Jacobian RBF DFE for a data-transmission target BER of 10^{-4} over the two-path Rayleigh fading channel of Table 10.4. The average LLR magnitude modem mode switching thresholds before decoding used for this scheme are listed in Table 10.8. The Jacobian RBF DFE had a feedforward order of $m = 2$, feedback order of $n = 1$ and decision delay of $\tau = 1$ symbol. The turbo coding parameters were given in Table 10.3 and the number of turbo decoder iterations was six. The BbB turbo interleaver size was fixed according to the modulation modes used, as shown in Table 10.5.

Referring to Figure 10.27, the coded BPS performance was better than that of the uncoded scheme for the channel SNR range of 0dB to 26dB, with a maximum SNR gain of 4dB at a channel SNR of 0dB. However, at high SNRs, the BPS performance is limited by the coding rate of the system to achieve a maximum BPS throughput of $\frac{26}{36} \cdot 6 = 4.33$. The turbo BCH(31,26) coded AQAM system also exhibited a superior BER performance, when compared to the uncoded system for the channel SNR range of 0dB to 16dB and for the range above 28dB. However, the coded AQAM system failed to achieve the target BER of 10^{-4} for the SNR range of 16dB to 28dB. This was because the spread nature of the short-term BER versus LLR magnitude curves observed in Figure 10.25 leads to inaccuracies in obtaining these LLR magnitude thresholds, especially for $\mathcal{L}_{16}^{\mathcal{M}}$ and $\mathcal{L}_{64}^{\mathcal{M}}$, as demonstrated in Figure 10.25. These inaccuracies affect the switching performance for the SNR range of 16dB to 28dB. The spread nature of the short-term BER versus LLR magnitude curves in Figure 10.25 is due to a number of factors and these investigations are set aside for future work.

Since the estimated short-term BER is a somewhat eratic function of the turbo decoder's

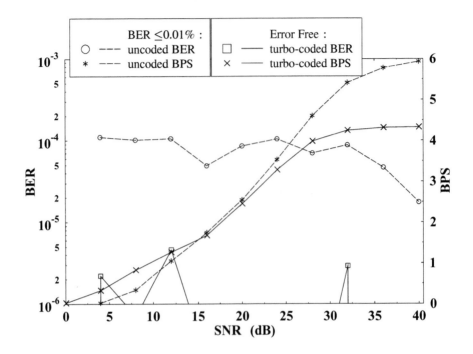

Figure 10.28: The BER and BPS performance of the turbo BCH(31,26) coded AQAM Jacobian RBF
DFE for targetted no error transmission over the two-path Rayleigh fading channel of
Table 10.4. The average LLR magnitude modem mode switching thresholds before decoding used for this scheme are listed in Table 10.9. The Jacobian RBF DFE had a
feedforward order of $m = 2$, feedback order of $n = 1$ and decision delay of $\tau = 1$
symbol. The turbo coding parameters were given in Table 10.3 and the number of turbo
decoder iterations is six. The BbB turbo interleaver size was fixed according to the modulation mode used as shown in Table 10.5.

	$\mathcal{L}_2^{\mathcal{M}}$	$\mathcal{L}_4^{\mathcal{M}}$	$\mathcal{L}_{16}^{\mathcal{M}}$	$\mathcal{L}_{64}^{\mathcal{M}}$
NO TX	10.0	30.0	280.0	1000.0
BPSK	10.0	30.0	280.0	1000.0
4-QAM	8.0	12.0	100.0	350.0
16-QAM	3.0	5.0	30.0	120.0
64-QAM	2.5	3.0	13.0	70.0

Table 10.9: The switching LLR magnitude thresholds $\mathcal{L}_i^{\mathcal{M}}$ before decoding of the RBF DFE BbB
AQAM scheme using turbo coding for the zero-error target performance over the two-path
Rayleigh fading channel of Table 10.4.

input LLR, the switching LLR values have to be conservative, if the target BER cannot be exceeded. For the BER = 10^{-4} scenario the switching LLR was adjusted experimentally to be near the upper end of the LLR-range observed in Figure 10.25. When aiming for virtually error-free communications, an even more conservative LLR threshold has to be chosen, in order not to precipitate a plethora of transmission errors, even at the cost of thereby reducing the achievable, BPS throughput of the system. Figure 10.28 shows the BER and BPS performance of the near-error-free, turbo-coded AQAM Jacobian RBF DFE scheme with the more conservative, increased LLR magnitude switching thresholds listed in Table 10.9. The BER and BPS performance of the uncoded AQAM RBF DFE system is also given in the figure for comparison. The BPS performance of the error-free coded system was better, than that of the uncoded AQAM system for the channel range of 0dB to 15dB, as evidenced by Figure 10.28. However, the BPS performance is limited by the coding rate of the system to a maximum value of 4.33 at high channel SNRs. This suggests that the best overall BER/BPS performance is achieved by our system, if we add the AQAM option of switching off the turbo BCH(31,26) code under high SNR conditions, namely around 25dB. This allows us to attain a BPS of 6 in this SNR region.

Wong [296] introduced the concept of variable rate turbo coding AQAM schemes with the aim of improving the throughput of turbo block coded AQAM scheme at high channel SNRs. Two types of variable code rate schemes were implemented:

1. **Partial turbo block coded adaptive modulation scheme**: The switching mechanism is capable of disabling and enabling the channel encoder for a chosen modulation mode.

2. **Variable rate turbo block coded adaptive modulation scheme**: The coding rate is varied by utilizing different BCH component codes for the different modulation modes. The higher-order modulation modes are assigned a higher code rate, in order to improve the effective data throughput at medium to high average channel SNRs and conversely, the lower-order modulation modes will be accompanied by lower code rates, in order to ensure maximum error protection at low average channel SNRs, where these modes have a high selection probability.

These methods can similarly be implemented for our turbo-coded AQAM RBF DFE system, in order to improve the throughput performance at high channel SNRs.

10.6.4   Switching Metric Selection

The choice of the switching metric depends on a variety of factors, which are discussed here with reference to Figures 10.21 – 10.27. The most reliable channel quality metric is the BER of a given transmitted burst, since this metric is capable of quantifying all channel impairments, irrespective of the effects of its source. Explicitly, the BER of the transmission burst quantifies the influence of reduced received signal strength or reduced SNR, that of increased ISI or co-channel interference, etc. The short-term BER of a transmission burst can be estimated for example with the aid of the RBF DFE using Equation 10.15.

In conjunction with turbo FEC coding also, the LLR of Equation 10.6 at the input or output of the turbo decoder can be used with the aid of Equation 10.14 and 10.15, in order to estimate the BER. Explicitly, the probability of a specific bit being in error is given by

Equation 10.14, which can be averaged according to Equation 10.15 for a transmission burst. The corresponding short-term BER versus transmission burst index was plotted using both the channel decoder's input and output SNRs in Figure 10.22 over the two-path equal-weight symbol-spaced Rayleigh channel of Table 10.4. Observe that due to the higher fluctuation of the FEC decoder's output LLRs the output BER fluctuates over a wider range. The corresponding turbo decoder LLRs both before and after turbo decoding are plotted in Figure 10.23 and 10.24, respectively, for 0dB, 8dB and 16dB channel SNRs. As expected, the evolution of the LLRs is similar, although the output LLR fluctuates over a wider dynamic range, since the turbo decoder typically improves the input LLRs upon each iteration, unless the LLR changes polarity several times, which is the sign of a low-reliability decision.

Since the BER curves of the turbo-coded constituent AQAM modem modes seen in Figure 10.9 are extremely steep, upon switching for example from the BPSK mode to 4-QAM the BER increases dramatically, by several orders of magnitude. Hence, for example the BPSK BER has to become significantly lower than 10^{-5} in Figure 10.9, before switching to 4-QAM can take place. This justifies the extreme BER differences observed in Table 10.6 and Table 10.7. In conclusion of our discussions on the choice of switching metric we infer from Figure 10.21 that whilst the BER of the AQAM switching regime using the LLRs before turbo-decoding attains a lower BER, this is not associated with any reduction of the BPS throughput, and hence this switching metric was deemed more beneficial to invoke. This is because due to the higher steepness of the turbo-decoded BER curve, the BER is more often misjudged on the basis of the output LLRs. This then often results in using an 'optimistic' high BPS AQAM mode, which increases the BER. When the channel quality is under-estimated, a reduced number of bits per symbol is used, however the associated BER reduction is insufficient for compensating for the increase of BER of the 'over-estimated' channel quality scenario. These under- and over-estimated BERs result in the high spread of the curves seen in Figure 10.25.

10.7   Review and Discussion

In this chapter, we have investigated the performance of the RBF equalizer using turbo coding. We have also demonstrated the application of turbo BCH coding in conjunction with AQAM in a wideband fading channel. The use of different switching criteria – namely the short-term BER and average burst LLR magnitude before and after decoding – was discussed. We observed that the performance of the switching mechanism depends on the fluctuation of the switching metric, since the AQAM scheme assumes that the channel quality is slowly varying. The turbo-coded AQAM RBF DFE system exhibited a better BPS performance, when compared to the uncoded system at low to medium channel SNRs, as evidenced by Figure 10.27. The same figure also showed an improved coded BER performance at higher channel SNRs. A virtually error-free turbo-coded AQAM scheme was also characterized in Figure 10.28.

In the next chapter we will explore the recently developed family of iterative equalization and channel decoding techniques, a scheme which is termed as turbo equalization. We will investigate the employment of RBF equalizer in the equalizer component.

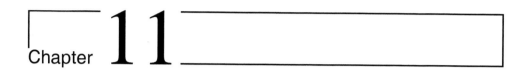

Chapter **11**

RBF Turbo Equalization

This chapter presents a novel turbo equalization scheme, which employs a RBF equaliser instead of the conventional trellis-based equaliser of Douillard *et al.* [153]. The basic principles of turbo equalization will be highlighted. Structural, computational cost and performance comparisons of the RBF-based and trellis-based turbo equalisers are provided. A novel element of our design is that in order to reduce the computational complexity of the RBF turbo equaliser (TEQ), we propose invoking further iterations only, if the decoded symbol has a high error probability. Otherwise we curtail the iterations, since a reliable decision can be taken. Let us now introduce the concept of turbo equalization.

11.1 Introduction to Turbo equalization

In the conventional RBF DFE based systems discussed in Chapter 10 equalization and channel decoding ensued independently. However, it is possible to improve the receiver's performance, if the equaliser is fed by the channel outputs plus the soft decisions provided by the channel decoder, invoking a number of iterative processing steps. This novel receiver scheme was first proposed by Douillard *et al.* [153] for a convolutional coded binary phase shift keying (BPSK) system, using a similar principle to that of turbo codes and hence it was termed *turbo equalization*. This scheme is illustrated in Figure 11.1, which will be detailed during our forthcoming discourse. Gertsman and Lodge [308] extended this work and showed that the iterative process of turbo equalization can compensate for the performance degradation due to imperfect channel estimation. Turbo equalization was implemented in conjunction with turbo coding, rather than conventional convolutional coding by Raphaeli and Zarai [309], demonstrating an increased performance gain due to turbo coding as well as with advent of enhanced ISI mitigation achieved by turbo equalization.

The principles of iterative turbo decoding [152] were modified appropriately for the coded $\mathcal{M} - QAM$ system of Figure 11.2. The channel encoder is fed with independent binary data d_n and every $\log_2(\mathcal{M})$ number of bits of the interleaved, channel encoded data c_k is mapped to an $\mathcal{M}$-ary symbol before transmission. In this scheme the channel is viewed as an 'inner encoder' of a serially concatenated arrangement, since it can be modelled with the aid of

453

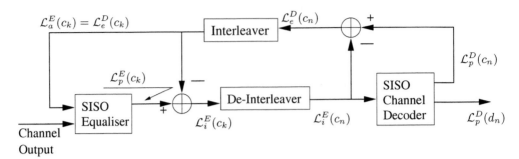

Figure 11.1: Iterative turbo equalization schematic

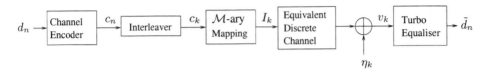

Figure 11.2: Serially concatenated coded $\mathcal{M}$-ary system using the turbo equaliser, which performs the equalization, demodulation and channel decoding iteratively.

a tapped delay line similar to that of a convolutional encoder [153, 310]. At the receiver the equaliser and decoder employ a Soft-In/Soft-Out (SISO) algorithm, such as the optimal Maximum *A Posteriori*(MAP) algorithm [162] or the Log-MAP algorithm [288]. The SISO equaliser processes the *a priori* information associated with the coded bits c_k transmitted over the channel and – in conjunction with the channel output values v_k – computes the *a posteriori* information concerning the coded bits. The soft values of the channel coded bits c_k are typically quantified in the form of the log-likelihood ratio defined in Equation 10.6. Note that in the context of turbo decoding – which was discussed in Chapter 10 – the SISO decoders compute the *a posteriori* information of the *source* bits only, while in turbo equalization the *a posteriori* information concerning all the *coded* bits is required.

In our description of the turbo equaliser depicted in Figure 11.1, we have used the notation $\mathcal{L}^E$ and $\mathcal{L}^D$ to indicate the LLR values output by the SISO equaliser and SISO decoder, respectively. The subscripts e, i, a and p were used to represent the extrinsic LLR, the combined channel and extrinsic LLR, the *a priori* LLR and the *a posteriori* LLR, respectively. Referring to Figure 11.1, the SISO equaliser processes the channel outputs and the *a priori* information $\mathcal{L}_a^E(c_k)$ of the coded bits, and generates the *a posteriori* LLR values $\mathcal{L}_p^E(c_k)$ of the interleaved coded bits c_k seen in Figure 11.2. Before passing the above *a posteriori* LLRs generated by the SISO equaliser to the SISO decoder of Figure 11.1, the contribution of the decoder — in the form of the *a priori* information $\mathcal{L}_a^E(c_k)$ — from the previous iteration must be removed, in order to yield the combined channel and extrinsic information $\mathcal{L}_i^E(c_k)$ seen in Figure 11.1. They are referred to as 'combined', since they are intrinsically bound and cannot be separated. However, note that at the initial iteration stage, no *a priori* information is available yet, hence we have $\mathcal{L}_a^E(c_k) = 0$. To elaborate further, the *a priori* information $\mathcal{L}_a^E(c_k)$ was removed at this stage, in order to prevent the decoder from processing its own output information, which would result in overwhelming the decoder's current reliability-estimation characterizing the coded bits, *i.e.* the extrinsic information. The combined channel and extrinsic LLR values are channel-deinterleaved – as seen in Figure 11.1 – in order to yield $\mathcal{L}_i^E(c_n)$, which is then passed to the SISO channel decoder. Subsequently, the channel decoder computes the *a posteriori* LLR values of the coded bits $\mathcal{L}_p^D(c_n)$. The *a posteriori* LLRs at the output of the channel decoder are constituted by the extrinsic LLR $\mathcal{L}_e^D(c_n)$ and the channel-deinterleaved combined channel and extrinsic LLR $\mathcal{L}_i^E(c_n)$ extracted from the equaliser's *a posteriori* LLR $\mathcal{L}_p^E(c_k)$. The extrinsic part can be interpreted as the incremental information concerning the current bit obtained through the decoding process from all the information available due to all other bits imposed by the code constraints, but excluding the information directly conveyed by the bit. This information can be calculated by subtracting bitwise the LLR values $\mathcal{L}_i^E(c_n)$ at the input of the decoder from the *a posteriori* LLR values $\mathcal{L}_p^D(c_n)$ at the channel decoder's output, as seen also in Figure 11.1, yielding:

$$\mathcal{L}_e^D(c_n) = \mathcal{L}_p^D(c_n) - \mathcal{L}_i^E(c_n). \tag{11.1}$$

The extrinsic information $\mathcal{L}_e^D(c_n)$ of the coded bits is then interleaved in Figure 11.1, in order to yield $\mathcal{L}_e^D(c_k)$, which is fed back in the required bit-order to the equaliser, where it is used as the *a priori* information $\mathcal{L}_a^E(c_k)$ in the next equalization iteration. This constitutes the first iteration. Again, it is important that only the channel-interleaved extrinsic part – *i.e.* $\mathcal{L}_e^D(c_k)$ of $\mathcal{L}_p^D(c_n)$ – is fed back to the equaliser, since the interdependence between the a priori information $\mathcal{L}_a^E(c_k) = \mathcal{L}_e^D(c_k)$ used by the equaliser and the previous decisions of the equaliser should be minimized. This independence assists in obtaining the equaliser's reliability-estimation of the coded bits for the current iteration, without being 'influenced' by its previous estimations. Ideally, the *a priori* information should be based on an independent estimation. As argued above, this is the reason that the *a priori* information $\mathcal{L}_a^E(c_k)$ is subtracted from the *a posteriori* LLR value $\mathcal{L}_p^E(c_k)$ at the output of the equaliser in Figure 11.1, before passing the LLR values to the channel decoder. In the final iteration, the *a posteriori* LLRs $\mathcal{L}_p^D(d_n)$ of the source bits are computed by the channel decoder. Subsequently, the transmitted bits are estimated by comparing $\mathcal{L}_p^D(d_n)$ to the threshold value of 0. For $\mathcal{L}_p^D(d_n) < 0$ the transmitted bit d_n is deemed to be a logical 0, while $d_n = +1$ or a logical 1 is output, when $\mathcal{L}_p^D(d_n) \geq 0$.

Previous turbo equalization research has implemented the SISO equaliser using the Soft-Output Viterbi Algorithm (SOVA) [153], the optimal MAP algorithm [311] and linear filters [172]. We will now introduce the proposed RBF based equaliser as the SISO equaliser in the context of turbo equalization. The following sections will discuss the implementational details and the performance of this scheme, benchmarked against the optimal MAP turbo equaliser scheme of [311].

11.2 RBF Assisted Turbo equalization

The RBF network based equaliser is capable of utilizing the *a priori* information $\mathcal{L}_a^E(c_k)$ provided by the channel decoder of Figure 11.1, in order to improve its performance. This *a priori* information can be assigned namely to the weights of the RBF network [312]. We will describe this in more detail in this section. For convenience, we will rewrite Equation 8.80, describing the conditional probability density function (PDF) of the *i*th symbol,

$i = 1, \ldots, \mathcal{M}$, associated with the ith subnet of the $\mathcal{M}$-ary RBF equaliser:

$$
\begin{aligned}
f^i_{RBF}(\mathbf{v}_k) &= \sum_{j=1}^{n^i_s} w^i_j \varphi(\|\mathbf{v}_k - \mathbf{c}^i_j\|), \\
\varphi(x) &= \exp\left(\frac{-x^2}{\rho}\right) \\
i &= 1, \ldots, \mathcal{M}, \qquad j = 1, \ldots, n^i_s
\end{aligned}
\tag{11.2}
$$

where $\mathbf{c}^i_j$, w^i_j, $\varphi(\cdot)$ and ρ are the RBF's centres, weights, activation function and width, respectively. In order to arrive at the Bayesian equalisation solution [85] – which was highlighted in Section 8.9 – the RBF centres are assigned the values of the channel states $\mathbf{r}^i_j$ defined in Equation 8.83, the RBF weights defined in Section 8.7.1 correspond to the a priori probability of the channel states $p^i_j = P(\mathbf{r}^i_j)$ and the RBF width introduced in Section 8.7.1 is given the value of $2\sigma^2_\eta$ where σ^2_η is the channel noise variance. The actual number of channel states n^i_s is determined by the specific design of the algorithm invoked, reducing the number of channel states from the optimum number of $\mathcal{M}^{m+L-1}$, where m is the equaliser feedforward order and $L + 1$ is the CIR duration [246, 286, 287]. The probability p^i_j of the channel states $\mathbf{r}^i_j$, and therefore the weights of the RBF equaliser can be derived from the LLR values of the transmitted bits, as estimated by the channel decoder.

Expounding further from Equation 8.2 and 8.10, the channel output can be defined as

$$
\mathbf{r}_j = \mathbf{F}\mathbf{s}_j,
\tag{11.3}
$$

where $\mathbf{F}$ is the CIR matrix defined in Equation 8.11 and $\mathbf{s}_j$ is the jth possible combination of the $(L+m)$ transmitted symbol sequence, $\mathbf{s}_j = \begin{bmatrix} s_{j1} & \cdots & s_{jp} & \cdots & s_{j(L+m)} \end{bmatrix}^T$. Hence – for a time-invariant CIR and assuming that the symbols in the sequence $\mathbf{s}_j$ are statistically independent of each other – the probability of the received channel output vector $\mathbf{r}_j$ is given by:

$$
\begin{aligned}
P(\mathbf{r}_j) &= P(\mathbf{s}_j) \\
&= P(s_{j1} \cap \ldots s_{jp} \cap \ldots s_{j(L+m)}) \\
&= \prod_{p=1}^{L+m} P(s_{jp}) \qquad j = 1, \ldots, n^i_s.
\end{aligned}
\tag{11.4}
$$

The transmitted symbol vector component s_{jp} – i.e. the pth symbol in the vector – is given by $m = \log_2 \mathcal{M}$ number of bits $c_{jp1}, c_{jp2}, \ldots, c_{jpm}$. Therefore,

$$
\begin{aligned}
P(s_{jp}) &= P(c_{jp1} \cap \ldots c_{jpq} \cap \ldots c_{jpm}) \\
&= \prod_{q=1}^{m} P(c_{jpq}) \qquad j = 1, \ldots, n^i_s, \qquad p = 1, \ldots, L+m.
\end{aligned}
\tag{11.5}
$$

We have to map the bits c_{jpq} representing the $\mathcal{M}$-ary symbol s_{jp} to the corresponding bit $\{c_k\}$. Note that the probability $P(\mathbf{r}_j)$ of the channel output states and therefore also the RBF weights defined in Equation 11.2 are time-variant, since the values of $\mathcal{L}_p(c_k)$ are time-variant.

Based on the definition of the bit LLR of Equation 10.6, the probability of bit c_k having the value of +1 or -1 can be obtained after a few steps from the *a priori* information $\mathcal{L}_a^E(c_k)$ provided by the channel decoder of Figure 11.1, according to:

$$P(c_k = \pm 1) = \frac{\exp(-\mathcal{L}_a^E(c_k)/2)}{1 + \exp(-\mathcal{L}_a^E(c_k))} \cdot \exp(\pm \mathcal{L}_a^E(c_k)/2). \tag{11.6}$$

Hence, referring to Equation 11.4, 11.5 and 11.6, the probability $P(\mathbf{r}_j)$ of the received channel output vector can be represented in terms of the bit LLRs $\mathcal{L}_a^E(c_{jpq})$ as follows:

$$
\begin{aligned}
P(\mathbf{r}_j) &= P(\mathbf{s}_j) \\
&= \prod_{p=1}^{L+m} P(s_{jp}) \\
&= \prod_{p=1}^{L+m} \prod_{q=1}^{m} P(c_{jpq}) \\
&= \prod_{p=1}^{L+m} \prod_{q=1}^{m} \frac{\exp(-\mathcal{L}_a^E(c_{jpq})/2)}{1 + \exp(-\mathcal{L}_a^E(c_{jpq}))} \cdot \exp\left(\frac{1}{2} \cdot c_{jpq} \cdot \mathcal{L}_a^E(c_{jpq})\right) \\
&= C_{\mathcal{L}_a^E(\mathbf{s}_j)} \cdot \prod_{p=1}^{L+m} \prod_{q=1}^{m} \exp\left(\frac{1}{2} \cdot c_{jpq} \cdot \mathcal{L}_a^E(c_{jpq})\right) \\
&= C_{\mathcal{L}_a^E(\mathbf{s}_j)} \cdot \exp\left(\frac{1}{2} \sum_{p=1}^{L+m} \sum_{q=1}^{m} c_{jpq} \cdot \mathcal{L}_a^E(c_{jpq})\right) \qquad j = 1, \ldots, n_s^i, \tag{11.7}
\end{aligned}
$$

where the constant $C_{\mathcal{L}_a^E(\mathbf{s}_j)} = \prod_{p=1}^{L+m} \prod_{q=1}^{m} \frac{\exp(-\mathcal{L}_a^E(c_{jpq})/2)}{1+\exp(-\mathcal{L}_a^E(c_{jpq}))}$ is independent of the bit c_{jpq}.

Therefore, we have demonstrated how the soft output $\mathcal{L}_a^E(c_k)$ of the channel decoder of Figure 11.1 can be utilized by the RBF equaliser. Another way of viewing this process is that the RBF equaliser is trained by the information generated by the channel decoder. The RBF equaliser provides the *a posteriori* LLR values of the bits c_k according to

$$\mathcal{L}_p^E(c_k) = \ln\left(\frac{\sum_{c_k=+1}^{i} f_{RBF}^i(\mathbf{v}_k)}{\sum_{c_k=-1}^{i} f_{RBF}^i(\mathbf{v}_k)}\right), \tag{11.8}$$

where $f_{RBF}^i(\mathbf{v}_k)$ was defined by Equation 11.2 and the received sequence $\mathbf{v}_k$ is shown in Figure 11.2. In the next section we will provide a comparative study of the RBF equaliser and the conventional MAP equaliser of [313].

11.3 Comparison of the RBF and MAP Equaliser

The *a posteriori* LLR value $\mathcal{L}_p^E$ of the coded bit c_k, given the received sequence $\mathbf{v}_k$ of Figure 11.2, can be calculated according to [311]:

$$\mathcal{L}_p^E(c_k) = \ln\left(\frac{\sum_{(s',s)\Rightarrow c_k=+1} p(s', s, \mathbf{v}_k)}{\sum_{(s',s)\Rightarrow c_k=-1} p(s', s, \mathbf{v}_k)}\right), \tag{11.9}$$

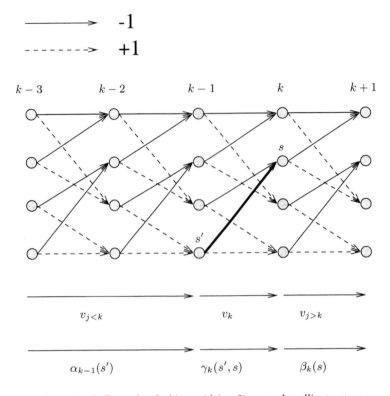

Figure 11.3: Example of a binary ($\mathcal{M} = 2$) system's trellis structure

where s' and s denote the states of the trellis seen in Figure 11.3 at trellis stages $k - 1$ and k, respectively. The joint probability $p(s', s, \mathbf{v}_k)$ is the product of three factors [311]:

$$p(s', s, \mathbf{v}_k) = \underbrace{p(s', v_{j<k})}_{\alpha_{k-1}(s')} \cdot \underbrace{P(s|s') \cdot p(v_k|s', s)}_{\gamma_k(s', s)} \cdot \underbrace{p(v_{j>k}|s)}_{\beta_k(s)}, \tag{11.10}$$

where the term $\alpha_{k-1}(s')$ and $\beta_k(s)$ are the so-called forward- and backward oriented transition probabilities, respectively, which can be obtained recursively, as follows [311]:

$$\alpha_k(s) = \sum_{s'} \gamma_k(s', s) \cdot \alpha_{k-1}(s') \tag{11.11}$$

$$\beta_{k-1}(s) = \sum_{s} \gamma_k(s', s) \cdot \beta_k(s). \tag{11.12}$$

Furthermore, $\gamma_k(s', s), k = 1, \ldots, \mathcal{F}$ represents the trellis transitions between the trellis stages $(k - 1)$ and k. The trellis has to be of finite length and for the case of MAP equalization, this corresponds to the length $\mathcal{F}$ of the received sequence or the transmission burst. The branch transition probability $\gamma_k(s', s)$ can be expressed as the product of the *a priori* probability $P(s|s') = P(c_k)$ and the transition probability $p(v_k|s', s)$:

$$\gamma_k(s', s) = P(c_k) \cdot p(v_k|s', s). \tag{11.13}$$

The transition probability is given by:

$$p(v_k|s', s) = \frac{1}{\sqrt{2\pi\sigma_\eta^2}} \exp(-\frac{(v_k - \tilde{v}_k)^2}{2\sigma_\eta^2}), \qquad (11.14)$$

where $\tilde{v}_k$ is the noiseless channel output, and the *a priori* probability of bit c_k being a logical 1 or a logical 0 can be expressed in terms of its LLR values according to Equation 11.6. Since the term $\frac{1}{\sqrt{2\pi\sigma_\eta^2}}$ in the transition probability expression of Equation 11.14 and the term $\frac{\exp(-\mathcal{L}_a^E(c_k)/2)}{1+\exp(-\mathcal{L}_a^E(c_k))}$ in the *a priori* probability formula of Equation 11.6 are constant over the summation in the numerator and denominator of Equation 11.9, they cancel out. Hence, the transition probability is calculated according to [311]:

$$\gamma_k(s', s) = w_k \cdot \gamma_k^*(s', s), \qquad (11.15)$$

$$\gamma^*(s', s) = \exp(-\frac{|v_k - \tilde{v}_k|^2}{2\sigma_\eta^2}) \qquad (11.16)$$

$$w_k = \exp(\frac{1}{2} \cdot c_k \cdot \mathcal{L}_a^E(c_k)). \qquad (11.17)$$

Note the similarity of the transition probability of Equation 11.15 with the PDF of the RBF equaliser's ith symbol described by Equation 10.3, where the terms w_k and $\gamma^*(s', s)$ are the RBF's weight and activation function, respectively, while the number of RBF nodes n_s^i is one. We also note that the computational complexity of both the MAP and the RBF equalisers can be reduced by representing the output of the equalisers in the logarithmic domain, utilizing the Jacobian logarithmic relationship [288] described in Equation 10.1. The RBF equaliser based on the Jacobian logarithm – highlighted in Section 10.2 – was hence termed as the Jacobian RBF equaliser.

The memory of the MAP equaliser is limited by the length of the trellis, provided that decisions about the kth transmitted symbol I_k are made in possession of the information related to all the received symbols of a transmission burst. In the MAP algorithm the recursive relationships of the forward and backward transition probabilities of Equation 11.11 and 11.12, respectively, allow us to avoid processing the entire received sequence $\mathbf{v}_k$ everytime the *a posteriori* LLR $\mathcal{L}_p^E(c_k)$ is evaluated from the joint probability $p(s', s, \mathbf{v}_k)$ according to Equation 11.9. This approach is different from that of the RBF based equaliser having a feedforward order of m, where the received sequence $\mathbf{v}_k$ of m-symbols is required each time the *a posteriori* LLR $\mathcal{L}_p^E(c_k)$ is evaluated using Equation 11.8. However, the MAP algorithm has to process the received sequence both in a forward and backward oriented fashion and store both the forward and backward recursively calculated transition probabilities $\alpha_k(s)$ and $\beta_k(s)$, before the LLR values $\mathcal{L}_p^E(c_k)$ can be calculated from Equation 11.9. The equaliser's delay facilitates invoking information from the 'future' samples $v_k, \ldots, v_{k-\tau+1}$ in the detection of the transmitted symbol $I_{k-\tau}$. In other words, the delayed decision of the MAP equaliser provides the necessary information concerning the 'future' samples $v_{j>k}$ – relative to the delayed kth decision – to be utilized and the information of the future samples is generated by the backward recursion of Equation 11.12.

The MAP equaliser exhibits optimum performance. However, if decision feedback is used in the RBF subset centre selection as in [246] or in the RBF space-translation as in

Section 8.11.2, the performance of the RBF DFE TEQ in conjunction with the idealistic assumption of *correct* decision feedback is better, than that of the MAP TEQ due to the increased Euclidean distance between channel states, as it will be demonstrated in Section 11.5. However, this is not so for the more practical RBF DFE feeding back the detected symbols, which may be erroneous.

11.4   Comparison of the Jacobian RBF and Log-MAP Equaliser

Building on Section 11.3, in this section the Jacobian logarithmic algorithm is invoked, in order to reduce the computational complexity of the MAP algorithm. We denote the forward, backward and transition probability in the logarithmic form as follows:

$$A_k(s) = \ln(\alpha_k(s)) \tag{11.18}$$
$$B_k(s) = \ln(\beta_k(s)) \tag{11.19}$$
$$\Gamma_k(s', s) = \ln(\gamma_k(s', s)), \tag{11.20}$$

which we also used in Section 11.3. Thus, we could rewrite Equation 11.11 as:

$$
\begin{aligned}
A_k(s) &= \ln\left(\sum_{s'} \gamma_k(s', s) \cdot \alpha_{k-1}(s')\right) \\
&= \ln\left(\sum_{s'} \exp\left(\Gamma_k(s', s) + A_{k-1}(s')\right)\right),
\end{aligned} \tag{11.21}
$$

and Equation 11.12 as:

$$
\begin{aligned}
B_{k-1}(s') &= \ln\left(\sum_{s} \gamma_k(s', s) \cdot \beta_k(s)\right) \\
&= \ln\left(\sum_{s} \exp\left(\Gamma_k(s', s) + B_k(s)\right)\right).
\end{aligned} \tag{11.22}
$$

¿From Equation 11.21 and 11.22, the logarithmic-domain forward and backward recursion can be evaluated, once $\Gamma_k(s', s)$ was obtained. In order to evaluate the logarithmic-domain branch metric $\Gamma_k(s', s)$, Equations 11.15–11.17 and 11.20 are utilized to yield:

$$\Gamma_k(s', s) = -\frac{|v_k - \tilde{v}_k|^2}{2\sigma_\eta^2} + \frac{1}{2} \cdot c_k \cdot \mathcal{L}_a^E(c_k). \tag{11.23}$$

By transforming $\alpha_k(s)$, $\gamma_k(s', s)$ and $\beta_k(s)$ into the logarithmic domain in the Log-MAP algorithm, the expression for the LLR, $\mathcal{L}_p^E(c_k)$ in Equation 11.9 is also modified to yield:

$$
\begin{aligned}
\mathcal{L}_p^E(c_k) &= \ln\left(\frac{\sum_{(s',s)\Rightarrow c_k=+1}\alpha_{k-1}(s')\cdot\gamma_k(s',s)\cdot\beta_k(s)}{\sum_{(s',s)\Rightarrow c_k=-1}\alpha_{k-1}(s')\cdot\gamma_k(s',s)\cdot\beta_k(s)}\right) \\
&= \ln\left(\frac{\sum_{(s',s)\Rightarrow c_k=+1}\exp\left(A_{k-1}(s')+\Gamma_k(s',s)+B_k(s)\right)}{\sum_{(s',s)\Rightarrow c_k=-1}\exp\left(A_{k-1}(s')+\Gamma_k(s',s)+B_k(s)\right)}\right) \\
&= \ln\left(\sum_{(s',s)\Rightarrow c_k=+1}\exp\left(A_{k-1}(s')+\Gamma_k(s',s)+B_k(s)\right)\right) \\
&\quad -\ln\left(\sum_{(s',s)\Rightarrow c_k=-1}\exp\left(A_{k-1}(s')+\Gamma_k(s',s)+B_k(s)\right)\right). \quad (11.24)
\end{aligned}
$$

In the trellis of Figure 11.3 there are $\mathcal{M}$ possible transitions from state s' to all possible states s or to state s from all possible states s'. Hence, there are $\mathcal{M}-1$ summations of the exponentials in the forward and backward recursion of Equation 11.21 and 11.22, respectively. Using the Jacobian logarithmic relationship of Equation 10.2, $\mathcal{M}-1$ summations of the exponentials requires $2(\mathcal{M}-1)$ additions/subtractions, $(\mathcal{M}-1)$ maximum search operations and $(\mathcal{M}-1)$ table look-up steps. Together with the $\mathcal{M}$ additions necessitated to evaluate the term $\Gamma_k(s',s)+A_{k-1}(s')$ and $\Gamma_k(s',s)+B_k(s)$ in Equation 11.21 and 11.22, respectively, the forward and backward recursion requires a total of $(6\mathcal{M}-4)$ additions/subtractions, $2(\mathcal{M}-1)$ maximum search operations and $2(\mathcal{M}-1)$ table look-up steps. Assuming that the term $\frac{1}{2}\cdot c_k\cdot\mathcal{L}_a^E(c_k)$ in Equation 11.23 is a known weighting coefficient, evaluating the branch metrics given by Equation 11.23 requires a total of 2 additions/subtractions, 1 multiplication and 1 division.

By considering a trellis having χ number of states at each trellis stage and $\mathcal{M}$ legitimate transitions leaving each state, there are $\frac{1}{2}\mathcal{M}\chi$ number of transitions due to the bit $c_k=+1$. Each of these transitions belongs to the set $(s',s)\Rightarrow c_k=+1$. Similarly, there will be $\frac{1}{2}\mathcal{M}\chi$ number of $c_k=-1$ transitions, which belong to the set $(s',s)\Rightarrow c_k=-1$. Evaluating $A_k(s)$, $B_{k-1}(s')$ and $\Gamma_k(s',s)$ of Equation 11.21, 11.22 and 11.23, respectively, at each trellis stage k associated with a total of $\mathcal{M}\chi$ transitions requires $\mathcal{M}\chi(6\mathcal{M}-2)$ additions/subtractions, $\mathcal{M}\chi(2\mathcal{M}-2)$ maximum search operations, $\mathcal{M}\chi(2\mathcal{M}-2)$ table look-up steps, plus $\mathcal{M}\chi$ multiplications and $\mathcal{M}\chi$ divisions. With the terms $A_k(s)$, $B_{k-1}(s')$ and $\Gamma_k(s',s)$ of Equations 11.21, 11.22 and 11.23 evaluated, computing the LLR $\mathcal{L}_p^E(c_k)$ of Equation 11.24 using the Jacobian logarithmic relationship of Equation 10.2 for the summation terms $\ln(\sum_{(s',s)\Rightarrow c_k=+1}\exp(\cdot))$ and $\ln(\sum_{(s',s)\Rightarrow c_k=+1}\exp(\cdot))$ requires a total of $4(\frac{1}{2}\mathcal{M}\chi-1)+2\mathcal{M}\chi+1$ additions/subtractions, $\mathcal{M}\chi-2$ maximum search operations and $\mathcal{M}\chi-2$ table look-up steps. The number of states at each trellis stage is given by $\chi=\mathcal{M}^L=n_{s,f}/\mathcal{M}$. Therefore, the total computational complexity associated with generating the a posteriori LLRs using the Jacobian logarithmic relationship for the Log-MAP equaliser is given in Table 11.1.

For the Jacobian RBF equaliser, the LLR expression of Equation 11.8 is rewritten in terms

	Log-MAP	Jacobian RBF
subtraction	$n_{s,f}(6\mathcal{M}+2)-3$	$n_{s,f}+$
and addition		$\mathcal{M}n_s^i(m+2)-4$
multiplication	$n_{s,f}$	$n_{s,f}$
division	$n_{s,f}$	$n_{s,f}$
max	$n_{s,f}(2\mathcal{M}-1)-2$	$\mathcal{M}n_s^i-2$
table look-up	$n_{s,f}(2\mathcal{M}-1)-2$	$\mathcal{M}n_s^i-2$

Table 11.1: Computational complexity of generating the *a posteriori* LLR $\mathcal{L}_p^E$ for the Log-MAP equaliser and the Jacobian RBF equaliser [314]. The RBF equaliser order is denoted by m and the number of RBF centres is n_s^i. The notation $n_{s,f} = \mathcal{M}^{L+1}$ indicates the number of trellis states for the Log-MAP equaliser and also the number of scalar channel states for the Jacobian RBF equaliser.

of the logarithmic form $\ln\left(f_{RBF}^i(\mathbf{v}_k)\right)$ to yield:

$$
\begin{aligned}
\mathcal{L}_p^E(c_k) &= \ln\left(\frac{\sum_{c_k=+1}^i f_{RBF}^i(\mathbf{v}_k)}{\sum_{c_k=-1}^i f_{RBF}^i(\mathbf{v}_k)}\right) \\
&= \ln\left(\frac{\sum_{c_k=+1}^i \exp\left(\ln(f_{RBF}^i(\mathbf{v}_k))\right)}{\sum_{c_k=-1}^i \exp\left(\ln(f_{RBF}^i(\mathbf{v}_k))\right)}\right) \\
&= \ln\left(\sum_{c_k=+1}^i \exp\left(\ln(f_{RBF}^i(\mathbf{v}_k))\right)\right) - \ln\left(\sum_{c_k=-1}^i \exp\left(\ln(f_{RBF}^i(\mathbf{v}_k))\right)\right).
\end{aligned}
\tag{11.25}
$$

The summation of the exponentials in Equation 11.25 requires $2(\mathcal{M}-2)$ additions/subtractions, $(\mathcal{M}-2)$ table look-up and $(\mathcal{M}-2)$ maximum search operations. The associated complexity of evaluating the conditional PDF of $\mathcal{M}$ symbols in logarithmic form according to Equation 10.4 was given in Table 10.1. Therefore, – similarly to the Log-MAP equaliser – the computational complexity associated with generating the *a posteriori* LLR $\mathcal{L}_p^E$ for the Jacobian RBF equaliser is given in Table 11.1. Figure 11.4 compares the number of additions/subtractions per turbo iteration involved in evaluating the *a posteriori* LLRs $\mathcal{L}_p^E$ for the Log-MAP equaliser and Jacobian RBF equaliser according to Table 11.1. More explicitly, the complexity is evaluated upon with varying the feedforward order m for different values of L, where $(L+1)$ is the CIR duration under the assumption that the feedback order $n = L$ and the number of RBF centres is $n_s^i = \mathcal{M}^{m+L-n}/\mathcal{M}$. Since the number of multiplications and divisions involved is similar, and by comparison, the number of maximum search and table look-up stages is insignificant, the number of additions/subtractions incurred in Figure 11.4 approximates the relative computational complexities involved. Figure 11.4 shows significant computational complexity reduction upon using Jacobian RBF equalisers of relatively low feedforward order, especially for higher-order modulation modes, such as $\mathcal{M} = 64$. The figure also shows an exponential increase of the computational complexity, as the CIR length increases. Observe in Figure 11.4 that as a rule of thumb, the feedforward order of the Jacobian RBF DFE must not exceed the CIR length $(L + 1)$ in order to achieve a computational complexity improvement relative to the Log-MAP equaliser, provided that we use the optimal number of RBF centres, namely $n_s^i = \mathcal{M}^{m+L-n}/\mathcal{M}$.

The length of the trellis determines the storage requirements of the Log-MAP equaliser, since the Log-MAP algorithm has to store both the forward- and backward-recursively calculated metrics $A_k(s)$ and $B_{k-1}(s')$ before the LLR values $\mathcal{L}_p^E(c_k)$ can be calculated. For the Jacobian RBF DFE, we have to store the value of the RBF centres and the storage requirements will depend on the CIR length $L + 1$ and on the modulation mode characterized by $\mathcal{M}$.

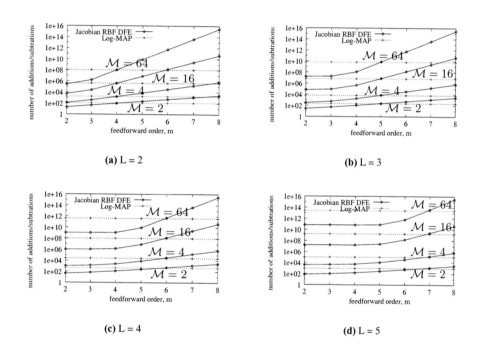

(a) L = 2   **(b)** L = 3

(c) L = 4   **(d)** L = 5

Figure 11.4: Number of additions/subtractions per iteration for the Jacobian RBF DFE of varying equaliser order m and the Log-MAP equaliser for various values of L, where $L + 1$ is the CIR length. The feedback order of the Jacobian RBF DFE is set to $n = L$ and the number of RBF centres is set to $n_s^i = \mathcal{M}^{m+L-n}/\mathcal{M}$.

11.5 RBF Turbo Equaliser Performance

The schematic of the entire system was shown in Figure 11.2, where the transmitted source bits are convolutionally encoded, channel-interleaved and mapped to an $\mathcal{M}$-ary modulated symbol. The encoder utilized a half-rate recursive systematic convolutional (RSC) code, having a constraint length of $K = 5$ and octal generator polynomials of $G_0 = 35$ and $G_1 = 23$. The transmission burst structure used in this system is the FMA1 non-spread speech burst, as specified in the Pan-European FRAMES proposal [151], which is seen in Figure 11.5.

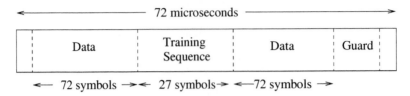

Figure 11.5: Transmission burst structure of the so-called FMA1 nonspread speech mode as specified in the FRAMES proposal [151].

11.5.1 Dispersive Gaussian Channels

The performance of the Jacobian RBF DFE TEQ was initially investigated over a dispersive Gaussian channel. A random channel interleaver of 4000-bit memory was invoked. We have assumed that perfect knowledge of the CIR was available, which implies that our results portray the best-case performance. Figure 11.6 provides the BER performance comparison of the Log-MAP and Jacobian RBF DFEs in the context of turbo equalization. Various equaliser orders were used over a three-path Gaussian channel having a z-domain transfer function of $F(z) = 0.5773 + 0.5773z^{-1} + 0.5773z^{-2}$ and employing BPSK. Figure 11.6(b) shows that when the feedback information is not error-free, the Log-MAP TEQ outperforms the Jacobian RBF DFE TEQ for the same number of iterations. The corresponding uncoded systems using the Log-MAP equaliser and the Jacobian RBF DFE exhibit similar performance trends. Comparing Figure 11.6(a) for the equaliser parameters of $m = 3$, $n = 2$ and $\tau = 2$, as well as Figure 11.6(b) for the equaliser parameters of $m = 4$, $n = 2$ and $\tau = 3$, we observe that the performance of the Jacobian RBF DFE TEQ improves, as the feedforward order and the decision delay of the equaliser increases. This is achieved at the expense of increased computational complexities as evidenced by Figure 11.4. The above trend is a consequence of the enhanced DFE performance in conjunction with increasing feedforward order and decision delay, as it was demonstrated and justified in Section 8.11. However, as seen in Table 11.1, the approximate number of additions/subtrations for the Jacobian RBF DFE increased from 44 to 100 for a feedforward order increase from $m = 3$ to $m = 4$. Both the Log-MAP and the Jacobian RBF DFE TEQs converge to a similar BER performance upon increasing the number of iterations. The Log-MAP TEQ performs better, than the Jacobian RBF DFE TEQ at a lower number of iterations, as shown in Figure 11.6. This is, because effectively the Log-MAP equaliser has a higher feedforward order, which is equivalent to the length of the trellis and also exhibits a longer decision delay, as discussed in Section 11.3. The performance of the Log-MAP TEQ in the zero-ISI – i.e. non-dispersive – Gaussian channel environment was also presented in Figure 11.6(b) for comparison. The Log-MAP TEQ, the Jacobian RBF DFE TEQ using $m = 4$, $n = 2$, $\tau = 3$ and the Jacobian RBF DFE TEQ employing $m = 3$, $n = 2$, $\tau = 2$ performed within approximately 0.2dB, 0.2dB and 0.5dB, respectively, from this zero-ISI, i.e. non-dispersive AWGN benchmarker at BER of 10^{-4}. The BER performance of the RBF DFE TEQ using *correct* decision fedback is also shown in Figure 11.6, which exhibits a better performance than the Log-MAP TEQ. This is possible – although the

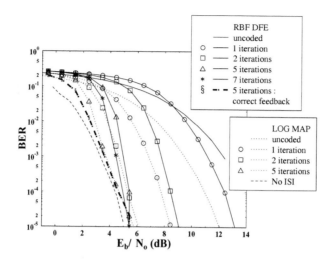

(a) The Jacobian RBF DFE has a feedforward order of $m = 3$, feedback order of $n = 2$ and decision delay of $\tau = 2$ symbols.

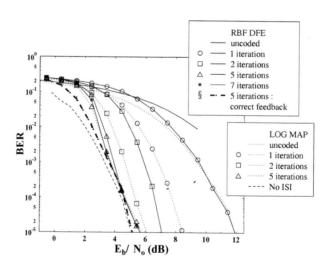

(b) The Jacobian RBF DFE has a feedforward order of $m = 4$, feedback order of $n = 2$ and decision delay of $\tau = 3$ symbols.

Figure 11.6: Performance of the Log-MAP TEQ and Jacobian RBF DFE TEQ over the three-path Gaussian channel having a z-domain transfer function of $F(z) = 0.5773 + 0.5773z^{-1} + 0.5773z^{-2}$ for BPSK.

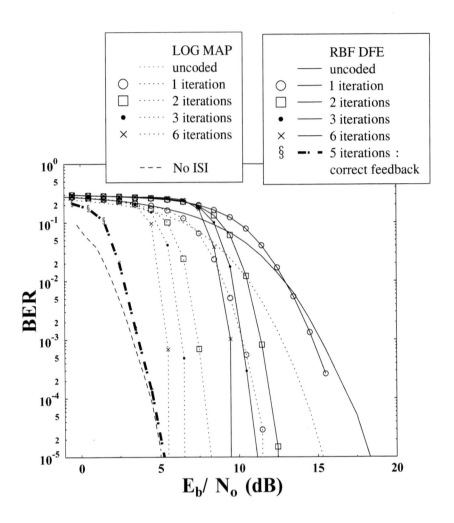

Figure 11.7: Performance of the Log-MAP TEQ and Jacobian RBF DFE TEQ over the five-path Gaussian channel having a z-domain transfer function of $F(z) = 0.227 + 0.46z^{-1} + 0.688z^{-2} + 0.46z^{-3} + 0.227z^{-4}$ for BPSK. The Jacobian RBF DFE has a feedforward order of $m = 5$, feedback order of $n = 4$ and decision delay of $\tau = 4$ symbols.

Log-MAP equaliser is known to approximate the optimal performance – because the RBF DFE's subset centre selection mechanism creates an increased Euclidean distance between the channel states [246] and effectively eliminates the postcursor ISI, which improves the performance of the Jacobian RBF DFE TEQ.

The performance of the TEQs was then investigated over a dispersive Gaussian channel having an increased CIR length. Figure 11.7 compares the performance of the Log-MAP TEQ and the Jacobian RBF DFE ($m = 5$, $n = 4$, $\tau = 4$) TEQ over the five-path Gaussian channel associated with the transfer function of $F(z) = 0.227 + 0.46z^{-1} + 0.688z^{-2} + 0.46z^{-3} + 0.227z^{-4}$. The performance of both the Log-MAP and Jacobian RBF DFE TEQs degrades with increasing CIR lengths, especially at lower SNRs, when we compare Figures 11.6 and 11.7. This is due to the increased number of multipath components to be resolved, when the CIR length is increased, a phenomenon which was also demonstrated in Figures 8.32 and 8.33 for an uncoded RBF DFE over the three-path and five-path channels, respectively. For the five-path channel, the Log-MAP TEQ and the Jacobian RBF DFE TEQ using $m = 5$, $n = 4$, $\tau = 4$ performed within about 1dB and 5dB, respectively, from the zero-ISI, non-dispersive Gaussian limit at a BER of 10^{-4}. We observed from Figures 11.6(b) and 11.7, that the coded BERs only start to decrease once the uncoded BERs reached approximately 2×10^{-1}.

11.5.2 Dispersive Rayleigh Fading Channels

Let us now investigate the performance of the TEQs in a dispersive Rayleigh fading chan-nel environment. In order to quntify the tolerable delay and hence on the required depth of the channel interleaver, we considered the maximum affordable delay of a Time Division Multiple Access/Time Division Duplex (TDMA/TDD) speech system, which employs eight uplink and eight downlink slots. Hence a certain user's transmission slot has the periodicity of sixteen TDMA slots. In our investigations the transmission delay of the BPSK, 4-QAM and 16-QAM system was limited to approximately 30ms. This corresponds to 3456 data symbols transmitted within 30ms and hence 3456-bit, 6912-bit and 13824-bit random chan-nel interleavers were utilized for BPSK, 4-QAM and 16-QAM, respectively. A three-path, symbol-spaced fading channel of equal weights was employed, where the Rayleigh fading statistics obeyed a normalized Doppler frequency of 1.5×10^{-4}. The CIR was assumed to be burst-invariant, in other words the fading envelope was time-invariant for the duration of a transmission burst and then it was faded at the end of the burst. The CIR was estimated iteratively using the LMS channel estimator. Specifically, during the first iteration only the midamble was used as the training sequence in conjunction with a fast learning rate of 0.1. By contrast, during all the forthcoming iterations all symbols of the channel decoded burst were used as a longer training sequence in conjunction with a finer learning rate of 0.01. Figures 11.8, 11.9 and 11.10 portray the performance of the Log-MAP TEQ and that of the Jacobian RBF DFE TEQ for BPSK, 4-QAM and 16-QAM, respectively. The Jacobian RBF DFE has a feedforward order of $m = 3$, feedback order of $n = 2$ and decision delay of $\tau = 2$ symbols. Figure 11.8 and Figure 11.9 show for BPSK and 4-QAM, that the Log-MAP TEQ and the Jacobian RBF DFE TEQ converge to a similar BER performance, but the Log-MAP TEQ requires a lower number of iterations. Specifically, two iterations are required for the Log MAP TEQ and three iterations for the Jacobian RBF DFE TEQ to achieve near-perfect convergence, since the Log-MAP TEQ exhibited a better BER performance for an uncoded system than the Jacobian RBF DFE. The performance of the Log-MAP TEQ at two iterations

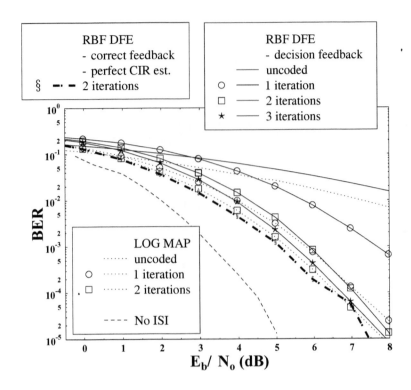

Figure 11.8: Performance of the Log-MAP TEQ and Jacobian RBF DFE TEQ over the three-path Rayleigh fading channel for **BPSK**. The Jacobian RBF DFE has a feedforward order of $m = 3$, feedback order of $n = 2$ and decision delay of $\tau = 2$ symbols.

and that of the Jacobian RBF DFE TEQ at three iterations is about 2dB and 2.5dB away from the zero-ISI Gaussian BER curve for BPSK and 4-QAM, respectively, at a BER of 10^{-4}. For 16-QAM, the effect of error propagation degrades the performance of the Jacobian RBF DFE TEQ by 4dB at BER of 10^{-4}, when we compare the Jacobian RBF DFE TEQ's correct feedback based and decision feedback assisted performance after the final iteration, as seen in Figure 11.10. Again, the performance can be improved by increasing the equaliser's feedforward order at the expense of a higher computational complexity, as discussed in Section 11.5.1.

The iteration gain of the Jacobian RBF DFE TEQ after the final iteration at a BER of 10^{-3} was 2dB, 3dB and more than 15dB for the modulation modes of BPSK, 4-QAM and 16-QAM, respectively. By contrast, for the Log-MAP TEQ the corresponding iteration gains were 0.5dB, 1dB and 2dB for the modulation modes of BPSK, 4-QAM and 16-QAM, respectively. Explicitly, the iteration gain was defined as the difference between the channel SNR required in order to achieve a certain BER after one iteration and the corresponding channel SNR required after n number of iterations. The iteration gain was higher for the higher-order

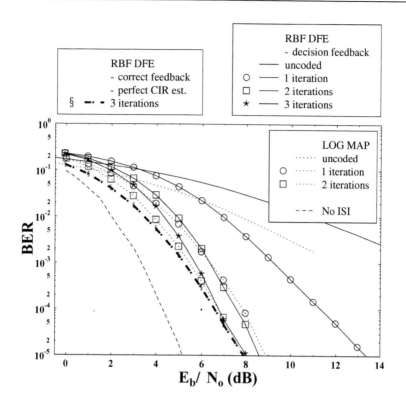

Figure 11.9: Performance of the Log-MAP TEQ and Jacobian RBF DFE TEQ over the three-path Rayleigh fading channel for **4-QAM**. The Jacobian RBF DFE has a feedforward order of $m = 3$, feedback order of $n = 2$ and decision delay of $\tau = 2$ symbols.

modulation modes, since the distance between two neighbouring points in the higher-order constellations was lower and hence it was more gravely affected by ISI and noise.

Since the computation of the associated implementational complexity summarised in Table 11.1 is quite elaborate, here we only give an estimate of the Log-MAP TEQ's and the Jacobian RBF DFE TEQ's complexity in the context of both BPSK and 4-QAM, employing the parameters used in our simulations. Specifically, in the BPSK scheme the approximate number of additions/subtractions and multiplications/divisions for the Log-MAP TEQ was 109 and 16 per iteration, respectively, whereas for the Jacobian RBF DFE TEQ ($m = 3$, $n = 2$, $\tau = 2$) the corresponding figures were 44 and 16, respectively. The 'per iteration' complexity of the Jacobian RBF DFE TEQ was approximately a factor of (109/44 =)2.5, 4.4 and 16.3 lower, than that of the Log-MAP TEQ, for BPSK, 4-QAM and 16-QAM, respectively.

Overall, due to the error propagation that gravely degrades the performance of the Jacobian RBF DFE TEQ when using 16-QAM, the Jacobian RBF DFE TEQ could only provide a practical performance versus complexity advantage for lower order modulation modes, such

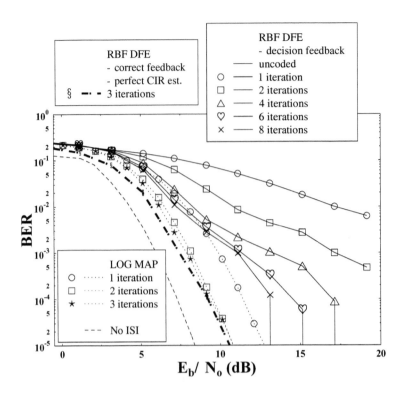

Figure 11.10: Performance of the Log-MAP TEQ and Jacobian RBF DFE TEQ over the three-path
Rayleigh fading channel for **16-QAM**. The Jacobian RBF DFE has a feedforward order
of $m = 3$, feedback order of $n = 2$ and decision delay of $\tau = 2$ symbols.

as BPSK and 4-QAM. It is worth noting here that we have attempted using the LLR values
output by the decoder in the previous iteration as the feedback information for the feedback
section of the RBF DFE. However, we attained an inferior performance compared to the scenario using the RBF DFE outputs as the feedback information. This is because the BER
improves on every iterations and the BER of the input of the equaliser fed back from the
decoder was improved after equalization. Therefore the output of the equaliser was more
reliable, than the output of the decoder in the previous iteration. Turbo equalization research
has been focused on developing reduced complexity equalisers, such as the receiver structure
proposed by Glavieux *et al.* [172], where the equaliser is constituted by two linear filters.
Motivated by this trend, Yeap, Wong and Hanzo [164, 315, 316] proposed a reduced complexity trellis-based equaliser scheme based on equalising the in-phase and quadrature-phase
component of the transmitted signal independently. This novel reduced complexity equaliser
is termed as the In-Phase/Quadrature-phase Equaliser (I/Q EQ). When a channel having a
memory of L symbol durations was encountered, the trellis-based equaliser must consider
$\mathcal{M}^{L+1}$ total number of transitions at each trellis stage, as discussed in Section 11.4. The

complexity of the complex-valued trellis-based equaliser increased rapidly with L. However, by removing the associated cross-coupling of the in-phase and quadrature-phase signal components and hence rendering the channel output to be only dependent on either quadrature component, the number of transitions considered was reduced to $(\sqrt{\mathcal{M}})^{L+1}$. Therefore, there will be an I/Q EQ for each I/Q component, substituting the original trellis-based equaliser and giving a complexity reduction factor of $\frac{\mathcal{M}^{L+1}}{2 \times \sqrt{\mathcal{M}}^{L+1}} = 0.5 \times \sqrt{\mathcal{M}^{L+1}}$. The TEQ using I/Q EQs was capable of achieving the same performance as the Log-MAP TEQ for 4-QAM and 16-QAM, while maintaining a complexity reduction factor of 2.67 and 16, respectively, over the equally-weighted three-path Rayleigh fading channel using a normalized Doppler frequency of 3.3×10^{-5} [164, 315, 316]. The complexity of the RBF DFE could be similarly reduced to that of the I/Q EQ by equalising the in-phase and quadrature-phase components of the transmitted signal separately. In the following section, we proposed another novel method of reducing the complexity of TEQ by making use of the fact that the RBF DFE evaluates its output on a symbol-by-symbol basis.

11.6   Reduced-complexity RBF Assisted Turbo equalization

The Log-MAP algorithm requires forward and backward recursions through the entire sequence of symbols in the received burst in order to evaluate the forward and backward transition probability of Equation 11.11 and 11.12, before calculating the a posteriori LLR values $\mathcal{L}_p(c_k)$. Therefore, effectively the computation of the a posteriori LLRs $\mathcal{L}_p(c_k)$ is performed on a burst-by-burst basis. The RBF based equaliser, however, performs the evaluation of the a posteriori LLRs $\mathcal{L}_p(c_k)$ on a symbol-by-symbol basis. Therefore, in order to reduce the associated computational complexity, the RBF based TEQ may skip evaluating the symbol LLRs according to Equation 11.8 in the current iteration, when the symbol has a low error probability or high a priori LLR magnitude $|\mathcal{L}_a^E(c_k)|$ after channel decoding in the previous iteration. If, however this is not the case, the equaliser invokes a further iteration and attempts to improve the decoder's reliablility estimation of the coded bits. The output $f_{RBF}^i(\mathbf{v}_k)$ of the RBF equaliser provides the likelihood of the ith symbol at instant k. The log-likelihood values of the ith symbol provided by the channel decoder in the previous iteration obey an approximately linear relationship versus the log-likelihood values from the equaliser in the current iteration, as demonstrated in Figure 11.11 for the BPSK mode over a three-path, symbol-spaced fading channel of equal CIR tap weights, where the Rayleigh fading statistics obeyed a normalized Doppler frequency of 1.5×10^{-4}. Therefore, the logarithmic domain output $\ln\left(\tilde{f}_{RBF}^i(\mathbf{v}_k)\right)$ of the RBF equaliser can be estimated based on this near-linear relationship portrayed in Figure 11.11 according to:

$$\ln\left(\tilde{f}_{RBF}^i(\mathbf{v}_k)\right) = g \cdot \ln(L_a(I_k = \mathcal{I}_i)) + c, \tag{11.26}$$

where $\ln(L_a(I_k = \mathcal{I}_i))$ is the log-likelihood of the transmitted symbol I_k being the ith QAM symbol $\mathcal{I}_i$ based on the decoder's soft output, g is the log-likelihood gradient and c is the log-likelihood intercept point. Both g and c can be inferred from Figure 11.11. As our next action, we have to set the LLR magnitude threshold $|\mathcal{L}|_{threshold}$, where the estimated coded bits c_k output by the decoder in the.previous iteration become sufficiently reliable for refraining from further iterations. Hence the symbols exhibiting an LLR value above this threshold are not fed back to the equaliser for futher iterations, since they can be considered sufficiently reliable for subjecting them to hard decision. The LLRs passed to the decoder from the equaliser are calculated from the symbols' log-likelihood values based on the linear relationship of Equation 11.26 instead of the more computationally demanding Equation 11.8, in order to reduce the computational complexity. We refer to this RBF based-TEQ as the *reduced-complexity RBF TEQ*.

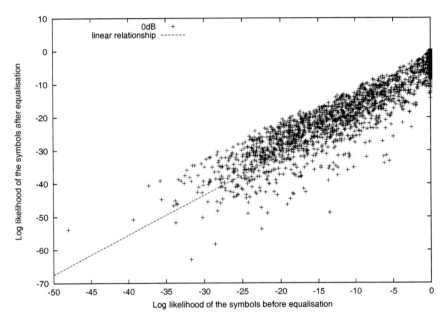

Figure 11.11: The log-likelihood of the RBF turbo equalised symbols before and after equalization over the three-path, symbol-spaced fading channel of equal CIR tap weights, where the Rayleigh fading statistics obeyed a normalised Doppler frequency of 1.5×10^{-4}, at an SNR of 0dB using BPSK.

In our experiments, the above mentioned log-likelihood gradient and the intercept point were found to be $g = 1.2$ and $c = -7.5$, respectively, according to the near-linear relationship of Figure 11.11. We set the LLR magnitude threshold $|\mathcal{L}|_{threshold}$ such that the symbols in the burst that were not fed back to the equaliser for further iterations became sufficiently reliable and hence exhibited a low probability of decoding error. The threshold was initially set to $|\mathcal{L}|_{threshold} = 10$ based on our experiments, such that the symbols that were not fed back to the decoder exhibited a probability of error below 5×10^{-5} according to Equation 10.14. Figures 11.12 and 11.13 compare the performance of the reduced-complexity Jacobian RBF DFE TEQ to that of the Jacobian RBF DFE TEQ of Section 10.2 over the three-path Gaussian channel having a transfer function of $F(z) = 0.5773 + 0.5773z^{-1} + 0.5773z^{-2}$. The reduced-complexity Jacobian RBF DFE TEQ provides an equivalent BER performance to that of the Jacobian RBF DFE TEQ of Section 10.2, while exhibiting a reduced computational complexity, which is proportional to the percentage of the BPSK symbols fed back for further iterations in Figure 11.12 and 11.13. We note that in our experiments the reduced-complexity Jacobian RBF DFE TEQ using the detected decision feedback – rather than error-free feed-

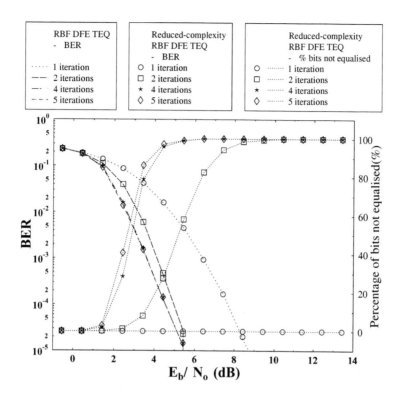

Figure 11.12: BER performance and the percentage of symbols not requiring equalization by the reduced-complexity RBF TEQ using **correct** decision feedback over the three-path Gaussian channel having a z-domain transfer function of $F(z) = 0.5773 + 0.5773z^{-1} + 0.5773z^{-2}$ for BPSK. The LLR magnitude threshold, the log-likelihood gradient and the log-likelihood intercept point were set to $|\mathcal{L}|_{threshold} = 10$, $g = 1.2$ and $c = -7.5$, respectively.

back – required a higher LLR magnitude threshold of $|\mathcal{L}|_{threshold} = 26$ (which guaranteed a probability of error of 5×10^{-12} according to Equation 10.14), in order to provide an equivalent BER performance to that of the Jacobian RBF DFE TEQ, since the decision feedback error propagation reduced the decoder's reliability estimation of the coded bits. The higher the LLR magnitude threshold, the higher the percentage of bits fed back, resulting in a higher complexity. According to Figure 11.12 depicting theperformance of the reduced-complexity Jacobian RBF DFE TEQ relying on correct decision feedback, the average percentage of bits not requiring further iterations for a channel SNR of 4dB was 20% for the second iteration, 70% for third iteration and approximately 90% for the consecutive iterations. This amounts to a total of approximately 54% computational complexity reduction at the SNR of 4dB. Referring to Figure 11.13, the reduced-complexity Jacobian RBF DFE TEQ relying on detected symbol-based – rather than perfect – decision feedback with its associated higher LLR magnitude threshold provides a total of approximately 21% computational reduction at an SNR

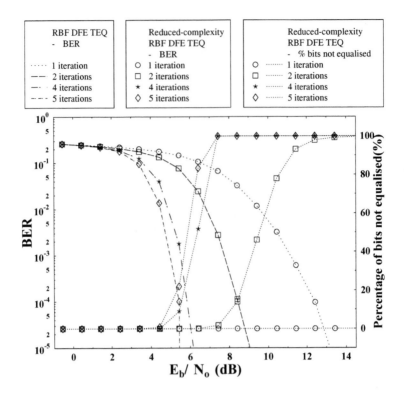

Figure 11.13: BER performance and the percentage of symbols not requiring equalization by the reduced-complexity RBF TEQ using **detected** decision feedback over the three-path Gaussian channel having a z-domain transfer function of $F(z) = 0.5773 + 0.5773z^{-1} + 0.5773z^{-2}$ for BPSK. The LLR magnitude threshold, the log-likelihood gradient and the log-likelihood intercept point were set to $|\mathcal{L}|_{threshold} = 26$, $g = 1.2$ and $c = -7.5$, respectively.

of 6dB. Figure 11.14 depicts the performance of the reduced-complexity Jacobian RBF DFE TEQ relying on detected decision feedback over the three-tap equal gain, symbol-spaced Rayleigh faded CIR obeying a Doppler frequency of 1.5×10^{-4}. A LLR magnitude threshold of 10 was sufficient for the reduced-complexity Jacobian RBF DFE TEQ in order to provide an equivalent BER performance to that of the Jacobian RBF DFE TEQ. The RBF DFE provided a better reliability-estimation over the dispersive burst-invariant Rayleigh fading channel compared to the dispersive Gaussian channel, since the uncoded BER performance was better over the Rayleigh fading channel, as it is seen from comparing Figures 11.13 and 11.14. [1] Hence less errors were propagated from the equaliser's decision feedback to future bits. Referring to Figure 11.14, the reduced-complexity Jacobian RBF DFE TEQ using de-

---

[1] The three-tap Rayleigh fading channel has a better BER performance than the three-path Gaussian channel, because the dispersive Gaussian channel has a bad spectral characteristic exhibiting spectral null. By contrast, for the Rayleigh fading channel, the CIR taps are faded and hence the frequency-domain transfer function does not exhibit a permanent null.

cision feedback provides approximately 35% computational complexity reduction at an SNR of 4dB.

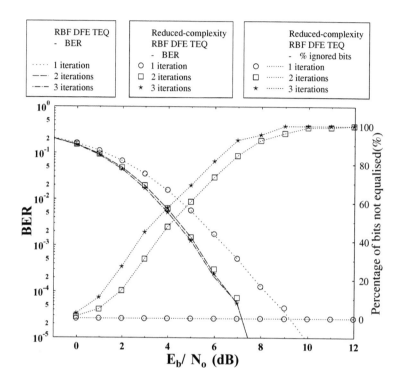

Figure 11.14: BER performance and the percentage of symbols not requiring equalization by the reduced-complexity RBF TEQ using **detected** decision feedback over the three-tap equal-gain Rayleigh fading channel for BPSK. The LLR magnitude threshold, the log-likelihood gradient and the log-likelihood interception were set to $|\mathcal{L}|_{threshold} = 10$, $g = 1.2$ and $c = -7.5$, respectively.

The reduced-complexity RBF DFE TEQ implementation can be used instead of the RBF DFE TEQ in order to provide substantial computational reductions without degrading the BER performance. Since the reliability of the symbols in the decoded burst is provided by the channel decoder in the previous iteration, we were capable of designing a system, where the percentage of bits not equalised in the decoded burst was set according to our design criteria for every iteration, such that each burst exhibited a predetermined fixed computational complexity reduction for the sake of practical, constant-complexity implementations.

In an effort to further reduce the associated computational complexity in the next section we will study the feasibility of separately equalizing the in-phase (I) and quadrature-phase (Q) components of the received signal. In general this receiver simplification can only be judiciously carried out, if the channel's impulse response is real-valued, since for transmissions over channels described by a complex channel impluse response there is 'cross-talk' between the I and Q components. However, in the next section we will show that **with the advent of using turbo equalization the BER degradation imposed by the gross simplification of assuming the absence of cross-talk between the received I and Q components can be completely eliminated, while maintaining a reduced complexity.**

11.7 Reduced Complexity In-Phase/Quadrature-Phase Turbo equalization Using Radial Basis Functions

M. S. Yee, B. L. Yeap and L. Hanzo

11.7.1 Introduction

As mentioned earlier in this section, the principle of turbo equalization [153] was contrived by Douillard *et al.*, which was shown to provide performance advantages in the context of a rate $R = 0.5$ convolutional coded system, where channel decoding and channel equalization was performed in unison. More specifically, the turbo equaliser exhibited a performance that was close to that achievable over non-dispersive channels, despite the presence of Inter-Symbol Interference (ISI), when performing the channel equalization and channel decoding by exchanging information between these two processing blocks. Due to complexity reasons, early turbo equalization investigations using the conventional trellis-based equaliser (CT-EQ) were constrained to applying Binary Phase Shift Keying (BPSK) and Quadrature Phase Shift Keying (QPSK) modulation schemes [308] and to limited Channel Impulse Response (CIR) durations, since the computational complexity incurred by the CT-EQ is dependent on both the maximum CIR duration and on the modulation mode utilized. Hence, turbo equalization research has been focused on developing reduced complexity equalisers, such as the low-complexity linear equaliser proposed by Glavieux *et al.* [172], the Radial Basis Function (RBF) equaliser advocated by Yee *et al.* [44] and the In-phase/Quadrature-phase turbo equaliser (I/Q-TEQ) introduced by Yeap [315]. Motivated by these trends, in this section we proposed a novel reduced complexity RBF channel equaliser based on the I/Q concept [315], which is referred to as the In-phase/Quadrature-phase RBF Equaliser (I/Q-RBF-EQ). We employ this concept here in the context of full response systems.

11.7.2 Principle of I/Q Equalisation

We denote the modulated signal by $s(t)$, which is transmitted over the dispersive channel characterized by the CIR $h(t)$. The signal is also contaminated by the zero-mean Additive White Gaussian Noise (AWGN) $n(t)$ exhibiting a variance of $\sigma^2 = N_o/2$, where N_o is the single-sided noise power spectral density. The received signal $r(t)$ is then formulated as:

$$
\begin{aligned}
r(t) &= s(t) * h(t) + n(t) \\
&= [s_I(t) + js_Q(t)] * [h_I(t) + jh_Q(t)] \\
&\quad + n_I(t) + jn_Q(t) \\
&= r_I(t) + jr_Q(t),
\end{aligned}
\tag{11.27}
$$

where

$$
\begin{aligned}
r_I(t) &= s_I(t) * h_I(t) - s_Q(t) * h_Q(t) + n_I(t) \\
r_Q(t) &= s_I(t) * h_Q(t) + s_Q(t) * h_I(t) + n_Q(t),
\end{aligned}
\tag{11.28}
$$

since the CIR $h(t)$ is complex and therefore consists of the I component $h_I(t)$ and Q component $h_Q(t)$.

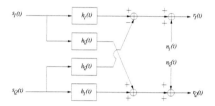

Figure 11.15: Model of the complex channel. After transmission over the complex channel $h(t)$, the received signal $r(t)$ becomes dependent on the in-phase component $s_I(t)$ and quadrature-phase component $s_Q(t)$ of the transmitted signal, as expressed in Equations 11.27 and 11.28.

On the same note, $s_I(t)$ and $s_Q(t)$ are the I and Q components of $s(t)$ in Figure 11.15, while $n_I(t)$ and $n_Q(t)$ denote the corresponding AWGN components. Both of the received I/Q signals, namely $r_I(t)$ and $r_Q(t)$ of Equation 11.28 become dependent on both $s_I(t)$ and $s_Q(t)$ due to the cross-coupling effect imposed by the complex channel. Hence a conventional channel equaliser - regardless, whether it is an iterative or non-iterative equaliser - would have to consider the effects of this cross-coupling.

In this contribution we propose a technique of reducing the complexity of the equaliser by initially neglecting the channel-induced cross-coupling of the received signal's quadrature components and then by compensating for this gross simplification with the aid of the proposed turbo equaliser. This simplification would result in an unacceptable performance degradation in the context of conventional non-iterative channel equalization, since the turbo iterations allow us to compensate for the above simplification. Then the I and Q components of the decoupled channel output $r'(t)$ are only dependent on $s_I(t)$ or $s_Q(t)$, as portrayed in Figure 11.16 in the context of the following equations:,

$$
\begin{aligned}
r'_I(t) &= s_I(t) * h(t) + n_I(t) \\
&= s_I(t) * h_I(t) + j[s_I(t) * h_Q(t)] + n(t) \\
r'_Q(t) &= -s_Q(t) * h(t) + n_Q(t) \\
&= -(s_Q(t) * h_I(t) + j[s_Q(t) * h_Q(t)]) + n(t).
\end{aligned}
\tag{11.29}
$$

More explicitly, the cross-coupling is facilitated by generating the estimates $\hat{s}_I(t)$ and $\hat{s}_Q(t)$ of the transmitted signal [315] with the aid of the reliability information generated by the channel decoder and then by cancelling the cross-coupling effects imposed by the channel, yielding $r'_I(t)$ and $r'_Q(t)$, respectively, in Figure 11.16. In the ideal scenario, where perfect knowledge of both the CIR and that of the transmitted signal is available, it is plausible that the channel-induced cross-coupling between the quadrature components can be removed.

However, when unreliable symbol estimates are generated due to the channel-impaired low-confidence reliability values, errors introduced in the decoupling operation. Nonetheless, we will show that the associated imperfect decoupling effects are compensated with the aid of the iterative turbo equalization process and the performance approaches that of the turbo equalizer utilizing the conventional trellis-based equalizer, where the cross-coupling is not neglected. As an added benefit, the complexity of the equalization process is susbtantially reduced. Following the above decoupling operation, the modified complex channel outputs,

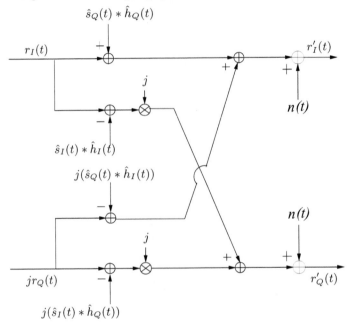

Figure 11.16: Removing the dependency of $r_I(t)$ and $r_Q(t)$ on the quadrature components of the transmitted signals, namely $s_I(t)$ and $s_Q(t)$, to give $r'_I(t)$ and $r'_Q(t)$, respectively. In this figure, it is assumed that the CIR estimation is perfect, i.e. $\hat{h}_I(t) = h_I(t)$ as well as $\hat{h}_Q(t) = h_Q(t)$ and that the transmitted signals are known, giving $\hat{s}_I(t) = s_I(t)$ and $\hat{s}_Q(t) = s_Q(t)$. In this case, perfect decoupling is achieved. However, in practice these estimates have to be generated at the receiver.

namely $r'_I(t)$ and $r'_Q(t)$, respectively, can be viewed as the result of convolving both quadrature components independently with the complex CIR on each quadrature arm. Consequently, we can equalise $s_I(t)$ and $s_Q(t)$ independently, hence reducing the number of channel states significantly. Again, note that in Equation 11.29 we have assumed that perfect signal regeneration and perfect decoupling is achieved at the receiver, in order to highlight the underlying principle of the reduced complexity equaliser.

11.7.3 RBF Assisted Turbo equalization

The RBF network based equaliser is capable of utilizing the *a priori* information provided by the channel decoder and in turn provide the decoder with the *a posteriori* information concerning the coded bits [312]. We will now provide a short description of the RBF based

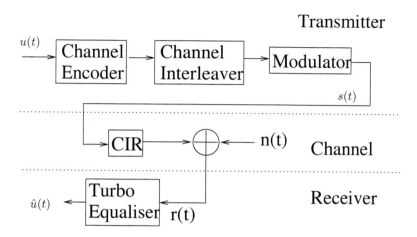

Figure 11.17: A coded M-QAM system employing a turbo equaliser at the receiver.

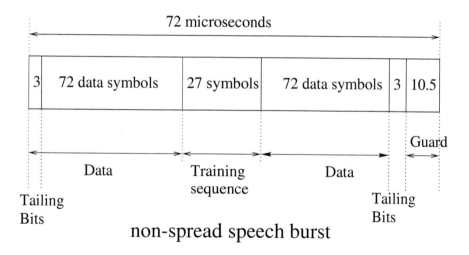

Figure 11.18: Transmission burst structure of the FMA1 non-spread speech burst of the FRAMES proposal [307].

equaliser and how it interacts with the channel decoder.

The conditional probability density function (PDF) of the ith symbol, $i = 1, \ldots, M$, associated with the ith subnet of the M-ary RBF equaliser having a feedforward order of m

is given by [312]:

$$f^i_{RBF}(\mathbf{r}(t)) = \sum_{j=1}^{n^i_s} w^i_j \varphi(|\mathbf{r}(t) - \mathbf{c}^i_j|), \tag{11.30}$$

$$w^i_j = p^i_j (2\pi\sigma^2_\eta)^{-m/2}, \tag{11.31}$$

$$\varphi(x) = \exp\left(\frac{-x^2}{2\sigma^2_\eta}\right) \tag{11.32}$$

$$i = 1, \ldots, \mathcal{M}, \qquad j = 1, \ldots, n^i_s$$

where $\mathbf{c}^i_j$, $\mathbf{w}^i_j$ and $\varphi(\cdot)$ are the RBF's centers, weights and activation function, respectively, and σ^2_η is the noise variance of the channel. The term

$$\mathbf{r}(t) = [r(t); r(t-1); \ldots ; r(t-m+1)]^T$$

is the m-dimensional channel output vector in the memory of the RBF-EQ. The RBF's centers $\mathbf{c}^i_j$ are assigned to the channel output states $\mathbf{r}^i_j$. The channel output state, which is the product of the CIR matrix $\mathbf{F}$ and the channel input state $\mathbf{s}_j$, is represented as follows [312]: $\mathbf{r}_j = \mathbf{H}\mathbf{s}_j$, where the z-transform of the CIR $h(t)$ having a memory of L symbols is represented by $H(z) = \sum_{n=0}^{L} h_n z^{-n}$ and $\mathbf{H}$ is an $m \times (m + L)$ matrix given by the CIR taps as follows:

$$\mathbf{H} = \begin{bmatrix} h_0 & h_1 & \cdots & h_L & \cdots & 0 \\ 0 & h_0 & \cdots & f_{L-1} & \cdots & 0 \\ \vdots & \vdots & & & & \vdots \\ 0 & 0 & f_0 & \cdots & f_{L-1} & f_L \end{bmatrix}. \tag{11.33}$$

The channel input state $\mathbf{s}_j$ is given by the jth combination of $(L + m)$ possible transmitted symbols, namely by $\mathbf{s}_j = \begin{bmatrix} s_j(t) & \cdots & s_j(t-1) & \cdots & s_j(t-L-m+1) \end{bmatrix}^T$. The term p^i_j in Equation 11.32 is the probability of occurance of the channel state $\mathbf{r}^i_j$ and it determines the values of the RBF weights w^i_j. The actual number of channel states n^i_s is determined by the design of the algorithm that reduces the number of channel states from the optimum number of $\mathcal{M}^{m+L-1}$ [246]. The probability of the channel states $\mathbf{r}^i_j$ and therefore the weights of the RBF equaliser can be derived from the LLR values of the transmitted bits, as estimated by the channel decoder.

In the I/Q-RBF-EQ we utilized the principle of I/Q equalization outlined in Section 11.7.2, where two separate RBF-EQ is used for the in-phase and quadrature component of the transmitted symbols. The in-phase-RBF-EQ has centers, which consist of the in-phase decoupled channel output $r'_I(t)$ of Equation 11.28 and vice-versa for the quadrature-RBF-EQ. The number of channel states is reduced, since the decoupled channel output $r'(t)$ is dependent on $\sqrt{M}$ possible in-phase or quadrature-phase transmitted symbols instead of the original M symbols. The following section describes, how the I/Q-RBF-DFE is incorporated into the schematic of the turbo equaliser.

11.7.4 System Overview

The schematic of the coded M-QAM system employing a TEQ at the receiver is shown in Figure 11.17. The transmitted source bits $u(t)$ are convolutionally encoded, interleaved

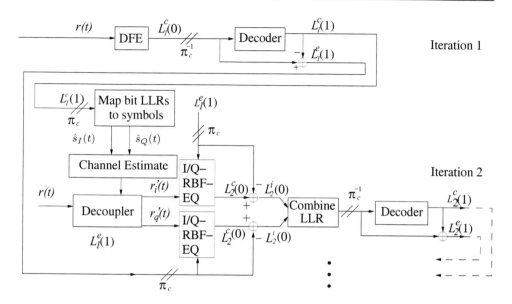

Figure 11.19: Schematic of the turbo equaliser employing a DFE and a SISO channel decoder in the first turbo equalization iteration. In subsequent iterations, two I/Q-EQs and one SISO channel decoder are employed. The notation π_c represents a channel interleaver, while π_c^{-1} is used to denote a channel deinterleaver.

and mapped to an M-QAM symbol. The encoder utilized an $\frac{1}{2}$-rate Recursive Systematic Convolutional (RSC) code having a constraint length of $K = 5$ and octal generator polynomials of $G_0 = 35$ and $G_1 = 23$. The transmission burst structure used in this system is the FMA1 non-spread speech burst specified by the Pan-European FRAMES proposal [151], which is shown in Figure 11.18. In order to decide on the tolerable delay and hence on the depth of the channel interleaver, we considered the maximum affordable delay of a speech system. In our investigations, the transmission delay of the 4-QAM, 16-QAM and 64-QAM systems [317] was limited to approximately 30 ms. This corresponds to 3456 symbols at a symbol rate of 13.9 Kbauds and hence 6912-bit, 13824-bit and 20736-bit random channel interleavers were utilized for 4-QAM, 16-QAM and 64-QAM, respectively. A three-path, symbol-spaced fading CIR of equal weights was used, which can be expressed as: $h(t) = 0.577 + 0.577z^{-1} + 0.577z^{-2}$, where the Rayleigh fading statistics obeyed a normalized Doppler frequency of 3.3615×10^{-5}. The fading magnitude and phase was kept constant for the duration of a transmission burst, a condition which we refer to as employing transmission burst-invariant fading.

Figure 11.19 illustrates the schematic of the turbo equaliser utilizing two reduced complexity I/Q-RBF-EQs. We express the LLRs of the equaliser and decoder using vector notations, according to the approach of [308], but using different specific notations. The superscript denotes the nature of the LLR, namely 'c' is used for the composite a *posteriori* [162] information, 'i' for the combined channel and extrinsic information and 'e' [162] for the extrinsic information. The subscripts in Figure 11.19 are used to represent the iteration index, while the argument within the brackets ( ) indicates the index of the receiver stage, where the equalisers are denoted as stage 0, while the channel decoder as stage 1.

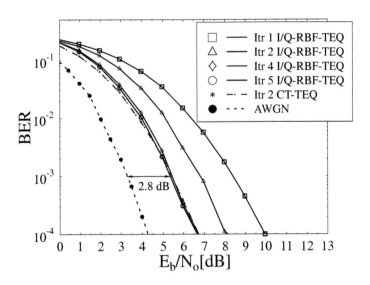

Figure 11.20: Performance of the I/Q-RBF-TEQ using iterative CIR estimation and the CT-TEQ having perfect channel estimation for a convolutional-coded 4QAM system possessing a channel interleaving depth of 69124 bits over the equally-weighted three-path Rayleigh fading CIR using a normalized Doppler frequency of 3.3×10^{-5}. The initial LMS CIR estimation step-size is 0.1 and the subsequent step-size is 0.01.

The conventional Decision Feedback Equaliser (DFE), as seen in Figure 11.19 is used for the first turbo equalization iteration for providing soft decisions in the form of the LLR $L_1^c(0)$ to the channel decoder. Invoking the DFE at the first iteration constitutes a low-complexity approach to providing an initial estimate of the transmittd symbols, as compared to the more complex I/Q-RBF-TEQ. The Soft-In/Soft-Out (SISO) channel decoder of Figure 11.19 generates the *a posteriori* LLR $L_1^c(1)$ and from that, the extrinsic information of the encoded bits $L_1^e(1)$ is extracted. In the next iteration, the *a posteriori* LLR $L_1^c(1)$ is used for regenerating estimates of the I and Q components of the transmitted signal, namely $\hat{s}_I(t)$ and $\hat{s}_Q(t)$, as seen in the 'MAP bit LLRs to symbols' block of Figure 11.19. The *a posteriori* information was transformed from the log domain to modulated symbols using the approach employed in [318]. The estimated transmitted quadrature components $\hat{s}_I(t)$ and $\hat{s}_Q(t)$ are then convolved with the estimate of the CIR $h(t)$. At the decoupler block of Figure 11.19, the resultant signal is used for removing the cross-coupling effect — seen in Equation 11.28 — according to Equation 11.29 from both quadrature components of the transmitted signal, yielding $r'_I(t)$ and $r'_Q(t)$.

After the decoupling operation, $r'_I(t)$ and $r'_Q(t)$ are passed to the I/Q-RBF-EQ in the schematic of Figure 11.19. In addition to these received quadrature signals, the I/Q-RBF-EQ also processes the *a priori* information received — which is constituted by the extrinsic LLRs $L_1^e(1)$ derived from the previous iteration — and generates the *a posteriori* information $L_2^c(0)$. Subsequently, the combined channel and extrinsic information $L_2^i(0)$ is extracted from both I/Q-RBF-EQs in Figure 11.19 and combined, before being passed to the Log-MAP channel decoder. As in the first turbo equalization iteration, the *a posteriori* and extrinsic information

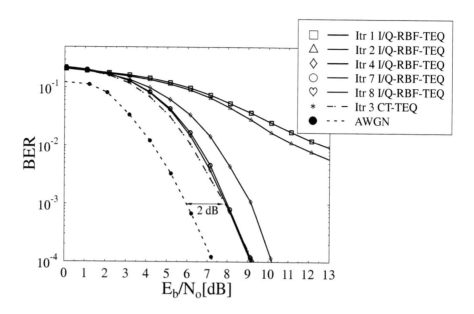

Figure 11.21: Performance of the I/Q-RBF-TEQ using iterative CIR estimation and the CT-TEQ having perfect channel estimation for a convolutional-coded 16QAM system possessing a channel interleaving depth of 13824 bits over the equally-weighted three-path Rayleigh fading CIR using a normalized Doppler frequency of 3.3×10^{-5}. The initial LMS CIR estimation step-size is 0.05 and the subsequent step-size is 0.01.

of the encoded bits, namely $L_2^c(1)$ and $L_2^e(1)$, respectively, are evaluated. Subsequent turbo equalization iterations obey the same sequence of operations, until the iteration termination criterion is met.

11.7.5 Results and Discussion

In our simulations, the Jacobian RBF DFE of [314] was employed, which reduced the complexity of the RBF-EQ by utilizing Jacobian logarithmic function and decision feedback for RBF-center selection [312]. Note that the I/Q-EQ scheme reduces the effect of error propagation, since the set of centers to be selected using the DFE mechanism is reduced from M^n to $M^{n/2}$ [312]. The feedforward and feedback order of the RBF DFE was three and two, respectively, while that of the conventional DFE was fifteen and four. We employed iterative LMS-based CIR estimation for both 4QAM and 16QAM in Figures 11.20 and 11.21, respectively. An initial step-size of 0.1 and 0.05 was set for 4QAM and 16QAM, respectively, in order to provide a rough estimate of the CIR. This initial CIR estimate was utilized by the conventional DFE employed during the first turbo equalization. In the subsequent iterations, the CIR was re-estimated and refined using a smaller step-size of 0.01. Figure 11.20 and 11.21 also present the performance of the CT-TEQ in conjunction with perfect CIR information. We defined the **critical number** of turbo equalization iterations, as the number, where no further significant performance improvement can be obtained upon invoking further iterations. We

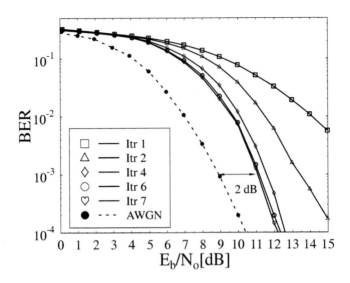

Figure 11.22: Performance of the I/Q-RBF-TEQ using perfect CIR estimation for a convolutional-coded 64QAM
system possessing a channel interleaving depth of 20736 bits over the equally-weighted three-path
Rayliegh fading CIR using a normalized Doppler frequency of 3.3×10^{-5}.

	Log-MAP	Jacobian RBF
sub. and add.	$n_{s,f}(6\mathcal{M}+2)-3$	$n_{s,f}+$ $\mathcal{M}n_s^i(m+2)-4$
mult. and div.	$2n_{s,f}$	$n_{s,f}$

Table 11.2: Computational complexity of generating the *a posteriori* LLR for the Log-MAP equaliser and the
Jacobian RBF equaliser [314]. The RBF equaliser order is denoted by m and the number of RBF
nodes is n_s^i. The notation $n_{s,f} = \mathcal{M}^{L+1}$ indicates the number of trellis transition for the Log-MAP
equaliser and also the number of scalar channel states for the Jacobian RBF equaliser.

found that both the CT-TEQ and the I/Q-RBF-TEQ provided a similar performance at the
corresponding critical number of iterations. The performance of the I/Q-RBF-TEQ at the
critical number of iteration using iterative CIR estimation at BER = 10^{-3} was about 2.8dB
and 2dB from the decoding performance curve recorded over the non-dispersive Gaussian
channel, as shown for 4QAM and 16QAM in Figures 11.20 and 11.21, respectively. Following the system complexity study of [45] and considering the number of critical iterations
needed, the complexity of the I/Q-RBF-TEQ imposed by the equaliser and decoder components was found to be a factor of 1.5 and 109.6 lower, than that of the CT-TEQ for 4-QAM
and 16-QAM, respectively, based on the general complexity expressions of Table 11.2.

The performance of our 64-QAM system for transmission over the same Rayleigh fading
channel but in conjunction with perfect channel estimation shows in Figure 11.22 that it is
only 2dB away from the decoding performance curve recorded for seven iterations over the
non-dispersive Gaussian channel at BER = 10^{-3}. We could not provide the performance
curve of the CT-TEQ, since this scheme was excessively complex. Assuming that the critical
number of iteration for the 64QAM CT-TEQ is two, the I/Q-RBF-TEQ provides a complexity

reduction by a factor of 3313.

The performance of the I/Q-RBF-EQ and that of the trellis-based I/Q-TEQ of [315] was found to be similar, although I/Q-RBF-EQ provided a complexity reduction factor of 1.2, 2.2 and 7.8 for 4QAM, 16QAM and 64QAM respectively.

In conclusion, our simulation results show significant complexity reductions for the I/Q-RBF-EQ when compared to the conventional CT-EQ, while achieving virtually the same performance. This is demonstrated in Figure 11.20 and 11.21, where the complexity reduction factor was 1.5 and 109.6 for 4QAM and 16QAM, respectively. The achievable complexity reduction increases for higher-dispersion channels and for high-order modulation schemes.

Throughout the book we have studied a host of adaptive transceiver schemes, which were invoked for mitigating the detrimental effects of the multipath-induced channel quality fluctuations. We also argued in the Prologue, namely in Chapter 1 that the same adaptive techniques can be used for combating the effects of the time-variant cochannel interference fluctuations imposed by the time-variant number of co-channel users. Furthermore, we suggested in the Prologue of the book, namely in Chapter 1 that the multipath-induced channel quality fluctuations may be mitigated also with the aid of multiple transmitter and multiple receiver assisted space-time coding arrangements [12], if the associated higher complexity is affordable. In the latter scenario the performance benefits of AQAM erode, since the the multipath-induced channel quality fluctuations are mitigated by the space-time coding schemes [12] used. Hence in the next section we consider a space-time trellis coded arrangement, which employs a sophisticated iterative turbo equalization based receiver. We note, however that fixed-mode modulation based space-time codecs are less efficient in terms of mitigating the effects of the time-variant co-channel interference fluctuations, than their adaptive counterparts.

11.8 Turbo Equalization of Convolutional Coded and Concatenated Space Time Trellis Coded Systems using Radial Basis Function Aided Equalizers

M. S. Yee, B. L. Yeap and L. Hanzo

11.8.1 Introduction

The family of transmission diversity techniques referred to as Space Time Trellis (STT) coding [217] also been introduced in order to provides a substantial diversity gain for mobile stations by upgrading the base stations, hence potentiallly increasing the achievable user capacity of the system. STT coding [217] jointly designs the channel coding, modulation, transmit diversity and the optional receiver diversity schemes invoked. Following the research by Tarokh et al. [217], Bauch et al. [319] proposed a joint equalization and STT decoding scheme, which yielded an improved performance with the advent of exploiting the soft-decision based feedback from the STT decoder's output to the channel equalizer's input. In [320] the performance of the STT encoded system was further improved by employing additional channel encoding in conjunction with turbo equalization. We refer to this turbo equalizer as the TEQ-STTC scheme. However, due to the associated computational complexity, the employment of this scheme was limited to low-order modulation modes, such as for example 4-level Quadrature Amplitude Modulation (4QAM).

Motivated by these trends, in this contribution we aim for reducing the complexity associated with the channel-coded and concatenated STT encoded system by using a reduced-complexity Jacobian Radial Basis Function (RBF) equalizer [314], which we will refer to as the RBF-TEQ-STTC scheme. We will investigate the BER performance achieved by the RBF-TEQ-STTC scheme and evaluate the achievable computational complexity reduction compared to the conventional trellis-based TEQ-STTC (CT-TEQ-STTC) arrangement of [320].

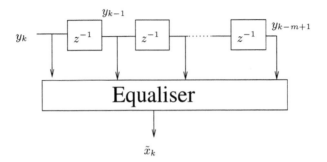

Figure 11.23: Schematic of m-tap equalizer

11.8.2   RBF aided channel equalizer for space-time-coding

In this section we will show that the channel equalization problem encountered in a space-time coded system can be considered as a geometric classification problem [245], namely that of classifying an $\mathcal{M}$-ary received phasor into one of $\mathcal{M}$ classes. Figure 11.23 shows an m-tap equalizer schematic, where the channel output observed by the equalizer can be written in vectorial form as

$$\mathbf{y}_k = \begin{bmatrix} y_k & y_{k-1} & \cdots & y_{k-m+1} \end{bmatrix}. \tag{11.34}$$

The baseband representation of the p-transmitter space-time coded system is shown in Figure 11.24, which transmits a sequence of p symbols $\mathbf{x}_k^p = \begin{bmatrix} x_{1,k} & \cdots & x_{p,k} \end{bmatrix}$ during each signalling instant k. The channel output at instant k is given by:

$$y_k = \sum_{i=1}^{p} \mathbf{h}_i \mathbf{x}_{i,k}^T + \eta_k, \tag{11.35}$$

where the i-th channel impulse response (CIR) $\mathbf{h}_i = \begin{bmatrix} h_{0,i} & h_{1,i} & \cdots & h_L \end{bmatrix}$, having a memory of L symbols, is convolved with a sequence of $L + 1$ transmitted symbols, namely with $\mathbf{x}_{i,k} = \begin{bmatrix} x_{i,k} & x_{i,k-1} & \cdots & x_{i,k-L} \end{bmatrix}$ and η_k is the additive Gaussian noise term having a variance of σ_η. For a p-transmitter system using an m-tap equalizer and communicating over a channel having a CIR memory of L (assuming that all of the p CIRs have the same memory), there are $n_s = \mathcal{M}^{(m+L) \cdot p}$ number of possible received phasor combinations due to the transmitted sequence, hence producing n_s number of different possible channel output vectors in the absence of channel noise:

$$\tilde{\mathbf{y}}_k = \begin{bmatrix} \tilde{y}_k & \tilde{y}_{k-1} & \cdots & \tilde{y}_{k-m+1} \end{bmatrix}, \tag{11.36}$$

where m is the length of the equalizer in Figure 11.23. Upon adding the noise we have: $\mathbf{y}_k = \tilde{\mathbf{y}}_k + \eta_k$. Expounding further, we denote each of the n_s number of different possible combinations of the channel's input sequence $\bar{\mathbf{x}}_k = \begin{bmatrix} \mathbf{x}_k^p & \cdots & \mathbf{x}_{k-m+1}^p \end{bmatrix}$ of length $(L + m) \times p$ symbols as $\mathbf{s}_i, i = 1, \ldots, n_s$, where the channel's input state $\mathbf{s}_i$ determines the desired channel output state $\mathbf{r}_i, i = 1, \ldots, n_s$. This is formulated as:

$$\tilde{\mathbf{y}}_{\mathbf{k}} = \mathbf{r}_i, \quad \text{if } \bar{\mathbf{x}}_k = \mathbf{s}_i, \quad i = 1, \ldots, n_s. \tag{11.37}$$

For an $\mathcal{M}$-level modulation scheme, the noisy channel output states $\mathbf{y}_k$ can be partitioned into $\mathcal{M}^p$ classes according to the sequence of p number of τ-delayed transmitted symbols, $\mathbf{x}_{k-\tau}^p$. The equalizer has to provide the associated non-linear decision boundaries for the classification strategy. The optimum equalizer is the so-called Bayesian equalizer [245], which has an excessive complexity. Hence here we advocate the reduced-complexity, but suboptimum Jacobian RBF equalizer, introduced in [314], which has N hidden nodes. The output of this Jacobian RBF equalizer can be represented mathematically as [314]:

$$\begin{aligned} f_{RBF}^{\ln}(\mathbf{y}_k) &= \ln\left(\sum_{i=1}^{N} w_i \exp(-\|\mathbf{y}_k - \mathbf{c}_i\|^2/\lambda)\right) \tag{11.38} \\ &= \ln\left(\sum_{i=1}^{N} \exp(\ln(w_i) - \|\mathbf{y}_k - \mathbf{c}_i\|^2/\lambda)\right) \\ &= J(\delta_{N,k}, J(\delta_{N-1,k}, \ldots J(\delta_{2,k}, \delta_{1,k}) \ldots)), \end{aligned}$$

where the terms w_i, $\mathbf{c}_i$ and λ are the weights, centers and width of the RBF nodes, respectively. Furthermore, we have $\delta_{i,k} = \exp(\ln(w_i) - \|\mathbf{y}_k - \mathbf{c}_i\|^2/\lambda)$ and $J(\delta_1, \delta_2)$ is the Jacobian logarithmic relationship defined in [163] as $J(\delta_1, \delta_2) \approx \max(\delta_1, \delta_2) + f_c(\|\delta_1 - \delta_2\|)$. The correction function $f_c(x) = \ln(1 + \exp(-x))$ is tabulated in a look-up table, in order to reduce the computational complexity [163].

The full-complexity RBF equalizer provides the so-called optimal Bayesian equalization solution [245] and generates the conditional probability density functions of $\mathcal{M}^p$ number of possible transmitted symbols $\mathbf{x}_{k-\tau}^p$ emitted by the transmitters at instant $k - \tau$ in the form of:

$$\begin{aligned} P(\mathbf{y}_k | \mathbf{x}_{k-\tau}^p = \mathbf{I}_j) &= \sum_{i=1}^{n_s^j} p_{i,j}(2\pi\sigma_\eta^2)^{-m/2} \cdot \\ &\quad \exp\left\{-\frac{1}{2\sigma_\eta^2}\|\mathbf{y}_k - \mathbf{r}_{i,j}\|^2\right\}, \\ &\quad j = 1, \ldots, \mathcal{M}^p, \tag{11.39} \end{aligned}$$

where the RBF parameters defined in the context of Equation 11.39 are assigned the values of $w_i = p_{i,j}(2\pi\sigma_\eta^2)^{-m/2}$, $\mathbf{c}_i = r_{i,j}$, $N = n_s^j$ and $\lambda = 2\sigma_\eta^2$. The term n_s^j is the number of possible channel states $\mathbf{r}_{i,j}$ corresponding to the jth transmitted symbol sequence $\mathbf{I}_j$ of the

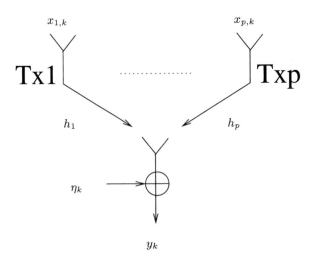

Figure 11.24: Baseband representation of p-transmitter space-time coded system using one receiver.

p-antenna SSTTC scheme that consist of p symbols, where we have $j = 1, \ldots, \mathcal{M}^p$. The term $p_{i,j}$ is the *a priori* probability of occurance of the channel state $r_{i,j}$. The *a posteriori* probability of the transmitted symbols $\mathbf{x}^p_{k-\tau}$ in Equation 11.39 provides the *a posteriori* Log-Likelihood Ratio (LLR) values of the convolutionally coded symbols, which can then be fed to the STT decoder, as shown in Figure 11.26. The *a priori* probability of occurance of the ith channel state $\mathbf{r}_{i,j}$ corresponding to the transmitted symbol sequence $\mathbf{I}_j$, $p_{i,j}$, can be evaluated from the LLRs generated by the STT decoder as described in Section 11.8.3.

11.8.3   System Overview

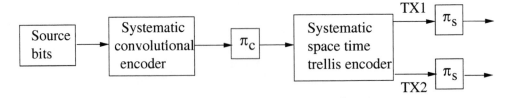

Figure 11.25: Transmitter of the serially concatenated systematic convolutional coded and systematic STT coded system.

In an effort to create a low-complexity, high-performance system, we employ the Jacobian RBF equalizer [314] in the context of a turbo equalizer in conjunction with a convolutional coded systematic STTC (CC-SSTTC) system employing two transmitters. Specifically, we use the decision feedback assisted Jacobian RBF equalizer (Jacobian RBF DFE) [314] for the sake of attaining a reduced computational complexity, where the detected symbol is fed back to the equalizer for selecting a reduced-size subset of RBF centers, which are then used

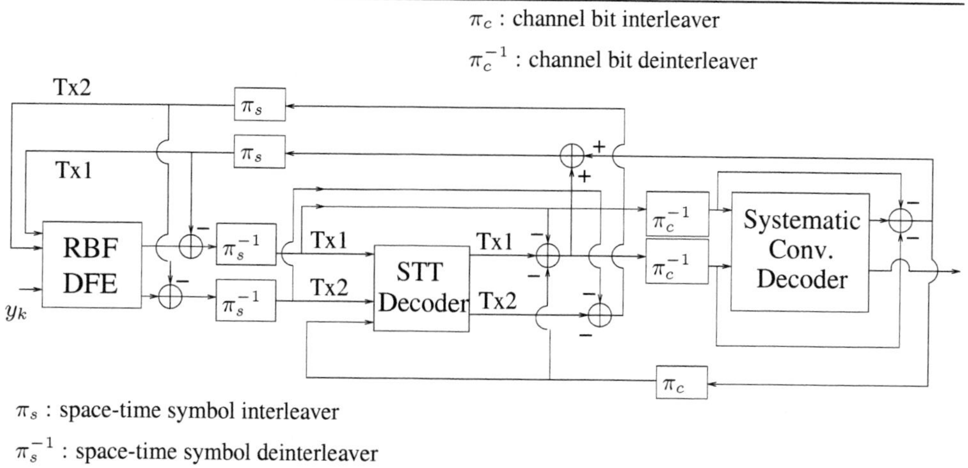

π_c : channel bit interleaver

π_c^{-1} : channel bit deinterleaver

π_s : space-time symbol interleaver

π_s^{-1} : space-time symbol deinterleaver

Figure 11.26: Receiver of the serially concatenated systematic convolutional coded and systematic STTC system using RBF DFE assisted turbo equalization.

for evaluating the *a posteriori* probability of the transmitted signal [245]. The schematic of the CC-SSTTC transmitter consists of a serially concatenated systematic convolutional encoder and a systematic STT encoder, as shown in Figure 11.25. The transmitted source bits are convolutionally encoded and directed to a random channel bit interleaver π_c. The convolutional encoder denoted as CC(2,1,3) is a $\frac{1}{2}$-rate Recursive Systematic Convolutional (RSC) coding scheme having a constraint length of $K = 3$ and octal generator polynomials of $G_0 = 7$ and $G_1 = 5$. The RSC codeword consists of a systematic bit and a parity bit.

Subsequently, the encoded bits are passed to a systematic STT encoder using two transmit antennas, as illustrated in Figure 11.25. We denote the systematic STT encoder used as the SSTTC($n = 4, m = 4$) scheme, since it is an $n = 4$-state, $m = 4$-PSK based STT code [217]. Upon receiving an input symbol, the SSTTC produces a symbol in each transmitter arm of Figure 11.25. Note that we have employed the simple SSTTC(4,4) code instead of more complex systematic STT codes using a higher number of encoder states, since our aim was to invoke the turbo equalization principle and 'invest' the affordable implementational complexity in a number of consecutive iterations, rather than in a high-complexity non-iterative decoder. The STT encoded symbols are interleaved by a random STT symbol interleaver represented as π_s in Figure 11.25.

The schematic of the receiver is shown in Figure 11.26. The channel equalizer of Figure 11.26 computes the *a posteriori* LLR values for the systematic STT coded symbols of both transmitter TX1 and TX2. Subsequently, these LLR values are deinterleaved by the STT deinterleaver π_s^{-1} of Figure 11.26 and passed to the SSTTC(4,4) decoder. In the first iteration, the channel equalizer only evaluates the received signal y_k, since there is no *a priori* feedback information from the output of the RSC decoder. However, in subsequent iterations the channel equalizer will receive additional *a priori* information concerning the STT codeword from the other decoding stages. In order to avoid passing the *a priori* information contributed by the other concatenated decoding states back to these stages in Figure 11.26, we subtract the *a priori* LLRs fed back to the input of the equalizer from the corresponding *a posteriori* LLRs output by the equalizer, in order to derive the combined channel and extrinsic information. Similar LLR subtraction stages can be seen at the output of the STT decoder and that of the convolutional decoder, again providing the extrinsic information for the next component of the receiver, as detailed in [320].

In our investigations the transmission burst structure consists of 100 data symbols. A two-path, symbol-spaced fading Channel Impulse Response (CIR) of equal weights was used, where the Rayleigh fading statistics obeyed a normalized Doppler frequency of 3.3615×10^{-5}. The fading magnitude and phase was kept constant for the duration of a transmission burst, a condition which we refer to as employing burst-invariant fading. Furthermore, in order to investigate the best-case performance of these systems, we have assumed that the CIR was perfectly estimated at the receiver. Our future research will characterize the ability of the proposed turbo scheme to compensate for the effects of CIR estimation errors. At the receiver, the systematic STT decoder and the RSC decoder employed the Log-MAP algorithm [163]. The Jacobian RBF DFE has a feedforward order of $m = 2$, feedback order of $n = 1$ and decision delay of $\tau = 1$.

11.8.4   Results and Discussion

Figure 11.27 shows the performance of the proposed RBF-TEQ-STTC and that of the CT-TEQ-STTC scheme [320], using various STTC interleaving sizes, namely 100, 400, 800, 1600, 3200 and 6400 symbols after eight turbo equalization iterations. It was observed in Figure 11.27 that by increasing the STTC interleaving size from 100 to 6400, the performance degradation of the RBF-TEQ-STTC scheme compared to the CT-TEQ-STTC arrangement expressed in terms of the excess SNR required for attaining a BER of 10^{-4} decreases from 3.8dB recorded for an STTC interleaver size of 100 symbols to 0dB, as observed for the STTC interleaver size of 6400 symbols. This is because the error propagation of the RBF DFE component decreases, as the BER performance improves, when using a longer STTC interleaver. The performance difference of the two schemes is less than 1dB at a STTC interleaver length of 400 symbols, although the RBF-TEQ-STTC scheme has a lower computational complexity, when the feedforward order m and feedback order n are set to $m = L + 1, n = L$. The interleaving gain attained by the RBF-TEQ-STTC scheme was approximately 9dB at a BER of 10^{-4}. Although higher interleaving gains can be achieved using longer STTC interleavers, the interleaver gain gradually saturates, when the STTC interleaver size is in excess of 1600 symbols.

Following the approach of our computational complexity study in [321], Table 11.3 summarises the computational complexity of generating the *a posteriori* LLRs for each received signal at instant k in the context of a p-transmitter space-time coded system. Figure 11.28 demonstrates the complexity reduction achieved by the Jacobian RBF DFE for various feedforward orders m, over the trellis-based equalizer. The feedback order n and decision delay τ of the RBF DFE was set to $n = L$ and $\tau = m$ for the sake of attaining the optimum performance, as stated in [245]. The performance of the RBF DFE improves, when increasing the feedforward order [245]. However, Figure 11.28 shows that the Jacobian RBF DFE only provides a significant complexity reduction compared to the trellis based equalizer, when the feedforward order is less than $L + 2$ and imposes a higher computational complexity for $m > L + 2$. Therefore, as a rule of thumb, the feedforward order of the Jacobian RBF DFE must not exceed $L + 1$ in order to achieve a computational complexity improvement relative

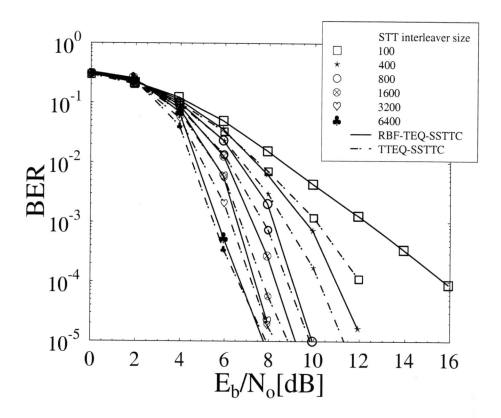

Figure 11.27: BER performance of the RBF DFE ($m = 2$, $n = 1$, $\tau = 1$) assisted turbo-equalized serially concatenated convolutional coded and STTC system using various STTC interleaver sizes, namely 100, 400, 800, 1600, 3200 and 6400 symbols, after eight turbo equalization iterations. The performance of the CT-TEQ-SSTTC system is also shown as a benchmarker.

	trellis-based	Jacobian RBF
sub. and add.	$n_{s,f}(6\mathcal{M}^p + 2)$ -3	$n_{s,f}+$ $\mathcal{M}^p n_s^i(m + 2) - 4$
mult. and div.	$2n_{s,f}$	$2n_{s,f}$

Table 11.3: Computational complexity of generating the *a posteriori* LLRs for the trellis-based equalizer and for the Jacobian RBF equalizer [321]. The RBF equalizer's feedforward and feedback order are denoted by m and n, respectively, and the number of RBF nodes is $n_s^i = \mathcal{M}^{(m+L-n)\cdot p}/\mathcal{M}, i = 1, \ldots, \mathcal{M}^p$, where L is the CIR memory and p is the number of STTC transmitters. The notation $n_{s,f} = \mathcal{M}^{(L+1)\cdot p}$ indicates the number of trellis transitions encountered in the trellis-based equalizer and also the number of possible different noise-free channel outputs $\tilde{y}_k$ of the Jacobian RBF equalizer.

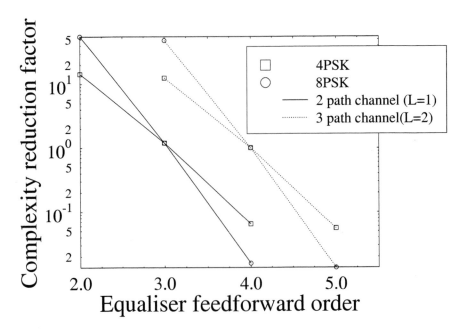

Figure 11.28: Complexity reduction factor achieved by the RBF DFE equalizer over the trellis based equalizer according to Table 11.3. The feedback order n was set to L and the number of transmitters was two.

to the trellis-based equalizer. The complexity imposed by the RBF-TEQ-STTC scheme using an equalizer feedforward order of 2 and a feedback order of 1 was found to be a factor of 14 lower, than that of the CT-TEQ-STTC scheme in the context of a two transmitter, one receiver system, based on the general complexity expressions of Table 11.3. For example, if we used a higher order modulation mode, such as 8PSK used in [217] along with the same number of transmitters, as well as equalizer and channel parameters, the achievable computational complexity reduction is a factor of 55, as shown in Figure 11.28.

In conclusion, a turbo equalization scheme using the Jacobian RBF equalizer principle of [314] was invoked in a serially concatenated systematic convolutional coded and systematic STT coded system. It was observed in Figure 11.27 that the BER performance degradation compared to the CT-TEQ-STTC system [320] was less than 1dB for a STTC interleaver length of 400 symbols, while achieving a computational complexity reduction factor 14. Hence, the Jacobian RBF equalizer based TEQ constitutes a better design choice in STTC systems, especially in the context of complex STTC schemes, having a high number of encoder states. Near-optimum performance was achieved, provided that a sufficiently high STTC interleaver length was affordable.

11.9 Review and Discussion

In conclusion, in this chapter the Jacobian RBF DFE TEQ has been proposed and analysed comparatively in conjunction with the well-known Log-MAP TEQ [288,311]. The associated performances and complexities have been compared in the context of BPSK, 4-QAM and 16-QAM. The computational complexity of the Jacobian RBF DFE TEQ is dependent on the number of RBF centres, the CIR length and modulation mode. The associated 'per iteration' implementational complexity of the Jacobian RBF DFE TEQ ($m = 3$, $n = 2$, $\tau = 2$) was approximately a factor 2.5, 4.4 and 16.3 lower in the context of BPSK, 4-QAM and 16-QAM, respectively, for the three-path channel considered. The performance degradation compared to the conventional Log-MAP TEQ [311] was negligible for BPSK and 4-QAM, but was approximately 4dB for 16-QAM, when communicating over the three-path, equal-weight, symbol-spaced burst-invariant Rayleigh fading channel environment considered. The large performance degradation for the 16-QAM scheme is due to the error propagation effect of the DFE, which becomes more grave in conjunction with higher order constellations. Therefore, the Jacobian RBF DFE TEQ could only provide a practical performance versus complexity advantage over the conventional Log-MAP TEQ [311] for lower modulation modes. Our proposed reduced-complexity Jacobian RBF DFE TEQ was shown to provide an equivalent BER performance to that of the RBF DFE TEQ at a reduced computational load. The reduced-complexity Jacobian RBF DFE TEQ using detected decision feedback provided approximately 21% (at SNR of 6dB) and 35% (at SNR of 4dB) computational reduction for dispersive Gaussian and Rayleigh channels, respectively.

In the last-but-one section which were aiming for reducing the computational complexity of the RBF turbo equaliser with the aid of separate I/Q based detection. Although when communicating over channels exhibiting complex-valued CIRs the assumption of having independent I and Q components constitutes a gross simplification and hence introduces a substantial error, the turbo equaliser is capable of gradually removing this error during its consecutive iterations. The performance of the I/Q-RBF-TEQ was characterized in a noise limited environment when communicating over an equally weighted, symbol-spaced three-path Rayleigh fading channel. The I/Q-RBF-TEQ maintained the same performance as the conventional turbo equaliser, while achieving a complexity reduction by a factor of 1.5 and 109.6 for 4-QAM and 16-QAM, respectively.

Finally, in the last section of this chapter the turbo equalization principle was employed for iteratively imporving the performance of convolutional coded and concatenated space-time coded transceivers. A two-path Rayleigh fading channel having a normalized Doppler frequency of 3.3615×10^{-5} was used. The BER performance of the RBF-CC-SSTTC(4,4) scheme employing a transmission burst consisting of 100 symbols using a space-time-trellis (STT) interleaver of at least 400 symbols and eight turbo equalization iterations was found to be similar to that of the CC-SSTTC system using a trellis-based TEQ, which attains the optimum performance. However, the Jacobian RBF based TEQ provided a complexity reduction factor of 14.

Part III

Near-Instantaneously Adaptive CDMA and Adaptive Space-Time Coded OFDM

Chapter 12

Burst-by-Burst Adaptive Multiuser Detection CDMA

E. L. Kuan and L. Hanzo[1]

12.1 Motivation

As argued throughout the previous chapters of the book, mobile propagation channels exhibit time-variant propagation properties [13]. Although apart from simple cordless telephone schemes most mobile radio systems employ power control for mitigating the effects of received power fluctuations, rapid channel quality fluctuations cannot be compensated by practical, finite reaction-time power control schemes. Furthermore, the ubiquitous phenomenon of signal dispersion due to the multiplicity of scattering and reflecting objects cannot be mitigated by power control. Similarly, other performance limiting factors, such as adjacent- and co-channel intereference as well as multi-user interference vary as a function of time. The ultimate channel quality metric is constituted by the bit error rate experienced, irrespective of the specific impairment encountered. The channel quality variations are typically higher near the fringes of the propagation cell or upon moving from an indoor scenario to an outdoor cell due to the high standard deviation of the shadow- and fast-fading [13] encountered, even in conjunction with agile power control. Furthermore, the bit errors typically occur in bursts due to the time-variant channel quality fluctuations and hence it is plausible that a fixed transceiver mode cannot achieve a high flexibility in such environments.

The design of powerful and flexible transceivers has to be based on finding the best compromise amongst a number of contradicting design factors. Some of these contradicting factors are low power consumption, high robustness against transmission errors amongst various channel conditions, high spectral efficiency, low-delay for the sake of supporting interactive real-time multimedia services, high-capacity networking and so forth [2]. In this chapter we

---

[1]This chapter is based on Kuan and Hanzo: Burst-by-Burst Adaptive Multiuser Detection CDMA:
A Framework for Existing and Future Wireless Standards, submitted to the Proceedings of the IEEE ©IEEE, 2001

will address a few of these issues in the context of Direct Sequence Code Division Multiple Access (DS-CDMA) systems. It was argued in [2] that the time-variant optimization criteria of a flexible multi-media system can only be met by an adaptive scheme, comprising the firmware of a suite of system components and invoking that particular combination of speech codecs, video codecs, embedded un-equal protection channel codecs, voice activity detector (VAD) and transceivers, which fulfils the currently prevalent set of transceiver optimization requirements.

These requirements lead to the concept of arbitrarily programmable, flexible so-called software radios [322], which is virtually synonymous to the so-called tool-box concept invoked to a degree in a range of existing systems at the time of writing [3]. This concept appears attractive also for third- and future fourth-generation wireless transceivers. A few examples of such optimization criteria are maximising the teletraffic carried or the robustness against channel errors, while in other cases minimization of the bandwidth occupancy or the power consumption is of prime concern.

Motivated by these requirements in the context of the CDMA-based third-generation wireless systems [13, 146], the outline of the chapter is as follows. In Section 12.2 we review the current state-of-the-art in multi-user detection with reference to the receiver family-tree of Figure 12.4. Section 12.4 is dedicated to adaptive CDMA schemes, which endeavour to guarantee a better performance than their fixed-mode counterparts. Burst-by-burst (BbB) adaptive quadrature amplitude modulation (AQAM) based and Variable Spreading Factor (VSF) assisted CDMA system proposals are studied comparatively in Section 12.5. Lastly our conclusions are offered in Section 12.6.

12.2 Multiuser Detection

12.2.1 Single-User Channel Equalisers

12.2.1.1 Zero-Forcing Principle

The fundamental approach of multiuser equalisers accrues from recognising the fact that the nature of the interference is similar, regardless, whether its source is dispersive multipath propagation or multiuser interference. In other words, the effects of imposing interference on the received signal by a K-path dispersive channel or by a K-user system are similar. Hence below we continue our discourse with a rudimentary overview of single-user equalisers, in order to pave the way for a more detailed discourse on multiuser equalisers.

The concept of zero-forcing (ZF) channel equalizers can be readily followed for example using the approach of [89]. Specifically, the zero-forcing criterion [89] constrains the signal component at the output of the equalizer to be free of intersymbol interference (ISI). More explicitly, this implies that the product of the transfer functions of the dispersive and hence frequency-selective channel and the channel equaliser results in a 'frequency-flat' constant, implying that the concatenated equaliser restores the perfect all-pass channel transfer function. This can be formulated as:

$$G(z) = F(z)B(z) = 1, \tag{12.1}$$

$$F(z) = \frac{1}{B(z)}, \tag{12.2}$$

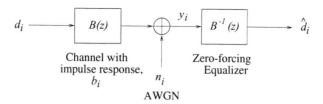

Figure 12.1: Block diagram of a simple transmission scheme using a zero-forcing equalizer.

where $F(z)$ and $B(z)$ are the z-transforms of the ZF-equaliser and of the dispersive channel, respectively. The impulse response corresponding to the concatenated system hence becomes a Dirac delta, implying that no ISI is inflicted. More explicitly, the zero-forcing equalizer is constituted by the inverse filter of the channel. Figure 12.1 shows the simplified block diagram of the corresponding system.

Upon denoting by $D(z)$ and $N(z)$ the z-transforms of the transmitted signal and the additive noise respectively, the z-transform of the received signal can be represented by $R(z)$, where

$$R(z) = D(z)B(z) + N(z). \tag{12.3}$$

The z-transform of the multiuser equalizer's output will be

$$\hat{D}(z) \;=\; F(z)R(z) \tag{12.4}$$

$$\;=\; \frac{R(z)}{B(z)} \tag{12.5}$$

$$\;=\; D(z) + \frac{N(z)}{B(z)}. \tag{12.6}$$

From Equation 12.6, it can be seen that the output signal is free of ISI. However, the noise component is enhanced by the inverse of the transfer function of the channel. This may have a disastrous effect on the output of the equalizer, in terms of noise amplification in the frequency domain at frequencies where the transfer function of the channel was severely attenuated. Hence a disadvantage of the ZF-equaliser is that in an effort to compensate for the effects of the dispersive and consequently frequency-selective channel and the associated ISI it substantially enhances the originally white noise spectrum by frequency-selectively amplifying it. This deficiency can be mitigated by invoking the so-called minimum mean square error linear equalizer, which is capable of jointly minimising the effects of noise and interference, rather than amplifying the effects of noise.

12.2.1.2 Minimum Mean Square Error Equalizer

Minimum mean square error (MMSE) equalizers have been considered in depth for example in [89] and a similar approach is followed here. Upon invoking the MMSE criterion [89], the equalizer tap coefficients are calculated in order to minimize the MSE at the output of the multiuser equalizer, where the MSE is defined as :

$$e_k^2 = E[|d_k - \hat{d}_k|^2], \tag{12.7}$$

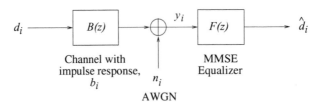

Figure 12.2: Block diagram of a simple transmission scheme employing an MMSE equalizer.

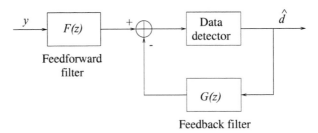

Figure 12.3: Block diagram of a decision feedback equalizer.

where the function $E[x]$ indicates the expected value of x. Figure 12.2 shows the system's schematic using an MMSE equalizer, where $B(z)$ is the channel's transfer function and $F(z)$ is the transfer function of the equalizer. The output of the equalizer is given by :

$$\hat{D}(z) = F(z)B(z)D(z) + F(z)N(z), \qquad (12.8)$$

where $D(z)$ is the z-transform of the data bits d_i, $\hat{D}(z)$ is the z-transform of the data estimates $\hat{d}_i$ and $N(z)$ is the z-transform of the noise samples n_i.

12.2.1.3 Decision Feedback Equalizers

The decision feedback equalizer (DFE) [89] can be separated into two components, a feed-forward filter and a feedback filter. The schematic of a general DFE is depicted in Figure 12.3. The philosophy of the DFE is two-fold. Firstly, it aims for reducing the filter-order of the ZFE, since with the aid of Equation 12.2 and Figure 12.1 it becomes plausible that the inverse filter of the channel, $B^{-1}(z)$, can only be implemented as an Infinite Impulse Response (IIR) filter, requiring a high implementational complexity. Secondly, provided that there are no transmission errors, the output of the hard-decision detector delivers the transmitted data bits, which can provide valuable explicit training data for the DFE. Hence a reduced-length feed-forward filter can be used, which however does not entirely eliminate the ISI. Instead, the feedback filter uses the data estimates at the output of the data detector in order to subtract the ISI from the output of the feed-forward filter, such that the input signal of the data detector has less ISI, than the signal at the output of the feed-forward filter. If it is assumed that the data estimates fed into the feedback filter are correct, then the DFE is superior to the linear equalizers, since the noise enhancement is reduced. One way of explaining this would be to say that if the data estimates are correct, then the noise has been eliminated and there is

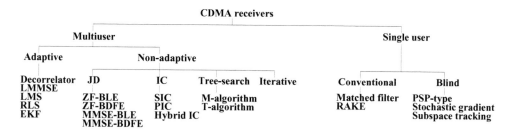

Figure 12.4: Classification of CDMA detectors.

no noise enhancement in the feedback loop. However, if the data estimates are incorrect, these errors will propagate through to future decisions and this problem is known as error propagation.

There are two basic DFEs, the ZF-DFE and the MMSE-DFE. Analogous to its linear counterpart, the coefficients of the feedback filter for the ZF-DFE are calculated so that the ISI at the output of the feed-forward filter is eliminated and the input signal of the data detector is free of ISI [76]. Let us now focus our attention on CDMA multiuser detection equalizers.

12.3   Multiuser Equaliser Concepts

DS-CDMA systems [323,324] support a multiplicity of users within the same bandwidth by assigning different - typically unique - codes to different users for their communications, in order to be able to distinguish their signals from each other. When the transmitted signal is subjected to hostile wireless propagation environments, the signals of different users interfere with each other and hence CDMA systems are interference-limited due to the multiple access interference (MAI) generated by the users transmitting within the same bandwidth simultaneously. The subject of this chapter is, how the MAI can be mitigated. A whole range of detectors have been proposed in the literature, which will be reviewed with reference to the family-tree of Figure 12.4 during our forthcoming discourse.

The conventional so-called single-user CDMA detectors of Figure 12.4 – such as the matched filter [280,325] and the RAKE combiner [76] – are optimized for detecting the signal of a single desired user. RAKE combiners exploit the inherent multi-path diversity in CDMA, since they essentially consist of matched filters for each resolvable path of the multipath channel. The outputs of these matched filters are then coherently combined according to a diversity combining technique, such as maximal ratio combining, equal gain combining or selection diversity combining [76]. These conventional single-user detectors are inefficient, since the interference is treated as noise and the knowledge of the channel impulse response (CIR) or the spreading sequences of the interferers is not exploited.

In order to mitigate the problem of MAI, Verdú [326] proposed and analysed the optimum multiuser detector for asynchronous Gaussian multiple access channels. The optimum detector invokes all the possible bit sequences, in order to find the sequence that maximizes the correlation metric given by [225] :

$$\Omega(\mathbf{r}, \mathbf{d}) = 2\mathbf{d}^T \mathbf{r} - \mathbf{d}^T \mathbf{R} \mathbf{d}, \tag{12.9}$$

where the elements of the vector $\mathbf{r}$ represent the cross-correlation of the spread, channel-impaired received signal with each of the users' spreading sequence, the vector $\mathbf{d}$ consists of the bits transmitted by all the users during the current signalling instant and the matrix $\mathbf{R}$ is the cross-correlation (CCL) matrix of the spreading sequences. This optimum detector significantly outperforms the conventional single-user detector and – in contrast to single user detectors – it is insensitive to power control errors, which is often termed as being near-far resistant. However, unfortunately its complexity grows exponentially in the order of $O(2^{NK})$, where N is the number of overlapping asynchronous bits considered in the detector's decision window and K is the number of interfering users. In order to reduce the complexity of the receiver and yet to provide an acceptable Bit Error Rate (BER) performance, significant research efforts have been invested in the field of sub-optimal CDMA multiuser receivers [225]. Multiuser detection exploits the base station's knowledge of the spreading sequences and that of the estimated (CIRs) in order to remove the MAI. These multiuser detectors can be categorized in a number of ways, such as linear versus non-linear, adaptive versus non-adaptive algorithms or burst transmission versus continuous transmission regimes. Excellent summaries of some of these sub-optimum detectors can be found in the monographs by Verú [225], Prasad [327], Glisic and Vucetic [328]. Other MAI-mitigating techniques include the employment of adaptive antenna arrays, which mitigate the level of MAI at the receiver by forming a beam in the direction of the wanted user and a null towards the interfering users. Research efforts invested in this area include, amongst others, the investigations carried out by Thompson, Grant and Mulgrew [329,330]; Naguib and Paulraj [331]; Godara [332]; as well as Kohno, Imai, Hatori and Pasupathy [333]. However, the area of adaptive antenna arrays is beyond the scope of this article and the reader is referred to the references cited for further discussions. In the forthcoming section, a brief survey of the sub-optimal multiuser receivers will be presented with reference to Figure 12.4, which constitutes an attractive compromise in terms of the achievable performance and the associated complexity.

12.3.1   Linear Receivers

Following the seminal work by Verdú [326], numerous sub-optimum multiuser detectors have been proposed for a variety of channels, data modulation schemes and transmission formats [334]. These CDMA detector schemes will be classified with reference to Figure 12.4, which will be referred to throughout our discussions. Lupas and Verdú [335] initially suggested a sub-optimum linear detector for symbol-synchronous transmissions and further developed it for asynchronous transmissions in a Gaussian channel [336]. This linear detector inverted the CCL matrix R seen in Equation 12.9, which was constructed from the CCLs of the spreading codes of the users and this receiver was termed the decorrelating detector. It was shown that this decorrelator exhibited the same degree of near-far resistance, as the optimum multiuser detector. A further sub-optimum multiuser detector investigated was the minimum mean square error (MMSE) detector, where a biased version of the CCL matrix was inverted and invoked, in order to optimize the receiver obeying the MMSE criterion.

Zvonar and Brady [337] proposed a multiuser detector for synchronous CDMA systems designed for a frequency-selective Rayleigh fading channel. Their approach also used a bank of matched filters followed by a so-called whitening filter, but maximal ratio combining was used to combine the resulting signals. The decorrelating detector of [336] was further developed for differentially-encoded coherent multiuser detection in flat fading channels by Zvonar *et al.* [338]. Zvonar also amalgamated the decorrelating detector with diversity combining, in order to achieve performance improvements in frequency selective fading channels [339]. A multiuser detector jointly performing decorrelating CIR estimation and data detection was investigated by Kawahara and Matsumoto [340]. Path-by-path decorrelators were employed for each user in order to obtain the input signals required for CIR estimation and the CIR estimates as well as the outputs of a matched filter bank were fed into a decorrelator for demodulating the data. A variant of this idea was also presented by Hosseinian, Fattouche and Sesay [341], where training sequences and a decorrelating scheme were used for determining the CIR estimate matrix. This matrix was then used in a decorrelating decision feedback scheme for obtaining the data estimates. Juntti, Aazhang and Lilleberg [342] proposed iterative schemes, in order to reduce the complexity. Sung and Chen [343] advocated using a sequential estimator for minimizing the mean square estimation error between the received signal and the signal after detection. The cross-correlations between the users' spreading codes and the estimates of the channel-impaired received signal of each user were needed, in order to obtain estimates of the transmitted data for each user. Duel-Hallen [344] proposed a decorrelating decision-feedback detector for removing the MAI from a synchronous system communicating over a Gaussian channel. The outputs from a bank of filters matched to the spreading codes of the users were passed through a whitening filter. This filter was obtained by decomposing the CCL matrix of the users' spreading codes with the aid of the Cholesky decomposition [233] technique. The results showed that MAI could be removed from each user's signal successively, assuming that there was no error propagation. However, estimates of the received signal strengths of the users were needed, since the users had to be ranked in order of decreasing signal strengths so that the more reliable estimates were obtained first. Duel-Hallen's decorrelating decision feedback detector [344] was improved by Wei and Schlegel [345] with the aid of a sub-optimum variant of the Viterbi algorithm, where the most likely paths were retained in the case of merging paths in the Viterbi algorithm. The decorrelating decision feedback detector [344] was also improved with the assistance of soft-decision convolutional coding by Hafeez and Stark [346]. Soft decisions from a Viterbi channel decoder were fed back into the filter for signal cancellation.

Having reviewed the range of linear receivers, let us now consider the class of joint detection schemes, which can be found in the family-tree of Figure 12.4 in the next section.

12.3.2 Joint Detection

12.3.2.1 Joint Detection Concept

As mentioned before in the context of single-user channel equalization, the effect of MAI on the desired signal is similar to the impact of multipath propagation-induced Inter-symbol Interference (ISI) on the same signal. Each user in a K-user system suffers from MAI due to the other $(K-1)$ users. This MAI can also be viewed as a single-user signal perturbed by ISI inflicted by $(K-1)$ paths in a multipath channel. Therefore, classic equalization techniques

[76, 103, 118, 280] used to mitigate the effects of ISI can be modified for multiuser detection and these types of multiuser detectors can be classified as joint detection receivers. The joint detection (JD) receivers were developed for burst-based, rather than continuous transmission. The concept of joint detection for the uplink was proposed by Klein and Baier [226] for synchronous burst transmissions, which is visualised with the aid of Figure 12.5.

In Figure 12.5 there are a total of K users in the system, where the information is transmitted in bursts. Each user transmits N data symbols per burst and the data vector for user k is represented as $\mathbf{d}^{(k)}$. Each data symbol is spread with a user-specific spreading sequence, $\mathbf{c}^{(k)}$, which has a length of Q chips. In the uplink, the signal of each user passes through a different mobile channel characterized by its time-varying complex impulse response, $\mathbf{h}^{(k)}$. By sampling at the chip rate of $1/T_c$, the impulse response can be represented by W complex samples. Following the approach of Klein $et\ al.$ [226], the received burst can be represented as $\mathbf{y} = \mathbf{Ad} + \mathbf{n}$, where $\mathbf{y}$ is the received vector and consists of the synchronous sum of the transmitted signals of all the K users, corrupted by a noise sequence, $\mathbf{n}$. The matrix $\mathbf{A}$ is referred to as the system matrix and it defines the system's response, representing the effects of MAI and the mobile channels. Each column in the matrix represents the combined impulse response obtained by convolving the spreading sequence of a user with its channel impulse response, $\mathbf{b}^{(k)} = \mathbf{c}^{(k)} * \mathbf{h}^{(k)}$. This is the impulse response experienced by a transmitted data symbol. Upon neglecting the effects of the noise the joint detection formulation is simply based on inverting the system matrix $\mathbf{A}$, in order to recover the data vector constituted by the superimposed transmitted information of all the K CDMA users. The dimensions of the matrix $\mathbf{A}$ are $(NQ + W - 1) \times KN$ and an example of it can be found in reference [226] by Klein $et\ al$, where the list of the symbols used is given as :

- K for the total number of users,

- N is the number of data symbols transmitted by each user in one transmission burst,

- Q represents the number of chips in each spreading sequence,

- W denotes the length of the wideband CIR, where W is assumed to be an integer multiple of the number of chip intervals, T_c.

- L indicates the number of multipath components or taps in the wideband CIR.

In order to introduce compact mathematical expressions, matrix notation will be employed. The transmitted data symbol sequence of the k-th user is represented by a vector as:

$$\mathbf{d}^{(k)} = (d_1^{(k)}, d_2^{(k)}, \dots, d_n^{(k)}, \dots, d_N^{(k)})^T, \qquad (12.10)$$
$$\text{for } k = 1, \dots, K; \quad n = 1, \dots, N,$$

where k is the user index and n is the symbol index. There are N data symbols per transmission burst and each data symbol is generated using an m-ary modulation scheme [76].

The Q-chip spreading sequence vector of the k-th user is expressed as :

$$\mathbf{c}^{(k)} = (c_1^{(k)}, c_2^{(k)}, \dots, c_q^{(k)}, \dots, c_Q^{(k)})^T, \qquad (12.11)$$
$$\text{for } k = 1, \dots, K; \quad q = 1, \dots, Q.$$

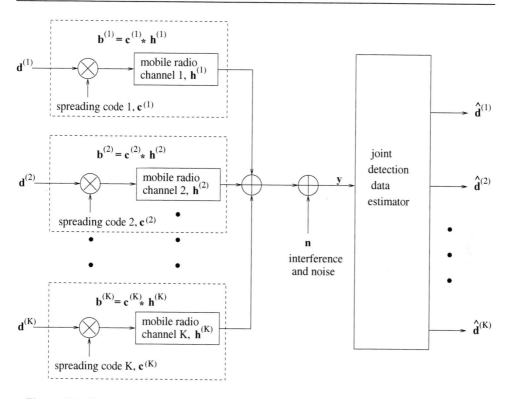

Figure 12.5: System model of a synchronous CDMA system on the up-link using joint detection.

The CIR for the n-th data symbol of the k-th user is represented as :

$$\mathbf{h}_n^{(k)} = (h_n^{(k)}(1), \ldots, h_n^{(k)}(w), \ldots, h_n^{(k)}(W))^T,$$
$$\text{for } k = 1, \ldots, K; \ w = 1, \ldots, W, \qquad (12.12)$$

consisting of W complex CIR samples $h_n^{(k)}(w)$ taken at the chip rate of $1/T_c$.

The combined impulse response, $\mathbf{b}_n^{(k)}$, due to the spreading sequence and the CIR is defined by the convolution of $\mathbf{c}^{(k)}$ and $\mathbf{h}_n^{(k)}$, which is represented as :

$$\begin{aligned}
\mathbf{b}_n^{(k)} &= (b_n^{(k)}(1), \ldots, b_n^{(k)}(l), \ldots, b_n^{(k)}(Q + W - 1))^T \\
&= \mathbf{c}^{(k)} * \mathbf{h}_n^{(k)}, \\
&\quad \text{for } k = 1 \ldots K; \ n = 1, \ldots N.
\end{aligned} \qquad (12.13)$$

In order to represent the ISI due to the N symbols and the dispersive combined impulse responses, the discretised received signal, $\mathbf{r}^{(k)}$, of user k can be expressed as the product of a matrix $\mathbf{A}^{(k)}$ and its data vector $\mathbf{d}^{(k)}$, where :

$$\mathbf{r}^{(k)} = \mathbf{A}^{(k)} \mathbf{d}^{(k)}. \qquad (12.14)$$

The i-th element of the received signal vector $\mathbf{r}^{(k)}$ is :

$$r_i^{(k)} = \sum_{n=1}^{N} [\mathbf{A}^{(k)}]_{in} d_n^{(k)}, \quad \text{for } i = 1, \dots, NQ + W - 1. \qquad (12.15)$$

Again, the matrix $\mathbf{A}^{(k)}$ is the so-called system matrix of the k-th user and it is constructed from the combined impulse responses of Equation 12.13. It represents the effect of the combined impulse responses on each data symbol $d_n^{(k)}$ in the data vector, $\mathbf{d}^{(k)}$. Each column in the matrix $\mathbf{A}$ indexed by n contains the combined impulse response, $\mathbf{b}_n^{(k)}$ that affects the n-th symbol of the data vector. However, since the data symbols are spread by the Q-chip spreading sequences, they are transmitted Q chips apart from each other. Hence the start of the combined impulse response, $\mathbf{b}_n^{(k)}$, for each column is offset by Q rows from the start of $\mathbf{b}_{n-1}^{(k)}$ in the preceding column. Therefore, the element in the $[(n-1)Q + l]$-th row and the n-th column of $\mathbf{A}^{(k)}$ is the l-th element of the combined impulse response, $\mathbf{b}_n^{(k)}$, for $l = 1, \dots, Q + W - 1$. All other elements in the column are zero-valued.

The pictorial representation of Equation 12.14 is shown in Figure 12.6, where $Q = 4$, $W = 2$ and $N = 3$. As it can be seen from the diagram, in each column of the matrix $\mathbf{A}^{(k)}$ – where a box with an asterisk marks a non-zero element – the vector $\mathbf{b}_n^{(k)}$ starts at an offset of $Q = 4$ rows below its preceding column, except for the first column, which starts at the first row. The total number of elements in the vector $\mathbf{b}_n^{(k)}$ is $(Q + W - 1) = 5$. The total number of columns in the matrix $\mathbf{A}^{(k)}$ equals the number of symbols in the data vector, $\mathbf{d}^{(k)}$, i.e. N. Finally, the received signal vector product, $\mathbf{r}^{(k)}$ in Equation 12.14, has a total of $(NQ + W - 1) = 13$ elements due to the ISI imposed by the multipath channel, as opposed to $NQ = 12$ elements in a narrowband channel.

The joint detection receiver aims for detecting the symbols of all the users jointly by utilizing the information available on the spreading sequences and CIR estimates of all the users. Therefore, as seen in Figure 12.7, the data symbols of all K users can be viewed as the transmitted data sequence of a single user, by concatenating all the data sequences. The overall transmitted sequence can be rewritten as :

$$\mathbf{d} = (\mathbf{d}^{(1)T}, \mathbf{d}^{(2)T}, \dots, \mathbf{d}^{(K)T})^T \qquad (12.16)$$

$$= (d_1, d_2, \dots, d_{KN})^T, \qquad (12.17)$$

where $d_j = d_n^{(k)}$ for $j = n + N.(k-1)$, $k = 1, 2, \dots, K$ and $n = 1, 2, \dots, N$.

The system matrix for the overall system can be constructed by appending the $\mathbf{A}^{(k)}$ matrix of each of the K users column-wise, whereby :

$$\mathbf{A} = (\mathbf{A}^{(1)}, \mathbf{A}^{(2)}, \dots, \mathbf{A}^{(k)}, \dots, \mathbf{A}^{(K)}). \qquad (12.18)$$

The construction of matrix $\mathbf{A}$ from the system matrices of the K users is depicted in Figure 12.7. Therefore, the discretised received composite signal can be represented in matrix form as :

$$\mathbf{y} = \mathbf{A}\mathbf{d} + \mathbf{n}, \qquad (12.19)$$

$$\mathbf{y} = (y_1, y_2, \dots, y_{NQ+W-1})^T,$$

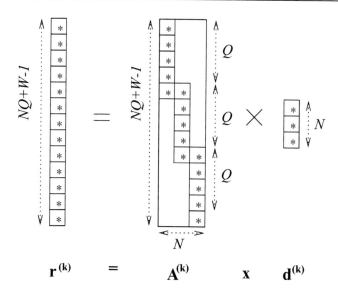

Figure 12.6: Stylized structure of Equation 12.14 representing the received signal vector of a wideband channel, where $Q = 4$, $W = 2$ and $N = 3$. The column vectors in the matrix $\mathbf{A}^{(k)}$ are the combined impulse response vectors, $\mathbf{b}_n^{(k)}$ of Equation 12.13. A box with an asterisk in it represents a non-zero element, and the remaining notation is as follows : K represents the total number of users, N denotes the number of data symbols transmitted by each user, Q represents the number of chips in each spreading sequence, and W indicates the length of the wideband CIR.

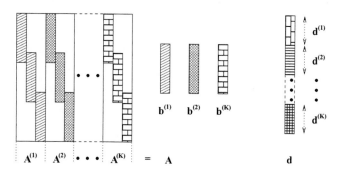

Figure 12.7: The construction of matrix $\mathbf{A}$ from the individual system matrices, $\mathbf{A}^{(k)}$ seen in Figure 12.6, and the data vector $\mathbf{d}$ from the concatenation of data vectors, $\mathbf{d}^{(k)}$, of all K users.

where $\mathbf{n} = (n_1, n_2, \ldots, n_{NQ+W-1})^T$, is the noise sequence, which has a covariance matrix of $\mathbf{R}_n = E[\mathbf{n}.\mathbf{n}^H]$. The composite signal vector $\mathbf{y}$ has $(NQ + W - 1)$ elements for a data burst of length N symbols. Upon multiplying the matrix $\mathbf{A}$ with the vector $\mathbf{d}$ seen in Figure 12.7, we obtain the MAI- and ISI-contaminated received symbols according to Equation 12.19.

Taken as a whole, the system matrix, $\mathbf{A}$, can be constructed from the combined response vectors, $\mathbf{b}_n^{(k)}$ of all the K users, in order to depict the effect of the system's response on the data vector of Equation 12.16. The dimensions of the matrix are $(NQ + W - 1) \times KN$. Figure 12.8 shows an example of the matrix, $\mathbf{A}$, for an N-bit long data burst. For ease of representation, we assumed that the channel length, W, for each user is the same and that it remains constant throughout the data burst. We have also assumed that the channel experiences slow fading and that the fading is almost constant across the data burst. Therefore, the combined response vector for each transmitted symbol of user k is represented by $\mathbf{b}^{(k)}$, where $\mathbf{b}^{(k)} = \mathbf{b}_1^{(k)} \mathbf{b}_2^{(k)} = \ldots = \mathbf{b}_N^{(k)}$. Focusing our attention on Figure 12.8, the elements in the j-th column of the matrix constitute the combined response vector that affects the j-th data symbol in the transmitted data vector $\mathbf{d}$. Therefore, columns $j = 1$ to N of matrix $\mathbf{A}$ correspond to symbols $m = 1$ to N of vector $\mathbf{d}$, which are also the data symbols of user $k = 1$. The next N columns correspond to the next N symbols of data vector $\mathbf{d}$, which are the data symbols of user $k = 2$ and so on.

For user k, each successive response vector, $\mathbf{b}^{(k)}$, is placed at an offset of Q rows from the preceding vector, as shown in Figure 12.8. For example, the combined response vector in column 1 of matrix $\mathbf{A}$ is $\mathbf{b}^{(1)}$ and it starts at row 1 of the matrix because that column corresponds to the first symbol of user $k = 1$. In column 2, the combined response vector is also $\mathbf{b}^{(1)}$, but it is offset from the start of the vector in column 1 by Q rows. This is because the data symbol corresponding to this matrix column is transmitted Q chips later. This is repeated until the columns $j = 1, \ldots, N$ contain the combined response vectors that affect all the data symbols of user $k = 1$. The next column of $j = N + 1$ in the matrix $\mathbf{A}$ contains the combined impulse response vector that affects the data symbol, $d_{N+1} = d_1^{(2)}$, which is the first data symbol of user $k = 2$. In this column, the combined response vector for user $k = 2$, $\mathbf{b}^{(2)}$, is used and the vector starts at row 1 of the matrix because it is the first symbol of this user. The response matrix, $\mathbf{b}^{(2)}$ is then placed into columns $j = N+1, \ldots, 2N$ of the matrix $\mathbf{A}$, with the same offsets for each successive vector, as was carried out for user 1. This process is repeated for all the other users until the system matrix is completely constructed.

The mathematical representation of matrix $\mathbf{A}$ in general can be written as :

$$[\mathbf{A}]_{ij} = \begin{cases} b_n^{(k)}(l) & \text{for } k = 1, \ldots, K; \; n = 1, \ldots, N; \\ & l = 1, \ldots, Q + W - 1 \\ 0 & \text{otherwise,} \end{cases} \tag{12.20}$$
$$\text{for } i = 1, \ldots, NQ + W - 1, \quad j = 1, \ldots, KN,$$

where $i = Q(n - 1) + l$ and $j = n + N(k - 1)$.

Figure 12.9 shows the stylized structure of Equation 12.19 for a specific example. In the figure, a system with $K = 2$ users is depicted. Each user transmits $N = 3$ symbols per transmission burst, and each symbol is spread with a signature sequence of length $Q = 3$ chips. The channel for each user has a dispersion length of $W = 3$ chips. The blocked segments in the figure represent the combination of elements that result in the element y_4, which is obtained from Equation 12.19 by :

$$y_4 = \sum_{i=1}^{KN=6} [\mathbf{A}]_{4,i} d_i + n_4 \tag{12.21}$$

$$= [\mathbf{A}]_{4,1} d_1 + [\mathbf{A}]_{4,2} d_2 + [\mathbf{A}]_{4,4} d_4 + [\mathbf{A}]_{4,5} d_5 + n_4 \tag{12.22}$$

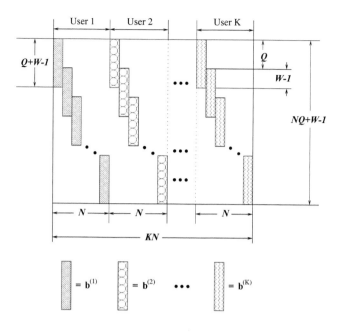

Figure 12.8: Stylized structure of the system matrix **A**, where $\mathbf{b}^{(1)}$, $\mathbf{b}^{(2)}$ and $\mathbf{b}^{(K)}$ are column vectors representing the combined impulse responses of users 1, 2 and K, respectively in Equation 12.13. The notation is as follows : K represents the total number of users, N denotes the number of data symbols transmitted by each user, Q represents the number of chips in each spreading sequence, and W indicates the length of the wideband CIR.

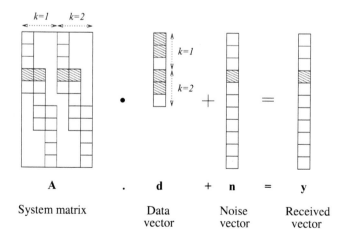

Figure 12.9: Stylized structure of the matrix equation $\mathbf{y} = \mathbf{Ad} + \mathbf{n}$ for a $K = 2$–user system. Each user transmits $N = 3$ symbols per transmission burst, and each symbol is spread with a signature sequence of length $Q = 3$ chips. The channel for each user has a dispersion length of $W = 3$ chips.

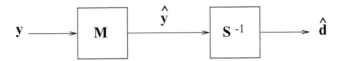

Figure 12.10: Structure of the receiver represented in Equation 12.23.

Given the above transmission regime, the basic concept of joint detection is centred around processing the received composite signal vector, **y**, in order to determine the transmitted data vector, **d**. This concept is encapsulated in the following set of equations :

$$\hat{\mathbf{y}} = \mathbf{S}\hat{\mathbf{d}} = \mathbf{M}\mathbf{y}, \tag{12.23}$$

where **S** is a square matrix with dimensions $(KN \times KN)$ and the matrix **M** is a $[KN \times (NQ + W - 1)]$-matrix. These two matrices determine the type of joint detection algorithm, as it will become explicit during our further discourse. The schematic in Figure 12.10 shows the receiver structure represented by this equation.

A range of joint detection schemes designed for uplink communications were proposed by Jung, Blanz, Nasshan, Steil, Baier and Klein, such as the minimum mean-square error block linear equalizer (MMSE-BLE) [208, 219, 227, 228], the zero-forcing block decision feedback equalizer (ZF-BDFE) [219, 228] and the minimum mean-square error block decision feedback equalizer (MMSE-BDFE) [219, 228].

These joint-detection receivers were also combined with coherent receiver antenna diversity (CRAD) techniques [219, 227, 228, 347] and turbo coding [348, 349] for performance improvement. Joint detection receivers were proposed also for downlink scenarios by Nasshan, Steil, Klein and Jung [350, 351]. CIR estimates were required for the joint detection receivers and CIR estimation algorithms were proposed by Steiner and Jung [352] for employment in conjunction with joint detection. Werner [353] extended the joint detection receiver by combining ZF-BLE and MMSE-BLE techniques with a multistage decision mechanism using soft inputs to a Viterbi decoder.

Having considered the family of JD receivers, which typically exhibit a high complexity, let us now highlight the state-of-the-art in the context of lower complexity interference cancellation schemes in the next section.

12.3.3   Interference Cancellation

Interference cancellation (IC) schemes constitute another variant of multiuser detection and they can be broadly divided into three categories, parallel interference cancellation (PIC), successive interference cancellation (SIC) and the hybrids of both, as seen in Figure 12.4. Varanasi and Aazhang [354] proposed a multistage detector for an asynchronous system, where the outputs from a matched filter bank were fed into a detector that performed MAI cancellation using a multistage algorithm. At each stage in the detector, the data estimates $\hat{\mathbf{d}}^{(1)}, \ldots, \hat{\mathbf{d}}^{(K-1)}$ of all the other $(K-1)$ users from the previous stage were used for reconstructing an estimate of the MAI and this estimate was then subtracted from the interfered received signal representing the wanted bit. The computational complexity of this detector was linear with respect to the number of users, K. Figure 12.11 depicts the schematic of

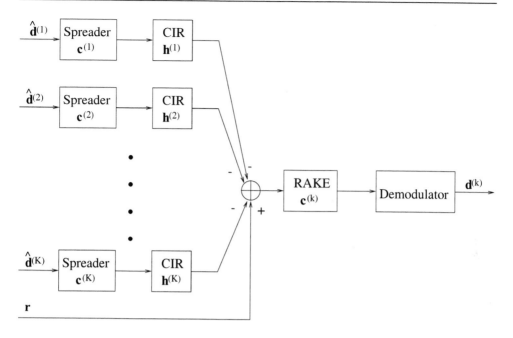

Figure 12.11: Schematic of a single cancellation stage for user k in the parallel interference cancellation (PIC) receiver for K users. The data estimates, $\hat{\mathbf{d}}^{(1)}, \ldots, \hat{\mathbf{d}}^{(K-1)}$ of the other
$(K-1)$ users were obtained from the previous cancellation stage and the received sig-
nal of each user other than the k-th one is reconstructed and cancelled from the received
signal, $\mathbf{r}$.

a single cancellation stage in the PIC receiver. Varanasi further modified the above parallel cancellation scheme, in order to create a parallel group detection scheme for Gaussian
channels [355] and later developed it further for frequency-selective slow Rayleigh fading
channels [356]. In this scheme, K users were divided into P groups and each group was demodulated in parallel using a group detector. Yoon, Kohno and Imai [357] then extended the
applicability of the multistage interference cancellation detector to a multipath, slowly fading
channel. At each cancellation stage, hard decisions generated by the previous cancellation
stage were used for reconstructing the signal of each user and for cancelling its contribution
from the composite signal. The effects of CIR estimation errors on the performance of the
cancellation scheme were also considered. A multiuser receiver that integrated MAI rejection and channel decoding was investigated by Giallorenzi and Wilson [358]. The MAI was
cancelled via a multistage cancellation scheme and soft-outputs were fed from the Viterbi
channel decoder of each user to each stage for improving the performance.

The PIC receiver of Figure 12.11 [354] was also modified for employment in multi-carrier
modulation [359] by Sanada and Nakagawa. Specifically, convolutional coding was used in
order to obtain improved estimates of the data for each user at the initial stage and these
estimates were then utilized for interference cancellation in the following stages. The employment of convolutional coding improved the performance by 1.5 dB. Latva-aho, Juntti
and Heikkilä [360] enhanced the performance of the parallel interference cancellation re-

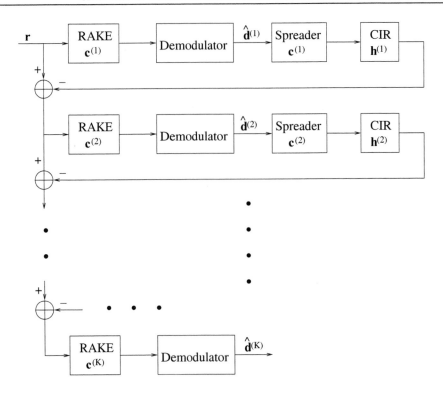

Figure 12.12: Schematic of the successive interference cancellation (SIC) receiver for K users. The users' signals have been ranked, where user 1's signal was received at the highest power, while user K's signal at the lowest power. In the order of ranking, the data estimates of each user are obtained and the received signal of each user is reconstructed and cancelled from the received composite signal, $\mathbf{r}$.

ceiver by feeding back CIR estimates to the signal reconstruction stage of the multistage receiver seen in Figure 12.11 and proposed an algorithm for mitigating error propagation. Dahlhaus, Jarosch, Fleury and Heddergott [361] combined multistage detection with CIR estimation techniques utilizing the outputs of antenna arrays. The CIR estimates obtained were fed back into the multistage detector in order to refine the data estimates. An advanced parallel cancellation receiver was also proposed by Divsalar, Simon and Raphaeli [362]. At each cancellation stage, only partial cancellation was carried out by weighting the regenerated signals with a less than unity scaling factor. At each consecutive stage, the weights were increased based on the assumption that the estimates became increasingly accurate.

Following the above brief notes on PIC receivers, let us now consider the family of reduced-complexity SIC receivers classified in Figure 12.4. A simple SIC scheme was analysed by Patel and Holtzman [363]. The received signals were ranked according to their correlation values, which were obtained by utilizing the correlations between the received signal and the spreading codes of the users. The transmitted information of the strongest user was estimated, enabling the transmitted signal to be reconstructed with the aid of the spreader as well as the CIR and subtracted from the received signal, as portrated in Figure 12.12. This was repeated for the next strongest user, where the reconstructed signal of this second user was cancelled from the composite signal remaining after the first cancellation. The interference cancellation was carried out successively for all the other users, until eventually only the signal of the weakest user remained. It was shown that the SIC receiver improved the BER and the system's user capacity over that of the conventional matched filter for the Gaussian, for narrowband Rayleigh and for dispersive Rayleigh channels. Multipath diversity was also exploited by combining the SIC receiver with the RAKE correlator [363]. Again, Figure 12.12 shows the schematic of the SIC receiver. Soong and Krzymien [364] extended the SIC receiver by using reference symbols in order to aid the CIR estimation. The performance of the receiver was investigated in flat and frequency-selective Rayleigh fading channels, as well as in multi-cell scenarios. A soft-decision based adaptive SIC scheme was proposed by Hui and Letaief [365], where soft decisions were used in the cancellation stage and if the decision statistic did not satisfy a certain threshold, no data estimation was carried out for that particular data bit, in order to reduce error propagation.

Hybrid SIC and PIC schemes were proposed by Oon, Li and Steele [366, 367], where SIC was first performed on the received signal, followed by a multistage PIC arrangement. This work was then extended to an adaptive hybrid scheme for flat Rayleigh fading channels [368]. In this scheme, successive cancellation was performed for a fraction of the users, while the remaining users' signals were processed via a parallel cancellation stage. Finally, multistage parallel cancellation was invoked. The number of serial and parallel cancellations performed was varied adaptively according to the BER estimates. Sawahashi, Miki, Andoh and Higuchi [369] proposed a pilot symbol-assisted multistage hybrid successive-parallel cancellation scheme. At each stage, data estimation was carried out successively for all the users, commencing with the user having the strongest signal and ending with the weakest signal. For each user, the interference inflicted by the other users was regenerated using the estimates of the current stage for the stronger users and the estimates of the previous stage for the weaker users. CIR estimates were obtained for each user by employing pilot symbols and a recursive estimation algorithm. Another hybrid successive and parallel interference cancellation receiver was proposed by Sun, Rasmussen, Sugimoto and Lim [370], where the users to be detected were split into a number of groups. Within each group, PIC was performed on the signals of these users belonging to the group. Between the separate groups, SIC was employed. This had the advantage of a reduced delay and improved performance compared to the SIC receiver. A further variant of the hybrid cancellation scheme was constituted by the combination of MMSE detectors with SIC receivers, as proposed by Cho and Lee [371]. Single-user MMSE detectors were used to obtain estimates of the data symbols, which were then fed back into the SIC stages. An adaptive interference cancellation scheme was investigated by Agashe and Woerner [372] for a multicellular scenario, where interference cancellation was performed for both in-cell interferers and out-of-cell interferers. It was shown that cancelling the estimated interference from users having weak signals actually degraded the performance, since the estimates were inaccurate. The adaptive scheme exercised interference cancellation in a discriminating manner, using only the data estimates of users having strong received signals. Therefore signal power estimation was needed and the threshold for signal cancellation was adapted accordingly. Following the above brief discourse on interference cancellation algorithms, let us now focus our attention on the tree-type detection techniques, which were also categorized in Figure 12.4.

12.3.4   Tree-Search Detection

Several tree-search detection [373–375] receivers have been proposed in the literature, in order to reduce the complexity of the original maximum likelihood detection scheme proposed by Verdú [326]. Specifically, Rasmussen, Lim and Aulin [373] investigated a tree-search detection algorithm, where a recursive, additive metric was developed in order to reduce the search complexity. Reduced tree-search algorithms, such as the well-known M-algorithms [376] and T-algorithms [376] were used by Wei, Rasmussen and Wyrwas [374] in order to reduce the complexity incurred by the optimum multiuser detector. According to the M-algorithm, at every node of the trellis search algorithm, only M surviving paths were retained, depending on certain criteria such as for example the highest-metric M number of paths. Alternatively, all the paths that were within a fixed threshold, T, compared to the highest metric were retained. At the decision node, the path having the highest metric was chosen as the most likely transmitted sequence. Maximal-ratio combining was also used in conjunction with the reduced tree-search algorithms and the combining detectors outperformed the "non-combining" detectors. The T-algorithm was combined with soft-input assisted Viterbi detectors for channel-coded CDMA multiuser detection in the work carried out by Nasiri-Kenari, Sylvester and Rushforth [375]. The recursive tree-search detector generated soft-outputs, which were fed into single-user Viterbi channel decoders, in order to generate the bit estimates.

The so-called multiuser projection based receivers were proposed by Schlegel, Roy, Alexander and Jiang [377] and by Alexander, Rasmussen and Schlegel [378]. These receivers reduced the MAI by projecting the received signal onto a space which was orthogonal to the unwanted MAI, where the wanted signal was separable from the MAI. Having reviewed the two most well-known tree-search type algorithms, we now concentrate on the family of intelligent adaptive detectors in the next section, which can be classified with the aid of Figure 12.4.

12.3.5   Adaptive Multiuser Detection

In all the multiuser receiver schemes discussed earlier, the required parameters - except for the transmitted data estimates - were assumed to be known at the receiver. In order to remove this constraint while reducing the complexity, adaptive receiver structures have been proposed [379]. An excellent summary of these adaptive receivers has been provided by Woodward and Vucetic [380]. Several adaptive algorithms have been introduced for approximating the performance of the MMSE receivers, such as the Least Mean Squares (LMS) [118] algorithm, the Recursive Least Squares (RLS) algorithm [118] and the Kalman filter [118]. Xie, Short and Rushforth [381] showed that the adaptive MMSE approach could be applied to multiuser receiver structures with a concomitant reduction in complexity. In the adaptive receivers employed for asynchronous transmission by Rapajic and Vucetic [379], training sequences were invoked, in order to obtain the estimates of the parameters required. Lim, Rasmussen and Sugimoto introduced a multiuser receiver for an asynchronous flat-fading channel based on the Kalman filter [382], which compared favourably with the finite impulse response MMSE detector. An adaptive decision feedback based joint detection scheme was investigated by Seite and Tardivel [383], where the least mean squares (LMS) algorithm was used to update the filter coefficients, in order to minimize the mean square error of the data estimates. New adaptive filter architectures for downlink DS-CDMA receivers were suggested by Spangenberg, Cruickshank, McLaughlin, Povey and Grant [66], where an adaptive algorithm was employed in order to estimate the CIR, and this estimated CIR was then used by a channel equalizer. The output of the channel equalizer was finally processed by a fixed multiuser detector in order to provide the data estimates of the desired user.

12.3.6 Blind Detection

The novel class of multiuser detectors, referred to as "blind" detectors, does not require explicit knowledge of the spreading codes and CIRs of the multiuser interferers. These detectors do not require the transmission of training sequences or parameter estimates for their operation. Instead, the parameters are estimated "blindly" according to certain criteria, hence the term "blind" detection. RAKE-type blind receivers have been proposed, for example by Povey, Grant and Pringle [384] for fast-fading mobile channels, where decision-directed CIR estimators were used for estimating the multipath components and the output of the RAKE fingers was combined employing various signal combining methods. Liu and Li [385] also proposed a RAKE-type receiver for frequency-selective fading channels. In [385], a weighting factor was utilized for each RAKE finger, which was calculated based on maximizing the signal-to-interference-plus-noise ratio (SINR) at the output of each RAKE finger.

Xie, Rushforth, Short and Moon [386] proposed an approximate Maximum Likelihood Sequence Estimation (MLSE) solution known as the per-survivor processing (PSP) type algorithm, which combined a tree-search algorithm for data detection with the Recursive Least Squares (RLS) adaptive algorithm used for channel amplitude and phase estimation. The PSP algorithm was first proposed by Seshadri [387]; as well as by Raheli, Polydoros and Tzou [388, 389] for blind equalization in single-user ISI-contaminated channels. Xie, Rushforth, Short and Moon extended their own earlier work [386], in order to include the estimation of user-delays along with channel- and data-estimation [390].

Iltis and Mailaender [391] combined the PSP algorithm with the Kalman filter, in order to adaptively estimate the amplitudes and delays of the CDMA users. In other blind detection schemes, Mitra and Poor compared the application of neural networks and LMS filters for obtaining data estimates of the CDMA users [392]. In contrast to other multiuser detectors, which required the knowledge of the spreading codes of all the users, only the spreading code of the desired user was needed for this adaptive receiver [392]. An adaptive decorrelating detector was also developed by Mitra and Poor [393], which was used to determine the spreading code of a new user entering the system.

Blind equalization was combined with multiuser detection for slowly fading channels in the work published by Wang and Poor [394]. Only the spreading sequence of the desired user was needed and a zero-forcing as well as an MMSE detector were developed for data detection. As a further solution, a so-called sub-space approach to blind multiuser detection was also proposed by Wang and Poor [395], where only the spreading sequence and the delay of the desired user were known at the receiver. Based on this knowledge, a blind sub-space tracking algorithm was developed for estimating the data of the desired user. Further blind adaptive algorithms were developed by Honig, Madhow and Verdú [396], Mandayam and Aazhang [397], as well as by Ulukus and Yates [398]. In [396], the applicability of two adaptive algorithms to the multiuser detection problem was investigated, namely that of the stochastic gradient algorithm and the least squares algorithm, while in [398] an adaptive detector that converged to the solution provided by the decorrelator was analysed.

The employment of the Kalman filter for adaptive data, CIR and delay estimation was carried out by Lim and Rasmussen [399]. They demonstrated that the Kalman filter gave a good performance and exhibited a high grade of flexibility. However, the Kalman filter required reliable initial delay estimates in order to initialize the algorithm. Miguez and Castedo [400] modified the well-known constant modulus approach [401, 402] to blind equalization for ISI-contaminated channels in the context of multiuser interference suppression. Fukawa and Suzuki [403] proposed an orthogonalizing matched filtering detector, which consisted of a bank of despreading filters and a signal combiner. One of the despreading filters was matched to the desired spreading sequence, while the other despreading sequences were arbitrarily chosen such that the impulse responses of the filters were linearly independent of each other. The filter outputs were adaptively weighted in the complex domain under the constraint that the average output power of the combiner was minimized. In another design, an iterative scheme used to maximize the so-called log-likelihood function was the basis of the research by Fawer and Aazhang [404]. RAKE correlators were employed for exploiting the multipath diversity and the outputs of the correlators were fed to an iterative scheme for joint CIR estimation and data detection using the Gauss-Seidel [297] algorithm.

12.3.7   Hybrid and Novel Multiuser Receivers

Several hybrid multiuser receiver structures have also been proposed recently [405–408]. Bar-Ness [405] advocated the hybrid multiuser detector that consisted of a decorrelator for detecting asynchronous users, followed by a data combiner maximising the Signal-to-noise Ratio (SNR), an adaptive canceller and another data combiner. The decorrelator matrix was adaptively determined.

A novel multiuser CDMA receiver based on genetic algorithms (GA) was considered by Yen *et al.* [406], where the transmitted symbols and the channel parameters of all the users were jointly estimated. The maximum likelihood receiver of synchronous CDMA systems exhibits a computational complexity that is exponentially increasing with the number of users, since at each signalling instant the corresponding data bit of all users has to be determined. Hence the employment of maximum likelihood detection invoking an exhaustive search is not a practical approach. GAs have been widely used for solving complex optimization problems in engineering, since they typically constitute an attractive compromise in performance versus complexity terms. Using the approach of [406] GAs can be invoked, in order to jointly estimate the users' channel parameters as well as the transmitted bit vector of all the users at the current signalling instant with the aid of a bank of matched filters at the receiver. It was shown in [406] that GA-based multi-user detectors can approach the single-user BER performance at a significantly lower complexity than that of the optimum ML multiuser detector without the employment of training sequences for channel estimation.

The essence of this GA-based technique [406] is that the search-space for the most likely data vector of all the users at a given signalling instant was limited to a certain population of vectors and the candidate vectors were updated at each iteration according to certain probabilistic genetic operations, known as *reproduction, crossover* or *mutation*. Commencing with a population of tentative decisions concerning the vector of all the users' received bits at the current signalling instant, the best n data vectors were selected as so-called "parent" vectors according to a certain "fitness" criterion - which can be also considered to be a cost-function

- based on the likelihood function [406] in order to generate the so-called "offspring" for the next generation of data vector estimates. The aim is that the off-spring should exhibit a better "fitness" or cost-function contribution, than the "parents", since then the algorithm will converge. The offspring of data vector estimates were generated by employing a so-called uniform "crossover process", where the bits between two parent or candidate data vectors were exchanged according to a random cross-over mask and a certain exchange probability. Finally, the so-called "mutation" was performed, where the value of a bit in the data vector was flipped according to a certain mutation probability. In order to prevent the loss of "high-fitness" parent sequences during the process of evolution of the estimated user data vectors, the "highest-merit" estimated user data vector that was initially excluded from the pool of parent vectors in creating a new generation of candidate data vectors was then used to replace the "lowest-merit" offspring.

Neural network-type multi-user equalizers have also been proposed as CDMA receivers [409,410]. Specifically, Tanner and Cruickshank proposed a non-linear receiver that exploited neural-network structures and employed pattern recognition techniques for data detection [409]. This work [409] was extended to a reduced complexity neural network receiver for the downlink scenario [410]. The advantage of the neural-network based receivers is that they are capable of 'learning' the optimum partitioning rules in the signal constellation space, even, when the received interference-contaminated constellation points linearly non-separable. In this scenario linear receivers would exhibit a residual BER even in the absence of channel noise.

Other novel techniques employed for mitigating the multipath fading effects inflicted upon multiple users include **joint transmitter-receiver optimization** proposed by Jang, Vojčić and Pickholtz [407, 408]. In these schemes, transmitter precoding was carried out, such that the mean squared errors of the signals at all the receivers were minimized. This required the knowledge of the CIRs of all the users and the assumption was made that the channel fading was sufficiently slow, such that CIR prediction could be employed reliably by the transmitter.

Recently, there has been significant interest in **iterative detection** schemes, where channel coding was exploited in conjunction with multiuser detection, in order to obtain a high BER performance. The spreading of the data and the convolutional channel coding was viewed as a serially concatenated code structure, where the CDMA channel was viewed as the inner code and the single user convolutional codes constituted the outer codes. After processing the received signal in a bank of matched filters - often referred to as orthogonalizing whitening matched filter - the matched filter outputs were processed using a so-called **turbo-style iterative decoding (TEQ)** [411] process. In this process, a multiuser decoder was used to produce bit confidence measures, which were used as soft inputs of the single-user channel decoders. These single-user decoders then provided similar confidence metrics, which were fed back to the multiuser detector. This iterative process continued, until no further performance improvement was recorded.

Giallorenzi and Wilson [412] presented the maximum likelihood solution for the asynchronous CDMA channel, where the user data was encoded with the aid of convolutional codes. Near-single-user performance was achieved for the two-user case in conjunction with fixed length spreading codes. The decoder was implemented using the Viterbi channel decoding algorithm, where the number of states increased exponentially with the product of the number of users and the constraint length of the convolutional codes. Later, a suboptimal modification of this technique was proposed [358], where the MAI was cancelled via multi-stage cancellation and the soft outputs of the Viterbi algorithm were supplied to each stage of the multistage canceller for improving the performance. Following this, several iterative multiuser detection schemes employing channel-coded signals have been presented [413–418]. For example, Alexander, Astenstorfer, Schlegel and Reed [415, 417] proposed the multiuser maximum a-posteriori (MAP) detectors for the decoding of the inner CDMA channel code and invoked single-user MAP decoders for the outer convolutional code. A reduced complexity solution employing the M-algorithm [376] was also suggested, which resulted in a complexity that increased linearly – rather than exponentially, as in [412] – with the number of users [416]. Wang and Poor [418] employed a soft-output multiuser detector for the inner channel code, which combined soft interference cancellation and instantaneous linear MMSE filtering, in order to reduce the complexity. These iterative receiver structures showed considerable promise and near-single-user performance was achieved at high SNRs.

Figure 12.4 portrays the classification of most of the CDMA detectors that have been discussed previously. All the acronyms for the detectors have been defined in the text. Examples of the different classes of detectors are also included. Having considered the family of various CDMA detectors, let us now turn our attention to adaptive rate CDMA schemes.

12.4 Adaptive CDMA Schemes

Mobile radio signals are subject to propagation path loss as well as slow fading and fast fading. Due to the nature of the fading channel, transmission errors occur in bursts, when the channel exhibits deep fades due to shadowing, obstructing vehicles, etc. or when there is a sudden surge of multiple access interference (MAI) or inter-symbol interference (ISI). In mobile communications systems power control techniques are used to mitigate the effects of path loss and slow fading [13]. However, in order to counteract the problem of fast fading and co-channel interference, agile power control algorithms are required [419]. Another technique that can be used to overcome the problems due to the time-variant fluctuations of the channel is Burst-by-Burst (BbB) adaptive transmission [1,75], where the information rate is varied according to the near-instantaneous quality of the channel, rather than according to user requirements. When the near-instantaneous channel quality is low, a lower information rate is supported, in order to reduce the number of errors. Conversely, when the near-instantaneous channel quality is high, a higher information rate is used, in order to increase the average throughput of the system. More explicitly, this method is similar to multi-rate transmission [65], except that in this case, the transmission rate is modified according to the near-instantaneous channel quality, instead of the service required by the mobile user. BbB-adaptive CDMA systems are also useful for employment in arbitrary propagation environments or in hand-over scenarios, such as those encountered when a mobile user moves from an indoor to an outdoor environment or in a so-called 'birth-death' scenario, where the number of transmitting CDMA users changes frequently [66], thereby changing the interference dramatically. Various methods of multi-rate transmission have been proposed in the research literature. Next we will briefly discuss some of the current research on multi-rate transmission schemes, before focusing our attention on BbB-adaptive systems.

Ottosson and Svensson compared various multi-rate systems [65], including multiple spreading factor (SF) based, multi-code and multi-level modulation schemes. According to the multi-code philosophy, the SF is kept constant for all users, but multiple spreading codes transmitted simultaneously are assigned to users requiring higher bit rates. In this case - unless the spreading codes's perfect orthogonality is retained after transmission over the channel - the multiple codes of a particular user interfere with each other. This inevitably reduces the system's performance.

Multiple data rates can also be supported by a variable SF scheme, where the chip rate is kept constant, but the data rates are varied, thereby effectively changing the SF of the spreading codes assigned to the users; at a fixed chip rate the lower the SF, the higher the supported data rate. Performance comparisons for both of these schemes have been carried out by Ottosson and Svensson [65], as well as by Ramakrishna and Holtzman [67], demonstrating that both schemes achieved a similar performance. Adachi, Ohno, Higashi, Dohi and Okumura proposed the employment of multi-code CDMA in conjunction with pilot symbol-assisted channel estimation, RAKE reception and antenna diversity for providing multi-rate capabilities [68, 69]. The employment of multi-level modulation schemes was also investigated by Ottosson and Svensson [65], where higher-rate users were assigned higher-order modulation modes, transmitting several bits per symbol. However, it was concluded that the performance experienced by users requiring higher rates was significantly worse than that experienced by the lower-rate users. The use of M-ary orthogonal modulation in providing variable rate transmission was investigated by Schotten, Elders-Boll and Busboom [70]. According to this method, each user was assigned an orthogonal sequence set, where the number of sequences, M, in the set was dependent on the data rate required – the higher the rate required, the larger the sequence set. Each sequence in the set was mapped to a particular combination of $b = (\log_2 M)$ bits to be transmitted. The M-ary sequence was then spread with the aid of a spreading code of a constant SF before transmission. It was found [70] that the performance of the system depended not only on the MAI, but also on the Hamming distance between the sequences in the M-ary sequence set.

Saquib and Yates [71] investigated the employment of the decorrelating detector in conjunction with the multiple-SF scheme and proposed a modified decorrelating detector, which utilized soft decisions and maximal ratio combining, in order to detect the bits of the different-rate users. Multi-rate transmission schemes involving interference cancellation receivers have previously been investigated amongst others by Johansson and Svensson [72, 73], as well as by Juntti [74]. Typically, multiple users transmitting at different bit rates are supported in the same CDMA system invoking multiple codes or different spreading factors. SIC schemes and multi-stage cancellation schemes were used at the receiver for mitigating the MAI [72–74], where the bit rate of the users was dictated by the user requirements. The performance comparison of various multiuser detectors in the context of a multiple-SF transmission scheme was presented for example by Juntti [74], where the detectors compared were the decorrelator, the PIC receiver and the so-called group serial interference cancellation (GSIC) receiver. It was concluded that the GSIC and the decorrelator performed better than the PIC receiver, but all the interference cancellation schemes including the GSIC, exhibited an error floor at high SNRs due to error propagation.

The bit rate of each user can also be adapted according to the near-instantaneous channel quality, in order to mitigate the effects of channel quality fluctuations. Kim [75] analysed the performance of two different methods of combating the near-instantaneous quality variations of the mobile channel. Specifically, Kim studied the adaptation of the transmitter power or the switching of the information rate, in order to suit the near-instantaneous channel conditions. Using a RAKE receiver [76], it was demonstrated that rate adaptation provided a higher average information rate than power adaptation for a given average transmit power and a given BER [75]. Abeta, Sampei and Morinaga [77] conducted investigations into an adaptive packet transmission based CDMA scheme, where the transmission rate was modified by varying the channel code rate and the processing gain of the CDMA user, employing the carrier to interference plus noise ratio (CINR) as the switching metric. When the channel quality was favourable, the instantaneous bit rate was increased and conversely, the instantaneous bit rate was reduced when the channel quality dropped. In order to maintain a constant overall bit rate, when a high instantaneous bit rate was employed, the duration of the transmission burst was reduced. Conversely, when the instantaneous bit rate was low, the duration of the burst was extended. This resulted in a decrease in interference power, which translated to an increase in system capacity. Hashimoto, Sampei and Morinaga [78] extended this work also to demonstrate that the proposed system was capable of achieving a higher user capacity with a reduced hand-off margin and lower average transmitter power. In these schemes the conventional RAKE receiver [76] was used for the detection of the data symbols. A variable-rate CDMA scheme – where the transmission rate was modified by varying the channel code rate and, correspondingly, the M-ary modulation constellations – was investigated by Lau and Maric [38]. As the channel code rate was increased, the bit-rate was increased by increasing M correspondingly in the M-ary modulation scheme. Another adaptive system was proposed by Tateesh, Atungsiri and Kondoz [79], where the rates of the speech and channel codecs were varied adaptively [79]. In their adaptive system, the gross transmitted bit rate was kept constant, but the speech codec and channel codec rates were varied according to the channel quality. When the channel quality was low, a lower rate speech codec was used, resulting in increased redundancy and thus a more powerful channel code could be employed. This resulted in an overall coding gain, although the speech quality dropped with decreasing speech rate. A variable rate data transmission scheme was proposed by Okumura and Adachi [80], where the fluctuating transmission rate was mapped to discontinuous transmission, in order to reduce the interference inflicted upon the other users, when there was no transmission. The transmission rate was detected blindly at the receiver with the help of cyclic redundancy check decoding and RAKE receivers were employed for coherent reception, where pilot-symbol-assisted channel estimation was performed.

The information rate can also be varied in accordance with the channel quality, as it will be demonstrated shortly. However, in comparison to conventional power control techniques - which again, may disadvantage other users in an effort to maintain the quality of the links considered - the proposed technique does not disadvantage other users and increases the network capacity [81]. The instantaneous channel quality can be estimated at the receiver and the chosen information rate can then be communicated to the transmitter via explicit signalling in a so-called closed-loop controlled scheme. Conversely, in an open-loop scheme - provided that the downlink and uplink channels exhibit a similar quality - the information rate for the downlink transmission can be chosen according to the channel quality estimate related to the uplink and vice versa. The validity of the above channel reciprocity issues in TDD-CDMA systems have been investigated by Miya *et al.* [82], Kato *et al.* [83] and Jeong *et al.* [84].

In the next section two different methods of varying the information rate are considered, namely the Adaptive Quadrature Amplitude Modulated (AQAM) scheme and the Variable Spreading Factor (VSF) scheme. AQAM is an adaptive-rate technique, whereby the data

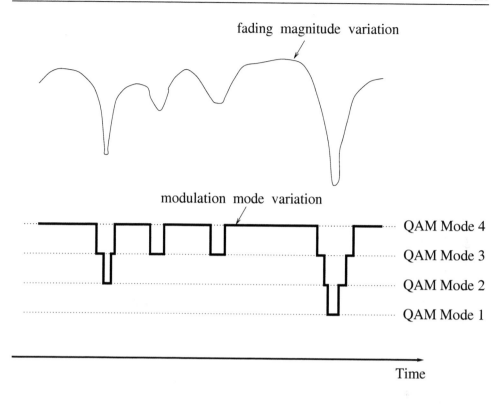

Figure 12.13: Basic concept of a four-mode AQAM transmission in a narrowband channel. The variation of the modulation mode follows the fading variation of the channel over time.

modulation mode is chosen according to some criterion related to the channel quality. On the other hand, in VSF transmission, the information rate is varied by adapting the spreading factor of the CDMA codes used, while keeping the chip rate constant. Further elaborations on these two methods will be given in subsequent sections.

12.5 Burst-by-Burst AQAM/CDMA

12.5.1 Burst-by-Burst AQAM/CDMA Philosophy

Burst-by-burst AQAM [76] is a technique that attempts to increase the average throughput of the system by switching between modulation modes depending on the instantaneous state or quality of the channel. When the channel quality is favourable, a modulation mode having a high number of constellation points is used to transmit as many bits per symbol as possible, in order to increase the throughput. Conversely, when the channel is hostile, the modulation mode is switched to using a low number of constellation points, in order to reduce the error probability and to maintain a certain adjustable target BER. Figure 12.13 shows the stylized quality variation of the fading channel and the switching of the modulation modes in a four-mode AQAM system, where both the BER and the throughput increase, when switching from

Mode 1 to 4.

In order to determine the best choice of modulation mode in terms of the required trade-off between the BER and the throughput, the near-instantaneous quality of the channel has to be estimated. The near-instantaneous channel quality can be estimated at the receiver and the chosen modulation mode is then communicated using explicit signalling to the transmitter in a closed-loop scheme. In other words, the receiver instructs the remote transmitter as to the choice of the transmitter's required modulation mode, in order to satisfy the receiver's prevalent BER and bits per symbol (BPS) throughput trade-off. This closed-loop scheme is depicted schematically in Figure 12.14(a). By contrast, in the open-loop scheme of Figure 12.14(b), the channel quality estimation can be carried out at the transmitter itself based on the co-located receiver's perception of the channel quality during the last received burst, provided that the uplink and downlink channel quality can be considered similar. Then the transmitter explicitly informs the remote receiver as to the modem mode used in the burst and modulation mode detection would then be performed on this basis at the receiver. This scheme performs most successfully in situations, where the channel fading varies slowly in comparison to the burst transmission rate. Channel quality estimation is inherently less accurate in a fast-fading channel and this lag in the quality estimation renders the choice of modulation mode less appropriate for the channel. Another approach to this issue would be for the receiver to detect the modulation mode used blindly [168, 420], as shown in Figure 12.14(c).

12.5.2   Channel Quality Metrics

As stated earlier, a metric corresponding to the near-instantaneous channel quality is required in order to adapt the AQAM modes. Some examples of these metrics include the carrier-to-interference (C/I) ratio of the channel [148], the SNR of the channel [1, 76], the received signal strength indicator's (RSSI) output [145], the mean square error (MSE) at the output of the receiver's channel equalizer and the BER of the system [421]. The most accurate metric is the BER of the system, since this metric corresponds directly to the system's performance, irrespective of the actual source of the channel impairment. However, the BER is dependent on the AQAM mode employed, and cannot be estimated directly for most receivers. For a system that incorporates channel coding, such as turbo coding [152], the so-called log-likelihood ratios (LLR) – which indicate the ratio of the probabilities of the estimated bit taking its two possible values – at the input and output of the turbo decoder can also be employed as the adaptation metric. AQAM systems were first proposed for narrowband channels and the research in this field includes work published by Webb and Steele [1, 76], Sampei, Komaki and Morinaga [148] Goldsmith and Chua [36]; as well as Torrance et al. [31]. Webb et al. [1, 76] employed Star QAM [76] and the channel quality was determined by measuring the received signal strength and the near-instantaneous BER. Sampei et al. [148] switched the modulation modes by estimating the signal to co-channel interference ratio and the expected delay spread of the channel. This work has been extended to wideband channels by Wong et al. [33], where the received signal also suffers from ISI in addition to amplitude and phase distortions due to the fading channel. In wideband AQAM systems the channel SNR, the Carrier-to-Interference (C/I) ratio or RSSI metrics cannot be readily estimated or predicted, amongst other factors due to the multipath nature of the channel or as a result of the so-called 'birth-death' processes associated with the sudden appearance or disappearance of commu-

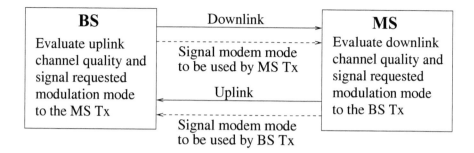

(a) Closed-loop modulation mode signalling from receiver to transmitter

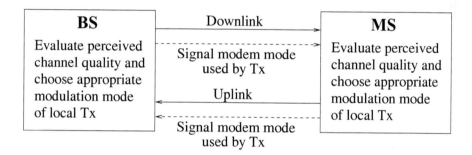

(b) Open-loop modulation mode signalling from transmitter to receiver

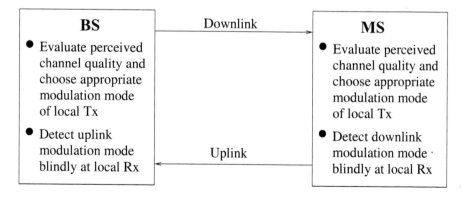

(c) Blind modulation mode detection at the receiver

Figure 12.14: Three different methods of modulation mode signalling for the adaptive schemes, where BS represents the Base Station, MS denotes the Mobile Station, the transmitter is represented by Tx and the receiver is denoted by Rx.

Switching criterion	Modulation mode
$\gamma_o(k) < t_1$	V_1
$t_1 \leq \gamma_o(k) < t_2$	V_2
.	.
.	.
.	.
$t_M \leq \gamma_o(k)$	V_M

Table 12.1: The general rules employed for switching the modulation modes in an AQAM system. The choice of modulation modes are denoted by V_m, where the total number of modulation modes is M and $m = 1, 2, \dots, M$. The modulation modes with the lowest and highest number of constellation points are V_1 and V_M, respectively. The SINR at the output of the multiuser receiver is represented by $\gamma_o(k)$ and the values $(t_1, \cdots, t_M)$ represent the switching thresholds, where $t_1 < t_2 < \cdots < t_M$.

nicating users. Additionally, the above simple metrics do not provide accurate measures of the system performance at the output of the receiver employed, since the effects of the CIR are not considered in the estimation of these metrics. Wong *et al.* [32] proposed a combined adaptive modulation and equalization scheme, where a Kalman-filtered DFE was used to mitigate the effects of ISI inflicted upon the signal. The CIR estimate was used to calculate the Signal to residual ISI plus Noise Ratio (SINR) at the output of the channel equalizer and this SINR value was used to switch the modulation modes. This was a more appropriate switching parameter than the received signal level, since it was a reliable indicator of the performance that could be achieved after equalization.

In AQAM/CDMA systems the SINR – $\gamma_o(k)$, for $k = 1, \dots, K$ – at the output of the multiuser receiver is estimated for all the K users by employing the estimated CIRs and spreading sequences of all the users [234]. After $\gamma_o(k)$ is calculated, the modulation mode is chosen accordingly and communicated to the transmitter. Let us designate the choice of modulation modes by V_m, where the total number of modulation modes is M and $m = 1, 2, \dots, M$. The modulation mode having the lowest number of modulation constellation points is V_1 and the one with the highest is V_M. The rules used to switch the modulation modes are tabulated in Table 12.1, where $\gamma_o(k)$ is the SINR of the k-th user at the output of the multiuser receiver and the values $(t_1, \cdots, t_M)$ represent the switching thresholds, where $t_1 < t_2 < \cdots < t_M$.

The schematic of the transmitter is shown in Figure 12.15. The data bits are mapped to their respective symbols according to the modulation mode chosen. The QAM symbols are then spread with the spreading code assigned to the user, modulated on to the carrier and transmitted. At the output of the multiuser receiver, the data estimates are demodulated according to the modulation mode used for transmission. Following the above brief protrayal of the associated AQAM/CDMA philosophy, let us now embark on the comparative performance study of various AQAM/CDMA schemes in the next section.

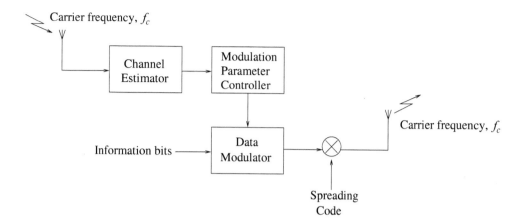

Figure 12.15: The schematic of the transmitter of an AQAM/CDMA system. In this transmitter, a TDD transmission scheme is assumed. Channel estimates are obtained by assuming close correlation between the uplink and downlink channels, which are used to measure the quality of the channel. This quality measure is passed to the modulation parameter controller, which selects the modulation mode according to the thresholds set. The data bits are mapped to QAM symbols according to the chosen modulation mode, spread with the spreading code and modulated on to the carrier.

Parameter	Value
Doppler frequency	80 Hz
Spreading ratio	64
Spreading sequence	Pseudo-random
Chip rate	2.167 MBaud
Burst structure	FMA1 Spread burst 1 [422]
Burst duration	577 μs

Table 12.2: Simulation parameters for the JD, SIC and PIC AQAM-CDMA systems.

12.5.3 Comparison of JD, SIC and PIC CDMA Receivers for AQAM Transmission

In this section a comparative performance study of JD-CDMA, SIC-CDMA and PIC-CDMA systems is provided in the context of AQAM transmissions. The switching criterion employed was the SINR at the output of the respective multiuser receivers. For JD-CDMA, the SINR was estimated by employing the SINR expression given by Klein *et al.* [208], whilst for the SIC and PIC receivers the SINR expression presented by Patel and Holtzman [363] was adopted. For the SIC and PIC receivers, the RAKE receiver was used to provide the initial data estimates required for the subsequent cancellation stages. The rest of the simulation parameters are summarized in Table 12.2. In our investigations the COST 207 [423] channel models developed for the Global System of Mobile Communications known as GSM were employed. Although due to the orthogonal spreading sequences and due to the resultant high

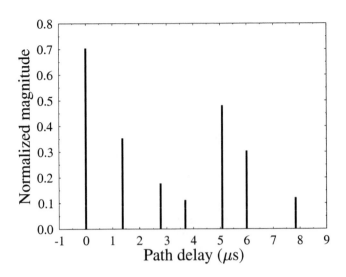

Figure 12.16: Normalized channel impulse response for the COST 207 protect [423] **seven path Bad Urban channel**.

chip-rate the number of resolvable multipath components is higher in our CDMA system than in GSM, nonetheless we used the COST 207 models, since these are widely used in the community. The CIR profile of the COST 207 [423] Bad Urban channel is shown in Figure 12.16.

Figure 12.17 compares the performance of the JD, the SIC receiver and the PIC receiver for a twin-mode, eight-user AQAM-CDMA system, switching between BPSK and 4-QAM. Here, the BER performance of all three receivers was kept as similar as possible and the performance comparison was evaluated on the basis of their BPS throughput. The BPS throughput of the JD was the highest, where approximately 1.9 BPS was achieved at $E_s/N_0 = 14$ dB. The PIC receiver outperformed the SIC receiver in BPS throughput terms, where the BPS throughput of the PIC receiver was approximately 1.55 BPS at $E_s/N_0 = 14$ dB compared to the approximately 1.02 BPS achieved by the SIC receiver. The two IC receivers suffered from MAI and they were unable to match the performance of the JD receiver. The PIC receiver outperformed the SIC receiver, since the received signal powers of all the users were similar on average and hence - as expected - the PIC receiver achieved a higher degree of interference cancellation than the SIC receiver.

The previous performance comparisons between the three multiuser receivers were then extended to triple-mode AQAM systems – switching between BPSK, 4-QAM and 16-QAM – that supported $K = 8$ users, as portrayed in Figure 12.18. For these systems, we can observe from the results that both the PIC and SIC receivers were unable to match the BPS performance of the JD receiver, when the multiuser receivers achieved similar BERs. This was because the increase in the number of users aggravated the MAI, thus degrading the

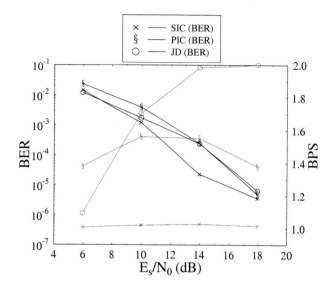

(a) BER comparisons are in bold, BPS comparisons are in grey.

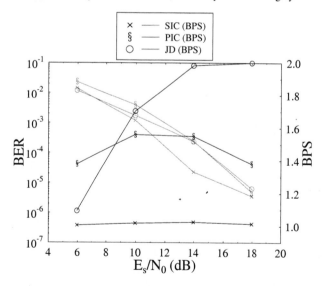

(b) BPS comparisons are in bold, BER comparisons are in grey.

Figure 12.17: Performance comparison of the JD, SIC and PIC CDMA receivers for **twin-mode** (BPSK, 4-QAM) AQAM transmission and $K = 8$ users over the **Bad Urban channel** of Figure 12.16. The rest of the simulation parameters are tabulated in Table 12.2.

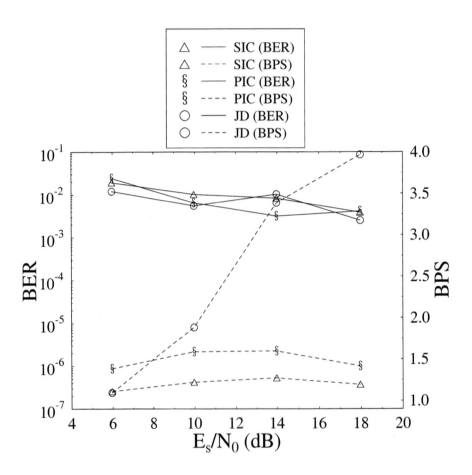

Figure 12.18: BER and BPS performance comparisons for **triple-mode** JD, SIC and PIC AQAM-CDMA schemes supporting $K = 8$ users over the **Bad Urban channel** of Figure 12.16. The modulation mode was chosen to be BPSK, 4-QAM or 16-QAM. The rest of the simulation parameters are tabulated in Table 12.2.

Switching criterion	Spreading code
$\gamma_o(k) < t_1$	$\mathbf{c}_1^{(k)}, \ Q_h$
$t_1 \leq \gamma_o(k) < t_2$	$\mathbf{c}_2^{(k)}, \ 2^{-1}Q_h$
.	.
.	.
.	.
$t_M \leq \gamma_o(k)$	$\mathbf{c}_M^{(k)}, \ 2^{-(M-1)}Q_h$

Table 12.3: The general switching rules for a VSF/CDMA system, where $\gamma_o(k)$ is the SINR of the k-th user at the output of the multiuser receiver. The values of $t_1, \cdots, t_M$ represent the switching thresholds for the different modulation modes, where $t_1 < t_2 < \cdots < t_M$.

ability of the RAKE receivers to provide reliable data estimates for interference cancellation. Here again, the PIC receiver outperformed the SIC receiver in BPS throughput terms.

Let us now provide performance comparisons for the JD, SIC and PIC receivers in the context of VSF-CDMA schemes, rather than AQAM-CDMA.

12.5.4 VSF-CDMA

Multi-rate transmission systems using spreading sequences having different processing gains have been proposed in the literature amongst others by Adachi, Sawahashi and Okawa [424]; Ottosson and Svensson [65]; Ramakrishna and Holtzman [67]; Saquib and Yates [71]; as well as by Johansson and Svensson [72]. In the FRAMES FMA2 Wideband CDMA proposal for UMTS [425], different bit rates are accommodated by supporting variable spreading factor (VSF) [424] and multi-code based operations. In this section, we discuss the employment of VSF codes in adaptive-rate CDMA systems, where the chip rate of the CDMA users is kept constant throughout the transmission, while the bit rate is varied by using spreading codes exhibiting different spreading factors over the course of transmission. For example, by keeping the chip rate constant, the number of bits transmitted in the same period of time upon using a spreading code of length $Q = 16$ is twice the number of bits transmitted upon using a spreading code of length $Q = 32$.

In the VSF-based scheme the rate adaptation is dependent on the channel quality. Generally, when the channel quality is favourable, a code with a low spreading factor is used, in order to increase the throughput. Conversely, when the channel is hostile, a code with a high spreading factor is employed, in order to minimize the number of errors inflicted and to maintain a given target BER performance. Each user in the VSF-CDMA system is assigned M number of legitimate spreading codes having different lengths, $Q_1, Q_2, \ldots, Q_M$. Analogously to the AQAM/CDMA system discussed in Section 12.5, the SINR at the output of the multiuser receiver is estimated and used as the metric for choosing the spreading code to be used for transmission. Generally, when the channel quality is favourable, a code with a low spreading factor will be used, in order to increase the throughput. Conversely, when the channel is hostile, a code with a high spreading factor will be used, in order to minimize the number of errors inflicted. In order to have a system that can accommodate a large number of spreading codes having different SFs, the simplest choice of spreading codes would

Parameter	Value
Doppler frequency	80 Hz
Modulation mode	4-QAM
Spreading sequence	Pseudo-random
Chip rate	2.167 MBaud
Burst structure	FMA1 Spread burst 1 [422]
Burst duration	577 μs

Table 12.4: Simulation parameters for the JD, SIC and PIC VSF-CDMA systems.

be those having SFs of $Q = 2^r$, where $r = 1, 2, 3, \cdots$. Let us denote the set of spreading codes assigned to the k-th user by $\{c_1^{(k)}, c_2^{(k)}, \cdots, c_M^{(k)}\}$, where $c_1^{(k)}$ is the spreading code having the highest spreading factor and the code $c_M^{(k)}$ has the lowest spreading factor. If the highest spreading factor is denoted by Q_h, then the rules used to choose the spreading code are tabulated in Table 12.3.

12.5.5   Comparison of JD, SIC and PIC CDMA Receivers for VSF Transmission

In this section, the above JD, PIC and SIC receivers are compared in the context of adaptive VSF/CDMA schemes. The spreading factor used was varied adaptively, opting for $Q_1 = 64$, $Q_2 = 32$ or $Q_3 = 16$, while the rest of the simulation parameters employed are summarized in Table 12.4.

Figure 12.19 portrays the associated BER and throughput comparisons for the adaptive VSF PIC-, SIC- and JD-CDMA schemes using 4-QAM. The minimum and maximum throughput values were approximately 68 kbits/s and 271 kbits/s, respectively, spanning a bit rate range of a factor of four according to $Q_1/Q_3 = 4$. The throughput performance values were normalized to the minimum throughput of 68 kbits/s, thus giving minimum and maximum normalized throughput values of 1 and 4, respectively. We compared the normalized throughput of the JD-CDMA system to that of the PIC and SIC based systems, under the condition of similar BER performances. At low E_s/N_0 values the PIC and SIC receivers outperformed the JD receiver in throughput terms, because the MAI was relatively low in a two-user scenario and the additive noise was the main impairment at low E_s/N_0 values. However, as E_s/N_0 increased, the JD gradually outperformed the two interference cancellation based receivers in both BER and BPS throughput performance terms, achieving a normalized throughput of 3.5 compared to 2.7 and 2 for the PIC and SIC receivers, respectively, at $E_s/N_0 = 16$ dB.

The above triple-mode VSF schemes using 4-QAM were then extended to systems supporting $K = 8$ users. From the results presented in Figure 12.20, it can be seen that the JD consistently outperformed the PIC and SIC receivers in both BER and throughput performance terms. The VSF/IC-CDMA systems were unable to accommodate the variability in the channel conditions and hence their normalized throughput performance remained in the range of 1 to 1.5, compared to the JD, which was capable of achieving a normalized throughput of 2.8 at $E_s/N_0 = 14$ dB.

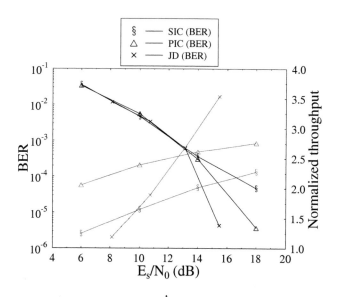

(a) BER comparisons are in bold, throughput comparisons are in grey.

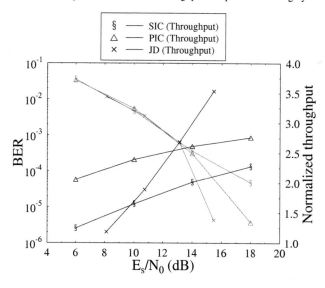

(b) Throughput comparisons are in bold, BER comparisons are in grey.

Figure 12.19: BER and throughput performance comparisons for triple-mode JD, SIC and PIC VSF-CDMA schemes supporting $K=2$ users over the Bad Urban channel of Figure 12.16. The spreading factor was adaptively varied using $Q_1 = 64$, $Q_2 = 32$ or $Q_3 = 16$; and 4-QAM was used as the data modulation mode. The rest of the simulation parameters are tabulated in Table 12.4.

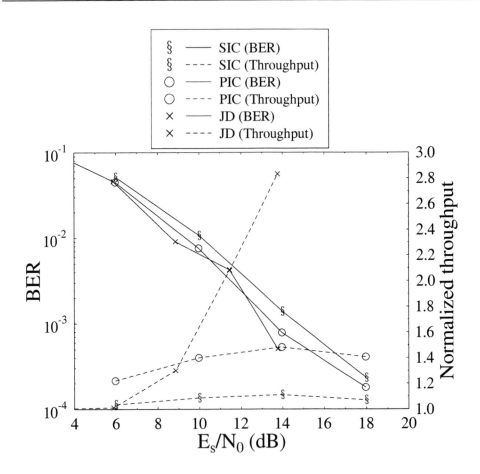

Figure 12.20: BER and throughput performance comparisons for triple-mode JD, SIC and PIC VSF-CDMA schemes supporting **K=8** users over the Bad Urban channel of Figure 12.16. The modulation mode was chosen to be 4-QAM. The rest of the simulation parameters are tabulated in Table 12.4.

	JD		SIC		PIC	
E_s/N_0	BPS	Cmplx	BPS	Cmplx	BPS	Cmplx
10	1.7	2641	1.02	1869	1.55	4605
14	1.98	2641	1.03	1869	1.38	4605

Table 12.5: Performance summary of the eight-user, **twin-mode AQAM-CDMA** results for the JD, SIC and PIC receivers, using a spreading factor of $Q = 64$ and transmitting over the seven-path Bad Urban channel of Figure 12.16. The E_s/N_0 values in the table are the values at which the AQAM-CDMA systems achieved the target BER of 1% or less. The complexity (Cmplx) values are in terms of the number of additions plus multiplications required per detected data symbol.

	JD		SIC		PIC	
E_s/N_0	BPS	Cmplx	BPS	Cmplx	BPS	Cmplx
10	1.88	2641	1.02	1869	1.99	4605
14	3.39	2641	1.03	1869	2	4605

Table 12.6: Performance summary of the eight-user, **triple-mode AQAM-CDMA** results for the JD, SIC and PIC receivers, using a spreading factor of $Q = 64$ and transmitting over the seven-path Bad Urban channel of Figure 12.16. The E_s/N_0 values in the table are the values at which the AQAM systems achieved the target BER of 1% or less. The complexity (Cmplx) values are in terms of the number of additions plus multiplications operations required per detected data symbol.

	JD		SIC		PIC	
E_s/N_0	Tp.	Cmplx	Tp.	Cmplx	Tp.	Cmplx
10	2.07	18298	1.08	1869	1.4	4605
14	2.83	18298	1.11	1869	1.48	4605

Table 12.7: Performance summary of **triple-mode VSF-CDMA** results for the JD, SIC and PIC receivers, where 4-QAM was employed as the modulation mode and transmission was conducted over the Bad Urban channel of Figure 12.16. The number of users in the system was $K = 8$. The E_s/N_0 in the table are the values at which the VSF systems achieved the target BER of 1% or less. The notation "Tp." denotes the throughput values normalized to the throughput of the fixed-rate scheme employing $Q = 64$ and 4-QAM. The complexity (Cmplx) values indicated are valid for the modem mode that incurred the highest complexity for each receiver.

12.6 Review and Discussion

The recent history of smart CDMA MUDs has been reviewed and the most promising schemes have been comparatively studied, in order to assist in the design of third- and fourth-generation receivers. All future transceivers are likely to become BbB-adaptive, in order to be able to accommodate the associated channel quality fluctuations without disadvantageously affecting the system's capacity. Hence the methods reviewed in this chapter are advantageous, since they often assist in avoiding powering up, which may inflict increased levels of co-channel interference and power consumption. Furthermore, the techniques characterized in this chapter support an increased throughput within a given bandwidth and will contribute towards reducing the constantly increasing demand for more bandwidth.

Both SIC and PIC receivers were investigated in the context of AQAM/CDMA schemes, which were outperformed by the JD based twin-mode AQAM/CDMA schemes for a spreading factor of $Q = 64$ and $K = 8$ users. Both IC receivers were unable to provide good performances in the triple-mode AQAM/CDMA arrangement, since the BER curves exhibited error floors. In the VSF/CDMA systems the employment of variable spreading factors of $Q = 32$ and $Q = 16$ enabled the PIC and SIC receivers to provide a reasonable BER and throughput performance, which was nonetheless inferior to that of the JD. When the number of users in the system was increased, the PIC and SIC receivers were unable to exploit the variability in channel conditions in order to provide a higher information throughput, as opposed to the JD scheme, which showed performance gains in both the adaptive-rate AQAM and VSF CDMA schemes. However, the complexity of the IC receivers increased only linearly with the number of CDMA users, K, compared to the joint detector, which exhibited a complexity proportional to $O(K^3)$. Tables 12.5, 12.6 and 12.7 summarize our performance comparisons for all three multiuser detectors in terms of E_s/N_0 required to achieve the target BER of 1% or less, the normalized throughput performance and the associated complexity in terms of the number of operations required per detected symbol. Both the third and future fourth generation standard work may benefit from this study. In conjunction with adaptive beam-steering [81] the order of the receiver performance may change.

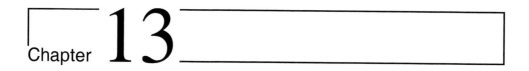

Chapter **13**

Adaptive Multicarrier Modulation

T. Keller and L. Hanzo[1]

13.1   Introduction

High data rate communications are limited not only by noise, but —- especially with increasing symbol rates — often more significantly by the Inter Symbol Interference (ISI) due to the memory of the dispersive wireless communications channel [317]. Explicitly, this channel memory is caused by the dispersive channel impulse response (CIR) due to the different–length propagation paths between the transmitting and the receiving antennae. This dispersion effect could theoretically be measured by transmitting an infinitely short impulse and "receiving" the CIR itself. On this basis, several measures of the effective duration of the impulse response can be calculated, one being the delay spread. The multipath propagation of the channel manifests itself by different echos of possibly different transmitted symbols overlapping at the receiver, which leads to error rate degradation.

   This effect occurs not only in wireless communications, but also over all types of electrical and optical wave-guides, although for these media the relative time differences are comparatively small, mostly due to multi–mode transmission or incorrect electrical or optical termination at interfaces.

   In wireless communications systems the duration and the shape of the CIR depend heavily on the propagation environment of the communications system in question. While indoor wireless networks typically exhibit only short relative delays, outdoor networks, like the Global System of Mobile communications (GSM) [13] can face delay spreads in the order of $15\mu s$.

   As a general rule, the effects of ISI on the transmission error statistics are negligible as long as the delay spread is significantly shorter than the duration of one transmitted symbol. This implies that the symbol rate of communications systems is practically limited by the

[1]This chapter is based on Keller and Hanzo: Adaptive Multicarrier modulation: A convenient framework for time-frequency processing in wireless communications, Proceedings of the IEEE, May 2000, Vol. 88, No. 5, pp 611-642 ©IEEE, 2000.

channel's memory. For higher symbol rates, there is typically significant deterioration of the system's error rate performance.

If symbol rates exceeding this limit are to be transmitted over the channel, mechanisms must be implemented in order to combat the effects of ISI. Channel equalization techniques [317] can be used to suppress the echoes caused by the channel. In order to perform this operation, the CIR must be estimated. Significant research efforts were invested into the development of such channel equalisers, and most wireless systems in operation use equalisers to combat ISI.

There is, however, an alternative approach towards transmitting data over a multipath channel. Instead of attempting to cancel the effects of the channel's echos, Orthogonal Frequency Division Multiplexing (OFDM) [317] modems employ a set of *subcarriers* in order to transmit information symbols in parallel - in so-called *subchannels* - over the channel. Since the system's data throughput is the sum of all the parallel channels' throughputs, the data rate per subchannel is only a fraction of the data rate of a conventional single–carrier system having the same throughput. This allows us to design a system supporting high data rates, while maintaining symbol durations much longer than the channel's memory, thus circumventing the need for channel equalization.

The outline of the chapter is as follows. Section 2 commences with a historical perspective on OFDM, highlighting the associated research issues with reference to the literature. Based on the above overview of the state-of-the-art, Section 3 characterizes the performance of OFDM over dispersive, wideband channels, while Section 4 quantifies the effects of synchronization errors on OFDM, leading on to Section 5, which highlights the range of synchronization solutions proposed by the research community at large. Again, commencing with a literature survey, the key topic of adaptive bit allocation over highly frequency-selective wireless channels is the subject of Section 6, while Section 7 is dedicated to the closely related subject of pre–equalization and channel coding. Our discourse is concluded in Section 8 with a wide-ranging throughput comparison of the schemes discussed in the chapter under the unified constraint of a fixed target bit error rate of 10^{-4}.

13.2   Orthogonal Frequency Division Multiplexing

13.2.1   Historical Perspective

Frequency Division Multiplexing (FDM) or multi–tone systems have been employed in military applications since the 1960s, for example by Bello [426], Zimmerman [427], Powers and Zimmerman [428], and others. Orthogonal Frequency Division Multiplexing (OFDM), which employs multiple carriers overlapping in the frequency domain, was pioneered by Chang [429,430]. Saltzberg [431] studied a multi–carrier system employing orthogonal time–staggered quadrature amplitude modulation (O-QAM) on the carriers.

The use of the discrete Fourier transform (DFT) to replace the banks of sinusoidal generators and the demodulators - suggested by Weinstein and Ebert [432] in 1971 - significantly reduces the implementation complexity of OFDM modems. This substantial implementational complexity reduction was attributable to the simple realization that the DFT uses a set of harmonically related sinusoidal and cosinusoidal basis functions, whose frequency is an integer multiple of the lowest non-zero frequency of the set, which is referred to as the basis frequency. These harmonically related frequencies can hence be used as the set of carriers required by the OFDM system. For a formal proof of this the interested reader is referred to [317].

In 1980, Hirosaki [433] suggested an equalization algorithm in order to suppress both inter–symbol and inter–subcarrier interference caused by the CIR or timing– and frequency–errors. Simplified OFDM modem implementations were studied by Peled [434] in 1980, while Hirosaki [435] introduced the DFT–based implementation of Saltzberg's O-QAM OFDM system. Kolb [436], Schüßler [437], Preuss [438] and Rückriem [439] conducted further research into the application of OFDM. Kalet [62] introduced the concept of allocating more bits to subcarriers, which were for example near the centre of the transmission frequency band and hence were less attenuated than those near the edge of the transmission band. However, since Kalet's discussions were cast in the context of slowly varying channels, the concept of near-instantaneously adaptive transmission was not introduced at this early stage of OFDM research. This concept was often referred to as 'water-filling' in the frequency domain. A few years later Cimini [440] provided early seminal results on the performance of OFDM modems in mobile communications channels.

More recent advances in OFDM transmission are presented in the impressive state-of-the-art collection of works edited by Fazel and Fettweis [441], including research by Fettweis *et al.*, Rohling *et al.*, Vandendorp, Huber *et al.*, Lindner *et al.*, Kammeyer *et al.*, Meyr *et al.* [442,443], but the impressive individual contributions are too numerous to mention.

While OFDM transmissions over mobile communications channels can alleviate the problem of multi–path propagation, recent research efforts have focussed on solving a set of inherent difficulties regarding OFDM, namely on reducing the associated peak–to–mean–power ratio fluctuation, on time– and frequency synchronization and on mitigating the effects of co-channel interference sensitivity in multi-user environments. These issues are addressed below in more depth.

13.2.1.1 Peak–to–Mean Power Ratio

It is plausible that the OFDM signal - which is the superposition of a high number of modulated subchannel signals - may exhibit a high instantaneous signal peak with respect to the average signal level. Furthermore, large signal amplitude swings are encountered, when the time-domain signal traverses from a low instantaneous power waveform to a high-power waveform. Similarly, the peak-to-mean power envelope fluctuates dramatically, when traversing the origin upon switching from one phasor to another. Both of these events may results in a high out-of-band (OOB) harmonic distortion power, unless the transmitter's power amplifier exhibits an extremely high linearity [317] across the entire signal dynamic range. This potentially contaminates the adjacent channels with adjacent channel interference. Practical amplifiers exhibit a finite amplitude range, in which they can be considered near-linear. In order to prevent severe clipping of the high OFDM signal peaks - which is the main source of OOB emissions - the power amplifier must not be driven into saturation and hence they are typically operated with a certain so-called backoff, creating a 'head-room' for the signal peaks, which reduces the risk of amplifier saturation and OOB emmission. Two different families of solutions have been suggested in the literature, in order to mitigate these problems, either reducing the peak–to–mean power ratio, or improving the amplification stage of the transmitter.

More explicitly, Shepherd [444], Jones [445], and Wulich [446] suggested different coding techniques which aim to minimise the peak power of the OFDM signal. According to their approach different data encoding or mapping schemes are employed before modulation. A simple example is concatenating a number of dummy bits to a string of information bits with the sole aim of mitigating the so-called Crest Factor (CF) or peak-to-mean signal envelope ratio. In a further attempt to mitigate the CF problem Müller [447], Pauli [448], May [449] and Wulich [450] suggested different algorithms for post–processing the time–domain OFDM signal prior to amplification, while Schmidt and Kammeyer [451] employed adaptive subcarrier allocation in order to reduce the Crest factor. Dinis and Gusmão [452–454] researched the use of two–branch amplifiers, while the so-called clustered OFDM technique introduced by Daneshrad, Cimini and Carloni [455] operates with a set of parallel partial FFT processors with associated transmitting chains. More explicitly, clustered OFDM allows a number of users to share a given bandwidth amongst a number of users on a demand basis, potentially supporting a peak data rate identical to that of a single-user OFDM system. The bandwidth assigned to a particular user is typically constituted by a number of subcarrier clusters, which are spread sufficiently far apart from each other, in order to provide frequency diversity. OFDM systems with increased robustness to nonlinear distortion have been proposed for example by Okada, Nishijima and Komaki [456] as well as by Dinis and Gusmão [457].

13.2.1.2   Synchronization

Time and frequency synchronization between the transmitter and receiver are of crucial importance in terms of the performance of an OFDM link [458–462]. A wide variety of techniques has been proposed for estimating and correcting both timing and carrier-frequency offsets at the OFDM receiver. Rough timing and frequency acquisition algorithms relying on known pilot symbols or pilot tones embedded into the OFDM symbols have been suggested by Claßen [442], Warner [463], Sari [464], Moose [465], as well as Brüninghaus and Rohling [466]. Fine frequency and timing tracking algorithms exploiting the OFDM signal's cyclic extension were published by Moose [465], Daffara [467] and Sandell [468].

13.2.1.3   OFDM / CDMA

Combining OFDM transmissions with Code Division Multiple Access (CDMA) allows us to exploit the wideband channel's inherent frequency diversity by spreading each symbol across multiple subcarriers. This technique has been pioneered by Yee, Linnartz and Fettweis [206], by Chouly, Brajal and Jourdan [469], as well as by Fettweis, Bahai and Anvari [470]. Fazel and Papke [207] investigated convolutional coding in conjunction with OFDM/CDMA. Prasad and Hara [471] compared various methods of combining the two techniques, identifying three different structures, namely multi–carrier CDMA (MC-CDMA), multi–carrier direct–sequence CDMA (MC-DS-CDMA) and multi–tone CDMA (MT-CDMA). Like non–spread OFDM transmission, OFDM/CDMA methods suffer from high peak–to–mean power ratios, which are dependent on the frequency–domain spreading scheme, as has been investigated by Choi, Kuan and Hanzo [220].

13.2.1.4   Adaptive Antennas

Combining adaptive antenna techniques with OFDM transmissions was shown to be advantageous in suppressing co–channel interference in cellular communications systems. Li, Cimini and Sollenberger [472–475], Kim, Choi and Cho [476] as well as Münster *et al.* [477] have investigated algorithms for multi–user channel estimation and interference suppression. The employment of adaptive antennas is always beneficial in terms of mitigating the effects of multi–user interference, since with the aid of beam–steering it becomes possible to focus the receiver's antenna beam on the served user, while attenuating the co–channel interferers. This is of particularly high importance in conjunction with OFDM, which exhibits a high sensitivity against co–channel interference, potentially hampering its application in co-channel interference limited multi–user scenarios.

13.2.1.5   OFDM Applications

Due to their implementational complexity, OFDM applications have been scarce until quite recently. Recently, however, OFDM has been adopted as the new European digital audio broadcasting (DAB) standard [478–482] as well as for Terrestrial Digital Video Broadcasting (DVB-T) system [464, 483]. The hostile propagation environment of the terrestrial system requires concatenated Reed-Solomon [13] (RS) and rate compatible punctured convolutional coding [13] (RCPCC) combined with OFDM. These schemes are capable of delivering high-definition video at bitrates of up to 20 Mbits/s in slowly time-varying broadcast-mode distributive wireless scenarios. Recently a range of DVB system performance studies were also published in the literature [484–487], portraying the DVB-T system.

For fixed–wire applications, OFDM is employed in the Asynchronous Digital Subscriber Line (ADSL) and High bit-rate Digital Subscriber Line (HDSL) systems [488–491] and it has also been suggested for power–line communications systems [492, 493] due to its resilience to time–dispersive channels and narrow–band interferers.

More recently, OFDM applications were studied within the European 4th Framework Advanced Communications Technologies and Services (ACTS) programme [494]. Specifically, the Pan-European Median project investigated a 155 Mbit/s (Mbps) Wireless Asynchronous Transfer Mode (WATM) network [495–498], while the Magic WAND group [499, 500] developed a wireless Local Area Network (LAN). Hallmann and Rohling [501] presented a range of different OFDM–based systems that were applicable to the European Telecommunication Standardization Institute's (ETSI) third-generation air interface [502].

Lastly, the recently standardized High PERformance Local Area Network standard known as HIPERLAN/2 was designed for providing convenient wireless networking in indoor environments and also invoked OFDM. The wireless provision of high bit rate services appears a more attractive alternative than installing wireline based networks. The HIPERLAN standard specifies the air interface and the physical layer, in order to ensure the compatibility of different manufacturers' equipment, while refraining from standardising the higher layer functions of the system. The HIPERLAN standard constitutes a member of the Broadband Radio Access Networks family often referred to as BRAN [503]- [504]. The BRAN family of recommendations is constituted by the HIPERLAN/1 and /2 systems operating in the 5GHz frequency band. Further members of the family include the so-called HIPERACCESS standard contrived for fixed wireless broadband Point-to-multipoint access and the HIPERLINK recommendation designed for wireless broadband communications in the 17 GHz frequency band. The system's parameters are summarised in Table 13.1.

No. of subcarriers (SC)	64
No. of active SCs	52, where 48 are used for data and 4 for pilots
Channel spacing	20 MHz
Sampling rate	20 Msample/s
Guard interval	800 ns or 16 time-domain samples
SC modulation	BPSK, QPSK, 16QAM and 64QAM
Demodulation	coherent
Channel coding	rate 1/2 convolutional coding with optional puncturing to rate 9/16 and 3/4
Data rates	6, 9, 12, 18, 27, 36, 54 Mbps
Interleaving	Block interleaving over one OFDM symbol

Table 13.1: HIPERLAN/2 physical layer parameters [505].

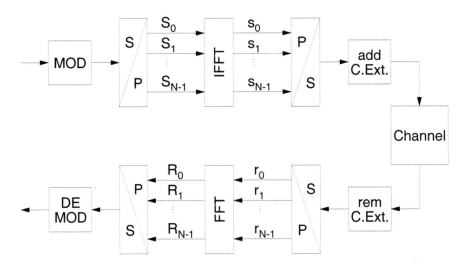

Figure 13.1: Schematic of N–subcarrier OFDM transmission system.

13.2.2   OFDM Modem Structure

The principle of any Frequency Division Multiplexing (FDM) system is to split the information to be transmitted into N parallel streams, each of which modulates a carrier using an arbitrary modulation technique. The frequency spacing between adjacent carriers is Δf, resulting in a total signal bandwidth of $N \cdot \Delta f$. The resulting N modulated and multiplexed signals are transmitted over the channel, and at the receiver N parallel receiver branches recover the information. A multiplexer then recombines the N parallel information streams

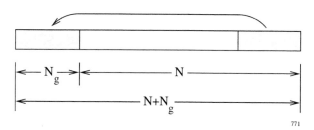

771

Figure 13.2: Stylized plot of N–subcarrier OFDM time domain signal with a cyclic extension of N_g samples.

into a high–rate serial stream.

The conceptually simplest implementation of an FDM modem is to employ N independent transmitter/receiver pairs, which is often prohibitive in terms of complexity and cost [435]. Weinstein [432] suggested the digital implementation of FDM subcarrier modulators/demodulators based on the Discrete Fourier Transform (DFT).

The DFT and its more efficient implementation, the Fast Fourier Transform (FFT) are employed for the base–band OFDM modulation/demodulation process, as it can be seen in the schematic shown in Figure 13.1. The associated harmonically related frequencies can hence be used as the set of subchannel carriers required by the OFDM system. However, instead of carrying out the modulation / demodulation on a subcarrier by subcarrier basis, as in Hirosaki's early proposal for example [433], all OFDM subchannels are modulated / demodulated in a single inverse DFT (IDFT) / DFT step. For more detailed explanations and signal waveforms the interested reader is referred to [317].

The serial data stream is mapped to data symbols with a symbol rate of $1/T_s$, employing a general phase and amplitude modulation scheme, and the resulting symbol stream is demultiplexed into a vector of N data symbols S_0 to S_{N-1}. The parallel data symbol rate is $1/N \cdot T_s$, i.e. the parallel symbol duration is N times longer than the serial symbol duration T_s. Hence the effects of the dispersive channel - which are imposed on the transmitted signal as the convolution of the signal with the CIR - become less damaging, affecting only a fraction of the extended signalling pulse duration. The inverse FFT (IFFT) of the data symbol vector is computed and the coefficients s_0 to s_{N-1} constitute an OFDM symbol, as seen in the figure. Since the harmonically related and modulated individual OFDM subcarriers can be conveniently visualised as the spectrum of the signal to be transmitted, it is the IFFT - rather than the FFT - which is invoked, in order to transform the signal's spectrum to the time-domain for transmission over the channel. The associated modulated signal samples s_n are the time–domain samples of the OFDM symbol and are transmitted sequentially over the channel at a symbol rate of $1/T_s$. At the receiver, a spectral decomposition of the received time–domain samples r_n is computed employing an N-tap FFT, and the recovered data symbols R_n are restored in serial order and demultiplexed, as seen in Figure 13.1.

The underlying assumption in the context of OFDM upon invoking the IFFT for modulation is that although N frequency–domain samples produce N time-domain samples, both signals are assumed to be periodically repeated over an infinite time–domain and frequency–domain interval, respectively. In practice, however, it is sufficient to repeat the time-domain signal periodically for the duration of the channel's memory, i.e. for a duration that is comparable to the length of the CIR. This is namely the time interval required for the channel's transient response to die down after exciting the channel with a time-domain OFDM symbol. Once the channel's transient response time has elapsed, its output is constituted by its steady-state response constituted by the received time-domain OFDM symbol. In order to ensure that the received time-domain OFDM symbol is demodulated from the channel's steady-state - rather than from its transient - response, each time–domain OFDM symbol is extended by the so–called cyclic extension (C. Ext. in Figure 13.1) or guard interval of N_g samples duration, in order to overcome the inter–OFDM symbol interference due to the channel's memory. The signal samples received during the guard interval are discarded at the receiver and the N-sample received time–domain OFDM symbol is deemed to follow the guard interval of N_g samples duration. The demodulated OFDM symbol is then generated from the remaining N samples upon invoking the IFFT. We note, however that since the transmitted time–domain signal was windowed to the finite duration of $N + N_g$ samples, the corresponding transmitted frequency–domain signal is convolved with the sinc–shaped frequency-domain transfer function of the rectangular time–domain window function. As a results of this frequency–domain convolution, the originally pure line–spectrum of the IFFT's output generates a sinc–shaped sub–channel spectrum centred on each OFDM sub–carrier.

The samples of the cyclic extension are copied from the end of the time–domain OFDM symbol, generating the transmitted time domain signal $(s_{N-N_g-1}, \cdots, s_{N-1}, s_0, \cdots, s_{N-1})$ depicted in Figure 13.2. At the receiver, the samples of the cyclic extension are discarded. Clearly, the need for a cyclic extension in time dispersive environments reduces the efficiency of OFDM transmissions by a factor of $N/(N + N_g)$. Since the duration N_g of the necessary cyclic extension depends only on the channel's memory, OFDM transmissions employing a high number of carriers N are desirable for efficient operation. Typically a guard interval length of not more than 10% of the OFDM symbol's duration is employed. Again, for further details concerning the operation of OFDM modems please refer to [205, 317, 506].

13.2.3 Modulation in the Frequency Domain

Modulation of the OFDM subcarriers is analogous to the modulation in conventional serial systems. The modulation schemes of the subcarriers are generally Quadrature Amplitude Modulation (QAM) or Phase Shift Keying (PSK) [317] in conjunction with both coherent and non–coherent detection. Differentially coded Star-QAM (DSQAM) [317] can also be employed. If coherently detected modulation schemes are employed, then the reference phase of the OFDM symbol must be known, which can be acquired with the aid of pilot tones [19] embedded in the spectrum of the OFDM symbol, as will be discussed in Section 13.3. For differential detection the knowledge of the absolute subcarrier phase is not necessary, and differentially coded signalling can be invoked either between neighbouring subcarriers or between the same subcarriers of consecutive OFDM symbols.

13.3 OFDM Transmission over Frequency Selective Channels

13.3.1 System Parameters

Based on the above advances in the field of OFDM modems, below we will characterize the expected performance of OFDM modems using the example of high-rate Wireless Asynchronous Transfer Mode (WATM) systems [495–497, 499, 500]. Specifically, the system parameters used in characterizing the performance of various OFDM algorithms closely followed the specifications of the Advanced Communications Technologies and Services (ACTS) Median system [495–498], which is a proposed wireless extension to fixed–wire ATM–type networks. In the Median system, the OFDM FFT length is 512, and each symbol is padded with a cyclic prefix of length 64. The sampling rate of the Median system is 225 Msamples/s, and the carrier frequency is 60 GHz. The uncoded target data rate of the Median system is 155Mbps.

OFDM modems were originally conceived in order to transmit data reliably in time–dispersive or frequency–selective channels without the need for a complex time–domain channel equaliser. In this chapter the techniques employed for the transmission of QAM OFDM signals over a time–dispersive channel are discussed and channel estimation methods are investigated [317].

13.3.2 The Channel Model

The channel model assumed in this chapter is that of a Finite Impulse Response (FIR) filter with time–varying tap values. Every propagation path i is characterized by a fixed delay τ_i and a time–varying amplitude $A_i(t) = a_i \cdot g_i(t)$, which is the product of a complex amplitude a_i and a Rayleigh fading process $g_i(t)$. The Rayleigh processes g_i are independent from each other, but they all exhibit the same normalized Doppler frequency f_d'.

The ensemble of the p propagation paths constitutes the impulse response

$$h(t, \tau) = \sum_{i=1}^{p} A_i(t) \cdot \delta(\tau - \tau_i) = \sum_{i=1}^{p} a_i \cdot g_i(t) \cdot \delta(\tau - \tau_i), \tag{13.1}$$

which is convolved with the transmitted signal.

The channel model employed in this chapter is the worst-case operating environment for an indoor wireless ATM network similar to that of the ACTS Median system [495–498]. We assumed a vehicular velocity of about 50 km/h or 13.9 m/s, resulting in a normalized Doppler frequency of $f_d' = 1.235 \cdot 10^{-5}$. We note here that the normalized Doppler frequency in this chapter was related to the OFDM symbol duration, rather than to the time–domain signal's sample duration. This relationship will be formally defined in Equation 13.5, hence suffice to say here that the normalized Doppler frequency in this sense is typically 512 times lower, than the conventional normalized Doppler frequency due to having 512 samples per PFDM symbol. The significance of this will become more clear in the context of adaptive OFDM schemes, where the predictability of the channel's frequency-domain transfer function between consecutive OFDM symbols depends explicitly on the duration of the symbol.

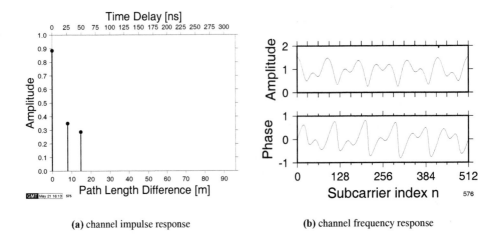

(a) channel impulse response   (b) channel frequency response

Figure 13.3: WATM channel: (a) impulse response (b) frequency domain channel transfer function $H(n)$ experienced by a specific OFDM symbol.

The vehicular velocity of 50 km/h constitutes the highest possible speed of for example an indoor fork-truck in a warehouse environment. Again, this worst-case speed was employed in order to provide performance results characterizing the worst possible scenario in the context of adaptive OFDM transceivers, which are sensitive to rapid CIR or transfer function variations. This issue will become more explicit during our further discourse. The impulse response was determined by simple ray-tracing in a warehouse–type environment, and is shown in Figure 13.3(a), where each CIR tap corresponds to a specifically delayed propagation path. We note that this indoor CIR is not particularly dispersive, however, at the 155 Mbps WATM rate, the dispersion corresponds to 11 sample periods, which would require a high-performance channel equaliser in a serial modem.

The last CIR path arrives at a delay of 48.9 ns due to the reflection with an excess path length of about 15 m with respect to the line–of–sight path, which again corresponds to 11 sample periods. The impulse response exhibits a Root Mean Squared (RMS) delay spread of $1.5276 \cdot 10^{-8}$ s, and is shown in Figure 13.3(a). The resulting frequency domain transfer function for this WATM impulse response is given in Figure 13.3(b), which exhibits an undulating behaviour across the 512 subcarriers. This suggests that the high-quality subcarrier may be able to use several bits per subcarrier, while others may have to be disabled. This issue will be further detailed during our later discourse.

13.3.3   Effects of Time–Dispersive Channels

The effects of the time–variant and time–dispersive channels on the data symbols transmitted in an OFDM symbol's subcarriers are diverse. Firstly, if the impulse response of the channel is longer than the duration of the OFDM guard interval, then energy will spill over between consecutive OFDM symbols, leading to inter–OFDM–symbol interference. We will not elaborate on these effects here, since the length of the guard interval is generally chosen to be longer than the longest anticipated CIR.

If the channel is changing only slowly compared to the duration of an OFDM symbol, then a near–time–invariant CIR can be associated with each transmitted OFDM symbol, which however slightly changes between consecutive OFDM symbols. In this case, the frequency-selective transfer function of the channel results in a frequency–dependent multiplicative distortion of the received frequency-domain OFDM symbols. This frequency-domain phenomenon is somewhat analogous to the time-domain effects of a time–domain fading channel envelope in a serial or single-carrier modem.

Let us now briefly view the system in the time-domain again. The role of the guard interval was discussed in depth before, hence suffice to state here that if the CIR duration is shorter than the OFDM guard interval, then no inter–OFDM–symbol interference is experienced. More explicitly, if the 'memory' or the 'echoes' of the dispersive CIR have died down during the guard interval i.e. before the commencement of the information-bearing OFDM symbol section, the consecutive OFDM symbols will not interfere with each other. This scenario is analogous to a narrow-band or non-dispersive fading channel in the context of a serial modem. This will be elaborated on in Section 13.3.3.1. A rapidly time–varying channel, however, will introduce inter–subcarrier interference due to the channel's time variant impulse response. The effects of this will be studied in Section 13.3.3.2.

13.3.3.1   Effects of the Slowly Time–Varying Time–Dispersive Channel

Here a channel is referred to as slowly time-varying, if the CIR does not vary significantly over the duration of one OFDM symbol, but it is time–variant over longer periods of time. In this case, the time–domain convolution of the transmitted time–domain signal with the CIR corresponds simply to the multiplication of the spectrum of the signal with the channel's frequency-domain transfer function $H(f)$, as seen below:

$$s(t) * h(t) \longleftrightarrow S(f) \cdot H(f), \tag{13.2}$$

where the channel's frequency–domain transfer function $H(f)$ is the Fourier transform of the impulse response $h(t)$:

$$h(t) \longleftrightarrow H(f). \tag{13.3}$$

Since the information symbols $S(n)$ are encoded into the amplitude of the transmitted spectrum at the subcarrier frequencies f_n, the received symbols $R(n)$ are the product of the transmitted symbol with the channel's frequency–domain transfer function $H(n)$ plus the additive complex Gaussian noise samples $N(n)$:

$$R(n) = S(n) \cdot H(n) + N(n). \tag{13.4}$$

We note here that the additive time-domain noise imposed by the channel becomes correlated due to the filtering effect of the demodulator's FFT operation. Let us now consider the effects of rapidly time-varying channels.

13.3.3.2   Rapidly Time–Varying Channel

A channel is classified here as rapidly time–varying, if the CIR changes significantly over the duration of an OFDM symbol. In this case, the frequency–domain transfer function is

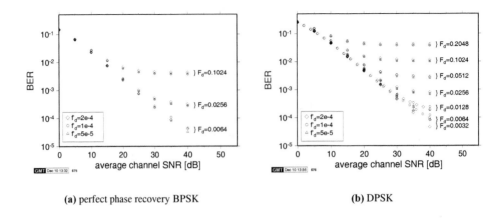

(a) perfect phase recovery BPSK   **(b)** DPSK

Figure 13.4: BPSK OFDM modem performance in a fading narrow-band channel for normalized Doppler frequencies of $f'_d = 5 \cdot 10^{-5}, 1 \cdot 10^{-4}$ and $2 \cdot 10^{-4}$ and FFT lengths between 16 and 4096, where $F_d = f_d \cdot NT_s = f'_d \cdot N$.

time–variant during the transmission of an OFDM symbol and this time–varying frequency–domain transfer function leads to the loss of orthogonality between the OFDM symbol's subcarriers. The amount of this inter–subcarrier interference depends on the rate of change in the impulse response.

The simplest environment to study the effects of rapidly time–varying channels is the narrow-band channel, whose impulse response consists of only one fading path. If the amplitude of this path is varying in time, then the received OFDM symbol's spectrum will be the original OFDM spectrum convolved with the spectrum of the channel variation during the transmission of the OFDM symbol. Since this short–term channel spectrum is varying between different OFDM transmission bursts, the effects of the time–varying narrow–band channel have to be averaged over a high number of transmission bursts for the sake of arriving at reliable performance estimates.

Since the interference is caused by the variation of the CIR during the transmission of each OFDM symbol, we introduce the "OFDM–symbol normalized" Doppler frequency F_d:

$$F_d = f_d \cdot NT_s = f'_d \cdot N, \tag{13.5}$$

where N is the FFT length, $1/T_s$ is the sampling rate, f_d is the Doppler frequency characterizing the fading channel and $f'_d = f_d \cdot T_s$ is the conventional normalized Doppler frequency.

The BER performance for an OFDM modem for a set of different FFT lengths and different channel Doppler frequencies was determined by simulation and the simulation results for BPSK are given in Figure 13.4. Figure 13.4(a) depicts the BER performance of an OFDM modem employing BPSK with perfect narrow-band fading channel estimation, where it can be observed that for any given value of F_d the different FFT lengths and channels behave similarly. For an F_d value of 0.0256, a residual bit error rate of about $2.8 \cdot 10^{-4}$ is observed, while for $F_d = 0.1024$ the residual BER is about 0.37%, where - again - $F_d = f_d \cdot NT_s = f'_d \cdot N$.

13.3.3.3 Transmission over Time–Dispersive OFDM Channels

Analogously to the case of serial modems in narrow-band fading channels, the amplitude–and phase variations inflicted by the channel's frequency–domain transfer function $H(n)$ upon the received symbols will severely affect the bit error probabilities, where different modulation schemes suffer to different extents from the effects of the channel transfer function. Coherent modulation schemes rely on the knowledge of the symbols' reference phase, which will be distorted by the phase of $H(n)$ and hence if such a modulation scheme is to be employed, then this phase distortion has to be estimated and corrected. For multi–level modulation schemes [317], where the magnitude of the received symbol also bears information, the magnitude of $H(n)$ will affect the demodulation. Clearly, the performance of such a system depends on the quality of the channel estimation.

A simpler approach to signalling over fading channels is to employ differential modulation, where the information is encoded in the difference between the individual modulated symbols mapped to consecutive subcarriers of the OFDM symbol. Differential Phase Shift Keying (DPSK) employs the phase of the previous modulated symbol conveyed by the previous subcarrier as phase reference, encoding information in the phase difference between consecutive modulated symbols. DPSK is thus only affected by the differential channel phase distortion between two consecutive symbols assigned to consecutive subcarriers, rather than by the channel phase distortion's absolute value. We note here that differential encoding between the corresponding identical-frequency subcarriers of consecutive OFDM symbols could also be invoked, although the associated channel phase change would be more subtantial and hence the former approach to differential encoding is more advantageous.

13.4 OFDM Performance with Frequency Errors and Timing Errors

In this section we will highlight the effects of time– and frequency domain synchronization errors on the performance of an OFDM system. Furthermore, a number of synchronization algorithms will be briefly highlighted for time–domain burst–based OFDM communications systems based on the recent advances in the literature.

The performance of the synchronization subsystem, in particular the accuracy of the frequency and timing error estimations, is of major influence on the overall OFDM system performance. In order to demonstrate the effects of carrier frequency and time–domain FFT window alignment errors, a series of results will be presented over different channels. For all the Additive White Gaussian Noise (AWGN) channel experiments rectangular time–domain pulse shaping was assumed.

13.4.1 Effects of Frequency Shift on OFDM

Carrier frequency errors result in a shift of the received signal's spectrum in the frequency domain. If the frequency error is an integer multiple I of the subcarrier spacing Δf, then the received frequency domain subcarriers are shifted by $I \cdot \Delta f$. The subcarriers are still mutually orthogonal, but the received data symbols, which were mapped to the OFDM spectrum, are in the wrong position in the demodulated spectrum, resulting in a bit error rate of 0.5.

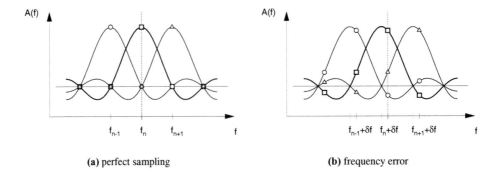

(a) perfect sampling   (b) frequency error

Figure 13.5: Stylized plot of OFDM symbol spectrum with sampling points for three subcarriers. The symbols on the curves signify the contributions of the three subcarriers to the sum at the sampling point. (a) – no frequency offset between transmitter and receiver, (b) – frequency error δf present.

If the carrier frequency error is not an integer multiple of the subcarrier spacing, then energy is spilling over between the subcarriers, resulting in loss of their mutual orthogonality. In other words, interference is observed between the subcarriers, which deteriorates the bit error rate of the system. The amount of this inter–subcarrier–interference can be evaluated by observing the spectrum of the OFDM symbol.

The spectrum of the OFDM signal is derived from its time domain representation transmitted over the channel. A single OFDM symbol in the time domain can be described as:

$$u(t) = \left[\sum_{n=0}^{N-1} a_n e^{j\omega_n \cdot t} \right] \times \text{rect}\left(\frac{t}{N \cdot T_s} \right),$$
(13.6)

which is the sum of N subcarriers $e^{\omega_n \cdot t}$, each modulated by a QAM symbol a_n and windowed by a rectangular window of the OFDM symbol duration T_s. The Fourier transform of this rectangular window is a frequency–domain sinc–function, which is convolved with the dirac–delta subcarriers, determining the spectrum of each of the windowed complex exponential functions, leading to the spectrum of the nth single subcarrier in the form of

$$A_n(\omega) = \frac{\sin(N \cdot T_s \cdot \omega/2)}{N \cdot T_s \cdot \omega/2} * \delta(\omega - \omega_n).$$
(13.7)

Replacing the radian frequencies ω by frequencies and using the relationship $N \cdot T_s = 1/\Delta f$, the spectrum of a subcarrier can be expressed as:

$$A_n(f) = \frac{\sin(\pi \frac{f-f_n}{\Delta f})}{\pi \frac{f-f_n}{\Delta f}} = \text{sinc}\left(\frac{f - f_n}{\Delta f} \right).$$
(13.8)

The OFDM receiver samples the received time–domain signal, demodulates it by invoking the FFT and - in case of a carrier frequency shift - generates the sub–channel signals in the frequency domain at the sampling points $f_n + \delta f$. These sampling points are spaced from each other by the subcarrier spacing Δf and misaligned by the frequency error δf. This scenario is shown in Figure 13.5. Figure 13.5(a) shows the sampling of the subcarrier at frequency f_n at the optimum frequency raster, resulting in a maximum signal amplitude and no inter–subcarrier interference. If the frequency reference of the receiver is offset with respect to that of the transmitter by a frequency error of δf, then the received symbols suffer from inter–subcarrier interference, as depicted in Figure 13.5(b).

The total amount of inter–subcarrier interference experienced by subcarrier n is the sum of the interference amplitude contributions of all the other subcarriers in the OFDM symbol:

$$I_n = \sum_{j,j \neq n} a_j \cdot A_j(f_n + \delta f). \tag{13.9}$$

Since the QAM symbols a_j are random variables, the interference amplitude in subcarrier n, I_n, is also a random variable, which cannot be calculated directly. If the number of interferers is high, however, then the power spectral density of I_n can be approximated with that of a Gaussian process, according to the central limit theorem. Therefore, the effects of the inter–subcarrier interference can be modelled by additional white Gaussian noise superimposed on the frequency domain data symbols.

The variance of this Gaussian process σ is the sum of the variances of the interference contributions,

$$\sigma^2 = \sum_{j,j \neq n} \sigma_{a_j}^2 \cdot |A_j(f_n + \delta f)|^2. \tag{13.10}$$

The quantities $\sigma_{a_j}^2$ are the variances of the data symbols, which are the same for all j in a system that is not varying the average symbol power across different subcarriers. Additionally, because of the constant subcarrier spacing of Δf, the interference amplitude contributions can be expressed more conveniently as:

$$A_j(f_n + \delta f) = A_j n(\delta f) = \text{sinc}((n - j) + \frac{\delta f}{\Delta f}). \tag{13.11}$$

The sum of the interferer powers leads to the inter–subcarrier interference variance expression of:

$$\sigma^2 = \sigma_a^2 \cdot \sum_{i=-N/2-1}^{N/2} \left| \text{sinc}(i + \frac{\delta f}{\Delta f}) \right|^2. \tag{13.12}$$

The value of the inter–subcarrier interference (ISCI) variance for FFT lengths of $N = 64$, 512 and 4096 and for a range of frequency errors δf is shown in Figure 13.6. It can be seen that the number of subcarriers does not influence the ISCI noise variance for OFDM symbol lengths of more than 64 subcarriers. This is due to the rapid decrease of the interference amplitude with increasing frequency separation, so that only the interference from close subcarriers contributes significantly to the interference load on the subcarriers.

In order to quantify the accuracy of the Gaussian approximation, histograms of the measured interference amplitude were produced for QPSK and 16QAM modulation of the subcarriers. The triangles in Figure 13.7 depict the histograms of ISCI noise magnitudes recorded for a 512–subcarrier OFDM modem employing QPSK and 16QAM in a system having a

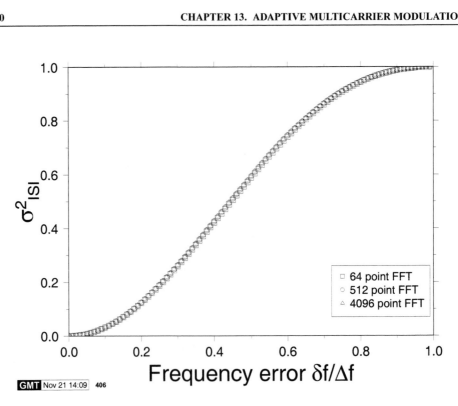

Figure 13.6: Inter–subcarrier interference variance due to a frequency shift δf FFT lengths of $N = 64$, 512, and 4096, for normalized frequency errors $\delta f / \Delta f$ between 0 and 1.

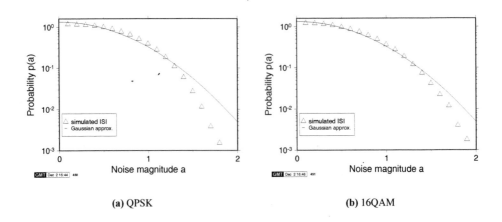

(a) QPSK   (b) 16QAM

Figure 13.7: Histogram of the ISCI magnitude for a simulated 512–subcarrier OFDM modem using QPSK or 16QAM for $\delta f = 0.3\Delta f$; the line represents the Gaussian approximation having the same variance.

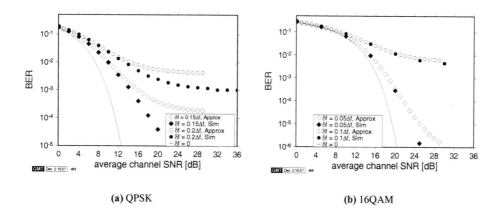

(a) QPSK　　　　　　　　　　　　　　　　(b) 16QAM

Figure 13.8: The effect of inter–subcarrier interference due to frequency synchronization error on the BER over AWGN channels: (a) Bit error probability versus channel SNR for frequency errors of $0.15\Delta f$ and $0.2\Delta f$ for a QPSK modem. (b) BER versus channel SNR for frequency errors of $0.05\Delta f$ and $0.1\Delta f$ for a 16QAM modem. In both graphs, the black markers are simulated BER results, while the white markers are the predicted BER curves using the Gaussian inter–subcarrier interference model.

frequency error of $\delta f = 0.3\Delta f$. The continuous line drawn on the same graph is the corresponding approximation of the histogram by a Gaussian probability density function (PDF) of the variance calculated using Equation 13.12. It can be observed that the Gaussian curve is a reasonable approximation for both histograms in the central region, but that for the tails of the distributions the Gaussian function exhibits high relative errors. The histogram of the interference caused by the 16QAM signal is, however, closer to the Gaussian curve than the QPSK interference histogram.

The frequency mismatch between the transmitter and receiver of an OFDM system not only results in inter–subcarrier interference, but it also reduces the useful signal amplitude at the frequency–domain sampling point by a factor of $f(\delta f) = \mathrm{sinc}(\delta f/\Delta f)$. Using this and σ^2, the theoretical influence of the inter–subcarrier interference, approximated by a Gaussian process, can be calculated for a given modulation scheme in a AWGN channel. In the case of coherently detected QPSK, the closed–form expression for the BER $P_e(\gamma)$ at a channel SNR γ is given by [280]:

$$P_e(\gamma) = Q(\sqrt{\gamma}), \tag{13.13}$$

where the Gaussian $Q()$–function is defined as [280]:

$$Q(y) = \frac{1}{\sqrt{2\pi}} \int_y^\infty e^{-x^2/2} dx = \frac{1}{2}\mathrm{erfc}\left(\frac{y}{\sqrt{2}}\right). \tag{13.14}$$

Assuming that the effects of the frequency error can be approximated by white Gaussian noise of variance σ^2 and taking into account the attenuated signal magnitude $f(\delta f) =$

$sinc(\pi\delta f/\Delta f)$, we can adjust the equivalent SNR to:

$$\gamma' = \frac{f(\delta f) \cdot \sigma_a^2}{\sigma^2 + \sigma_a^2/\gamma}, \tag{13.15}$$

where σ_a^2 is the average symbol power and γ is the real channel SNR. Comparison between the theoretical BER calculated using γ' and QPSK simulation results for frequency errors of $\delta f = 0.15\Delta f$ and $0.2\Delta f$ are shown in Figure 13.8(a). While for both frequency errors the theoretical BER using the Gaussian approximation fits the simulation results well for channel SNR values of up to 12 dB, the predictions and the simulation results diverge for higher values of SNR. The pessimistic BER prediction is due to the pronounced discrepancy between the histogram and the Gaussian curve in Figure 13.7 at the tail ends of the amplitude histograms, since for high noise amplitudes the Gaussian model is a poor approximation for the inter–subcarrier interference.

The equivalent experiment - conducted for coherently detected 16QAM - results in the simulated and predicted bit error rates depicted in Figure 13.8(b). For 16QAM transmission, the noise resilience is much lower than for QPSK, hence for these experiments smaller values of $\delta f = 0.05\Delta f$ and $0.1\ \Delta f$ have been chosen. It can be observed that the Gaussian noise approximation is a much better fit for the simulated BER in a 16QAM system than for a QPSK modem. This is in accordance with Figure 13.7, where the histograms of the interference magnitudes were depicted.

13.4.2   Effect of Time–Domain Synchronization Errors on OFDM

Unlike frequency mismatch, as discussed above, time synchronization errors do not result in inter–subcarrier interference. Instead, if the receiver's FFT window spans samples from two consecutive OFDM symbols, inter–OFDM–symbol interference occurs.

Additionally, even small time-domain misalignments of the FFT window result in an evolving phase shift in the frequency domain symbols, leading to BER degradation. Initially, we will concentrate on these phase errors.

If the receiver's FFT window is shifted with respect to that of the transmitter, then the time shift property of the Fourier transform, formulated as:

$$
\begin{aligned}
f(t) &\longleftrightarrow F(\omega) \\
f(t - \tau) &\longleftrightarrow e^{-j\omega\tau} F(\omega)
\end{aligned}
$$

describes its effects on the received symbols. Any misalignment τ of the receiver's FFT window will introduce a phase error of $2\pi\Delta f\tau/T_s$ between two adjacent subcarriers. If the time shift is an integer multiple m of the sampling time T_s, then the phase shift introduced between two consecutive subcarriers is $\delta\phi = 2\pi m/N$, where N is the FFT length employed. This evolving phase error has a considerable influence on the BER performance of the OFDM system, clearly depending on the modulation scheme used.

13.4.2.1   Coherent Modulation

Coherent modulation schemes suffer the most from FFT window misalignments, since the reference phase evolves by 2π throughout the frequency range for every sampling time misalignment $n \cdot T_s$. Clearly, this results in a total loss of the reference phase, and hence coherent modulation cannot be employed without phase correction mechanisms, if imperfect time synchronization is to be expected.

13.4.2.2   Pilot Symbol Assisted Modulation

Pilot-symbol-assisted-modulation (PSAM) schemes [19, 317] can be employed in order to mitigate the effects of spectral attenuation and the phase rotation throughout the FFT bandwidth. Pilots are interspersed with the data symbols in the frequency domain and the receiver can estimate the evolving phase error from the received pilots' phases.

The number of pilot subcarriers necessary for correctly estimating the channel transfer function depends on the maximum anticipated time shift τ. Following the notion of the frequency domain channel transfer function $H(n)$ introduced in Section 13.3, the effects of phase errors can be written as:

$$H(f) = e^{-j2\pi f \tau}.\tag{13.16}$$

Replacing the frequency variable f by the subcarrier index n, where $f = n\Delta f = n/(NT_s)$ and normalizing the time misalignment τ to the sampling time T_s, so that $\tau = m \cdot T_s$, the frequency domain channel transfer function can be expressed as:

$$H(n) = e^{-j2\pi \frac{n\,m}{N}}.\tag{13.17}$$

The number of pilots necessary for correctly estimating this frequency domain channel transfer function $H(n)$ is dependent on the normalized time delay m. Following the Nyquist sampling theorem, the distance Δp between two pilot tones in the OFDM spectrum must be less than or equal to half the period of $H(n)$, so that

$$\Delta p \leq \frac{N}{2m}.\tag{13.18}$$

The simulated performance of a 512–subcarrier 16QAM PSAM modem in the presence of a constant timing error of $\tau = 10T_s$ in an AWGN channel is depicted in Figure 13.9 for both ideal low-pass and for simple linear interpolation. Following Equation 13.18, the maximum acceptable pilot subcarrier distance required for resolving a normalized FFT–window misalignment of $m = \tau/T_s = 10$ is $\Delta p = N/20 = 512/20 = 25.6$, requiring at least 20 pilot subcarriers equidistantly spaced in the OFDM symbol. We can see in both graphs of Figure 13.9 that the bit error rate is 0.5 for both schemes, if less than 20 pilot subcarriers are employed in the OFDM symbol. For pilot numbers above the required minimum of 20, however, the performance of the ideal low-pass interpolated PSAM scheme does not vary with the number of pilots employed, while the linearly interpolated PSAM scheme needs higher numbers of pilot subcarriers in order to achieve a similar performance. The continuous lines in the graphs show the BER for a coherently detected 16QAM OFDM modem in the absence of timing errors, while utilizing no PSAM. Observe in the figure that there is a BER penalty for PSAM in a narrow-band AWGN channel, since the pilots are affected by noise, which is interpreted by the PSAM schemes as a channel induced fluctuation, which has to be compensated.

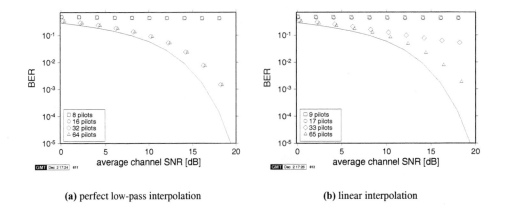

(a) perfect low-pass interpolation   (b) linear interpolation

Figure 13.9: Bit error rate versus channel SNR performance for 16–level PSA-QAM in an AWGN
channel for different pilot subcarrier spacings in the presence of a fixed FFT window
misalignment of $\tau = 10T_s$. The OFDM FFT length is 512. (a) PSAM interpolation
using ideal low-pass interpolator, (b) PSAM using linear interpolator. In both graphs, the
line marks the coherently detected 16QAM performance in absence of both FFT window
misalignment and PSAM.

13.4.2.3   Differential Modulation

As stated before, differential encoding [317] can be implemented both between corresponding subcarriers of consecutive OFDM symbols or between adjacent subcarriers of the same
OFDM symbol. The latter was found more advantageous, since there is less channel-induced
- rather than modulation-induced - phase rotation between consecutive subcarriers of an
OFDM symbol than between the identical-frequency subcarriers of consecutive OFDM symbols. Hence differential encoding between adjacent subcarriers was employed here. Simulations have been performed for a 512–subcarrier OFDM system, employing DBPSK and
DQPSK for different FFT window misalignment values. The BER performance curves for
timing errors of up to six positive and negative sampling intervals are displayed in Figure
13.10. The figure suggests that in case of time-advanced data (bold markers in the figure)
or time-delayed FFT windows the BER degrades due to including samples of the previous
OFDM symbol in the current FFT window, while neglecting some of samples belonging to
the current OFDM symbol. This data-dependent error is the reason for the fluctuating BER in
the figure. Note that one sample interval misalignment represents a phase error of $2\pi/512$ between two consecutive samples, which explains why the BER effects of the simulated positive
timing misalignments marked by the hollow symbols are negligible for DBPSK. Specifically,
a maximum SNR degradation of 0.5 dB was observed for DQPSK.

Positive FFT window time shifts correspond to a delayed received data stream and hence
all samples in the receiver's FFT window belong to the same quasi–periodically extended
OFDM symbol. In the case of negative time shifts, however, the effects on the bit error rate
are much more severe due to inter–OFDM–symbol interference. Since the data is received
prematurely, the receiver's FFT window contains samples of the forthcoming OFDM symbol,
not from the cyclic extension of the wanted symbol. This scenario can only be encountered

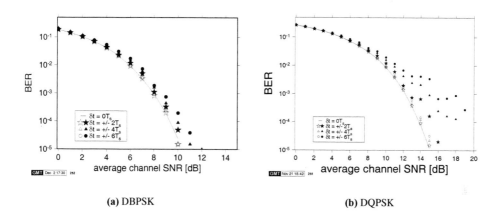

(a) DBPSK   (b) DQPSK

Figure 13.10: Bit error rate versus SNR over AWGN channels for a 512–subcarrier OFDM modem
employing DBPSK and DQPSK, respectively. Positive time shifts imply time–advanced
FFT window or delayed received data.

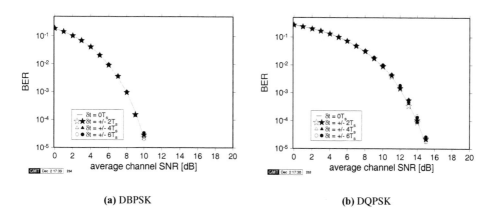

(a) DBPSK   (b) DQPSK

Figure 13.11: Bit error rate versus SNR over AWGN channels for a 512–subcarrier OFDM modem
employing a post–amble of 10 symbols for DBPSK and DQPSK, respectively. Positive
time shifts correspond to time–advanced FFT window or delayed received data.

in conjunction with imperfect OFDM symbol synchronization, when the OFDM symbols are received prematurely.

This non-symmetrical behaviour of the OFDM receiver with respect to positive and negative relative timing errors can be mitigated by adding a short post–amble, consisting of copies of the OFDM symbol's first samples. Figure 13.11 shows the BER versus SNR curves for the same offsets, while using a 10–sample post–amble. Now, the behaviour for positive and negative timing errors becomes symmetrical. Clearly, the required length of this post–amble depends on the largest anticipated timing error, which adds further redundancy to the system. This post–amble can be usefully employed, however, to make an OFDM system more robust to time misalignments and thus to simplify the task of the time–domain FFT window synchronization system.

13.5 Synchronization Algorithms

The results of Section 13.4 indicate that the accuracy of a modem's time– and frequency–domain synchronization system dramatically influences the overall BER performance. We have seen that carrier frequency differences between the transmitter and the receiver of an OFDM system will introduce additional impairments in the frequency domain caused by inter–subcarrier–interference, while FFT window misalignments in the time–domain will lead to phase errors between the subcarriers. Both of these effects will degrade the system's performance and have to be kept to a minimum by the synchronization system.

In a TDMA based OFDM system, the frame synchronization between a master station — in cellular systems generally the base station — and the portable stations has to be also maintained. For these systems, a so-called reference symbol marking the beginning of a new time frame is commonly used. This added redundancy can be exploited for both frequency synchronization and FFT–window alignment, if the reference symbol is correctly chosen.

In order to achieve synchronization with a minimal amount of computational effort at the receiver, while also minimising the amount of redundant information added to the data signal, the synchronization process is normally split into a coarse acquisition phase and a fine tracking phase, if the characteristics of the random frequency– and timing–errors are known. In the acquisition phase, an initial estimate of the errors is acquired, using more complex algorithms and possibly a higher amount of synchronization information in the data signal, whereas later the tracking algorithms only have to correct for small short–term deviations.

At the commencement of the synchronization process neither the frequency error nor the timing misalignment are known, hence synchronization algorithms must be found that are sufficiently robust to initial timing and frequency errors. In the forthcoming sections we will briefly review the associated literature, before providing some performance figures for the sake of illustration.

13.5.1 Coarse Transmission Frame and OFDM Symbol Synchronization

Coarse frame and symbol synchronization algorithms presented in the literature all rely on additional redundancy inserted in the transmitted data stream. The Pan–European DVB system uses a so–called Null–symbol as the first OFDM symbol in the time frame, during whose duration no energy is transmitted [507], and which is detected by monitoring the received baseband power in the time domain, without invoking FFT processing. Claßen [442] proposed an OFDM synchronization burst of at least three OFDM symbols per time frame. Two of the OFDM symbols in the burst would contain synchronization subcarriers bearing known symbols along with normal data transmission carriers, but one of the OFDM symbols would be the exact copy of one of the other two. This results in more than one OFDM symbol synchronization overhead per synchronization burst. For the so-called ALOHA environment, Warner [463] proposed the employment of a power detector and subsequent correlation–based detection of a set of received synchronization subcarriers embedded in the data symbols. The received synchronization tones are extracted from the received time–domain signal using an iterative algorithm for updating the synchronization tone values once per sampling interval. For a more detailed discussion on these techniques the interested reader is referred to the literature [442, 463].

13.5.2 Fine Symbol Tracking Overview

Fine symbol tracking algorithms are generally based on correlation operations either in the time– or in the frequency–domain. Warner [463] and Bingham [508] employed frequency–domain correlation of the received synchronization pilot tones with known synchronization sequences, while de Couasnon [509] utilized the redundancy of the cyclic prefix by integrating over the magnitude of the difference between the data and the cyclic extension samples. Sandell [468] proposed to exploit the autocorrelation properties of the received time–domain samples imposed by the cyclic extension for fine time–domain tracking.

13.5.3 Frequency Acquisition Overview

The frequency acquisition algorithm has to provide an initial frequency error estimate, which is sufficiently accurate for the subsequent frequency tracking algorithm to operate reliably. Generally the initial estimate must be accurate to half a subcarrier spacing. Sari [464] proposed the use of a pilot tone embedded into the data symbol, surrounded by zero–valued virtual subcarriers, so that the frequency–shifted pilot can be located easily by the receiver. Moose [465] suggested a shortened repeated OFDM symbol pair, analogous to his frequency tracking algorithm to be highlighted in the next section. By using a shorter DFT for this reference symbol pair, the subcarrier distance is increased and thus the frequency error estimation range is extended. Claßen [442, 443] proposed to use binary pseudo-noise (PN) or so–called CAZAC training sequences carried by synchronization subcarriers, which are also employed for the frequency tracking. The frequency acquisition, however, is performed by a search for the training sequence in the frequency domain. This is achieved by means of frequency–domain correlation of the received symbol with the training sequence.

13.5.4 Frequency Tracking Overview

Frequency tracking generally relies on an already established coarse frequency estimation having a frequency error of less than half a subcarrier spacing. Moose [465] suggested the use of the phase difference between subcarriers of repeated OFDM symbols in order to estimate frequency deviations of up to one half of the subcarrier spacing, while Claßen [442] employed frequency–domain synchronization subcarriers embedded into the data symbols, for which the phase shift between consecutive OFDM symbols can be measured. Daffara [467] and Sandell [468] used the phase of the received signal's autocorrelation function, which represents a phase shift between the received data samples and their repeated copies in the cyclic extension of the OFDM symbols.

13.5.5   The Effects of Oscillator Phase Noise

In practice a carrier recovery loop has to be employed, in order to synchronize the local oscillator with the remote oscillator and once the synchronization loop is locked, there is no carrier frequency offset. However, the synchronization loop is prone to oscillator phase noise or phase jitter. This becomes a particularly grave problem in high-frequency, high-bandwidth applications found for example in 155 Mbps WATM systems operating at 60 GHz, such as the applications considered in this contribution. The 60 GHz band is attractive in terms of having a relatively high propagation pathloss due to vapour attenuation and the phenomenon of oxygen-absorption and hence it conveniently curtails co-channel interferences [13]. Furthermore, there is sufficient spectrum available for the 200 MHz bandwidth required by our 155 Mbps WATM system. However, at this extremely high frequency there is a paucity of high-quality oscillators, since no standard systems operate in this frequency band at the time of writing. Hence in this section we consider briefly the issue of phase noise.

The presence of phase noise is an important limiting factor for an OFDM system's performance [458, 459, 510], and depends on the quality and the operating conditions of the system's RF hardware. In conventional mobile radio systems around a carrier frequency of 2GHz the phase noise constitutes typically no severe limitation, however in the 60GHz carrier frequency, 225MHz bandwidth WATM system considered here its effects were less negligible and hence had to be investigated in more depth. Oscillator noise stems from oscillator inaccuracies in both the transmitter and receiver and manifests itself in the baseband as additional phase– and amplitude modulation of the received samples [511]. The oscillator noise influence on the signal depends on the noise characteristics of the oscillators in the system and on the signal bandwidth. It is generally split in amplitude noise $A(t)$ and phase noise $\Phi(t)$, and the influence of the amplitude noise $A(t)$ on the data samples is often neglected. The time domain functions $A(t)$ and $\Phi(t)$ have Gaussian histograms, and their time domain correlation is determined by their respective long–term power spectra through the Wiener–Khintchine theorem.

If the amplitude noise is neglected, imperfect oscillators are characterized by the long–term power spectral density (PSD) $N_p(f')$ of the oscillator output signal's phase noise, which is also referred to as the phase noise mask. The variable f' represents the frequency distance from the oscillator's nominal carrier frequency in a band–pass model, or equivalently, the absolute frequency in the base–band. An example of this phase noise mask for a practical oscillator is given in Figure 13.12(a). If the phase–noise PSD $N_p(f')$ of a specific oscillator is known, then the variance of the phase error $\Phi(t)$ for noise components in a frequency band $[f_1, f_2]$ is the integral of the phase noise spectral density over this frequency band as in [511]:

$$\bar{\Phi}^2 = \int_{f_1}^{f_2} \left(\frac{2N_p(f')}{C} \right) df', \tag{13.19}$$

where C is the carrier power and the factor 2 represents the double sided spectrum of the

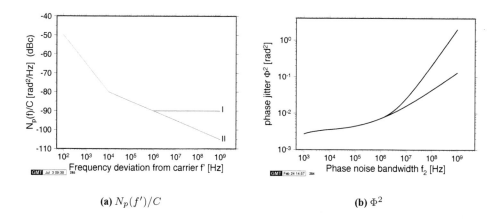

(a) $N_p(f')/C$ **(b)** Φ^2

Figure 13.12: Phase noise characterization: (a) spectral phase noise density (phase noise mask), (b) integrated phase jitter for two different phase noise masks.

phase noise. The phase noise variance $\bar{\Phi}^2$ is also referred to as the integrated phase jitter, which is depicted in Figure 13.12(b).

The phase noise contribution of both the transmitter and receiver can be viewed as an additional multiplicative effect of the radio channel, like fast and slow fading. The performance of the carrier recovery is affected by the phase noise, which in turn degrades the performance of a coherently detected scheme.

For OFDM schemes, multiplication of the received time–domain signal with a time–varying channel transfer function is equivalent to convolving the frequency–domain spectrum of the OFDM signal with the frequency domain channel transfer function. Since the phase noise spectrum's bandwidth is wider than the subcarrier spacing, this results in energy spillage into other subchannels and therefore in inter–subcarrier interference, an effect which will be quantified below. Let us now consider the phase noise model employed in our performance study.

13.5.5.1 Coloured Phase Noise Model

The integral $\bar{\Phi}^2$ of Equation 13.19 characterizes the long–term statistical properties of the oscillator's phase and frequency errors due to phase noise. In order to create a time–domain function satisfying the standard deviation $\bar{\Phi}^2$, a white Gaussian noise spectrum was filtered with the phase noise mask $N_p(f')$ depicted in Figure 13.12(a), which was transformed into the time domain. A frequency resolution of about 50 Hz was assumed in order to model the shape of the phase noise mask at low frequencies, which led to a FFT transform length of $2^{22} = 4194304$ samples for the frequency range of Figure 13.12(a).

The resulting time–domain phase noise channel data is a stream of phase error samples, which were used to distort the incoming signal at the receiver. The double–sided phase noise mask used for the simulations is given in Table 13.2. Between the points given in Table 13.2, a log–linear interpolation is assumed, as shown in Figure 13.12(a). As the commercial oscillator's phase noise mask used in our investigations was not specified for frequencies beyond

f'[Hz]	100	1k	10k	100k	1M
N_p/C[dB]	-50	-65	-80	-85	-90

Table 13.2: Two–sided phase noise mask used for simulations. f' — frequency distance from carrier, N_p/C — normalized phase noise density.

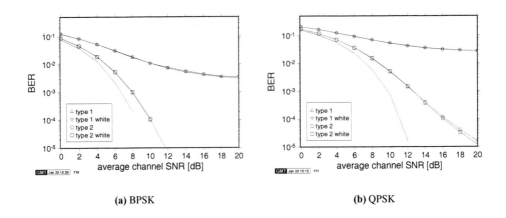

(a) BPSK   (b) QPSK

Figure 13.13: Bit error rate versus channel SNR for a 512–subcarrier OFDM modem in the presence of phase noise. Type 1 represents the coloured phase noise channel with the phase noise mask depicted in Figure 13.12(a) assuming a noise floor of 90rad^2/Hz, while Type 2 is the channel without phase noise floor. The curves designated "white" are the corresponding white phase noise results. The lines without markers give the corresponding results in the absence of phase noise.

1MHz, two different cases were considered for frequencies beyond 1MHz: (I) a phase noise floor at -90dB, and (II) a $f^{-1/2}$ law. Both of these extended phase noise masks are shown in Figure 13.12(a). The integrated phase jitter has been calculated using Equation 13.19 for both scenarios, and the value of the integral for different noise bandwidths is depicted in Figure 13.12(b).

For the investigated 155 Mbits/s Wireless ATM (WATM) system's [495–498] double–sided bandwidth of 225 MHz, the integration of the phase noise masks results in phase jitter values of $\bar{\Phi}^2 = 0.2303$rad^2 and $\bar{\Phi}^2 = 0.04533$rad^2 for the phase noise mask with and without noise floor, respectively.

The simulated BER performance of a 512–subcarrier OFDM system with a subcarrier distance $\Delta f = 440$kHz over the two different phase noise channels is depicted in Figure 13.13 for coherently detected BPSK and QPSK. In addition to the BER graphs corresponding to the coloured phase noise channels described above, graphs of the modems' BER performance over white phase noise channels with the equivalent integrated phase jitter values was also plotted in the figures. It can be observed that the BER performance for both modulation schemes and for both phase noise masks is very similar for the coloured and the white phase noise models.

The simulated BER results shown in Figure 13.13 show virtually indistinguishable performance for the modems in both the coloured and the white phase noise channels, when using BPSK. By contrast, a slight BER difference can be observed for QPSK between the Type 1 and Type 2 phase noise masks, where the corresponding white phase noise results in a better performance than the coloured noise. This difference can be explained with the interference being caused by fewer dominant interfering subcarriers compared to the white phase noise scenario, resulting in a non–Gaussian error histogram.

In summary, phase noise, like all time–varying channel conditions experienced by the time–domain signal, results in inter–subcarrier interference in OFDM transmissions. If the bandwidth of the phase noise is high compared to the OFDM subcarrier spacing, then this interference is caused by a high number of contributions from different subcarriers, resulting in a Gaussian noise–like interference. In addition to this noise inflicted upon the received symbols, the signal level in the subcarriers drops by the amount of energy spread over the adjacent subcarriers. The integral over the phase noise mask, termed as phase jitter, is a measure of the signal–to–interference ratio that can be expected in the received subcarriers, if the phase noise has a wide bandwidth and is predominantly white. For narrow–band phase noise this estimation is pessimistic.

Following the above overview of the associated synchronization issues, we will investigate two different synchronization algorithms, both making use of a reference symbol marking the beginning of a new time frame. This limits the use of both algorithms to systems whose channel access scheme is based on Time Division Multiple Access (TDMA) frames.

13.5.6 BER Performance with Frequency Synchronization

Here we refrain from characterizing the performance of all the previously reviewed synchronization algorithms and refer the interested reader to [317] for implementation specific details. However, as a representative example, Figure 13.14(a) depicts the BER versus channel SNR for BPSK, QPSK and 16QAM in an AWGN channel with a frequency error of $\delta f = 0.3\Delta f$. The white symbols in the graph portray the BER performance of an OFDM modem employing no frequency synchronization. It can be seen that the uncorrected frequency errors result in heavy inter–subcarrier interference, which manifests itself as a high residual bit error rate of about 5% for BPSK and QPSK and about 20% for 16QAM. The lines in the graph characterize the performance of the modem in the absence of frequency errors. The black markers correspond to the BER recorded with the frequency synchronization algorithm in operation. It can be seen that the performance of the modem employing the frequency synchronization algorithm of [512] is nearly indistinguishable from the perfectly synchronised case. In Figure 13.14(b), the modem's BER curves for an AWGN channel at a frequency error of $7.5\Delta f$ are depicted. Since the synchronization algorithm's accuracy does not vary with varying frequency errors, the modem's BER performance employing the synchronization algorithm considered at $\delta f = 7.5\Delta f$ is the same as at $\delta f = 0.3\Delta f$. The BER for the non–synchronised modem is, however, 50% and the corresponding markers are off the graph.

The synchronised modem's BER performance in wideband channels is characterized in Figures 13.14(c) and 13.14(d). The impulse response used was the WATM impulse response of Figure 13.3(a). Perfect knowledge of the CIR was assumed for perfect phase– and amplitude–correction of the data symbols with coherent detection. Again, the BER curves for both the non–fading and the fading channels show a remarkable correspondence between

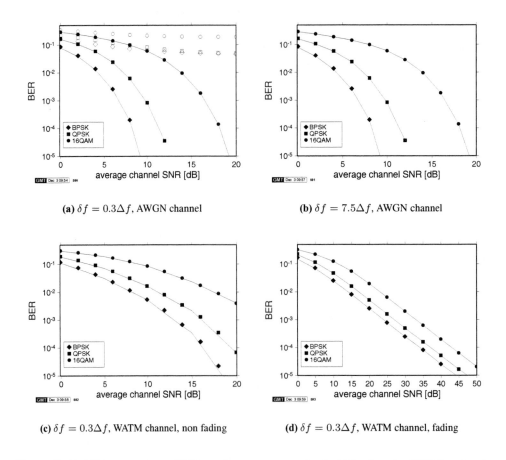

(a) $\delta f = 0.3\Delta f$, AWGN channel

(b) $\delta f = 7.5\Delta f$, AWGN channel

(c) $\delta f = 0.3\Delta f$, WATM channel, non fading

(d) $\delta f = 0.3\Delta f$, WATM channel, fading

Figure 13.14: BER versus channel SNR performance curves for the 512 subcarrier OFDM system in the presence of fixed frequency errors. The lines indicate the performance for perfectly corrected frequency error, and the white symbols show the performance for uncorrected frequency errors. The black symbols indicate simulations based on frequency error estimation using the time–domain correlation technique. The WATM CIR is shown in Figure 13.3(a).

the ideal performance lines and the performance of the synchronised modems. The modem's performance was unaffected by the estimation accuracy of the time–domain reference symbol synchronization algorithm in all the investigated environments.

In summary, the effects of frequency and timing errors in OFDM transmissions have been characterized. While frequency errors result in frequency–domain inter–subcarrier interference, timing errors lead to time–domain inter–OFDM symbol interference and to frequency–domain phase rotations. In order to overcome the effects of moderate timing errors, a cyclic post–amble and the use of pilot–symbol assisted modulation or differential detection was recommended. Different frequency and timing error estimation algorithms were reviewed and characterized. Let us now in the next section consider the recent advances in the field of sophisticated adaptive OFDM schemes.

13.6 Adaptive OFDM

13.6.1 Survey and Motivation

Steele and Webb [1, 16] proposed adaptive modulation for exploiting the time-variant Shannonian channel capacity of fading narrow-band channels, which stimulated further research by Sampei et al. [34], Goldsmith et al. [513], by Pearce, Burr and Tozer [51], Lau and Mcleod [52], as well as Torrance and Hanzo [60, 61]. The associated principles can also be invoked in the context of parallel modems, as it has been demonstrated by Kalet [62], Czylwik et al. [63] as well as by Chow, Cioffi and Bingham [64].

Based on the philosophy of the above contributions, below we summarise the ideas behind adaptive modulation. We have seen in Section 13.3 that the bit error probability of different OFDM subcarriers transmitted in time dispersive channels depends on the frequency domain channel transfer function. The occurrence of bit errors is normally concentrated in a set of severely faded subcarriers, while in the rest of the OFDM spectrum, often no bit errors are observed. If the subcarriers that will exhibit high bit error probabilities in the OFDM symbol to be transmitted can be identified and excluded from data transmission, the overall BER can be improved in exchange for a slight loss of system throughput. Since the frequency domain fading deteriorates the SNR of certain subcarriers, but improves others' above the average SNR value, the potential loss of throughput due to the exclusion of faded subcarriers can be mitigated by employing higher-order modulation modes on the subcarriers exhibiting high SNR values.

As a further conceptual augmentation of the above ideas let as consider the following example. The associated channel SNR of an adaptive OFDM modem is shown in a three-dimensional form in Figure 13.15, which was generated with the aid of the FFT of the Rayleigh-faded CIR of Figure 13.3. Observe that the instantaneous channel SNR is a function of both time and frequency. An example of the associated time and frequency–dependent modulation scheme allocation for an adaptive OFDM modem carrying 578 data bits per OFDM symbol at an average channel SNR of 5dB is given in Figure 13.16(a) for 100 consecutive OFDM symbols. The unused sub–bands with indices 15 and 16 contain the virtual carriers, and therefore do not transmit any useful data. It can be seen that the adaptation algorithm allocates data to the better quality subcarriers on a symbol–by–symbol basis, while keeping the total number of bits per OFDM symbol constant. As a comparison, Figure 13.16(b) shows the equivalent overview of the modulation schemes employed for the

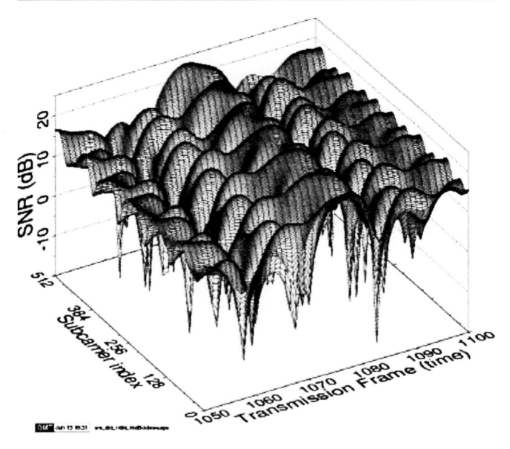

Figure 13.15: Instanteous Channel SNR for all 512 subcarriers versus time, for an average channel
SNR of 16dB over the channel characterized by the impulse response of Figure 13.3.

substantially higher - nearly tripled - fixed bit rate of 1458 bits per OFDM symbol. It can be
seen that in order to achieve the throughput target, hardly any sub–bands are in "no transmission" mode, and overall predominantly higher order modulation schemes, such as QPSK and
16QAM, have to be employed.

In addition to excluding sets of faded subcarriers and varying the modulation modes employed, other parameters such as the coding rate of error correction coding schemes can be
also adapted at the transmitter according to the perceived channel transfer function.

Adaptation of the transmission parameters may be based on the transmitter's perception
of the channel conditions in the forthcoming TDMA / TDD duplex time-slot. Clearly, this
estimation of future channel parameters can only be obtained by extrapolation of previous
channel estimations, which are acquired upon detecting each received OFDM symbol. The
channel characteristics therefore have to be varying sufficiently slowly compared to the channel estimation interval.

Adapting the transmission technique to the channel conditions on a timeslot–by–timeslot
basis for serial modems in narrow-band fading channels has been shown to considerably

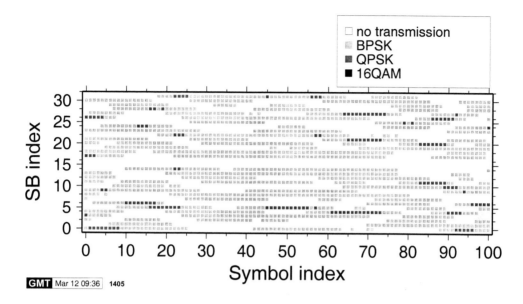

(a) 578 data bits per OFDM symbol

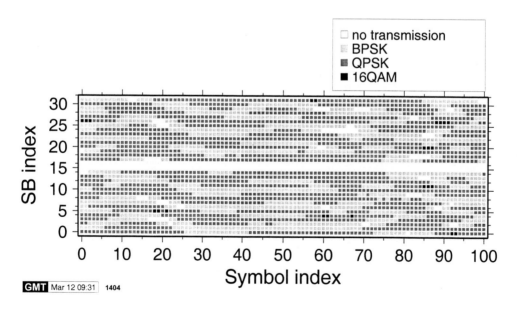

(b) 1458 data bits per OFDM symbol

Figure 13.16: An example of modem mode allocation for a 578-bit and 1458-bit fixed throughput adaptive OFDM modem over fading time–dispersive channels at 5dB average channel SNR.

improve the BER performance [145] for Time Division Duplex (TDD) systems assuming duplex reciprocal channels. However, the Doppler fading rate of the narrow-band channel has a strong effect on the achievable system performance. If the fading is rapid, then the prediction of the channel conditions for the next transmit timeslot is inaccurate, and therefore the wrong set of transmission parameters may be chosen. If however the channel varies slowly, then the data throughput of the system is varying dramatically over time, and large data buffers are required at the transmitters in order to smoothen the bit rate fluctuation. For time–critical applications, such as interactive speech transmission, the potential delays can become problematic. A given single–carrier adaptive system in narrow-band channels will therefore operate efficiently only in a limited range of channel conditions.

Adaptive OFDM modems have the potential of mitigating the problem of slowly time–varying channels, since the variation of the signal quality can be exploited in both the time– and the frequency–domain. The channel conditions still have to be monitored based on the received OFDM symbols, and relatively slowly varying channels have to be assumed, since we have seen in Section 13.3.3.2 that OFDM transmissions are not well suited to rapidly varying channel conditions.

13.6.2   Adaptive Techniques

Adaptive modulation is only suitable for duplex communication between two stations, since the transmission parameters have to be adapted using some form of two–way transmission in order to allow channel measurements and signalling to take place. These issues are studied below.

Transmission parameter adaptation is a response of the transmitter to time–varying channel conditions. In order to efficiently react to the changes in channel quality, the following steps have to be taken:

- *Channel quality estimation:* In order to appropriately select the transmission parameters to be employed for the next transmission, a reliable estimation of the channel transfer function during the next active transmit timeslot is necessary.

- *Choice of the appropriate parameters for the next transmission:* Based on the prediction of the channel conditions for the next timeslot, the transmitter has to select the appropriate modulation modes for the subcarriers.

- *Signalling or blind detection of the employed parameters:* The receiver has to be informed, as to which demodulator parameters to employ for the received packet. This information can either be conveyed within the OFDM symbol itself, at the cost of a loss of effective data throughput, or the receiver can attempt to estimate the parameters employed by the remote transmitter by means of blind detection mechanisms. These issues will be made more explicit in the context of Figure 13.17.

13.6.2.1   Channel Quality Estimation

The transmitter requires an estimate of the expected channel conditions for the time instant, when the next OFDM symbol is to be transmitted. Since this knowledge can only be gained by prediction from past channel quality estimations, the adaptive system can only operate efficiently in an environment exhibiting relatively slowly varying channel conditions.

The channel quality estimation can be acquired from a range of different sources. If the communication between the two stations is bidirectional and the channel can be considered reciprocal, then each station can estimate the channel quality on the basis of the received OFDM symbols, and adapt the parameters of the local transmitter to this estimation. We may refer to such a regime as *open–loop adaptation*, since there is no feedback between the receiver of a given OFDM symbol and the choice of the modulation parameters. An indoor Time Division Duplex (TDD) system in the absence of interference is an example of such a system, and hence a TDD regime is assumed for generating the performance results below. Channel reciprocity issues were addressed, for example, in [82, 83].

If the channel is not reciprocal, as far as the up- and down-link are concerned, as in a Frequency Division Duplex (FDD) system, then the stations cannot determine the parameters for the next OFDM symbol's transmission from the received symbols. In this case, the receiver has to estimate the channel quality and explicitly signal this perceived channel quality information to the transmitter in the reverse link. Since in this case the receiver explicitly instructs the remote transmitter as to which modem modes to invoke, this regime is referred to as *closed–loop adaptation*. The adaptation algorithms can — with the aid of this technique — take into account effects such as interference as well as non–reciprocal channels. If the communication between the stations is essentially unidirectional, then a low–rate signalling channel must be implemented from the receiver to the transmitter. If such a channel exists, then the same technique as for non-reciprocal channels can be employed.

Different techniques can be employed to estimate the channel quality. For OFDM modems, the bit error probability in each subcarrier is determined by the fluctuations of the channel's current frequency domain channel transfer function H_n with the aid of the channel transfer function estimates provided by the pilot symbols, provided that no interference is present. The estimate of the channel transfer function $\hat{H}_n$ can be acquired by means of pilot–tone based channel estimation. More accurate measures of the channel transfer function can be gained by means of decision–directed or time–domain training sequence based techniques. The estimate of the channel transfer function $\hat{H}_n$ does not take into account effects, such as co–channel or inter–subcarrier interference. Alternative channel quality measures including interference effects can be devised on the basis of the error correction decoder's soft output information or by means of decision-feedback local SNR estimations.

The delay between the channel quality estimation and the actual transmission of the OFDM symbol in relation to the maximal Doppler frequency of the channel is crucial to the adaptive system's performance. If the channel estimate is obsolete at the time of transmission, then poor system performance will result. For a closed–loop adaptive system the delays between channel estimation and transmission of the packet are generally longer than for an open–loop adaptive system, and therefore the Doppler frequency of the channel is a more critical parameter for the system's performance than in open–loop adaptive systems.

13.6.2.2 Parameter Adaptation

Different transmission parameters - such as the modulation and coding modes - can be adapted to the anticipated channel conditions. Adapting the number of modulation levels in response to the anticipated local SNR encountered in each subcarrier can be employed, in order to achieve a wide range of different trade–offs between the received data integrity and throughput. Corrupted subcarriers can be excluded from data transmission and left blank or used, for example, for Crest–factor reduction. A range of different algorithms for selecting the appropriate modulation modes have been proposed in the literature [168, 420].

The adaptive channel coding parameters include code rate, adaptive interleaving and puncturing for convolutional and turbo codes, or varying block lengths for block codes [317]. These techniques can be combined with adaptive modulation mode selection.

Based on the estimated frequency–domain channel transfer function, spectral pre–equalization at the transmitter of one or both communicating stations can be invoked, in order to partially or fully counteract the frequency–selective fading of the time–dispersive channel. Unlike frequency–domain equalization at the receiver — which corrects for the amplitude– and phase–errors inflicted upon the subcarriers by the channel — spectral pre–equalization at the OFDM transmitter can deliver near–constant signal–to–noise levels for all subcarriers. Hence the above concept can be interpreted as power control on a subcarrier–by–subcarrier basis.

In addition to improving the system's BER performance in time–dispersive channels, spectral pre–equalization can be employed in order to perform all channel estimation and equalization functions at only one of the two communicating duplex stations. Low–cost, low–power consumption mobile stations can communicate with a base station that performs the channel estimation and frequency–domain equalization of the up–link, and uses the estimated channel transfer function for pre–equalizing the down–link OFDM symbol. This setup would lead to different overall channel quality on the up–link and down–link, and the superior down–link channel quality could be exploited by using a computationally less complex channel decoder having weaker error correction capabilities in the mobile station than in the base station.

If the channel's frequency–domain transfer function is to be fully counteracted by the spectral pre-equalization upon adapting the subcarrier power to the inverse of the channel transfer function, then the output power of the transmitter can become excessive, if heavily faded subcarriers are present in the system's frequency range. In order to limit the transmitter's maximal output power, hybrid channel pre–equalization and adaptive modulation schemes can be devised, which would deactivate transmission in deeply faded subchannels, while retaining the benefits of pre–equalization in the remaining subcarriers.

13.6.2.3 Signalling the Parameters

Signalling plays an important role in adaptive systems and the range of signalling options is summarised in Figure 13.17 for both open–loop and closed–loop signalling, as well as for blind detection. If the channel quality estimation and parameter adaptation have been performed at the transmitter of a particular link, based on open–loop adaptation, then the resulting set of parameters has to be communicated to the receiver in order to successfully demodulate and decode the OFDM symbol. If the receiver itself determines the requested parameter set to be used by the remote transmitter — the closed–loop scenario — then the same amount of information has to be transported to the remote transmitter in the reverse link. If this signalling information is corrupted, then the receiver is generally unable to correctly decode the OFDM symbol corresponding to the incorrect signalling information.

Unlike adaptive serial systems, which employ the same set of parameters for all data symbols in a transmission packet [60, 61], adaptive OFDM systems have to react to the frequency-selective nature of the channel, by adapting the modem parameters across the subcarriers. The

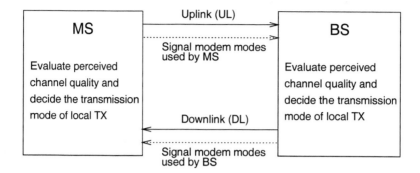

(a) Reciprocal channel, open–loop control

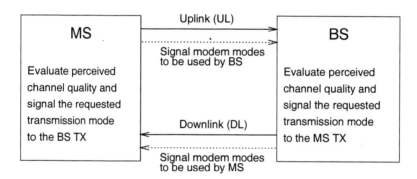

(b) Non–reciprocal channel, closed–loop signalling

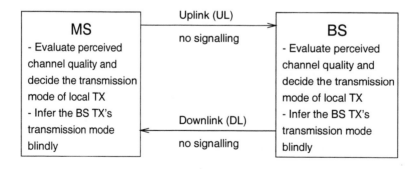

(c) Reciprocal channel, blind modem–mode detection

Figure 13.17: Signalling scenarios in adaptive modems.

resulting signalling overhead could become significantly higher than that for serial modems, and can be prohibitive, for example, for subcarrier–by–subcarrier modulation mode adaptation. In order to overcome these limitations, efficient and reliable signalling techniques have to be employed for practical implementation of adaptive OFDM modems.

If some flexibility in choosing the transmission parameters is sacrificed in an adaptation scheme, like in the sub–band adaptive OFDM schemes described below [168, 420], then the amount of signalling can be reduced. Alternatively, blind parameter detection schemes can be devised, which require little or no signalling information, respectively. A simple blind modulation scheme detection algorithms will be highlighted in Section 13.6.4.2 [168, 420].

The effects of transmission parameter adaptation for OFDM systems on the overall communication system have to be appraised in at least the following areas: data buffering and latency due to varying data throughput [60], the effects of co–channel interference and bandwidth efficiency [61].

13.6.3   Choice of the Modulation Modes

In [168, 420] the two communicating stations use the open–loop predicted channel transfer function acquired from the most recent received OFDM symbol, in order to allocate the appropriate modulation modes to the subcarriers. The modulation modes were chosen from the set of Binary Phase Shift Keying (BPSK), Quadrature Phase Shift Keying (QPSK), 16-level Quadrature Amplitude Modulation (16QAM), as well as "No Transmission", for which no signal was transmitted. These modulation modes are denoted by M_n, where $n \in (0, 1, 2, 4)$ is the number of data bits associated with one data symbol of each mode.

In order to keep the system complexity low, the modulation mode was not varied on a subcarrier–by–subcarrier basis, but instead the total OFDM bandwidth of 512 subcarriers was split into blocks of adjacent subcarriers, referred to as sub–bands, and the same modulation scheme was employed for all subcarriers of the same sub–band. This substantially simplified the task of signalling the modem mode and rendered the employment of alternative blind detection mechanisms feasible, which will be discussed in Section 13.6.4.

Three sub–band modulation mode allocation algorithms were investigated in the literature [168, 420]: a fixed threshold controlled algorithm, an upper–bound BER estimator and a fixed–throughput adaptation algorithm.

13.6.3.1   Fixed Threshold Adaptation Algorithm

The fixed threshold algorithm was derived from the adaptation algorithm proposed by Torrance for serial modems [37]. In the case of a serial modem, the channel quality was assumed to be constant for all symbols in the time slot, and hence the channel had to be slowly varying, in order to allow accurate channel quality prediction. Under these circumstances, all data symbols in the transmit time slot employed the same modulation mode, chosen according to the predicted SNR. The SNR thresholds for a given long–term target BER were determined by Powell–optimization [37]. Torrance assumed two uncoded target bit error rates: 1% for a high data rate "speech" system, and 10^{-4} for a higher integrity, lower data rate "data" system. The resulting SNR thresholds l_n invoked for activating a given modulation mode M_n in a slowly Rayleigh fading narrow-band channel for both systems are given in Table 13.3. Specifically, the modulation mode M_n is selected if the instantaneous channel SNR exceeds

	l_0	l_1	l_2	l_4
speech system	$-\infty$	3.31	6.48	11.61
data system	$-\infty$	7.98	10.42	16.76

Table 13.3: Optimized switching levels for adaptive modulation over Rayleigh fading channels for the "speech" and "data" system, shown in instantaneous channel SNR [dB] (from [37]).

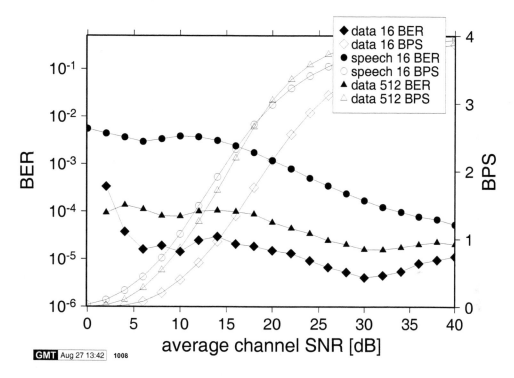

Figure 13.18: BER and Bit Per Symbol (BPS) throughput performance of the 16 sub–band 512–subcarrier switching level adaptive OFDM modem employing BPSK, QPSK, 16QAM and "no transmission" over the Rayleigh fading time dispersive channel of Figure 13.3 using the switching thresholds of Table 13.3.

the switching level l_n.

This adaptation algorithm originally assumed a constant instantaneous SNR over all of the transmission burst's symbols, but in the case of an OFDM system in a frequency-selective channel the channel quality varies across the different subcarriers. For sub–band adaptive OFDM transmission, this implies that if the the sub–band width is wider than the channel's coherence bandwidth [317], then the original switching algorithm cannot be employed. In the performance evaluations the lowest quality subcarrier in the sub–band was employed for the adaptation algorithm based on the thresholds given in Table 13.3. The performance of the 16 sub–band adaptive system over the WATM Rayleigh fading channel of Figure 13.3 is shown in Figure 13.18.

Adjacent or consecutive timeslots have been used for the up–link and down–link slots in these simulations, so that the delay between channel estimation and transmission was rendered as short as possible. Figure 13.18 shows the long–term average BER and throughput of the studied modem for the "speech" and "data" switching levels of Table 13.3 as well as for a subcarrier–by–subcarrier adaptive modem employing the "data" switching levels. The results show the typical behaviour of a variable–throughput Adaptive OFDM (AOFDM) system, which constitutes a tradeoff between the best BER and best throughput performance. For low SNR values, the system achieves a low BER by transmitting very few bits and only when the channel conditions allow. With increasing long–term SNR, the throughput increases, without significant change in the BER. For high SNR values the BER drops as the throughput approaches its maximum of 4 bits per symbol, since the highest–order constellation was 16QAM.

It can be seen from the figure that the adaptive system performs better than its target bit error rates of 10^{-2} and 10^{-4} for the "speech" and "data" system, respectively, resulting in measured bit error rates lower than the targets. This can be explained by the adaptation regime, which was based on the conservative principle of using the lowest quality subcarrier in each sub–band for channel quality estimation, leading to a pessimistic channel quality estimate for the entire sub–band. For low values of SNR, the throughput in bits per data symbol is low and exceeds the fixed BPSK throughput of 1 bit/symbol only for SNR values in excess of 9.5 dB and 14 dB for the "speech" and "data" systems, respectively.

The upper–bound performance of the system with subcarrier–by–subcarrier adaptation is also portrayed in the figure, shown as 512 independent sub–bands, for the "data" optimized set of threshold values. It can be seen that in this case the target BER of 10^{-4} is closely met over a wide range of SNR values from about 2 dB to 20 dB, and that the throughput is considerably higher than in the case of the 16 sub–band modem. This is the result of more accurate subcarrier-by-subcarrier channel quality estimation and fine–grained adaptation, leading to better exploitation of the available channel capacity.

Figure 13.19 shows the long–term modulation mode histograms for a range of channel SNR values for the "data" switching levels in both the 16 sub–band and the subcarrier-by-subcarrier adaptive modems using the switching thresholds of Table 13.3. Comparison of the graphs shows that higher order modulation modes are used more frequently by the subcarrier–by–subcarrier adaptation algorithm, which is in accordance with the overall throughput performance of the two modems in Figure 13.18.

The throughput penalty of employing sub–band adaptation depends on the frequency–domain variation of the channel transfer function. If the sub–band bandwidth is lower than the channel's coherence bandwidth, then the assumption of constant channel quality per sub–band is closely met, and the system performance is equivalent to that of a subcarrier–by–subcarrier adaptive scheme.

13.6.3.2   Sub–Band BER Estimator Adaptation Algorithm

We have seen above that the fixed switching level based algorithm leads to a throughput performance penalty, if used in a sub–band adaptive OFDM modem, when the channel quality is not constant throughout each sub–band. This is due to the conservative adaptation based on the subcarrier experiencing the most hostile channel in each sub–band.

An alternative scheme taking into account the non–constant SNR values γ_j across the N_s

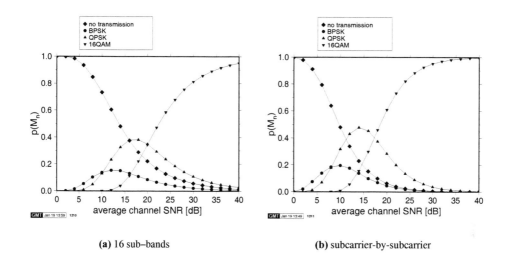

(a) 16 sub-bands   (b) subcarrier-by-subcarrier

Figure 13.19: Histograms of modulation modes versus channel SNR for the "data" switching level adaptive 512–subcarrier 16 sub–band OFDM modem over the Rayleigh fading time dispersive channel of Figure 13.3 using the switching thresholds of Table 13.3.

subcarriers in the j-th sub–band can be devised by calculating the expected overall bit error probability for all available modulation modes M_n in each sub–band, which is denoted by $\bar{p}_e(n) = 1/N_s \sum_j p_e(\gamma_j, M_n)$. For each sub–band, the mode having the highest throughput, whose estimated BER is lower than a given threshold, is then chosen. While the adaptation granularity is still limited to the sub–band width, the channel quality estimation includes not only the most corrupted subcarrier, which leads to an improved throughput.

Figure 13.20 shows the BER and throughput performance for the 16 sub–band adaptive OFDM modem employing the BER estimator adaptation algorithm in the Rayleigh fading time dispersive channel of Figure 13.3. The two sets of curves in the figure correspond to target bit error rates of 10^{-2} and 10^{-1}, respectively. Comparing the modem's performance for a target BER of 10^{-2} with that of the "speech" modem in Figure 13.18, it can be seen that the BER estimator algorithm results in significantly higher throughput, while meeting the BER requirements. The BER estimator algorithm is readily adjustable to different target bit error rates, which is demonstrated in the figure for a target BER of 10^{-1}. Such adjustability is beneficial, when combining adaptive modulation with channel coding, as will be discussed in Section 13.6.5.

13.6.4 Signalling and Blind Detection

The adaptive OFDM receiver has to be informed of the modulation modes used for the different sub–bands. This information can either be conveyed using signalling subcarriers in the OFDM symbol itself, or the receiver can employ blind detection techniques in order to estimate the transmitted symbols' modulation modes, as seen in Figure 13.17 [168,420].

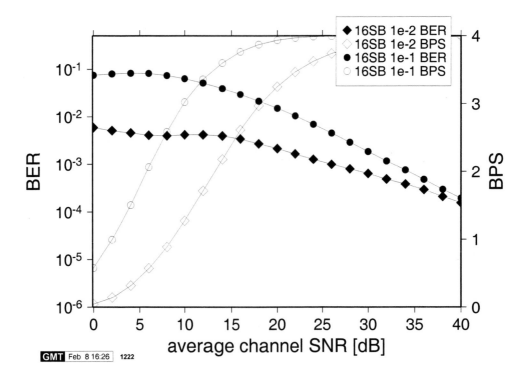

Figure 13.20: BER and BPS throughput performance of the 16 sub–band 512–subcarrier **BER estima-**
tor adaptive OFDM modem employing BPSK, QPSK, 16QAM and "no transmission"
over the Rayleigh fading time dispersive channel of Figure 13.3.

13.6.4.1 Signalling

The simplest way of signalling the modulation mode employed in a sub–band is to replace
one data symbol by an M-PSK symbol, where M is the number of possible modulation modes.
In this case, reception of each of the constellation points directly signals a particular modulation mode in the current sub–band. For four modulation modes - and assuming perfect
phase recovery - the probability of a signalling error $p_s(\gamma)$, when employing one signalling
symbol, is the symbol error probability of QPSK. Then the correct sub–band mode signalling
probability is:

$$(1 - p_s(\gamma)) = (1 - p_{b,QPSK}(\gamma))^2, \qquad (13.20)$$

where $p_{b,QPSK}$ is the bit error probability for QPSK:

$$p_{b,QPSK}(\gamma) = Q(\sqrt{\gamma}) = \frac{1}{2} \cdot \mathrm{erfc}\left(\sqrt{\frac{\gamma}{2}}\right), \qquad (13.21)$$

which leads to the expression for the modulation mode signalling error probability of

$$p_s(\gamma) = 1 - \left(1 - \frac{1}{2} \cdot \mathrm{erfc}\left(\sqrt{\frac{\gamma}{2}}\right)\right)^2. \qquad (13.22)$$

The modem mode signalling error probability can be reduced by employing multiple signalling symbols and maximum ratio combining of the received signalling symbols $R_{s,n}$, in order to generate the decision variable R'_s prior to decision:

$$R'_s = \sum_{n=1}^{N_s} R_{s,n} \cdot \hat{H}^*_{s,n}, \tag{13.23}$$

where N_s is the number of signalling symbols per sub–band, the quantities $R_{s,n}$ are the received symbols in the signalling subcarriers, and $\hat{H}_{s,n}$ represents the estimated values of the frequency domain channel transfer function at the signalling subcarriers. Assuming perfect channel estimation and constant values of the channel transfer function across the group of signalling subcarriers, the signalling error probability for N_s signalling symbols can be expressed as:

$$p'_s(\gamma, N_s) = 1 - \left(1 - \frac{1}{2} \cdot \text{erfc}\left(\sqrt{\frac{N_s \gamma}{2}}\right)\right)^2. \tag{13.24}$$

Figure 13.21 shows the signalling error rate in an AWGN channel for 1, 2, 4 and 8 signalling symbols per sub–band, respectively. It can be seen that doubling the number of signalling subcarriers improves the performance by 3 dB. Modem mode detection error ratios (DER) below 10^{-5} can be achieved at 10 dB SNR over AWGN channels if two signalling symbols are used. The signalling symbols for a given sub–band can be interleaved across the entire OFDM symbol bandwidth, in order benefit from frequency diversity in fading wide-band channels.

As seen in Figure 13.17, blind detection algorithms aim to estimate the employed modulation mode directly from the received data symbols, therefore avoiding the loss of data capacity due to signalling subcarriers. Two algorithms have been proposed in the literature [168,420], one based on SNR estimation, and another one incorporating error correction coding. Let us briefly highlight the conceptually more simple one in the next section.

13.6.4.2   Blind Detection by SNR Estimation

The receiver has no *a priori* knowledge of the modulation mode employed in a particular received sub–band and estimates this parameter by quantising the de–faded received data symbols $R_n/\hat{H}_n$ in the sub–band to the closest symbol $\hat{R}_{n,m}$ for all possible modulation modes M_m for each subcarrier index n in the current sub–band. The decision–directed error energy e_m between the ideal constellation phasor positions and the received phasors is calculated for each modulation mode according to:

$$e_m = \sum_n \left(R_n/\hat{H}_n - \hat{R}_{n,m}\right)^2 \tag{13.25}$$

and then the modulation mode M_m which minimises e_m is chosen for the demodulation of the sub–band.

The DER of the blind modulation mode detection algorithm described in this section for a 512–subcarrier OFDM modem in an AWGN channel is depicted in Figure 13.22. It can be

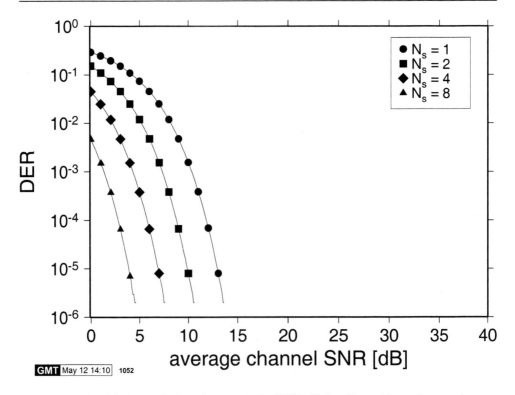

GMT May 12 14:10 | 1052

Figure 13.21: Modulation mode detection error ratio (DER), if signalling with maximum ratio combining is employed for QPSK symbols in an AWGN channel for 1, 2, 4 and 8 signalling symbols per sub–band, evaluated from Equation 13.24.

seen that the detection performance depends on the number of symbols per sub–band, with fewer sub–bands and therefore longer symbol sequences per sub–band leading to a better detection performance. It is apparent, however, that the number of available modulation modes has a more significant effect on the detection reliability than the block length. If all four legitimate modem modes are employed, then reliable detection of the modulation mode is only guaranteed for AWGN SNR values of more than 15–18 dB, depending on the number of sub–bands per OFDM symbol. If only M_0 and M_1 are employed, however, the estimation accuracy is dramatically improved. In this case, AWGN SNR values above 5–7 dB are sufficient to ensure reliable detection.

Figure 13.23 shows the BER performance of the fixed–threshold "data"–type 16 sub–band adaptive system in the fading wideband channel of Figure 13.3 for both sets of modulation modes, namely for (M_0, M_1) and (M_0, M_1, M_2, M_4) with blind modulation mode detection. Erroneous modulation mode decisions were assumed to yield a BER of 50% in the received block. This is optimistic, since in a realistic scenario the receiver would have no knowledge of the number of bits actually transmitted, leading to loss of synchronization in the data stream. This problem is faced by all systems having a variable throughput and not employing an ideal reliable signalling channel. This impediment must be mitigated by data

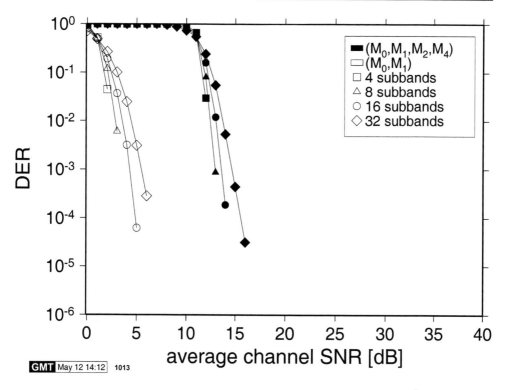

$$\text{(legend: } \blacksquare\ (M_0,M_1,M_2,M_4);\ \square\ (M_0,M_1);\ \square\ \text{4 subbands};\ \triangle\ \text{8 subbands};\ \circ\ \text{16 subbands};\ \diamond\ \text{32 subbands)}$$

Figure 13.22: Blind modulation mode detection error ratio (DER) for 512–subcarrier OFDM systems employing (M_0, M_1) as well as for (M_0, M_1, M_2, M_4) for different numbers of sub–bands in an AWGN channel.

synchronization measures.

It can be seen from Figure 13.23 that while blind modulation mode detection yields poor performance for the quadruple–mode adaptive scheme, the twin–mode scheme exhibits BER results consistently better than 10^{-4}.

13.6.5 Sub–Band Adaptive OFDM and Channel Coding

Adaptive modulation can reduce the BER to a level, where channel decoders can perform well. Figure 13.24 shows both the uncoded and coded BER performance of a 512–subcarrier OFDM modem in the fading wideband channel of Figure 13.3, assuming perfect channel estimation. The channel coding employed in this set of experiments was a turbo coder [152, 155] with a data block length of 1000 bits, employing a random interleaver and 8 decoder iterations. The log–MAP decoding algorithm was used [162]. The constituent half–rate convolutional encoders were of constraint length 3, with octally represented generator polynomials of $(7, 5)$ [317]. It can be seen that the turbo decoder provides a considerable coding gain for the different fixed modulation schemes, with a BER of 10^{-4} for SNR values of 13.8 dB, 17.3 dB and 23.2 dB for BPSK, QPSK and 16QAM transmission, respectively.

Figure 13.25 depicts the BER and throughput performance of the same decoder employed

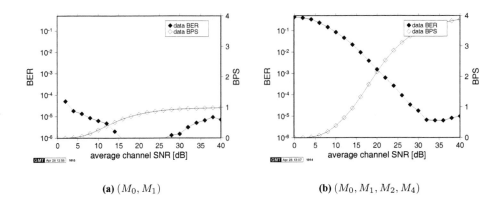

(a) (M_0, M_1) **(b)** (M_0, M_1, M_2, M_4)

Figure 13.23: BER and BPS throughput performance of a 16 sub–band 512–subcarrier adaptive OFDM modem employing (a) No Transmission (M_0) and BPSK (M_1) or (b) (M_0, M_1, M_2, M_4), both using the data–type switching levels of Table 13.3 and the SNR–based blind modulation mode detection of Section 13.6.4.2 over the Rayleigh fading time–dispersive channel of Figure 13.3.

in conjunction with the adaptive OFDM modem for different adaptation algorithms. Figure 13.25(a) shows the performance for the "speech" system employing the switching levels listed in Table 13.3. As expected, the half–rate channel coding results in a halved throughput compared to the uncoded case, but offers low–BER transmission over the channel of Figure 13.3 for SNR values of down to 0 dB, maintaining a BER below 10^{-6}.

Further tuning of the adaptation parameters can ensure a better average throughput, while retaining error–free data transmission. The switching level based adaptation algorithm of Table 13.3 is difficult to control for arbitrary bit error rates, since the set of switching levels was determined by an optimization process for uncoded transmission. Since the turbo–codec has a non–linear BER versus SNR characteristic, direct switching–level optimization is an arduous task. The sub–band BER predictor of Section 13.6.3.2 is easier to adapt to a channel codec, and Figure 13.25(b) shows the performance for the same decoder, with the adaptation algorithm employing the BER–prediction method having an upper BER–bound of 1%. It can be seen that the less stringent uncoded BER constraints when compared to Figure 13.25(a) lead to a significantly higher throughput for low SNR values. The turbo–decoded data bits are error–free, hence a further increase in throughput is possible while maintaining a high degree of coded data integrity.

The second set of curves in Figure 13.25(b) show the system's performance, if an uncoded target BER of 10% is assumed. In this case, the turbo decoder's output BER is below 10^{-5} for all the SNR values plotted, and shows a slow decrease for increasing values of SNR. The throughput of the system, however, exceeds 0.5 data bits per symbol for SNR values of more than 2 dB. In closing, we note that it is an attractive adaptive OFDM property that the adaptive modulation mode selection results in less bursty error statistics than fixed modulation, which improves the achievable performance of the channel codec.

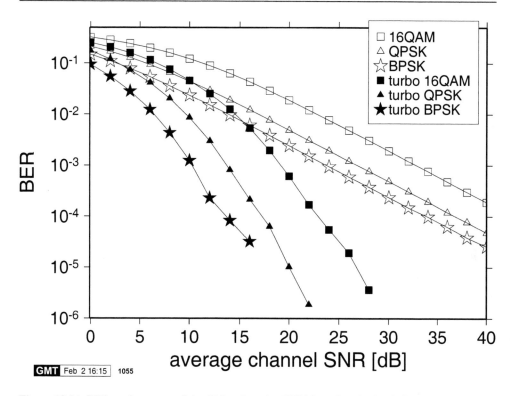

Figure 13.24: BER performance of the 512–subcarrier OFDM modem in the fading time–dispersive channel of Figure 13.3 for both uncoded and half–rate turbo–coded transmission, using 8–iteration log–MAP turbo decoding, 1000–bit random interleaver, and a constraint length of 3.

13.7 Pre–Equalization

We have seen above how the receiver's estimate of the channel transfer function can be employed by the transmitter in order to dramatically improve the performance of an OFDM system by adapting the subcarrier modulation modes to the channel conditions. For subchannels exhibiting a low signal–to–noise ratio, robust modulation modes are used, while for subcarriers having a high SNR, high throughput multi–level modulation modes can be employed. An alternative approach to combating the frequency-selective channel behaviour was proposed in [514], applying pre–equalization to the OFDM symbol prior to transmission on the basis of the anticipated channel transfer function. We will highlight a range of related topics in this section.

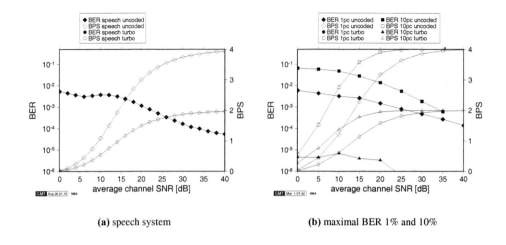

(a) speech system   (b) maximal BER 1% and 10%

Figure 13.25: BER and BPS throughput performance of 16 sub–band 512–subcarrier adaptive turbo coded and uncoded OFDM modem employing (M_0, M_1, M_2, M_4) for (a) speech type switching levels of Table 13.3 and (b) a maximal estimated sub–band BER of 1% and 10% over the channel of Figure 13.3. The turbo coded transmission over the speech system and the 1% maximal BER system are error free for all examined SNR values and therefore the corresponding BER curves are omitted from the graphs.

13.7.1   Motivation

As discussed above, the received data symbol R_n of subcarrier n over a slowly time–varying time–dispersive channel can be characterized by:

$$R_n = S_n \cdot H_n + n_n, \tag{13.26}$$

where S_n is the transmitted data symbol, H_n is the channel transfer function of subcarrier n, and n_n is a noise sample.

The pilot-based frequency–domain equalization at the receiver — which is necessary for non–differential detection of the data symbols — corrects the phase and amplitude of the received data symbols using the estimate of the channel transfer function $\hat{H}_n$ as follows:

$$R'_n = R_n/\hat{H}_n = S_n \cdot H_n/\hat{H}_n + n_n/\hat{H}_n. \tag{13.27}$$

If the estimate $\hat{H}_n$ is accurate, this operation de–fades the constellation points before decision. However, upon de–fading the noise sample n_n is amplified by the same amount as the signal, therefore failing to improve the SNR of the received sample.

Pre–equalization for the OFDM modem operates by scaling the data symbol of subcarrier n, S_n, by a pre–equalization function E_n, computed from the inverse of the anticipated channel transfer function, prior to transmission. At the receiver, no equalization is performed, hence the received symbols can be expressed as:

$$R_n = S_n \cdot E_n \cdot H_n + n_n. \tag{13.28}$$

Since no equalization is performed, there is no noise amplification at the receiver. Similarly to the adaptive modulation techniques illustrated above, pre–equalization is only applicable to a duplex link, since the transmitted signal is adapted to the specific channel conditions perceived by the receiver. Like for other adaptive schemes, the transmitter needs an estimate of the current frequency–domain channel transfer function, which can be obtained from the received signal in the reverse link, as seen in Figure 13.17.

13.7.2   Pre–Equalization with Sub–Band Blocking

Direct channel inversion at the transmitter is not practical, as the output power fluctuations are prohibitive; excluding those subcarriers that are faded too low can be used to limit the necessary transmit power. Analogously to the adaptive modulation schemes above, the transmitter decides for all subcarriers in each sub–band, whether to transmit data or not. If pre–equalization is possible under the power constraints, then the subcarriers are modulated with the pre–equalized data symbols. The information whether a sub–band is used for transmission or not is signalled to the receiver.

Since no attempt is made to transmit in the sub–bands that cannot be pre–equalized, the power not employed in the blank subcarriers can be used for 'boosting' the data–bearing sub–bands. This scheme allows for a more flexible pre–equalization algorithm than the fixed threshold based method described above, which is summarised as follows:

- Calculate the necessary transmit power p_n for each sub–band n, assuming perfect pre–equalization.

- Sort sub–bands according to their required transmit power p_n.

- Select sub–band n with the lowest power p_n, and add p_n to the total transmit power. Repeat this procedure with the next–lowest power, until no further sub–bands can be added without the total power $\sum p_j$ exceeding the power limit l.

Figure 13.26 depicts the 16–QAM BER performance over the WATM channel of Figure 13.3. Three different blocking thresholds, namely 0, 6 and 12 dB, were used, where sub–bands requiring more power than the allowable maximum were blocked from transmission. The BER floor stems from the channel variability, due to the time delay between channel estimation and transmission. The average throughput figures for the 6 dB and 12 dB symbol power limits are 3.54 and 3.92 bits per data symbol, respectively. It can be noted that the BER floor is lower for $l = 6$dB than for $l = 12$dB. This is because the effects of the channel variation due to the delay between the instants of channel estimation and reception in the faded subcarriers on the equalization function is much more dramatic for low-quality sub–bands requiring more 'power boosting' than in the higher–quality subcarriers. The lower the total symbol power limit l, the fewer the number of low–quality subcarriers used for transmission. Although not explicitly shown in the figure, we note that for both $l = 6$dB and $l = 12$dB, the BER performance of the blocking modem is better than that of a modem employing full channel inversion, provided that 1 time slot delay is assumed. Again, the reason for this is the exclusion of the deeply faded corrupted subcarriers. If the OFDM symbol power is limited to 0 dB, then the BER floor drops to $1.5 \cdot 10^{-6}$ at the expense of the throughput, which attains 2.5 BPS. Figure 13.27 depicts the mean OFDM symbol power

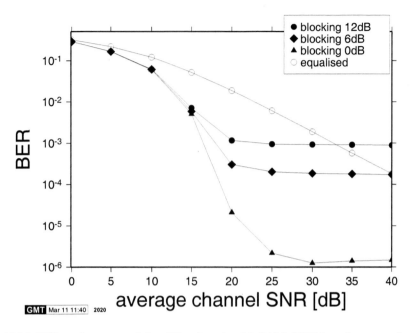

Figure 13.26: BER performance of the 512–subcarrier 16–QAM OFDM modem over the fading
WATM channel of Figure 13.3 employing 16 sub–band pre–equalization with blocking and a delay of 1 timeslot between the instants of perfect channel estimation and
reception.

histogram for this scenario. It can be seen that - as expected - the 6 dB-limited scheme has a
more compact symbol power PDF, which is associated with less dramatic power fluctuation.

13.7.3   Adaptive Modulation with Spectral Pre–Equalization

The pre–equalization algorithm discussed above inverts the channel's anticipated transfer
function, in order to transform the resulting channel into a Gaussian–like non–fading channel, whose SNR is dependent only on the path–loss. Sub–band blocking has been introduced
above, in order to limit the transmitter's output power, while maintaining the near–constant–
SNR across the used subcarriers. The pre–equalization algorithms discussed above do not
cancel out the channel's path loss, but rely on the receiver's gain control algorithm to automatically account for the channel's average path–loss.

   We have seen in Section 13.6 on adaptive modulation algorithms that maintaining Gaussian channel characteristics is not the most efficient way of exploiting the channel's time–
variant capacity. If maintaining a constant data throughput is not required by the rest of the
communications system, then a fixed BER scheme in conjunction with error correction coding can assist in maximising the system's throughput. The results presented for the target–
BER adaptive modulation scheme in Figure 13.25(b) showed that for the particular turbo
coding scheme used an uncoded BER of 1% resulted in error–free channel coded data transmission, and that for an uncoded target BER of 10% the turbo decoded data BER was below

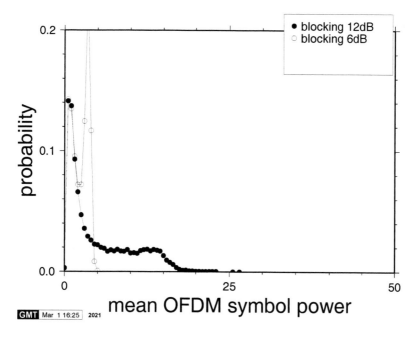

Figure 13.27: OFDM symbol power histogram for 512–subcarrier 16 sub–band pre–equalization with blocking over the WATM channel of Figure 13.3 using 16QAM. The corresponding BER curves are given in Figure 13.26.

10^{-5}. We have seen that it was impossible to exactly reach the anticipated uncoded target BER with the adaptive modulation algorithm, since the adaptation algorithm operated in discrete steps between modulation modes.

Combining the target–BER adaptive modulation scheme and spectral pre–equalization allows the transmitter to react to the channel's time– and frequency–variant nature, in order to fine–tune the behaviour of the adaptive modem in fading channels. It also allows the transmitter to invest the energy that is not used in "no transmission" sub–bands into the other sub–bands without affecting the equalization at the receiver.

The combined algorithm for adaptive modulation with spectral pre–equalization described here does not intend to invert the channel's transfer function across the OFDM symbol's range of subcarriers, it is therefore not a pure pre–equalization algorithm. Instead, the aim is to transmit a sub–band's data symbols at a power level which ensures a given target SNR at the receiver, that is constant for all subcarriers in the sub–band, which in turn results in the required BER. Clearly, the receiver has to anticipate the different relative power levels for the different modulation modes, so that error–free demodulation of the multi–level modulation modes employed can be ensured.

The joint adaptation algorithm requires the estimates of the noise floor level at the receiver as well as the channel transfer function, which includes the path–loss. On the basis of these values, the necessary amplitude of E_n required to transmit a data symbol over the subcarrier

target BER	10^{-4}	10^{-2}	10^{-1}
SNR(BPSK)[dB]	8.4	4.33	−0.85
SNR(QPSK)[dB]	11.42	7.34	2.16
SNR(16QAM)[dB]	18.23	13.91	7.91

Table 13.4: Required target SNR levels for 1% and 10% target BER for the different modulation schemes over an AWGN channel.

n for a given subcarrier SNR of γ_n can be calculated as follows:

$$|E_n| = \frac{\sqrt{N_0 \cdot \gamma_n}}{\left|\hat{H}_n\right|}, \tag{13.29}$$

where N_0 is the noise floor at the receiver. The phase of E_n is used for the pre–equalization, and hence:

$$\angle E_n = -\angle \hat{H}_n. \tag{13.30}$$

The target SNR of subcarrier n, γ_n, is dependent on the modulation mode that is signalled over the subcarrier, and determines the system's target BER. We have identified three sets of target SNR values for the modulation modes, with uncoded target BER values of 1% and 10% for use in conjunction with channel coders, as well as 10^{-4} for transmission without channel coding. Table 13.4 gives an overview of these levels, which have been read from the BER performance curves of the different modulation modes in a Gaussian channel.

Figure 13.28 shows the performance of the joint pre–equalization and adaptive modulation algorithm over the fading time–dispersive WATM channel of Figure 13.3 for the set of different target BER values of Table 13.4, as well as the comparison curves of the perfectly equalised 16QAM modem under the same channel conditions. It can be seen that the BER achieved by the system is close to the BER targets. Specifically, for a target BER of 10%, no perceptible deviation from the target has been recorded, while for the lower BER targets the deviations increase for higher channel SNRs. For a target BER of 1%, the highest measured deviation is at the SNR of 40 dB, where the recorded BER is 1.36%. For the target BER of 10^{-4}, the BER deviation is small at 0 dB SNR, but at an SNR of 40 dB the experimental BER is $2.2 \cdot 10^{-3}$. This increase of the BER with increasing SNR is due to the rapid channel variations in the deeply faded subcarriers, which are increasingly used at higher SNR values. The half–tone curve in the figure denotes the system's performance, if no delay is present between the channel estimation and the transmission. In this case, the simulated BER shows only very little deviation from the target BER value. This is consistent with the behaviour of the full channel inversion pre–equalizing modem.

13.8   Review and Discussion

We commenced with a historical perspective on OFDM transmissions with reference to the literature of the past 30 years. The advantages and disadvantages of various OFDM techniques were considered briefly and the expected performance was characterized for the sake

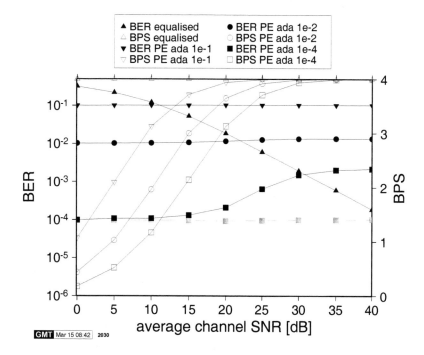

Figure 13.28: BER performance and BPS throughput of the 512–subcarrier 16 sub–band adaptive OFDM modem with spectral pre–equalization over the Rayleigh fading time dispersive WATM channel of Figure 13.3, and that of the perfectly equalised 16QAM modem. The half–tone BER curve gives the performance of the adaptive modem for a target BER of 10^{-4} with no delay between channel estimation and transmission, while the other results assume 1 timeslot delay between up–link and down–link.

of illustration in the context of WATM systems. Our discussions deepened as we approached the subject of adaptive subcarrier modem mode allocation and channel coding. Here we would like to close with a brief discussion comparing the various coded and uncoded, fixed and adaptive OFDM modems and identify future research issues.

Specifically, Figure 13.29 compares the different adaptive modulation schemes discussed. The comparison graph is split into two sets of curves, depicting the achievable data throughput for a data BER of 10^{-4} highlighted for the fixed throughput systems in Figure 13.29(a), and for the time–variant–throughput systems in Figure 13.29(b).

The fixed throughput systems — highlighted in black in Figure 13.29(a) — comprise the non–adaptive BPSK, QPSK and 16QAM modems, as well as the fixed–throughput adaptive scheme, both for coded and uncoded applications. The non–adaptive modems' performance is marked on the graph as diamonds, and it can be seen that the uncoded fixed schemes require the highest channel SNR of all reviewed transmission methods to achieve a data BER of 10^{-4}. Channel coding employing the turbo coding schemes considered dramatically improved the SNR requirements, at the expense of half the data throughput. The uncoded fixed–throughput (FT) adaptive scheme, marked by the filled triangles, yielded consistently

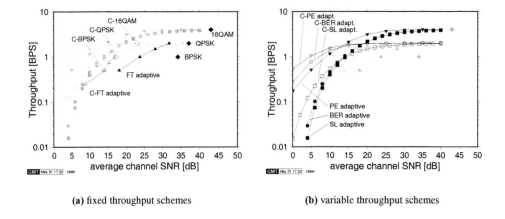

(a) fixed throughput schemes   **(b)** variable throughput schemes

Figure 13.29: BPS throughput versus average channel SNR for non–adaptive and adaptive modulation as well as for pre–equalized adaptive techniques, for a data bit error rate of 10^{-4}.
Note that for the coded schemes the achieved BER values are lower than 10^{-4}. (a)
Fixed throughput systems: coded (C-) and uncoded BPSK, QPSK, 16QAM, and fixed
throughput (FT) adaptive modulation. (b) Variable throughput systems: coded (C-) and
uncoded switching level adaptive (SL), target–BER adaptive (BER) and pre–equalized
adaptive (PE) systems. Note that the separately plotted variable–throughput graph also
shows the lightly shaded benchmark curves of the complementary fixed–rate schemes
and vice versa.

worse data throughput than the coded (C-) fixed modulation schemes C-BPSK, C-QPSK and
C-16QAM, with its throughput being about half the coded fixed scheme's at the same SNR
values. The coded FT–adaptive (C-FT) system, however, delivered a fairly similar throughput
to the C-BPSK and C-QPSK transmissions, and were capable of delivering a BER of 10^{-4}
for SNR values down to about 9 dB.

The variable throughput schemes, highlighted in Figure 13.29(b), outperformed the comparable fixed throughput algorithms [420]. For high SNR values, all uncoded schemes' performance curves converged to a throughput of 4bits/symbol, which was equivalent to 16QAM
transmission. The coded schemes reached a maximal throughput of 2BPS. Of the uncoded
schemes, the "data" switching–level (SL) and target–BER adaptive modems delivered a similar BPS performance, with the target–BER scheme exhibiting slightly better throughput than
the SL adaptive modem. The adaptive modem employing pre–equalization (PE) significantly
outperformed the other uncoded adaptive schemes and offered a throughput of 0.18 BPS at
an SNR of 0 dB, although its crest-factor PDF is slightly less attractive.

The coded transmission schemes suffered from limited throughput at high SNR values,
since the half–rate channel coding limited the data throughput to 2 BPS. For low SNR values,
however, the coded schemes offered better performance than the uncoded schemes, with the
exception of the "speech" SL–adaptive coded scheme, which was outperformed by the uncoded PE-adaptive modem. The poor performance of the coded SL–scheme can be explained
by the lower uncoded target BER of the "speech" scenario, which was 1%, in contrast to the
10% uncoded target BER for the coded BER– and PE–adaptive schemes. The coded PE–

adaptive modem outperformed the target–BER adaptive scheme, thanks to its more accurate control of the uncoded BER, leading to a higher throughput for low SNR values.

It is interesting to observe that for the given set of four modulation modes the uncoded PE–adaptive scheme was close in performance to the coded adaptive schemes, and that for SNR values of more than 14 dB it outperformed all other studied schemes. It is clear, however, that the coded schemes would benefit from higher order modulation modes or higher-rate channel codecs at high SNR, which would allow these modems to increase the data throughput further when the channel conditions allow.

Based on these findings, adaptive-rate channel coding is an interesting area for future research in conjunction with adaptive OFDM, adaptive beam-stearing and interference cancellation. Provided that these OFDM enhancements reach a similar state of maturity to the OFDM components used in the current standard digital audio and video broadcasting, WATM, ADSL and power-line communications proposals, OFDM is likely to find a range of further attractive applications in wireless and wireline communications both for businesses and in the home.

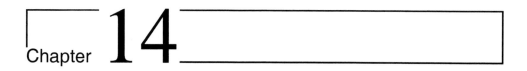
Space-Time Trellis Coding versus Adaptive Modulation

T.H. Liew and L. Hanzo[1]

14.1 Introduction

In [515] the encoding and decoding processes as well as the various design trade-offs of *space-time block codes* [7,218,516] were reviewed. More explicitly, various previously proposed space-time block codes [218,516] have been discussed and their performance was investigated over perfectly interleaved, non-dispersive Rayleigh fading channels. A range of systems consisting of space-time block codes and different channel codecs were investigated. The performance versus estimated complexity trade-off of the different systems was investigated and compared.

In an effort to provide as comprehensive a technology road-map as possible and to identify the most promising schemes in the light of their performance versus estimated complexity, in this chapter we shall explore the family of *space-time trellis codes* [217,517–521], which were proposed by Tarokh *et al*. Space-time trellis codes incorporate jointly designed channel coding, modulation, transmit diversity and optional receiver diversity. The performance criteria for designing space-time trellis codes were outlined in [217], under the assumption that the channel is fading slowly and that the fading is frequency non-selective. It was shown in [217] that the system's performance is determined by matrices constructed from pairs of distinct code sequences. Both the *diversity gain* and *coding gain* of the codes are determined by the minimum rank and the minimum determinant [217,522] of the matrices, respectively. The results were then also extended to fast fading channels. The space-time trellis codes proposed in [217] provide the best tradeoff between data rate, diversity advantage and trellis

[1]Space-Time Trellis Coding and Space-Time Block Coding versus Adaptive Modulation: An Overview and Comparative Study for Transmission over Wideband Channels, submitted to IEEE Tr. on Vehicular Technology, 2001 ©IEEE

complexity.

The performance of both space-time trellis and block codes over narrowband Rayleigh fading channels was investigated by numerous researchers [7, 181, 217, 218, 520]. The investigation of space-time codes was then also extended to the class of practical wideband fading channels. The effect of multiple paths on the performance of space-time trellis codes was studied in [521] for transmission over slowly varying Rayleigh fading channels. It was shown in [521] that the presence of multiple paths does not decrease the diversity order guaranteed by the design criteria used to construct the space-time trellis codes. The evidence provided in [521] was then also extended to rapidly fading dispersive and non-dispersive channels. As a further performance improvement, turbo equalization was employed in [319] in order to mitigate the effects dispersive channels. However space-time coded turbo equalization involved an enormous complexity. In addressing the complexity issues, Bauch *et al.* [523] derived finite-length multi-input multi-output (MIMO) channel filters and used them as prefilters for turbo equalizers. These prefilters significantly reduce the number of turbo equalizer states and hence mitigate the decoding complexity. As an alternative solution, the effect of Inter Symbol Interference (ISI) could be eliminated by employing Orthogonal Frequency Division Multiplexing (OFDM) [4]. A system using space-time trellis coded OFDM is attractive, since the decoding complexity reduced, as demonstrated by the recent surge of research interests [181, 524–526]. In [181, 524, 526], non-binary Reed-Solomon (RS) codes were employed in the space-time trellis coded OFDM systems for improving its performance.

Similarly, the performance of space-time block codes was also investigated over frequency selective Rayleigh fading channels. In [527], a multiple input multiple output equalizer was utilized for equalising the dispersive multipath channels. Furthermore, the advantages of OFDM were also exploited in space-time block coded systems [181, 528, 529].

We commence our discussion with a detailed description of the encoding and decoding processes of the space-time trellis codes in Section 14.2. The state diagrams of a range of other space-time trellis codes are also given in Section 14.2.2. In Section 14.3, a specific system was introduced, which enables the comparison of space-time trellis codes and space-time block codes over wideband channels. Our simulation results are then given in Section 14.4. We continue our investigations by employing space-time coded adaptive modulation based OFDM in Section 14.5. Finally, we conclude in Section 14.6.

14.2   Space-Time Trellis Codes

In this section, we will detail the encoding and decoding processes of space-time trellis codes. Space-time trellis codes are defined by the number of transmitters p, by the associated state diagram and the modulation scheme employed. For ease of explanation, as an example we shall use the simplest 4-state, 4-level Phase Shift Keying (4PSK) space-time trellis code, which has $p = 2$ two transmit antennas.

14.2.1   The 4-State, 4PSK Space-Time Trellis Encoder

At any time instant k, the 4-state 4PSK space-time trellis encoder transmits symbols $x_{k,1}$ and $x_{k,2}$ over the transmit antennas Tx 1 and Tx 2, respectively. The output symbols at time

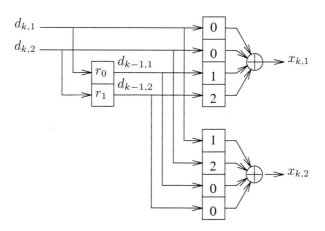

Figure 14.1: The 4-state, 4PSK space-time trellis encoder.

instant k are given by [217]:

$$x_{k,1} = 0.d_{k,1} + 0.d_{k,2} + 1.d_{k-1,1} + 2.d_{k-1,2} \tag{14.1}$$
$$x_{k,2} = 1.d_{k,1} + 2.d_{k,2} + 0.d_{k-1,1} + 0.d_{k-1,2} \tag{14.2}$$

where $d_{k,i}$ represents the current input bits, whereas $d_{k-1,i}$ the previous input bits and $i = 1, 2$. More explicitly, we can represent Equation 14.2 with the aid of a shift register, as shown in Figure 14.1, where $\oplus$ represents modulo 4 addition. Let us explain the operation of the shift register encoder for the random input data bits 01111000. The shift register stages r_0 and r_1 must be reset to zero before the encoding of a transmission frame starts. They represent the state of the encoder. The operational steps are summarised in Table 14.1. Again, given the register stages $d_{k-1,1}$ and $d_{k-1,2}$ as well as the input bits $d_{k,1}$ and $d_{k,2}$, the output symbols seen in the table are determined according to Equation 14.2 or Figure 14.1. Note that the

Input queue	Instant k	Input bits $(d_{k,1}; d_{k,2})$	Shift register $(d_{k-1,1}; d_{k-1,2})$	State S_k	Transmitted symbols $(x_{k,1}; x_{k,2})$
00011110	0	- -	0 0	0	- -
000111	1	0 1(2)	0 0	0	0 2
0001	2	1 1(3)	0 1	2	2 3
00	3	1 0(1)	1 1	3	3 1
	4	0 0(2)	1 0	1	1 0
	5	- -	0 0	0	- -

Table 14.1: Operation of the space-time encoder of Figure 14.1.

last two binary data bits in Table 14.1 are intentionally set to zero in order to force the 4-state 4PSK trellis encoder back to the zero state which is common practice at the end of a transmission frame. Therefore, the transmit antenna $Tx\ 1$ will transmit symbols $0, 2, 3, 1$. By contrast, symbols $2, 3, 1, 0$ are then transmitted by the antenna $Tx\ 2$.

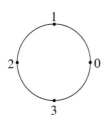

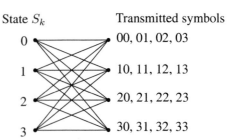

Figure 14.2: The 4PSK constellation points.

Figure 14.3: The 4-state, 4PSK space-time trellis code.

According to the shift register encoder shown in Figure 14.1, we can find all the legitimate subsequent states, which result in transmitting the various symbols $x_{k,1}$ and $x_{k,2}$, depending on a particular state of the shift register. This enables us to construct the state diagram for the encoder. The 4PSK constellation points are seen in Figure 14.2, while the corresponding state diagram of the 4-state 4PSK space-time trellis code [217] is shown in Figure 14.3. In Figure 14.3, we can see that for each current state there are four possible trellis transitions to the states $0, 1, 2$ and 3, which correspond to the legitimate input symbols of $0(d_{k,1} = 0, d_{k,2} = 0), 1(d_{k,1} = 1, d_{k,2} = 0), 2(d_{k,1} = 0, d_{k,2} = 1)$ and $3(d_{k,1} = 1, d_{k,2} = 1)$, respectively. Correspondingly, there are four sets of possible transmitted symbols associated with the four trellis transitions, shown at right of the state diagram. Each trellis transition is associated with two transmitted symbols, namely with x_1 and x_2, which are transmitted by the antennas Tx 1 and Tx 2, respectively. In Figure 14.4, we have highlighted the trellis transitions from state zero $S_k = 0$ to various states. The associated input symbols and the transmitted symbols of each trellis transitions are shown on top of each trellis transition. If

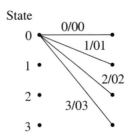

Figure 14.4: The trellis transitions from state $S_k = 0$ to various states.

the input symbol is 0, then the symbol $x_1 = 0$ will be sent by the transmit antenna Tx 1, and symbol $x_2 = 0$ by the transmit antenna Tx 2 as seen in Figure 14.4 or Figure 14.3. The next state remains $S_{k+1} = 0$. However, if the input symbol is 2 associated with $d_{k,1} = 0$, $d_{k,2} = 1$ in Table 14.1 then, the trellis traverses from state $S_k = 0$ to state $S_{k+1} = 2$ and the symbols $x_1 = 0$ and $x_2 = 2$ are transmitted over the antennas Tx 1 and Tx 2, respectively. Again, the encoder is required to be in the zero state both at the beginning and at the end of the encoding process.

14.2.1.1 The 4-State, 4PSK Space-Time Trellis Decoder

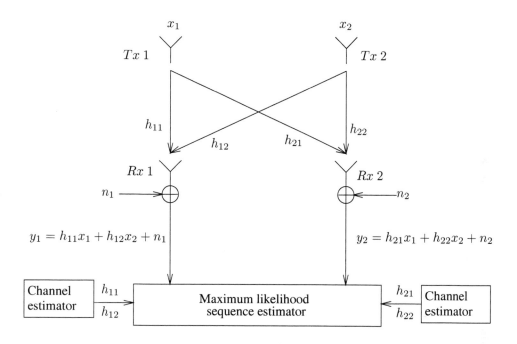

Figure 14.5: Baseband representation of the 4-state, 4PSK space-time trellis code using two receivers.

In Figure 14.5 we show the baseband representation of the 4-state, 4PSK space-time trellis code using two receivers. At any transmission instant, we have symbols x_1 and x_2 transmitted by the antennas Tx 1 and Tx 2, respectively. At the receivers Rx 1 and Rx 2, we would have:

$$y_1 = h_{11}x_1 + h_{12}x_2 + n_1 \tag{14.3}$$
$$y_2 = h_{21}x_1 + h_{22}x_2 + n_2 , \tag{14.4}$$

where h_{11}, h_{12}, h_{21} and h_{22} represent the corresponding complex time-domain channel transfer factors. Aided by the channel estimator, the Viterbi Algorithm based maximum likelihood sequence estimator [217] first finds the branch metric associated with every transition in the decoding trellis diagram, which is identical to the state diagram shown in Figure 14.3. For each trellis transition, we have two estimated transmit symbols, namely $\tilde{x}_1$ and $\tilde{x}_2$, for which the branch metric BM is given by:

$$
\begin{aligned}
BM &= |y_1 - h_{11}\tilde{x}_1 - h_{12}\tilde{x}_2 + y_2 - h_{21}\tilde{x}_1 - h_{22}\tilde{x}_2|^2 \\
&= \sum_{i=1}^{2} |y_i - h_{i1}\tilde{x}_1 - h_{i2}\tilde{x}_2|^2 \\
&= \sum_{i=1}^{2} \left| y_i - \sum_{j=1}^{2} h_{ij}\tilde{x}_j \right|^2 .
\end{aligned}
\tag{14.5}
$$

We can however generalise Equation 14.5 to p transmitters and q receivers, as follows:

$$
BM = \sum_{i=1}^{p} \left| y_i - \sum_{j=1}^{q} h_{ij}\tilde{x}_j \right|^2 .
\tag{14.6}
$$

When all the transmitted symbols were received and the branch metric of each legitimate transition was calculated, the maximum likelihood sequence estimator invokes the Viterbi Algorithm (VA) in order to find the maximum likelihood path associated with the best accumulated metric.

14.2.2   Other Space-Time Trellis Codes

In Section 14.2.1, we have shown the encoding and decoding process of the simple 4-state, 4PSK space-time trellis code. More sophisticated 4PSK space-time trellis codes were designed by increasing the number of trellis states [217], which are reproduced in Figures 14.6 to 14.8. With an increasing number of trellis states the number of tailing symbols required for terminating the trellis at the end of a transmitted frame is also increased. Two zero-symbols are needed to force the trellis back to state zero for the space-time trellis codes shown in Figures 14.6 and 14.7. By contrast, three zero-symbols are required for the space-time trellis code shown in Figure 14.8.

Space-time trellis codes designed for the higher-order modulation scheme of 8PSK were also proposed in [217]. In Figure 14.9, we showed the constellation points employed in [217]. The trellises of the 8-state, 16-state and 32-state 8PSK space-time trellis codes were reproduced from [217] in Figures 14.10, 14.11 and 14.12, respectively. One zero-symbol is required to terminate the 8-state, 8PSK space-time trellis code, whereas two zero-symbols are needed for both the 16-state and 32-state 8PSK space-time trellis codes.

14.3   Space-Time Coded Transmission Over Wideband Channels

In Section 14.2, we have detailed the concept of space-time trellis codes. Let us now elaborate further by investigating the performance of space-time codes over dispersive wideband fading channels. As mentioned in Section 14.1, Bauch's approach [319, 523] of using turbo equalization for mitigating the ISI exhibits a considerable complexity. Hence we argued that using space-time coded OFDM constitutes a more favourable approach to transmission over

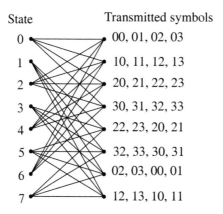

Figure 14.6: The 8-state, 4PSK space-time trellis code ©IEEE [217].

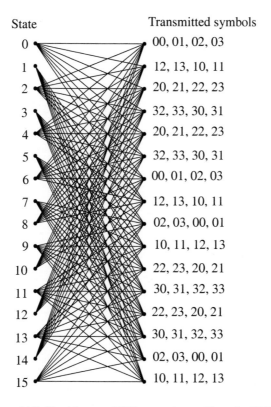

Figure 14.7: The 16-state, 4PSK space-time trellis code ©IEEE [217].

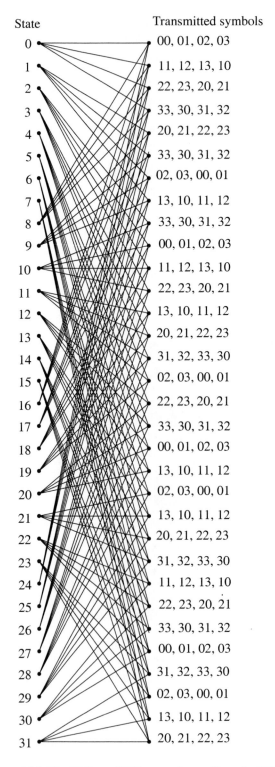

Figure 14.8: The 32-State, 4PSK space-time trellis code ©IEEE [217].

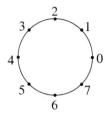

Figure 14.9: The 8PSK constellation points ©IEEE [217].

State   Transmitted symbols

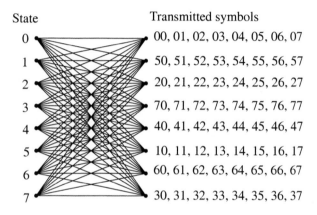

Figure 14.10: The 8-state, 8PSK space-time trellis code ©IEEE [217].

dispersive wireless channels, since the associated decoding complexity is significantly lower. Therefore, in this chapter OFDM is employed for mitigating the effects of dispersive channels.

It is widely recognised that space-time trellis codes [217] perform well at the cost of high complexity. However, Alamouti's G_2 space-time block code [7] could be invoked instead of space-time trellis codes. The space-time block code G_2 is appealing in terms of its simplicity, although there is a slight loss in performance. Therefore, we concatenate the space-time block code G_2 with Turbo Convolutional (TC) codes in order to improve the performance of the system. The family of TC codes was favoured, because it was shown in [530, 531] that TC codes achieve an enormous coding gain at a moderate complexity, when compared to convolutional codes, turbo BCH codes, trellis coded modulation and turbo trellis coded modulation. The performance of concatenated space-time block codes and TC codes will then be compared to that of space-time trellis codes. Conventionally, Reed-Solomon (RS) codes have been employed in conjunction with the space-time trellis codes [181,524,526] for improving the performance of the system. In our forthcoming discussion, we will concentrate on comparing the performance of space-time block and trellis codes in conjunction with various channel coders.

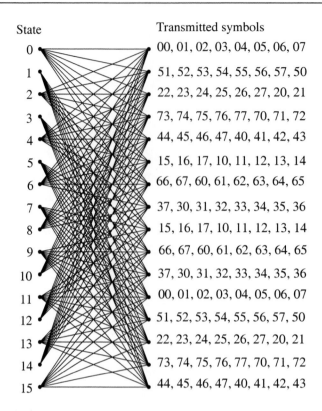

State   Transmitted symbols

0   00, 01, 02, 03, 04, 05, 06, 07

1   51, 52, 53, 54, 55, 56, 57, 50

2   22, 23, 24, 25, 26, 27, 20, 21

3   73, 74, 75, 76, 77, 70, 71, 72

4   44, 45, 46, 47, 40, 41, 42, 43

5   15, 16, 17, 10, 11, 12, 13, 14

6   66, 67, 60, 61, 62, 63, 64, 65

7   37, 30, 31, 32, 33, 34, 35, 36

8   15, 16, 17, 10, 11, 12, 13, 14

9   66, 67, 60, 61, 62, 63, 64, 65

10   37, 30, 31, 32, 33, 34, 35, 36

11   00, 01, 02, 03, 04, 05, 06, 07

12   51, 52, 53, 54, 55, 56, 57, 50

13   22, 23, 24, 25, 26, 27, 20, 21

14   73, 74, 75, 76, 77, 70, 71, 72

15   44, 45, 46, 47, 40, 41, 42, 43

Figure 14.11: The 16-State, 8PSK space-time trellis code ©IEEE [217].

14.3.1 System Overview

Figure 14.13 shows the schematic of the system employed in our performance study. At the transmitter, the information source generates random information data bits. The information bits are then encoded by TC codes, RS codes or left uncoded. The coded or uncoded bits are then channel interleaved, as shown in Figure 14.13. The output bits of the channel interleaver are then passed to the Space-Time Trellis (STT) or Space-Time Block (STB) encoder. We will investigate all the previously mentioned space-time trellis codes proposed by Tarokh, Seshadri and Calderbank in [217], where the associated state diagrams are shown in Figures 14.3, 14.6, 14.7, 14.10, 14.11 and 14.12. The modulation schemes employed are 4PSK as well as 8PSK and the corresponding trellis diagrams were shown in Figures 14.2 and 14.9, respectively. On the other hand, from the family of space-time block codes only Alamouti's G_2 code is employed in the system, since it was shown in [531] that the best performance is achieved by concatenating the space-time block code G_2 with TC codes. For convenience, the transmission matrix of the space-time block code G_2 is reproduced here as follows:

$$G_2 = \begin{pmatrix} x_1 & x_2 \\ -\bar{x}_2 & \bar{x}_1 \end{pmatrix} .$$ (14.7)

State   Transmitted symbols

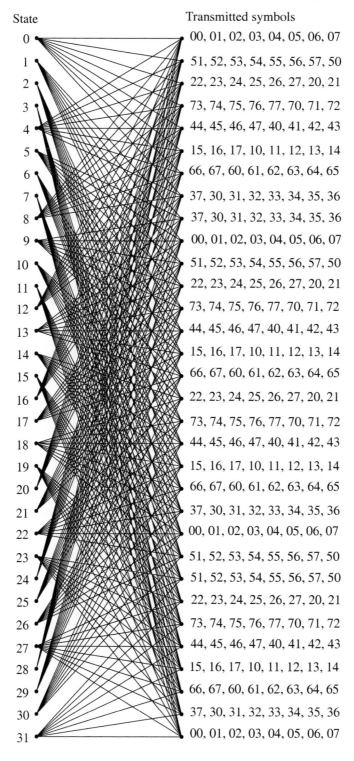

State	Transmitted symbols
0	00, 01, 02, 03, 04, 05, 06, 07
1	51, 52, 53, 54, 55, 56, 57, 50
2	22, 23, 24, 25, 26, 27, 20, 21
3	73, 74, 75, 76, 77, 70, 71, 72
4	44, 45, 46, 47, 40, 41, 42, 43
5	15, 16, 17, 10, 11, 12, 13, 14
6	66, 67, 60, 61, 62, 63, 64, 65
7	37, 30, 31, 32, 33, 34, 35, 36
8	37, 30, 31, 32, 33, 34, 35, 36
9	00, 01, 02, 03, 04, 05, 06, 07
10	51, 52, 53, 54, 55, 56, 57, 50
11	22, 23, 24, 25, 26, 27, 20, 21
12	73, 74, 75, 76, 77, 70, 71, 72
13	44, 45, 46, 47, 40, 41, 42, 43
14	15, 16, 17, 10, 11, 12, 13, 14
15	66, 67, 60, 61, 62, 63, 64, 65
16	22, 23, 24, 25, 26, 27, 20, 21
17	73, 74, 75, 76, 77, 70, 71, 72
18	44, 45, 46, 47, 40, 41, 42, 43
19	15, 16, 17, 10, 11, 12, 13, 14
20	66, 67, 60, 61, 62, 63, 64, 65
21	37, 30, 31, 32, 33, 34, 35, 36
22	00, 01, 02, 03, 04, 05, 06, 07
23	51, 52, 53, 54, 55, 56, 57, 50
24	51, 52, 53, 54, 55, 56, 57, 50
25	22, 23, 24, 25, 26, 27, 20, 21
26	73, 74, 75, 76, 77, 70, 71, 72
27	44, 45, 46, 47, 40, 41, 42, 43
28	15, 16, 17, 10, 11, 12, 13, 14
29	66, 67, 60, 61, 62, 63, 64, 65
30	37, 30, 31, 32, 33, 34, 35, 36
31	00, 01, 02, 03, 04, 05, 06, 07

Figure 14.12: The 32-State, 8PSK space-time trellis code ©IEEE [217].

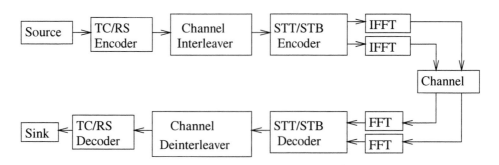

Figure 14.13: System overview.

Different modulation schemes could be employed [13], such as Binary Phase Shift Keying (BPSK), Quadrature Phase Shift Keying (QPSK), 16-level Quadrature Amplitude Modulation (16QAM) and 64-level Quadrature Amplitude Modulation (64QAM). Gray-mapping of the bits to symbols was applied and this resulted in different protection classes in higher-order modulation schemes [4]. The mapping of the data bits and parity bits of the TC encoder was chosen such that it yielded the best achievable performance along with the application of the random separation channel interleaver [530]. The output of the space-time encoder was then OFDM [4] modulated and transmitted by the corresponding antenna. The number of transmit antennas was fixed to two, while the number of receive antennas constituted a design parameter. Dispersive wideband channels were used and the associated channels' profiles will be discussed later.

At the receiver the signal of each receive antenna is OFDM demodulated. The demodulated signals of the receiver antennas are then fed to the space-time trellis or space-time block decoder. The space-time decoders apply the MAP [162] or Log-MAP [163, 532] decoding algorithms for providing soft outputs for the channel decoders. If no channel codecs are employed in the system, the space-time decoders apply the VA [217,533], which gives similar performance to the MAP decoder at a lower complexity. The decoded bits are finally passed to the sink for the calculation of the Bit Error Rate (BER) or Frame Error Rate (FER).

14.3.2   Space-Time and Channel Codec Parameters

In Figure 14.13, we have given an overview of the system studied. In this section, we present the parameters of the space-time codes and the channel codecs employed in the system. We will employ the set of various space-time trellis codes shown in Figures 14.3, 14.6, 14.7, 14.8, 14.10, 14.11 and 14.12. The associated space-time trellis coding parameters are summarised in Table 14.2. On the other hand, from the family of space-time block codes only Alamouti's G_2 code is employed, since we have shown in [531] that the best performance in the set of investigated schemes was yielded by concatenating the space-time block code G_2 with TC codes. The transmission matrix of the code is shown in Equation 14.7, while the number of transmitters used by the space-time block code G_2 is two, which is identical to the number of transmitters in the space-time trellis codes shown in Table 14.2.

Let us now briefly consider the TC channel codes used. In this chapter we will con-

Modulation scheme	BPS	Decoding algorithm	No. of states	No. of transmitters	No. of termination symbols
4PSK	2	VA	4	2	1
			8	2	2
			16	2	2
			32	2	3
8PSK	3	VA	8	2	1
			16	2	2
			32	2	2

Table 14.2: Parameters of the space-time trellis codes shown in Figures 14.3, 14.6, 14.7, 14.8, 14.10, 14.11 and 14.12.

centrate on using the simple half-rate TC(2, 1, 3) code. Its associated parameters are shown in Table 14.3. As seen in Table 14.4, different modulation schemes are employed in con-

Code	Octal generator polynomial	No. of states	Decoding algorithm	Puncturing pattern	No. of iterations
TC(2, 1, 3)	7,5	4	Log-MAP	10,01	8

Table 14.3: The associated parameters of the TC(2, 1, 3) code.

junction with the concatenated space-time block code G_2 and the TC(2,1,3) code. Since the half-rate TC(2,1,3) code is employed, higher-order modulation schemes such as 16QAM and 64QAM were chosen, so that the throughput of the system remained the same as that of the system employing the space-time trellis codes without channel coding. It is widely recognised that the performance of TC codes improves upon increasing the turbo interleaver size and near-optimum performance can be achieved using large interleaver sizes exceeding 10,000 bits. However, this performance gain is achieved at the cost of high latency, which is impractical for a delay-sensitive real-time system. On the other hand, space-time trellis codes offer impressive coding gains [217] at low latency. The decoding of the space-time trellis codes is carried out on a transmission burst-by-burst basis. In order to make a fair comparison between the systems investigated, the turbo interleaver size was chosen such that all the coded bits were hosted by one transmission burst. This enables burst-by-burst turbo decoding at the receiver. Since we employ an OFDM modem, latency may also be imposed by a high number of subcarriers in an OFDM symbol. Therefore, the turbo interleaver size was increased, as the number of sub-carriers increased in our investigations. In Table 14.4, we summarised the modulation schemes and interleaver sizes used for different number of OFDM subcarriers in the system. The random separation based channel interleaver of [530] was used. The mapping of the data bits and parity bits into different protection classes of the higher-order modulation scheme was carried out such that the best possible performance was attained. This issue was addressed in [530].

Reed-Solomon codes were employed in conjunction with the space-time trellis codes.

Code	Code Rate R	Modula- tion Mode	BPS	Random turbo interleaver depth	Random separation interleaver depth
			128 carriers		
TC(2, 1, 3)	0.50	16QAM	2	256	512
		64QAM	3	384	768
			512 carriers		
		QPSK	1	512	1024
		16QAM	2	1024	2048

Table 14.4: The simulation parameters associated with the TC(2, 1, 3) code.

Code	Galois Field	Rate	Correctable symbol errors
RS(105,51)	2^{10}	0.49	27
RS(153,102)	2^{10}	0.67	25

Table 14.5: The coding parameters of the Reed-Solomon codes employed.

Hard decision decoding was utilized and the coding parameters of the Reed-Solomon codes employed are summarised in Table 14.5.

14.3.3 Complexity Issues

In this section, we will address the implementational complexity issues of the systems studied. We will however focus mainly on the relative complexity of the systems, rather than attempting to quantify their exact complexity. In order to simplify our comparative study, several assumptions were stipulated. In our simplified approach, the estimated complexity of the system is deemed to depend only on that of the space-time trellis decoder and turbo decoder. In other words, the complexity associated with the modulator, demodulator, space-time block encoder and decoder as well as that of the space-time trellis encoder and turbo encoder are assumed to be insignificant compared to the complexity of space-time trellis decoder and turbo decoder.

In [531], we have detailed our complexity estimates for the TC decoder and the reader is referred to the paper for further details. The estimated complexity of the TC decoder is assumed to depend purely on the number of trellis transitions per information data bit and this simple estimated complexity measure was also used in [531] as the basis of our comparisons. Here, we adopt the same approach and evaluate the estimated complexity of the space-time trellis decoder on the basis of the number of trellis transitions per information data bit.

In Figures 14.3, 14.6, 14.7, 14.8, 14.10, 14.11 and 14.12, we have shown the state diagrams of the 4PSK and 8PSK space-time trellis codes. From these state diagrams, we can see that the number of trellis transitions leaving each state is equivalent to 2^{BPS}, where BPS denotes the number of transmitted bits per modulation symbol. Since the number of information bits is equal to BPS, we can approximate the complexity of the space-time trellis decoder as:

$$comp\{STT\} \quad = \quad \frac{2^{BPS} \times \text{No. of States}}{BPS}$$

$$= \quad 2^{BPS-1} \times \text{No. of States} . \qquad (14.8)$$

Applying Equation 14.8 and assuming that the Viterbi decoding algorithm was employed, we tabulated the approximated complexities of the space-time trellis decoder in Table 14.6.

Modulation scheme	BPS	No. of states	Complexity
4PSK	2	4	8
		8	16
		16	32
		32	64
8PSK	3	8	21.33
		16	42.67
		32	85.33

Table 14.6: Estimated complexity of the space-time trellis decoders shown in Figures 14.3, 14.6, 14.7, 14.8, 14.10, 14.11 and 14.12.

14.4 Simulation Results

In this section, we will present our simulation results characterizing the OFDM-based system investigated. As mentioned earlier, we will investigate the system's performance over dispersive wideband Rayleigh fading channels. We will commence our investigations using a simple two-ray channel impulse response (CIR) having equal tap weights, followed by a more realistic Wireless Asynchronous Transfer Mode (WATM) channel [4]. The CIR of the two-ray model is shown in Figure 14.14. From the figure we can see that the reflected path has the same amplitude as the Line Of Sight (LOS) path, although arriving $5\mu s$ later. However, in our simulations we also present results over two-ray channels separated by various delay spreads, up to $40\mu s$. Jakes' model [199] was adapted for modelling the fading channels. In Figure 14.15, we portray the 128-subcarrier OFDM symbol employed, having a guard period of $40\mu s$. The guard period of $40\mu s$ or cyclic extension of 32 samples was employed to overcome the inter-OFDM symbol interference due to the channel's memory.

In order to obtain our simulation results, several assumptions were stipulated:

- The average signal power received from each transmitter antenna was the same;

- All multipath components undergo independent Rayleigh fading;

- The receiver has a perfect knowledge of the CIR.

We note that the above assumptions are unrealistic, yielding the best-case performance, nonetheless, facilitating the performance comparison of the various techniques under identical circumstances.

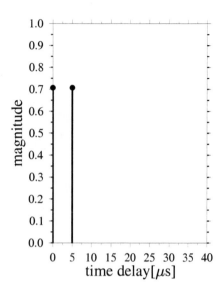

Figure 14.14: Two-ray channel impulse response having equal amplitudes.

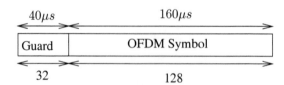

Figure 14.15: Stylised plot of 128-subcarrier OFDM time-domain signal using a cyclic extension of 32 samples.

14.4.1   Space-Time Coding Comparison − Throughput of 2 BPS

In Figure 14.16, we show our frame error rate (FER) performance comparison between 4PSK space-time trellis codes and the space-time block code G_2 concatenated with the TC(2,1,3) code using one receiver and the 128-subcarrier OFDM modem. The CIR had two equal-power rays separated by a delay spread of $5\mu s$ and the maximum Doppler frequency was 200 Hz. The TC$(2, 1, 3)$ code is a half-rate code and hence 16QAM was employed, in order to support the same 2 BPS throughput, as the 4PSK space-time trellis codes using no channel codes. We can clearly see that at FER=10^{-3} the performance of the concatenated scheme is at least 7 dB better, than that of the space-time trellis codes.

The performance of the space-time block code G_2 without TC$(2, 1, 3)$ channel coding is also shown in Figure 14.16. It can be seen in the figure that the space-time block code G_2

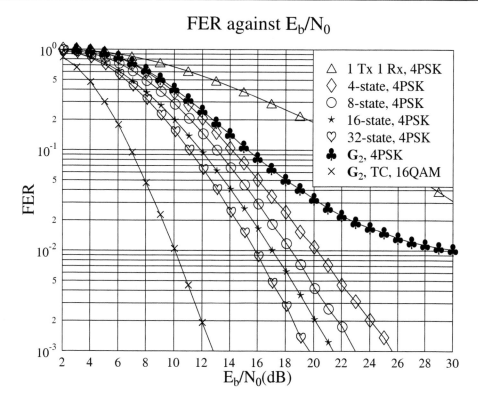

Figure 14.16: FER performance comparison between various 4PSK space-time trellis codes and the
space-time block code $\mathbf{G}_2$ concatenated with the TC(2,1,3) code using **one receiver** and
the 128-subcarrier OFDM modem over a channel having a CIR characterised by two
equal-power rays separated by a delay spread of $5\mu s$. The maximum Doppler frequency
was 200 Hz. The effective throughput was **2 BPS** and the coding parameters are shown
in Tables 14.2, 14.3 and 14.4.

does not perform well, exhibiting a residual BER. Moreover, at high E_b/N_0 values, the performance of the single-transmitter, single-receiver system is better than that of the space-time
block code $\mathbf{G}_2$. This is because the assumption that the fading is constant over the two consecutive transmission instants is no longer valid in this situation. Here, the two consecutive
transmission instants are associated with two adjacent subcarriers in the OFDM symbol and
the fading variation is relatively fast in the frequency domain. Therefore, the orthogonality
of the space-time code has been destroyed by the frequency-domain variation of the fading
envelope. At the receiver, the combiner can no longer separate the two different transmitted
signals, namely x_1 and x_2. More explicitly, the signals interfere with each other. The increase in SNR does not improve the performance of the space-time block code $\mathbf{G}_2$, since this
also increases the power of the interfering signal. We will address this issue more explicitly
in Section 14.4.4. By contrast, the TC(2, 1, 3) channel codec succeeds in overcoming this
problem. However, we will show later in Section 14.4.4 that the concatenated channel coded

scheme exhibits the same residual BER problem, if the channel's variation becomes more rapid.

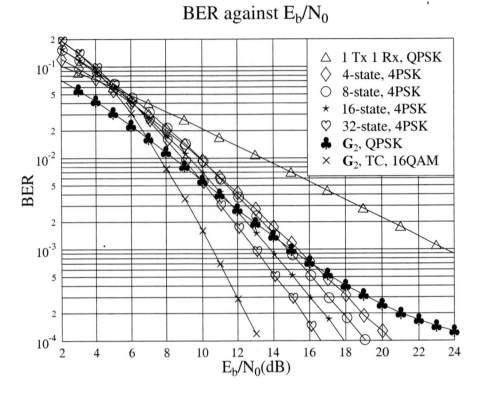

Figure 14.17: **BER** performance comparison between various 4PSK space-time trellis codes and the space-time block code G_2 concatenated with the TC(2,1,3) code using **one receiver** and the 128-subcarrier OFDM modem over a channel having a CIR characterised by two equal-power rays separated by a delay spread of $5\mu s$. The maximum Doppler frequency was 200 Hz. The effective throughput was **2 BPS** and the coding parameters are shown in Tables 14.2, 14.3 and 14.4.

In Figure 14.17, we provide the corresponding BER performance comparison between the 4PSK space-time trellis codes and the space-time block code G_2 concatenated with the TC(2,1,3) code using one receiver and the 128-subcarrier OFDM modem over a channel characterised by two equal-power rays separated by a delay spread of $5\mu s$ and having a maximum Doppler frequency of 200 Hz. Again, we show in the figure that the 2 BPS throughput concatenated G_2/TC(2,1,3) scheme outperforms the 2 BPS space-time trellis codes using no channel coding. At a BER of 10^{-4}, the concatenated channel coded scheme is at least 2 dB superior in SNR terms to the space-time trellis codes using no channel codes. At high E_b/N_0 values, the space-time block code G_2, again exhibits a residual BER. On the other hand, at low E_b/N_0 values the latter outperforms the concatenated G_2/TC(2,1,3) channel coded scheme as well as the space-time trellis codes using no channel coding.

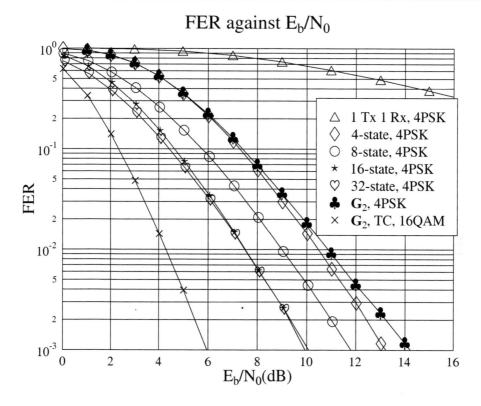

Figure 14.18: FER performance comparison between various 4PSK space-time trellis codes and the
space-time block code G_2 concatenated with the TC(2,1,3) code using **two receivers**
and the 128-subcarrier OFDM modem over a channel having a CIR characterised by two
equal-power rays separated by a delay spread of $5\mu s$. The maximum Doppler frequency
was 200 Hz. The effective throughput was **2 BPS** and the coding parameters are shown
in Tables 14.2, 14.3 and 14.4.

Following the above investigations, the number of receivers was increased to two. In
Figure 14.18, we show our FER performance comparison between the various 4PSK space-
time trellis codes and the space-time block code G_2 concatenated with the TC(2,1,3) code
using two receivers and the 128-subcarrier OFDM modem. As before, the CIR had two
equal-power rays separated by a delay spread of $5\mu s$. Again, we can see that the concatenated G_2/TC(2,1,3) channel coded scheme outperforms the space-time trellis codes using
no channel coding. However, the associated difference is lower and at a FER of 10^{-3} the
concatenated channel coded scheme is about 4 dB better in E_b/N_0 terms than the space-time
trellis codes using no channel codes. On the other hand, by employing two receivers the performance of the space-time block code G_2 improved and the performance flattening effect
happens at a lower FER.

In [531] and 14.3.3, we have derived the complexity estimates of the TC decoders and
space-time trellis decoders, respectively. By employing Equation 14.8 and equations in [531],

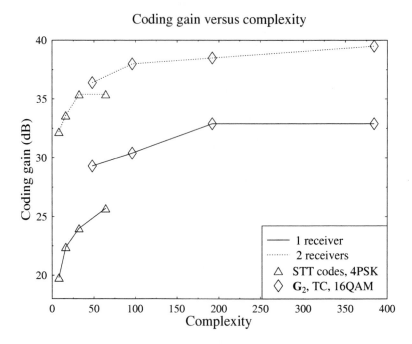

Figure 14.19: Coding gain versus estimated complexity for the various 4PSK space-time trellis codes and the space-time block code G_2 concatenated with the TC(2,1,3) code **using one as well as two receivers** and the 128-subcarrier OFDM modem over a channel having a CIR characterised by two equal-power rays separated by a delay spread of $5\mu s$. The maximum Doppler frequency was 200 Hz. The effective throughput was **2 BPS** and the coding parameters are shown in Tables 14.2, 14.3 and 14.4.

we compare the performance of the schemes studied, while considering their approximate complexity. Our performance comparison of the various schemes was carried out on the basis of the coding gain defined as the E_b/N_0 difference, expressed in decibels (dB), at FER$= 10^{-3}$ between the schemes studied and the uncoded single-transmitter, single-receiver system having the same throughput of 2 BPS. In Figure 14.19, we show our coding gain versus estimated complexity comparison for the various 4PSK space-time trellis codes and the space-time block code G_2 concatenated with the TC(2,1,3) code using one as well as two receivers. The 128-subcarrier OFDM modem transmitted over the channel having a CIR of two equal-power rays separated by a delay spread of $5\mu s$ and a maximum Doppler frequency of 200 Hz. The estimated complexity of the space-time trellis codes was increased by increasing the number of trellis states. By constrast, the estimated complexity of the TC$(2, 1, 3)$ code was increased by increasing the number of turbo iterations. The coding gain of the concatenated G_2/TC(2,1,3) scheme using one, two, four and eight iterations is shown in Figure 14.19. It can be seen that the concatenated scheme outperforms the space-time trellis codes using no channel coding, even though the number of turbo iterations was only one. Moreover, the improvement in coding gain was obtained, at an estimated complexity comparable to that of the 32-state 4PSK space-time trellis code using no channel coding.

From the figure we can also see that the performance gain of the concatenated G_2/TC(2,1,3) channel coded scheme over the space-time trellis codes becomes lower, when the number of receivers is increased to two.

14.4.2 Space-Time Coding Comparison − Throughput of 3 BPS

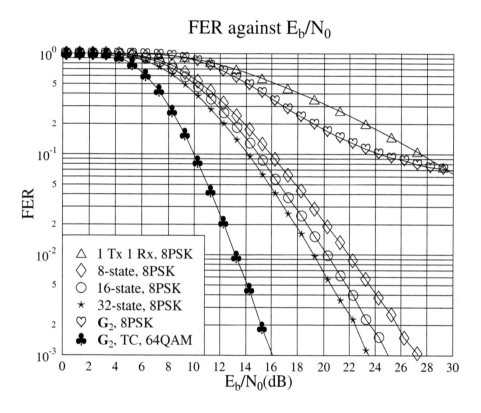

Figure 14.20: **FER** performance comparison between various 8PSK space-time trellis codes and the space-time block code G_2 concatenated with the TC(2,1,3) code using **one receiver** and the 128-subcarrier OFDM modem over a channel having a CIR characterised by two equal-power rays separated by a delay spread of $5\mu s$. The maximum Doppler frequency was 200 Hz. The effective throughput was **3 BPS** and the coding parameters are shown in Tables 14.2, 14.3 and 14.4.

In Figure 14.20, we show our FER performance comparison between the various 8PSK space-time trellis codes of Table 14.2 and space-time block code G_2 concatenated with the TC(2,1,3) code using one receiver and the 128-subcarrier OFDM modem. The CIR exhibited two equal-power rays separated by a delay spread of $5\mu s$ and a maximum Doppler frequency of 200 Hz. Since the TC(2, 1, 3) scheme is a half-rate code, 64QAM was employed in order to ensure the same 3 BPS throughput, as the 8PSK space-time trellis codes using no channel coding. We can clearly see that at FER=10^{-3} the performance of the concatenated channel coded scheme is at least 7 dB better in terms of the required E_b/N_0 than that of the space-time trellis codes. The performance of the space-time block code $\mathbf{G}_2$ without the concatenated TC(2, 1, 3) code is also shown in the figure. In Table 14.4, we can see that although there is an increase in the turbo interleaver size, due to employing a higher-order modulation scheme, nonetheless, no performance gain is observed for the concatenated TC(2, 1, 3)-$\mathbf{G}_2$ scheme over the space-time trellis codes using no channel coding. We speculate that this is because the potential gain due to the increased interleaver size has been offset by the vulnerable 64QAM scheme.

We also show in Figure 14.20 that the performance of the 3 BPS 8PSK space-time block code $\mathbf{G}_2$ without the concatenated TC(2, 1, 3) scheme is worse, than that of the other schemes investigated. It exhibits the previously noted flattening effect, which becomes more pronounced near FER= 10^{-1}. The same phenomenon was observed near FER= 10^{-2} for the corresponding $\mathbf{G}_2$-coded 4PSK scheme, which has a throughput of 2 BPS.

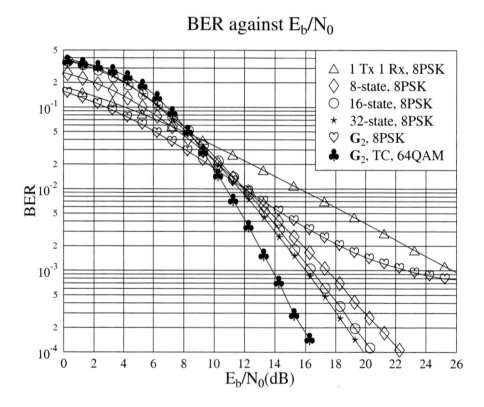

Figure 14.21: **BER** performance comparison between various 8PSK space-time trellis codes and the space-time block code $\mathbf{G}_2$ concatenated with the TC(2,1,3) code using **one receiver** and the 128-subcarrier OFDM modem over a channel having a CIR characterised by two equal-power rays separated by a delay spread of $5\mu s$. The maximum Doppler frequency was 200 Hz. The effective throughput was **3 BPS** and the coding parameters are shown in Tables 14.2, 14.3 and 14.4.

In Figure 14.21, we portray our BER performance comparison between the various 8PSK space-time trellis codes and the space-time block code G_2 concatenated with the TC(2,1,3) scheme using one receiver and the 128-subcarrier OFDM modem. The CIR exhibited two equal-power rays separated by a delay spread of $5\mu s$ and the maximum Doppler frequency was 200 Hz. Again, we observe in the figure that the concatenated G_2/TC(2,1,3)-coded scheme outperforms the space-time trellis codes using no channel coding. At a BER of 10^{-4}, the concatenated scheme is at least 2 dB better in terms of its required E_b/N_0 value, than the space-time trellis codes. The performance of the space-time block code G_2 without TC(2,1,3) channel coding is also shown in Figure 14.21. As before, at high E_b/N_0 values, the space-time block code G_2 exhibits a flattening effect. On the other hand, at low E_b/N_0 values it outperforms the concatenated G_2/TC(2,1,3) scheme as well as the space-time trellis codes.

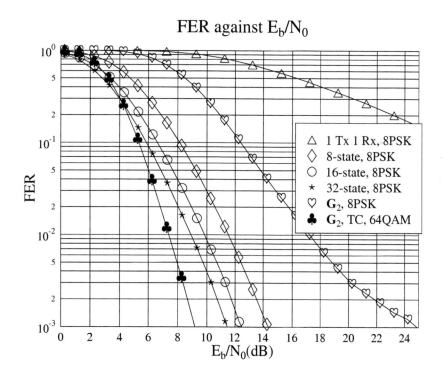

Figure 14.22: FER performance comparison between various 8PSK space-time trellis codes and the space-time block code G_2 concatenated with the TC(2,1,3) code using **two receivers** and the 128-subcarrier OFDM modem over a channel having a CIR characterised by two equal-power rays separated by a delay spread of $5\mu s$. The maximum Doppler frequency was 200 Hz. The effective throughput was **3 BPS** and the coding parameters are shown in Tables 14.2, 14.3 and 14.4.

In Figure 14.22, we compare the FER performance of the 8PSK space-time trellis codes and the space-time block code G_2 concatenated with the TC(2,1,3) channel codec using two receivers and the 128-subcarrier OFDM modem. As before, the CIR has two equal-power rays separated by a delay spread of $5\mu s$ and exhibits maximum Doppler frequency of 200 Hz. Again, with the increase in the number of receivers the performance gap between the concatenated channel coded scheme and the space-time trellis codes using no channel coding becomes smaller. At a FER of 10^{-3} the concatenated channel coded scheme is only about 2 dB better in terms of its required E_b/N_0, than the space-time trellis codes using no channel coding.

With the increase in the number of receivers, the previously observed flattening effect of the space-time block code G_2 has been substantially mitigated, dipping to values below FER= 10^{-3}. However, it can be seen in Figure 14.22 that its performance is about 10 dB worse, than that of the 8-state 8PSK space-time trellis code. In the previous system characterised in Figure 14.18, which had an effective throughput of 2 BPS, the performance of the space-time block code G_2 was only about 1 dB worse in E_b/N_0 terms, than that of the 4-state 4PSK space-time trellis code, when the number of receivers was increased to two. This observation clearly shows that higher-order modulation schemes have a tendency to saturate the channel's capacity and hence result in a poorer performance, than the identical-throughput space-time trellis codes using no channel coding.

Similarly to the 2 BPS schemes of Figure 14.19, we compare the performance of the 3 BPS throughput schemes by considering their approximate decoding complexity. The derivation of the estimated complexity has been detailed in [531] and 14.3.3. As mentioned earlier, the performance comparison of the various schemes was made on the basis of the coding gain defined as the E_b/N_0 difference, expressed in decibels, at a FER= 10^{-3} between the schemes investigated and the uncoded single-transmitter, single-receiver system having a throughput of 3 BPS. In Figure 14.23, we show the associated coding gain versus estimated complexity curves for the 8PSK space-time trellis codes using no channel coding and the space-time block code G_2 concatenated with the TC(2,1,3) code using one and two receivers and the 128-subcarrier OFDM modem. For the sake of consistency, the CIR, again, exhibited two equal-power rays separated by a delay spread of $5\mu s$ and a maximum Doppler frequency of 200 Hz. Again, the estimated complexity of the space-time trellis codes was increased by increasing the number of states. On the other hand, the estimated complexity of the TC(2, 1, 3) code was increased by increasing the number of iterations. The coding gain of the concatenated channel coded scheme invoking one, two, four and eight iterations is shown in Figure 14.23. Previously in Figure 14.19 we have shown that the concatenated TC(2,1,3)-coded scheme using one iteration outperformed the space-time trellis codes using no channel coding. However, in Figure 14.23 the concatenated scheme does not exhibit the same performance trend. For the case of one receiver, the concatenated scheme using one iteration has a negative coding gain and exhibits a saturation effect. This is again, due to the employment of the high-order 64QAM scheme, which has a preponderance to exceed the channel's capacity. Again, we can also see that the performance gain of the concatenated G_2/TC(2,1,3)-coded scheme over the space-time trellis codes using no channel coding becomes smaller, when the number of receivers is increased to two. Having studied the performance of the various schemes over the channel characterised by the two-path, 5μ-dispersion CIR at a fixed Doppler frequency of 200Hz, let us in the next section study the effects of varying the Doppler frequency.

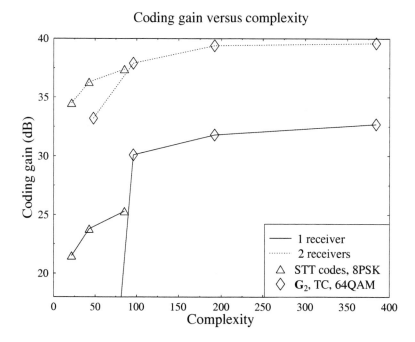

Figure 14.23: Coding gain versus estimated complexity for the various 8PSK space-time trellis codes and the space-time block code G_2 concatenated with the TC(2,1,3) code **using one and two receivers** and the 128-subcarrier OFDM modem over a channel having a CIR characterised by two equal-power rays separated by a delay spread of $5\mu s$. The maximum Doppler frequency was 200 Hz. The effective throughput was **3 BPS** and the coding parameters are shown in Tables 14.2, 14.3 and 14.4.

14.4.3 The Effect of Maximum Doppler Frequency

In our further investigations we have generated the FER versus E_b/N_0 curves similar to those in Figure 14.16, when the Doppler frequency was fixed to 5, 10, 20, 50 and 100 Hz. In order to present these results in a compact form, we then extracted the required E_b/N_0 values for maintaining a FER of 10^{-3}. In Figure 14.24, we show the E_b/N_0 crossing point at FER=10^{-3} versus the maximum Doppler frequency for the 32-state 4PSK space-time trellis code using no channel coding and for the space-time block code G_2 concatenated with the TC(2,1,3) code using one receiver and the 128-subcarrier OFDM modem. As before, the CIR exhibited two equal-power rays separated by a delay spread of $5\mu s$. We conclude from the near-horizontal curves shown in the figure that the maximum Doppler frequency does not significantly affect the performance of the space-time trellis codes and the concatenated scheme. Furthermore, the performance of the concatenated scheme is always better, than that of the space-time trellis codes using no channel coding. Having studied the effects of various Doppler frequencies, let us now consider the impact of varying the delay spread.

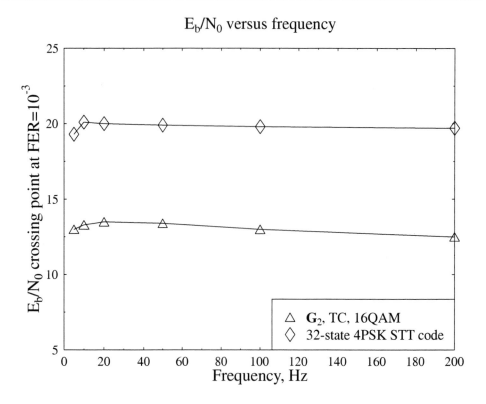

Figure 14.24: The E_b/N_0 value required for maintaining FER=10^{-3} versus the maximum Doppler frequency for the 32-state 4PSK space-time trellis code and for the space-time block code G_2 concatenated with the TC(2,1,3) code using **one receiver** and the 128-subcarrier OFDM modem. The CIR exhibited two equal-power rays separated by a delay spread of $5\mu s$. The effective throughput was **3 BPS** and the coding parameters are shown in Tables 14.2, 14.3 and 14.4.

14.4.4   The Effect of Delay Spreads

In this section, we will study how the variation of the delay spread between the two paths of the channel affects the system performance. By varying the delay spread, the channel's frequency-domain response varies as well. In Figure 14.25, we show the fading amplitude variation of the 128 subcarriers in an OFDM symbol for a delay spread of (a) $5\mu s$, (b) $10\mu s$, (c) $20\mu s$ and (d) $40\mu s$. It can be seen from the figure that the fading amplitudes vary more rapidly, when the delay spread is increased. For the space-time block code G_2 the fading envelopes of the two consecutive transmission instants of antennas Tx 1 and Tx 2 are assumed to be constant [7]. In Figure 14.25 (d), we can see that the variation of the frequency-domain fading amplitudes is so dramatic that we can no longer assume that the fading envelopes are constant for two consecutive transmission instants. The variation of the frequency-domain fading envelope will eventually destroy the orthogonality of the space-time block code G_2.

We show in Figure 14.26 that the two transmission instants are no longer assumed to be

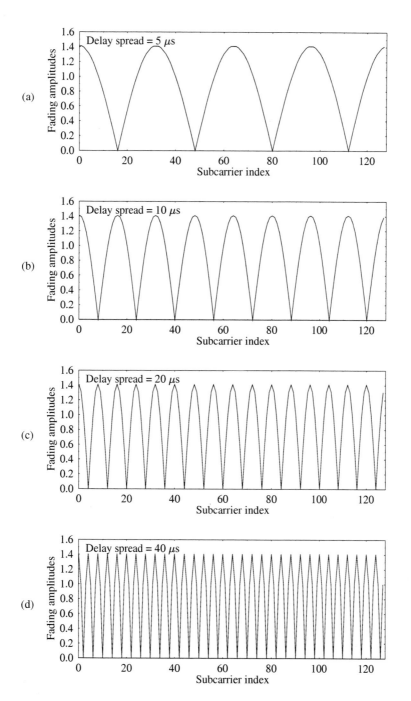

Figure 14.25: Frequency-domain fading amplitudes of the 128 subcarriers in an OFDM symbol for a delay spread of (a) $5\mu s$, (b) $10\mu s$, (d) $20\mu s$ and (c) $40\mu s$.

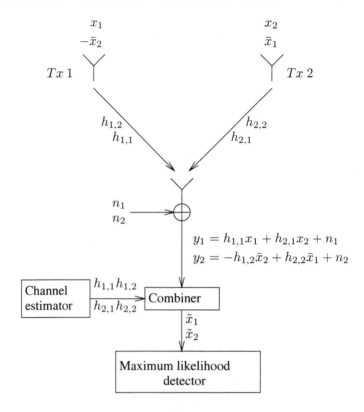

Figure 14.26: Baseband representation of the simple twin-transmitter space-time block code $\mathbf{G}_2$ of Equation 14.7 using one receiver over varying fading conditions.

associated with the same complex transfer function values. The figure shows the baseband representation of the simple twin-transmitter space-time block code $\mathbf{G}_2$ of Equation 14.7 using one receiver. At the receiver, we have

$$y_1 = h_{1,1}x_1 + h_{2,1}x_2 + n_1 \tag{14.9}$$
$$y_2 = -h_{1,2}\bar{x}_2 + h_{2,2}\bar{x}_1 + n_2 \,, \tag{14.10}$$

where y_1 is the first received signal and y_2 is the second. Both signals y_1 and y_2 are passed to the combiner in order to extract the signals x_1 and x_2. Aided by the channel estimator, which in this example provides perfect estimation of the diversity channels' frequency-domain transfer functions, the combiner performs simple signal processing in order to separate the signals x_1 and x_2. Specifically, in order to extract the signal x_1, it combines signals y_1 and y_2 as follows:

$$\tilde{x}_1 = \bar{h}_{1,1}y_1 + h_{2,2}\bar{y}_2 \tag{14.11}$$
$$= \bar{h}_{1,1}h_{1,1}x_1 + \bar{h}_{1,1}h_{2,1}x_2 + \bar{h}_{1,1}n_1 - h_{2,2}\bar{h}_{1,2}x_2 + h_{2,2}\bar{h}_{2,2}x_1 + h_{2,2}\bar{n}_2$$
$$= \left(|h_{1,1}|^2 + |h_{2,2}|^2\right)x_1 + \left(\bar{h}_{1,1}h_{2,1} - h_{2,2}\bar{h}_{1,2}\right)x_2 + \bar{h}_{1,1}n_1 + h_{2,2}\bar{n}_2.$$

Similarly, for signal x_2 the combiner generates:

$$
\begin{aligned}
\tilde{x}_2 &= \bar{h}_{2,1}y_1 - h_{1,2}\bar{y}_2 \quad\quad\quad\quad\quad\quad\quad\quad\quad\quad\quad\quad\quad (14.12)\\
&= \bar{h}_{2,1}h_{1,1}x_1 + \bar{h}_{2,1}h_{2,1}x_2 + \bar{h}_{2,1}n_1 + h_{1,2}\bar{h}_{1,1}x_2 - h_{1,2}\bar{h}_{2,2}x_1 - h_{1,2}\bar{n}_2\\
&= \left(|h_{2,1}|^2 + |h_{1,2}|^2\right)x_2 + \left(\bar{h}_{2,1}h_{1,1} - h_{1,2}\bar{h}_{2,2}\right)x_1 + \bar{h}_{2,1}n_1 - h_{1,2}\bar{n}_2.
\end{aligned}
$$

In contrast to the prefect cancellation scenario of [7], we can see from Equations 14.12 and 14.13 that the signals x_1 and x_2 now interfere with each other. We can no longer cancel the cross-coupling of signals x_2 and x_1 in Equations 14.12 and 14.13, respectively, unless the fading envelopes satisfy the condition of $h_{1,1} = h_{1,2}$ and $h_{2,1} = h_{2,2}$.

At high SNRs the noise power is insignificant compared to the transmitted power of the signals x_1 and x_2. Therefore, we can ignore the noise terms n in Equations 14.12 and 14.13. However, the interference signals' power increases, as we increase the transmission power. Assuming that both the signals x_1 and x_2 have an equivalent signal power, we can then express the signal to interference ratio (SIR) for signal x_1 as:

$$
SIR = \frac{|h_{1,1}|^2 + |h_{2,2}|^2}{\bar{h}_{1,1}h_{2,1} - h_{2,2}\bar{h}_{1,2}}, \quad\quad\quad\quad (14.13)
$$

and similarly for signal x_2 as:

$$
SIR = \frac{|h_{2,1}|^2 + |h_{1,2}|^2}{\bar{h}_{2,1}h_{1,1} - h_{1,2}\bar{h}_{2,2}}. \quad\quad\quad\quad (14.14)
$$

In Figure 14.27, we show the FER performance of the space-time block code G_2 concatenated with the TC(2,1,3) code using one receiver and the 128-subcarrier 16QAM OFDM modem. The CIR has two equal-power rays separated by various delay spreads and a maximum Doppler frequency of 200 Hz. As we can see in Equations 14.13 and 14.14, we have $SIR \rightarrow \infty$, if $h_{1,1} = h_{1,2}$ and $h_{2,1} = h_{2,2}$. On the other hand, we encounter $SIR \rightarrow 1$, if $h_{1,1} = \delta h_{1,2}$ and $h_{2,1} = \delta h_{2,2}$, where $\delta \rightarrow \infty$. Since the SIR decreases, when the delay spread increases due to the rapidly fluctuating frequency-domain fading envelopes, as shown in Figure 14.25, we can see in Figure 14.27 that the performance of the concatenated scheme degrades, when increasing the delay spread. When the delay spread is more than $15\mu s$, we can see from the figure that the concatenated scheme exhibits the previously observed flattening effect. Furthermore, the error floor of the concatenated scheme becomes higher, as the delay spread is increased.

Similarly to Figure 14.24, where the Doppler frequency was varied, we show in Figure 14.28 the E_b/N_0 value required for maintaining FER=10^{-3} versus the delay spread for the 32-state 4PSK space-time trellis code and for the space-time block code G_2 concatenated with the TC(2,1,3) code using one receiver and the 128-subcarrier OFDM modem. The CIR exhibited two equal-power rays separated by various delay spreads. The maximum Doppler frequency was 200 Hz. We can see in the figure that the performance of the 32-state 4PSK space-time trellis code does not vary significantly with the delay spread. However, the concatenated TC(2,1,3)-coded scheme suffers severe performance degradation upon increasing the delay spread, as evidenced by the associated error floors shown in Figure 14.27. The SIR associated with the various delay spreads was obtained using computer simulations and the

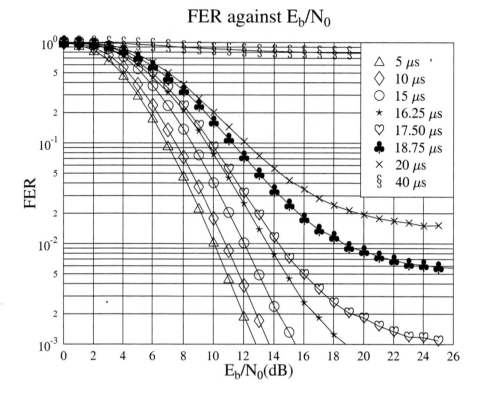

Figure 14.27: FER performance of the space-time block code $\mathbf{G}_2$ concatenated with the TC(2,1,3) code using one receiver, the 128-subcarrier OFDM modem and 16QAM. The CIR exhibits two equal-power rays separated by various delay spreads and a maximum Doppler frequency of 200 Hz. The coding parameters are shown in Tables 14.2, 14.3 and 14.4.

associated SIR values are also shown in Figure 14.28, denoted by the hearts. As we have expected, the calculated SIR decreases with the delay spread. We can see in the figure that the performance of the concatenated $\mathbf{G}_2$/TC(2,1,3) scheme suffers severe degradation, when the delay spread is in excess of $15\mu s$, as indicated by the near-vertical curve marked by triangles. If we relate this curve to the SIR curve marked by the hearts, we can see from the figure that the SIR is approximately 10 dB. Hence the SIR of the concatenated $\mathbf{G}_2$/TC(2,1,3) scheme has to be more than 10 dB, in order for it to outperform the space-time trellis codes using no channel coding.

14.4.5 Delay Non-sensitive System

Previously, we have provided simulation results for a delay-sensitive, OFDM symbol-by-symbol decoded system. More explicitly, the received OFDM symbol had to be demodulated and decoded on a symbol-by-symbol basis, in order to provide decoded bits for example for a low-delay source decoder. Therefore, the two transmission instants of the space-time block

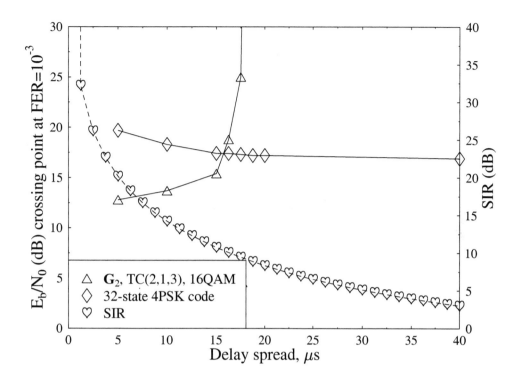

Figure 14.28: The E_b/N_0 values required for maintaining FER=10^{-3} versus delay spreads for the 32-state 4PSK space-time trellis code and for the space-time block code $\mathbf{G}_2$ concatenated with the TC(2,1,3) code using **one receiver** and the 128-subcarrier OFDM modem. The CIR exhibited two equal-power rays separated by various delay spreads and a maximum Doppler frequency of 200 Hz. The effective throughput was **2 BPS** and the coding parameters are shown in Tables 14.2, 14.3 and 14.4. The SIR of various delay spreads are shown as well.

code $\mathbf{G}_2$ had to be in the same OFDM symbol. They were allocated to the adjacent subcarriers in our previous studies. Moreover, we have shown in Figure 14.25 that the variation of the frequency-domain fading amplitudes along the subcarriers becomes more severe, as we increase the delay spread of the two rays. In Figure 14.29 we show both the frequency-domain and time-domain fading amplitudes of the channels' fading amplitudes for a fraction of the subcarriers in the 128-subcarrier OFDM symbols over the previously used two-path channel having two equal-power rays separated by a delay spread of $40\mu s$. The maximum Doppler frequency was set here to 100 Hz. It can be clearly seen from the figure that the fading amplitude variation versus time is slower, than that versus the subcarrier index within the OFDM symbols. This implies that the SIR attained would be higher, if we were to allocate the two transmission instants of the space-time block code $\mathbf{G}_2$ to the same subcarrier of consecutive OFDM symbols. This increase in SIR is achieved by doubling the delay of the system, since in this scenario two consecutive OFDM symbols have to be decoded, before all the received

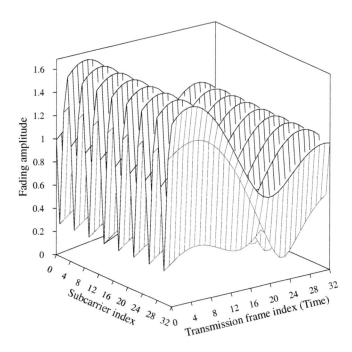

Figure 14.29: The fading amplitude versus time and frequency for various 128-subcarrier OFDM symbols over the two-path channel exhibiting two equal-power rays separated by a delay spread of $40\mu s$ and maximum Doppler frequency of 100 Hz.

data becomes available.

In Figure 14.30, we show our FER performance comparison for the above two scenarios, namely using two adjacent subcarriers and the same subcarrier in two consecutive OFDM symbols for the space-time block code G_2 concatenated with the TC(2,1,3) code using one 128-subcarrier 16QAM OFDM receiver. As before, the CIR exhibited two equal-power rays separated by a delay spread of $40\mu s$ and a maximum Doppler frequency of 100 Hz. It can be seen from the figure that there is a severe performance degradation, if the two transmission instants of the space-time block G_2 are allocated to two adjacent subcarriers. This is evidenced by the near-horizontal curve marked by diamonds across the figure. On the other hand, upon assuming that having a delay of two OFDM-symbol durations does not pose any problems in terms of real-time interactive communications, we can allocate the two transmission instants of the space-time block code G_2 to the same subcarrier of two consecutive OFDM symbols. From Figure 14.30, we can observe a dramatic improvement over the previous allocation method. Furthermore, the figure also indicates that by tolerating a two OFDM-symbol delay, the concatenated G_2/TC(2,1,3) scheme outperforms the 32-state 4PSK space-time trellis code by approximately 2 dB in terms of the required E_b/N_0 value at a FER of 10^{-3}.

Since the two transmission instants of the space-time block code G_2 are allocated to the same subcarrier of two consecutive OFDM symbols, it is the maximum Doppler fre-

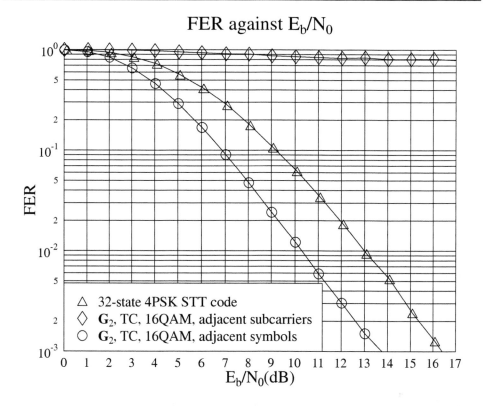

Figure 14.30: FER performance comparison between adjacent subcarriers and adjacent OFDM symbols allocation for the space-time block code G_2 concatenated with the TC(2,1,3) code using one receiver, the 128-subcarrier OFDM modem and 16QAM over a channel having a CIR characterised by two equal-power rays separated by a delay spread of $40\mu s$. The maximum Doppler frequency was 100 Hz. The coding parameters are shown in Tables 14.2, 14.3 and 14.4.

quency that would affect the performance of the concatenated scheme more gravely, rather than the delay spread. Hence we extended our studies to consider the effects of the maximum Doppler frequency on the performance of the concatenated G_2/TC(2,1,3) scheme. Specifically, Figure 14.31 shows the E_b/N_0 values required for maintaining FER=10^{-3} versus the Doppler frequency for the 32-state 4PSK space-time trellis code, and for the space-time block code G_2 concatenated with the TC(2,1,3) code using one 128-subcarrier 16QAM OFDM receiver, when mapping the two transmission instants to the same subcarrier of two consecutive OFDM symbols. The channel exhibited two equal-power rays separated by a delay spread of $40\mu s$ and various maximum Doppler frequencies. The SIR achievable at various maximum Doppler frequencies is also shown in Figure 14.31. Again, we can see that the performance of the concatenated G_2/TC(2,1,3) scheme suffers severely, if the maximum Doppler frequency is above 160 Hz. More precisely, we can surmise that in order for the concatenated scheme to outperform the 32-state 4PSK space-time trellis code, the SIR should be at least 15 dB,

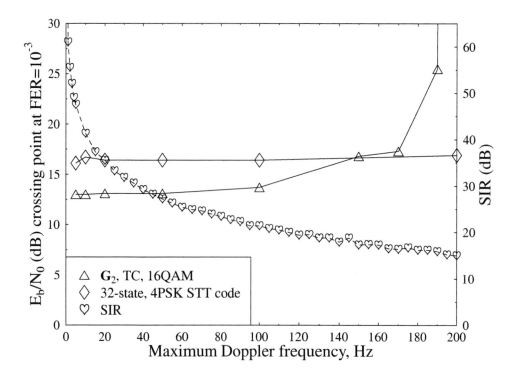

Figure 14.31: The E_b/N_0 value required for maintaining FER=10^{-3} versus the maximum Doppler
frequency for the 32-state 4PSK space-time trellis code and for the adjacent OFDM
symbols allocation of the space-time block code $\mathbf{G}_2$ concatenated with the TC(2,1,3)
code using **one receiver** and the 128-subcarrier OFDM modem. The CIR exhibited two
equal-power rays separated by a delay spread of $40\mu s$. The effective throughput was **2
BPS** and the coding parameters are shown in Tables 14.2, 14.3 and 14.4. The SIR of
various maximum Doppler frequencies are shown as well.

which is about the same as the required SIR in Figure 14.28. From Figure 14.28 and 14.31,
we can conclude that the concatenated $\mathbf{G}_2$/TC(2,1,3) scheme performs better, if the SIR is
in excess of about 10-15 dB.

14.4.6 The Wireless Asynchronous Transfer Mode System

We have previously investigated the performance of different schemes over two-path channels
having two equal-power rays. In this section, we investigate the performance of the various
systems over indoor Wireless Asynchronous Transfer Mode (WATM) channels. The WATM
system used 512 subcarriers and each OFDM symbol was extended with a cyclic prefix of
length 64. The sampling rate was 225 Msamples/s and the carrier frequency was 60 GHz.
In [4] two WATM CIRs were used, namely a five-path and a three-path model, where the latter
one was referred to as the shortened WATM CIR. This CIR was used also in our investigations

here.

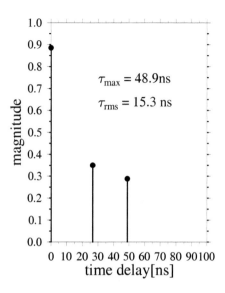

Figure 14.32: Short WATM channel impulse response.

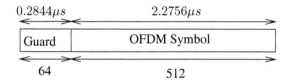

Figure 14.33: Short WATM plot of 512-subcarrier OFDM time domain signal with a cyclic extension of 64 samples.

The shortened WATM channel's impulse response is depicted in Figure 14.32, where the longest-delay path arrived at a delay of 48.9 ns, which corresponds to 11 sample periods at the 225 Msamples/s sampling rate. The 512-subcarrier OFDM time-domain transmission frame having a cyclic extension of 64 samples is shown in Figure 14.33.

14.4.6.1   Channel Coded Space-Time Codes – Throughput of 1 BPS

Previously, we have compared the performance of the space-time trellis codes to that of the TC(2,1,3)-coded space-time block code G_2. We now extend our comparisons to Reed-Solomon (RS) coded space-time trellis codes, which were used in [181, 524, 526]. In Fig-

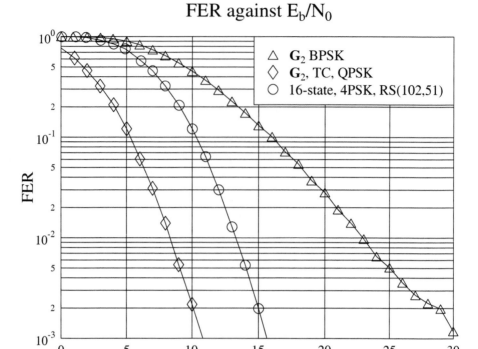

Figure 14.34: FER performance comparison between the TC(2,1,3) coded space-time block code G_2 and the RS(102,51) GF(2^{10}) coded 16-state 4PSK space-time trellis code using one 512-subcarrier OFDM receiver over the shortened WATM channel at an effective throughput of **1 BPS**. The coding parameters are shown in Tables 14.2, 14.3, 14.4 and 14.5.

ure 14.34 we show our FER performance comparison between the TC(2,1,3) coded space-time block code G_2 and the RS(102,51) GF(2^{10}) coded 16-state 4PSK space-time trellis code using one 512-subcarrier OFDM receiver over the shortened WATM CIR of Figure 14.32 at an effective throughput of 1 BPS. We can see from the figure that the TC(2,1,3) coded space-time block code G_2 outperforms the RS(102,51) GF(2^{10}) coded 16-state 4PSK space-time trellis code by approximately 5 dB in E_b/N_0 terms at a FER of 10^{-3}. The performance of the RS(102,51) GF(2^{10}) coded 16-state 4PSK space-time trellis code would be improved by about 2 dB, if the additional complexity of maximum likelihood decoding were affordable. However, even assuming this improvement, the TC(2,1,3) coded space-time block code G_2 would outperform the RS(102,51) GF(2^{10}) coded 16-state 4PSK space-time trellis code.

14.4.6.2 Channel Coded Space-Time Codes — Throughput of 2 BPS

In our next experiment, the throughput of the system was increased to 2 BPS by employing a higher-order modulation scheme. In Figure 14.35 we show our FER performance comparison between the TC(2,1,3) coded space-time block code G_2 and the RS(153,102) GF(2^{10})

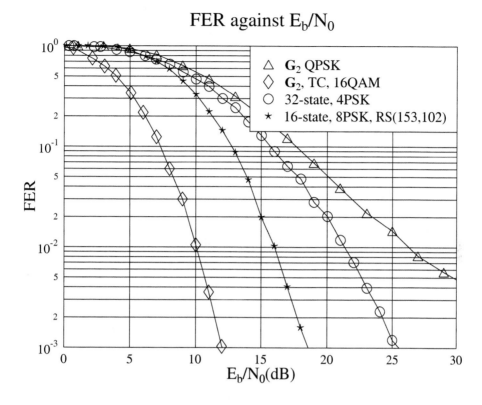

Figure 14.35: FER performance comparison between the TC(2,1,3) coded space-time block code G_2 and the RS(153,102) GF(2^{10}) coded 16-state 8PSK space-time trellis code using one 512-subcarrier OFDM receiver over the shortened WATM channel at an effective throughput of 2 BPS. The coding parameters are shown in Tables 14.2, 14.3, 14.4 and 14.5.

coded 16-state 8PSK space-time trellis code using one 512-subcarrier OFDM receiver over the shortened WATM channel of Figure 14.32 at an effective throughput of 2 BPS. Again, we can see that the TC(2,1,3) coded space-time block code G_2 outperforms the RS(153,102) GF(2^{10}) coded 16-state 8PSK space-time trellis code by approximately 5 dB in terms of E_b/N_0 at a FER of 10^{-3}. The corresponding performance of the 32-state 4PSK space-time trellis code is also shown in the figure. It can be seen that its performance is about 13 dB worse in E_b/N_0 terms, than that of the TC(2,1,3) coded space-time block code G_2. Let us now continue our investigations by considering, whether channel-quality controlled adaptive space-time coded OFDM is capable of providing further performance benefits.

14.5   Space-Time Coded Adaptive Modulation for OFDM

14.5.1   Introduction

Adaptive modulation was proposed by Steele and Webb [1, 16], in order to combat the time-variant fading of mobile channels. The main idea of adaptive modulation is that when the channel quality is favourable, higher-order modulation modes are employed, in order to increase the throughput of the system. On the other hand, more robust but lower-throughput modulation modes are employed, if the channel quality is low. This simple but elegant idea has motivated a number of researchers to probe further [4, 52, 60, 61, 168, 420, 513, 534, 535].

Recently adaptive modulation was also invoked in the context of OFDM, which was termed adaptive OFDM (AOFDM) [4, 62, 168, 420]. AOFDM exploits the variation of the signal quality both in the time domain as well as in the frequency domain. In what is known as sub-band adaptive OFDM transmission, all subcarriers in an AOFDM symbol are split into blocks of adjacent subcarriers, referred to as sub-bands. The same modulation scheme is employed for all subcarriers of the same sub-band. This substantially simplifies the task of signalling the modulation modes, since there are typically four modes and for example 32 sub-bands, requiring a total of 64 AOFDM mode signalling bits.

14.5.2   Turbo-Coded and Space-Time-Coded Adaptive OFDM

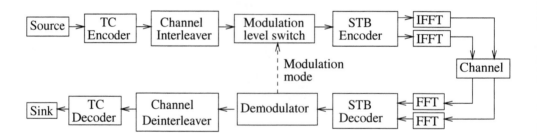

Figure 14.36: System overview of the turbo-coded and space-time-coded adaptive OFDM.

In this section, the adaptive OFDM philosophy portrayed for example in [4, 168, 204, 420] is extended, in order to exploit the advantages of multiple transmit and receive antennas. Additionally, turbo coding is employed for improving the performance of the system. In Figure 14.36, we show the system schematic of the turbo-coded and space-time-coded adaptive OFDM system. Similarly to Figure 14.13, random data bits are generated and encoded by the TC(2,1,3) encoder using an octal generator polynomial of (7, 5). Various TC(2,1,3) coding rates were used for the different modulation schemes. The encoded bits were channel interleaved and passed to the modulator. The choice of the modulation scheme to be used by the transmitter for its next OFDM symbol is determined by the channel quality estimate of the receiver based on the current OFDM symbol. In this study, we assumed perfect channel quality estimation and perfect signalling of the required modem mode of each sub-band based on the channel quality estimate acquired during the current OFDM symbol. Aided by the perfect channel quality estimator, the receiver determines the highest-throughput modulation mode to be employed by the transmitter for its next transmission while maintaining the system's target BER. Five possible transmission modes were employed in our investigations, which are no transmission (NoTx), BPSK, QPSK, 16QAM and 64QAM. In order to simplify the task of signalling the required modulation modes, we employed the sub-band adaptive OFDM transmission scheme detailed for example in [4,168,204,420]. The modulated signals were then passed to the encoder of the space-time block code G_2. The space-time encoded signals were OFDM modulated and transmitted by the corresponding antennas. The shortened WATM channel was used, where the CIR profile and the OFDM transmission frame are shown in Figures 14.32 and 14.33, respectively.

The number of receivers invoked constitutes a design parameter. The received signals were OFDM demodulated and passed to the space-time decoders. Log-MAP [532] decoding of the received space-time signals was performed, in order to provide soft-outputs for the TC(2,1,3) decoder. Assuming that the demodulator of the receiver has perfect knowledge of the instantaneous channel quality, this information is passed to the transmitter in order to determine its next AOFDM modulation mode allocation. The received bits were then channel deinterleaved and passed to the TC decoder, which again, employs the Log-MAP decoding algorithm [163]. The decoded bits were finally passed to the sink for calculation of the BER.

14.5.3 Simulation Results

As mentioned earlier, all the AOFDM based simulation results were obtained over the shortened WATM channel. The channels' profile and the OFDM transmission frame structure are shown in Figures 14.32 and 14.33, respectively. Again, Jakes' model [199] was adopted for modelling the fading channels.

In order to obtain our simulation results, several assumptions were stipulated:

- The average signal power received from each transmitter antenna was the same;

- All multipath components undergo independent Rayleigh fading;

- The receiver has a perfect knowledge of the CIR;

- Perfect signalling of the AOFDM modulation modes.

Again, we note that the above assumptions are unrealistic, yielding the best-case performance, nonetheless, they facilitate the performance comparison of the various techniques under identical circumstances.

14.5.3.1 Space-Time Coded Adaptive OFDM

In this section, we employ the fixed threshold based modem mode selection algorithm, which was also used in [4], adapting the techniques of [26,534,536] for serial modems. It was assumed that the channel quality was constant for all the symbols in a transmission burst, i.e. that the channel's fading envelope varied slowly across the transmission burst. Under these conditions, all the transmitted symbols are modulated using the same modulation mode, chosen according to the predicted SNR. Torrance optimized the modem mode switching thresholds [26,534,536] for the target BERs of 10^{-2} and 10^{-4}, which will be appropriate for a

System	NoTx	BPSK	QPSK	16QAM	64QAM
Speech	$-\infty$	3.31	6.48	11.61	17.64
Data	$-\infty$	7.98	10.42	16.76	26.33

Table 14.7: Optimized switching levels quoted from [534] for adaptive modulation over Rayleigh fading channels for the speech and data systems, shown in instantaneous channel SNR (dB).

high-BER speech system and for a low-BER data system, respectively. The resulting SNR switching thresholds for activating a given modulation mode in a slowly Rayleigh fading narrowband channel are given in Table 14.7 for both systems. Assuming perfect channel quality estimation, the instantaneous channel SNR is measured by the receiver and the information is passed to the modulation mode selection switch at the transmitter, as shown in Figure 14.36 using the system's control channel. This side-information signalling does not constitute a problem, since state-of-the-art wireless systems, such as for example IMT-2000 [13] have a high-rate, low-delay signalling channel. This modem mode signalling feedback information is utilized by the transmitter for selecting the next modulation mode. Specifically, a given modulation mode is selected, if the instantaneous channel SNR perceived by the receiver exceeds the corresponding switching levels shown in Figure 14.7, depending on the target BER.

As mentioned earlier, the adaptation algorithm of [26, 534, 536] assumes constant instantaneous channel SNR over the whole transmission burst. However, in the case of an OFDM system transmitting over frequency selective channels, the channels' quality varies across the different subcarriers. In [4, 204] employing the lowest-quality sub-carrier in the sub-band for controlling the adaptation algorithm based on the switching thresholds given in Table 14.7. Again, this approach significantly simplifies the signalling and therefore it was also adopted in our investigations.

In Figure 14.37, we show the BER and BPS performance of the 16 sub-band AOFDM scheme employing the space-time block code G_2 in conjunction with multiple receivers and a target BER of 10^{-4} over the shortened WATM channel shown in Figure 14.32. The switching thresholds are shown in Table 14.7. The performance of the conventional AOFDM scheme using no diversity [4] is also shown in the figure. From Figure 14.37, we can see that the BPS performance of the space-time coded AOFDM scheme using one receiver is better, than that of the conventional AOFDM scheme. The associated performance gain improves, as the throughput increases. At a throughput of 6 BPS, the space-time coded scheme outperforms the conventional scheme by at least 10 dB in E_b/N_0 terms. However, we notice in Figure 14.37 that as a secondary effect, the BER performance of the space-time coded AOFDM scheme using one receiver degrades, as we increase the average channel SNR. Again, this problem is due to the interference of signals x_1 and x_2 caused by the rapidly varying frequency-domain fading envelope across the subcarriers. At high SNRs, predominantly 64QAM was employed. Since the constellation points in 64QAM are densely packed, this modulation mode is more sensitive to the 'cross-talk' of the signals x_1 and x_2. This limited the BER performance to 10^{-3} even at high SNRs. However, at SNRs lower than 30 dB typically more robust modulation modes were employed and hence the target BER of 10^{-4} was readily met. We will show in the next section that this problem can be overcome by employing turbo channel coding in the system.

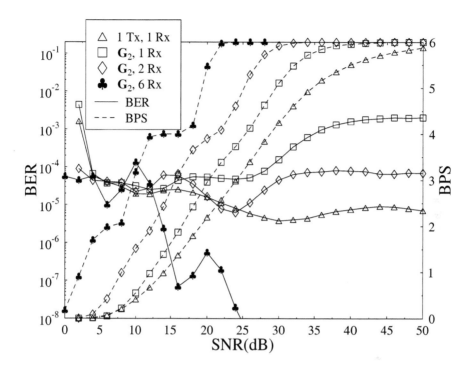

Figure 14.37: BER and BPS performance of 16 sub-band AOFDM employing the space-time block code G_2 using multiple receivers for a target BER of 10^{-4} over the shortened WATM channel shown in Figure 14.32 and the transmission format of Figure 14.33. The switching thresholds are shown in Table 14.7.

In Figure 14.37 we also observe that the BER and BPS performance improves, as we increase the number of AOFDM receivers, since the interference between the signals x_1 and x_2 is eliminated. Upon having six AOFDM receivers, the BER of the system drops below 10^{-8}, when the average channel SNR exceeds 25 dB and there is no sign of the BER flattening effect. At a throughput of 6 BPS, the space-time coded AOFDM scheme using six receivers outperforms the conventional system by more than 30 dB.

Figure 14.38 shows the probability of each AOFDM sub-band modulation mode for (a) conventional AOFDM and for space-time coded AOFDM using (b) 1, (c) 2 and (d) 6 receivers over the shortened WATM channel shown in Figure 14.32. The transmission format obeyed Figure 14.33. The switching thresholds were optimized for the data system having a target BER of 10^{-4} and they are shown in Table 14.7. By employing multiple transmitters and receivers, we increase the diversity gain and we can see in the figure that this increases the probability of the most appropriate modulation mode at a certain average channel SNR. This is clearly shown by the increased peaks of each modulation mode at different average channel

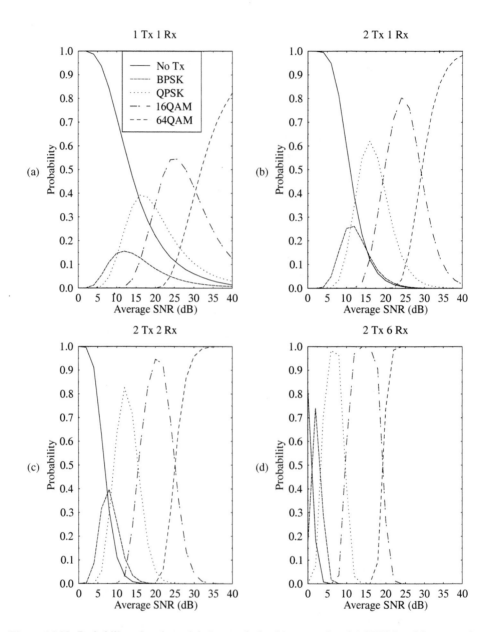

Figure 14.38: Probability of each modulation mode for (a) conventional AOFDM and for space-time coded AOFDM using (b) 1, (c) 2 and (d) 6 receivers over the shortened WATM channel shown in Figure 14.32 and using the transmission frame of Figure 14.33. The thresholds were optimized are for the data system and they are shown in Table 14.7. All sub-figures share the legends seen in Figure 14.38 (a).

SNRs. As an example, in Figure 14.38 (d) we can see that there is an almost 100% probability of transmitting in the QPSK and 16QAM modes at average channel SNR of approximately 6 dB and 15 dB, respectively. This strongly suggests that it is a better solution, if fixed modulation based transmission is employed in space-time coded OFDM, provided that we can afford the associated complexity of using six receivers. We shall investigate these issues in more depth at a later stage.

On the other hand, the increased probability of a particular modulation mode at a certain average channel SNR also means that there is less frequent switching amongst the various modulation modes. For example, we can see in Figure 14.38 (b) that the probability of employing 16QAM increased to 0.8 at an average channel SNR of 25 dB compared to 0.5 in Figure 14.38 (a). Furthermore, there are almost no BPSK transmissions at SNR=25 dB in Figure 14.38 (b). This situation might be an advantage in the context of the AOFDM system, since most of the time the system will employ 16QAM and only occasionally switches to the QPSK and 64QAM modulation modes. This can be potentially exploited to reduce AOFDM modem mode the signalling traffic and to simplify the system.

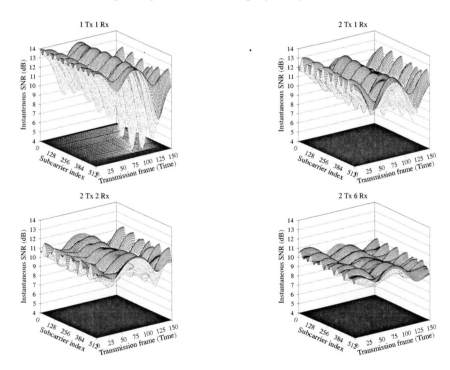

Figure 14.39: Instantaneous channel SNR of 512-subcarrier OFDM symbols for single-transmitter single-receiver and for the space-time block code G_2 using one, two and six receivers over the shortened WATM channel shown in Figure 14.32 and using the transmission format of Figure 14.33. The average channel SNR is 10 dB.

The characteristics of the modem mode probability density functions in Figure 14.38 in conjunction with multiple transmit antennas can be further explained with the aid of Figure 14.39. In Figure 14.39 we show the instantaneous channel SNR experienced by the 512-subcarrier OFDM symbols for a single-transmitter, single-receiver scheme and for the space-time block code G_2 using one, two and six receivers over the shortened WATM channel. The average channel SNR is 10 dB. We can see in Figure 14.39 that the variation of the instantaneous channel SNR for a single transmitter and single receiver is fast and severe. The instantaneous channel SNR may become as low as 4 dB due to deep fades of the channel. On the other hand, we can see that for the space-time block code G_2 using one receiver the variation in the instantaneous channel SNR is slower and less severe. Explicitly, by employing multiple transmit antennas as shown in Figure 14.39, we have reduced the effect of the channels' deep fades significantly. This is advantageous in the context of adaptive modulation schemes, since higher-order modulation modes can be employed, in order to increase the throughput of the system. However, as we increase the number of receivers, i.e. the diversity order, we observe that the variation of the channel becomes slower. Effectively, by employing higher-order diversity, the fading channels have been converted to AWGN-like channels, as evidenced by the space-time block code G_2 using six receivers. Since adaptive modulation only offers advantages over fading channels, we argue that using adaptive modulation might become unnecessary, as the diversity order is increased.

To elaborate a little further, from Figure 14.38 and 14.39 we surmise that fixed modulation schemes might become more attractive, when the diversity order increases, which is achieved in this case by employing more receivers. This is because for a certain average channel SNR, the probability of a particular modulation mode increases. In other words, the fading channel has become an AWGN-like channel, as the diversity order is increased. In Figure 14.40 we show our throughput performance comparison between AOFDM and fixed modulation based OFDM in conjunction with the space-time block code G_2 employing (a) one receiver and (b) two receivers over the shortened WATM channel. The throughput of fixed OFDM was 1, 2, 4 and 6 BPS and the corresponding E_b/N_0 values were extracted from the associated BER versus E_b/N_0 curves of the individual fixed-mode OFDM schemes. It can be seen from Figure 14.40 (a) that the throughput performance of the adaptive and fixed OFDM schemes is similar for a 10^{-2} target BER system. However, for a 10^{-4} target BER system, there is an improvement of 5-10 dB in E_b/N_0 terms at various throughputs for the adaptive OFDM scheme over the fixed OFDM scheme. At high average channel SNRs the throughput performance of both schemes converged, since 64QAM became the dominant modulation mode for AOFDM.

On the other hand, if the number of receivers is increased to two, we can see in Figure 14.40 (b) that the throughput performance of both adaptive and fixed OFDM is similar for both the 10^{-4} and 10^{-2} target BER systems. We would expect similar trends, as the number of receivers is increased, since the fading channels become AWGN-like channels. From Figure 14.40, we conclude that AOFDM is only beneficial for the space-time block code G_2 using one receiver in the context of the 10^{-2} target BER system.

14.5.3.2   Turbo and Space-Time Coded Adaptive OFDM

In the previous section we have discussed the performance of space-time coded adaptive OFDM. Here we extend our study by concatenating turbo coding with the space-time coded AOFDM scheme in order to improve both the BER and BPS performance of the system. As earlier, the turbo convolutional code TC(2,1,3) having a constraint length of 3 and octal gen-

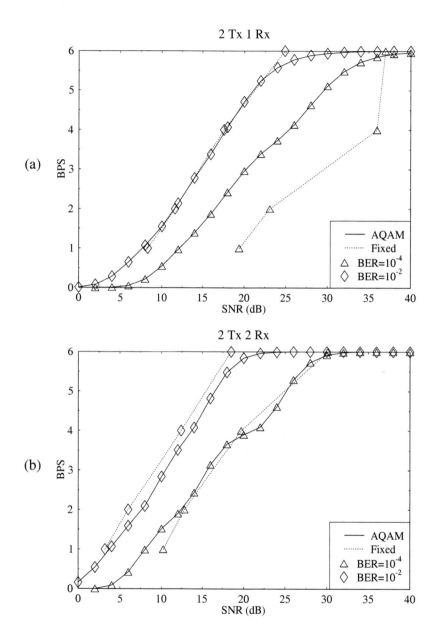

Figure 14.40: BPS throughput performance comparison between adaptive OFDM and fixed modulation based OFDM using the space-time block code $\mathbf{G}_2$ employing (a) one receiver and (b) two receivers over the shortened WATM channel shown in Figure 14.32 and the transmission format of Figure 14.33.

erator polynomial of $(7, 5)$ was employed. Since the system was designed for high-integrity, low-BER data transmission, it was delay non-sensitive. Hence a random turbo interleaver size of approximately 10,000 bits was employed. The random separation channel interleaver [530] was utilized in order to disperse the bursty channel errors. The Log-MAP [163] decoding algorithm was employed, using eight iterations.

We proposed two TC coded schemes for the space-time coded AOFDM system. The first scheme is a fixed half-rate turbo and space-time coded adaptive OFDM system. It achieves a high BER performance, but at the cost of a maximum throughput limited to 3 BPS due to half-rate channel coding. The second one is a variable-rate turbo and space-time coded adaptive OFDM system. This scheme sacrifices the BER performance in exchange for an increased system throughput. Different puncturing patterns are employed for the various code rates R. The puncturing patterns were optimized experimentally by simulations. The design procedure for punctured turbo codes was proposed by Acikel *et al.* [537] in the context of BPSK and QPSK. The optimum AOFDM mode switching thresholds were obtained by computer simulations over the shortened WATM channel of Figure 14.32 and they are shown in Table 14.8.

	NoTx	BPSK	QPSK	16QAM	64QAM
Half-rate TC(2,1,3)					
Rate	–	0.50	0.50	0.50	0.50
Thresholds (dB)	$-\infty$	-4.0	-1.3	5.4	9.8
Variable-rate TC(2,1,3)					
Rate	–	0.50	0.67	0.75	0.90
Thresholds (dB)	$-\infty$	-4.0	2.0	9.70	21.50

Table 14.8: Coding rates and switching levels (dB) for TC(2,1,3) and space-time coded adaptive OFDM over the shortened WATM channel of Figure 14.32 for a target BER of 10^{-4}.

In Figure 14.41, we show the BER and BPS performance of 16 sub-band AOFDM employing the space-time block code G_2 concatenated with both half-rate and variable-rate TC(2,1,3) coding at a target BER of 10^{-4} over the shortened WATM channel of Figure 14.32. We can see in the figure that by concatenating fixed half-rate turbo coding with the space-time coded adaptive OFDM scheme, the BER performance of the system improves tremendously, indicated by a steep dip of the associated BER curve marked by the solid line and diamonds. There is an improvement in the BPS performance as well, exhibiting an E_b/N_0 gain of approximately 5 dB and 10 dB at an effective throughput of 1 BPS, compared to space-time coded AOFDM and conventional AOFDM, respectively. However, again, the maximum throughput of the system is limited to 3 BPS, since half-rate channel coding was employed. In Figure 14.41, we can see that at an E_b/N_0 value of about 30 dB the maximum throughput of the turbo coded and space-time adaptive OFDM system is increased from 3.0 BPS to 5.4 BPS by employing the variable-rate TC(2,1,3) code. Furthermore, the BPS performance of the variable-rate turbo coded scheme is similar to that of the half-rate turbo coded scheme at an average channel SNR below 15 dB. The BER curve marked by the solid line and clubs drops, as the average channel SNR is increased from 0 dB to 15 dB. Due to the increased probability of the 64QAM transmission mode, the variable-rate turbo coded scheme was overloaded by the plethora of channel errors introduced by the 64QAM

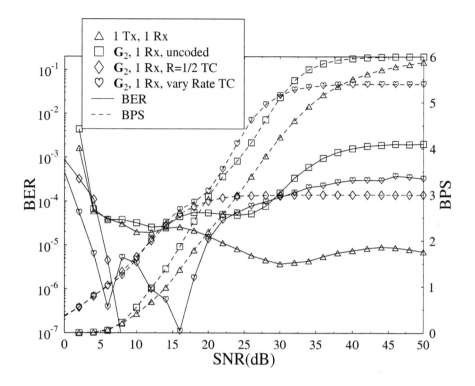

Figure 14.41: BER and BPS performance of 16 sub-band AOFDM employing the space-time block code $\mathbf{G}_2$ concatenated with both half-rate and variable-rate TC(2,1,3) at a target BER of 10^{-4} over the shortened WATM channel shown in Figure 14.32 and using the transmission format of Figure 14.33. The switching thresholds and coding rates are shown in Table 14.7.

mode. Therefore, we can see in Figure 14.41 that the BER increases and stabilises at 10^{-4}. Again, the interference of the signals x_1 and x_2 in the context of the space-time block code $\mathbf{G}_2$ prohibits further improvements in the BER performance, as the average channel SNR is increased. However, employing the variable-rate turbo codec has lowered the BER floor, as demonstrated by the curve marked by the solid line and squares.

14.6 Review and Discussion

Space-time trellis codes [217, 517–521] and space-time block codes [7, 218, 516] constitute state-of-the-art transmission schemes based on multiple transmitters and receivers. Both codes have been introduced in Section 14.1. Space-time trellis codes were introduced in Section 14.2 by utilizing the simplest possible 4-state, 4PSK space-time trellis code as an example. The state diagrams for other 4PSK and 8PSK space-time trellis codes were also given. The branch metric of each trellis transition was derived, in order to facilitate their maximum likelihood (ML) decoding.

In Section 14.3, we proposed to employ an OFDM modem for mitigating the effects of dispersive multipath channels due to its simplicity compared to other approachs [319, 523]. Turbo codes and Reed-Solomon codes were invoked in Section 14.3.1 for concatenation with the space-time block code G_2 and the various space-time trellis codes, respectively. The estimated complexity of the various space-time trellis codes was derived in Section 14.3.3.

We presented our simulation results for the proposed schemes in Section 14.4. The first scheme studied was the TC(2,1,3) coded space-time block code G_2, whereas the second one was based on the family of space-time trellis codes. It was found that the FER and BER performance of the TC(2,1,3) coded space-time block G_2 was better than that of the investigated space-time trellis codes at a throughput of 2 and 3 BPS over the channel exhibiting two equal-power rays separated by a delay spread of $5\mu s$ and having a maximum Doppler frequency of 200 Hz. Our comparison between the two schemes was performed by also considering the estimated complexity of both schemes. It was found that the concatenated G_2/TC(2,1,3) scheme still outperformed the space-time trellis codes using no channel coding, even though both schemes exhibited a similar complexity.

The effect of the maximum Doppler frequency on both schemes was also investigated in Section 14.4.3. It was found that the maximum Doppler frequency had no significant impact on the performance of both schemes. By contrast, in Section 14.4.4 we investigated the effect of the delay spread on the system. Initially, the delay-spread dependent SIR of the space-time block code G_2 was quantified. It was found that the performance of the concatenated TC(2,1,3)-G_2 scheme degrades, as the delay spread increases due to the decrease in the associated SIR. However, varying the delay spread had no significant effect on the space-time trellis codes. We proposed in Section 14.4.5 an alternative mapping of the two transmission instants of the space-time block code G_2 to the same subcarrier of two consecutive OFDM symbols, a solution which was applicable to a delay non-sensitive system. By employing this approach, the performance of the concatenated scheme was no longer limited by the delay spread, but by the maximum Doppler frequency. We concluded that a certain minimum SIR has to be maintained for attaining the best possible performance of the concatenated scheme.

The shortened WATM channel was introduced in Section 14.4.6. In this section, space-time trellis codes were concatenated with Reed-Solomon codes, in order to improve the performance of the system. Once again, both channel coded space-time block and trellis codes were compared at a throughput of 1 and 2 BPS. It was also found that the TC(2,1,3) coded space-time block code G_2 outperforms the RS coded space-time trellis codes.

Space-time block coded AOFDM was studied in the Section 14.5, which is the last section of this chapter. It was shown in Section 14.5.3.1 that only the space-time block code G_2 using one AOFDM receiver was capable of outperforming the conventional single-transmitter, single-receiver AOFDM system designed for a data transmission target BER of 10^{-4} over the shortened WATM channel. We also confirmed that upon increasing the diversity order, the fading channels become AWGN-like channels. This explains, why fixed-mode OFDM transmission constitutes a better trade-off, than AOFDM, when the diversity order is increased. In Section 14.5.3.2, we continued our investigations into AOFDM by concatenating turbo coding with the system. Two schemes were proposed: half-rate turbo and space-time coded AOFDM as well as variable-rate turbo and space-time coded AOFDM. Despite the impressive BER performance of the half-rate turbo and space-time coded scheme, the maximum throughput of the system was limited to 3 BPS. However, by employing the variable-rate turbo and space-time coded scheme, the BPS performance improved, achieving a maximum throughput of 5.4 BPS. However, the improvement in BPS performance was achieved at the cost of a poorer BER performance. In conclusion, a burgeoning research area, which is expected to achieve further advances in the field of turbo space-time coding.

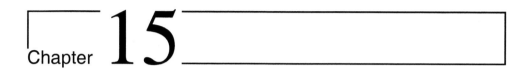

Conclusions and Suggestions for Further Research

In this brief chapter a summary of this monograph is presented and the corresponding conclusions that can be drawn are presented. This will be followed by a range of ideas set aside for future research.

15.1 Book Summary and Conclusions

In Part I of this monograph initially the performance of the DFE was analysed using multilevel modulation modes over static multi-path Gaussian channels, as shown in Figure 2.10. These discussions were further developed in the context of a multi-path fading channel environment, where the recursive Kalman algorithm was invoked in order to track and equalize the received linearly distorted data, as evidenced by Figure 3.16. Explicitly, an adaptive CIR estimator and DFE were implemented in two different receiver structures, as shown in Figure 3.14, while their performances were compared in Figure 2.10. In this respect, Structure 1, which utilized the adaptive CIR estimator provided a better performance, when compared to that of Structure 2, which involved the adaptive DFE, as evidenced by Table 3.10. Furthermore, the complexity of Structure 2 was higher than that of Structure 1, which was studied in Section 3.4. However, these experiments were conducted in a fast start-up environment, where adaptation was restricted to the duration of the training sequence length. By contrast, if the adaptation was invoked over the entire transmission frame using a decision directed scheme, the complexity advantage of Structure 1 was eroded, as discussed in Section 3.5. The application of these fast adapting and accurate CIR estimators was crucial in a wideband AQAM scheme, where the CIR variation across the transmission frame was slow. In these experiments valuable insights were obtained with regards to the design of the equalizer and to the characteristics of the adaptive algorithm itself. This provided a firm foundation for the further investigation of the proposed wideband AQAM scheme.

Following our introductory chapters, in **Chapter 4** the concept of adaptive modulation cast in the context of a narrow-band environment was introduced in conjunction with the application of threshold-based power control. In this respect, power control was applied in the vicinity of the switching thresholds of the AQAM scheme. The associated performance was recorded in Table 4.4, where the trade-off between the BER and BPS performance was highlighted. The relative frequency of modulation mode switching was also reduced at the cost of a slight BER degradation. However, the complexity of the scheme increased due to the implementation of the power control regime. Moreover, the performance gains portrayed at this stage represented an upper-bound estimate, since perfect power control was applied. Consequently, the introduction of threshold-based power control in a narrow-band AQAM did not offer an attractive complexity versus performance gain trade-off.

The concept of AQAM was subsequently invoked in the context of a wideband channel, where the DFE was utilized in conjunction with the AQAM modem mode switching regime. Due to the dispersive multi-path characteristics of the wideband channel, a metric based on the output SNR of the DFE was proposed in order to quantify the channel's quality. This ensured that the wideband channel effects were mitigated by the employment of AQAM and equalization techniques. Subsequently a numerical model based on this criterion was established for the wideband AQAM scheme, as evidenced by Figures 4.16 and 4.17. The wideband AQAM switching thresholds were optimised for maintaining a certain target BER and BPS performance, as shown in Figure 4.3.5. The wideband AQAM BPS throughput performance was then compared to that of the constituent fixed modulation modes, where BPS/SNR gains of approximately $1 - 3$dB and $7 - 9$dB were observed for target BERs of 1% and 0.01%, respectively. However, as a result of the assumption made in Section 4.3.1, these gains constituted an upper bound estimate. Nevertheless, the considerable gains achieved provided further motivation for the research of wideband AQAM schemes.

The concept of wideband coded AQAM was presented in **Chapter 5**, where turbo block coding was invoked in the switching regime for different wideband AQAM schemes. The key characteristics of these four schemes, namely those of the **FCFI-TBCH-AQAM, FCVI-TBCH-AQAM, P-TBCH-AQAM** and **VR-TBCH-AQAM** arrangements, were highlighted in Table 5.10 in terms of the respective turbo interleaver size and the coding rate utilized. The general aim of using turbo block coding in conjunction with a high code rates was to increase the effective BPS transmission throughput, which was achieved, as shown in Table 5.11 for the arrangement that we referred to as the **Low-BER** scheme. In this respect, all the schemes produced gains in terms of their BER and BPS performance, when compared to the uncoded AQAM scheme, which was optimised for a BER of 0.01%. This comparison was recorded in Table 5.11 for the four different turbo coded AQAM schemes studied.

The **FCFI-TBCH-AQAM** scheme exhibited a better throughput gain, when compared to the other schemes. This was achieved as a result of the larger turbo interleaver used in this scheme, which also incurred a higher delay. The size of the turbo interleaver was then varied, while retaining identical coding rate for each modulation mode, resulting in the **FCVI-TBCH-AQAM** scheme, where burst-by-burst decoding was achieved at the receiver. The BPS throughput performance of this scheme was also compared to that of the constituent fixed modulation modes, which utilized different channel interleaver sizes, as shown in Figure 5.11. SNR gains of approximately 1.5 and 5.0dB were achieved by the adaptive scheme for a target BERs of 0.01%, when compared to the fixed modulation modes using the small- and large-channel interleavers, respectively. By contrast, for a target BER of 1% only modest gains were achieved by the wideband AQAM scheme. These apparently low gains were the consequence of an 'unfair' comparison, since sibnificantly larger turbo interleaver and channel interleaver sizes were utilized by the fixed modulation modes. Naturally, his resulted in a high transmission delay for the fixed modulation modes. By contrast, the **FCVI-TBCH-AQAM** scheme employed low-latency instantaneous burst-by-burst decoding, which is important in real-time interactive communications.

The size of the turbo interleaver and the coding rate was then varied according to the modulation mode, in order to ensure burst-by-burst decoding at the receiver. This resulted in the **P-TBCH-AQAM** scheme, which also incorporated un-coded modes for the sake of increasing the achievable throughput. Finally, the **VR-TBCH-AQAM** scheme activated different code rates in conjunction with the different modulation modes. These schemes produced a higher maximum throughput due to the utilization of higher code rates. However, the SNR gains in terms of both the BER and BPS performance degraded, when compared to the **FCFI-TBCH-AQAM** scheme as a result of the reduced-size turbo interleaver used. Furthermore, the utilization of higher code rates for the **VR-TBCH-AQAM** arrangement resulted in a higher decoding complexity. Once again, these comparisons are recorded in Table 5.7. Similar characteristics were also observed in the context of the **High-BER** candidate scheme and in conjunction with the near-error-free schemes. However, the performance gains of the **High-BER** schemes were less than those of the **Low-BER** schemes. This was primarily due to the lower channel coding gain achieved at higher BERs and due to the smaller turbo interleaver size used.

The advantages of burst-by-burst decoding were also exploited in the context of blindly detecting the modulation modes. In this respect, the channel coding information and the mean square phasor error was utilized in the hybrid SD-MSE modulation mode detection algorithm of Section 5.6.2 characterized by Equation 5.6. Furthermore, concatenated m-sequences [169] were used in order to detect the NO TX mode while also estimating the channel's quality. The performance of this algorithm was shown in Figure 5.16, where a modulation mode detection error rate (DER) of 10^{-4} was achieved at an average channel SNR of 15dB. However, the complexity incurred by this algorithm was high due to the multiple channel decoding processes required for each individual modulation mode.

Turbo convolutional coding was then introduced and its performance using fixed modulation modes was compared to that of turbo block coding, as shown in Figure 5.23. A BER versus SNR degradation of approximately $1 - 2$dB was observed for the turbo convolutional coded performance at a BER of 10^{-4}. However, the complexity was significantly reduced, namely by a factor of seven, when compared to the previously studied turbo block coded schemes. Turbo convolutional coding was then incorporated in our wideband AQAM scheme and its performance was compared to that of the turbo block coded AQAM schemes, where the results were similar, as evidenced by Figure 5.24. Consequently the complexity versus performance gain trade-off was more attractive for our turbo convolutional coded AQAM schemes.

In our continued investigations of coded AQAM schemes, turbo equalization was invoked where BPS/SNR gains of approximately $1 - 2$dB were achieved by our AQAM scheme. In achieving this performance, iterative CIR estimation was implemented based on the LMS algorithm, which approached the perfect CIR estimation assisted AQAM performance, as shown in Figure 5.31. However, the implementation of this scheme was severely hindered by the high complexity incurred, which increased exponentially in conjunction with higher-order modulation modes and longer CIR memory.

The chapter was concluded with a system design example cast in the context of TCM,

TTCM and BICM based AQAM schemes, which were studied under the constraint of a similar implementational complexity. The BbB adaptive TCM and TTCM schemes were investigated when communicating over wideband fading channels both with and without channel interleaving and they were characterised in performance terms over the COST 207 TU fading channel. When observing the associated BPS curves, adaptive TTCM exhibited up to 2.5 dB SNR-gain for a channel interleaver length of four transmission bursts in comparison to the non-interleaved scenario, as it was evidenced in Figure 5.40. Upon comparing the associated BPS curves, adaptive TTCM also exhibited up to 0.7 dB SNR-gain compared to adaptive TCM of the same complexity in the context of **System II**, while maintaining a target BER of less than 0.01 %, as it was shown in Figure 5.44. Finally, adaptive TCM performed better, than the adaptive BICM benchmarker in the context of **System I**, while the adaptive BICM-ID scheme performed marginally worse, than adaptive TTCM in the context of **System II**, as it was discussed in Section 5.11.5.

In **Chapter 6** following a brief introduction to several fading counter-measures, a general model was used for describing various adaptive modulation schemes employing various constituent modulation modes, such as PSK, Star QAM and Square QAM, as one of the attractive fading counter-measures. In Section 6.3.3.1, the closed form expressions were derived for the average BER, the average BPS throughput and the mode selection probability of the adaptive modulation schemes, which were shown to be dependent on the mode-switching levels as well as on the average SNR. In Sections 6.4.1, 6.4.2 and 6.4.3 we reviewed the existing techniques devised for determining the mode-switching levels. Furthermore, in Section 6.4.4 the optimum switching levels achieving the highest possible BPS throughput were studied, while maintaining the average target BER. These switching levels were developed based on the Lagrangian optimization method.

Then, in Section 6.5.1 the performance of uncoded adaptive PSK, Star QAM and Square QAM was characterised, when the underlying channel was a Nakagami fading channel. It was found that an adaptive scheme employing a k-BPS fixed-mode as the highest throughput constituent modulation mode was sufficient for attaining all the benefits of adaptive modulation, while achieving an average throughput of up to $k - 1$ BPS. For example, a three-mode adaptive PSK scheme employing No-Tx, 1-BPS BPSK and 2-BPS QPSK modes attained the maximum possible average BPS throughput of 1 BPS and hence adding higher-throughput modes, such as 3-BPS 8-PSK to the three-mode adaptive PSK scheme resulting in a four-mode adaptive PSK scheme did not achieved a better performance across the 1 BPS throughput range. Instead, this four-mode adaptive PSK scheme extended the maximum BPS throughput achievable by any adaptive PSK scheme to 2 BPS, while asymptotically achieving a throughput of 3 BPS, as the average SNR increases.

On the other hand, the relative SNR advantage of adaptive schemes in comparison to fixed-mode schemes increased as the target average BER became lower and decreased as the fading became less severe. More explicitly, less severe fading corresponds to an increased Nakagami fading parameter m, to an increased number of diversity antennas, or to an increased number of multi-path components encountered in wide-band fading channels. As the fading becomes less severe, the average BPS throughput curves of our adaptive Square QAM schemes exhibit undulations owing to the absence of 3-BPS, 5-BPS and 7-BPS square QAM modes.

The comparisons between fixed-mode MC-CDMA and adaptive OFDM (AOFDM) were made based on different channel models. In Section 6.5.4 it was found that fixed-mode MC-

CDMA might outperform adaptive OFDM, when the underlying channel provides sufficient diversity. However, a definite conclusion could not be drawn since in practice MC-CDMA might suffer from MUI and AOFDM might suffer from imperfect channel quality estimation and feedback delays.

Concatenated space-time block coded and turbo convolutional-coded adaptive multi-carrier systems were investigated in Section 6.5.5. The coded schemes reduced the required average SNR by about 6dB-7dB at a throughput of 1 BPS, achieving near error-free transmission. It was also observed in Section 6.5.5 that increasing the number of transmit antennas in adaptive schemes was not very effective, achieving less than 1dB SNR gain, due the fact that the transmit power per antenna had to be reduced in order to limit the total transmit power for the sake of a fair comparison.

The practical issues regarding the implementation of the advocated wideband AQAM scheme was analysed in **Chapter 7**. The impact of error propagation in the DFE was highlighted in Figure 7.2, where the BER degradation was minimal and the target BERs were achieved without any degradation to the transmission throughput performance. The impact of channel quality estimation latency was also studied, where the sub frame based TDD/TDMA system of Section 7.2.1 was implemented. In this system, a channel quality estimation delay of 2.3075ms was imposed and the channel quality estimates were predicted using a linear prediction technique. In this practical wideband AQAM scheme, SNR gains of approximately 1.4dB and 6.4dB were achieved for target BERs of 1% and 0.01%, when compared to the performance of the constituent fixed modulation modes. This was shown graphically in Figure 7.11.

CCI was then subsequently introduced in Section 7.3, where in terms of channel quality estimation, the minimum average SIR that can be tolerated by the wideband AQAM scheme was approximately 10dB, as evidenced by Figure 7.14. In order to mitigate the impact of CCI on the demodulation process, the JD-MMSE-BDFE scheme using an embedded convolutional encoder was invoked, where the performance was shown in Figure 7.21 and 7.22 for the fixed modulation modes and for the wideband AQAM scheme, respectively. The performance gains achieved by the wideband AQAM scheme were approximately $2 - 4dB$ and $7 - 9dB$ for the target BERs of 1% and 0.01%, when compared to the performance of the associated fixed modulation modes. However, these gains constituted an upper bound estimate, since perfect channel estimation of the reference user and the interferer was assumed.

The concept of segmented wideband AQAM was then introduced, in order to reduce the impact of CCI. In this scheme, the inner and outer switching thresholds were developed based on a noise and interference limited environment, respectively, for any given average channel SNR and average SIR. A channel quality estimation delay of 2.3075ms was imposed in estimating the instantaneous SIR and the associated performance was displayed in Figure 7.24. By employing this segmented AQAM scheme, accurate estimation of the instantaneous SIR was needed, which was provided by the Kalman-filter based mid-amble assisted CIR estimation. However, information regarding the interferer was not required for reducing the impact of CCI, which was a substantial advantage.

In Part II of the book we investigated the application of neural networks in the context of channel equalization. As an introduction, the family of established neural network based equalizer structures was reviewed. We opted for studying RBF network based equalizers in detail and investigated their implementation in conjunction with adaptive modulation and turbo channel coding, in order to improve the performance of the transceivers investigated.

More explicitly, **Chapter 8** provided a brief overview of neural networks and augmented, why channel equalization can also be viewed as a classification problem, namely that of classifying the received phasor into one of the M phasors associated with an M-ary modulation scheme. We studied the performance of the RBF equalizer assisted QAM schemes and their adaptive convergence performance in conjunction with both clustering algorithms and LMS channel estimators. The RBF equalizer provided superior performance compared to the linear MSE equalizer using an equivalent equalizer order at the expense of a higher computational complexity, as it was shown in Figure 8.28 and 8.29. According to Figure 8.28 and 8.29, the RBF equalizer ($m = 9$) provided performance improvements of 10dB and 20dB over the linear MSE equalizer for transmission over two-path and three-path Gaussian channels, respectively, at a BER of 10^{-3}. We note that both the linear MSE equalizer and the RBF equalizer exhibited residual BER characteristics, if the channel states corresponding to different transmitted symbols are inseparable in the channel's output observation space, as it was shown in Figure 8.30. The adaptive performance of the RBF equalizer employing the LMS channel estimator of Section 8.9.4, the vector centre clustering algorithm of Section 8.9.5 and the scalar centre clustering algorithm of Section 8.10 was compared. The convergence rate of the clustering algorithm depends on the number of channel coefficients to be adapted and therefore also on the modulation scheme used as well as on the CIR length. However, the convergence of the LMS channel estimation technique only depends on the CIR length and therefore this technique is preferred for high-order modulation schemes and high CIR lengths.

In Section 8.11 decision feedback was introduced into the RBF equalizer, in order to reduce its computational complexity and to improve its performance, since due to its employment the Euclidean distance between the channel states corresponding to different transmitted symbols was increased. The performance degradation due to decision error propagation increased as the BER increased, which became more significant for higher-order QAM constellations, as it was shown in Figure 8.42. The performance degradation for higher-order modulation schemes was higher for fading channel conditions, since they are more sensitive to fades due to the reduced Euclidean distance between the neighbouring channel states.

We note that even for relatively slow fading channels, the channel states value can change significantly on a symbol-by-symbol basis in a transmission burst duration. Inseparable channel state clusters were observed for symbol-invariant - but burst-variant - fading, as it was shown in Figure 8.48(b), which is due to the fading effects manifesting themselves across the burst duration. These phenomena, together with the non-ideal learnt channel states, explain the residual BERs present in our simulations.

Chapter 9 introduced the concept of adaptive modulation invoked for improving the throughput of the system, while maintaining a certain target BER performance. The RBF DFE's 'on-line' BER estimation of the received data burst was used as the AQAM modem mode switching metric in order to quantify the channel's quality. Our simulation results of Section 9.4.5 showed that the proposed RBF DFE-assisted BbB adaptive modem outperformed the individual constituent fixed modulation modes in terms of the mean BER and BPS. The AQAM scheme employing RBF DFE was compared to the AQAM scheme using a conventional DFE, in terms of mitigating the effects of the dispersive wideband channel. Our results in Section 9.4.5 showed that the AQAM RBF DFE scheme was capable of performing as well as the conventional AQAM DFE at a lower decision delay and lower feedforward as well as feedback order. It is worth noting futhermore that the performance of the AQAM

RBF DFE can be improved by increasing both the decision delay τ and the feedforward order m, at the expense of increased computational complexity, while the performance of the conventional AQAM DFE cannot be improved significantly by increasing its equalizer order. However, the computational complexity of the RBF DFE is dependent on the AQAM mode and increases significantly for higher-order modulation modes. This is not so in the context of the conventional DFE, where the computational complexity is only dependent on the feedforward and feedback order.

A practical method of obtaining the switching BER thresholds of the joint AQAM RBF DFE scheme was proposed in Section 9.5, which was shown to provide a near-identical performance in comparison to the achievable best-case performance for the target BER of 10^{-2}. However, for the lower target BER of 10^{-4}, the BER performance degradation in comparison to the best-case performance was more significant, since the RBF DFE was unable to provide a BER estimate of such high accuracy and also because of the spread nature of the BER estimates seen for example in Figure 9.14.

Overall, we have shown that our proposed AQAM scheme improved the throughput performance compared to the fixed modulation modes. On the whole, the RBF DFE provides a reliable channel quality measure for the AQAM scheme, which quantifies all channel impairments, irrespective of their source and at the same time improves the BER performance.

Chapter 10 proposed the Jacobian RBF equalizer that invoked the Jacobian logarithmic approximation, in order to reduce the computational complexity of the original RBF equalizer discussed in Section 8.9.1, while providing a similar BER performance. For example, the total complexity reduction was by a factor of about 2.1, when we considered a 16-QAM RBF DFE in conjunction with the equalizer parameters of $m = 3$, $n = 1$ and $\tau = 2$. The performance of the RBF DFE was investigated using turbo coding and it was compared to the turbo-coded conventional DFE scheme in Section 10.4. Introducing BCH(31,26) turbo coding into the system improved the SNR-performance by 9.5dB for BPSK and by about 8dB for 4-QAM, 16-QAM and 64-QAM at a BER of 10^{-4}. The performance of the conventional DFE and RBF DFE schemes depends on their uncoded performance.

We have also investigated the application of turbo BCH coding in conjunction with AQAM in a wideband fading channel. We observed in Section 10.6.2 that the performance of the switching mechanism depends on the fluctuation of the switching metric since the AQAM switching regime assumed that the channel quality was slowly varying. This was demonstrated in Section 10.6.2, when we compared the performance of the AQAM scheme using the short-term BER before and after turbo decoding, as the switching metric. The spurious nature of the short-term BER after turbo decoding was shown in Figure 10.22, which degraded the performance of the AQAM scheme, as it assumed that the channel quality was slowly varying.

The turbo-coded AQAM RBF DFE system exhibited a better BPS performance, when compared to the uncoded system at low to medium channel SNRs – in the range of 0dB to 26 dB – as evidenced by Figure 10.27. The same figure also showed an improved coded BER performance at higher channel SNRs – in the range above 30dB. A virtually error-free turbo-coded AQAM scheme was also characterized in Figure 10.28. The BPS performance of the error-free coded system was better, than that of the uncoded AQAM system for the channel SNR range of 0dB to 15dB, as evidenced by Figure 10.28. Overall, we have presented the advantageous interactions of RBF-aided DFE and BbB AQAM in conjunction with turbo FEC.

Chapter 11 presented the Jacobian RBF DFE TEQ and comparatively analysed its associated performance and complexity in conjunction with the well-known Log-MAP TEQ in the context of BPSK, 4-QAM and 16-QAM. The computational complexity of the Jacobian RBF DFE TEQ was shown in Section 11.4 to be dependent on the number of RBF centres, on the CIR length and on the modulation mode. The associated 'per iteration' implementational complexity of the Jacobian RBF DFE TEQ ($m = 3$, $n = 2$, $\tau = 2$) was approximately a factor 2.5, 4.4 and 16.3 lower in the context of BPSK, 4-QAM and 16-QAM, respectively, for transmission over the three-path channel considered as seen in Table 11.1. The associated performance degradation compared to the Log-MAP TEQ was shown in Figures 11.8, 11.9 and 11.10 to be approximately 0.2dB, 0.2dB and 10dB for BPSK, 4-QAM and 16-QAM, respectively over the three-path, equal-weight, symbol-spaced Rayleigh fading channel environment considered. The large performance degradation for the 16-QAM scheme was due to the error propagation effect of the DFE, which became more grave in conjunction with higher-order constellations. Therefore, the Jacobian RBF DFE TEQ of Section 11.2 could only provide a practical performance versus complexity advantage for lower modulation modes. In terms of storage requirements, the Jacobian RBF DFE is less demanding, as it only has to store the values of the RBF centres, while the Log-MAP equalizer has to store both the forward- and backward-recursively calculated metrics. Our proposed reduced-complexity RBF DFE TEQ – where the RBF DFE skips evaluating the symbol LLRs in the current iteration when the symbol is sufficiently reliable after channel decoding in the previous iteration – was shown in Section 11.6 to give significant computational complexity reductions, while providing an equivalent BER performance to the RBF DFE TEQ. The complexity reduction was approximately 21% (at an SNR of 6dB) and 35% (at an SNR of 4dB) for dispersive Gaussian and Rayleigh channels, respectively.

In Part III of the book adaptive CDMA, OFDM and space-time coded systems have been investigated by invoking the previously detailed burst-by-burst adaptive principles.

Specifically, in **Chapter 12** we commenced our discussions with the recent history of smart CDMA MUDs and the most promising schemes have been comparatively studied, in order to assist in the design of powerful third- and fourth-generation receivers. Future transceivers may become BbB-adaptive, in order to be able to accommodate the associated channel quality fluctuations without disadvantageously affecting the system's capacity. Hence the methods studied in this monograph are advantageous, since they often assist in avoiding having to power up at the transmitter, which would result in inflicting increased levels of co-channel interference and power consumption. Furthermore, the techniques characterized in this chapter support an increased throughput within a given bandwidth and will contribute towards reducing the constantly increasing demand for more bandwidth.

Both SIC and PIC receivers were investigated in the context of AQAM/CDMA schemes, which were outperformed by the JD based twin-mode AQAM/CDMA schemes for a spreading factor of $Q = 64$ and $K = 8$ users. Both IC receivers were unable to provide good performances in the triple-mode AQAM/CDMA arrangement, since the BER curves exhibited error floors. In the VSF/CDMA systems the employment of variable spreading factors of $Q = 32$ and $Q = 16$ enabled the PIC and SIC receivers to provide a reasonable BER and throughput performance, which was nonetheless inferior to that of the JD. When the number of users in the system was increased, the PIC and SIC receivers were unable to exploit the variability in channel conditions in order to provide a higher information throughput, as opposed to the JD scheme, which showed performance gains in both the adaptive-rate AQAM and VSF CDMA schemes. However, the complexity of the IC receivers increased only linearly with the number of CDMA users, K, compared to the joint detector, which exhibited a complexity proportional to $O(K^3)$. Tables 12.5, 12.6 and 12.7 summarize our performance comparisons for all three multiuser detectors in terms of the E_s/N_0 required for achieving the target BER of 1% or less, the normalized throughput performance and the associated complexity in terms of the number of operations required per detected symbol, respectively. Both the third and future fourth generation standardisation activities may benefit from this study. We note, however that in conjunction with adaptive beam-steering [81] and space-time coding the preference order of the various receivers' performance may change.

In **Chapter 13** we commenced our discussions with a historical perspective on OFDM transmissions with reference to the literature of the past 30 years. The advantages and disadvantages of various OFDM techniques were considered briefly and the expected performance was characterized for the sake of illustration in the context of WATM systems. Our discussions deepened, as we approached the subject of adaptive OFDM subcarrier modulation mode allocation and channel coding. Here we would like conclude with a brief discussion comparing the various coded and uncoded, fixed and adaptive OFDM modems and identify future research issues.

Specifically, Figure 13.29 compared the different adaptive modulation schemes discussed. The comparison graph was split into two sets of curves, depicting the achievable data throughput for a data BER of 10^{-4} highlighted for the fixed throughput systems in Figure 13.29(a), and for the time–variant–throughput systems in Figure 13.29(b).

The fixed throughput systems — highlighted in black in Figure 13.29(a) — comprise the non–adaptive BPSK, QPSK and 16QAM modems, as well as the fixed–throughput adaptive scheme, both for coded and uncoded applications. The non–adaptive modems' performance is marked on the graph as diamonds, and it can be seen that the uncoded fixed schemes require the highest channel SNR of all reviewed transmission methods to achieve a data BER of 10^{-4}. Channel coding employing the turbo coding schemes considered dramatically improved the SNR requirements, at the expense of half the data throughput. The uncoded fixed–throughput (FT) adaptive scheme, marked by the filled triangles, yielded consistently worse data throughput than the coded (C-) fixed modulation schemes C-BPSK, C-QPSK and C-16QAM, with its throughput being about half the coded fixed scheme's at the same SNR values. The coded FT–adaptive (C-FT) system, however, delivered a fairly similar throughput to the C-BPSK and C-QPSK transmissions, and were capable of delivering a BER of 10^{-4} for SNR values down to about 9 dB.

The variable throughput schemes, highlighted in Figure 13.29(b), outperformed the comparable fixed throughput algorithms [420]. For high SNR values, all uncoded schemes' performance curves converged to a throughput of 4bits/symbol, which was equivalent to 16QAM transmission. The coded schemes reached a maximal throughput of 2BPS. Of the uncoded schemes, the "data" switching–level (SL) and target–BER adaptive modems delivered a similar BPS performance, with the target–BER scheme exhibiting slightly better throughput than the SL adaptive modem. The adaptive modem employing pre–equalization (PE) significantly outperformed the other uncoded adaptive schemes and offered a throughput of 0.18 BPS at an SNR of 0 dB, although its crest-factor PDF is slightly less attractive.

The coded transmission schemes suffered from limited throughput at high SNR values, since the half–rate channel coding limited the data throughput to 2 BPS. For low SNR values, however, the coded schemes offered better performance than the uncoded schemes, with the exception of the "speech" SL–adaptive coded scheme, which was outperformed by the un-coded PE-adaptive modem. The poor performance of the coded SL–scheme can be explained by the lower uncoded target BER of the "speech" scenario, which was 1%, in contrast to the 10% uncoded target BER for the coded BER– and PE–adaptive schemes. The coded PE–adaptive modem outperformed the target–BER adaptive scheme, thanks to its more accurate control of the uncoded BER, leading to a higher throughput for low SNR values.

It is interesting to observe that for the given set of four modulation modes the uncoded PE–adaptive scheme was close in performance to the coded adaptive schemes, and that for SNR values of more than 14 dB it outperformed all other studied schemes. It is clear, how-ever, that the coded schemes would benefit from higher order modulation modes or higher-rate channel codecs at high SNR, which would allow these modems to increase the data throughput further when the channel conditions allow.

Based on these findings, adaptive-rate channel coding is an interesting area for future research in conjunction with adaptive OFDM, adaptive beam-stearing and interference can-cellation. Provided that these OFDM enhancements reach a similar state of maturity to the OFDM components used in the current standard digital audio and video broadcasting, WATM, ADSL and power-line communications proposals, OFDM is likely to find a range of further attractive applications in wireless and wireline communications both for businesses and in the home.

In our closing chapter, namely in **Chapter 14** our discussions were centred around the various trade-offs in the context of adaptive transmission and space-time coding. Space-time trellis codes and space-time block codes constitute state-of-the-art transmission schemes based on multiple transmitters and receivers. Space-time trellis codes were introduced in Section 14.2 by utilising the simplest possible 4-state, 4PSK space-time trellis code as an example. The state diagrams for other 4PSK and 8PSK space-time trellis codes were also given. The branch metric of each trellis transition was derived, in order to facilitate their maximum likelihood (ML) decoding.

In Section 14.3, we proposed to employ an OFDM modem for mitigating the effects of dispersive multipath channels due to its simplicity compared to other approachs [319, 523]. Turbo codes and Reed-Solomon codes were invoked in Section 14.3.1 for concatenation with the space-time block code G_2 and the various space-time trellis codes, respectively. The estimated complexity of the various space-time trellis codes was derived in Section 14.3.3.

We presented our simulation results for the proposed schemes in Section 14.4. The first scheme studied was the TC(2,1,3) coded space-time block code G_2, whereas the second one was based on the family of space-time trellis codes. It was found that the FER and BER per-formance of the TC(2,1,3) coded space-time block G_2 was better than that of the investigated space-time trellis codes at a throughput of 2 and 3 BPS over the channel exhibiting two equal-power rays separated by a delay spread of $5\mu s$ and having a maximum Doppler frequency of 200 Hz. Our comparison between the two schemes was performed by also considering the estimated complexity of both schemes. It was found that the concatenated G_2/TC(2,1,3) scheme still outperformed the space-time trellis codes using no channel coding, even though both schemes exhibited a similar complexity.

The effect of the maximum Doppler frequency on both schemes was also investigated in Section 14.4.3. It was found that the maximum Doppler frequency had no significant impact on the performance of both schemes. By contrast, in Section 14.4.4 we investigated the effect of the delay spread on the system. Initially, the delay-spread dependent SIR of the space-time block code G_2 was quantified. It was found that the performance of the concatenated TC(2,1,3)-G_2 scheme degrades, as the delay spread increases due to the decrease in the associated SIR. However, varying the delay spread had no significant effect on the space-time trellis codes. We proposed in Section 14.4.5 an alternative mapping of the two transmission instants of the space-time block code G_2 to the same subcarrier of two consecutive OFDM symbols, a solution which was applicable to a delay non-sensitive system. By employing this approach, the performance of the concatenated scheme was no longer limited by the delay spread, but by the maximum Doppler frequency. We concluded that a certain minimum SIR has to be maintained for attaining the best possible performance of the concatenated scheme.

The shortened WATM channel was introduced in Section 14.4.6. In this section, space-time trellis codes were concatenated with Reed-Solomon codes, in order to improve the performance of the system. Once again, both channel coded space-time block and trellis codes were compared at a throughput of 1 and 2 BPS. It was also found that the TC(2,1,3) coded space-time block code G_2 outperforms the RS coded space-time trellis codes.

Space-time block coded AOFDM was studied in the Section 14.5, which is the last section of this treatise. It was shown in Section 14.5.3.1 that only the space-time block code G_2 using one AOFDM receiver was capable of outperforming the conventional single-transmitter, single-receiver AOFDM system designed for a data transmission target BER of 10^{-4} over the shortened WATM channel. We also confirmed that upon increasing the diversity order, the fading channels become AWGN-like channels. This explains, why fixed-mode OFDM transmission constitutes a better trade-off, than AOFDM, when the diversity order is increased. In Section 14.5.3.2, we continued our investigations into AOFDM by concatenating turbo coding with the system. Two schemes were proposed: half-rate turbo and space-time coded AOFDM as well as variable-rate turbo and space-time coded AOFDM. Despite the impressive BER performance of the half-rate turbo and space-time coded scheme, the maximum throughput of the system was limited to 3 BPS. However, by employing the variable-rate turbo and space-time coded scheme, the BPS performance improved, achieving a maximum throughput of 5.4 BPS. However, the improvement in BPS performance was achieved at the cost of a poorer BER performance. In conclusion, a burgeoning research area, which is expected to achieve further advances in the field of turbo space-time coding.

15.2 Suggestions for Future Research

The impact of latency on the performance of the wideband AQAM scheme was reduced due to the implementation of a channel quality predictor using a simple linear interpolation technique. It is therefore feasible to formulate a more sophisticated prediction algorithm in order to cater for schemes having longer channel quality estimation delays or for a faster varying CIR. In this respect, channel prediction algorithms have been proposed by amongst others Lau *et al.* [52] and Eyceoz *et al.* [538], and Duell-Hallen *et al.* [14], which can be applied in an AQAM scheme. Equally important is the development of intelligent learning schemes for the appropriate adjustment of the AQAM switching thresholds, following for example the approach proposed by Tang [30].

The AQAM mode switching criterion can be further refined by exploiting the channel coding schemes described in Chapter 5, where the decoder's soft output information can be used, in order to deduce the estimated BER on a burst-by-burst basis. Consequently the instantaneous burst-BER can be used as an additional channel quality measure. Since the impact of CCI was severe in a wideband AQAM scheme - as discussed in Section 7.3 - adaptive beam-forming techniques [332, 539] can be employed in this scenario in order to reduce the impact of CCI on the wideband AQAM scheme.

The entire cellular network's performance utilizing AQAM schemes can be studied with respect to the network capacity, interference levels as well as the cell-cluster configuration. In this respect, it is interesting to observe and analyse the impact of AQAM on the network and subsequently optimize the network, in order to achieve a better network capacity. These network-layer issues were documented by Blogh and Hanzo [540, 541], where the network capacity of both TDMA and CDMA systems was shown to increase substantially, as a result of employing AQAM and beam-forming. The concept of Trellis-Coded Modulation (TCM) - where the channel coding and modulation constellation mapping was jointly optimized - was introduced into an AQAM scheme by Chua *et al.* [154]. This scheme can be extended to incorporate iterative channel decoding, where TCM codes are used as the component codes, which is termed as Turbo Trellis Coded Modulation (TTCM) [12, 177].

One of the most important requirements in the implementation of adaptive modulation is to be able to anticipate the channel's variation before transmission. This requirement is also critical to the concept of pre-equalization or pre-coding at the transmitter. This idea was formulated by Tomlinson [112] and Harashima *et al* [113] independently at about the same time for pulse amplitude modulation (PAM). Like the DFE with its feedback filter, this pre-coding technique compensates for the ISI caused by the past data symbols. However, the error propagation phenomenon of the DFE is circumvented by pre-equalizing the signal at the transmitter with the aid of the Tomlinson-Harashima (TH) pre-coding. Assuming that the CIR is known at the transmitter, controlled intersymbol interference is introduced into the transmitted signal by the TH pre-coding technique based on the perceived CIR at that moment. This is expected to be cancelled out by the actual intersymbol interference induced by the channel. However, careful considerations have to be given to limit and control the peak transmitted power of the pre-equalizer, which was addressed to a certain extent by the utilization of an inverse modulo filter [4, 112].

The BER degradation due to carrier recovery, clock recovery and amplifier distortion imperfections was beyond the scope of this research. However, the issues regarding the phase and timing recovery have to be addressed. Similarly, the power efficiency and out of band emission properties associated with the non-linear amplification of QAM schemes must be considered.

Research has also been conducted in employing neural network based equalizers in code-division multiple-access (CDMA) environments [409, 542–548] and it appears promising to explore further this area of research in conjunction with the coded AQAM schemes portrayed in this monograph.

Chen *et. al.* proposed a strategy for designing DFE-based support vector machines (SVM) [293]. The SVM design in conjunction with low-complexity conventional DFE structures achieves asymptotically the minimum BER (MBER) solution [290], which provides a performance close to that of the optimal Bayesian DFE. Unlike the exact MBER solution, the SVM solution can be implemented significantly more efficiently. The low-complexity structure of the SVM DFE provides an attractive alternative to the optimal Bayesian-based RBF DFE and to its implementation in conjunction with AQAM schemes and turbo coding.

15.3 Closing Remarks

Throughout the book we reviewed the 30-year history of adaptive wireless transmission schemes. Most of the material was presented in the context of a *novel design paradigm, namely that of 'just' maintaining the required target integrity typically expressed in terms of the required BER, rather than aiming for the highest possible transmission integrity. This allowed us to maximize the achievable throughput.* Given the increasing price and scarcity of spectrum, which became more apparent during the past decade, the associated research efforts have substantially intensified in recent years, which motivated us to collate the research advances of the past decade.

We attempted to portray the range of contradictory system design trade-offs unique to a variety of applications in an unbiased fashion and sufficiently richly illustrated. In a nutshell, the BbB-adaptive transceivers studied are capable of substantially mitigating the effects of the near-instantaneous channel quality fluctuations experience in wireless channels. Naturally, this is also true for the space-time coding assisted transmit diversity schemes employing multiple transmitters and receivers, provided that the associated higher complexity is affordable. This argument was explicitly supported by Figure 1.1 of the Preface. The future is expected to witness the co-exitence of different-complexity BbB-adaptive and space-time coded arrangements.

Our sincere hope is that you will find the book useful, assisting you in solving your own particular communications problem. In this rapidly evolving field it is a real challenge to complete a comprehensive treatise, since new advances are contrived at an ever increasing pace, which one would like to report on. Hence this edition of the book was released in the hope that we will be able to raise your interest in this fascinating research field and that you will be able to dedicate some of your research efforts to solving the outstanding problems, ultimately leading to more innovative wireless systems in the near-future!

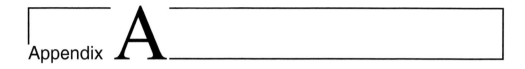

Appendix **A**

Appendices

A.1 Turbo Decoding and Equalization Algorithms

The derivation of the Maximum A Posteriori (MAP) [162] algorithm is presented here and subsequently the algorithm is extended to the Log-MAP [163] algorithm, which can be utilized in the context of an equalizer or a channel decoder. The main objective of these algorithms is to provide the Log Likelihood Ratio (LLR) for each input bit u_n, which will be defined in the next section. Depending on the implementation of these algorithms, u_n can be considered as an encoded bit in an equalizer or a source bit in a channel decoder. However, the derivation of these algorithms is generic and can be applied to an equalizer or a channel decoder.

A.1.1 MAP Algorithm

The so-called a posteriori LLRs, $L(u_n|r)$ can be defined as the ratio of the probability of the transmitted coded bit taking its two possible values, given that r was the symbol sequence received. This can be written as follows:

$$L(u_n|r) = \ln \left(\frac{P(u_n = +1|r)}{P(u_n = -1|r)} \right).$$
(A.1)

By exploiting Bayes' theorem [549] of $P(a \wedge b) = P(a|b) \cdot P(b)$, where the symbol $\wedge$ represents the logical 'and' operation, Equation A.1 can be rewritten as:

$$L(u_n|r) = \ln \left(\frac{P(u_n = +1 \wedge r)}{P(u_n = -1 \wedge r)} \right).$$
(A.2)

For a binary system's trellis with χ number of states at each trellis stage, there will be χ sets of $u_n = +1$ and $u_n = -1$ bit transitions, which are mutually exclusive. This is because only one transition could have occurred at the encoder, depending on the value of u_n. Therefore, the probability that u_n was transmitted can also be expressed as the sum of

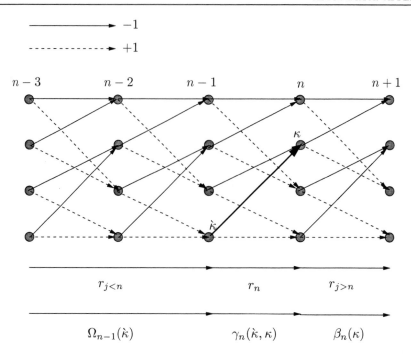

Figure A.1: Example of a binary $M = 2$ system's trellis structure.

the individual probabilities of the associated χ number of transitions. We can then express the LLR in Equation A.2 as:

$$L(u_n|r) = \ln \left(\frac{\displaystyle\sum_{(\check{k},\kappa) \Rightarrow u_n = +1} P(\check{k} \wedge \kappa \wedge r)}{\displaystyle\sum_{(\check{k},\kappa) \Rightarrow u_n = -1} P(\check{k} \wedge \kappa \wedge r)} \right), \tag{A.3}$$

where $(\check{k}, \kappa) \Rightarrow u_n = +1$ denotes the set of transitions from state $\check{k}$ to κ caused by the bit $u_n = +1$ and similarly, $(\check{k}, \kappa) \Rightarrow u_n = -1$ for the bit $u_n = -1$. In Figure A.1 we show an example of a binary system, which has $\chi = 4$ states at each stage n in the trellis. Since this is an example of a binary system's trellis, at each trellis transition interval n, there are $M = 2$ transitions leaving each state and $\chi = 4$ transitions are due to the bit $u_n = +1$. Each of these transitions belongs to the set $(\check{k}, \kappa) \Rightarrow u_n = +1$. Similarly, there will be $\chi = 4$ number of $u_n = -1$ transitions, which belong to the set $(\check{k}, \kappa) \Rightarrow u_n = -1$. In our forthcoming discussions, we will introduce the remaining notations of Figure A.1, with emphasis on $\Omega_{n-1}(\check{k})$, $\beta_n(\kappa)$, and $\gamma_n(\check{k}, \kappa)$.

The received symbol sequence r can be viewed in different time periods, where $r_{j<n}$, r_n and $r_{j>n}$ represents the past, present and future received symbol sequence. Consequently, by exploiting Bayes' theorem [549], Equation A.3 can be rewritten as [162]:

$$L(u_n|r) = \ln \left(\frac{\displaystyle\sum_{(\check{k},\kappa)\Rightarrow u_n=+1} P(r_{j<n}\wedge\check{k})\cdot P(r_n\wedge\kappa|\check{k})\cdot P(r_{j>n}|\kappa)}{\displaystyle\sum_{(\check{k},\kappa)\Rightarrow u_n=-1} P(r_{j<n}\wedge\check{k})\cdot P(r_n\wedge\kappa|\check{k})\cdot P(r_{j>n}|\kappa)} \right)$$

(A.4)

$$= \ln \left(\frac{\displaystyle\sum_{(\check{k},\kappa)\Rightarrow u_n=+1} \Omega_{n-1}(\check{k}) \cdot \gamma_n(\check{k},\kappa) \cdot \beta_n(\kappa)}{\displaystyle\sum_{(\check{k},\kappa)\Rightarrow u_n=+1} \Omega_{n-1}(\check{k}) \cdot \gamma_n(\check{k},\kappa) \cdot \beta_n(\kappa)} \right).$$

In Equation A.4 we have introduced the notation $\Omega_{n-1}(\check{k})$, which is defined as follows:

$$\Omega_{n-1}(\check{k}) = P(r_{j<n}\wedge\check{k}),$$ (A.5)

representing the probability that state $\check{k}$ is attained at instant $n-1$ and that the previous received channel output is $r_{j<n}$. Similarly, the notation $\gamma_n(\check{k},\kappa)$ can be written as:

$$\gamma_n(\check{k},\kappa) = P(r_n\wedge\kappa|\check{k}),$$ (A.6)

which indicates the probability that given state $\check{k}$ is attained at the trellis stage of $n-1$ and that the channel output r_n is received at instant n, the resulting transition is to state κ. Lastly, the notation $\beta_n(\kappa)$ is given by :

$$\beta_n(\kappa) = P(r_{j>n}|\kappa),$$ (A.7)

where $\beta_n(\kappa)$ is the probability that at an instant n we will receive the future channel sequence $r_{j>n}$, provided that the present state at instant n is κ. Hence, by using Equations A.4, A.5, A.6 and A.7 we can calculate the LLR for each input bit, u_n. In the following subsection, we will describe how each of these three variables, namely $\Omega_{n-1}(\check{k})$, $\gamma_n(\check{k},\kappa)$ and $\beta_n(\kappa)$ can be determined recursively.

A.1.1.1 The Calculation of the Log Likelihood Ratio

By considering the calculation of $\Omega_n(\kappa)$, which will become $\Omega_n(\check{k})$ at instant $n+1$, Equation A.5 can be rewritten as:

$$\Omega_n(\kappa) = P(\kappa\wedge r_{j<n+1})$$
$$= P(\kappa\wedge r_{j<n}\wedge r_n).$$ (A.8)

It was shown in Reference [550] that by using Bayes' theorem and writing the probability $P(\kappa\wedge r_{j<n}\wedge r_n)$ as the sum of the joint probabilities $P(\kappa\wedge\check{k}\wedge r_{j<n}\wedge r_n)$ over all possible states $\check{k}$, Equation A.8 becomes:

$$\Omega_n(\kappa) = \sum_{\text{all } \check{k}} P(\kappa\wedge\check{k} \wedge r_{j<n}\wedge r_n)$$
$$= \sum_{\text{all } \check{k}} P([\kappa\wedge r_n]|[\check{k} \wedge r_{j<n}])\cdot P(\check{k} \wedge r_{j<n}).$$ (A.9)

By assuming that the channel is memoryless, we can rewrite $P([\kappa \wedge r_n]|[\check{k} \wedge r_{j<n}])$ as $P([\kappa \wedge r_n]|[\check{k}])$. Consequently the probability that r_n was received and that state κ was reached at instant n, is only dependent on the previous state $\check{k}$, yielding:

$$\Omega_n(\kappa) = \sum_{\text{all } \check{k}} P([\kappa \wedge r_n]|\check{k}) \cdot P(\check{k} \wedge r_{j<n}). \tag{A.10}$$

Upon substituting Equations A.6 and A.5 into Equation A.10, we can write:

$$\Omega_n(\kappa) = \sum_{\text{all } \check{k}} \Omega_{n-1}(\check{k}) \cdot \gamma_n(\check{k}, \kappa). \tag{A.11}$$

Hence by determining $\gamma_{(}\check{k}, \kappa)$, $\Omega_n(\kappa)$ can be evaluated recursively, where the initial value of $\Omega_n(\kappa)$, $\Omega_0(\kappa)$ is set as follows:

$$\Omega_0(\kappa) = \begin{cases} 1 & \text{for } \kappa = S_{\text{start}} \\ 0 & \text{for } \kappa \neq S_{\text{start}}, \end{cases} \tag{A.12}$$

where S_{start} is the starting state.

The derivation of $\beta_n(\kappa)$ utilizes a similar approach to that of $\Omega_n(\kappa)$, where we can rewrite Equation A.7 as:

$$\beta_{n-1}(\check{k}) = P(r_{j>n-1}|\check{k}), \tag{A.13}$$

Consequently, by expressing $P(r_{j>n-1}|\check{k})$ as the sum of joint probabilities over all states κ, we have:

$$\begin{aligned} \beta_{n-1}(\check{k}) &= P(r_{j>n-1}|\check{k}) \\ &= \sum_{\text{all } \kappa} P([\kappa \wedge r_{j>n-1}]|\check{k}) \\ &= \sum_{\text{all } \kappa} P([\kappa \wedge r_{j>n} \wedge r_n]|\check{k}) \end{aligned} \tag{A.14}$$

and by employing Bayes' theorem, we can write $\beta_{n-1}(\check{k})$ as:

$$\begin{aligned} \beta_{n-1}(\check{k}) &= \sum_{\text{all } \kappa} \frac{P(\kappa \wedge r_{j>n} \wedge r_n \wedge \check{k})}{P(\check{k})} \\ &= \sum_{\text{all } \kappa} P(r_{j>n}|[r_n \wedge \check{k} \wedge \kappa]) \cdot \left[\frac{P(r_n \wedge \check{k} \wedge \kappa)}{P(\check{k})} \right] \\ &= \sum_{\text{all } \kappa} P(r_{j>n}|[r_n \wedge \check{k} \wedge \kappa]) \cdot P([r_n \wedge \kappa]|\check{k}). \end{aligned} \tag{A.15}$$

By assuming that the channel is memoryless, the term $P(r_{j>n}|[r_n \wedge \check{k} \wedge \kappa])$ can be written as $P(r_{j>n}|\kappa)$, since the probability of $r_{j>n}$ occurring depends only on the current state κ, yielding:

$$\beta_{n-1}(\check{k}) = \sum_{\text{all } \kappa} P(r_{j>n}|\kappa) \cdot P([r_n \wedge \kappa]|\check{k}). \tag{A.16}$$

Lastly, upon invoking Equations A.6 and A.7, Equation A.16 can be rewritten as:

$$\beta_{n-1}(\grave{k}) = \sum_{\text{all } \kappa} \beta_n(\kappa) \cdot \gamma_n(\grave{k}, \kappa). \tag{A.17}$$

Hence $\beta_{n-1}(\grave{k})$ can be evaluated recursively, given the value of $\gamma_n(\grave{k}.\kappa)$.

In order to determine the initial conditions for $\beta_N(\kappa)$, where N is the length of the received symbol sequence, Equations A.17 and A.7 are employed, in order to yield the following equations [551]:

$$\beta_{N-1}(\grave{k}) = P(r_N|\grave{k})$$
$$= \sum_{\text{all } \kappa} P([r_N \wedge \kappa]|\grave{k}), \tag{A.18}$$

$$\beta_{N-1}(\grave{k}) = \sum_{\text{all } \kappa} \gamma_N(\grave{k}, \kappa). \tag{A.19}$$

Consequently, by observing Equations A.19 and A.17, the only condition that satisfies both these equations for all states κ is as follows:

$$\beta_N(\kappa) = 1 \tag{A.20}$$

In deriving the mathematical expression for $\gamma_n(\kappa)$, we will rearrange Equation A.6 by exploiting Bayes' theorem, yielding:

$$\gamma_n(\grave{k}, \kappa) = P(r_n \wedge \kappa|\grave{k})$$
$$= \frac{P(r_n \wedge \kappa \wedge \grave{k})}{P(\grave{k})}$$
$$= P(r_n|\kappa \wedge \grave{k}) \cdot \left[\frac{P(\kappa \wedge \grave{k})}{P(\grave{k})}\right] \tag{A.21}$$
$$= P(r_n|\kappa \wedge \grave{k}) \cdot P(\kappa|\grave{k})$$
$$= P(r_n|s_i) \cdot P(u_n),$$

where the notation s_i represents the estimate of the transmitted signal arising from a transition from state $\grave{k}$ to κ, while $P(u_n)$ is the so-called a priori probability for bit u_n. At the beginning, the channel decoder or the equalizer have no prior knowledge concerning bit u_n. However, for subsequent iterations, $P(u_n)$ can be obtained from independent sources. In turbo equalization $P(u_n)$ is produced by the SISO decoder, whereas in turbo decoding, $P(u_n)$ is obtained from the other independent decoder component. Hence this can be used in evaluating $\gamma_n(\grave{k}, \kappa)$ for subsequent iterations in the context of turbo decoding and turbo equalization. By assuming that s_i was transmitted over a Gaussian channel, we can express $P(r_n|s_i)$ as:

$$P(r_n|s_i) = \frac{1}{\sqrt{\pi N_o}} \exp\left[-\frac{1}{N_o}(r_n - s_i)^2\right], \tag{A.22}$$

where $\frac{N_o}{2}$ is the double sided power spectral density of the Gaussian noise and the notations r_n and s_i represent the received signal and transmitted signal, respectively. Therefore, $\gamma_n(\check{k}, \kappa)$ of the MAP equalizer is given by:

$$
\begin{aligned}
\gamma_n(\check{k}, \kappa) &= P(r_n|s_i) \cdot P(u_n) \\
&= \frac{1}{\sqrt{\pi N_o}} \exp\left[-\frac{1}{N_o}(r_n - s_i)^2\right] \cdot P(u_n).
\end{aligned}
\tag{A.23}
$$

However, for the MAP channel decoder, the $\gamma_n(\check{k}, \kappa)$ can be obtained as follows [311]:

$$
\gamma_n(\check{k}, \kappa) = \exp\left[\frac{1}{2}\sum_{i=1}^{J} L_{(i,n)}^{in} \cdot c_{(i,n)}\right] \cdot P(u_n),
\tag{A.24}
$$

where J denotes the inverse code rate and $c_{(i,n)}$ represents ith coded bit at a particular trellis transition interval n. In our context of an equalized system, the input LLR of the ith coded bit at the instant n for the channel decoder $l_{(i,n)}^{in}$, is derived from the output of the equalizer. In this respect, the input LLR consists of the composite channel output information and the contribution of the equalizer.

A.1.1.2 Summary of the MAP algorithm

Let us now summarise the operations required in order to calculate the LLR, which is needed by the turbo decoder or the turbo equalizer. Initially, as shown in Figure A.2, we evaluate $\gamma_n(\check{k}, \kappa)$ for the entire received symbol sequence r using Equations A.21 and A.22. Subsequently, $\Omega_{n-1}(\check{k})$ and $\beta_n(\kappa)$ is calculated recursively using Equations A.11 and A.17, respectively. Having evaluated these three terms, namely $\Omega_{n-1}(\check{k})$, $\beta_n(\kappa)$ and $\gamma_n(\check{k}, \kappa)$, we can calculate the LLR $L(u_n|r)$ for each coded bit u_n, using Equation A.4, which is passed to the turbo decoder or the turbo equalizer. The complexity involved in implementing the MAP algorithm is high, since we have to consider all possible paths in the trellis, unlike in the Viterbi algorithm, which only considers the survivor paths. Consequently, the Log-MAP algorithm was proposed by Robertson *et al.* [552], which provides an identical performance to that of the MAP algorithm at a lower complexity.

A.1.2 The Log-MAP Algorithm

In the MAP algorithm, the LLR, $L(u_n|r)$ at instant n was calculated by evaluating $\Omega_{n-1}(\check{k})$, $\beta_n(\kappa)$ and $\gamma_n(\check{k}, \kappa)$ in Equations A.11, A.17 and A.21. The calculation was performed exhaustively, which involved multiplications and evaluating the natural logarithms as well as the exponential terms, in order to determine these three variables. However, in utilizing the Log-MAP algorithm, the complexity of these recursive calculations is reduced by transforming $\Omega_{n-1}(\check{k})$, $\gamma_n(\check{k}, \kappa)$ and $\beta_n(\kappa)$ into the logarithmic domain and subsequently invoking the so-called Jacobian logarithm relationship, which is defined as [289, 553]:

$$
\begin{aligned}
\ln(e^{x_1} + e^{x_2}) &= \max(x_1, x_2) + \ln(1 + e^{-|x_1 - x_2|}) \\
&= \max(x_1, x_2) + f_c(|x_1 - x_2|) \\
&= \max(x_1, x_2) + f_c(\delta) \\
&= J(x_1, x_2),
\end{aligned}
\tag{A.25}
$$

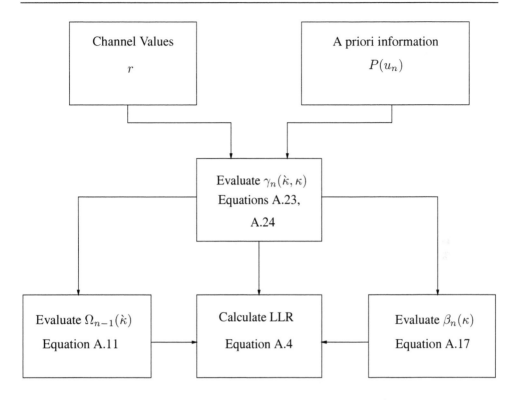

Figure A.2: Summary of key operations in the MAP algorithm, in order to evaluate the LLR for bit u_n at instant n.

where $\delta = |x_1 - x_2|$ and $f_c(\delta)$ can be viewed as a correction term and $\max(x_1, x_2)$ is defined as follows:

$$\max(x_1, x_2) = \begin{cases} x_1 & \text{if } x_1 > x_2 \\ x_2 & \text{if } x_2 > x_1. \end{cases} \quad \text{(A.26)}$$

The transformation of the three variables, $\Omega_n(\kappa)$, $\beta_n(\kappa)$ and $\gamma_n(\check{k}, \kappa)$ to the logarithmic domain is represented as follows:

$$A_n(\kappa) = \ln(\Omega_n(\kappa)), \quad \text{(A.27)}$$
$$B_n(\kappa) = \ln(\beta_n(\kappa)), \quad \text{(A.28)}$$
$$\Gamma_n(\kappa) = \ln(\gamma_n(\check{k}, \kappa)). \quad \text{(A.29)}$$

In deriving the Log-MAP algorithm, the Jacobian logarithmic relationship described by Equation A.25 is exploited. Consequently, the LLR can be evaluated in terms of additions and table look-up operations, instead of the more complex multiplication, natural logarithm $\ln(\cdot)$ and exponential calculations.

The forward recursive calculations for $\Omega_n(s)$ can be transformed into the logarithmic domain by utilizing Equations A.27, A.29 and A.11, yielding the term $A_n(\kappa)$ as follows:

$$
\begin{aligned}
A_n(\kappa) &= \ln(\Omega_n(\kappa)) \\
&= \ln\left(\sum_{\text{all } \check{\kappa}} \Omega_{n-1}(\check{\kappa}) \cdot \gamma_n(\check{\kappa}, \kappa)\right) \\
&= \ln\left(\sum_{\text{all } \check{\kappa}} \exp\left[A_{n-1}(\check{\kappa}) + \Gamma_n(\check{\kappa}, \kappa)\right]\right) \\
&= \ln\left(\sum_{\text{all } \check{\kappa}} \exp\left[\Upsilon_f(\check{\kappa}, \kappa)\right]\right),
\end{aligned}
\tag{A.30}
$$

where $\Upsilon_f(\check{\kappa}, \kappa)$ is defined as:

$$
\Upsilon_f(\check{\kappa}, \kappa) = A_{n-1}(\check{\kappa}) + \Gamma_n(\check{\kappa}, \kappa).
\tag{A.31}
$$

The term $\Upsilon_f(\check{\kappa}, \kappa)$ represents the accumulated metric for a forward transition from state $\check{\kappa}$ to κ as shown in Figure A.1. Explicitly, for each transition from state $\check{\kappa}$ to κ, the branch metric $\Gamma_n(\check{\kappa}, \kappa)$ is added to the previous value $A_{n-1}(\check{\kappa})$, in order to obtain the current value of $\Upsilon_f(\check{\kappa}, \kappa)$ for that particular path. In a binary trellis, there will be two branches leaving each state, and consequently there will be two possible transitions merging to every state, as illustrated in Figure A.1. Therefore, at state κ, there exists two values of $\Upsilon_f(\check{\kappa}, \kappa)$ arising from the two branches, which is referred to as $\Upsilon_{f1}(\check{\kappa}, \kappa)$ and $\Upsilon_{f2}(\check{\kappa}, \kappa)$. By exploiting this property inherent in a binary system, Equation A.30 can be written as:

$$
\begin{aligned}
A_n(\kappa) &= \ln\left(\sum_{\text{all } \check{\kappa}} \exp\left[\Upsilon(\check{\kappa}, \kappa)\right]\right) \\
&= \ln\left(\exp\left[\Upsilon_{f1}(\check{\kappa}, \kappa) + \Upsilon_{f2}(\check{\kappa}, \kappa)\right]\right).
\end{aligned}
\tag{A.32}
$$

Subsequently, by using the Jacobian logarithmic relationship in Equation A.25, we can rewrite Equation A.32 as:

$$
\begin{aligned}
A_n(\kappa) &= \ln\left(\exp\left[\Upsilon_{f1}(\check{\kappa}, \kappa) + \Upsilon_{f2}(\check{\kappa}, \kappa)\right]\right) \\
&= \max(\Upsilon_{f1}(\check{\kappa}, \kappa), \Upsilon_{f2}(\check{\kappa}, \kappa)) + \ln\left(1 + \exp[-|\Upsilon_{f1}(\check{\kappa}, \kappa) - \Upsilon_{f2}(\check{\kappa}, \kappa)|]\right).
\end{aligned}
\tag{A.33}
$$

This representation of $A_n(\kappa)$ enabled the calculation to be performed based on recursive additions, rather than employing the more complex multiplication operations. However, the natural logarithm term, $\ln\left(1 + \exp[-|\Upsilon_{f1}(\check{\kappa}, \kappa) - \Upsilon_{f2}(\check{\kappa}, \kappa)|]\right)$ or $f_c(\delta_f)$ has to be computed at the cost of a higher complexity. Fortunately, the term $f_c(\delta_f)$ can be stored as a look-up table in order to avoid the need for computing it repeatedly. Furthermore, it was shown by Robertson et al. [552] that it is sufficient to store only eight values of δ_f, ranging from 0 to 5, in order to evaluate the term $f_c(\delta_f)$. Therefore, in the Log-MAP algorithm, the evaluation of $A_n(\kappa)$, which is the logarithmic domain representation of $\Omega_n(\kappa)$, is less complex, when compared to that of the MAP algorithm, since it only involves additions and a look-up table search.

By employing the same approach for $B_{n-1}(\check{k})$, and using Equations A.17, A.28 and A.29, we obtain the following:

$$B_{n-1}(\check{k}) = \ln\left(\sum_{\text{all }\kappa} \beta_n(\kappa)\cdot\gamma_n(\check{k},\kappa)\right)$$

$$= \ln\left(\sum_{\text{all }\kappa} \exp[B_n(\kappa) + \Gamma_n(\check{k},\kappa)]\right) \tag{A.34}$$

$$= \ln\left(\sum_{\text{all }\kappa} \exp[\Upsilon_b(\check{k},\kappa)]\right),$$

where $\Upsilon_b(\check{k},\kappa)$ is defined as:

$$\Upsilon_b(\check{k},\kappa) = B_n(\kappa) + \Gamma_n(\check{k},\kappa). \tag{A.35}$$

The term $\Upsilon_b(\check{k},\kappa)$ represents the accumulated metric for the backward transition from state κ to $\check{k}$. We observe in Equation A.35 that $\Upsilon_b(\check{k},\kappa)$ for each path from state κ to $\check{k}$ is evaluated by adding $\Gamma(\check{k},\kappa)$ to $B_n(\kappa)$. In considering a binary system, there exists only two possible transitions or branches from state κ to states $\check{k}$. As before, by representing the accumulated metric from these two branches as $\Upsilon_{b1}(\check{k},\kappa)$ and $\Upsilon_{b2}(\check{k},\kappa)$, we can rewrite Equation A.34 as:

$$B_{n-1}(\kappa) = \ln\left(\sum_{\text{all }\kappa} \exp[\Upsilon_b(\check{k},\kappa)]\right) \tag{A.36}$$

$$= \ln\left(\exp[\Upsilon_{b1}(\check{k},\kappa) + \Upsilon_{b2}(\check{k},\kappa)]\right).$$

By exploiting the Jacobian logarithmic relationship in Equation A.25, Equation A.36 is transformed as follows:

$$\begin{aligned} B_{n-1}(\kappa) &= \max(\Upsilon_{b1}(\check{k},\kappa), \Upsilon_{b2}(\check{k},\kappa)) + \ln\left(1 + \exp[-|\Upsilon_{b1}(\check{k},\kappa) - \Upsilon_{b2}(\check{k},\kappa)|]\right) \\ &= \max(\Upsilon_{b1}(\check{k},\kappa), \Upsilon_{b2}(\check{k},\kappa)) + f_c(\delta_b), \end{aligned} \tag{A.37}$$

where $\delta_b = \Upsilon_{b1}(\check{k},\kappa) - \Upsilon_{b2}(\check{k},\kappa)$. Similarly to $A_n(\kappa)$, the recursive calculations for $B_{n-1}(\check{k})$ are also based on recursive additions and the evaluation of $f_c(\delta_b)$ can be stored in a look-up table, in order to reduce the computational complexity.

From Equations A.34 and A.30, the variables $A_n(\kappa)$ and $B_{n-1}(\check{k})$ can be evaluated, once $\Gamma_n(\check{k},\kappa)$ was obtained. In order to evaluate the branch metric $\Gamma_n(\check{k},\kappa)$, Equations A.29 and A.23 are utilized to yield:

$$\begin{aligned} \Gamma_n(\check{k},\kappa) &= \ln\left(\gamma_n(\check{k},\kappa)\right) \\ &= \ln[P(u_n)\cdot P(r_n|s_i)] \\ &= \ln\left[\left(\frac{1}{2\sqrt{\pi N_o}} \exp\left[-\frac{1}{N_o}(r_n - s_i)^2\right]\right)\cdot P(u_n)\right] \\ &= \ln\left(\frac{1}{2\sqrt{\pi N_o}}\right) - \frac{1}{N_o}(r_n - s_i)^2 + \ln(P(u_n)). \end{aligned} \tag{A.38}$$

The term $\ln\left(\frac{1}{2\sqrt{\pi N_o}}\right)$ on the right hand side of Equation A.38 is independent of the bit u_n and therefore can be considered as a constant. Consequently, this term is neglected and Equation A.38 for the Log-MAP equalizer can be rewritten as:

$$\Gamma_n(\acute{k},\kappa) = -\frac{1}{N_o}(r_n - s_i)^2 + \ln(P(u_n)). \tag{A.39}$$

and similarly for the channel decoder, the branch metric $\Gamma_n(\acute{k},\kappa)$ is given by :

$$\Gamma_n(\acute{k},\kappa) = \frac{1}{2}\sum_{i=1}^{J} L_{(i,n)}^{in} \cdot c_{(i,n)} + \ln[P(u_n)], \tag{A.40}$$

where the notations are identical to those in Equation A.24.

By transforming $\Omega_n(\kappa)$, $\gamma_n(\acute{k},\kappa)$ and $\beta_n(\kappa)$ into the logarithm domain in the Log-MAP algorithm, the expression for the LLR, $L(u_n|r)$ in Equation A.4 is also modified to yield:

$$L(u_n|r) = \ln\left(\frac{\displaystyle\sum_{(\acute{k},\kappa)\Rightarrow u_n=+1} \Omega_{n-1}(\acute{k})\cdot\gamma_n(\acute{k},\kappa)\cdot\beta_n(\kappa)}{\displaystyle\sum_{(\acute{k},\kappa)\Rightarrow u_n=-1} \Omega_{n-1}(\acute{k})\cdot\gamma_n(\acute{k},\kappa)\cdot\beta_n(\kappa)}\right)$$

$$= \ln\left(\frac{\displaystyle\sum_{(\acute{k},\kappa)\Rightarrow u_n=+1} \exp\left(A_{n-1}(\acute{k})+\Gamma_n(\acute{k},\kappa)+B_n(\kappa)\right)}{\displaystyle\sum_{(\acute{k},\kappa)\Rightarrow u_n=-1} \exp\left(A_{n-1}(\acute{k})+\Gamma_n(\acute{k},\kappa)+B_n(\kappa)\right)}\right) \tag{A.41}$$

$$= \ln\left(\sum_{(\acute{k},\kappa)\Rightarrow u_n=+1} \exp\left(A_{n-1}(\acute{k})+\Gamma_n(\acute{k},\kappa)+B_n(\kappa)\right)\right)$$

$$- \ln\left(\sum_{(\acute{k},\kappa)\Rightarrow u_n=-1} \exp\left(A_{n-1}(\acute{k})+\Gamma_n(\acute{k},\kappa)+B_n(\kappa)\right)\right).$$

By considering a trellis with χ number of states at each trellis stage, there exist χ transitions, which belong to the set $(\acute{k},\kappa) \Rightarrow u_n = +1$, and similarly, χ number of $u_n = -1$ transitions, belonging to $(\acute{k},\kappa) \Rightarrow u_n = -1$. Therefore, we must evaluate the natural logarithm $\ln(\cdot)$ of the sum of χ number of exponential terms. This evaluation can be simplified by extending and generalising the Jacobian logarithmic relationship of Equation A.25 in order to cope with higher number of exponential summations. This can be achieved by nesting the $J(x_1, x_2)$ operations as follows [550]:

$$\ln\left(\sum_{k=1}^{V} e^{x_k}\right) = J(x_V, J(x_{V-1}, \ldots J(x_3, J(x_2, x_1)))), \tag{A.42}$$

where $V = \chi$ for our χ-state trellis.

Hence by using this relationship and Equation A.41, the LLR values for each input bit u_n - where x_k is equal to the sum of $A_{n-1}(\acute{k})$, $\Gamma_n(\acute{k},\kappa)$ and $B_n(\kappa)$, for the kth path at instant n - can be evaluated.

In summary, by transforming $\Omega_n(\kappa)$, $\beta_n(\kappa)$ and $\gamma_n(\check{k}, \kappa)$ to the logarithmic domain terms of $A_n(\kappa)$, $B_n(\kappa)$ and $\Gamma_n(\check{k}, \kappa)$ for the Log-MAP algorithm, the LLR values can be calculated through additions and a $f_c(\delta)$ look-up table, instead of multiplications and $\ln(\cdot)$ as well as exponential evaluations. This reduced the computational complexity, when compared to that of the MAP algorithm.

A.1.3 Calculation of the Source and Parity Log Likelihood Ratio for Turbo Equalization

In the implementation of turbo equalization, the SISO decoder block of Figure 5.26 accepts soft inputs from the SISO equaliser, and produces the LLR of not only the source bits, but also the parity bits in order to implement turbo equalisation. The LLR computation of the source and parity bits involves two important steps. Firstly, the forward and backward-recursion values, $A_{n-1}(\check{k})$ and $B_n(\kappa)$, and $\Gamma_n(\check{k}, \kappa)$, which are the logarithmic domain counterparts of $\Omega_{n-1}(\kappa)$, $\beta_n(\kappa)$ and $\gamma_n(\check{k}, \kappa)$, are computed by using Equations A.30, A.34 and A.38.

Secondly, we have to modify Equation A.41 for the SISO decoder in order to give the LLR of the source and parity bits. Recalling Equation A.41 and representing the input LLR sequence to the decoder as $L^{in}_{(i,n)}$ we have :

$$L(u_n | l^{in}_{(i,n)}) = \ln \left(\frac{\sum\limits_{(\check{k},\kappa) \Rightarrow u_n = +1} \exp\left(A_{n-1}(\check{k}) + \Gamma_n(\check{k}, \kappa) + B_n(\kappa)\right)}{\sum\limits_{(\check{k},\kappa) \Rightarrow u_n = -1} \exp\left(A_{n-1}(\check{k}) + \Gamma_n(\check{k}, \kappa) + B_n(\kappa)\right)} \right), \qquad (A.43)$$

where we note that at the nth instant, we perform the summation of $e^{(A_{n-1}(\check{k}) + \Gamma_n(\check{k}, \kappa) + B_n(\kappa))}$ for all branches, which were due to the source bit $u_n = +1$ and $u_n = -1$. In order to determine the LLR for each coded bit $c_{i,n}$, where $c_{i,n}$ is the ith code bit at instant n, we must sum $e^{(A_{n-1}(\check{k}) + \Gamma_n(\check{k}, \kappa) + B_n(\kappa))}$ for all branches, which gives $c_{i,n} = +1$ and also for branches resulting in $c_{i,n} = -1$. Therefore, Equation A.43 can be rewritten as:

$$L(c_{i,n} | L^{in}_{(i,n)}) = \ln \left(\frac{\sum\limits_{(\check{k},\kappa) \Rightarrow c_{i,n} = +1} \exp\left(A_{n-1}(\check{k}) + \Gamma_n(\check{k}, \kappa) + B_n(\kappa)\right)}{\sum\limits_{(\check{k},\kappa) \Rightarrow c_{i,n} = -1} \exp\left(A_{n-1}(\check{k}) + \Gamma_n(\check{k}, \kappa) + B_n(\kappa)\right)} \right)$$

$$= \ln \left(\sum\limits_{(\check{k},\kappa) \Rightarrow c_{i,n} = +1} \exp\left(A_{n-1}(\check{k}) + \Gamma_n(\check{k}, \kappa) + B_n(\kappa)\right) \right) \qquad (A.44)$$

$$- \ln \left(\sum\limits_{(\check{k},\kappa) \Rightarrow c_{i,n} = -1} \exp\left(A_{n-1}(\check{k}) + \Gamma_n(\check{k}, \kappa) + B_n(\kappa)\right) \right).$$

Let us consider an example of a SISO convolutional decoder trellis for a rate $R = \frac{1}{2}$, constraint length $K = 3$ encoder, as illustrated in Figure A.3. Since $R = \frac{1}{2}$, the output codeword consists of two coded bits, namely $c_{1,n}$ and $c_{2,n}$. The output codeword for each transition is given by as $(c_{1,n}, c_{2,n})$ in Figure A.3. The encoder is specified by the generator polynomials, G_0 and G_1, which can be represented in octal form as 7 and 5, respectively. In order to determine the LLR for $c_{1,n}$, all paths which give the code bit $c_{1,n} = +1$ have to

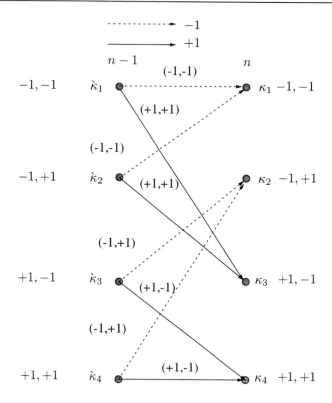

Figure A.3: Trellis of a $K = 3$, $R = \frac{1}{2}$ convolutional decoder, where the octal representation of the generator polynomials, G_0 and G_1 is 7 and 5, respectively, at the nth coding instant.

be considered. Referring to Figure A.3, we observe that there are four branches, which fulfil this requirement, namely branches from state $\check{k}_1$ to κ_3, $\check{k}_2$ to κ_3, $\check{k}_3$ to κ_4 and $\check{k}_4$ to κ_4. By comparing Equations A.3 and A.4, and by substituting Equations A.27, A.28 and A.29 into Equations A.4, we note that the probability that a particular transition from state $\check{k}$ to κ occurs and that the received LLR information $L_{(1,n)}^{in}$ was obtained is given by $P(\check{k} \wedge \kappa \wedge L_{(1,n)}^{in})$, which can be written as :

$$P(\check{k} \wedge \kappa \wedge L_{(1,n)}^{in}) = \exp\left(A_{n-1}(\check{k}) + \Gamma_n(\check{k},\kappa) + B_n(\kappa)\right). \tag{A.45}$$

Therefore, the individual branch probabilities, which provide the coded bit $c_{1,n} = +1$ can be written as:

$$\begin{aligned}
P(\check{k}_1 \wedge \kappa_3 \wedge L_{(1,n)}^{in}) &= \exp\left(A_{n-1}(\check{k}_1) + \Gamma_n(\check{k}_1,\kappa_3) + B_n(\kappa_3)\right) \\
P(\check{k}_2 \wedge \kappa_3 \wedge L_{(1,n)}^{in}) &= \exp\left(A_{n-1}(\check{k}_2) + \Gamma_n(\check{k}_2,\kappa_3) + B_n(\kappa_3)\right) \\
P(\check{k}_3 \wedge \kappa_4 \wedge L_{(1,n)}^{in}) &= \exp\left(A_{n-1}(\check{k}_3) + \Gamma_n(\check{k}_3,\kappa_4) + B_n(\kappa_4)\right) \\
P(\check{k}_4 \wedge \kappa_4 \wedge L_{(1,n)}^{in}) &= \exp\left(A_{n-1}(\check{k}_4) + \Gamma_n(\check{k}_4,\kappa_4) + B_n(\kappa_4)\right).
\end{aligned} \tag{A.46}$$

The probability of each of these transitions occurring can be calculated using $A_{n-1}(\check{k})$ and

$B_n(\kappa)$, and $\Gamma_n(\grave{k}, \kappa)$, which was evaluated by using Equations A.30, A.34 and A.38. Therefore, the probability that $c_{1,n} = +1$ was transmitted at instant n and that $L_{(1,n)}^{in}$ was generated, is determined by the sum of the individual branch probabilities $P(\grave{k}_1 \wedge \kappa_3 \wedge L_{(1,n)}^{in})$, $P(\grave{k}_2 \wedge \kappa_3 \wedge L_{(1,n)}^{in})$, $P(\grave{k}_3 \wedge \kappa_4 \wedge L_{(1,n)}^{in})$, and $P(\grave{k}_4 \wedge \kappa_4 \wedge L_{(1,n)}^{in})$.

Similarly, for the paths which give $c_{1,n} = -1$, namely branches from state $\grave{k}_1$ to κ_1, $\grave{k}_2$ to κ_1, $\grave{k}_3$ to κ_2 and $\grave{k}_4$ to κ_2, the individual branch probabilities can be calculated as:

$$
\begin{aligned}
P(\grave{k}_1 \wedge \kappa_1 \wedge L_{(1,n)}^{in}) &= \exp\left(A_{n-1}(\grave{k}_1) + \Gamma_n(\grave{k}_1, \kappa_1) + B_n(\kappa_1)\right) \\
P(\grave{k}_2 \wedge \kappa_1 \wedge L_{(1,n)}^{in}) &= \exp\left(A_{n-1}(\grave{k}_2) + \Gamma_n(\grave{k}_2, \kappa_1) + B_n(\kappa_1)\right) \\
P(\grave{k}_3 \wedge \kappa_2 \wedge L_{(1,n)}^{in}) &= \exp\left(A_{n-1}(\grave{k}_3) + \Gamma_n(\grave{k}_3, \kappa_2) + B_n(\kappa_2)\right) \\
P(\grave{k}_4 \wedge \kappa_2 \wedge L_{(1,n)}^{in}) &= \exp\left(A_{n-1}(\grave{k}_4) + \Gamma_n(\grave{k}_4, \kappa_2) + B_n(\kappa_2)\right).
\end{aligned}
\tag{A.47}
$$

As before, the probability that $c_{1,n} = -1$ was transmitted at instant n and that $L_{(1,n)}^{in}$ was generated is evaluated by taking the sum of the individual branch probabilities $P(\grave{k}_1 \wedge \kappa_1 \wedge L_{(1,n)}^{in})$, $P(\grave{k}_2 \wedge \kappa_1 \wedge L_{(1,n)}^{in})$, $P(\grave{k}_3 \wedge \kappa_2 \wedge L_{(1,n)}^{in})$, and $P(\grave{k}_4 \wedge \kappa_2 \wedge L_{(1,n)}^{in})$.

With the aid of Equation A.44 and using Equations A.45, A.46 and A.47, the LLR for the first coded bit $c_{1,n}$ can be written as:

$$
\begin{aligned}
&L(c_{1,n} | L_{(1,n)}^{in}) \\
&= \ln\left(\frac{\displaystyle\sum_{(\grave{k},\kappa) \Rightarrow c_{1,n}=+1} P(\grave{k} \wedge \kappa \wedge L_{(1,n)}^{in})}{\displaystyle\sum_{(\grave{k},\kappa) \Rightarrow c_{1,n}=-1} P(\grave{k} \wedge \kappa \wedge L_{(1,n)}^{in})} \right) \\
&= \ln\left(\frac{P(\grave{k}_1 \wedge \kappa_3 \wedge L_{(1,n)}^{in}) + P(\grave{k}_2 \wedge \kappa_3 \wedge L_{(1,n)}^{in}) + P(\grave{k}_3 \wedge \kappa_4 \wedge L_{(1,n)}^{in}) + P(\grave{k}_4 \wedge \kappa_4 \wedge L_{(1,n)}^{in})}{P(\grave{k}_1 \wedge \kappa_1 \wedge L_{(1,n)}^{in}) + P(\grave{k}_2 \wedge \kappa_1 \wedge L_{(1,n)}^{in}) + P(\grave{k}_3 \wedge \kappa_2 \wedge L_{(1,n)}^{in}) + P(\grave{k}_4 \wedge \kappa_2 \wedge L_{(1,n)}^{in})} \right) \\
&= \ln\left(e^{(A_{n-1}(\grave{k}_1) + \Gamma_n(\grave{k}_1, \kappa_3) + B_n(\kappa_3))} + e^{(A_{n-1}(\grave{k}_2) + \Gamma_n(\grave{k}_2, \kappa_3) + B_n(\kappa_3))} + \right. \\
&\qquad \left. e^{(A_{n-1}(\grave{k}_3) + \Gamma_n(\grave{k}_3, \kappa_4) + B_n(\kappa_4))} + e^{(A_{n-1}(\grave{k}_4) + \Gamma_n(\grave{k}_4, \kappa_4) + B_n(\kappa_4))} \right) \\
&\quad - \ln\left(e^{(A_{n-1}(\grave{k}_1) + \Gamma_n(\grave{k}_1, \kappa_1) + B_n(\kappa_1))} + e^{(A_{n-1}(\grave{k}_2) + \Gamma_n(\grave{k}_2, \kappa_1) + B_n(\kappa_1))} + \right. \\
&\qquad \left. e^{(A_{n-1}(\grave{k}_3) + \Gamma_n(\grave{k}_3, \kappa_2) + B_n(\kappa_2))} + e^{(A_{n-1}(\grave{k}_4) + \Gamma_n(\grave{k}_4, \kappa_2) + B_n(\kappa_2))} \right).
\end{aligned}
\tag{A.48}
$$

By employing the same approach for determining the LLR of $c_{1,n}$, the LLR of the second coded bit $c_{2,n}$ can be derived as:

$$L(c_{2,n}|L_{(2,n)}^{in})$$

$$= \ln \left(\frac{\displaystyle\sum_{(\hat{k},\kappa) \Rightarrow c_{2,n}=+1} P(\hat{k} \wedge \kappa \wedge L_{(2,n)}^{in})}{\displaystyle\sum_{(\hat{k},\kappa) \Rightarrow c_{2,n}=-1} P(\hat{k} \wedge \kappa \wedge L_{(2,n)}^{in})} \right)$$

$$= \ln \left(\frac{P(\hat{k}_1 \wedge \kappa_3 \wedge L_{(2,n)}^{in}) + P(\hat{k}_2 \wedge \kappa_3 \wedge L_{(2,n)}^{in}) + P(\hat{k}_3 \wedge \kappa_2 \wedge L_{(2,n)}^{in}) + P(\hat{k}_4 \wedge \kappa_2 \wedge L}{P(\hat{k}_1 \wedge \kappa_1 \wedge L_{(2,n)}^{in}) + P(\hat{k}_2 \wedge \kappa_1 \wedge L_{(2,n)}^{in}) + P(\hat{k}_3 \wedge \kappa_4 \wedge L_{(2,n)}^{in}) + P(\hat{k}_4 \wedge \kappa_4 \wedge L} \right)$$

$$= \ln \left(e^{(A_{n-1}(\hat{k}_1)+\Gamma_n(\hat{k}_1,\kappa_3)+B_n(\kappa_3))} + e^{(A_{n-1}(\hat{k}_2)+\Gamma_n(\hat{k}_2,\kappa_3)+B_n(\kappa_3))} + \right.$$

$$\left. e^{(A_{n-1}(\hat{k}_3)+\Gamma_n(\hat{k}_3,\kappa_2)+B_n(\kappa_2))} + e^{(A_{n-1}(\hat{k}_4)+\Gamma_n(\hat{k}_4,\kappa_2)+B_n(\kappa_2))} \right)$$

$$- \ln \left(e^{(A_{n-1}(\hat{k}_1)+\Gamma_n(\hat{k}_1,\kappa_1)+B_n(\kappa_1))} + e^{(A_{n-1}(\hat{k}_2)+\Gamma_n(\hat{k}_2,\kappa_1)+B_n(\kappa_1))} + \right.$$

$$\left. e^{(A_{n-1}(\hat{k}_3)+\Gamma_n(\hat{k}_3,\kappa_4)+B_n(\kappa_4))} + e^{(A_{n-1}(\hat{k}_4)+\Gamma_n(\hat{k}_4,\kappa_4)+B_n(\kappa_4))} \right).$$

Subsequently, the generalised Jacobian relationship in Equation A.42 is then used to evaluate the sum of exponential terms in Equations A.48 and A.49.

A.2 Least Mean Square Algorithm

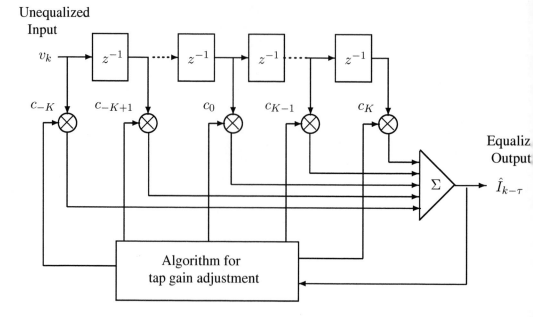

Figure A.4: Linear equaliser schematic.

For a linear transversal filter shown in Figure A.4, the optimum coefficients according to the mean square error (MSE) criterion [280], are determined from the solution of a set of linear equations, which can be expressed in matrix form as:

$$\mathbf{\Gamma} C = \boldsymbol{\xi}, \tag{A.50}$$

where $\mathbf{\Gamma}$ is the $(2K + 1) \times (2K + 1)$ covariance matrix of the input signal samples $\{v_k\}$, C is the column vector of $(2K+1)$ equaliser tap weights $\{c_k\}$ and $\boldsymbol{\xi}$ is a set of $(2K+1)$ cross-correlations between the unequalised input samples v_k and the equalised desired response $\{I_k\}$.

In order to avoid the direct matrix inversion in obtaining C_{opt}, we can minimize the MSE J by iteratively descending on the associated MSE versus the equaliser coefficient surface via gradient methods [119, 280]. Each equaliser tap weight is changed in the direction opposite to its corresponding gradient component $\delta J/\delta c_k, k = -K, \ldots, -1, 0, -1, \ldots, K$ at the currently encountered point of the MSE surface. The iteratively updated values of the coefficient vector C are given by [280]:

$$C_{k+1} = \mathbf{\Gamma} C_k - \boldsymbol{\xi} = C_k + \mu E(\varepsilon_k V_k^*), \tag{A.51}$$

where the vector C_k is the set of equaliser coefficients at the kth iteration, $\varepsilon_k = I_k - \hat{I}_k$ is the equalisation error at kth iteration, V_k is the vector of the equaliser input signal samples that generate the equaliser output $\hat{I}_k$, i.e., $V_k = [v_{k+K} \ldots v_k \ldots v_{k-K}]^T$ and μ is the associated step-size. The difficulty with the gradient descent method is in determining the covariance matrix $\mathbf{\Gamma}$ and the vector $\boldsymbol{\xi}$ of cross correlations, which will need a collection of unequalised data $\{v_k\}$. An alternative is to estimate the MSE surface gradient and adjust the tap weights according to the relation [119, 280]:

$$C_{k+1} = C_k + \mu \varepsilon V_k^*. \tag{A.52}$$

This is the LMS algorithm, which is implementationally simple. In order to guarantee convergence of the recursive relation in Equation A.52, the step-size μ must satisfy the inequality [280]:

$$0 < \mu < 2/\lambda_{\max}, \tag{A.53}$$

where $\lambda_{\max}$ is the largest so-called eigenvalue of $\mathbf{\Gamma}$ [280]. Note that $\lambda_{\max}$ cannot be greater than the trace of $\mathbf{\Gamma}$, tr$[\mathbf{\Gamma}]$, which can be expressed as [280] $(2K + 1)E(v_k^2)$ for a linear transversal filter. Thus convergence of the coefficient vector is assured by [119]

$$
\begin{aligned}
\text{In general:} \quad & 0 < \mu < \frac{2}{tr[\mathbf{\Gamma}]} \\
\text{Transversal filter:} \quad & 0 < \mu < \frac{2}{(2K+1)(\text{received signal power})}.
\end{aligned} \tag{A.54}
$$

In practical applications, the LMS algorithm employs noisy estimates of the MSE surface gradient. The noise in these estimates causes the coefficients to fluctuate randomly around the optimal values. The final MSE in steady state is $J_{\min} + J_{\text{excess}}$. The *excess mean square error* term J_{excess} is defined in a simplified form by Proakis [280] and Widrow [119] as:

$$J_{\text{excess}} \approx \frac{1}{2}\mu J_{\min} tr[\mathbf{\Gamma}]. \tag{A.55}$$

From Equation A.55, we can see that the value of μ has to be as small as possible, in order to reduce the excess MSE. However, at the same time the step size μ is proportional to the speed of convergence. A fast convergence is important, if the statistical time variations of the signal occur rapidly. Therefore a compromise is necessary for ensuring good tracking of the time-variant signal statistics without undue degradation of the associated performance. To overcome this problem, in the LMS algorithm the step-size is often made time-varying. A few of the time-varying forms found in the literature are:

$$\text{Stochastic approximation schedule [554]}: \quad \mu(k) = \frac{a}{k} \tag{A.56}$$

$$\text{Search-then-converge schedule [555]}: \quad \mu(k) = \frac{\mu_0}{1 + (k/\kappa)} \tag{A.57}$$

where a, μ_0 and κ are constant.

A.3 Minimal Feedforward Order of the RBF DFE [Proof] [246]

The RBF DFE has a feedforward order of m, feedback order of n and a decision delay of τ. We denote the $(m + L)$-symbol length channel input sequence that determines the values of the noiseless channel state $\mathbf{r}_j, j = 1, \dots, n_s$ by $\tilde{\mathbf{I}}_{k-\tau}$, where the CIR length is $L + 1$. Let $\tilde{\mathbf{I}}_{k-\tau} = \mathbf{s}_j, j = 1, \dots, n_s$, where $\mathbf{s}_j$ represents the n_s possible states of $\tilde{\mathbf{I}}_{k-\tau}$. Referring to Equation 8.105, we consider $\mathbf{r}_{j,l}^i \in V_{m,\tau,j}^i, j = 1, \dots, n_f$ for $I_{k-\tau} = \mathcal{I}_i, i = 1, \dots, M$, where

$$\mathbf{r}_{j,l}^i = \begin{bmatrix} r_{j,l,0}^i & \cdots & r_{j,l,m-1}^i \end{bmatrix}^T, \quad 1 \leq i \leq M, \quad 1 \leq j \leq n_s, \quad 1 \leq l \leq n_f.$$

Assuming $m > \tau + 1$, the squared distance between the channel output vector $\mathbf{v}_k$ of Equation 8.2 and $\mathbf{r}_{j,l}^i$ is

$$
\begin{aligned}
\omega_{m,j,l}^i(k) &= \|\mathbf{v}_k - \mathbf{r}_{j,l}^i\|^2 \\
&= \sum_{u=0}^{\tau} (v_{k-u} - r_{j,l,u}^i)^2 + \sum_{u=\tau+1}^{m-1} (v_{k-u} - r_{j,l,u}^i)^2 \\
&= \omega_{\tau+1,j,l}^i(k) + \sum_{u=\tau+1}^{m-1} (v_{k-u} - r_{j,l,u}^i)^2, \\
&\qquad\qquad 1 \leq i \leq M.
\end{aligned}
\tag{A.58}
$$

The feedback symbols are assumed to be correct, that is,

$$\tilde{\mathbf{I}}_{k-\tau} = \begin{bmatrix} I_{k-\tau-1} & \cdots & I_{k-\tau-n} \end{bmatrix}^T, \tag{A.59}$$

where $n = L + m - 1 - \tau$. For any $\mathbf{r}_{j,l}^i \in V_{m,\tau,j}^i$ and $1 \leq i \leq M$, we have:

$$r_{j,l,u}^i = \sum_{n=0}^{L} f_n I_{k-u-n} \quad \tau + 1 \leq u \leq m - 1. \tag{A.60}$$

Upon introducing

$$\tilde{\omega}(k) = \sum_{u=\tau+1}^{m-1} (v_{k-u} - r_{j,l,u}^i)^2. \tag{A.61}$$

We have:

$$\omega_{m,j,l}^i(k) = \omega_{\tau+1,j,l}^i(k) + \tilde{\omega}(k). \tag{A.62}$$

The conditional Bayesian decision variables, given that $\tilde{\mathbf{I}}_{k-\tau} = \mathbf{s}_{f,j}$, are as follows:

$$\zeta_i(k|\tilde{\mathbf{I}}_{k-\tau} = \mathbf{s}_{f,j}) = \sum_{l=1}^{n_{s,j}^i} \alpha \cdot exp(-\omega_{m,j,l}^i/\rho), \qquad 1 \leq i \leq \mathcal{M}, \tag{A.63}$$

where α is an arbitrary positive scalar, $\rho = 2\sigma_\eta^2$ and $n_{s,j}^i$ is the number of states in $V_{m,\tau,j}^i$. Substituting Equation A.62 into Equation A.63 yields:

$$\begin{aligned}
\zeta_i(k|\tilde{\mathbf{I}}_{k-\tau} = \mathbf{s}_{f,j}) &= \sum_{l=1}^{n_{s,j}^i} \alpha \cdot exp\left(-\tilde{\omega}(k)/\rho\right) exp\left(-\omega_{\tau+1,j,l}^i(k)/\rho\right) \\
&= \sum_{l=1}^{\tilde{n}_{s,j}^i} \tilde{\alpha} \cdot exp(-\omega_{\tau+1,j,l}^i(k)/\rho) \qquad 1 \leq i \leq \mathcal{M}, \tag{A.64}
\end{aligned}$$

where $\tilde{n}_{s,j}^i$ is the number of states in $V_{\tau+1,\tau,j}^i$ and $\tilde{\alpha}$ is a positive scalar, since α and $\tilde{\omega}(k)$ are positive scalars. This proves that the RBF DFE based on the Bayesian solution [246] having a feedforward order of $m = \tau + 1$ has the same conditional decision variables, as those of arbitrary higher feedforward orders of $m > \tau + 1$.

In the above proof, the number of states in $V_{\tau+1,\tau,j}^i$, $\tilde{n}_{s,j}^i$ has first implicitly been multiplied by a factor of $\mathcal{M}^{m-\tau-1}$ so as to match the number of states in $V_{m,\tau,j}^i$, n_s, and then reduced to the original $\tilde{n}_{s,j}^i$. This is allowed, since $\tilde{\alpha}$ is an arbitrary positive scalar.

A.4 BER Analysis of Type-I Star-QAM

The Star Quadrature Amplitude Modulation (SQAM) technique [4], also known as Amplitude-modulated Phase Shift Keying (APSK), employs circular constellations, rather than rectangular constellation as in Square QAM [556]. Although Square QAM has the maximum possible minimum Euclidean distance amongst its phasors given a constant average symbol power, in some situations Star QAM may be preferred due to its relatively simple detector and for its low Peak-to-Average Power Ratio (PAPR) compared to Square QAM [556]. Since differentially detected non-coherent Star QAM signals are robust against fading effects, many researchers analysed its Bit Error Ratio (BER) performance for transmission over AWGN channels [557], Rayleigh fading channels [557, 558] as well as Rician fading channels [559]. The effects diversity reception on its BER were also studied when communicating over Rayleigh fading channels [560, 561]. The BER of coherent 16 Star QAM was also analysed for transmission over AWGN channels [557] as well as when communicating over *Nakagami-m* fading channels [562]. However, the BER of Star-QAM schemes other than 16-level Star QAM, such as 8, 32 and 64-level Star QAM, has not been studied.

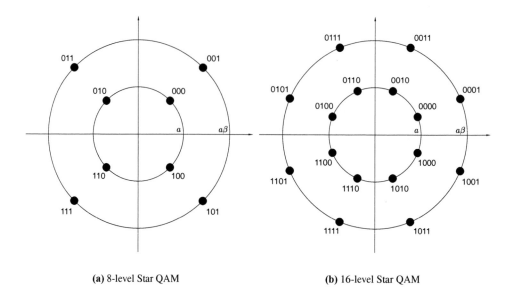

(a) 8-level Star QAM   **(b)** 16-level Star QAM

Figure A.5: Type-I constellations of Star QAM using two constellation rings.

A.4.1   Coherent Detection

The BER of coherent star QAM schemes employing Type-I constellations [556] when communicating over AWGN channels can be analysed using the signal-space method [196, 557, 562, 563]. The phasor constellations of the various Type-I Star QAM schemes are illustrated in Figure A.5 and A.10. Let us first consider 8-level Star QAM, which is also referred to as 2-level QPSK [556]. In Figure A.5(a), a is the radius of the inner ring, while $a\beta$ is the radius of the outer ring. The ring ratio is given by $a\beta/a = \beta$. The three bits, namely b_1, b_2 and b_3, are assigned as shown in Figure A.5(a), representing Gray coding for each ring using the bits $b_1 b_2$. The third bit, namely b_3 indicates, which ring of the constellation is encountered. The average symbol power is given as:

$$E_s = \frac{4a^2 + 4a^2\beta^2}{8} = \frac{1}{2}a^2(1 + \beta^2) .$$
(A.65)

In order to nomalise the constellations so that the average symbol power becomes unity, a should be given as:

$$a = \sqrt{\frac{2}{1 + \beta^2}} .$$
(A.66)

In terms of the signal space, the modulation scheme with respect to b_3 is an Amplitude Shift Keying (ASK) scheme. The decision rule related to bit b_3 is specified in Figure A.6(a). The

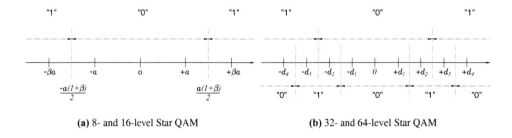

(a) 8- and 16-level Star QAM   (b) 32- and 64-level Star QAM

Figure A.6: Magnitude-bit decision regions for various Type-I Star QAM constellations.

BER of bit b_3 can be expressed as:

$$P_{b_3} = \frac{1}{2}\left[Q\left(\frac{a(\beta-1)}{2}\sqrt{2\gamma}\right) + Q\left(\frac{a(\beta+3)}{2}\sqrt{2\gamma}\right)\right]$$

$$+ \frac{1}{2}\left[Q\left(\frac{a(\beta-1)}{2}\sqrt{2\gamma}\right) - Q\left(a(\beta+1)\sqrt{2\gamma}\right)\right] \tag{A.67}$$

$$\simeq Q\left(\frac{a(\beta-1)}{2}\sqrt{2\gamma}\right) \tag{A.68}$$

$$= Q\left(\sqrt{\frac{(\beta-1)^2}{1+\beta^2}\gamma}\right), \tag{A.69}$$

where the Gaussian Q-function is defined as $Q(x) = \frac{1}{\sqrt{2\pi}}\int_x^\infty e^{-y^2/2}dy$ and γ is the SNR per symbol. Since bits b_1 and b_2 corresponds to Gray coded QPSK signals, their BER can be expressed as:

$$P_{b_1} = P_{b_2} \simeq \frac{1}{2}Q\left(\frac{a}{\sqrt{2}}\sqrt{2\gamma}\right) + \frac{1}{2}Q\left(\frac{a\beta}{\sqrt{2}}\sqrt{2\gamma}\right) \tag{A.70}$$

$$= \frac{1}{2}Q\left(\sqrt{\frac{2}{1+\beta^2}\gamma}\right) + \frac{1}{2}Q\left(\sqrt{\frac{2\beta^2}{1+\beta^2}\gamma}\right). \tag{A.71}$$

Hence, the average BER of an 8-level Star QAM scheme communicating over an AWGN channel can be expressed as:

$$P_8 = \frac{1}{3}P_{b_1} + \frac{1}{3}P_{b_2} + \frac{1}{3}P_{b_3} \tag{A.72}$$

$$\simeq \frac{1}{3}\left[Q\left(\sqrt{\frac{2}{1+\beta^2}\gamma}\right) + Q\left(\sqrt{\frac{2\beta^2}{1+\beta^2}\gamma}\right) + Q\left(\sqrt{\frac{(\beta-1)^2}{1+\beta^2}\gamma}\right)\right]. \tag{A.73}$$

The BER of (A.73) is plotted in Figure A.7(a) as a function of the ring ratio β for various values of the SNR per symbol γ. We can observe that the BER of 8-level Star QAM reaches its minimum, when the ring ratio is $\beta \simeq 2.4$. This is not surprising, considering that the ring ratio should be $\beta = 1 + \sqrt{2}$ in order to make the Euclidean distances between an inner ring

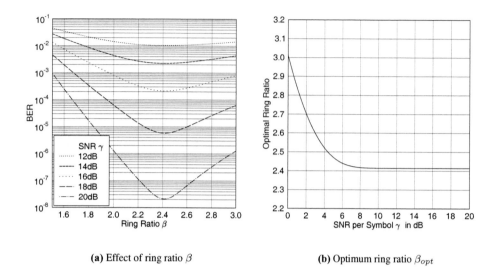

(a) Effect of ring ratio β **(b)** Optimum ring ratio β_{opt}

Figure A.7: BER of Gray-mapped 8-level Star QAM for transmission over AWGN channels.

constellation point and its three adjacent constellation points the same. However, the optimum ring ratio β_{opt}, where the BER reaches its minimum is SNR dependent. The optimum ring ratio versus the SNR per symbol is plotted in Figure A.7(b). It can be observed that when the SNR is lower than 8dB, the optimum ring ratio increases sharply. Since the corresponding BER improvement was however less than 0.1dB even at SNRs near 0dB, the fixed ring ratio of $\beta = 1 + \sqrt{2}$ can be used for all SNR values. Figure A.8(a) compares the BER of 8-level Star QAM and 8-PSK. Observe that 8-level Star QAM exhibits an approximately 1dB SNR performance gain, when the SNR is below 2dB, but above this SNR the SNR gain becomes marginal..

Let us now consider 16-level Star QAM. The corresponding phasor constellation is given in Figure A.5(b). Since the average symbol power is the same as that of 8-level Star QAM, Equation A.66 can be used for determining a. The BER analysis for the fourth bit, namely for b_4 is exactly the same as that of 8-level Star QAM and the corresponding value of P_{b_4} is given in (A.69). Since the first three bits, namely b_1, b_2 and b_3, are 8-PSK modulated, their BER can be expressed as:

$$P_{b_1} = P_{b_2} = P_{b_3} = \frac{1}{2} P_{8PSK}(a^2 \gamma) + \frac{1}{2} P_{8PSK}(a^2 \beta^2 \gamma) \,. \tag{A.74}$$

Lu, Letaief, Chuang and *Liou* found an accurate approximation of the BER of Gray-coded MPSK, which is given by [196]:

$$P_{MPSK} \simeq \frac{2}{\log_2 M} \sum_{i=1}^{2} Q\left(\sqrt{2 \sin^2 \left(\frac{2i-1}{M} \pi\right) \gamma}\right), \tag{A.75}$$

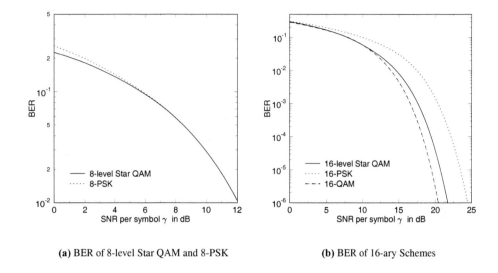

(a) BER of 8-level Star QAM and 8-PSK

(b) BER of 16-ary Schemes

Figure A.8: BER comparison of 8-ary and 16-ary modulation schemes for transmission over AWGN channels.

where γ is the SNR per symbol. Hence, the BER of (A.74) can be expressed as:

$$P_{b_1} = P_{b_2} = P_{b_3} \tag{A.76}$$

$$\simeq \frac{1}{3} \left[Q\left(\sqrt{\frac{4\sin^2(\pi/8)}{1+\beta^2}}\gamma \right) + Q\left(\sqrt{\frac{4\sin^2(3\pi/8)}{1+\beta^2}}\gamma, \right) \right]$$

$$+ \frac{1}{3} \left[Q\left(\sqrt{\frac{4\beta^2\sin^2(\pi/8)}{1+\beta^2}}\gamma \right) + Q\left(\sqrt{\frac{4\beta^2\sin^2(3\pi/8)}{1+\beta^2}}\gamma \right) \right]. \tag{A.77}$$

Now, the average BER of a 16-level Star QAM scheme for transmission over AWGN channel can be expressed as:

$$P_{16} = \frac{1}{4}P_{b_1} + \frac{1}{4}P_{b_2} + \frac{1}{4}P_{b_3} + \frac{1}{4}P_{b_4} \tag{A.78}$$

$$\simeq \frac{1}{4} \left[Q\left(\sqrt{\frac{4\sin^2(\pi/8)}{1+\beta^2}}\gamma \right) + Q\left(\sqrt{\frac{4\sin^2(3\pi/8)}{1+\beta^2}}\gamma, \right) + Q\left(\sqrt{\frac{4\beta^2\sin^2(\pi/8)}{1+\beta^2}}\gamma \right) \right.$$

$$\left. + Q\left(\sqrt{\frac{4\beta^2\sin^2(3\pi/8)}{1+\beta^2}}\gamma \right) + Q\left(\sqrt{\frac{(\beta-1)^2}{1+\beta^2}}\gamma \right) \right]. \tag{A.79}$$

The BER of (A.79) is plotted in Figure A.9(a) as a function of the ring ratio β for the various values of the SNR per symbol γ. We can observe that the ring ratio of $\beta \simeq 1.8$ minimises

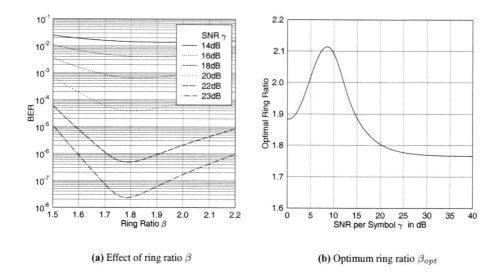

(a) Effect of ring ratio β (b) Optimum ring ratio β_{opt}

Figure A.9: BER of Gray mapped 16-level Star QAM for transmission over AWGN channels.

the BER of 16-level Star QAM, when communicating over AWGN channels [557]. This is also expected, since the ring ratio should be $\beta = 1 + 2\cos(3\pi/8) = 1.7654$ in order to render the Euclidean distances between an inner-ring constellation point and its three adjacent constellation points the same. The actual optimum ring ratio β_{opt}, where the BER reaches its minimum is plotted in Figure A.9(b). As for 8-level Star QAM, even though the optimum ratio is SNR dependent, the difference between the BER corresponding to the optimum ring ratio and that corresponding to the constant ring ratio of $\beta = 1 + 2\cos(3\pi/8)$ is negligible. Figure A.8(b) compares the BER of 16-level Star QAM, 16-PSK and 16-level Square QAM. We found that the BER performance of 16-level Star QAM is inferior to that of 16-level Square QAM. Viewing the corresponding performance from the perspective of the required SNR per symbol, the former requires an approxmately 1.3dB high SNR for maintaining the BER of 10^{-6}. By contrast, it requires a 2.7dB lower symbol-SNR, than 16-PSK.

Having considered the family of twin-ring constellations, let us focus our attention on two four-ring constellations. The Type-I constellations of 32-level and 64-level Star QAM scheme are depicted in Figure A.10. The last two bits of a symbol are Gray coded in the 'radial direction' and they are four-level ASK modulated. Let us assume that the Gray coding scheme for the bits "b_4b_5" of 32-level Star QAM in the 'radial direction' is given as "00", "01", "11" and "10", when viewing it from the inner most ring to the outer rings. The decision regions of these bits were illustrated in Figure A.6(b). The remaining bits are also Gray coded along each of the four rings and PSK modulated. Let us denote the radius of each ring as d_1, $d_2 = d_1\beta_1$, $d_3 = d_1\beta_2$ and $d_4 = d_1\beta_3$, where β_1, β_2 and β_3 are the corresponding ring ratios of each ring. Since the average power per symbol E_s is given as:

$$E_s = \frac{d_1^2 + d_2^2 + d_3^2 + d_4^2}{4} = \frac{d_1^2}{4}(1 + \beta_1^2 + \beta_2^2 + \beta_3^2), \qquad (A.80)$$

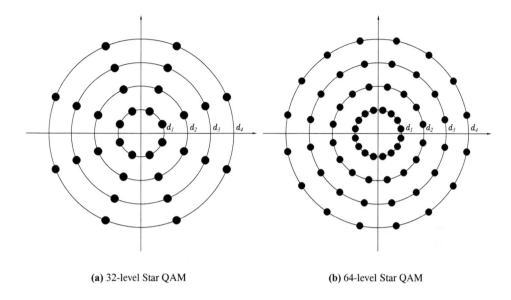

(a) 32-level Star QAM   **(b)** 64-level Star QAM

Figure A.10: Type-I constellations of Star QAM schemes having four rings.

the value of d_1 required for normalising the average power to unity can be expressed as:

$$d_1 = \sqrt{\frac{4}{1 + \beta_1^2 + \beta_2^2 + \beta_3^2}} \cdot \tag{A.81}$$

Inspecting Figure A.6(b), the BER of the fourth bit of 32-level Star QAM can be formulated as:

$$
P_{b_4} = \frac{1}{4}\left[Q\left(\frac{d_2 + d_3 - 2d_1}{2}\sqrt{2\gamma}\right) + Q\left(\frac{d_2 + d_3 + 2d_1}{2}\sqrt{2\gamma}\right)\right]
$$
$$
+ \frac{1}{4}\left[Q\left(\frac{d_3 - d_2}{2}\sqrt{2\gamma}\right) + Q\left(\frac{d_3 + 3d_2}{2}\sqrt{2\gamma}\right)\right]
$$
$$
+ \frac{1}{4}\left[Q\left(\frac{d_3 - d_2}{2}\sqrt{2\gamma}\right) - Q\left(\frac{3d_3 + d_2}{2}\sqrt{2\gamma}\right)\right]
$$
$$
+ \frac{1}{4}\left[Q\left(\frac{2d_4 - d_2 - d_3}{2}\sqrt{2\gamma}\right) - Q\left(\frac{2d_4 + d_2 + d_3}{2}\sqrt{2\gamma}\right)\right] \tag{A.82}
$$
$$
\simeq \frac{1}{4}\left[Q\left(\sqrt{\frac{2(\beta_1 + \beta_2 - 2)^2}{1 + \beta_1^2 + \beta_2^2 + \beta_3^2}\gamma}\right) + 2Q\left(\sqrt{\frac{2(\beta_2 - \beta_1)^2}{1 + \beta_1^2 + \beta_2^2 + \beta_3^2}\gamma}\right)\right.
$$
$$
\left. + Q\left(\sqrt{\frac{2(2\beta_3 - \beta_1 - \beta_2)^2}{1 + \beta_1^2 + \beta_2^2 + \beta_3^2}\gamma}\right)\right] \cdot \tag{A.83}
$$

The decision regions depicted in the lower part of Figure A.6(b) are valid for the fifth bit, namely b_5. The BER of b_5 can be expressed as:

$$
\begin{aligned}
P_{b_5} = \frac{1}{4} & \left[Q\left(\frac{d_2 - d_1}{2} \sqrt{2\gamma} \right) + Q\left(\frac{d_2 + 3 d_1}{2} \sqrt{2\gamma} \right) \right.\\
& \left. - Q\left(\frac{d_3 + d_4 - 2 d_1}{2} \sqrt{2\gamma} \right) - Q\left(\frac{d_3 + d_4 + 2 d_1}{2} \sqrt{2\gamma} \right) \right] \\
+ \frac{1}{4} & \left[Q\left(\frac{d_2 - d_1}{2} \sqrt{2\gamma} \right) + Q\left(\frac{d_3 + d_4 - 2 d_2}{2} \sqrt{2\gamma} \right) \right.\\
& \left. - Q\left(\frac{d_1 + 3 d_2}{2} \sqrt{2\gamma} \right) + Q\left(\frac{d_3 + d_4 + 2 d_2}{2} \sqrt{2\gamma} \right) \right] \\
+ \frac{1}{4} & \left[Q\left(\frac{2 d_3 - d_1 - d_2}{2} \sqrt{2\gamma} \right) + Q\left(\frac{d_4 - d_3}{2} \sqrt{2\gamma} \right) \right.\\
& \left. - Q\left(\frac{d_1 + d_2 + 2 d_3}{2} \sqrt{2\gamma} \right) + Q\left(\frac{3 d_3 + d_4}{2} \sqrt{2\gamma} \right) \right] \\
+ \frac{1}{4} & \left[Q\left(\frac{d_4 - d_3}{2} \sqrt{2\gamma} \right) - Q\left(\frac{2 d_4 - d_1 - d_2}{2} \sqrt{2\gamma} \right) \right.\\
& \left. + Q\left(\frac{d_1 + d_2 + 2 d_4}{2} \sqrt{2\gamma} \right) - Q\left(\frac{d_3 + 3 d_4}{2} \sqrt{2\gamma} \right) \right] .
\end{aligned}
\tag{A.84}
$$

The expression of the BER P_{b_5} can be accurately approximated as ;

$$
\begin{aligned}
P_{b_5} \simeq \frac{1}{4} & \left[2 Q\left(\sqrt{ \frac{2(\beta_1 - 1)^2}{1 + \beta_1^2 + \beta_2^2 + \beta_3^2} \gamma } \right) + Q\left(\sqrt{ \frac{2(\beta_2 + \beta_3 - 2\beta_1)^2}{1 + \beta_1^2 + \beta_2^2 + \beta_3^2} \gamma } \right) \right.\\
& \left. + Q\left(\sqrt{ \frac{2(2\beta_2 - 1 - \beta_1)^2}{1 + \beta_1^2 + \beta_2^2 + \beta_3^2} \gamma } \right) + 2 Q\left(\sqrt{ \frac{2(\beta_3 - \beta_2)^2}{1 + \beta_1^2 + \beta_2^2 + \beta_3^2} \gamma } \right) \right] .
\end{aligned}
\tag{A.85}
$$

Let us now find the BER of the PSK modulated bits b_1, b_2 and b_3. Since they are 8-PSK modulated, the BER can be expressed using the results of [196] as:

$$P_{b_1} = P_{b_2} = P_{b_3} = \frac{1}{4}P_{8PSK}(d_1^2\gamma) + \frac{1}{4}P_{8PSK}(d_2^2\gamma) + \frac{1}{4}P_{8PSK}(d_3^2\gamma) + \frac{1}{4}P_{8PSK}(d_4^2\gamma)$$

(A.86)

$$\simeq \frac{1}{6}\left[Q\left(\sqrt{\frac{8\sin^2(\pi/8)}{1+\beta_1^2+\beta_2^2+\beta_3^2}\gamma}\right) + Q\left(\sqrt{\frac{8\sin^2(3\pi/8)}{1+\beta_1^2+\beta_2^2+\beta_3^2}\gamma}\right)\right]$$

$$+ \frac{1}{6}\left[Q\left(\sqrt{\frac{8\beta_1^2\sin^2(\pi/8)}{1+\beta_1^2+\beta_2^2+\beta_3^2}\gamma}\right) + Q\left(\sqrt{\frac{8\beta_1^2\sin^2(3\pi/8)}{1+\beta_1^2+\beta_2^2+\beta_3^2}\gamma}\right)\right]$$

$$+ \frac{1}{6}\left[Q\left(\sqrt{\frac{8\beta_2^2\sin^2(\pi/8)}{1+\beta_1^2+\beta_2^2+\beta_3^2}\gamma}\right) + Q\left(\sqrt{\frac{8\beta_2^2\sin^2(3\pi/8)}{1+\beta_1^2+\beta_2^2+\beta_3^2}\gamma}\right)\right]$$

$$+ \frac{1}{6}\left[Q\left(\sqrt{\frac{8\beta_3^2\sin^2(\pi/8)}{1+\beta_1^2+\beta_2^2+\beta_3^2}\gamma}\right) + Q\left(\sqrt{\frac{8\beta_3^2\sin^2(3\pi/8)}{1+\beta_1^2+\beta_2^2+\beta_3^2}\gamma}\right)\right] . \quad \text{(A.87)}$$

Hence, the BER of 32-level Star QAM can be expressed as:

$$P_{32} = \frac{1}{5}(3P_{b_1} + P_{b_4} + P_{b_5}), \quad \text{(A.88)}$$

where P_{b_1}, P_{b_4} and P_{b_5} are given in Equations (A.87), (A.83) and (A.85), respectively. The optimum ring ratios of 32-level Star QAM are depicted in Figure A.11(a). The optimum ring ratios converge to $\beta_1 = 1.77$, $\beta_2 = 2.541$ and $\beta_3 = 3.318$. Note that the first optimum ring ratio is the same as the optimum ring ratio of 16-level Star QAM and the corresponding distances between the second, the third and the fourth rings are approximately equal, as one would expect in an effort to maintain an identical distance amongst the constellations points. Figure A.11(b) compares the BER of 32-level Star QAM and 32-PSK. We found that the SNR gain of 32-level Star QAM over 32-PSK is 4.6dB at a BER of 10^{-6}.

The BER of 64-level Star QAM can be obtained using the same procedure employed for determing the BER of 32-level Star QAM, considering that now the bits b_1, b_2, b_3 and b_4 are 16-PSK modulated on each ring. The BER of the last two bits, P_{b_5} and P_{b_6} are the same as those given in (A.83) and (A.85), respectively. On the other hand, the BER of the 16-PSK

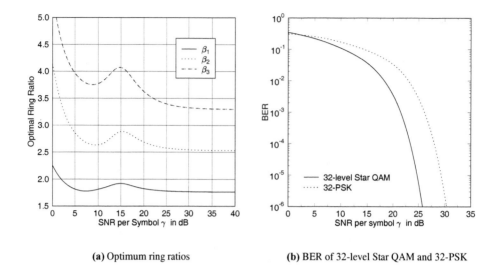

(a) Optimum ring ratios

(b) BER of 32-level Star QAM and 32-PSK

Figure A.11: BER of 32 Star QAM over AWGN channel.

modulated bits of 64-level Star QAM can be expressed as:

$$P_{b_1} = P_{b_2} = P_{b_3} = P_{b_4}$$

$$= \frac{1}{4}P_{16PSK}(d_1^2\gamma) + \frac{1}{4}P_{16PSK}(d_2^2\gamma) + \frac{1}{4}P_{16PSK}(d_3^2\gamma) + \frac{1}{4}P_{16PSK}(d_4^2\gamma) \quad \text{(A.89)}$$

$$\simeq \frac{1}{8}\left[Q\left(\sqrt{\frac{8\sin^2(\pi/16)}{1+\beta_1^2+\beta_2^2+\beta_3^2}\gamma}\right) + Q\left(\sqrt{\frac{8\sin^2(3\pi/16)}{1+\beta_1^2+\beta_2^2+\beta_3^2}\gamma}\right)\right]$$

$$+ \frac{1}{8}\left[Q\left(\sqrt{\frac{8\beta_1^2\sin^2(\pi/16)}{1+\beta_1^2+\beta_2^2+\beta_3^2}\gamma}\right) + Q\left(\sqrt{\frac{8\beta_1^2\sin^2(3\pi/16)}{1+\beta_1^2+\beta_2^2+\beta_3^2}\gamma}\right)\right]$$

$$+ \frac{1}{8}\left[Q\left(\sqrt{\frac{8\beta_2^2\sin^2(\pi/16)}{1+\beta_1^2+\beta_2^2+\beta_3^2}\gamma}\right) + Q\left(\sqrt{\frac{8\beta_2^2\sin^2(3\pi/16)}{1+\beta_1^2+\beta_2^2+\beta_3^2}\gamma}\right)\right]$$

$$+ \frac{1}{8}\left[Q\left(\sqrt{\frac{8\beta_3^2\sin^2(\pi/16)}{1+\beta_1^2+\beta_2^2+\beta_3^2}\gamma}\right) + Q\left(\sqrt{\frac{8\beta_3^2\sin^2(3\pi/16)}{1+\beta_1^2+\beta_2^2+\beta_3^2}\gamma}\right)\right] . \quad \text{(A.90)}$$

Hence, the average BER of 64-level Star QAM can be expressed as:

$$P_{64} = \frac{1}{6}(4P_{b_1} + P_{b_5} + P_{b_6}), \quad \text{(A.91)}$$

where P_{b_1} is given in (A.90), where P_{b_5} and P_{b_6} are given in (A.83) and (A.85), respectively. The optimum ring ratios of 64-level Star QAM are depicted in Figure A.12(a). The optimum

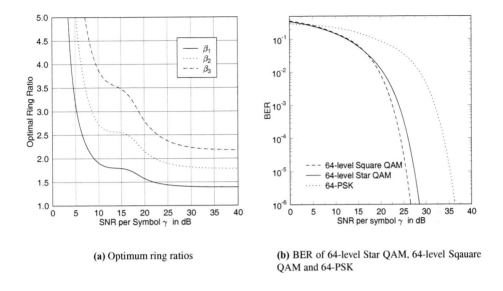

(a) Optimum ring ratios

(b) BER of 64-level Star QAM, 64-level Sqauare QAM and 64-PSK

Figure A.12: BER of 64-level Star QAM for transmission over AWGN channels.

ring ratios converge to $\beta_1 = 1.4$, $\beta_2 = 1.81$ and $\beta_3 = 2.23$. It was observed that the SNR difference between the optimised BER and that employing the asymptotic ring ratio is at most 1dB in the SNR range of 5dB to 15dB. Figure A.12(b) compares the BER of 64-level Star QAM, 64-level Square QAM and 64-PSK. We found that the SNR gain of 64-level Star QAM over 64-PSK is 7.7dB at the BER of 10^{-6}. The 64-level Square QAM arrangement is the most power-efficient scheme, which exhibits 2dB SNR gain over 64-level Star QAM at the BER of 10^{-6}.

A.5 Two-Dimensional Rake Receiver

A.5.1 System Model

The schematic of our Rake-receiver and D-antenna diversity assisted adaptive Square QAM (AQAM) system is illustrated in Figure A.13. A band-limited equivalent low-pass m-ary QAM signal $s(t)$, having a spectrum of $S(f)=0$ for $|f| > 1/2\,W$, is transmitted over time variant frequency selective fading channels and received by a set of D RAKE-receivers. Each Rake-receiver [87, 201] combines all the resolvable multi-path components using Maximal Ratio Combining (MRC). The combined signals of the D-antenna assisted Rake-receivers are summed and demodulated using the estimated channel quality information. The estimated signal-to-noise ratio is fed back to the transmitter and it is used for deciding upon the most appropriate m-ary square QAM modulation mode to be used during the next transmission burst. We assume that the channel quality is estimated perfectly and it is available at the transmitter immediately. The effects of channel estimation error and feedback delay on the

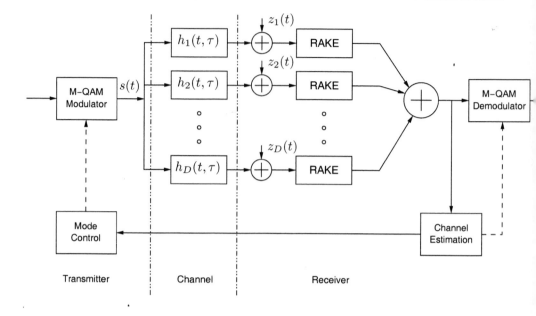

Figure A.13: Equivalent low-pass model of a D-th order antenna diversity based RAKE-receiver assisted AQAM system.

performance of AQAM were studied for example by *Goldsmith* and *Chua* [25].

The low-pass equivalent impulse response of the channel between the transmitter and the d-th antenna may be represented as [87] :

$$h_d(t, \tau) = \sum_{n=1}^{N} h_{d,n}(t) \, \delta \left(\tau - \frac{n}{W} \right) , \tag{A.92}$$

where $\{h_{d,n}(t)\}$ is a set of independent complex valued stationary random Gaussian processes. The maximum number of resolvable multi-path components N is given by $\lfloor T_m W \rfloor + 1$, where T_m is the multi-path delay spread of the channel [87]. Hence, the low-pass equivalent received signal $r_d(t)$ at the d-th antenna can be formulated as :

$$r_d(t) = \sum_{n=1}^{N} h_{d,n}(t) s \left(t - \frac{n}{W} \right) + z_d(t) , \tag{A.93}$$

where $z_d(t)$ is a zero mean Gaussian random process having a two-sided power spectral density of $N_o/2$. Let us assume that the fading is sufficiently slow or $(\Delta t)_c \ll T$, where $(\Delta t)_c$ is the channel's coherence time [13] and T is the signaling period. Then, $h_{d,n}(t)$ can be simplified to $h_{d,n}(t) = \alpha_{d,n} e^{j\phi_{d,n}}$ for the duration of signalling period T, where the fading magnitude $\alpha_{d,n}$ is assumed to be Rayleigh distributed and the phase $\phi_{d,n}$ is assumed to be uniformly distributed.

A.5.2 BER Analysis of Fixed-mode Square QAM

An ideal RAKE receiver [201] combines all the signal powers scattered over N paths in an optimal manner, so that the instantaneous Signal-to-Noise Ratio (SNR) per symbol at the RAKE receiver's output can be maximised [87]. The noise at the RAKE receiver's output is known to be Gaussian [87]. The SNR, γ_d, at the d-th ideal RAKE receiver's output is given as [87] :

$$\gamma_d = \sum_{n=1}^{N} \gamma_{d,n} \, , \tag{A.94}$$

where $\gamma_{d,n} = E/N_o \, \alpha_{d,n}^2$ and $\{\alpha_{d,n}\}$ is assumed to be normalised, such that $\sum_{n=1}^{N} \alpha_{d,n}^2$ becomes unity. Since we assumed that each multi-path component has an independent Rayleigh distribution, the characteristic function of γ_d can be represented as [87, pp 802] :

$$\psi_{\gamma_d}(j\upsilon) = \prod_{n=1}^{N} \frac{1}{1 - j\upsilon\bar{\gamma}_{d,n}} \, , \tag{A.95}$$

where $\bar{\gamma}_{d,n} = E/N_o \mathrm{E}[\alpha_{d,n}^2]$. Let us assume furthermore that each of the D diversity channels has the same multi-path intensity profile (MIP), although in practical systems each antenna may experience a different MIP. Under this assumption, $\bar{\gamma}_{d,n}$ in (A.95) can be written as $\bar{\gamma}_n$. The total SNR per symbol, γ, at the output of the demodulator depicted in Figure A.13 is given as :

$$\gamma = \sum_{d=1}^{D} \gamma_d \, , \tag{A.96}$$

while the characteristic function of the SNR per symbol γ, under the assumption of independent identical diversity channels, can be formulated as :

$$\psi_{\gamma}(j\upsilon) = \prod_{n=1}^{N} \frac{1}{(1 - j\upsilon\bar{\gamma}_n)^D} \, . \tag{A.97}$$

Applying the technique of Partial Fraction Expansion (PFE) [195], $\psi_{\gamma}(j\upsilon)$ can be expressed as :

$$\psi_{\gamma}(j\upsilon) = \sum_{d=1}^{D} \sum_{n=1}^{N} \Lambda_{D-d+1,n} \frac{1}{(1 - j\upsilon\bar{\gamma}_n)^d} \, . \tag{A.98}$$

Let us now determine the constant coefficients $\Lambda_{d,n}$. Equating (A.97) with (A.98) and substituting $j\upsilon = -p$, we have

$$\prod_{i=1}^{N} \frac{1}{(1 + p\bar{\gamma}_i)^D} = \sum_{d=1}^{D} \sum_{i=1}^{N} \Lambda_{D-d+1,i} \frac{1}{(1 + p\bar{\gamma}_i)^d} \, . \tag{A.99}$$

Multiplying by $(1 + p\bar{\gamma}_n)^D$ at both sides, (A.99) becomes :

$$\prod_{\substack{i=1 \\ i \neq n}}^{N} \frac{1}{(1 + p\bar{\gamma}_i)^D} = \sum_{d=1}^{D} \sum_{\substack{i=1 \\ i \neq n}}^{N} \Lambda_{D-d+1,i} \frac{1}{(1 + p\bar{\gamma}_i)^d} + \sum_{d=1}^{D} \Lambda_{d,n} (1 + p\bar{\gamma}_n)^{d-1} . \quad (A.100)$$

Setting the $(d-1)$th derivatives with respect to p and substituting $p = -1/\bar{\gamma}_n$, we have :

$$\frac{d^{d-1}}{dp^{d-1}} \left[\prod_{\substack{i=1 \\ i \neq n}}^{N} (p\,\bar{\gamma}_i + 1)^{-D} \right]_{p = -1/\bar{\gamma}_n} = (d-1)! \, \bar{\gamma}_n^{(d-1)} \Lambda_{d,n} . \quad (A.101)$$

Hence, $\Lambda_{d,n}$ is given as :

$$\Lambda_{d,n} \triangleq \frac{1}{(d-1)! \, \bar{\gamma}_n^{(d-1)}} \varphi_{d,n} (-1/\bar{\gamma}_n) , \quad (A.102)$$

where $\varphi_{d,n}(x)$ is defined as :

$$\varphi_{d,n}(x) \triangleq \frac{d^{d-1}}{dp^{d-1}} \left[\prod_{\substack{i=1 \\ i \neq n}}^{N} (p\,\bar{\gamma}_i + 1)^{-D} \right]_{p = x} . \quad (A.103)$$

Upon setting the derivatives directly, $\varphi_{d,n}(-1/\bar{\gamma}_n)$ of (A.103) can be represented recursively as :

$$\varphi_{1,n}(-1/\bar{\gamma}_n) = \pi_n^D$$

$$\varphi_{d,n}(-1/\bar{\gamma}_n) = \sum_{i=1}^{d-1} \left[C_{d,i} \, D \, \varphi_{d-i,n}(-1/\bar{\gamma}_n) \sum_{\substack{j=1 \\ j \neq n}}^{N} \left(\frac{\bar{\gamma}_n \bar{\gamma}_j}{\bar{\gamma}_n - \bar{\gamma}_j} \right)^i \right] , \quad (A.104)$$

where π_n is defined as :

$$\pi_n \triangleq \prod_{\substack{i=1 \\ i \neq n}}^{N} \frac{\bar{\gamma}_n}{\bar{\gamma}_n - \bar{\gamma}_i} \quad (A.105)$$

and the doubly indexed coefficient $C_{d,i}$ of (A.104) can also be expressed recursively as :

$$\begin{array}{llll} C_{d,1} & = & -1 & \text{for all } d \\ C_{d,d} & = & 0 & \text{for } d > 1 \\ C_{d,i} & = & -(i-1) \, C_{d-1,i-1} + C_{d-1,i} & \text{for } d > i . \end{array} \quad (A.106)$$

The PDF of γ, $f_{\bar{\gamma}}(\gamma)$, can be found by applying the inverse Fourier transform to $\psi_\gamma(jv)$ in (A.98), which is given by [87, pp 781, (14-4-13)] :

$$f_{\bar{\gamma}}(\gamma) =. \sum_{d=1}^{D} \sum_{n=1}^{N} \Lambda_{D-d+1,n} \frac{1}{(d-1)! \, \bar{\gamma}_n^d} \gamma^{d-1} e^{-\gamma/\bar{\gamma}_n} . \quad (A.107)$$

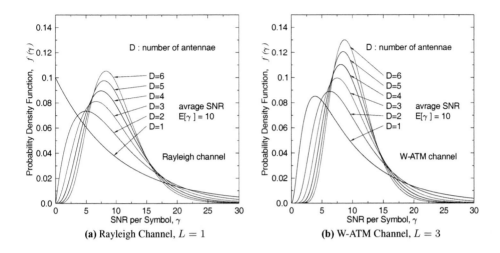

Figure A.14: PDF of γ given in (A.107) for an average SNR per symbol of $E[\gamma] = 10$dB.

Figure A.14 shows the PDF of the SNR per symbol over both a narrow-band Rayleigh channel and the dispersive Wireless Asynchronous Transfer Mode (W-ATM) channel of [4]. Specifically, the W-ATM channel is a 3-path indoor channel, where the average SNR for each path is given as $\bar{\gamma}_1 = 0.79192\bar{\gamma}$, $\bar{\gamma}_2 = 0.12424\bar{\gamma}$ and $\bar{\gamma}_3 = 0.08384\bar{\gamma}$.

Since we now have the PDF $f_{\bar{\gamma}}(\gamma)$ of the channel SNR, let us calculate the average BEP of m-ary square QAM employing Gray mapping. The average BEP P_e can be expressed as [4, 87]:

$$P_e = \int_0^\infty p_m(\gamma) f(\gamma) d\gamma \,, \tag{A.108}$$

where $p_m(\gamma)$ is the BER of m-ary square QAM employing Gray mapping over Gaussian channels [4]:

$$p_m(\gamma) = \sum_i A_i Q(\sqrt{a_i \gamma}) \,, \tag{A.109}$$

where $Q(x)$ is the Gaussian Q-function defined as $Q(x) \triangleq \frac{1}{\sqrt{2\pi}} \int_x^\infty e^{-t^2/2} dt$ and $\{A_i, a_i\}$ is a set of modulation mode dependent constants. For the modulation modes associated with $m = 2, 4, 16$ and 64, the sets $\{A_i, a_i\}$ are given as [4, 191]:

$$
\begin{array}{lll}
m = 2, & \text{BPSK} & \{(1, 2)\} \\
m = 4, & \text{QPSK} & \{(1, 1)\} \\
m = 16, & \text{16-QAM} & \left\{ \left(\frac{3}{4}, \frac{1}{5}\right), \left(\frac{2}{4}, \frac{3^2}{5}\right), \left(-\frac{1}{4}, \frac{5^2}{5}\right) \right\} \\
m = 64, & \text{64-QAM} & \left\{ \left(\frac{7}{12}, \frac{1}{21}\right), \left(\frac{6}{12}, \frac{3^2}{21}\right), \left(-\frac{1}{12}, \frac{5^2}{21}\right), \left(\frac{1}{12}, \frac{9^2}{21}\right), \left(-\frac{1}{12}, \frac{13^2}{21}\right) \right\} \,.
\end{array}
$$

$$\tag{A.110}$$

The average BEP of m-ary QAM in our scenario can be calculated by substituting $p_m(\gamma)$ of (A.109) and $f_{\bar{\gamma}}(\gamma)$ of (A.107) into (A.108) :

$$P_{e,m}(\bar{\gamma}) = \int_0^\infty \sum_i A_i Q(\sqrt{a_i \gamma}) f_{\bar{\gamma}}(\gamma) d\gamma \qquad (A.111)$$

$$= \sum_i A_i P_e(\bar{\gamma}; a_i) , \qquad (A.112)$$

where each constituent BEP $P_e(\bar{\gamma}; a_i)$ is defined as :

$$P_e(\bar{\gamma}; a_i) = \int_0^\infty Q(\sqrt{a_i \gamma}) f_{\bar{\gamma}}(\gamma) d\gamma . \qquad (A.113)$$

Using the similarity of $f_{\bar{\gamma}}(\gamma)$ in (A.107) and the PDF of the SNR of a D-antenna diversity-assisted Maximal Ratio Combining (MRC) system transmitting over flat Rayleigh channels [87, pp 781], the closed form solution for the component BEP $P_e(\bar{\gamma}; a_i)$ can be expressed as :

$$P_e(\bar{\gamma}; a_i) = \sum_{d=1}^D \sum_{n=1}^N \frac{1}{\sqrt{2\pi}} \int_0^\infty \int_{\sqrt{2\gamma}}^\infty e^{-x^2/2} \Lambda_{D-d+1,n} \frac{1}{(d-1)! \bar{\gamma}_n^d} \gamma^{d-1} e^{-\gamma/\bar{\gamma}_n} dx \, d\gamma \qquad (A.114)$$

$$= \sum_{d=1}^D \sum_{n=1}^N \left[\Lambda_{D-d+1,n} \left\{ \tfrac{1}{2}(1-\mu_n) \right\}^d \sum_{i=0}^{d-1} \binom{d-1+i}{i} \left\{ \tfrac{1}{2}(1+\mu_n) \right\}^i \right], \qquad (A.115)$$

where $\mu_n \triangleq \sqrt{\frac{a_i \bar{\gamma}_n}{2 + a_i \bar{\gamma}_n}}$ and the average SNR per symbol is $\bar{\gamma} = D \sum_{n=1}^N \bar{\gamma}_n$. Substituting $P_e(\bar{\gamma}; a_i)$ of (A.115) into (A.112), the average BEP of an m-ary QAM Rake receiver using antenna diversity can be expressed in a closed form.

Let us consider the performance of BPSK by setting $D=1$ or $N=1$. When the number of antennae is one, *i.e.* $D=1$, $P_{e,2}(\bar{\gamma})$ is reduced to

$$P_{e,BPSK} = \sum_{n=1}^N \Lambda_{1,n} \left\{ \tfrac{1}{2}(1-\mu_n) \right\} \qquad (A.116)$$

$$= \sum_{n=1}^N \pi_n \left\{ \tfrac{1}{2}(1-\mu_n) \right\} , \qquad (A.117)$$

which is identical to the result given in [87, (14-5-28), pp 802] . On the other hand, when the channels exhibit flat fading, *i.e.* $L=1$, our system is reduced to a D-antenna diversity-based MRC system transmitting over D number of flat Rayleigh channels. In this case, $\Lambda_{D-d+1,n}$ of (A.102) becomes zero for all values of d, except for $\Lambda_{1,1}=1$ when $d=D$, and the average

BPSK BEP in this scenario becomes :

$$P_{e,BPSK} = \sum_{d=1}^{D} \left[\Lambda_{D-d+1,1} \left\{ \tfrac{1}{2}(1-\mu_1) \right\}^d \sum_{i=0}^{d-1} \binom{d-1+i}{i} \left\{ \tfrac{1}{2}(1+\mu_1) \right\}^i \right] \qquad (A.118)$$

$$= \left\{ \tfrac{1}{2}(1-\mu_1) \right\}^D \sum_{i=0}^{D-1} \binom{D-1+i}{i} \left\{ \tfrac{1}{2}(1+\mu_1) \right\}^i , \qquad (A.119)$$

which is also given in [87, (14-4-15), pp781] .

A.6 Mode Specific Average BEP of an Adaptive Modulation Scheme

A closed form solution for the 'mode-specific average BER' of a Maximal Ratio Combining (MRC) receiver using Dth order antenna diversity over independent Rayleigh channels is derived, where the 'mode-specific average BER' refers to the BER of the adaptive modulation scheme, while activating one of its specific constituent modem modes. The PDF $f_{\bar{\gamma}}(\gamma)$ of the channel SNR γ is given as [87, (14-4-13)]

$$f_{\bar{\gamma}}(\gamma) = \frac{1}{(D-1)!\,\bar{\gamma}^D} \gamma^{D-1} e^{-\gamma/\bar{\gamma}} , \quad \gamma \geq 0 , \qquad (A.120)$$

where $\bar{\gamma}$ is the average channel SNR. Since the PDF of the instantaneous channel SNR γ over a Nakagami fading channel is given as :

$$f_{\bar{\gamma}}(\gamma) = \left(\frac{m}{\bar{\gamma}}\right)^m \frac{\gamma^{m-1}}{\Gamma(m)} e^{-m\gamma/\bar{\gamma}} , \quad \gamma \geq 0 , \qquad (A.121)$$

the following results can also be applied to a Nakagami fading channel with a simple change of variable given as $D = m$ and $\bar{\gamma} = \bar{\gamma}/m$.

The mode-specific average BEP is defined as

$$P_r(\alpha, \beta; \bar{\gamma}, D, a) \triangleq \int_{\alpha}^{\beta} Q(\sqrt{a\gamma}) \, f_{\bar{\gamma}}(\gamma) \, d\gamma \qquad (A.122)$$

$$= \int_{\alpha}^{\beta} Q(\sqrt{a\gamma}) \frac{1}{(D-1)!\,\bar{\gamma}^D} \gamma^{D-1} e^{-\gamma/\bar{\gamma}} \, d\gamma , \qquad (A.123)$$

where $Q(x) \triangleq \frac{1}{\sqrt{2\pi}} \int_x^{\infty} e^{-t^2/2} \, dt$. Applying integration-by-part or $\int u \, dv = uv - \int v \, du$ and noting that

$$u = Q(\sqrt{a\gamma})$$

$$du = -\frac{\sqrt{a}}{2\sqrt{2\pi\gamma}} e^{-a\gamma/2}$$

$$dv = \frac{1}{(D-1)!\,\bar{\gamma}^D} \gamma^{D-1} e^{-\gamma/\bar{\gamma}}$$

$$v = -e^{-\gamma/\bar{\gamma}} \sum_{d=0}^{D-1} (\gamma/\bar{\gamma})^d \frac{1}{d!}$$

(A.122) becomes

$$P_r(\alpha, \beta; \bar{\gamma}, D, a) = \left[e^{-\gamma/\bar{\gamma}} Q(\sqrt{a\gamma}) \sum_{d=0}^{D-1} (\gamma/\bar{\gamma})^d \frac{1}{d!} \right]_{\beta}^{\alpha} - \sum_{d=0}^{D-1} I_d(\alpha, \beta), \qquad (A.124)$$

where

$$I_d(\alpha, \beta) = \int_{\alpha}^{\beta} \frac{\sqrt{a}}{2\sqrt{2\pi}} \frac{1}{d!} (\gamma/\bar{\gamma})^d \frac{1}{\sqrt{\gamma}} e^{-a\gamma/(2\mu^2)} d\gamma \qquad (A.125)$$

and $\mu = \sqrt{\frac{a\bar{\gamma}}{a\bar{\gamma}+2}}$. Let us consider I_d for the case of $d = 0$:

$$I_0(\alpha, \beta) = \int_{\alpha}^{\beta} \frac{\sqrt{a}}{2\sqrt{2\pi}} \frac{1}{\sqrt{\gamma}} e^{-a\gamma/(2\mu^2)} d\gamma . \qquad (A.126)$$

Upon introducing the variable $t^2 = a\gamma/\mu^2$ and exploiting that $d\gamma = 2\mu\sqrt{\gamma/a}\, ds$, we have :

$$I_0(\alpha, \beta) = [\mu Q(\sqrt{a\gamma}/\mu)]_{\beta}^{\alpha} . \qquad (A.127)$$

Applying integration-by-part once to (A.125) yields

$$I_d(\alpha, \beta) = \left[\frac{\mu^2}{\sqrt{2a\pi}} \frac{1}{d!} (\gamma/\bar{\gamma})^d \frac{1}{\sqrt{\gamma}} e^{-a\gamma/(2\mu^2)} \right]_{\beta}^{\alpha} + \frac{2d-1}{a\bar{\gamma}d} \mu^2 I_{d-1}, \qquad (A.128)$$

which is a recursive form with the initial value given in (A.127). For this recursive form of (A.128), a non-recursive form of I_d can be expressed as :

$$I_d(\alpha, \beta) = \left[\frac{\mu^2}{\sqrt{2a\pi}} \frac{\Gamma(d+\frac{1}{2})}{\bar{\gamma}^d \Gamma(d+1)} \sum_{i=1}^{d} \left(\frac{2\mu^2}{a} \right)^{d-i} \frac{\gamma^{i-\frac{1}{2}}}{\Gamma(i+\frac{1}{2})} e^{-a\gamma/(2\mu^2)} \right]_{\beta}^{\alpha}$$

$$+ \left[\left(\frac{2\mu^2}{a\bar{\gamma}} \right)^d \frac{1}{\sqrt{\pi}} \frac{\Gamma(d+\frac{1}{2})}{\Gamma(d+1)} \mu Q(\sqrt{a\gamma}/\mu) \right]_{\beta}^{\alpha} . \qquad (A.129)$$

By substituting $I_d(\alpha, \beta)$ of (A.129) into (A.124), the regional BER $P_r(\bar{\gamma}; a, \alpha, \beta)$ can be represented in a closed form.

Bibliography

[1] W. Webb and R. Steele, "Variable rate QAM for mobile radio," *IEEE Transactions on Communications*, vol. 43, pp. 2223–2230, July 1995.

[2] L. Hanzo, "Bandwidth-efficient wireless multimedia communications," *Proceedings of the IEEE*, vol. 86, pp. 1342–1382, July 1998.

[3] S. Nanda, K. Balachandran, and S. Kumar, "Adaptation techniques in wireless packet data services," *IEEE Communications Magazine*, vol. 38, pp. 54–64, January 2000.

[4] L. Hanzo, W. Webb, and T. Keller, *Single- and Multi-carrier Quadrature Amplitude Modulation*. New York, USA: IEEE Press-John Wiley, April 2000.

[5] L. Hanzo, F. Somerville, and J. Woodard, *Voice Compression and Communications: Principles and Applications for Fixed and Wireless Channels*. IEEE Press and John Wiley, 2001. (For detailed contents and sample chapters please refer to http://www-mobile.ecs.soton.ac.uk.).

[6] L. Hanzo, P. Cherriman, and J. Streit, "Wireless video communications: From second to third generation systems, WLANs and beyond." IEEE Press, 2001. (For detailed contents please refer to http://www-mobile.ecs.soton.ac.uk.).

[7] S. M. Alamouti, "A simple transmit diversity technique for wireless communications," *IEEE Journal on Selected Areas in Communications*, vol. 16, pp. 1451–1458, October 1998.

[8] T. Liew and L. Hanzo, "Space-time block coded adaptive modulation aided ofdm," in *Proceeding of GLOBECOM'2001*, (San Antonio, USA), IEEE, 26-29 November 2001.

[9] J. K. Cavers, "Variable rate transmission for rayleigh fading channels," *IEEE Transactions on Communications Technology*, vol. COM-20, pp. 15–22, February 1972.

[10] W. Tuttlebee, ed., *Software Defined Radio, Volumes I and II*. John Wiley, 2002.

[11] L. Hanzo, C. Wong, and M. Yee, *Adaptive Wireless Transceivers*. John Wiley, IEEE Press, 2002. (For detailed contents, please refer to http://www-mobile.ecs.soton.ac.uk.).

[12] L. Hanzo, T. Liew, and B. Yeap, *Turbo Coding, Turbo Equalisation and Space-Time Coding*. John Wiley, IEEE Press, 2002. (For detailed contents, please refer to http://www-mobile.ecs.soton.ac.uk.).

[13] R. Steele and L. Hanzo, eds., *Mobile Radio Communications*. New York, USA: IEEE Press - John Wiley & Sons, 2nd ed., 1999.

[14] A. Duel-Hallen, S. Hu, and H. Hallen, "Long range prediction of fading signals," *IEEE Signal Processing Magazine*, vol. 17, pp. 62–75, May 2000.

[15] J. F. Hayes, "Adaptive feedback communications," *IEEE Transactions on Communication Technology*, vol. 16, no. 1, pp. 29–34, 1968.

[16] R. Steele and W. Webb, "Variable rate QAM for data transmission over Rayleigh fading channels," in *Proceeedings of Wireless '91*, (Calgary, Alberta), pp. 1–14, IEEE, 1991.

[17] W. T. Webb and R. Steele, "Variable rate QAM for mobile radio," *IEEE Transactions on Communications*, vol. 43, no. 7, pp. 2223–2230, 1995.

[18] M. Moher and J. Lodge, "TCMP—a modulation and coding strategy for rician fading channels," *IEEE Journal on Selected Areas in Communications*, vol. 7, pp. 1347–1355, December 1989.

[19] J. K. Cavers, "An Analysis of Pilot Symbol Assisted Modulation for Rayleigh Fading Channels," *IEEE Transactions on Vehicular Technology*, vol. 40, pp. 686–693, November 1991.

[20] S. Sampei and T. Sunaga, "Rayleigh fading compensation for QAM in land mobile radio communications," *IEEE Transactions on Vehicular Technology*, vol. 42, pp. 137–147, May 1993.

[21] S. Otsuki, S. Sampei, and N. Morinaga, "Square QAM adaptive modulation/TDMA/TDD systems using modulation level estimation with Walsh function," *Electronics Letters*, vol. 31, pp. 169–171, February 1995.

[22] W. Lee, "Estimate of channel capacity in Rayleigh fading environment," *IEEE Transactions on Vehicular Technology*, vol. 39, pp. 187–189, August 1990.

[23] A. Goldsmith and P. Varaiya, "Capacity of fading channels with channel side information," *IEEE Transactions on Information Theory*, vol. 43, pp. 1986–1992, November 1997.

[24] M. S. Alouini and A. J. Goldsmith, "Capacity of Rayleigh fading channels under different adaptive transmission and diversity-combining technique," *IEEE Transactions on Vehicular Technology*, vol. 48, pp. 1165–1181, July 1999.

[25] A. Goldsmith and S. Chua, "Variable rate variable power MQAM for fading channels," *IEEE Transactions on Communications*, vol. 45, pp. 1218–1230, October 1997.

[26] J. Torrance and L. Hanzo, "Optimisation of switching levels for adaptive modulation in a slow Rayleigh fading channel," *Electronics Letters*, vol. 32, pp. 1167–1169, 20 June 1996.

[27] B. J. Choi and L. Hanzo, "Optimum mode-switching levels for adaptive modulation systems," in *Submitted to IEEE GLOBECOM 2001*, 2001.

[28] B. J. Choi, M. Münster, L. L. Yang, and L. Hanzo, "Performance of Rake receiver assisted adaptive-modulation based CDMA over frequency selective slow Rayleigh fading channel," *Electronics Letters*, vol. 37, pp. 247–249, February 2001.

[29] W. H. Press, S. A. Teukolsky, W. T. Vetterling, and B. P. Flannery, *Numerical Recipies in C*. Cambridge University Press, 1992.

[30] C. Tang, "An intelligent learning scheme for adaptive modulation," in *Proceedings of the IEEE Vehicular Technology Conference*, (Atlantic City, USA), pp. 144–148, 7-10 October 2001.

[31] J. Torrance and L. Hanzo, "Upper bound performance of adaptive modulation in a slow Rayleigh fading channel," *Electronics Letters*, vol. 32, pp. 718–719, 11 April 1996.

[32] C. Wong and L. Hanzo, "Upper-bound of a wideband burst-by-burst adaptive modem," in *Proceedings of VTC'99 (Spring)*, (Houston, TX, USA), pp. 1851–1855, IEEE, 16–20 May 1999.

[33] C. Wong and L. Hanzo, "Upper-bound performance of a wideband burst-by-burst adaptive modem," *IEEE Transactions on Communications*, vol. 48, pp. 367–369, March 2000.

[34] H. Matsuoka, S. Sampei, N. Morinaga, and Y. Kamio, "Adaptive modulation system with variable coding rate concatenated code for high quality multi-media communications systems," in *Proceedings of IEEE VTC'96*, vol. 1, (Atlanta, GA, USA), pp. 487–491, IEEE, 28 April–1 May 1996.

[35] A. J. Goldsmith and S. G. Chua, "Adaptive coded modulation for fading channels," in *Proceedings of IEEE International Conference on Communications*, vol. 3, (Montreal, Canada), pp. 1488–1492, 8–12 June 1997.

[36] A. Goldsmith and S. Chua, "Variable-rate variable-power MQAM for fading channels," *IEEE Transactions on Communications*, vol. 45, pp. 1218–1230, October 1997.

[37] J. Torrance and L. Hanzo, "Demodulation level selection in adaptive modulation," *Electronics Letters*, vol. 32, pp. 1751–1752, 12 September 1996.

[38] V. Lau and S. Maric, "Variable rate adaptive modulation for DS-CDMA," *IEEE Transactions on Communications*, vol. 47, pp. 577–589, April 1999.

[39] S. Sampei, N. Morinaga, and Y. Kamio, "Adaptive modulation/TDMA with a BDDFE for 2 mbit/s multi-media wireless communication systems," in *Proceedings of IEEE Vehicular Technology Conference (VTC'95)*, vol. 1, (Chicago, USA), pp. 311–315, IEEE, 15–28 July 1995.

[40] J. Torrance and L. Hanzo, "Latency considerations for adaptive modulation in a slow Rayleigh fading channel," in *Proceedings of IEEE VTC'97*, vol. 2, (Phoenix, AZ, USA), pp. 1204–1209, IEEE, 4–7 May 1997.

[41] J. Torrance and L. Hanzo, "Statistical multiplexing for mitigating latency in adaptive modems," in *Proceedings of IEEE International Symposium on Personal, Indoor and Mobile Radio Communications, PIMRC'97*, (Marina Congress Centre, Helsinki, Finland), pp. 938–942, IEEE, 1–4 September 1997.

[42] T. Ue, S. Sampei, and N. Morinaga, "Symbol rate controlled adaptive modulation/TDMA/TDD for wireless personal communication systems," *IEICE Transactions on Communications*, vol. E78-B, pp. 1117–1124, August 1995.

[43] M. Yee and L. Hanzo, "Radial Basis Function decision feedback equaliser assisted burst-by-burst adaptive modulation," in *Proceedings of IEEE Global Telecommunications Conference (GLOBECOM)*, (Rio de Janeiro, Brazil), pp. 2183–2187, 5–9 December 1999.

[44] M. Yee, T. Liew, and L. Hanzo, "Radial basis function decision feedback equalisation assisted block turbo burst-by-burst adaptive modems," in *Proceedings of VTC '99 Fall*, (Amsterdam, Holland), pp. 1600–1604, 19-22 September 1999.

[45] M. S. Yee, B. L. Yeap, and L. Hanzo, "Radial basis function assisted turbo equalisation," in *Proceedings of IEEE Vehicular Technology Conference*, (Japan, Tokyo), pp. 640–644, IEEE, 15-18 May 2000.

[46] M. S. Yee and T. H. Liew and L. Hanzo, "Burst-by-burst adaptive turbo-coded radial basis function-assisted decision feedback equalization," *IEEE Transactions on Communications*, pp. 1935–1945, Nov. 2001.

[47] M. S. Yee and B. L. Yeap and L. Hanzo, "RBF-based decision feedback aided turbo equalisation of convolutional and space-time trellis coded systems," *IEE Electronics Letters*, pp. 1298–1299, October 2001.

[48] M. S. Yee and B. L. Yeap and L. Hanzo, "Turbo Equalization of Convolutional Coded and Concatenated Space Time Trellis Coded Systems using Radial Basis Function Aided Equalizers," in *Proceedings of the IEEE Vehicular Technology Conference*, (Atlantic City, USA), pp. 882–886, 7-10 October 2001.

[49] D. Goeckel, "Adaptive coding for fading channels using outdated fading estimates," *IEEE Transactions on Communications*, vol. 47, pp. 844–855, June 1999.

[50] K. J. Hole, H. Holm, and G. E. Oien, "Adaptive multidimensional coded modulation over flat fading channels," *IEEE Journal on Selected Areas in Communications*, vol. 18, pp. 1153–1158, July 2000.

[51] D. Pearce, A. Burr, and T. Tozer, "Comparison of counter-measures against slow Rayleigh fading for TDMA systems," in *IEE Colloquium on Advanced TDMA Techniques and Applications*, (London, UK), pp. 9/1–9/6, IEE, 28 October 1996. digest 1996/234.

[52] V. Lau and M. Macleod, "Variable rate adaptive trellis coded QAM for high bandwidth efficiency applications in Rayleigh fading channels," in *Proceedings of IEEE Vehicular Technology Conference (VTC'98)*, (Ottawa, Canada), pp. 348–352, IEEE, 18–21 May 1998.

[53] S. X. Ng, C. H. Wong and L. Hanzo, "Burst-by-Burst Adaptive Decision Feedback Equalized TCM, TTCM, BICM and BICM-ID," *International Conference on Communications (ICC)*, pp. 3031–3035, June 2001.

[54] T. Suzuki, S. Sampei, and N. Morinaga, "Space and path diversity combining technique for 10 Mbits/s adaptive modulation/TDMA in wireless communications systems," in *Proceedings of IEEE VTC'96*, (Atlanta, GA, USA), pp. 1003–1007, IEEE, 28 April–1 May 1996.

[55] K. Arimochi, S. Sampei, and N. Morinaga, "Adaptive modulation system with discrete power control and predistortion-type non-linear compensation for high spectral efficient and high power efficient wireless communication systems," in *Proceedings of the IEEE International Symposium on Personal, Indoor and Mobile Radio Communications (PIMRC)*, (Helsinki, Finland), pp. 472–477, 1–4 September 1997.

[56] T. Ikeda, S. Sampei, and N. Morinaga, "TDMA-based adaptive modulation with dynamic channel assignment (AMDCA) for high capacity multi-media microcellular systems," in *Proceedings of IEEE Vehicular Technology Conference*, (Phoenix, USA), pp. 1479–1483, May 1997.

[57] T. Ue, S. Sampei, and N. Morinaga, "Adaptive modulation packet radio communication system using NP-CSMA/TDD scheme," in *Proceedings of IEEE VTC'96*, (Atlanta, GA, USA), pp. 416–421, IEEE, 28 April–1 May 1996.

[58] M. Naijoh, S. Sampei, N. Morinaga, and Y. Kamio, "ARQ schemes with adaptive modulation/TDMA/TDD systems for wireless multimedia communication systems," in *Proceedings of the IEEE International Symposium on Personal, Indoor and Mobile Radio Communications (PIMRC)*, (Helsinki, Finland), pp. 709–713, 1–4 September 1997.

[59] S. Sampei, T. Ue, N. Morinaga, and K. Hamaguchi, "Laboratory experimental results of an adaptive modulation TDMA/TDD for wireless multimedia communication systems," in *Proceedings of IEEE International Symposium on Personal, Indoor and Mobile Radio Communications, PIMRC'97*, (Marina Congress Centre, Helsinki, Finland), pp. 467–471, IEEE, 1–4 September 1997.

[60] J. Torrance and L. Hanzo, "Latency and networking aspects of adaptive modems over slow indoors Rayleigh fading channels," *IEEE Transactions on Vehicular Technology*, vol. 48, no. 4, pp. 1237–1251, 1998.

[61] J. Torrance, L. Hanzo, and T. Keller, "Interference aspects of adaptive modems over slow Rayleigh fading channels," *IEEE Transactions on Vehicular Technology*, vol. 48, pp. 1527–1545, September 1999.

[62] I. Kalet, "The multitone channel," *IEEE Transactions on Communications*, vol. 37, pp. 119–124, February 1989.

[63] A. Czylwik, "Adaptive OFDM for wideband radio channels," in *Proceeding of IEEE Global Telecommunications Conference, Globecom 96*, (London, UK), pp. 713–718, IEEE, 18–22 November 1996.

[64] P. Chow, J. Cioffi, and J. Bingham, "A practical discrete multitone transceiver loading algorithm for data transmission over spectrally shaped channels," *IEEE Transactions on Communications*, vol. 48, pp. 772–775, 1995.

[65] T. Ottosson and A. Svensson, "On schemes for multirate support in DS-CDMA systems," *Wireless Personal Communications (Kluwer)*, vol. 6, pp. 265–287, March 1998.

[66] S. Spangenberg, D. Cruickshank, S. McLaughlin, G. Povey, and P. Grant, "Advanced multiuser detection techniques for downlink CDMA, version 2.0," tech. rep., Virtual Centre of Excellence in Mobile and Personal Communications Ltd (Mobile VCE), July 1999.

[67] S. Ramakrishna and J. Holtzman, "A comparison between single code and multiple code transmission schemes in a CDMA system," in *Proceedings of IEEE Vehicular Technology Conference (VTC'98)*, (Ottawa, Canada), pp. 791–795, IEEE, 18–21 May 1998.

[68] F. Adachi, K. Ohno, A. Higashi, T. Dohi, and Y. Okumura, "Coherent multicode DS-CDMA mobile Radio Access," *IEICE Transactions on Communications*, vol. E79-B, pp. 1316–1324, September 1996.

[69] T. Dohi, Y. Okumura, A. Higashi, K. Ohno, and F. Adachi, "Experiments on coherent multicode DS-CDMA," *IEICE Transactions on Communications*, vol. E79-B, pp. 1326–1332, September 1996.

[70] H. Schotten, H. Elders-Boll, and A. Busboom, "Adaptive multi-rate multi-code CDMA systems," in *Proceedings of the IEEE Vehicular Technology Conference (VTC)*, (Ottawa, Canada), pp. 782–785, 18–21 May 1998.

[71] M. Saquib and R. Yates, "Decorrelating detectors for a dual rate synchronous DS/CDMAchannel," *Wireless Personal Communications (Kluwer)*, vol. 9, pp. 197–216, May 1999.

[72] A.-L. Johansson and A. Svensson, "Successive interference cancellation schemes in multi-rateDS/CDMA systems," in *Wireless Information Networks (Baltzer)*, pp. 265–279, 1996.

[73] A. Johansson and A. Svensson, "Multistage interference cancellation in multirate DS/CDMA on a mobile radio channel," in *Proceedings of the IEEE Vehicular Technology Conference (VTC)*, (Atlanta, GA, USA), pp. 666–670, 28 April–1 May 1996.

[74] M. Juntti, "Multiuser detector performance comparisons inmultirate CDMA systems," in *Proceedings of the IEEE Vehicular Technology Conference (VTC)*, (Ottawa, Canada), pp. 36–40, 18–21 May 1998.

[75] S. Kim, "Adaptive rate and power DS/CDMA communications in fading channels," *IEEE Communications Letters*, vol. 3, pp. 85–87, April 1999.

[76] L. Hanzo, W. Webb, and T. Keller, *Single- and Multi-Carrier Quadrature Amplitude Modulation: Principles and Applications for Personal Communications, WLANs and Broadcasting*. IEEE Press, 2000.

[77] S. Abeta, S. Sampei, and N. Morinaga, "Channel activation with adaptive coding rate and processing gain control for cellular DS/CDMA systems," in *Proceedings of IEEE VTC'96*, (Atlanta, GA, USA), pp. 1115–1119, IEEE, 28 April–1 May 1996.

[78] M. Hashimoto, S. Sampei, and N. Morinaga, "Forward and reverse link capacity enhancement of DS/CDMA cellular system using channel activation and soft power control techniques," in *Proceedings of the IEEE International Symposium on Personal, Indoor and Mobile Radio Communications (PIMRC)*, (Helsinki, Finland), pp. 246–250, 1–4 September 1997.

[79] S. Tateesh, S. Atungsiri, and A. Kondoz, "Link adaptive multi-rate coding verification system for CDMA mobile communications," in *Proceedings of the IEEE Global Telecommunications Conference (GLOBECOM)*, (London, UK), pp. 1969–1973, 18–22 November 1996.

[80] Y. Okumura and F. Adachi, "Variable-rate data transmission with blind rate detection for coherent DS-CDMA mobile radio," *IEICE Transactions on Communications*, vol. E81B, pp. 1365–1373, July 1998.

[81] J. Blogh, P. Cherriman, and L. Hanzo, "Adaptive beamforming assisted dynamic channel allocation," in *Proceeding of VTC'99 (Spring)*, (Houston, TX, USA), pp. 199–203, IEEE, 16–20 May 1999.

[82] K. Miya, O. Kato, K. Homma, T. Kitade, M. Hayashi, and T. Ue, "Wideband CDMA systems in TDD-mode operation for IMT-2000," *IEICE Transactions on Communications*, vol. E81-B, pp. 1317–1326, July 1998.

[83] O. Kato, K. Miya, K. Homma, T. Kitade, M. Hayashi, and M. Watanabe, "Experimental performance results of coherent wideband DS-CDMA with TDD scheme," *IEICE Transactions on Communications.*, vol. E81-B, pp. 1337–1344, July 1998.

[84] I. Jeong and M. Nakagawa, "A novel transmission diversity system in TDD-CDMA," *IEICE Transactions on Communications*, vol. E81-B, pp. 1409–1416, July 1998.

[85] S. Chen, B. Mulgrew, and P. M. Grant, "A clustering technique for digital communications channel equalization using radial basis function networks," *IEEE Transactions on Neural Networks*, vol. 4, pp. 570–579, July 1993.

[86] R. Kalman and R. Bucy, "New results in linear filtering and prediction theory," *Transactions ASME. Journal Basic Engineering*, vol. 83-D, pp. 85–108, 1961.

[87] J. G. Proakis, *Digital Communications*. Mc-Graw Hill International Editions, 3rd ed., 1995.

[88] T. Rappaport, *Wireless Communications Principles and Practice*. Englewood Cliffs, NJ, USA: Prentice-Hall, 1996.

[89] E. Lee and D. Messerschmitt, *Digital Communication*. Dordrecht: Kluwer Academic Publishers, 1988.

[90] I. S. Gradshteyn and I. M. Ryzhik, *Table of Integrals, Series and Products*. New York, USA: Academic Press, 1980.

[91] D. Tufts, "Nyquist's problem - the joint optimisation of the transmitter and receiver in pulse amplitude modulation," *Proceedings of the IEEE*, vol. 53, pp. 248–260, March 1965.

[92] J. Smith, "The joint optimization of transmitted signal and receiving filter for data transmission filters," *Bell Systems Technical Journal*, vol. 44, pp. 2363–2392, December 1965.

[93] E. Hänsler, "Some properties of transmission systems with minimum mean square error," *IEEE Transactions on Communications Technology (Corresp)*, vol. COM-19, pp. 576–579, August 1971.

[94] T. Ericson, "Structure of optimum receiving filters in data transmission systems," *IEEE Transactions on Information Theory (Corresp)*, vol. IT-17, pp. 352–353, May 1971.

[95] G. Forney Jr, "Maximum likelihood sequence estimation of digital sequences in the presence of intersymbol interference," *IEEE Transactions on Information Theory*, vol. IT-18, pp. 363–378, May 1972.

[96] M. Austin, "Decision feedback equalization for fading dispersive channels," Tech. Rep. 461, M.I.T Research Lab. Electron, August 1971.

[97] P. Monsen, "Feedback equalization for fading dispersive channels," *IEEE Transactions on Information Theory*, vol. IT-17, pp. 1144–1153, January 1971.

[98] J. Salz, "Optimum mean square decision feedback equalization," *Bell Systems Techncial Journal*, vol. 52, pp. 1341–1373, October 1973.

[99] D. Falconer and G. Foschini, "Theory of mmse qam system employing decision feedback equalization," *Bell Systems Technical Journal*, vol. 52, pp. 1821–1849, November 1973.

[100] R. Price, "Non-linearly feedback equalized pam versus capacity for noisy filter channels," in *Rec. Int. Conf. Communication*, pp. 12–17, 1972.

[101] R. Lucky, "A survey of the communication theory literature : 1968–1973," *IEEE Transactions on Information Theory*, vol. IT-19, pp. 725–739, July 1973.

[102] C. Belfiore and J. Park Jr, "Decision feedback equalization," *Proceedings of the IEEE*, vol. 67, pp. 1143–1156, August 1979.

[103] S. Qureshi, "Adaptive equalization," in *Advanced Digital Communications Systems and Signal Processing Techniques* (K.Feher, ed.), pp. 640–713, Englewood Cliffs NJ, USA: Prentice-Hall, 1987.

[104] J. Cheung, *Adaptive Equalisers for Wideband TDMA Mobile Radio*. PhD thesis, Department of Electronics and Computer Science, University of Southampton, UK, 1991.

[105] J. Cheung and R. Steele, "Soft-decision feedback equalizer for continuous-phase modulated signals in wide-band mobile radio channels," *IEEE Transactions on Communications*, vol. 42, pp. 1628–1638, February/March/April 1994.

[106] J. Wu, A. Aghvami, and J. Pearson, "A reduced state soft decision feedback viterbi equaliser for mobile radio communications," in *Proceedings of IEEE International Symposium on Personal, Indoor and Mobile Radio Communications*, (Stockholm, Sweden), pp. 234–242, June 1994.

[107] J. Wu and A. Aghvami, "A new adaptive equalizer with channel estimator for mobile radio communications," *IEEE Transactions on Vehicular Technology*, vol. 45, pp. 467–474, August 1996.

[108] Y. Gu and T. Le-Ngoc, "Adaptive combined DFE/MLSE techniques for ISI channels," *IEEE Transactions on Communications*, vol. 44, pp. 847–857, July 1996.

[109] D. Duttweiler, J. Mazo, and D. Messerschmitt, "An upper bound on the error probability on decision feedback equalization," *IEEE Transactions on Information Theory*, vol. IT-20, pp. 490–497, July 1974.

[110] J. Smee and N. Beaulieu, "Error-rate evaluating of linear equalization and decision feedback equalization with error rate performance," *IEEE Transactions On Communications*, vol. 46, pp. 656–665, May 1998.

[111] S. Altekar and N. Beaulieu, "Upper bounds to the error probability of decision feedback equalization," *IEEE Transactions on Communications*, vol. 39, pp. 145–157, January 1993.

[112] M. Tomlinson, "New automatic equalizer employing modulo arithmetic," *IEE Electronics Letters*, vol. 7, pp. 138–139, March 1971.

[113] H. Harashima and H. Miyakawa, "Matched transmission technique for channels with intersymbol interference," *IEEE Transactions on Communications*, vol. COM-20, pp. 774–780, August 1972.

[114] M. Russell and J. Bergmans, "A technique to reduce error propagation in M-ary decision feedback equalization," *IEEE Transactions on Communications*, vol. 43, pp. 2878–2881, December 1995.

[115] M. Chiani, "Introducing erasures in decision feedback equalization to reduce error propagation," *IEEE Transactions on Communications*, vol. 45, pp. 757–760, July 1997.

[116] J. Proakis and D. Manolakis, *Digital Signal Processing — Principles, Algorithms and Applications*. Macmillan, 1992.

[117] D. Mix, *Random Signal Processing*. Englewood Cliffs NJ, USA: Prentice-Hall, 1995.

[118] S. Haykin, *Adaptive Filter Theory*. Englewood Cliffs, NJ, USA: Prentice-Hall, 1996.

[119] B. Widrow and S. Stearns, *Adaptive Signal Processing*. Englewood Cliffs, NJ, USA: Prentice-Hall, 1985.

[120] B. Widrow, "On the statistical on the efficiency of the LMS algorithm with non stationary inputs," *IEEE Transactions on Information Theory*, vol. IT 30, pp. 211–221, March 1984.

[121] B. Widrow, J. McCool, M. Larrimore, and C. Johnson Jr, "Stationary and non-stationary learning characteristics of the LMS adaptive filter," *Proceedings of the IEEE*, vol. 64, pp. 1151–1162, August 1976.

[122] D. Falconer and L. Ljung, "Application of fast kalman estimation to adaptive equalization," *IEEE Transactions on Communications*, vol. COM-26, pp. 1439–1446, October 1978.

[123] D. Godard, "Channel equalization using a kalman filter for fast data transmission," *IBM Journal on Research and Development*, pp. 267–273, May 1974.

[124] A. Clark and R. Harun, "Assessment of kalman-filter channel estimators for an HF radio link," *Proceedings of the IEE*, vol. 133, pp. 513–521, October 1986.

[125] P. Shukla and L. Turner, "Channel estimation-based adaptive dfe for fading multipath radio channels," *IEE Proceedings - I*, vol. 138, pp. 525–543, October 1991.

[126] R. Lucky, "Automatic equalization for digital communication," *Bell Systems Technical Journal*, vol. 44, pp. 547–588, April 1965.

[127] R. Lucky, "Techniques for adaptive equalization of digital communication," *Bell Systems Technical Journal*, vol. 45, pp. 255–286, February 1966.

[128] R. Lawrence and H. Kaufman, "The kalman filter for equalization of a digital communications channel," *IEEE Transactions on Communications*, vol. COM-19, pp. 1137–1141, 1971.

[129] J. Mark, "A note on the modified kalman filter for channel equalization," *Proceedings of the IEEE*, vol. 61, pp. 481–482, 1973.

[130] S. Kleibano, V. Privalkii, and I. Time, "Kalman filter for equalization of digital communications channel," *Automation and Remote Control*, vol. 2, pp. 1097–1102, 1974.

[131] A. Luvinson and G. Pirani, "Design and performance of an adaptive receiver for synchronous data transmission," *IEEE Transactions on AES*, vol. AES-15x, pp. 635–648, 1979.

[132] M. Kumar, M. Yurtseven, and A. Rahrooh, "Design and performance analysis of an adaptive fir kalman equalizer," *Journal of the Franklin Institute*, vol. 330, no. 5, pp. 929–938, 1993.

[133] T. Lee and D. Cunningham, "Kalman filter equalization for QPSK communications," *IEEE Transactions on Communications*, vol. COM-24, pp. 361–364, 1976.

[134] E. D. Messe and G. Corsini, "Adaptive kalman filter equalizer," *IEE Electronicss Letters*, vol. 16, no. 8, pp. 547–549, 1980.

[135] F. Hsu, "Square root kalman filtering for high speed data received over fading dispersive hf channels," *IEEE Transactions on Information Theory*, vol. IT-28, pp. 753–763, 1982.

[136] G. Richards, "Implementation of kalman filters for process identification," *GEC Journal Research*, vol. 1, pp. 100–107, 1983.

[137] M. Mueller, "Least squares algorithms for adaptive equalizers," *Bell Systems Technical Journal*, vol. 60, pp. 1905–1925, October 1981.

[138] A. Sayed and T. Kailath, "A state-space approach to adaptive rls filtering," *IEEE Signal Processing Magazine*, vol. 11, pp. 571–577, July 1994.

[139] S. Bozic, *Digital and Kalman Filtering*. Edward Arnold, 1994.

[140] A. Milewski, "Periodic sequences with optimal properties for channel estimation and fast start-up equalization," *IBM Journal Research and Development*, vol. 54, pp. 425–528, September 1983.

[141] L. Scharf, *Statistical signal processing : Detection, estimation and time signal analysis*. Addison-Wesley Publishing Company, Inc., 1991.

[142] R. Brown and P. Hwang, *Introduction to Random Signals and Applied Kalman Filtering*. New York, USA: John Wiley and Sons, 1997.

[143] W. Press, S. Teukolsky, W. Vetterling, and B. Flannery, "Minimization or maximization of functions," in *Numerical Recipes in C*, ch. 10, pp. 394–455, Cambridge: Cambridge University Press, 1992.

[144] K. Narayanan and L. Cimini, "Equalizer adaptation algorithms for high speed wireless communications," in *Proceedings of IEEE VTC'96*, (Atlanta, GA, USA), pp. 681–685, IEEE, 28 April–1 May 1996.

[145] J. Torrance, *Adaptive Full Response Digital Modulation for Wireless Communications Systems*. PhD thesis, Department of Electronics and Computer Science, University of Southampton, UK, 1997.

[146] T. Ojanperä and R. Prasad, *Wideband CDMA for Third Generation Mobile Communications*. London, UK: Artech House, 1998.

[147] J. Hayes, "Adaptive feedback communications," *IEEE Communication Technology*, vol. 16, pp. 29–34, February 1968.

[148] S. Sampei, S. Komaki, and N. Morinaga, "Adaptive modulation/TDMA scheme for large capacity personal multi-media communication systems," *IEICE Transactions on Communications (Japan)*, vol. E77-B, pp. 1096–1103, September 1994.

[149] P. Monsen, "Theoretical and measured performance of a dfe modem on a fading multipath channel," *IEEE Transactions on Communications*, vol. COM-25, pp. 1144–1153, October 1977.

[150] "COST 207: Digital land mobile radio communications, final report." Office for Official Publications of the European Communities, 1989. Luxembourg.

[151] A. Klein, R. Pirhonen, J. Skoeld, and R. Suoranta, "FRAMES multiple access mode 1 — wideband TDMA with and without spreading," in *Proceedings of the IEEE International Symposium on Personal, Indoor and Mobile Radio Communications (PIMRC)*, vol. 1, (Helsinki, Finland), pp. 37–41, 1–4 September 1997.

[152] C. Berrou and A. Glavieux, "Near optimum error correcting coding and decoding: Turbo codes," *IEEE Transactions on Communications*, vol. 44, pp. 1261–1271, October 1996.

[153] C. Douillard, A. Picart, M. Jézéquel, P. Didier, C. Berrou, and A. Glavieux, "Iterative correction of inter-symbol interference: Turbo-equalization," *European Transactions on Communications*, vol. 6, pp. 507–511, 1995.

[154] A. Goldsmith and S. Chua, "Adaptive coded modulation for fading channels," *IEEE Transactions on Communications*, vol. 46, pp. 595–602, May 1998.

[155] C. Berrou, A. Glavieux, and P. Thitimajshima, "Near shannon limit error-correcting coding and decoding: Turbo codes," in *Proceedings of the International Conference on Communications*, (Geneva, Switzerland), pp. 1064–1070, May 1993.

[156] P. Robertson, "Illuminating the structure of code and decoder of parallel concatenated recursive systematic (turbo) codes," *IEEE Globecom*, pp. 1298–1303, 1994.

[157] B. Sklar, *Digital Communications—Fundamentals and Applications*. Englewood Cliffs, NJ, USA: Prentice-Hall, 1988.

[158] S. Lin and D. Constello Jr., *Error Control Coding: Fundamentals and Applications*. Englewood Cliffs, NJ, USA: Prentice-Hall, October 1982. ISBN: 013283796X.

[159] R. Blahut, *Theory and Practice of Error Control Codes*. Reading, MA, USA: Addison-Wesley, 1983. ISBN 0-201-10102-5.

[160] J. Hagenauer, E. Offer, and L. Papke, "Iterative decoding of binary block and convolutional codes," *IEEE Transactions on Information Theory*, vol. 42, pp. 429–445, March 1996.

[161] R. Pyndiah, "Iterative decoding of product codes: Block turbo codes," in *Proceedings of the International Symposium on Turbo Codes & Related Topics*, (Brest, France), pp. 71–79, 3–5 September 1997.

[162] L. Bahl, J. Cocke, F. Jelinek, and J. Raviv, "Optimal decoding of linear codes for minimising symbol error rate," *IEEE Transactions on Information Theory*, vol. 20, pp. 284–287, March 1974.

[163] P. Robertson, E. Villebrun, and P. Hoeher, "A comparison of optimal and sub-optimal MAP decoding algorithms operating in the log domain," in *Proceedings of the International Conference on Communications*, (Seattle, United States), pp. 1009–1013, June 1995.

[164] B. Yeap, *Turbo Equalisation Algorithms for Full and Partial Response Modulation*. PhD thesis, University of Southampton, UK, 1999.

[165] A. Barbulescu and S. Pietrobon, "Interleaver design for turbo codes," *IEE Electronics Letters*, pp. 2107–2108, December 1994.

[166] T. Liew, C. Wong, and L. Hanzo, "Block turbo coded burst-by-burst adaptive modems," in *Proceedings of Microcoll'99, Budapest, Hungary*, pp. 59–62, 21–24 March 1999.

[167] C. Wong, T. Liew, and L. Hanzo, "Turbo coded burst by burst adaptive wideband modulation with blind modem mode detection," in *Proceeding of ACTS Mobile Communication Summit '99*, (Sorrento, Italy), pp. 303–308, ACTS, 8–11 June 1999.

[168] T. Keller and L. Hanzo, "Adaptive orthogonal frequency division multiplexing schemes," in *Proceeding of ACTS Mobile Communication Summit '98*, (Rhodes, Greece), pp. 794–799, ACTS, 8–11 June 1998.

[169] M. Simon, J. Omura, R. Scholtz, and B. Levitt, *Spread Spectrum Communications Handbook*. New York, USA: McGraw-Hill, 1994.

[170] C. Wong, T. Liew, and L. Hanzo, "Burst-by-burst turbo coded wodeband adaptive modulation," *Submitted to the IEEE Journal on Selected Areas in Communications*.

[171] U. Wachsmann and J. Huber, "Power and bandwidth efficient digital communications using turbo codes in multilevel codes," *European Transactions on Telecommunications*, vol. 6, pp. 557–567, September–October 1995.

[172] A. Glavieux, C. Laot, and J. Labat, "Turbo equalization over a frequency selective channel," in *Proceedings of the International Symposium on Turbo Codes*, (Brest, France), pp. 96–102, 3-5 September 1997.

[173] C. Wong, B. Yeap, and L. Hanzo, "Wideband burst-by-burst adaptive modulation with turbo equalization and iterative channel estimation," in *Accepted for the Proceedings of the IEEE Vehicular Technology Conference 2000*, 2000.

[174] G. Ungerboeck, "Channel coding with multilevel/phase signals," *IEEE Transactions on Information Theory*, vol. IT-28, pp. 55–67, January 1982.

[175] D. Divsalar and M. K. Simon, "The design of trellis coded mpsk for fading channel: Performance criteria," *IEEE Transactions on Communications*, vol. 36, pp. 1004–1012, September 1988.

[176] D. Divsalar and M. K. Simon, "The design of trellis coded MPSK for fading channel: set partitioning for optimum code design," *IEEE Transactions on Communications*, vol. 36, pp. 1013–1021, September 1988.

[177] P. Robertson and T. Wörz, "Bandwidth efficient turbo trellis-coded modulation using punctured component codes," *IEEE Journal on Selected Area on Communications*, vol. 16, pp. 206–218, February 1998.

[178] L. Piazzo and L. Hanzo, "TTCM-OFDM over Dispersive Fading Channels," *IEEE Vehicular Technology Conference*, vol. 1, pp. 66–70, May 2000.

[179] G. Caire, G. Taricco, and E. Biglieri, "Capacity of bit-interleaved channels," *Electronics Letters*, pp. 1060–1061, 1996.

[180] J. K. Cavers and P. Ho, "Analysis of the Error Performance of Trellis-Coded Modulations in Rayleigh-Fading Channels," *IEEE Transactions on Communications*, vol. 40, pp. 74–83, January 1992.

[181] A. Naguib, N. Seshdri, and A. Calderbank, "Increasing data rate over wireless channels," *IEEE Signal Processing Magazine*, vol. 17, pp. 76–92, May 2000.

[182] S. M. Alamouti and S. Kallel, "Adaptive trellis-coded multiple-phased-shift keying Rayleigh fading channels," *IEEE Transactions on Communications*, vol. 42, pp. 2305–2341, June 1994.

[183] E. Zehavi, "8-PSK trellis codes for a Rayleigh fading channel," *IEEE Transactions on Communications*, vol. 40, pp. 873–883, May 1992.

[184] X. Li and J. Ritcey, "Bit-interleaved coded modulation with iterative decoding," *IEEE Communications Letters*, vol. 1, November 1997.

[185] X. Li and J. Ritcey, "Bit-interleaved coded modulation with iterative decoding — approaching turbo-tcm performance without code concatenation," in *Proceedings of CISS 1998*, (Princeton University, USA), March 1998.

[186] X. Li and J. Ritcey, "Trellis-coded modulation with bit interleaving and iterative decoding," *IEEE Journal on Selected Areas in Communications*, vol. 17, April 1999.

[187] X. Li and J. Ritcey, "Bit-interleaved coded modulation with iterative decoding using soft feedback," *IEE Electronics Letters*, vol. 34, pp. 942–943, May 1998.

[188] J. Hagenauer, "Rate-compatible puncture convolutional codes (RCPC) and their application," *IEEE Transactions on Communications*, vol. 36, pp. 389–400, April 1988.

[189] C. Wong, P. Cherriman, and L. Hanzo, "Burst-by-burst adaptive wireless video telephony over dispersive channels," in *Proceeding of Globecom '99*, vol. 1a, (Rio de Janeiro, Brazil), pp. 204–208, IEEE, 5–9 December 1999.

[190] P. Cherriman, C. Wong, and L. Hanzo, "Turbo- and BCH-coded wide-band burst-by-burst adaptive H.263-assisted wireless video telephony," *IEEE Transactions on Circuits and Systems for Video Technology*, vol. 10, pp. 1355–1363, December 2000.

[191] M. S. Alouini, X. Tand, and A. J. Goldsmith, "An adaptive modulation scheme for simultaneous voice and data transmission over fading channels," *IEEE Journal on Selected Areas in Communications*, vol. 17, pp. 837–850, May 1999.

[192] D. Yoon, K. Cho, and J. Lee, "Bit error probability of M-ary Quadrature Amplitude Modulation," in *Proc. IEEE VTC 2000-Fall*, vol. 5, pp. 2422–2427, IEEE, September 2000.

[193] E. L. Kuan, C. H. Wong, and L. Hanzo, "Burst-by-burst adaptive joint-detection CDMA," in *Proc. of IEEE VTC'99 Fall*, vol. 2, (Amsterdam, Netherland), pp. 1628–1632, September 1999.

[194] M. Nakagami, "The m-distribution - A general formula of intensity distribution of rapid fading," in *Statistical Methods in Radio Wave Propagation* (W. C. Hoffman, ed.), pp. 3–36, Pergamon Press, 1960.

[195] J. Kreyszig, *Advanced engineering mathematics*. Wiley, 7th edition ed., 1993.

[196] J. Lu, K. B. Letaief, C. I. J. Chuang, and M. L. Lio, "M-PSK and M-QAM BER computation using signal-space concepts," *IEEE Transactions on Communications*, vol. 47, no. 2, pp. 181–184, 1999.

[197] T. Keller and L. Hanzo, "Adaptive modulation technique for duplex OFDM transmission," *IEEE Transactions on Vehicular Technology*, vol. 49, pp. 1893–1906, September 2000.

[198] G. S. G. Beveridge and R. S. Schechter, *Optimization: Theory and Practice*. McGraw-Hill, 1970.

[199] W. Jakes Jr., ed., *Microwave Mobile Communications*. New York, USA: John Wiley & Sons, 1974.

[200] "COST 207 : Digital land mobile radio communications, final report," tech. rep., Luxembourg, 1989.

[201] R. Price and E. Green Jr., "A communication technique for multipath channels," *Proceedings of the IRE*, vol. 46, pp. 555–570, March 1958.

[202] M. K. Simon and M. S. Alouini, *Digital Communication over Fading Channels: A Unified Approach to Performance Analysis*. John Wiley & Sons, Inc., 2000. ISBN 0471317799.

[203] C. Y. Wong, R. S. Cheng, K. B. Letaief, and R. D. Murch, "Multiuser OFDM with adaptive subcarrier, bit, and power allocation," *IEEE Journal on Selected Areas in Communications*, vol. 17, pp. 1747–1758, October 1999.

[204] T. Keller and L. Hanzo, "Adaptive multicarrier modulation: A convenient framework for time-frequency processing in wireless communications," *Proceedings of the IEEE*, vol. 88, pp. 611–642, May 2000.

[205] J. Bingham, "Multicarrier modulation for data transmission: an idea whose time has come," *IEEE Communications Magazine*, pp. 5–14, May 1990.

[206] N. Yee, J.-P. Linnartz, and G. Fettweis, "Multicarrier CDMA in indoor wireless radio networks," in *PIMRC'93*, pp. 109–113, 1993.

[207] K. Fazel and L. Papke, "On the performance of convolutionally-coded CDMA/OFDM for mobile communication system," in *PIMRC'93*, pp. 468–472, 1993.

[208] A. Klein, G. Kaleh, and P. Baier, "Zero forcing and minimum mean square error equalization for multiuser detection in code division multiple access channels," *IEEE Transactions on Vehicular Technology*, vol. 45, pp. 276–287, May 1996.

[209] B. J. Choi, T. H. Liew, and L. Hanzo, "Concatenated space-time block coded and turbo coded symbol-by-symbol adaptive OFDM and multi-carrier CDMA systems," in *Proceedings of IEEE VTC 2001-Spring*, p. P.528, IEEE, May 2001.

[210] B. Vucetic, "An adaptive coding scheme for time-varying channels," *IEEE Transactions on Communications*, vol. 39, no. 5, pp. 653–663, 1991.

[211] H. Imai and S. Hirakawa, "A new multi-level coding method using error correcting codes," *IEEE Transactions on Information Theory*, vol. 23, pp. 371–377, May 1977.

[212] S. Chua and A. Goldsmith, "Adaptive coded modulation for fading channels," *IEEE Transactions on Communications*, vol. 46, pp. 595–602, May 1998.

[213] T. Keller, T. Liew, and L. Hanzo, "Adaptive rate RRNS coded OFDM transmission for mobile communication channels," in *Proceedings of VTC 2000 Spring*, (Tokyo, Japan), pp. 230–234, 15-18 May 2000.

[214] T. Keller, T. H. Liew, and L. Hanzo, "Adaptive redundant residue number system coded multicarrier modulation," *IEEE Journal on Selected Areas in Communications*, vol. 18, pp. 1292–2301, November 2000.

[215] T. Liew, C. Wong, and L. Hanzo, "Block turbo coded burst-by-burst adaptive modems," in *Proceedings of Microcoll'99*, (Budapest, Hungary), pp. 59–62, 21-24 March 1999.

[216] C. Wong, T. Liew, and L. Hanzo, "Turbo coded burst by burst adaptive wideband modulation with blind modem mode detection," in *ACTS Mobile Communications Summit*, (Sorrento, Italy), pp. 303–308, 8-11 June 1999.

[217] V. Tarokh, N. Seshadri, and A. R. Calderbank, "Space-Time Codes for High Data Rate Wireless Communication: Performance Criterion and Code Construction," *IEEE Transactions on Information Theory*, vol. 44, pp. 744–765, March 1998.

[218] V. Tarokh, H. Jafarkhani, and A. R. Calderbank, "Space-time block coding for wireless communications: Performance results," *IEEE Journal on Selected Areas in Communications*, vol. 17, pp. 451–460, March 1999.

[219] P. Jung and J. Blanz, "Joint detection with coherent receiver antenna diversity in CDMA mobile radio systems," *IEEE Transactions on Vehicular Technology*, vol. 44, pp. 76–88, February 1995.

[220] B.-J. Choi, E.-L. Kuan, and L. Hanzo, "Crest–factor study of MC-CDMA and OFDM," in *Proceeding of VTC'99 (Fall)*, vol. 1, (Amsterdam, Netherlands), pp. 233–237, IEEE, 19–22 September 1999.

[221] 3GPP, *3rd Generation Patnership Project (3GPP); Technical Specification Group Radio Access Network Physical Channels and Mapping of Transport Channels onto Physical Channels (TDD) (3G TS 25.221 version 3.0.0) (www.3gpp.org)*, October 1999.

[222] W. Lee, *Mobile Cellular Telecommunications: Analog and Digital Systems*. New York, USA: McGraw-Hill, 1995.

[223] W. Webb, "Spectrum efficiency of multilevel modulation schemes in mobile radio communications," *IEEE Transactions on Communications*, vol. 43, no. 8, pp. 2344–2349, 1995.

[224] C. Lee and R. Steele, "Signal-to-interference calculations for modern TDMA cellular communication systems," *IEE Proceedings on Communication*, vol. 142, pp. 21–30, February 1995.

[225] S. Verdú, *Multiuser Detection*. Cambridge, UK: Cambridge University Press, 1998.

[226] A. Klein and P. Baier, "Linear unbiased data estimation in mobile radio sytems applying CDMA," *IEEE Journal on Selected Areas in Communications*, vol. 11, pp. 1058–1066, September 1993.

[227] J. Blanz, A. Klein, M. Nasshan, and A. Steil, "Performance of a cellular hybrid C/TDMA mobile radio system applying joint detection and coherent receiver antenna diversity," *IEEE Journal on Selected Areas in Communications*, vol. 12, pp. 568–579, May 1994.

[228] P. Jung, J. Blanz, M. Nasshan, and P. Baier, "Simulation of the uplink of the JD-CDMA mobile radio systems with coherent receiver antenna diversity," *Wireless Personal Communications (Kluwer)*, vol. 1, no. 1, pp. 61–89, 1994.

[229] H. Yoshino, K. Fukawa, and H. Suzuki, "Interference cancelling equalizer (ice) for mobile radio communication," *IEEE Transactions on Vehicular Technology*, vol. 46, pp. 849–61, November 1997.

[230] M. Valenti and B. Woerner, "Combined multiuser reception and channel decoding for TDMA cellular systems," in *Proceedings of IEEE Vehicular Technology Conference*, (Ottawa, Canada), pp. 1915–1919, May 1998.

[231] J. Joung and G. Stuber, "Performance of truncated co-channel interference cancelling MLSE for TDMA systems," in *Proceedings of IEEE Vehicular Technology Conference*, (Ottawa, Canada), pp. 1710–1714, May 1998.

[232] E. Kuan, *Burst-by-burst Adaptive Multiuser Detection CDMATechniques*. PhD thesis, University of Southampton, UK, 1999.

[233] G. Golub and C. van Loan, *Matrix Computations*. North Oxford Academic, 1983.

[234] C. Wong, E. Kuan, and L. Hanzo, "Joint detection based wideband burst-by-burst adaptive modem with co-channel interference," in *Proceedings of the IEEE Vehicular Technology Conference*, (Amsterdam, The Netherlands), pp. 613–617, September 1999.

[235] S. Mak and A. Aghvami, "Detection of trellis-coded modulation on time-dispersive channels," in *Proceeding of IEEE Global Telecommunications Conference, Globecom 96*, (London, UK), pp. 1825–1829, IEEE, 18–22 November 1996.

[236] S. Siu and C. F. N. Cowan, "Performance analysis of the l_p norm back propagation algorithm for adaptive equalisation," *IEE Proceedings*, vol. 140, pp. 43–47, February 1993.

[237] G. J. Gibson, S. Siu, and C. F. N. Cowan, "The application of nonlinear structures to the reconstruction of binary signals," *IEEE Transactions on Signal Processing*, vol. 39, pp. 1877–1884, August 1991.

[238] G. J. Gibson, S. Siu, and C. F. N. Cowan, "Multi-layer perceptron structures applied to adaptive equalizers for data communications," in *ICASSP, IEEE International Conference on Acoustics, Speech and Signal Processing*, vol. 2, (IEEE Service Center, Piscataway, NJ, USA), pp. 1183–1186, IEEE, May 1989.

[239] G. J. Gibson, S. Siu, S. Chen, P. M. Grant, and C. F. N. Cowan, "The application of nonlinear architectures to adaptive channel equalisation," in *IEEE International Conference on Communications*, vol. 2, (IEEE Service Center, Piscataway, NJ, USA), pp. 649–653, IEEE, April 1990.

[240] S. Siu, G. J. Gibson, and C. F. N. Cowan, "Decision feedback equalisation using neural network structures and performance comparison with standard architecture," *IEE Proceedings*, vol. 137, pp. 221–225, August 1990.

[241] S. Chen, G. J. Gibson, and C. F. N. Cowan, "Adaptive channel equalisation using a polynomial-perceptron structure," *IEE Proceedings*, vol. 137, pp. 257–264, October 1990.

[242] C.-H. Chang, S. Siu, and C.-H. Wei, "A polynomial-perceptron based decision feedback equalizer with a robust learning algorithm," *EURASIP Signal Processing*, vol. 47, pp. 145–158, November 1995.

[243] Z.-J. Xiang and G.-G. Bi, "A new lattice polynomial perceptron and its applications to frequency-selective fading channel equalization and ACI suppression," *IEEE Transactions on Communications*, vol. 44, pp. 761–767, July 1996.

[244] Z.-J. Xiang, G.-G. Bi, and T. Le-Ngoc, "Polynomial perceptrons and their applications to fading channel equalization and co-channel interference suppression," *IEEE Transactions on Signal Processing*, vol. 42, pp. 2470–2480, September 1994.

[245] S. Chen, S. McLaughlin, and B. Mulgrew, "Complex-valued radial basis function network, Part II: Application to digital communications channel equalisation," *EURASIP Signal Processing*, vol. 36, pp. 175–188, March 1994.

[246] S. Chen, B. Mulgrew, and S. McLaughlin, "Adaptive Bayesian equalizer with decision feedback," *IEEE Transactions on Signal Processing*, vol. 41, pp. 2918–2927, September 1993.

[247] S. Chen, G. J. Gibson, C. F. N. Cowan, and P. M. Grant, "Reconstruction of binary signals using an adaptive radial basis function equaliser," *EURASIP Signal Processing*, vol. 22, pp. 77–93, January 1991.

[248] B. Mulgrew, "Applying radial basis functions," *IEEE Signal Processing Magazine*, vol. 13, pp. 50–65, March 1996.

[249] C. A. Micchelli, "Interpolation of scatter data: Distance matrices and conditionally positive definite functions," *Constructive Approximation*, vol. 2, pp. 11–22, 1986.

[250] M. J. D. Powell, *Algorithms for Approximation*, ch. Radial Basis Functions for Multivariable Interpolation: a review, pp. 143–167. Oxford: Clarendon Press, 1987.

[251] H. L. V. Trees, *Detection, Estimation and Modulation Theory, Part 1*. New York: John Wiley and Sons, 1968.

[252] E. R. Kandel, *Principles of Neural Science*, ch. Nerve cells and behavior, pp. 13–24. Elsevier, New York, 2nd. ed., 1985.

[253] S. Haykin, *Neural Networks : A Comprehensive Foundation*. Macmillan Publishing Company, 1994.

[254] C. M. Bishop, *Neural Networks for Pattern Recognition*. Oxford University Press, 1995.

[255] D. E. Rumelhart, G. E. Hinton, and R. J. Williams, *Parallel Distributed Processing: Explorations in the Microstructure of Cognition*, ch. Learning internal representations by error propagation, pp. 318–362. Cambridge, Mass. : MIT Press, 1986.

[256] D. E. Rumelhart, G. E. Hinton, and R. J. Williams, "Learning representations by back-propagating errors," *Nature (London)*, vol. 323, pp. 533–536, 1986.

[257] J. B. Gomm and D. L. Yu, "Selecting radial basis function network centers with recursive orthogonal least squares training," *IEEE Transactions on Neural Networks*, vol. 11, pp. 306–314, March 2000.

[258] R. P. Lippmann, "An introduction to computing with neural nets," *IEEE ASSP Magazine*, pp. 4–22, April 1987.

[259] J. C. Sueiro, A. A. Rodriguez, and A. R. F. Vidal, "Recurrent radial basis function networks for optimal symbol-by-symbol equalization," *EURASIP Signal Processing*, vol. 40, pp. 53–63, October 1994.

[260] W. S. Gan, J. J. Soraghan, and T. S. Durrani, "New functional-link based equaliser," *Electronics Letter*, vol. 28, pp. 1643–1645, August 1992.

[261] A. Hussain, J. J. Soraghan, and T. S. Durrani, "A new artificial neural network based adaptive non-linear equalizer for overcoming co-channel interferenc," in *IEEE Global Telecommunications Conference*, pp. 1422–1426, November 1996.

[262] A. Hussain, J. J. Soraghan, and T. S. Durrani, "A new adaptive functional-link neural-network-based DFE for overcoming co-channel interference," *IEEE Transaction on Communications*, vol. 45, pp. 1358–1362, November 1997.

[263] T. Kohonen, O. Simula, A. Visa, and J. Kangas, "Engineering applications of the self-organizing map," *Proceedings of the IEEE*, vol. 84, pp. 1358–1384, October 1996.

[264] S. Bouchired, M. Ibnkahla, D. Roviras, and F. Castanié, "Neural network equalization of satellite mobile communication channels," *ICASSP, IEEE International Conference on Acoustics, Speech and Signal Processing - Proceedings, 1998*, vol. 6, pp. 3377–3379, May 1998.

[265] S. Bouchired, M. Ibnkahla, D. Roviras, and F. Castanié, "Equalization of satellite UMTS channels using neural network devices," in *ICASSP, IEEE International Conference on Acoustics, Speech and Signal Processing - Proceedings, 1999*, vol. 5, (Phoenix, Arizona), pp. 2563–2566, 15-19 March 1999.

[266] S. Bouchired, D. Roviras, and F. Castanié, "Equalization of satellite mobile channels with neural network techniques," *Space Communications*, vol. 15, no. 4, pp. 209–220, 1998-1999.

[267] G. J. Gibson and C. F. N. Cowan, "On the decision regions of multilayer perceptrons," *Proceedings of the IEEE*, vol. 78, pp. 1590–1594, October 1990.

[268] N. E. Cotter, "The Stone-Weierstrass theorem and its application to neural networks," *IEEE Transactions on Neural Networks*, vol. 1, pp. 290–295, December 1990.

[269] T. M. Cover, "Geometrical and statistical properties of systems of linear inequalities with applications in pattern recognition," *IEEE Transactions on Electronic Computers*, vol. 14, pp. 326–334, 1965.

[270] W. A. Light, "Some aspects of radial basis function approximation," *Approximation Theory, Spline Functions and Applications, NATO ASI Series*, vol. 256, pp. 163–190, 1992.

[271] V. A. Morozov, *Regularization Methods for Ill-Posed Problems*. Boca Raton, FL, USA: CRC Press, 1993.

[272] A. N. Tikhonov and V. Y. Arsenin, *Solutions of Ill-posed Problems*. Washington, DC:W. H. Winston, 1977.

[273] T. Poggio and F. Girosi, "Networks for approximation and learning," *Proceedings of the IEEE*, vol. 78, pp. 1481–1497, 1990.

[274] S. Chen, C. F. N. Cowan, and P. M. Grant, "Orthogonal least squares learning algorithm for radial basis function networks," *IEEE Transactions on Neural Networks*, vol. 2, pp. 302–309, March 1991.

[275] S. Chen, E.-S. Chng, and K. Alkadhimi, "Regularised orthogonal least squares algorithm for constructing radial basis function networks," *International Journal of Control*, vol. 64, pp. 829–837, July 1996.

[276] J. MacQueen, "Some methods for classsification and analysis of multivariate observation," in *Proceedings of the 5th Berkeley Symposium on Mathematical Statistics and Probability* (L. M. LeCun and J. Neyman, eds.), vol. 1, (Berkeley), pp. 281–297, University of California Press, 1967.

[277] D. E. Rumelhart and D. Zipser, "Feature discovery by competitive learning," in *Parallel Distributed Processing: Explorations in the Microstructure of Cognition*, vol. 1, Cambridge, MA: Bradford Books, 1986.

[278] D. DeSieno, "Adding a conscience to competitive learning," in *Proceedings 2nd IEEE International Conference on Neural Networks*, vol. 1, pp. 117–124, July 1988.

[279] C. Chinrungrueng and C. H. Sëquin, "Optimal adaptive k-means algorithm with dynamic adjustment of learning rate," *IEEE Transactions on Neural Networks*, vol. 6, pp. 157–169, January 1995.

[280] J. Proakis, *Digital communications*. McGraw-Hill, 1995.

[281] J. C. Sueiro and A. R. F. Vidal, "Recurrent radial basis function networks for optimal blind equalisation," in *Proceedings IEEE Workshop on Neural Networks for Signal Processing*, pp. 562–571, IEEE, June 1993.

[282] S. Chen, S. McLaughlin, B. Mulgrew, and P. M. Grant, "Adaptive Bayesian decision feedback equalizer for dispersive mobile radio channels," *IEEE Transactions on Communications*, vol. 43, pp. 1937–1945, May 1995.

[283] S. N. Crozier, D. D. Falconer, and S. A. Mahmoud, "Least sum of squared errors (LSSE) channel estimation," *IEE Proceedings Part F: Radar and Signal Processing*, vol. 138, pp. 371–378, August 1991.

[284] A. Urie, M. Streeton, and C. Mourot, "An advance TDMA mobile access system for UMTS," in *Proceedings of IEEE International Symposium on Personal, Indoor and Mobile Radio Communications*, (The Hague, The Netherlands), pp. 685–690, 1994.

[285] J. L. Valenzuela and F. Casadevall, "Performance of adaptive Bayesian equalizers in outdoor environment," in *Proceedings of IEEE Vehicular Technology Conference*, vol. 3, pp. 2143–2147, IEEE, May 1997.

[286] E.-S. Chng, H. Yang, and W. Skarbek, "Reduced complexity implementation of Bayesian equaliser using local RBF network for channel equalisation problem," *Electronics Letters*, vol. 32, pp. 17–19, January 1996.

[287] S. K. Patra and B. Mulgrew, "Computational aspects of adaptive radial basis function equalizer design," in *IEEE International Symposium on Circuits and Systems, ISCAS'97*, vol. 1, pp. 521–524, IEEE, Piscataway, NJ, USA, June 1997.

[288] P. Robertson, E. Villebrun, and P. Höher, "A comparison of optimal and sub-optimal map decoding algorithms operating in the log domain," *IEEE International Conference on Communications*, pp. 1009–1013, 1995.

[289] J. Erfanian, S. Pasupathy, and G. Gulak, "Reduced complexity symbol dectectors with parallel structures for ISI channels," *IEEE Transactions on Communications*, vol. 42, pp. 1661–1671, 1994.

[290] S. Chen, B. Mulgrew, E.-S. Chng, and G. J. Gibson, "Space translation properties and the minimum-BER linear-combiner DFE," *IEE Proceedings on Communications*, vol. 145, pp. 316–322, October 1998.

[291] S. Chen, "Importance sampling simulation for evaluation the lower-bound BER of the bayesian DFE." submitted to IEEE Transactions on Communications.

[292] S. Chen, E.-S. Chng, B. Mulgrew, and G. J. Gibson, "Minimum-BER linear-combiner DFE," in *Proceedings of ICC'96*, (Dallas, Texas), pp. 1173–1177, 1996.

[293] S. Chen, S. Gunn, and C. J. Harris, "Decision feedback equalizer design using support vector machines," *IEE Proceedings Vision, Image and Signal Processing*, 2000. Accepted for publication.

[294] S. Chen and C. J. Harris, "Design of the optimal separating hyperplane for the decision feedback equalizer using support vector machine," in *Proceedings of IEEE International Conference on Accoustic, Speech and Signal Processing*, pp. 2701–2704, IEEE, June 5-9 2000.

[295] B. Sklar, "Rayleigh fading channels in mobile digital communication systems, part 1: Characterization," *IEEE Communications Magazine*, vol. 35, pp. 136–146, September 1997.

[296] C. H. Wong, *Wideband Adaptive Full Response Multilevel Transceivers and Equalizers*. PhD thesis, University of Southampton, United Kingdom, November 1999.

[297] W. Press, S. Teukolsky, W. Vetterling, and B. Flannery, *Numerical Recipes in C*. Cambridge: Cambridge University Press, 1992.

[298] R. Iltis, J. Shynk, and K. Giridhar, "Baysian algorithms for blind equalization using parallel adaptive filtering," *IEEE Transactions on Communications*, vol. 43, pp. 1017–1032, February–April 1994.

[299] J. Rapeli, "UMTS:targets, system concept, and standardization in a global framework," *IEEE Personal Communications*, vol. 2, pp. 20–28, February 1995.

[300] E. Berruto, M. Gudmundson, R. Menolascino, W. Mohr, and M. Pizarroso, "Research activities on UMTS radio interface, network architectures, and planning," *IEEE Communications Magazine*, vol. 36, pp. 82–95, February 1998.

[301] T. Ojanperä and R. Prasad, "An overview of air interface multiple access for IMT-2000/UMTS," *IEEE Communications Magazine*, vol. 36, pp. 82–95, September 1998.

[302] J. Hagenauer and P. Hoeher, "A Viterbi algorithm with soft-decision outputs and its applications," in *IEEE Globecom*, pp. 1680–1686, 1989.

[303] M. Breiling and L. Hanzo, "Non-iterative optimum super-trellis decoding of turbo codes," *Electronics Letters*, vol. 33, pp. 848–849, May 1997.

[304] M. Breiling and L. Hanzo, "Optimum non-iterative turbo-decoding," in *IEEE International Symposium on Personal, Indoor and Mobile Radio Communications, PIMRC, 1997*, vol. 2, pp. 714–718, IEEE, 1997.

[305] P. Robertson, P. Hoeher, and E. Villebrun, "Optimal and sub-optimal maximum a posteriori algorithms suitable for turbo decoding," *European Transactions on Telecommunications*, vol. 8, pp. 119–125, March/April 1997.

[306] J. Hagenauer, E. E. Offer, and L. Papke, "Iterative decoding of binary block and convolutional codes," *IEEE Transactions on Information Theory*, pp. 429–437, 1996.

[307] A. Klein, R. Pirhonen, J. Sköld, and R. Suoranta, "FRAMES Multiple Access Mode1 - Wideband TDMA with and without spreading," in *Proceedings of the IEEE International Symposium on Personal, Indoor and Mobile Radio Communications 1997*, (Helsinki, Finland), pp. 37–41, 1-4 September 1997.

[308] M. Gertsman and J. Lodge, "Symbol-by-symbol MAP demodulation of CPM and PSK signals on Rayleigh flat-fading channels," *IEEE Transactions on Communications*, vol. 45, pp. 788–799, July 1997.

[309] D. Raphaeli and Y. Zarai, "Combined turbo equalization and turbo decoding," *IEEE Communications Letters*, vol. 2, pp. 107–109, April 1998.

[310] A. Knickenberg, B. L. Yeap, J. Hamorsky, M. Breiling, and L. Hanzo, "Non-iterative joint channel equalisation and channel decoding," in *Proceedings of Globecom'99*, (Rio de Janeiro, Brazil), pp. 442–446, 5-9 December 1999.

[311] G. Bauch, H. Khorram, and J. Hagenauer, "Iterative equalization and decoding in mobile communications systems," in *European Personal Mobile Communications Conference*, (Bonn, Germany), pp. 301–312, 30 September - 2 October 1997.

[312] M. S. Yee and L. Hanzo, "Multi-level radial basis function network based equalisers for Rayleigh channels," in *Proceedings of IEEE Vehicular Technology Conference*, (Houston, USA), pp. 707–711, IEEE, 16-19 May 1999.

[313] K. Abend, T. J. Harley Jr, B. D. Fritchman, and C. Gumacos, "On optimum receivers for channels having memory," *IEEE Transactions Information Theory*, vol. IT-14, pp. 818–819, November 1968.

[314] M. S. Yee, T. H. Liew, and L. Hanzo, "Block turbo coded burst-by-burst adaptive radial basis function decision feedback equaliser assisted modems," in *Proceedings of IEEE Vehicular Technology Conference*, vol. 3, (Amsterdam, Netherlands), pp. 1600–1604, September 1999.

[315] B. L. Yeap, C. H. Wong, and L. Hanzo, "Reduced complexity in-phase/quadrature-phase turbo equalisation," in *IEEE International Communications Conference*, (Helsinki, Finland), pp. 393–397, 11 - 14 June 2001.

[316] B. L. Yeap, C. H. Wong, and L. Hanzo, "Reduced complexity in-phase/quadrature-phase turbo equalisation." submitted to IEEE Journal on Selected Areas in Communication, 2000.

[317] L. Hanzo, W. Webb, and T. Keller, *Single- and Multi-carrier Quadrature Amplitude Modulation*. New York: John Wiley-IEEE Press, April 2000.

[318] A. Glavieux, C. Laot, and J. Labat, "Turbo Equalization over a frequency selective channel," in *Proceedings of the International Symposium on Turbo Codes & Related Topics*, (Brest, France), pp. 96–102, 3-5 September 1997.

[319] G. Bauch, A. Naguib, and N. Seshadri, "MAP equalization of space-time coded signals over frequency selective channels," in *Proceedings of Wireless Communications and Networking Conference*, (New Orleans, USA), pp. 261–265, September 1999.

[320] B. L. Yeap, T. H. Liew, and L. Hanzo, "Turbo equalization of serially concatenated systematic convolutional codes and systematic space time trellis codes," in *Proceedings of the IEEE Vehicular Technology Conference*, (Rhodes, Greece), pp. 1464–1468, 6 - 9 May 2001.

[321] M. S. Yee, B. L. Yeap, and L. Hanzo, "Radial Basis Function Assisted Turbo Equalisation," in *Proceedings of the IEEE Vehicular Technology Conference*, (Tokyo, Japan), pp. 640–644, 15-18 May 2000.

[322] "Feature topic: Software radios," *IEEE Communications Magazine*, vol. 33, pp. 24–68, May 1995.

[323] A. Viterbi, *CDMA: Principles of Spread Spectrum Communication*. Reading MA, USA: Addison-Wesley, June 1995. ISBN 0201633744.

[324] L. Miller and J. Lee, *CDMA Systems Engineering Handbook*. London, UK: Artech House, 1998.

[325] A. Whalen, *Detection of signals in noise*. New York, USA: Academic Press, 1971.

[326] S. Verdú, "Minimum probability of error for asynchronous Gaussian multiple-access channel," *IEEE Transactions on Communications*, vol. 32, pp. 85–96, January 1986.

[327] R. Prasad, *Universal Wireless Personal Communications*. London, UK: Artech House Publishers, 1998.

[328] S. Glisic and B. Vucetic, *Spread Spectrum CDMA Systems for Wireless Communications*. London, UK: Artech House, April 1997. ISBN 0890068585.

[329] J. Thompson, P. Grant, and B. Mulgrew, "Smart antenna arrays for CDMA systems," *IEEE Personal Communications Magazine*, vol. 3, pp. 16–25, October 1996.

[330] J. Thompson, P. Grant, and B. Mulgrew, "Performance of antenna array receiver algorithms for CDMA," in *Proceedings of the IEEE Global Telecommunications Conference (GLOBECOM)*, (London, UK), pp. 570–574, 18–22 November 1996.

[331] A. Naguib and A. Paulraj, "Performance of wireless CDMA with m-ary orthogonal modulation and cell site antenna arrays," *IEEE Journal on Selected Areas in Communications*, vol. 14, pp. 1770–1783, December 1996.

[332] L. Godara, "Applications of antenna arrays to mobile communications, part I: Performance improvement, feasibility, and system considerations," *Proceedings of the IEEE*, vol. 85, pp. 1029–1060, July 1997.

[333] R. Kohno, H. Imai, M. Hatori, and S. Pasupathy, "Combination of adaptive array antenna and a canceller of interference for direct-sequence spread-spectrum multiple-access system," *IEEE Journal on Selected Areas in Communications*, vol. 8, pp. 675–681, May 1998.

[334] S. Moshavi, "Multi-user detection for DS-CDMA communications," *IEEE Communications Magazine*, vol. 34, pp. 124–136, October 1996.

[335] R. Lupas and S. Verdú, "Linear multiuser detectors for synchronous code divison multiple access channels," *IEEE Transactions on Information Theory*, vol. 35, pp. 123–136, January 1989.

[336] R. Lupas and S. Verdú, "Near-far resistance of multiuser detectors in asynchronous channels," *IEEE Transactions on Communications*, vol. 38, pp. 509–519, April 1990.

[337] Z. Zvonar and D. Brady, "Suboptimal multiuser detector for frequency selective Rayleigh fading synchronous CDMA channels," *IEEE Transactions on Communications*, vol. 43, pp. 154–157, February–April. 1995.

[338] Z. Zvonar and D. Brady, "Differentially coherent multiuser detection in asynchronous CDMA flat Rayleigh fading channels," *IEEE Transactions on Communications*, vol. 43, pp. 1252–1255, February–April 1995.

[339] Z. Zvonar, "Combined multiuser detection and diversity reception for wireless CDMA systems," *IEEE Transactions on Vehicular Technology*, vol. 45, pp. 205–211, February 1996.

[340] T. Kawahara and T. Matsumoto, "Joint decorrelating multiuser detection and channel estimation in asynchronous cdma mobile communications channels," *IEEE Transactions on Vehicular Technology*, vol. 44, pp. 506–515, August 1995.

[341] M. Hosseinian, M. Fattouche, and A. Sesay, "A multiuser detection scheme with pilot symbol-aided channel estimation for synchronous CDMA systems," in *Proceedings of the IEEE Vehicular Technology Conference (VTC)*, (Ottawa, Canada), pp. 796–800, 18–21 May 1998.

[342] M. Juntti, B. Aazhang, and J. Lilleberg, "Iterative implementation of linear multiuser detection for dynamic asynchronous CDMA systems," *IEEE Transactions on Communications*, vol. 46, pp. 503–508, April 1998.

[343] P.-A. Sung and K.-C. Chen, "A linear minimum mean square error multiuser receiver in Rayleigh fading channels," *IEEE Journal on Selected Areas in Communications*, vol. 14, pp. 1583–1594, October 1996.

[344] A. Duel-Hallen, "Decorrelating decision-feedback multiuser detector for synchronous code-division multiple-access channel," *IEEE Transactions on Communications*, vol. 41, pp. 285–290, February 1993.

[345] L. Wei and C. Schlegel, "Synchronous DS-SSMA system with improved decorrelating decision-feedback multiuser detection," *IEEE Transactions on Vehicular Technology*, vol. 43, pp. 767–772, August 1994.

[346] A. Hafeez and W. Stark, "Combined decision-feedback multiuser detection/soft-decision decoding for CDMA channels," in *Proceedings of the IEEE Vehicular Technology Conference (VTC)*, (Atlanta, GA, USA), pp. 382–386, 28 April–1 May 1996.

[347] A. Steil and J. Blanz, "Spectral efficiency of JD-CDMA mobile radio systems applying coherent receiver antenna diversity," in *Proceedings of the International Symposium on Spread Spectrum Techniques and Applications (ISSSTA)*, (Mainz, Germany), pp. 313–319, 22–25 September 1996.

[348] P. Jung, M. Nasshan, and J. Blanz, "Application of turbo codes to a CDMA mobile radio system using joint detection and antenna diversity," in *Proceedings of the IEEE Vehicular Technology Conference (VTC)*, (Stockholm, Sweden), pp. 770–774, 8–10 June 1994.

[349] P. Jung and M. Nasshan, "Results on turbo-codes for speech transmission in a joint detection CDMA mobile radio system with coherent receiver antenna diversity," *IEEE Transactions on Vehicular Technology*, vol. 46, pp. 862–870, November 1997.

[350] M. Nasshan, A. Steil, A. Klein, and P. Jung, "Downlink cellular radio capacity of a joint detection CDMA mobile radio system," in *Proceedings of the 45th IEEE Vehicular Technology Conference (VTC)*, (Chicago, USA), pp. 474–478, 25–28 July 1995.

[351] A. Klein, "Data detection algorithms specially designed for the downlink of CDMA mobile radio systems," in *Proceedings of the IEEE Vehicular Technology Conference (VTC)*, (Phoenix, USA), pp. 203–207, 4–7 May 1997.

[352] B. Steiner and P. Jung, "Optimum and suboptimum channel estimation for the uplink of CDMA mobile radio systems with joint detection," *European Transactions on Telecommunications*, vol. 5, pp. 39–50, 1994.

[353] M. Werner, "Multistage joint detection with decision feedback for CDMA mobile radio applications," in *Proceedings of the IEEE International Symposium on Personal, Indoor and Mobile Radio Communications (PIMRC)*, pp. 178–183, 1994.

[354] M. Varanasi and B. Aazhang, "Multistage detection in asynchronous code-division multiple-access communications," *IEEE Transactions on Communications*, vol. 38, pp. 509–519, April 1990.

[355] M. Varanasi, "Group detection for synchronous Gaussian code-division multiple-access channels," *IEEE Transactions on Information Theory*, vol. 41, pp. 1083–1096, July 1995.

[356] M. Varanasi, "Parallel group detection for synchronous CDMA communication over frequency-selective Rayleigh fading channels," *IEEE Transactions on Information Theory*, vol. 42, pp. 116–128, January 1996.

[357] Y. Yoon, R. Kohno, and H. Imai, "A SSMA system with cochannel interference cancellation with multipath fading channels," *IEEE Journal on Selected Areas in Communications*, vol. 11, pp. 1067–1075, September 1993.

[358] T. Giallorenzi and S. Wilson, "Suboptimum multiuser receivers for convolutionally coded asynchronous DS-CDMA systems," *IEEE Transactions on Communications*, vol. 44, pp. 1183–1196, September 1996.

[359] Y. Sanada and M. Nakagawa, "A multiuser interference cancellation technique utilizing convolutional codes and multicarrier modulation for wireless indoor communications," *IEEE Journal on Selected Areas in Communications*, vol. 14, pp. 1500–1509, October 1996.

[360] M. Latva-aho, M. Juntti, and M. Heikkilä, "Parallel interference cancellation receiver for DS-CDMA systems in fading channels," in *Proceedings of the IEEE International Symposium on Personal, Indoor and Mobile Radio Communications (PIMRC)*, (Helsinki, Finland), pp. 559–564, 1–4 September 1997.

[361] D. Dahlhaus, A. Jarosch, B. Fleury, and R. Heddergott, "Joint demodulation in DS/CDMA systems exploiting the space and time diversity of the mobile radio channel," in *Proceedings of the IEEE International Symposium on Personal, Indoor and Mobile Radio Communications (PIMRC)*, (Helsinki, Finland), pp. 47–52, 1–4 September 1997.

[362] D. Divsalar, M. Simon, and D. Raphaeli, "Improved parallel interference cancellation for CDMA," *IEEE Transactions on Communications*, vol. 46, pp. 258–267, February 1998.

[363] P. Patel and J. Holtzman, "Analysis of a simple successive interference cancellation scheme in a DS/CDMA system," *IEEE Journal on Selected Areas in Communications*, vol. 12, pp. 796–807, June 1994.

[364] A. Soong and W. Krzymien, "A novel CDMA multi-user interference cancellation receiver with reference symbol aided estimation of channel parameters," *IEEE Journal on Selected Areas in Communications*, vol. 14, pp. 1536–1547, October 1996.

[365] A. Hui and K. Letaief, "Successive interference cancellation for multiuser asynchronous DS/CDMA detectors in multipath fading links," *IEEE Transactions on Communications*, vol. 46, pp. 384–391, March 1998.

[366] Y. Li and R. Steele, "Serial interference cancellation method for CDMA," *Electronics Letters*, vol. 30, pp. 1581–1583, September 1994.

[367] T. Oon, R. Steele, and Y. Li, "Cancellation frame size for a quasi-single-bit detector in asynchronous CDMA channel," *Electronics Letters*, vol. 33, pp. 258–259, February 1997.

[368] T.-B. Oon, R. Steele, and Y. Li, "Performance of an adaptive successive serial-parallel CDMA cancellation scheme in flat Rayleigh fading channels," in *Proceedings of the IEEE Vehicular Technology Conference (VTC)*, (Phoenix, USA), pp. 193–197, 4–7 May 1997.

[369] M. Sawahashi, Y. Miki, H. Andoh, and K. Higuchi, "Pilot symbol-assisted coherent multistage interference canceller using recursive channel estimation for DS-CDMA mobile radio," *IEICE Transactions on Communications*, vol. E79-B, pp. 1262–1269, September 1996.

[370] S. Sun, L. Rasmussen, H. Sugimoto, and T. Lim, "A hybrid interference canceller in CDMA," in *Proceedings of the IEEE International Symposium on Spread Spectrum Techniques and Applications (ISSSTA)*, (Sun City, South Africa), pp. 150–154, 2–4 September 1998.

[371] Y. Cho and J. Lee, "Analysis of an adaptive SIC for near-far resistant DS-CDMA," *IEEE Transactions on Communications*, vol. 46, pp. 1429–1432, November 1998.

[372] P. Agashe and B. Woerner, "Interference cancellation for a multicellular CDMA environment," *Wireless Personal Communications (Kluwer)*, vol. 3, no. 1–2, pp. 1–14, 1996.

[373] L. Rasmussen, T. Lim, and T. Aulin, "Breadth-first maximum likelihood detection in multiuser CDMA," *IEEE Transactions on Communications*, vol. 45, pp. 1176–1178, October 1997.

[374] L. Wei, L. Rasmussen, and R. Wyrwas, "Near optimum tree-search detection schemes for bit-synchronous multiuser CDMA systems over Gaussian and two-path Rayleigh fading channels," *IEEE Transactions on Communications*, vol. 45, pp. 691–700, June 1997.

[375] M. Nasiri-Kenari, R. Sylvester, and C. Rushforth, "Efficient soft-in-soft-out multiuser detector for synchronous CDMA with error-control coding," *IEEE Transactions on Vehicular Technology*, vol. 47, pp. 947–953, August 1998.

[376] J. Anderson and S. Mohan, "Sequential coding algorithms: a survey and cost analysis," *IEEE Transactions on Communications*, vol. 32, pp. 169–176, February 1984.

[377] C. Schlegel, S. Roy, P. Alexander, and Z.-J. Xiang, "Multiuser projection receivers," *IEEE Journal on Selected Areas in Communications*, vol. 14, pp. 1610–1618, October 1996.

[378] P. Alexander, L. Rasmussen, and C. Schlegel, "A linear receiver for coded multiuser CDMA," *IEEE Transactions on Communications*, vol. 45, pp. 605–610, May 1997.

[379] P. Rapajic and B. Vucetic, "Adaptive receiver structures for asynchornous CDMA systems," *IEEE Journal on Selected Areas in Communications*, vol. 12, pp. 685–697, May 1994.

[380] G. Woodward and B. Vucetic, "Adaptive detection for DS-CDMA," *Proceedings of the IEEE*, vol. 86, pp. 1413–1434, July 1998.

[381] Z. Xie, R. Short, and C. Rushforth, "Family of suboptimum detectors for coherent multiuser communications," *IEEE Journal on Selected Areas in Communications*, vol. 8, pp. 683–690, May 1990.

[382] T. Lim, L. Rasmussen, and H. Sugimoto, "An asynchronous multiuser CDMA detector based on the kalman filter," *IEEE Journal on Selected Areas in Communications*, vol. 16, pp. 1711–1722, December 1998.

[383] P. Seite and J. Tardivel, "Adaptive equalizers for joint detection in an indoor CDMA channel," in *Proceedings of the IEEE Vehicular Technology Conference (VTC)*, (Chicago, USA), pp. 484–488, 25–28 July 1995.

[384] G. Povey, P. Grant, and R. Pringle, "A decision-directed spread-spectrum RAKE receiver for fast-fading mobile channels," *IEEE Transactions on Vehicular Technology*, vol. 45, pp. 491–502, August 1996.

[385] H. Liu and K. Li, "A decorrelating RAKE receiver for CDMA communications over frequency-selective fading channels," *IEEE Transactions on Communications*, vol. 47, pp. 1036–1045, July 1999.

[386] Z. Xie, C. Rushforth, R. Short, and T. Moon, "Joint signal detection and parameter estimation in multiuser communications," *IEEE Transactions on Communications*, vol. 41, pp. 1208–1216, August 1993.

[387] N. Seshadri, "Joint data and channel estimation using blind trellis search techniques," *IEEE Transactions on Communications*, vol. 42, pp. 1000–1011, February/March/April 1994.

[388] A. Polydoros, R. Raheli, and C. Tzou, "Per–survivor processing: a general approach to MLSE in uncertain environments," *IEEE Transactions on Communications*, vol. COM–43, pp. 354–364, February–April 1995.

[389] R. Raheli, G. Marino, and P. Castoldi, "Per-survivor processing and tentative decisions: What is in between?," *IEEE Transactions on Communications*, vol. 44, pp. 127–129, February 1998.

[390] T. Moon, Z. Xie, C. Rushforth, and R. Short, "Parameter estimation in a multi-user communication system," *IEEE Transactions on Communications*, vol. 42, pp. 2553–2560, August 1994.

[391] R. Iltis and L. Mailaender, "Adaptive multiuser detector with joint amplitude and delay estimation," *IEEE Journal on Selected Areas in Communications*, vol. 12, pp. 774–785, June 1994.

[392] U. Mitra and H. Poor, "Adaptive receiver algorithms for near-far resistant CDMA," *IEEE Transactions on Communications*, vol. 43, pp. 1713–1724, February–April 1995.

[393] U. Mitra and H. Poor, "Analysis of an adaptive decorrelating detector for synchronous CDMA," *IEEE Transactions on Communications*, vol. 44, pp. 257–268, February 1996.

[394] X. Wang and H. Poor, "Blind equalization and multiuser detection in dispersive CDMA channels," *IEEE Transactions on Communications*, vol. 46, pp. 91–103, January 1998.

[395] X. Wang and H. Poor, "Blind multiuser detection: a subspace approach," *IEEE Transactions on Information Theory*, vol. 44, pp. 677–690, March 1998.

[396] M. Honig, U. Madhow, and S. Verdú, "Blind adaptive multiuser detection," *IEEE Transactions on Information Theory*, vol. 41, pp. 944–960, July 1995.

[397] N. Mandayam and B. Aazhang, "Gradient estimation for sensitivity analysis and adaptive multiuser interference rejection in code division multiple access systems," *IEEE Transactions on Communications*, vol. 45, pp. 848–858, July 1997.

[398] S. Ulukus and R. Yates, "A blind adaptive decorrelating detector for CDMA systems," *IEEE Journal on Selected Areas in Communications*, vol. 16, no. 8, pp. 1530–1541, 1998.

[399] T. Lim and L. Rasmussen, "Adaptive symbol and parameter estimation in asynchronous multiuser CDMA detectors," *IEEE Transactions on Communications*, vol. 45, pp. 213–220, February 1997.

[400] J. Míguez and L. Castedo, "A linearly constrained constant modulus approach to blind adaptive multiuser interference suppression," *IEEE Communications Letters*, vol. 2, pp. 217–219, August 1998.

[401] D. Godard, "Self–recovering equalization and carrier tracking in two–dimensional data communication systems," *IEEE Transactions on Communications*, vol. COM–28, pp. 1867–1875, November 1980.

[402] K. Wesolowsky, "Analysis and properties of the modified constant modulus algorithm for blind equalization," *European Transactions on Telecommunication*, vol. 3, pp. 225–230, May–June 1992.

[403] K. Fukawa and H. Suzuki, "Orthogonalizing matched filtering (OMF) detector for DS-CDMA mobile communication systems," *IEEE Transactions on Vehicular Technology*, vol. 48, pp. 188–197, January 1999.

[404] U. Fawer and B. Aazhang, "Multiuser receiver for code division multiple access communications over multipath channels," *IEEE Transactions on Communications*, vol. 43, pp. 1556–1565, February–April 1995.

[405] Y. Bar-Ness, "Asynchronous multiuser CDMA detector made simpler: Novel decorrelator, combiner, canceller, combiner (DC^3) structure," *IEEE Transactions on Communications*, vol. 47, pp. 115–122, January 1999.

[406] K. Yen and L. Hanzo, "Hybrid genetic algorithm based multi-user detection schemes for synchronous CDMA systems," in *submitted to the IEEE Vehicular Technology Conference (VTC)*, (Tokyo, Japan), 2000.

[407] W. Jang, B. Vojčić, and R. Pickholtz, "Joint transmitter-receiver optimization in synchronous multiuser communications over multipath channels," *IEEE Transactions on Communications*, vol. 46, pp. 269–278, February 1998.

[408] B. Vojčić and W. Jang, "Transmitter precoding in synchronous multiuser communications," *IEEE Transactions on Communications*, vol. 46, pp. 1346–1355, October 1998.

[409] R. Tanner and D. Cruickshank, "Receivers for nonlinearly separable scenarios in DS-CDMA," *Electronics Letters*, vol. 33, pp. 2103–2105, December 1997.

[410] R. Tanner and D. Cruickshank, "RBF based receivers for DS-CDMA with reduced complexity," in *Proceedings of the IEEE International Symposium on Spread Spectrum Techniques and Applications (ISSSTA)*, (Sun City, South Africa), pp. 647–651, 2–4 September 1998.

[411] C. Berrou, P. Adde, E. Angui, and S. Faudeil, "A low complexity soft-output viterbi decoder architecture," in *Proceedings of the International Conference on Communications*, pp. 737–740, May 1993.

[412] T. Giallorenzi and S. Wilson, "Multiuser ML sequence estimator for convolutionally coded asynchronous DS-CDMA systems," *IEEE Transactions on Communications*, vol. 44, pp. 997–1008, August 1996.

[413] M. Moher, "An iterative multiuser decoder for near-capacity communications," *IEEE Transactions on Communications*, vol. 46, pp. 870–880, July 1998.

[414] M. Moher and P. Guinaud, "An iterative algorithm for asynchronous coded multiuser detection," *IEEE Communications Letters*, vol. 2, pp. 229–231, August 1998.

[415] P. Alexander, A. Grant, and M. Reed, "Iterative detection in code-division multiple-access with error control coding," *European Transactions on Telecommunications*, vol. 9, pp. 419–426, September–October 1998.

[416] P. Alexander, M. Reed, J. Asenstorfer, and C. Schlegel, "Iterative multiuser interference reduction: Turbo CDMA," *IEEE Transactions on Communications*, vol. 47, pp. 1008–1014, July 1999.

[417] M. Reed, C. Schlegel, P. Alexander, and J. Asenstorfer, "Iterative multiuser detection for CDMA with FEC: Near-single-user performance," *IEEE Transactions on Communications*, vol. 46, pp. 1693–1699, December 1998.

[418] X. Wang and H. Poor, "Iterative (turbo) soft interference cancellation and decoding for coded CDMA," *IEEE Transactions on Communications*, vol. 47, pp. 1046–1061, July 1999.

[419] K. Gilhousen, I. Jacobs, R. Padovani, A. Viterbi, L. Weaver Jr., and C. Wheatley III, "On the capacity of a cellular CDMA system," *IEEE Transactions on Vehicular Technology*, vol. 40, pp. 303–312, May 1991.

[420] T. Keller and L. Hanzo, "Blind-detection assisted sub-band adaptive turbo-coded OFDM schemes," in *Proceeding of VTC'99 (Spring)*, (Houston, TX, USA), pp. 489–493, IEEE, 16–20 May 1999.

[421] M. Yee and L. Hanzo, "Multi-level Radial Basis Function network based equalisers for Rayleigh channel," in *Proceeding of VTC'99 (Spring)*, (Houston, TX, USA), pp. 707–711, IEEE, 16–20 May 1999.

[422] T. Ojanperä, A. Klein, and P.-O. Anderson, "FRAMES multiple access for UMTS," *IEE Colloquium (Digest)*, pp. 7/1–7/8, May 1997.

[423] M. Failli, "Digital land mobile radio communications COST 207," tech. rep., European Commission, 1989.

[424] F. Adachi, M. Sawahashi, and K. Okawa, "Tree-structured generation of orthogonal spreading codes with different lengths for forward link of DS-CDMA mobile radio," *Electronics Letters*, vol. 33, pp. 27–28, January 1997.

[425] A. Toskala, J. Castro, E. Dahlman, M. Latva-aho, and T. Ojanperä, "FRAMES FMA2 wideband-CDMA for UMTS," *European Transactions on Telecommunications*, vol. 9, pp. 325–336, July–August 1998.

[426] P. Bello, "Selective fading limitations of the KATHRYN modem and some system design considerations," *IEEE Trabsactions on Communications Technology*, vol. COM–13, pp. 320–333, September 1965.

[427] M. Zimmermann and A. Kirsch, "The AN/GSC-10/KATHRYN/variable rate data modem for HF radio," *IEEE Transactions on Communication Technology*, vol. CCM–15, pp. 197–205, April 1967.

[428] E. Powers and M. Zimmermann, "A digital implementation of a multichannel data modem," in *Proceedings of the IEEE International Conference on Communications*, (Philadelphia, USA), 1968.

[429] R. W. Chang, "Synthesis of band-limited orthogonal signals for multichannel data transmission," *Bell Systems Technical Journal*, vol. 46, pp. 1775–1796, December 1966.

[430] R. Chang and R. Gibby, "A theoretical study of performance of an orthogonal multiplexing data transmission scheme," *IEEE Transactions on Communication Technology*, vol. COM–16, pp. 529–540, August 1968.

[431] B. R. Saltzberg, "Performance of an efficient parallel data transmission system," *IEEE Transactions on Communication Technology*, pp. 805–813, December 1967.

[432] S. B. Weinstein and P. M. Ebert, "Data transmission by frequency division multiplexing using the discrete fourier transform," *IEEE Transactions on Communication Technology*, vol. COM–19, pp. 628–634, October 1971.

[433] B. Hirosaki, "An analysis of automatic equalizers for orthogonally multiplexed QAM systems," *IEEE Transactions on Communications*, vol. COM-28, pp. 73–83, January 1980.

[434] A. Peled and A. Ruiz, "Frequency domain data transmission using reduced computational complexity algorithms," in *Proceedings of International Conference on Acoustics, Speech, and Signal Processing, ICASSP'80*, vol. 3, (Denver, CO, USA), pp. 964–967, IEEE, 9–11 April 1980.

[435] B. Hirosaki, "An orthogonally multiplexed QAM system using the discrete fourier transform," *IEEE Transactions on Communications*, vol. COM-29, pp. 983–989, July 1981.

[436] H. Kolb, "Untersuchungen über ein digitales mehrfrequenzverfahren zur datenübertragung," in *Ausgewählte Arbeiten über Nachrichtensysteme*, no. 50, Universität Erlangen-Nürnberg, 1982.

[437] H. Schüssler, "Ein digitales Mehrfrequenzverfahren zur Datenübertragung," in *Professoren-Konferenz, Stand und Entwicklungsaussichten der Daten und Telekommunikation*, (Darmstadt, Germany), pp. 179–196, 1983.

[438] K. Preuss, "Ein Parallelverfahren zur schnellen Datenübertragung Im Ortsnetz," in *Ausgewählte Arbeiten über Nachrichtensysteme*, no. 56, Universität Erlangen-Nürnberg, 1984.

[439] R. Rückriem, "Realisierung und messtechnische Untersuchung an einem digitalen Parallelverfahren zur Datenübertragung im Fernsprechkanal," in *Ausgewählte Arbeiten über Nachrichtensysteme*, no. 59, Universität Erlangen-Nürnberg, 1985.

[440] L. Cimini, "Analysis and simulation of a digital mobile channel using orthogonal frequency division multiplexing," *IEEE Transactions on Communications*, vol. 33, pp. 665–675, July 1985.

[441] K. Fazel and G. Fettweis, eds., *Multi-Carrier Spread-Spectrum*. Dordrecht: Kluwer, 1997. ISBN 0-7923-9973-0.

[442] F. Classen and H. Meyr, "Synchronisation algorithms for an OFDM system for mobile communications," in *Codierung für Quelle, Kanal und Übertragung*, no. 130 in ITG Fachbericht, (Berlin), pp. 105–113, VDE–Verlag, 1994.

[443] F. Classen and H. Meyr, "Frequency synchronisation algorithms for OFDM systems suitable for communication over frequency selective fading channels," in *Proceedings of IEEE VTC '94*, (Stockholm, Sweden), pp. 1655–1659, IEEE, 8–10 June 1994.

[444] S. Shepherd, P. van Eetvelt, C. Wyatt-Millington, and S. Barton, "Simple coding scheme to reduce peak factor in QPSK multicarrier modulation," *Electronics Letters*, vol. 31, pp. 1131–1132, July 1995.

[445] A. E. Jones, T. A. Wilkinson, and S. K. Barton, "Block coding scheme for reduction of peak to mean envelope power ratio of multicarrier transmission schemes," *Electronics Letters*, vol. 30, pp. 2098–2099, December 1994.

[446] D. Wulich, "Reduction of peak to mean ratio of multicarrier modulation by cyclic coding," *Electronics Letters*, vol. 32, pp. 432–433, 1996.

[447] S. Müller and J. Huber, "Vergleich von OFDM–Verfahren mit reduzierter Spitzenleistung," in *2. OFDM–Fachgespräch in Braunschweig*, 1997.

[448] M. Pauli and H.-P. Kuchenbecker, "Neue Aspekte zur Reduzierung der durch Nichtlinearitäten hervorgerufenen Außerbandstrahlung eines OFDM–Signals," in *2. OFDM–Fachgespräch in Braunschweig*, 1997.

[449] T. May and H. Rohling, "Reduktion von Nachbarkanalstörungen in OFDM–Funkübertragungssystemen," in *2. OFDM–Fachgespräch in Braunschweig*, 1997.

[450] D. Wulich, "Peak factor in orthogonal multicarrier modulation with variable levels," *Electronics Letters*, vol. 32, no. 20, pp. 1859–1861, 1996.

[451] H. Schmidt and K. Kammeyer, "Adaptive Subträgerselektion zur Reduktion des Crest faktors bei OFDM," in *3. OFDM Fachgespräch in Braunschweig*, 1998.

[452] R. Dinis and A. G. ao, "Performance evaluation of OFDM transmission with conventional and 2-branch combining power amplification schemes," in *Proceeding of IEEE Global Telecommunications Conference, Globecom 96*, (London, UK), pp. 734–739, IEEE, 18–22 November 1996.

[453] R. Dinis, P. Montezuma, and A. G. ao, "Performance trade-offs with quasi-linearly amplified OFDM through a 2-branch combining technique," in *Proceedings of IEEE VTC'96*, (Atlanta, GA, USA), pp. 899–903, IEEE, 28 April–1 May 1996.

[454] R. Dinis, A. G. ao, and J. Fernandes, "Adaptive transmission techniques for the mobile broadband system," in *Proceeding of ACTS Mobile Communication Summit '97*, (Aalborg, Denmark), pp. 757–762, ACTS, 7–10 October 1997.

[455] B. Daneshrad, L. Cimini Jr., and M. Carloni, "Clustered-OFDM transmitter implementation," in *Proceedings of IEEE International Symposium on Personal, Indoor, and Mobile Radio Communications (PIMRC'96)*, (Taipei, Taiwan), pp. 1064–1068, IEEE, 15–18 October 1996.

[456] M. Okada, H. Nishijima, and S. Komaki, "A maximum likelihood decision based nonlinear distortion compensator for multi-carrier modulated signals," *IEICE Transactions on Communications*, vol. E81B, no. 4, pp. 737–744, 1998.

[457] R. Dinis and A. G. ao, "Performance evaluation of a multicarrier modulation technique allowing strongly nonlinear amplification," in *Proceedings of ICC 1998*, pp. 791–796, IEEE, 1998.

[458] T. Pollet, M. van Bladel, and M. Moeneclaey, "BER sensitivity of OFDM systems to carrier frequency offset and wiener phase noise," *IEEE Transactions on Communications*, vol. 43, pp. 191–193, February/March/April 1995.

[459] H. Nikookar and R. Prasad, "On the sensitivity of multicarrier transmission over multipath channels to phase noise and frequency offset," in *Proceedings of IEEE International Symposium on Personal, Indoor, and Mobile Radio Communications (PIMRC'96)*, (Taipei, Taiwan), pp. 68–72, IEEE, 15–18 October 1996.

[460] J. Kuronen, V.-P. Kaasila, and A. Mammela, "An all-digital symbol tracking algorithm in an OFDM system by using the cyclic prefix," in *Proc. ACTS Summit '96*, (Granada, Spain), pp. 340–345, 27–29 November 1996.

[461] M. Kiviranta and A. Mammela, "Coarse frame synchronization structures in OFDM," in *Proc. ACTS Summit '96*, (Granada, Spain), pp. 464–470, 27–29 November 1996.

[462] Z. Li and A. Mammela, "An all digital frequency synchronization scheme for OFDM systems," in *Proceedings of the IEEE International Symposium on Personal, Indoor and Mobile Radio Communications (PIMRC)*, (Helsinki, Finland), pp. 327–331, 1–4 September 1997.

[463] W. Warner and C. Leung, "OFDM/FM frame synchronization for mobile radio data communication," *IEEE Transactions on Vehicular Technology*, vol. 42, pp. 302–313, August 1993.

[464] H. Sari, G. Karam, and I. Jeanclaude, "Transmission techniques for digital terrestrial TV broadcasting," *IEEE Communications Magazine*, pp. 100–109, February 1995.

[465] P. Moose, "A technique for orthogonal frequency division multiplexing frequency offset correction," *IEEE Transactions on Communications*, vol. 42, pp. 2908–2914, October 1994.

[466] K. Brüninghaus and H. Rohling, "Verfahren zur Rahmensynchronisation in einem OFDM-System," in *3. OFDM Fachgespräch in Braunschweig*, 1998.

[467] F. Daffara and O. Adami, "A new frequency detector for orthogonal multicarrier transmission techniques," in *Proceedings of IEEE Vehicular Technology Conference (VTC'95)*, (Chicago, USA), pp. 804–809, IEEE, 15–28 July 1995.

[468] M. Sandell, J.-J. van de Beek, and P. Börjesson, "Timing and frequency synchronisation in OFDM systems using the cyclic prefix," in *Proceedings of International Symposium on Synchronisation*, (Essen, Germany), pp. 16–19, 14–15 December 1995.

[469] A. Chouly, A. Brajal, and S. Jourdan, "Orthogonal multicarrier techniques applied to direct sequence spread spectrum CDMA systems," in *Proceedings of the IEEE Global Telecommunications Conference 1993*, (Houston, TX, USA), pp. 1723–1728, 29 November – 2 December 1993.

[470] G. Fettweis, A. Bahai, and K. Anvari, "On multi-carrier code division multiple access (MC-CDMA) modem design," in *Proceedings of IEEE VTC '94*, (Stockholm, Sweden), pp. 1670–1674, IEEE, 8–10 June 1994.

[471] R. Prasad and S. Hara, "Overview of multicarrier CDMA," *IEEE Communications Magazine*, pp. 126–133, December 1997.

[472] Y. Li and N. Sollenberger, "Interference suppression in OFDM systems using adaptive antenna arrays," in *Proceeding of Globecom'98*, (Sydney, Australia), pp. 213–218, IEEE, 8–12 November 1998.

[473] Y. Li and N. Sollenberger, "Adaptive antenna arrays for OFDM systems with cochannel interference," *IEEE Transactions on Communications*, vol. 47, pp. 217–229, February 1999.

[474] Y. Li, L. Cimini, and N. Sollenberger, "Robust channel estimation for OFDM systems with rapid dispersive fading channels," *IEEE Transactions on Communications*, vol. 46, pp. 902–915, April 1998.

[475] L. Lin, L. Cimini Jr., and J.-I. Chuang, "Turbo codes for OFDM with antenna diversity," in *Proceeding of VTC'99 (Spring)*, (Houston, TX, USA), IEEE, 16–20 May 1999.

[476] C. Kim, S. Choi, and Y. Cho, "Adaptive beamforming for an OFDM sytem," in *Proceeding of VTC'99 (Spring)*, (Houston, TX, USA), IEEE, 16–20 May 1999.

[477] M. Münster, T. Keller, and L. Hanzo, "Co–channel interference suppression assisted adaptive OFDM in interference limited environments," in *Proceeding of VTC'99 (Fall)*, vol. 1, (Amsterdam, Netherlands), pp. 284–288, IEEE, 19–22 September 1999.

[478] F. Mueller-Roemer, "Directions in audio broadcasting," *Journal Audio Engineering Society*, vol. 41, pp. 158–173, March 1993.

[479] G. Plenge, "DAB — a new radio broadcasting system — state of development and ways for its introduction," *Rundfunktech. Mitt.*, vol. 35, no. 2, 1991.

[480] M. Alard and R. Lassalle, "Principles of modulation and channel coding for digital broadcasting for mobile receivers," *EBU Review, Technical No. 224*, pp. 47–69, August 1987.

[481] *Proceedings of 1st International Symposium,DAB*, (Montreux, Switzerland), June 1992.

[482] ETSI, *Digital Audio Broadcasting (DAB)*, 2nd ed., May 1997. ETS 300 401.

[483] ETSI, *Digital Video Broadcasting (DVB)*, 1.1.2 ed., August 1997. EN 300 744.

[484] S. O'Leary and D. Priestly, "Mobile broadcasting of DVB-T signals," *IEEE Transactions on Broadcasting*, vol. 44, pp. 346–352, September 1998.

[485] W.-C. Lee, H.-M. Park, K.-J. Kang, and K.-B. Kim, "Performance analysis of viterbi decoder using channel state information in COFDM system," *IEEE Transactions on Broadcasting*, vol. 44, pp. 488–496, December 1998.

[486] S. O'Leary, "Hierarchical transmission and COFDM systems," *IEEE Transactions on Broadcasting*, vol. 43, pp. 166–174, June 1997.

[487] L. Thibault and M. Le, "Performance evaluation of COFDM for digital audoo broadcasting Part I: parametric study," *IEEE Transactions on Broadcasting*, vol. 43, pp. 64–75, March 1997.

[488] P. Chow, J. Tu, and J. Cioffi, "A discrete multitone transceiver system for HDSL applications," *IEEE journal on selected areas in communications*, vol. 9, pp. 895–908, August 1991.

[489] P. Chow, J. Tu, and J. Cioffi, "Performance evaluation of a multichannel transceiver system for ADSL and VHDSL services," *IEEE journal on selected areas in communications*, vol. 9, pp. 909–919, August 1991.

[490] K. Sistanizadeh, P. Chow, and J. Cioffi, "Multi-tone transmission for asymmetric digital subscriber lines (ADSL)," in *Proceedings of ICC'93*, pp. 756–760, IEEE, 1993.

[491] ANSI, *ANSI/T1E1.4/94-007, Asymmetric Digital Subscriber Line (ADSL) Metallic Interface.*, August 1997.

[492] A. Burr and P. Brown, "Application of OFDM to powerline telecommunications," in *3rd International Symposium On Power-Line Communications*, (Lancaster, UK), 30 March – 1 April 1999.

[493] M. Deinzer and M. Stoger, "Integrated PLC-modem based on OFDM," in *3rd International Symposium On Power-Line Communications*, (Lancaster, UK), 30 March – 1 April 1999.

[494] R. Prasad and H. Harada, "A novel OFDM based wireless ATM system for future broadband multimedia communications," in *Proceeding of ACTS Mobile Communication Summit '97*, (Aalborg, Denmark), pp. 757–762, ACTS, 7–10 October 1997.

[495] C. Ciotti and J. Borowski, "The AC006 MEDIAN project — overview and state–of–the–art," in *Proc. ACTS Summit '96*, (Granada, Spain), pp. 362–367, 27–29 November 1996.

[496] J. Borowski, S. Zeisberg, J. Hübner, K. Koora, E. Bogenfeld, and B. Kull, "Performance of OFDM and comparable single carrier system in MEDIAN demonstrator 60GHz channel," in *Proceeding of ACTS Mobile Communication Summit '97*, (Aalborg, Denmark), pp. 653–658, ACTS, 7–10 October 1997.

[497] M. D. Benedetto, P. Mandarini, and L. Piazzo, "Effects of a mismatch in the in–phase and in–quadrature paths, and of phase noise, in QDCPSK-OFDM modems," in *Proceeding of ACTS Mobile Communication Summit '97*, (Aalborg, Denmark), pp. 769–774, ACTS, 7–10 October 1997.

[498] T. Rautio, M. Pietikainen, J. Niemi, J. Rautio, K. Rautiola, and A. Mammela, "Architecture and implementation of the 150 Mbit/s OFDM modem (invited paper)," in *IEEE Benelux Joint Chapter on Communications and Vehicular Technology, 6th Symposium on Vehicular Technology and Communications*, (Helsinki, Finland), p. 11, 12–13 October 1998.

[499] J. Ala-Laurila and G. Awater, "The magic WAND — wireless ATM network demondtrator system," in *Proceeding of ACTS Mobile Communication Summit '97*, (Aalborg, Denmark), pp. 356–362, ACTS, 7–10 October 1997.

[500] J. Aldis, E. Busking, T. Kleijne, R. Kopmeiners, R. van Nee, R. Mann-Pelz, and T. Mark, "Magic into reality, building the WAND modem," in *Proceeding of ACTS Mobile Communication Summit '97*, (Aalborg, Denmark), pp. 775–780, ACTS, 7–10 October 1997.

[501] E. Hallmann and H. Rohling, "OFDM-Vorschläge für UMTS," in *3. OFDM Fachgespräch in Braunschweig*, 1998.

[502] "Universal mobile telecommunications system (UMTS); UMTS terrestrial radio access (UTRA); concept evaluation," tech. rep., ETSI, 1997. TR 101 146.

[503] ETSI, *CEPT Recommendation TIR 22-06, Harmonized radio frequency bands for HIPERLAN systems.* European Telecommunications Standards Institute, Sophia Antipolis, Cedex, France, January 1998.

[504] ETSI, *ETSI TR 101 683, vO.1.2, Broadband Radio Access networks (BRAN); High Performance Radio Local Area Networks (HIPERLAN) Type 2, System Overview.* European Telecommunications Standards Institute, Sophia Antipolis, Cedex, France, 1999.

[505] N. Prasad and H. Teunissen, "A state-of-the-art of HIPERLAN/2," in *Proceeding of VTC'99 (Fall)*, (Amsterdam, Netherlands), pp. 2661–2665, IEEE, 19–22 September 1999.

[506] R. van Nee and R. Prasad, *OFDM for wireless multimedia communications.* London: Artech House Publishers, 2000.

[507] K. Fazel, S. Kaiser, P. Robertson, and M. Ruf, "A concept of digital terrestrial television broadcasting," *Wireless Personal Communications*, vol. 2, pp. 9–27, 1995.

[508] J. Bingham, "Method and apparatus for correcting for clock and carrier frequency offset, and phase jitter in multicarrier modems." U.S. Patent No. 5206886, 27 April 1993.

[509] T. de Couasnon, R. Monnier, and J. Rault, "OFDM for digital TV broadcasting," *Signal Processing*, vol. 39, pp. 1–32, 1994.

[510] A. Garcia and M. Calvo, "Phase noise and sub–carrier spacing effects on the performance of an OFDM communications system," *IEEE Communications Letters*, vol. 2, pp. 11–13, January 1998.

[511] W. Robins, *Phase Noise in signal sources*, vol. 9 of *IEE Telecommunication series.* Peter Peregrinus Ltd., 1982.

[512] T. Keller, "Orthogonal frequency division multiplex techniques for wireless local area networks," 1996. Internal Report.

[513] S.-G. Chua and A. Goldsmith, "Variable-rate variable-power mQAM for fading channels," in *Proceedings of IEEE VTC'96*, (Atlanta, GA, USA), pp. 815–819, IEEE, 28 April–1 May 1996.

[514] T. Keller and L. Hanzo, "Sub–band adaptive pre–equalised OFDM transmission," in *Proceeding of VTC'99 (Fall)*, (Amsterdam, Netherlands), pp. 334–338, IEEE, 19–22 September 1999.

[515] T. Liew and L. Hanzo, "Space-time block codes and concatenated channel codes for wireless communications," *Proceedings of the IEEE*, vol. 90, pp. 187–219, February 2002.

[516] V. Tarokh, H. Jafarkhani, and A. Calderbank, "Space-time block codes from orthogonal designs," *IEEE Transactions on Information Theory*, vol. 45, pp. 1456–1467, May 1999.

[517] N. Seshadri, V. Tarokh, and A. Calderbank, "Space-time codes for high data rate wireless communications: Code construction," in *Proceedings of IEEE Vehicular Technology Conference '97*, (Phoenix, Arizona), pp. 637–641, 1997.

[518] V. Tarokh, N. Seshadri, and A. Calderbank, "Space-time codes for high data rate wireless communications: Performance criterion and code construction," in *Proc IEEE International Conference on Communications '97*, (Montreal, Canada), pp. 299–303, 1997.

[519] V. Tarokh, A. Naguib, N. Seshadri, and A. Calderbank, "Space-time codes for high data rate wireless communications: Mismatch analysis," in *Proc IEEE International Conference on Communications '97*, (Montreal, Canada), pp. 309–313, 1997.

[520] A. F. Naguib, V. Tarokh, N. Seshadri, and A. R. Calderbank, "A space-time coding modem for high-data-rate wireless communications," *IEEE Journal on Selected Areas in Communications*, vol. 16, pp. 1459–1478, October 1998.

[521] V. Tarokh, A. Naguib, N. Seshadri, and A. R. Calderbank, "Space-time codes for high data rate wireless communication: Performance criteria in the presence of channel estimation errors, mobility, and multile paths," *IEEE Transactions on Communications*, vol. 47, pp. 199–207, February 1999.

[522] R. Horn and C. Johnson, *Matrix Analysis.* New York: Cambridge University Press, 1988.

[523] G. Bauch and N. Al-Dhahir, "Reduced-complexity turbo equalization with multiple transmit and receive antennas over multipath fading channels," in *Proceedings of Information Sciences and Systems*, (Princeton, USA), pp. WP3 13–18, March 2000.

[524] D. Agrawal, V. Tarokh, A. Naguib, and N. Seshadri, "Space-time coded OFDM for high data-rate wireless communication over wideband channels," in *Proceedings of IEEE Vehicular Technology Conference*, (Ottawa, Canada), pp. 2232–2236, May 1998.

[525] Y. Li, N. Seshadri, and S. Ariyavisitakul, "Channel estimation for OFDM systems with transmitter diversity in mobile wireless channels," *IEEE Journal on Selected Areas in Communications*, vol. 17, pp. 461–471, March 1999.

[526] Y. Li, J. Chuang, and N. Sollenberger, "Transmitter diversity for OFDM systems and its impact on high-rate data wireless networks," *IEEE Journal on Selected Areas in Communications*, vol. 17, pp. 1233–1243, July 1999.

[527] W.-J. Choi and J. Cioffi, "Space-time block codes over frequency selective fading channels," in *Proceedings of VTC 1999 Fall*, (Amsterdam, Holland), pp. 2541–2545, 19-22 September 1999.

[528] Z. Liu, G. Giannakis, A. Scaglione, and S. Barbarossa, "Block precoding and transmit-antenna diversity for decoding and equalization of unknown multipath channels," in *Proc 33rd Asilomar Conference Signals, Systems and Computers*, (Pacific Grove, Canada), pp. 1557–1561, 1-4 November 1999.

[529] Z. Liu and G. Giannakis, "Space-time coding with transmit antennas for multiple access regardless of frequency-selective multipath," in *Proc 1st Sensor Array and Multichannel SP Workshop*, (Boston, USA), 15-17 March 2000.

[530] T. Liew, J. Pliquett, B. Yeap, L.-L. Yang, and L. Hanzo, "Comparative study of space time block codes and various concatenated turbo coding schemes," in *PIMRC 2000*, (London, UK), pp. 741–745, 18-21 September 2000.

[531] T. Liew, J. Pliquett, B. Yeap, L.-L. Yang, and L. Hanzo, "Concatenated space time block codes and TCM, turbo TCM, convolutional as well as turbo codes," in *GLOBECOM 2000*, (San Francisco, USA), 27 Nov -1 Dec 2000.

[532] G. Bauch, "Concatenation of space-time block codes and Turbo-TCM," in *Proceedings of IEEE International Conference on Communications*, (Vancouver, Canada), pp. 1202–1206, June 1999.

[533] G. Forney, "The Viterbi algorithm," *Proceedings of the IEEE*, vol. 61, pp. 268–278, March 1973.

[534] J. Torrance and L. Hanzo, "Performance upper bound of adaptive QAM in slow Rayleigh-fading environments," in *Proceedings of IEEE ICCS'96/ISPACS'96*, (Singapore), pp. 1653–1657, IEEE, 25–29 November 1996.

[535] H. Matsuako, S. Sampei, N. Morinaga, and Y. Kamio, "Adaptive modulation systems with variable coding rate concatenated code for high quality multi-media communication systems," in *Proceedings of IEEE Vehicular Technology Conference*, (Atlanta, USA), pp. 487–491, April 1996.

[536] J. Torrance and L. Hanzo, "On the upper bound performance of adaptive QAM in a slow Rayleigh fading," *IEE Electronics Letters*, pp. 169–171, April 1996.

[537] O. Acikel and W. Ryan, "Punctured turbo-codes for BPSK/QPSK channels," *IEEE Transactions on Communications*, vol. 47, pp. 1315–1323, September 1999.

[538] T. Eyceoz, A. Duel-Hallen, and H. Hallen, "Determinsitic channel modeling and long range prediction of fast fading mobile radio channels," *IEEE Communications Letters*, vol. 2, pp. 254–256, September 1998.

[539] L. Godara, "Applications of antenna arrays to mobile communications, part II: Beam-forming and direction-of-arrival considerations," *Proceedings of the IEEE*, vol. 85, pp. 1193–1245, August 1997.

[540] J. Blogh and L. Hanzo, *3G Systems and Intelligent Networking*. John Wiley and IEEE Press, 2002. (For detailed contents, please refer to http://www-mobile.ecs.soton.ac.uk.).

[541] J. Blogh, P. Cherriman, and L. Hanzo, "Dynamic channel allocation using adaptive modulation and adaptive antennas," in *Proceeding of VTC'99 (Fall)*, vol. 4, (Amsterdam, Netherlands), pp. 2348–2352, IEEE, 19–22 September 1999.

[542] B. Aazhang, B.-P. Paris, and G. C. Orsak, "Neural networks for multiuser detection in code-division multiple-access communications," *IEEE Transactions on Communications*, vol. 40, pp. 1212–1222, July 1992.

[543] D. G. M. Cruickshank, "Radial basis function receivers for DS-CDMA," *Electronics Letters*, vol. 32, pp. 188–190, February 1996.

[544] T. Miyajima, T. Hasegawa, and M. Haneishi, "On the multiuser detection using a neural network in code-division multiple-access communications," *IEICE Transactions on Communications*, vol. E76-B, pp. 961–968, August 1993.

[545] T. Miyajima, "Adaptive multiuser receiver using a Hopfield network," *IEICE Transactions on Fundamentals of Electronics, Communications and Computer Sciences*, vol. E79-A, pp. 652–654, May 1996.

[546] T. Nagaosa, T. Miyajima, and T. Hasegawa, "Multiuser detection using a Hopfield network in asynchronous M-ary/SSMA communications," in *Proceedings of the 1996 IEEE 4th International Symposium on Spread Spectrum Techniques and Applications*, vol. 2, pp. 837–841, IEE, September 1996.

[547] T. Miyajima and T. Hasegawa, "Multiuser detection using a Hopfield network for asynchronous code-division multiple-access systems," *IEICE Transactions on Fundamentals of Electronics, Communications and Computer Sciences*, vol. E79-A, pp. 1963–1971, December 1996.

[548] G. I. Kechriotis and E. S. Manolakos, "Comparison of a neural network based receiver to the optimal and multistage CDMA multiuser detectors," in *Proceedings of the 5th IEEE Workshop on Neural Networks for Signal Processing (NNSP'95)*, (Piscataway, NJ, USA), pp. 613–622, IEEE, August 1995.

[549] A. Papoulis, *Probability, Random Variables, and Stochastic Processes*. McGraw-Hill, 3 ed., 1991.

[550] J. Woodard, "Turbo coding for Gaussian and fading channels," tech. rep., Southampton University, UK, 1998.

[551] M. Breiling, "Turbo coding simulation results," tech. rep., Universität Karlsruhe, Germany and Southampton University, UK, 1997.

[552] P. Robertson and T. Wörz, "Coded modulation scheme employing turbo codes," *IEE Electronics Letters*, vol. 31, pp. 1546–1547, 31st August 1995.

[553] W. Koch and A. Baier, "Optimum and sub-optimum detection of coded data disturbed by time-varying intersymbol interference," *IEEE Globecom*, pp. 1679–1684, December 1990.

[554] H. Robbins and S. Monroe, "A stochastic approximation method," *Annals of Mathematical Statistics*, vol. 22, pp. 400–407, 1951.

[555] C. Darken and J. Moody, "Towards faster stochastic gradient search," in *Advances in Neural Information Processing System 4* (J. E. Moody, S. J. Hanson, and R. P. Lippmann, eds.), pp. 1009–1016, San Mateo, CA: Morgan Kaufmann, 1992.

[556] W. T. Webb and L. Hanzo, *Modern Quadrature Amplitude Modulation : Principles and Applications for Fixed and Wireless Channels*. London: IEEE Press, and John Wiley & Sons, 1994.

[557] Y. Chow, A. Nix, and J. McGeehan, "Analysis of 16-APSK modulation in AWGN and rayleigh fading channel," *Electronics Letters*, vol. 28, pp. 1608–1610, November 1992.

[558] F. Adachi and M. Sawahashi, "Performance analysis of various 16 level modulation schemes under Rrayleigh fading," *Electronics Letters*, vol. 28, pp. 1579–1581, November 1992.

[559] F. Adachi, "Error rate analysis of differentially encoded and detected 16APSK under rician fading," *IEEE Transactions on Vehicular Technology*, vol. 45, pp. 1–12, February 1996.

[560] Y. C. Chow, A. R. Nix, and J. P. McGeehan, "Diversity improvement for 16-DAPSK in Rayleigh fading channel," *Electronics Letters*, vol. 29, pp. 387–389, February 1993.

[561] Y. C. Chow, A. R. Nix, and J. P. McGeehan, "Error analysis for circular 16-DAPSK in frquency-selective Rayleigh fading channels with diversity reception," *Electronics Letters*, vol. 30, pp. 2006–2007, November 1994.

[562] C. M. Lo and W. H. Lam, "Performance analysis of bandwidth efficient coherent modulation schems with L-fold MRC and SC in Nakagami-m fading channels," in *Proceedings of IEEE PIMRC 2000*, vol. 1, pp. 572–576, September 2000.

[563] S. Benedetto, E. Biglierri, and V. Castellani, *Digital Transmission Theory*. Prentice-Hall, 1987.

Author Index

H